Win the complete library of RSMeans Online data!

AF717529

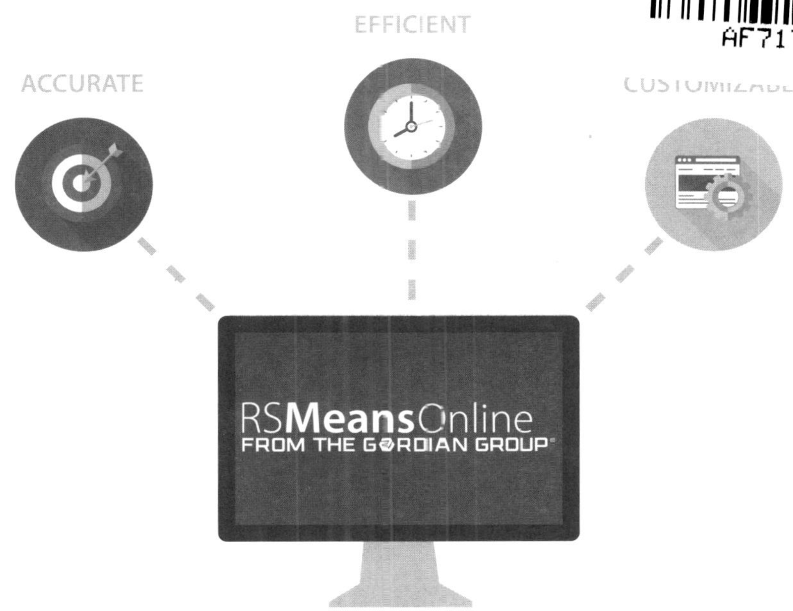

Register your book below to receive your **free** quarterly updates. **Plus** you will be entered into a quarterly drawing to win the **complete** RSMeans Online library of 2016 data!

Be sure to keep up-to-date in 2016!

Fill out the card below for RSMeans' free quarterly updates, as well as a chance to win the complete RSMeans Online library. Please provide your name, address, and email below and return this card by mail, or register online:

info.thegordiangroup.com/2016updates.html

Name ______________________________

Email ____________________ Title ____________

Company ______________________________

Street ______________________________

City/Town ____________ State/Prov. ______ Zip/Postal Code ____________

RSMeans
FROM THE GORDIAN GROUP

RSMeans
FROM THE GORDIAN GROUP®

Assemblies Cost Data

2016

41st annual edition

Marilyn Phelan, AIA Senior Editor

Engineering Director
Bob Mewis, CCP *(1, 2, 4, 13, 14, 31, 32, 33, 34, 35, 41, 44, 46)*

Contributing Editors
Christopher Babbitt
Adrian C. Charest, PE
Cheryl Elsmore
Wafaa Hamitou
Joseph Kelble
Charles Kibbee
Robert J. Kuchta *(8)*
Michael Landry
Thomas Lane (6, 7)
Genevieve Medeiros
Elisa Mello
Chris Morris *(26, 27, 28, 48)*
Melville J. Mossman, PE *(21, 22, 23)*
Marilyn Phelan, AIA *(9, 10, 11, 12)*
Stephen C. Plotner *(3, 5)*
Stephen Rosenberg
Kevin Souza
Keegan Spraker
Tim Tonello
David Yazbek

Vice President Data & Engineering
Chris Anderson

Product Manager
Andrea Sillah

Production Manager
Debbie Panarelli

Production
Sharon Larsen
Jonathan Forgit
Sheryl Rose
Mary Lou Geary

Technical Support
Gary L. Hoitt
Kathryn S. Rodriguez

Cover Design
Blaire Gaddis

**Numbers in italics are the divisional responsibilities for each editor. Please contact the designated editor directly with any questions.*

RSMeans
Construction Publishers & Consultants
1099 Hingham Street, Suite 201
Rockland, MA 02370
United States of America
1-877-669-1338
www.RSMeans.com

Printed in the United States of America
ISSN 1524-2153
ISBN 978-1-943215-00-3

0066 $349.99 per copy (in United States)
Price is subject to change without prior notice.

Related RSMeans Products and Services

This data set is aimed primarily at estimating commercial and industrial projects or large multi-family housing projects costing $3,500,000 and up. For civil engineering structures such as bridges, dams, highways, or the like, please refer to *RSMeans Heavy Construction Cost Data*. For procurement construction cost data such as Job Order Contracting and Change Order Management please refer to The Gordian Group's procurement solutions.

The engineers at RSMeans suggest the following products and services as companion information resources to *RSMeans Assemblies Cost Data:*

Construction Cost Data

Building Construction Cost Data 2016
Facilities Construction Cost Data 2016
Square Foot Costs 2016

Reference Books

Building Security: Strategies & Costs
Designing & Building with the IBC
Estimating Building Costs
RSMeans Estimating Handbook
Green Building: Project Planning & Estimating
How to Estimate with RSMeans Data and CostWorks
Plan Reading & Material Takeoff
Square Foot & Assemblies Estimating Methods

Seminars and In-House Training

RSMeans Online® Training
Square Foot & Assemblies Cost Estimating
Scheduling with MS Project for Construction Professionals
Scope of Work for Facilities Estimating
Unit Price Estimating

RSMeans Online Store

Visit RSMeans at www.RSMeans.com for the most reliable and current resources available on the market. Learn more about our more than 20 data sets available in Online, Book, eBook, and CD formats. Our library of reference books is also available, along with professional development seminars aimed at improving cost estimating, project management, administration, and facilities management skills.

RSMeans Electronic Data

Receive the most up-to-date cost information with RSMeans Online. This web-based service is quick, intuitive, and easy to use, giving you instant access to RSMeans' comprehensive database. Learn more at: **www.RSMeans.com/Online**.

RSMeans Custom Solutions

Building owners, facility managers, building product manufacturers, attorneys, and even insurance firms across the public and private sectors have engaged in RSMeans' custom solutions to solve their estimating needs including:

Custom Cost Engineering Solutions

Knowing the lifetime maintenance cost of your building, how much it will cost to build a specific building type in different locations across the country and globe, and if the estimate from which you are basing a major decision is accurate are all imperative for building owners and their facility managers.

Market & Custom Analytics

Taking a construction product to market requires accurate market research in order to make critical business development decisions. Once the product goes to market, one way to differentiate your product from the competition is to make any cost saving claims you have to offer.

Third-Party Legal Resources

Natural disasters can lead to reconstruction and repairs just as new developments and expansions can lead to issues of eminent domain. In these cases where one party must pay, it is plausible for legal disputes over costs and estimates to arise.

Construction Costs for Software Applications

More than 25 unit price and assemblies cost databases are available through a number of leading estimating and facilities management software partners. For more information, see "Other RSMeans Products and Services" at the back of this publication.

RSMeans data is also available to federal, state, and local government agencies as multi-year, multi-seat licenses.

For information on our current partners, call 1-877-669-1338.

Table of Contents

Foreword

Who We Are

Since 1942, RSMeans has delivered construction cost estimating information and consulting throughout North America. In 2014, RSMeans was acquired by The Gordian Group, combining two industry-leading construction cost databases. Through the RSMeans line of products and services, The Gordian Group provides innovative construction cost estimating data to organizations pursuing efficient and effective construction planning, estimating, procurement, and information solutions.

Our Offerings

With RSMeans' construction cost estimating data, contractors, architects, engineers, facility owners, and managers can utilize the most up-to-date data available in many formats to meet their specific planning and estimating needs.

When you purchase information from RSMeans, you are, in effect, hiring the services of a full-time staff of construction and engineering professionals.

Our thoroughly-experienced and highly-qualified staff works daily to collect, analyze, and disseminate comprehensive cost information to meet your needs.

These staff members have years of practical construction experience and engineering training prior to joining the firm. As a result, you can count on them not only for accurate cost figures, but also for additional background reference information that will help you create a realistic estimate.

The RSMeans organization is equipped to help you solve construction problems through its variety of data solutions.

Access our comprehensive database electronically with RSMeans Online. Quick, intuitive, easy to use, and updated continuously throughout the year, RSMeans Online is the most accurate and up-to-date cost estimating data.

This up-to-date, accurate, and localized data can be leveraged for a broad range of applications, such as Custom Cost Engineering solutions, Market and Custom Analytics, and Third-Party Legal Resources.

To ensure you are getting the most from your cost estimating data, we also offer a myriad of reference guides, training, and professional seminars (learn more at **www.RSMeans.com/Learn**).

In short, RSMeans can provide you with the tools and expertise for developing accurate and dependable construction estimates and budgets in a variety of ways.

Our Commitment

Today at RSMeans, we do more than talk about the quality of our data and the usefulness of the information. We stand behind all of our data—from historical cost indexes to construction materials and techniques—to craft current costs and predict future trends.

If you have any questions about our products or services, please call us toll-free at 1-877-669-1338. You can also visit our website at: **www.RSMeans.com**.

How the Cost Data Is Built: An Overview

A Powerful Construction Tool

You now have one of the most powerful construction tools available today. A successful project is built on the foundation of an accurate and dependable estimate. This tool will enable you to construct such an estimate.

For the casual user the information is designed to be:

- quickly and easily understood so you can get right to your estimate.
- filled with valuable information so you can understand the necessary factors that go into a cost estimate.

For the professional user, the information is designed to be:

- a handy reference that can be quickly referred to for key costs.
- a comprehensive, fully reliable source of current construction costs and productivity rates so you'll be prepared to estimate any project.
- a source for preliminary project costs, product selections, and alternate materials and methods.

To meet all of these requirements, we have organized the information into the following clearly defined sections.

Estimating with RSMeans Unit Price Cost Data*

Please refer to these steps for guidance on completing an estimate using RSMeans unit price cost data.

How to Use the Information: The Details

This section contains an in-depth explanation of how the information is arranged and how you can use it to determine a reliable construction cost estimate. It includes how we develop our cost figures and how to prepare your estimate.

Unit Prices*

All cost data has been divided into 50 divisions according to the MasterFormat system of classification and numbering.

Assemblies*

The cost data in this section has been organized in an "Assemblies" format. These assemblies are the functional elements of a building and are arranged according to the 7 elements of the UNIFORMAT II classification system. For a complete explanation of a typical "Assembly", see "How RSMeans Assemblies Data Works."

*Residential Models**

Model buildings for four classes of construction—economy, average, custom, and luxury—are developed and shown with complete costs per square foot.

*Commercial/Industrial/ Institutional Models**

This section contains complete costs for 77 typical model buildings expressed as costs per square foot.

*Green Commercial/Industrial/ Institutional Models**

This section contains complete costs for 25 green model buildings expressed as costs per square foot.

*References**

This section includes information on Equipment Rental Costs, Crew Listings, Historical Cost Indexes, City Cost Indexes, Location Factors, Reference Tables, and Change Orders, as well as a listing of abbreviations.

- **Equipment Rental Costs:** Included are the average costs to rent and operate hundreds of pieces of construction equipment.
- **Crew Listings:** This section lists all the crews referenced in the cost data. A crew is composed of more than one trade classification and/or the addition of power equipment to any trade classification. Power equipment is included in the cost of the crew. Costs are shown both with bare labor rates and with the installing contractor's overhead and profit added. For each, the total crew cost per eight-hour day and the composite cost per labor-hour are listed.
- **Historical Cost Indexes:** These indexes provide you with data to adjust construction costs over time.
- **City Cost Indexes:** All costs in this data set are U.S. national averages. Costs vary by region. You can adjust for this by CSI Division to over 700 locations throughout the U.S. and Canada by using this data.
- **Location Factors:** You can adjust total project costs to over 900 locations throughout the U.S. and Canada by using the weighted number, which applies across all divisions.
- **Reference Tables:** At the beginning of selected major classifications in the Unit Prices are reference numbers indicators. These numbers refer you to related information in the Reference Section. In this section, you'll find reference tables, explanations, and estimating information that support how we develop the unit price data, technical data, and estimating procedures.
- **Change Orders:** This section includes information on the factors that influence the pricing of change orders.
- **Abbreviations:** A listing of abbreviations used throughout this information, along with the terms they represent, is included.

Index (printed versions only)

A comprehensive listing of all terms and subjects will help you quickly find what you need when you are not sure where it occurs in MasterFormat.

Conclusion

This information is designed to be as comprehensive and easy to use as possible.

The Construction Specifications Institute (CSI) and Construction Specifications Canada (CSC) have produced the 2014 edition of MasterFormat®, a system of titles and numbers used extensively to organize construction information.

All unit prices in the RSMeans cost data are now arranged in the 50-division MasterFormat® 2014 system.

* Not all information is available in all data sets

Note: The material prices in RSMeans cost data are "contractor's prices." They are the prices that contractors can expect to pay at the lumberyards, suppliers'/distributors' warehouses, etc. Small orders of specialty items would be higher than the costs shown, while very large orders, such as truckload lots, would be less. The variation would depend on the size, timing, and negotiating power of the contractor. The labor costs are primarily for new construction or major renovation rather than repairs or minor alterations. With reasonable exercise of judgment, the figures can be used for any building work.

How to Use the Cost Data: The Details

What's Behind the Numbers? The Development of Cost Data

The staff at RSMeans continually monitors developments in the construction industry in order to ensure reliable, thorough, and up-to-date cost information. While overall construction costs may vary relative to general economic conditions, price fluctuations within the industry are dependent upon many factors. Individual price variations may, in fact, be opposite to overall economic trends. Therefore, costs are constantly tracked, and complete updates are performed yearly. Also, new items are frequently added in response to changes in materials and methods.

Costs—$ (U.S.)

All costs represent U.S. national averages and are given in U.S. dollars. The RSMeans City Cost Indexes (CCI) can be used to adjust costs to a particular location. The CCI for Canada can be used to adjust U.S. national averages to local costs in Canadian dollars. No exchange rate conversion is necessary because it has already been factored in.

G The processes or products identified by the green symbol in our publications have been determined to be environmentally responsible and/or resource-efficient solely by the RSMeans engineering staff. The inclusion of the green symbol does not represent compliance with any specific industry association or standard.

Material Costs

The RSMeans staff contacts manufacturers, dealers, distributors, and contractors all across the U.S. and Canada to determine national average material costs. If you have access to current material costs for your specific location, you may wish to make adjustments to reflect differences from the national average. Included within material costs are fasteners for a normal installation. RSMeans engineers use manufacturers' recommendations, written specifications, and/or standard construction practices for size and spacing of fasteners. Adjustments to material costs may be required for your specific application or location. The manufacturer's warranty is assumed. Extended warranties are not included in the material costs. **Material costs do not include sales tax.**

Labor Costs

Labor costs are based upon a mathematical average of trade-specific wages in 30 major U.S. cities. The type of wage (union, open shop, or residential) is identified on the inside back cover of printed publications or is selected by the estimator when using the electronic products. Markups for the wages can also be found on the inside back cover of printed publications and/or under the labor references found in the electronic products.

- If wage rates in your area vary from those used, or if rate increases are expected within a given year, labor costs should be adjusted accordingly.

Labor costs reflect productivity based on actual working conditions. In addition to actual installation, these figures include time spent during a normal weekday on tasks, such as material receiving and handling, mobilization at site, site movement, breaks, and cleanup.

Productivity data is developed over an extended period so as not to be influenced by abnormal variations and reflects a typical average.

Equipment Costs

Equipment costs include not only rental, but also operating costs for equipment under normal use. The operating costs include parts and labor for routine servicing, such as repair and replacement of pumps, filters, and worn lines. Normal operating expendables, such as fuel, lubricants, tires, and electricity (where applicable), are also included. Extraordinary operating expendables with highly variable wear patterns, such as diamond bits and blades, are excluded. These costs are included under materials. Equipment rental rates are obtained from industry sources throughout North America—contractors, suppliers, dealers, manufacturers, and distributors.

Rental rates can also be treated as reimbursement costs for contractor-owned equipment. Owned equipment costs include depreciation, loan payments, interest, taxes, insurance, storage, and major repairs.

Equipment costs do not include operators' wages.

Equipment Cost/Day—The cost of equipment required for each crew is included in the Crew Listings in the Reference Section (small tools that are considered essential everyday tools are not listed out separately). The Crew Listings itemize specialized tools and heavy equipment along with labor trades. The daily cost of itemized equipment included in a crew is based on dividing the weekly bare rental rate by 5 (number of working days per week), then adding the hourly operating cost times 8 (the number of hours per day). This Equipment Cost/Day is shown in the last column of the Equipment Rental Costs in the Reference Section.

Mobilization/Demobilization—The cost to move construction equipment from an equipment yard or rental company to the job site and back again is not included in equipment costs. Mobilization (to the site) and demobilization (from the site) costs can be found in the Unit Price Section. If a piece of equipment is already at the job site, it is not appropriate to utilize mob./demob. costs again in an estimate.

Overhead and Profit

Total Cost including O&P for the installing contractor is shown in the last column of the Unit Price and/or Assemblies. This figure is the sum of the bare material cost plus 10% for profit, the bare labor cost plus total overhead and profit, and the bare equipment cost plus 10% for profit. Details for the calculation of overhead and profit on labor are shown on the inside back cover of the printed product and in the Reference Section of the electronic product.

General Conditions

Cost data in this data set is presented in two ways: Bare Costs and Total Cost including O&P (Overhead and Profit). General Conditions, or General Requirements, of the contract should also be added to the Total Cost including O&P when applicable. Costs for General Conditions are listed in Division 1 of the Unit Price Section and in the Reference Section.

General Conditions for the installing contractor may range from 0% to 10% of the Total Cost including O&P. For the general or prime contractor, costs for General Conditions may range from 5% to 15% of the Total Cost including O&P, with a figure of 10% as the most typical allowance. If applicable, the Assemblies and Models sections use costs that include the installing contractor's overhead and profit (O&P).

Factors Affecting Costs

Costs can vary depending upon a number of variables. Here's a listing of some factors that affect costs and points to consider.

Quality—The prices for materials and the workmanship upon which productivity is based represent sound construction work. They are also in line with industry standard and manufacturer specifications and are frequently used by federal, state, and local governments.

Overtime—We have made no allowance for overtime. If you anticipate premium time or work beyond normal working hours, be sure to make an appropriate adjustment to your labor costs.

Productivity—The productivity, daily output, and labor-hour figures for each line item are based on working an eight-hour day in daylight hours in moderate temperatures. For work that extends beyond normal work hours or is performed under adverse conditions, productivity may decrease.

Size of Project—The size, scope of work, and type of construction project will have a significant impact on cost. Economies of scale can reduce costs for large projects. Unit costs can often run higher for small projects.

Location—Material prices are for metropolitan areas. However, in dense urban areas, traffic and site storage limitations may increase costs. Beyond a 20-mile radius of metropolitan areas, extra trucking or transportation charges may also increase the material costs slightly. On the other hand, lower wage rates may be in effect. Be sure to consider both of these factors when preparing an estimate, particularly if the job site is located in a central city or remote rural location. In addition, highly specialized subcontract items may require travel and per-diem expenses for mechanics.

Other Factors—

- season of year
- contractor management
- weather conditions
- local union restrictions
- building code requirements
- availability of:
  - adequate energy
  - skilled labor
  - building materials
- owner's special requirements/restrictions
- safety requirements
- environmental considerations
- access

Unpredictable Factors—General business conditions influence "in-place" costs of all items. Substitute materials and construction methods may have to be employed. These may affect the installed cost and/or life cycle costs. Such factors may be difficult to evaluate and cannot necessarily be predicted on the basis of the job's location in a particular section of the country. Thus, where these factors apply, you may find significant but unavoidable cost variations for which you will have to apply a measure of judgment to your estimate.

Rounding of Costs

In printed publications only, all unit prices in excess of $5.00 have been rounded to make them easier to use and still maintain adequate precision of the results.

How Subcontracted Items Affect Costs

A considerable portion of all large construction jobs is usually subcontracted. In fact, the percentage done by subcontractors is constantly increasing and may run over 90%. Since the workers employed by these companies do nothing else but install their particular product, they soon become experts in that line. The result is, installation by these firms is accomplished so efficiently that the total in-place cost, even adding the general contractor's overhead and profit, is no more, and often less, than if the principal contractor had handled the installation. Companies that deal with construction specialties are anxious to have their product perform well and, consequently, the installation will be the best possible.

Contingencies

The allowance for contingencies generally provides for unforeseen construction difficulties. On alterations or repair jobs, 20% is not too much. If drawings are final and only field contingencies are being considered, 2% or 3% is probably sufficient, and often nothing needs to be added. Contractually, changes in plans will be covered by extras. The contractor should consider inflationary price trends and possible material shortages during the course of the job. These escalation factors are dependent upon both economic conditions and the anticipated time between the estimate and actual construction. If drawings are not complete or approved, or a budget cost is wanted, it is wise to add 5% to 10%. Contingencies, then, are a matter of judgment.

Important Estimating Considerations

The productivity, or daily output, of each craftsman or crew assumes a well-managed job where tradesmen with the proper tools and equipment, along with the appropriate construction materials, are present. Included are daily set-up and cleanup time, break time, and plan layout time. Unless otherwise indicated, time for material movement on site (for items that can be transported by hand) of up to 200' into the building and to the first or second floor is also included. If material has to be transported by other means, over greater distances, or to higher floors, an additional allowance should be considered by the estimator.

While horizontal movement is typically a sole function of distances, vertical transport introduces other variables that can significantly impact productivity. In an occupied building, the use of elevators (assuming access, size, and required protective measures are acceptable) must be understood at the time of the estimate. For new construction, hoist wait and cycle times can easily be 15 minutes and may result in scheduled access extending beyond the normal work day. Finally, all vertical transport will impose strict weight limits likely to preclude the use of any motorized material handling.

The productivity, or daily output, also assumes installation that meets manufacturer/designer/standard specifications. A time allowance for quality control checks, minor adjustments, and any task required to ensure the proper function or operation is also included. For items that require connections to services, time is included for positioning, leveling, securing the unit, and for making all the necessary connections (and start up where applicable), ensuring a complete installation. Estimating of the services themselves (electrical, plumbing, water, steam, hydraulics, dust collection, etc.) is separate.

In some cases, the estimator must consider the use of a crane and an appropriate crew for the installation of large or heavy items. For those situations where a crane is not included in the assigned crew and as part of the line item cost, equipment rental costs, mobilization and demobilization costs, and operator and support personnel costs must be considered.

Labor-Hours

The labor-hours expressed in this publication are derived by dividing the total daily labor-hours for the crew by the daily output. Based on average installation time and the assumptions listed above, the labor-hours include: direct labor, indirect labor, and nonproductive time. A typical day for a craftsman might include, but is not limited to:

- Direct Work
  - Measuring and layout
  - Preparing materials
  - Actual installation
  - Quality assurance/quality control
- Indirect Work
  - Reading plans or specifications
  - Preparing space
  - Receiving materials
  - Material movement
  - Giving or receiving instruction
  - Miscellaneous
- Non-Work
  - Chatting
  - Personal issues
  - Breaks
  - Interruptions (i.e., sickness, weather, material or equipment shortages, etc.)

If any of the items for a typical day do not apply to the particular work or project situation, the estimator should make any necessary adjustments.

Final Checklist

Estimating can be a straightforward process provided you remember the basics. Here's a checklist of some of the steps you should remember to complete before finalizing your estimate.

Did you remember to:

- factor in the City Cost Index for your locale?
- take into consideration which items have been marked up and by how much?
- mark up the entire estimate sufficiently for your purposes?
- read the background information on techniques and technical matters that could impact your project time span and cost?
- include all components of your project in the final estimate?
- double check your figures for accuracy?
- call RSMeans if you have any questions about your estimate or the data you've used? Remember, RSMeans stands behind its information. If you have any questions about your estimate, about the costs you've used from our data, or even about the technical aspects of the job that may affect your estimate, feel free to call the RSMeans editors at 1-877-669-1338.

Free Quarterly Updates

Stay up-to-date throughout the year with RSMeans' free cost data updates four times a year. Sign up online to make sure you have access to the newest data. Every quarter we provide the city cost adjustment factors for hundreds of cities and key materials.

Register at: **http://info.TheGordianGroup.com/2016Updates.html**.

Assembly Section

Table of Contents

Table of Contents

Table of Contents

Table of Contents

Table of Contents

How RSMeans Assemblies Data Works

Assemblies estimating provides a fast and reasonably accurate way to develop construction costs. An assembly is the grouping of individual work items, with appropriate quantities, to provide a cost for a major construction component in a convenient unit of measure.

An assemblies estimate is often used during early stages of design development to compare the cost impact of various design alternatives on total building cost.

Assemblies estimates are also used as an efficient tool to verify construction estimates.

Assemblies estimates do not require a completed design or detailed drawings. Instead, they are based on the general size of the structure and other known parameters of the project. The degree of accuracy of an assemblies estimate is generally within +/- 15%.

Most RSMeans assemblies consist of three major elements: a graphic, the system components, and the cost data itself. The **Graphic** is a visual representation showing

1 Unique 12-character Identifier

RSMeans assemblies are identified by a **unique 12-character identifier**. The assemblies are numbered using UNIFORMAT II, ASTM Standard E1557. The first 5 characters represent this system to Level 3. The last 7 characters represent further breakdown by RSMeans in order to arrange items in understandable groups of similar tasks. Line numbers are consistent across all RSMeans publications, so a line number in any RSMeans assemblies data set will always refer to the same work.

2 Reference Box

Information is available in the Reference Section to assist the estimator with estimating procedures, alternate pricing methods, and additional technical information.

The **Reference Box** indicates the exact location of this information in the Reference Section. The "R" stands for "reference," and the remaining characters are the RSMeans line numbers.

3 Narrative Descriptions

RSMeans assemblies descriptions appear in two formats: narrative and table. **Narrative descriptions** are shown in a hierarchical structure to make them readable. In order to read a complete description, read up through the indents to the top of the section. Include everything that is above and to the left that is not contradicted by information below.

Narrative Format

C20 Stairs

C2010 Stair Construction

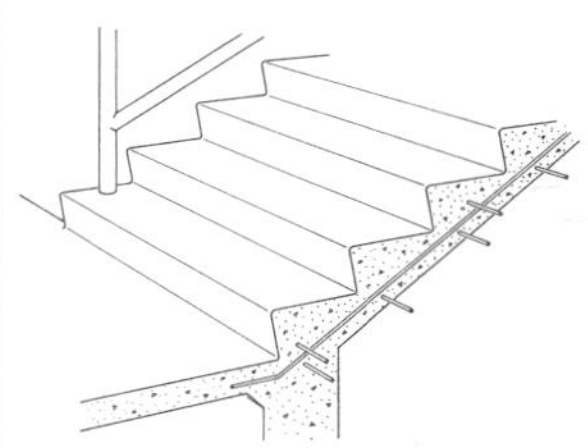

General Design: See reference section for code requirements. Maximum height between landings is 12′; usual stair angle is 20° to 50° with 30° to 35° best. Usual relation of riser to treads is:
Riser + tread = 17.5.
2x (Riser) + tread = 25.
Riser x tread = 70 or 75.
Maximum riser height is 7″ for commercial, 8-1/4″ for residential.
Usual riser height is 6-1/2″ to 7-1/4″.

Minimum tread width is 11″ for commercial and 9″ for residential.

For additional information please see reference section.

Cost Per Flight: Table below lists the cost per flight for 4′-0″ wide stairs. Side walls are not included. Railings are included.

System Components	QUANTITY	UNIT	COST PER FLIGHT MAT.	INST.	TOTAL
SYSTEM C2010 110 0560 (1)					
STAIRS, C.I.P. CONCRETE WITH LANDING, 12 RISERS					
Concrete in place, free standing stairs not incl. safety treads	48.000	L.F.	302.40	2,036.16	2,338.56
Concrete in place, free standing stair landing	32.000	S.F.	161.60	560.64	722.24
Cast alum nosing insert, abr surface, pre-drilled, 3″ wide x 4′ long	12.000	Ea.	966	209.40	1,175.40
Industrial railing, welded, 2 rail 3′-6″ high 1-1/2″ pipe	18.000	L.F.	540	216.72	756.72
Wall railing with returns, steel pipe	17.000	L.F.	289.85	204.68	494.53
TOTAL			2,259.85	3,227.60	5,487.45

C2010 110	Stairs	COST PER FLIGHT MAT.	INST.	TOTAL
0470	Stairs, C.I.P. concrete, w/o landing, 12 risers, w/o nosing (2) RC2010 -100	1,125	2,475	3,600
0480	With nosing	2,100	2,675	4,775
0550	W/landing, 12 risers, w/o nosing	1,300	3,025	4,325
0560	With nosing	2,250	3,225	5,475
0570	16 risers, w/o nosing	1,575	3,775	5,350
0580	With nosing	2,875	4,075	6,950
0590	20 risers, w/o nosing	1,875	4,575	6,450
0600	With nosing (3)	3,475	4,925	8,400
0610	24 risers, w/o nosing	2,150	5,325	7,475
0620	With nosing	4,100	5,750	9,850
0630	Steel, grate type w/nosing & rails, 12 risers, w/o landing	5,700	1,250	6,950
0640	With landing	7,975	1,700	9,675
0660	16 risers, with landing	9,875	2,100	11,975
0680	20 risers, with landing	11,800	2,500	14,300
0700	24 risers, with landing	13,700	2,925	16,625
0710	Concrete fill metal pan & picket rail, 12 risers, w/o landing	7,375	1,250	8,625
0720	With landing	9,725	1,850	11,575
0740	16 risers, with landing	12,200	2,250	14,450
0760	20 risers, with landing	14,600	2,675	17,275
0780	24 risers, with landing	17,100	3,050	20,150
0790	Cast iron tread & pipe rail, 12 risers, w/o landing	7,375	1,250	8,625
0800	With landing	9,725	1,850	11,575
0820	16 risers, with landing	12,200	2,250	14,450
0840	20 risers, with landing	14,600	2,675	17,275

For supplemental customizable square foot estimating forms, visit: **www.RSMeans.com/2016extras**

the typical appearance of the assembly in question, frequently accompanied by additional explanatory technical information describing the class of items. The **System Components** is a listing of the individual tasks that make up the assembly, including the quantity and unit of measure for each item, along with the cost of material and installation.

The **Assemblies Data** below lists prices for other similar systems with dimensional and/or size variations.

All RSMeans assemblies costs represent the cost for the installing contractor. An allowance for profit has been added to all material, labor, and equipment rental costs. A markup for labor burdens, including workers' compensation, fixed overhead, and business overhead, is included with installation costs.

The information in RSMeans cost data represents a "national average" cost. This data should be modified to the project location using the **City Cost Indexes** or **Location Factors** tables found in the Reference Section.

Table Format

A20 Basement Construction

A2020 Basement Walls

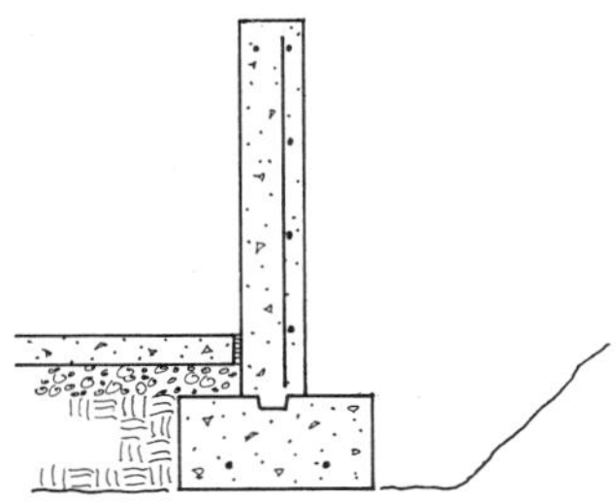

The Foundation Bearing Wall System includes: forms up to 16′ high (four uses); 3,000 p.s.i. concrete placed and vibrated; and form removal with breaking form ties and patching walls. The wall systems list walls from 6″ to 16″ thick and are designed with minimum reinforcement.

Excavation and backfill are not included.

Please see the reference section for further design and cost information.

System Components 5	QUANTITY	UNIT	COST PER L.F. 4 MAT.	INST.	TOTAL
SYSTEM A2020 110 1500					
FOUNDATION WALL, CAST IN PLACE, DIRECT CHUTE, 4′ HIGH, 6″ THICK					
Formwork	8.000	SFCA	6.32	46	52.32
Reinforcing	3.300	Lb.	1.73	1.44	3.17
Unloading & sorting reinforcing	3.300	Lb.		.08	.08
Concrete, 3,000 psi	.074	C.Y.	8.73		8.73
Place concrete, direct chute	.074	C.Y.		2.43	2.43
Finish walls, break ties and patch voids, one side	4.000	S.F.	.20	4	4.20
TOTAL			16.98	53.95	70.93

A2020 110 — Walls, Cast in Place

	WALL HEIGHT (FT.)	PLACING METHOD	CONCRETE (C.Y. per L.F.)	REINFORCING (LBS. per L.F.)	WALL THICKNESS (IN.)	COST PER L.F. MAT.	INST.	TOTAL
1500	4′	direct chute	.074	3.3	6	17	54	71
1520	6		.099	4.8	8	20.50	55.50	76
1540			.123	6.0	10	24	56.50	80.50
1560	RA2020 -210		.148	7.2	12	28	57.50	85.50
1580			.173	8.1	14	31	58.50	89.50
1600			.197	9.44	16	34.50	60	94.50
1700	4′	pumped	.074	3.3	6	17	55	72
1720			.099	4.8	8	20.50	57	77.50
1740			.123	6.0	10	24	58	82
1760			.148	7.2	12	28	59.50	87.50
1780			.173	8.1	14	31	61	92
1800			.197	9.44	16	34.50	62.50	97
3000	6′	direct chute	.111	4.95	6	25.50	81	106.50
3020			.149	7.20	8	31	83	114
3040			.184	9.00	10	36	84.50	120.50
3060			.222	10.8	12	41.50	86.50	128
3080			.260	12.15	14	47	87.50	134.50
3100			.300	14.39	16	52.50	90.50	143

4 Unit of Measure

All RSMeans assemblies data includes a typical **Unit of Measure** used for estimating that item. For instance, for continuous footings or foundation walls the unit is linear feet (L.F.). For spread footings the unit is each (Ea.). The estimator needs to take special care that the unit in the data matches the unit in the takeoff. Abbreviations and unit conversions can be found in the Reference Section.

5 System Components

System components are listed separately to detail what is included in the development of the total system price.

6 Table Descriptions

Table descriptions work in a similar fashion, except that if there is a blank in the column at a particular line number, read up to the description above in the same column.

How RSMeans Assemblies Data Works (Continued)

Sample Estimate

This sample demonstrates the elements of an estimate, including a tally of the RSMeans data lines. Published assemblies costs include all markups for labor burden and profit for the installing contractor. This estimate adds a summary of the markups applied by a general contractor on the installing contractor's work. These figures represent the total cost to the owner. The RSMeans location factor is applied at the bottom of the estimate to adjust the cost of the work to a specific location.

Project Name:	Interior Fit-out, ABC Office				
Location:	Anywhere, USA		Date: 1/1/2016		STD
Assembly Number	Description		Qty.	Unit	Subtotal
C1010 124 1200	Wood partition, 2 x 4 @ 16" OC w/5/8" FR gypsum board		560.000	S.F.	$2,738.40
C1020 114 1800	Metal door & frame, flush hollow core, 3'-0" x 7'-0"		2.000	Ea.	$2,470.00
C3010 230 0080	Painting, brushwork, primer & 2 coats	1	1,120.000	S.F.	$1,344.00
C3020 410 0140	Carpet, tufted, nylon, roll goods, 12' wide, 26 oz		240.000	S.F.	$813.60
C3030 210 6000	Acoustic ceilings, 24" x 48" tile, tee grid suspension		200.000	S.F.	$1,156.00
D5020 125 0560	Receptacles incl plate, box, conduit, wire, 20 A duplex		8.000	Ea.	$2,168.00
D5020 125 0720	Light switch incl plate, box, conduit, wire, 20 A single pole		2.000	Ea.	$532.00
D5020 210 0560	Fluorescent fixtures, recess mounted, 20 per 1000 SF		200.000	S.F.	$2,186.00
	Assembly Subtotal				**$13,408.00**
		Sales Tax @ 2	**5 %**		**$ 335.20**
		General Requirements @ 3	**7 %**		**$ 938.56**
	Subtotal A				**$14,681.76**
		GC Overhead @ 4	**5 %**		**$ 734.09**
	Subtotal B				**$15,415.85**
		GC Profit @ 5	**5 %**		**$ 770.79**
	Subtotal C				**$16,186.64**
	Adjusted by Location Factor 6		**118.1**		**$ 19,116.42**
		Architects Fee @ 7	**8 %**		**$ 1,529.31**
		Contingency @ 8	**15 %**		**$ 2,867.46**
	Project Total Cost				**$ 23,513.20**

This estimate is based on an interactive spreadsheet.
A copy of this spreadsheet is located on the RSMeans website at **www.RSMeans.com/2016extras**.
You are free to download it and adjust it to your methodology.

1 Work Performed

The body of the estimate shows the RSMeans data selected, including line numbers, a brief description of each item, its takeoff quantity and unit, and the total installed cost, including the installing contractor's overhead and profit.

2 Sales Tax

If the work is subject to state or local sales taxes, the amount must be added to the estimate. In a conceptual estimate it can be assumed that one half of the total represents material costs. Therefore, apply the sales tax rate to 50% of the assembly subtotal.

3 General Requirements

This item covers project-wide needs provided by the general contractor. These items vary by project but may include temporary facilities and utilities, security, testing, project cleanup, etc. In assemblies estimates a percentage is used, typically between 5% and 15% of project cost.

4 General Contractor Overhead

This entry represents the general contractor's markup on all work to cover project administration costs.

5 General Contractor Profit

This entry represents the GC's profit on all work performed. The value included here can vary widely by project, and is influenced by the GC's perception of the project's financial risk and market conditions.

6 Location Factor

RSMeans published data is based on national average costs. If necessary, adjust the total cost of the project using a location factor from the "Location Factor" table or the "City Cost Indexes" table found in the Reference Section. Use location factors if the work is general, covering the work of multiple trades. If the work is by a single trade (e.g. masonry) use the more specific data found in the City Cost Indexes.

To adjust costs by location factors, multiply the base cost by the factor and divide by 100.

7 Architect's Fee

If appropriate, add the design cost to the project estimate. These fees vary based on project complexity and size. Typical design and engineering fees can be found in the reference section.

8 Contingency

A factor for contingency may be added to any estimate to represent the cost of unknowns that may occur between the time that the estimate is performed and the time the project is constructed The amount of the allowance will depend on the stage of design at which the estimate is done, and the contractor's assessment of the risk involved.

A1010 Standard Foundations

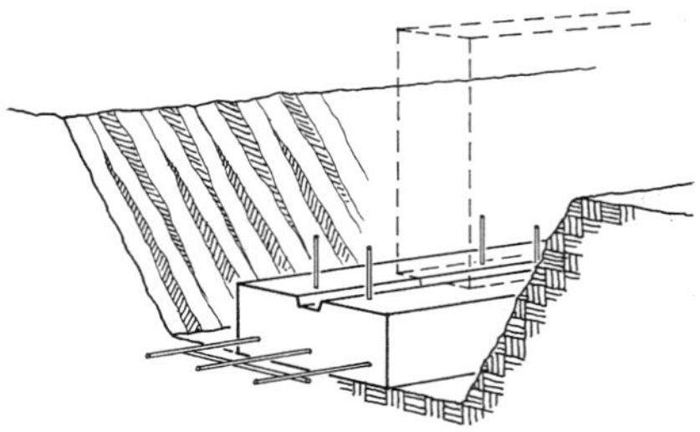

The Strip Footing System includes: excavation; hand trim; all forms needed for footing placement; forms for 2″ x 6″ keyway (four uses); dowels; and 3,000 p.s.i. concrete.

The footing size required varies for different soils. Soil bearing capacities are listed for 3 KSF and 6 KSF. Depths of the system range from 8″ and deeper. Widths range from 16″ and wider. Smaller strip footings may not require reinforcement.

Please see the reference section for further design and cost information.

System Components	QUANTITY	UNIT	COST PER L.F. MAT.	INST.	TOTAL
SYSTEM A1010 110 2500					
STRIP FOOTING, LOAD 5.1 KLF, SOIL CAP. 3 KSF, 24″ WIDE X 12″ DEEP, REINF.					
Trench excavation	.148	C.Y.		1.47	1.47
Hand trim	2.000	S.F.		1.96	1.96
Compacted backfill	.074	C.Y.		.30	.30
Formwork, 4 uses	2.000	S.F.	4.88	9.28	14.16
Keyway form, 4 uses	1.000	L.F.	.33	1.19	1.52
Reinforcing, fy = 60000 psi	3.000	Lb.	1.65	1.89	3.54
Dowels	2.000	Ea.	1.56	5.48	7.04
Concrete, f'c = 3000 psi	.074	C.Y.	8.73		8.73
Place concrete, direct chute	.074	C.Y.		1.82	1.82
Screed finish	2.000	S.F.		.78	.78
TOTAL			17.15	24.17	41.32

A1010 110	Strip Footings	COST PER L.F. MAT.	INST.	TOTAL
2100	Strip footing, load 2.6 KLF, soil capacity 3 KSF, 16″ wide x 8″ deep, plain	7.50	11.15	18.65
2300	Load 3.9 KLF, soil capacity 3 KSF, 24″ wide x 8″ deep, plain	9.50	12.55	22.05
2500	Load 5.1 KLF, soil capacity 3 KSF, 24″ wide x 12″ deep, reinf. RA1010 -140	17.15	24	41.15
2700	Load 11.1 KLF, soil capacity 6 KSF, 24″ wide x 12″ deep, reinf.	17.15	24	41.15
2900	Load 6.8 KLF, soil capacity 3 KSF, 32″ wide x 12″ deep, reinf.	20.50	26	46.50
3100	Load 14.8 KLF, soil capacity 6 KSF, 32″ wide x 12″ deep, reinf.	20.50	26	46.50
3300	Load 9.3 KLF, soil capacity 3 KSF, 40″ wide x 12″ deep, reinf.	24	28.50	52.50
3500	Load 18.4 KLF, soil capacity 6 KSF, 40″ wide x 12″ deep, reinf.	24.50	28.50	53
3700	Load 10.1 KLF, soil capacity 3 KSF, 48″ wide x 12″ deep, reinf.	27	30.50	57.50
3900	Load 22.1 KLF, soil capacity 6 KSF, 48″ wide x 12″ deep, reinf.	28.50	32.50	61
4100	Load 11.8 KLF, soil capacity 3 KSF, 56″ wide x 12″ deep, reinf.	31.50	34	65.50
4300	Load 25.8 KLF, soil capacity 6 KSF, 56″ wide x 12″ deep, reinf.	34	36.50	70.50
4500	Load 10 KLF, soil capacity 3 KSF, 48″ wide x 16″ deep, reinf.	34.50	36	70.50
4700	Load 22 KLF, soil capacity 6 KSF, 48″ wide, 16″ deep, reinf.	35	37	72
4900	Load 11.6 KLF, soil capacity 3 KSF, 56″ wide x 16″ deep, reinf.	39	51.50	90.50
5100	Load 25.6 KLF, soil capacity 6 KSF, 56″ wide x 16″ deep, reinf.	41	53.50	94.50
5300	Load 13.3 KLF, soil capacity 3 KSF, 64″ wide x 16″ deep, reinf.	44.50	42.50	87
5500	Load 29.3 KLF, soil capacity 6 KSF, 64″ wide x 16″ deep, reinf.	47	45.50	92.50
5700	Load 15 KLF, soil capacity 3 KSF, 72″ wide x 20″ deep, reinf.	59	51	110
5900	Load 33 KLF, soil capacity 6 KSF, 72″ wide x 20″ deep, reinf.	62.50	54.50	117
6100	Load 18.3 KLF, soil capacity 3 KSF, 88″ wide x 24″ deep, reinf.	83.50	65	148.50
6300	Load 40.3 KLF, soil capacity 6 KSF, 88″ wide x 24″ deep, reinf.	89.50	72	161.50
6500	Load 20 KLF, soil capacity 3 KSF, 96″ wide x 24″ deep, reinf.	90.50	68.50	159
6700	Load 44 KLF, soil capacity 6 KSF, 96″ wide x 24″ deep, reinf.	95	74.50	169.50

A1010 Standard Foundations

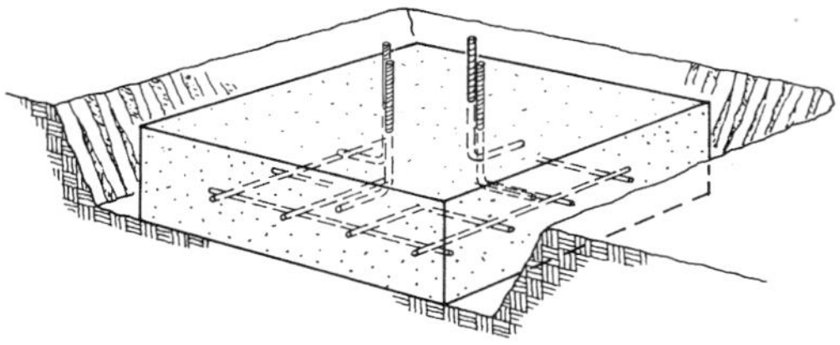

The Spread Footing System includes: excavation; backfill; forms (four uses); all reinforcement; 3,000 p.s.i. concrete (chute placed); and float finish.

Footing systems are priced per individual unit. The Expanded System Listing at the bottom shows various footing sizes. It is assumed that excavation is done by a truck mounted hydraulic excavator with an operator and oiler.

Backfill is with a dozer, and compaction by air tamp. The excavation and backfill equipment is assumed to operate at 30 C.Y. per hour.

Please see the reference section for further design and cost information.

System Components	QUANTITY	UNIT	COST EACH		
			MAT.	INST.	TOTAL
SYSTEM A1010 210 7100					
SPREAD FOOTINGS, LOAD 25K, SOIL CAPACITY 3 KSF, 3′ SQ X 12″ DEEP					
Bulk excavation	.590	C.Y.		5.08	5.08
Hand trim	9.000	S.F.		8.82	8.82
Compacted backfill	.260	C.Y.		1.06	1.06
Formwork, 4 uses	12.000	S.F.	9.24	65.40	74.64
Reinforcing, fy = 60,000 psi	.006	Ton	6.30	7.50	13.80
Dowel or anchor bolt templates	6.000	L.F.	5.82	27	32.82
Concrete, f′c = 3,000 psi	.330	C.Y.	38.94		38.94
Place concrete, direct chute	.330	C.Y.		8.11	8.11
Float finish	9.000	S.F.		3.51	3.51
TOTAL			60.30	126.48	186.78

A1010 210	Spread Footings	COST EACH		
		MAT.	INST.	TOTAL
7090	Spread footings, 3000 psi concrete, chute delivered			
7100	Load 25K, soil capacity 3 KSF, 3′-0″ sq. x 12″ deep	60.50	127	187.50
7150	Load 50K, soil capacity 3 KSF, 4′-6″ sq. x 12″ deep RA1010 -120	129	218	347
7200	Load 50K, soil capacity 6 KSF, 3′-0″ sq. x 12″ deep	60.50	127	187.50
7250	Load 75K, soil capacity 3 KSF, 5′-6″ sq. x 13″ deep	205	305	510
7300	Load 75K, soil capacity 6 KSF, 4′-0″ sq. x 12″ deep	104	187	291
7350	Load 100K, soil capacity 3 KSF, 6′-0″ sq. x 14″ deep	259	370	629
7410	Load 100K, soil capacity 6 KSF, 4′-6″ sq. x 15″ deep	159	256	415
7450	Load 125K, soil capacity 3 KSF, 7′-0″ sq. x 17″ deep	410	525	935
7500	Load 125K, soil capacity 6 KSF, 5′-0″ sq. x 16″ deep	205	305	510
7550	Load 150K, soil capacity 3 KSF 7′-6″ sq. x 18″ deep	500	610	1,110
7610	Load 150K, soil capacity 6 KSF, 5′-6″ sq. x 18″ deep	273	385	658
7650	Load 200K, soil capacity 3 KSF, 8′-6″ sq. x 20″ deep	710	810	1,520
7700	Load 200K, soil capacity 6 KSF, 6′-0″ sq. x 20″ deep	360	475	835
7750	Load 300K, soil capacity 3 KSF, 10′-6″ sq. x 25″ deep	1,300	1,325	2,625
7810	Load 300K, soil capacity 6 KSF, 7′-6″ sq. x 25″ deep	680	795	1,475
7850	Load 400K, soil capacity 3 KSF, 12′-6″ sq. x 28″ deep	2,075	1,950	4,025
7900	Load 400K, soil capacity 6 KSF, 8′-6″ sq. x 27″ deep	945	1,025	1,970
7950	Load 500K, soil capacity 3 KSF, 14′-0″ sq. x 31″ deep	2,875	2,550	5,425
8010	Load 500K, soil capacity 6 KSF, 9′-6″ sq. x 30″ deep	1,300	1,350	2,650
8050	Load 600K, soil capacity 3 KSF, 16′-0″ sq. x 35″ deep	4,225	3,500	7,725
8100	Load 600K, soil capacity 6 KSF, 10′-6″ sq. x 33″ deep	1,750	1,725	3,475

A1010 Standard Foundations

A1010 210	Spread Footings	COST EACH		
		MAT.	INST.	TOTAL
8150	Load 700K, soil capacity 3 KSF, 17'-0" sq. x 37" deep	4,975	4,000	8,975
8200	Load 700K, soil capacity 6 KSF, 11'-6" sq. x 36" deep	2,250	2,100	4,350
8250	Load 800K, soil capacity 3 KSF, 18'-0" sq. x 39" deep	5,875	4,650	10,525
8300	Load 800K, soil capacity 6 KSF, 12'-0" sq. x 37" deep	2,525	2,325	4,850
8350	Load 900K, soil capacity 3 KSF, 19'-0" sq. x 40" deep	6,825	5,325	12,150
8400	Load 900K, soil capacity 6 KSF, 13'-0" sq. x 39" deep	3,125	2,775	5,900
8450	Load 1000K, soil capacity 3 KSF, 20'-0" sq. x 42" deep	7,875	5,975	13,850
8500	Load 1000K, soil capacity 6 KSF, 13'-6" sq. x 41" deep	3,550	3,075	6,625
8550	Load 1200K, soil capacity 6 KSF, 15'-0" sq. x 48" deep	4,700	3,900	8,600
8600	Load 1400K, soil capacity 6 KSF, 16'-0" sq. x 47" deep	5,725	4,625	10,350
8650	Load 1600K, soil capacity 6 KSF, 18'-0" sq. x 52" deep	7,975	6,125	14,100
8700				

A1010 Standard Foundations

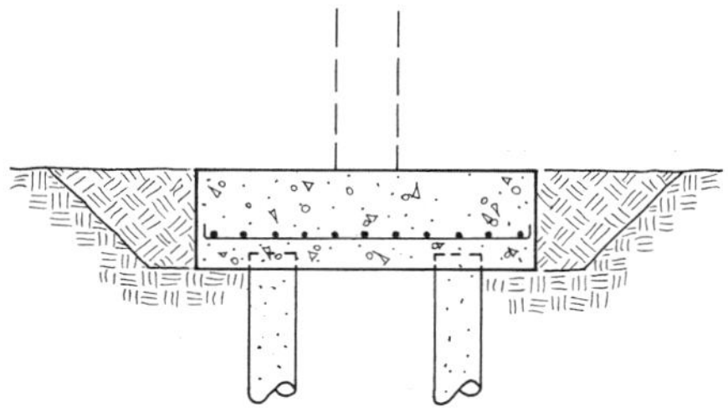

These pile cap systems include excavation with a truck mounted hydraulic excavator, hand trimming, compacted backfill, forms for concrete, templates for dowels or anchor bolts, reinforcing steel and concrete placed and floated.

Pile embedment is assumed as 6″. Design is consistent with the Concrete Reinforcing Steel Institute Handbook f'c = 3000 psi, fy = 60,000.

Please see the reference section for further design and cost information.

System Components	QUANTITY	UNIT	COST EACH		
			MAT.	INST.	TOTAL
SYSTEM A1010 250 5100					
CAP FOR 2 PILES, 6′-6″X3′-6″X20″, 15 TON PILE, 8″ MIN. COL., 45K COL. LOAD					
Excavation, bulk, hyd excavator, truck mtd. 30″ bucket 1/2 CY	2.890	C.Y.		24.89	24.89
Trim sides and bottom of trench, regular soil	23.000	S.F.		22.54	22.54
Dozer backfill & roller compaction	1.500	C.Y.		6.08	6.08
Forms in place pile cap, square or rectangular, 4 uses	33.000	SFCA	33.66	193.05	226.71
Templates for dowels or anchor bolts	8.000	Ea.	7.76	36	43.76
Reinforcing in place footings, #8 to #14	.025	Ton	26.25	18.25	44.50
Concrete ready mix, regular weight, 3000 psi	1.400	C.Y.	165.20		165.20
Place and vibrate concrete for pile caps, under 5 CY, direct chute	1.400	C.Y.		45.88	45.88
Float finish	23.000	S.F.		8.97	8.97
TOTAL			232.87	355.66	588.53

A1010 250	Pile Caps							
	NO. PILES	SIZE FT-IN X FT-IN X IN	PILE CAPACITY (TON)	COLUMN SIZE (IN)	COLUMN LOAD (K)	COST EACH		
						MAT.	INST.	TOTAL
5100	2	6-6x3-6x20	15	8	45	233	360	593
5150	RA1010 -330	26	40	8	155	286	435	721
5200		34	80	11	314	390	560	950
5250		37	120	14	473	415	595	1,010
5300	3	5-6x5-1x23	15	8	75	277	410	687
5350		28	40	10	232	310	465	775
5400		32	80	14	471	360	525	885
5450		38	120	17	709	415	600	1,015
5500	4	5-6x5-6x18	15	10	103	315	410	725
5550		30	40	11	308	460	590	1,050
5600		36	80	16	626	540	685	1,225
5650		38	120	19	945	570	720	1,290
5700	6	8-6x5-6x18	15	12	156	525	590	1,115
5750		37	40	14	458	850	875	1,725
5800		40	80	19	936	970	975	1,945
5850		45	120	24	1413	1,075	1,075	2,150
5900	8	8-6x7-9x19	15	12	205	790	800	1,590
5950		36	40	16	610	1,150	1,050	2,200
6000		44	80	22	1243	1,425	1,275	2,700
6050		47	120	27	1881	1,525	1,375	2,900

A1010 Standard Foundations

A1010 250 Pile Caps

	NO. PILES	SIZE FT-IN X FT-IN X IN	PILE CAPACITY (TON)	COLUMN SIZE (IN)	COLUMN LOAD (K)	COST EACH		
						MAT.	INST.	TOTAL
6100	10	11-6x7-9x21	15	14	250	1,125	980	2,105
6150		39	40	17	756	1,700	1,400	3,100
6200		47	80	25	1547	2,075	1,675	3,750
6250		49	120	31	2345	2,200	1,775	3,975
6300	12	11-6x8-6x22	15	15	316	1,425	1,175	2,600
6350		49	40	19	900	2,250	1,800	4,050
6400		52	80	27	1856	2,500	1,950	4,450
6450		55	120	34	2812	2,700	2,100	4,800
6500	14	11-6x10-9x24	15	16	345	1,825	1,475	3,300
6550		41	40	21	1056	2,425	1,825	4,250
6600		55	80	29	2155	3,175	2,325	5,500
6700	16	11-6x11-6x26	15	18	400	2,225	1,675	3,900
6750		48	40	22	1200	2,975	2,175	5,150
6800		60	80	31	2460	3,725	2,650	6,375
6900	18	13-0x11-6x28	15	20	450	2,525	1,850	4,375
6950		49	40	23	1349	3,475	2,400	5,875
7000		56	80	33	2776	4,100	2,850	6,950
7100	20	14-6x11-6x30	15	20	510	3,100	2,225	5,325
7150		52	40	24	1491	4,175	2,825	7,000

A1010 Standard Foundations

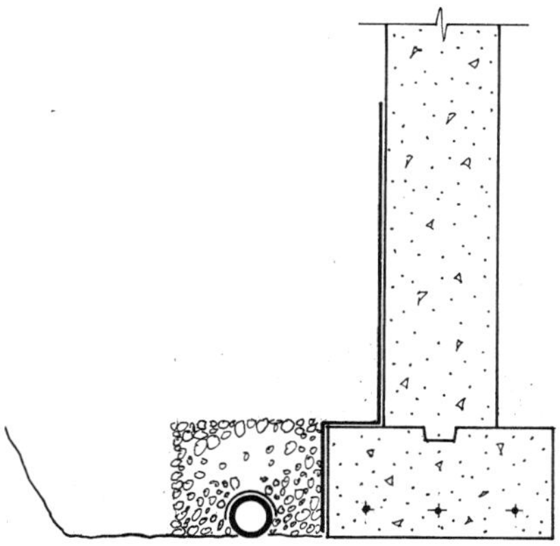

General: Footing drains can be placed either inside or outside of foundation walls depending upon the source of water to be intercepted. If the source of subsurface water is principally from grade or a subsurface stream above the bottom of the footing, outside drains should be used. For high water tables, use inside drains or both inside and outside.

The effectiveness of underdrains depends on good waterproofing. This must be carefully installed and protected during construction.

Costs below include the labor and materials for the pipe and 6″ of crushed stone around pipe. Excavation and backfill are not included.

System Components	QUANTITY	UNIT	COST PER L.F. MAT.	INST.	TOTAL
SYSTEM A1010 310 1000					
FOUNDATION UNDERDRAIN, OUTSIDE ONLY, PVC, 4″ DIAM.					
PVC pipe 4″ diam. S.D.R. 35	1.000	L.F.	1.64	4.19	5.83
Crushed stone 3/4″ to 1/2″	.070	C.Y.	1.86	.90	2.76
TOTAL			3.50	5.09	8.59

A1010 310	Foundation Underdrain	COST PER L.F. MAT.	INST.	TOTAL
1000	Foundation underdrain, outside only, PVC, 4″ diameter	3.50	5.10	8.60
1100	6″ diameter	6	5.65	11.65
1400	Perforated HDPE, 6″ diameter	4.13	2.18	6.31
1450	8″ diameter	6.40	2.74	9.14
1500	12″ diameter	11.30	6.70	18
1600	Corrugated metal, 16 ga. asphalt coated, 6″ diameter	9.55	5.30	14.85
1650	8″ diameter	12.30	5.65	17.95
1700	10″ diameter	15.20	7.35	22.55
3000	Outside and inside, PVC, 4″ diameter	7	10.20	17.20
3100	6″ diameter	12	11.30	23.30
3400	Perforated HDPE, 6″ diameter	8.25	4.37	12.62
3450	8″ diameter	12.75	5.50	18.25
3500	12″ diameter	22.50	13.35	35.85
3600	Corrugated metal, 16 ga., asphalt coated, 6″ diameter	19.05	10.60	29.65
3650	8″ diameter	24.50	11.30	35.80
3700	10″ diameter	30.50	14.70	45.20

A1010 Standard Foundations

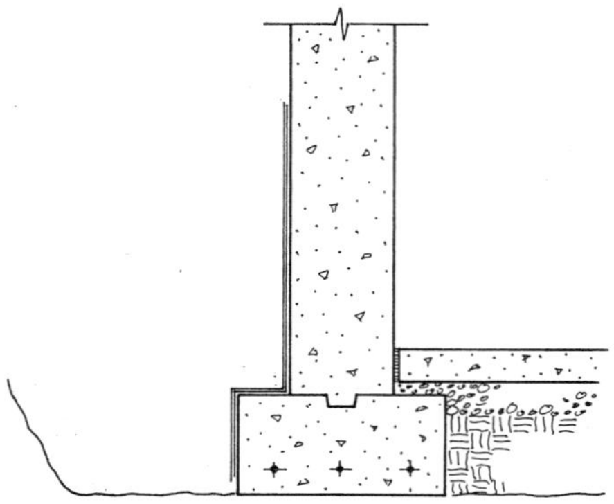

General: Apply foundation wall dampproofing over clean concrete giving particular attention to the joint between the wall and the footing. Use care in backfilling to prevent damage to the dampproofing.

Costs for four types of dampproofing are listed below.

System Components	QUANTITY	UNIT	COST PER L.F. MAT.	COST PER L.F. INST.	COST PER L.F. TOTAL
SYSTEM A1010 320 1000					
FOUNDATION DAMPPROOFING, BITUMINOUS, 1 COAT, 4′ HIGH					
Bituminous asphalt dampproofing brushed on below grade, 1 coat	4.000	S.F.	1	3.40	4.40
Labor for protection of dampproofing during backfilling	4.000	S.F.		1.40	1.40
TOTAL			1	4.80	5.80

A1010 320	Foundation Dampproofing	COST PER L.F. MAT.	COST PER L.F. INST.	COST PER L.F. TOTAL
1000	Foundation dampproofing, bituminous, 1 coat, 4′ high	1	4.80	5.80
1400	8′ high	2	9.60	11.60
1800	12′ high	3	14.85	17.85
2000	2 coats, 4′ high	1.96	5.95	7.91
2400	8′ high	3.92	11.90	15.82
2800	12′ high	5.90	18.35	24.25
3000	Asphalt with fibers, 1/16″ thick, 4′ high	1.56	5.95	7.51
3400	8′ high	3.12	11.90	15.02
3800	12′ high	4.68	18.35	23.03
4000	1/8″ thick, 4′ high	2.80	7.10	9.90
4400	8′ high	5.60	14.15	19.75
4800	12′ high	8.40	21.50	29.90
5000	Asphalt coated board and mastic, 1/4″ thick, 4′ high	5.10	6.45	11.55
5400	8′ high	10.25	12.85	23.10
5800	12′ high	15.35	19.75	35.10
6000	1/2″ thick, 4′ high	7.70	8.95	16.65
6400	8′ high	15.35	17.90	33.25
6800	12′ high	23	27.50	50.50
7000	Cementitious coating, on walls, 1/8″ thick coating, 4′ high	3.24	8.95	12.19
7400	8′ high	6.50	17.95	24.45
7800	12′ high	9.70	27	36.70
8000	Cementitious/metallic slurry, 4 coat, 1/2″thick, 2′ high	.77	9	9.77
8400	4′ high	1.54	18	19.54
8800	6′ high	2.31	27	29.31

A1020 Special Foundations

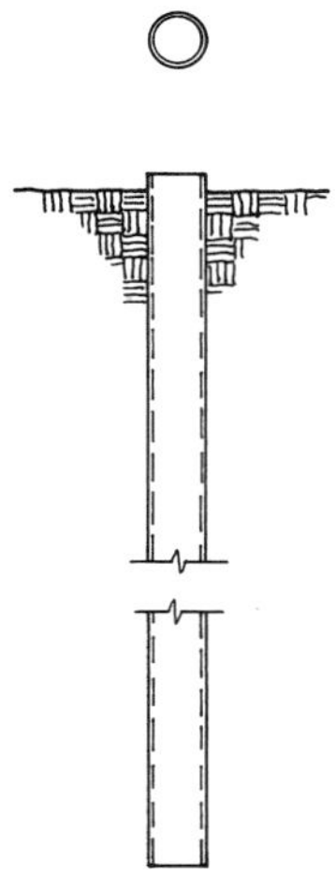

The Cast-in-Place Concrete Pile System includes: a defined number of 4,000 p.s.i. concrete piles with thin-wall, straight-sided, steel shells that have a standard steel plate driving point. An allowance for cutoffs is included.

The Expanded System Listing shows costs per cluster of piles. Clusters range from one pile to twenty piles. Loads vary from 50 Kips to 1,600 Kips. Both end-bearing and friction-type piles are shown.

Please see the reference section for cost of mobilization of the pile driving equipment and other design and cost information.

System Components

System Components	QUANTITY	UNIT	COST EACH MAT.	COST EACH INST.	COST EACH TOTAL
SYSTEM A1020 110 2220					
CIP SHELL CONCRETE PILE, 25′ LONG, 50K LOAD, END BEARING, 1 PILE					
7 Ga. shell, 12″ diam.	27.000	V.L.F.	918	305.91	1,223.91
Steel pipe pile standard point, 12″ or 14″ diameter pile	1.000	Ea.	118.50	89	207.50
Pile cutoff, conc. pile with thin steel shell	1.000	Ea.		16.15	16.15
TOTAL			1,036.50	411.06	1,447.56

A1020 110	C.I.P. Concrete Piles	COST EACH MAT.	COST EACH INST.	COST EACH TOTAL
2220	CIP shell concrete pile, 25′ long, 50K load, end bearing, 1 pile	1,025	410	1,435
2240	100K load, end bearing, 2 pile cluster	2,075	825	2,900
2260	200K load, end bearing, 4 pile cluster RA1020-100	4,150	1,650	5,800
2280	400K load, end bearing, 7 pile cluster	7,250	2,875	10,125
2300	10 pile cluster	10,400	4,125	14,525
2320	800K load, end bearing, 13 pile cluster	21,900	6,775	28,675
2340	17 pile cluster	28,600	8,850	37,450
2360	1200K load, end bearing, 14 pile cluster	23,600	7,275	30,875
2380	19 pile cluster	32,000	9,875	41,875
2400	1600K load, end bearing, 19 pile cluster	32,000	9,875	41,875
2420	50′ long, 50K load, end bearing, 1 pile	1,925	705	2,630
2440	Friction type, 2 pile cluster	3,675	1,400	5,075
2460	3 pile cluster	5,525	2,125	7,650
2480	100K load, end bearing, 2 pile cluster	3,850	1,400	5,250
2500	Friction type, 4 pile cluster	7,375	2,825	10,200
2520	6 pile cluster	11,000	4,250	15,250
2540	200K load, end bearing, 4 pile cluster	7,675	2,825	10,500
2560	Friction type, 8 pile cluster	14,700	5,650	20,350
2580	10 pile cluster	18,400	7,050	25,450
2600	400K load, end bearing, 7 pile cluster	13,400	4,925	18,325
2620	Friction type, 16 pile cluster	29,500	11,300	40,800
2640	19 pile cluster	35,000	13,500	48,500
2660	800K load, end bearing, 14 pile cluster	43,400	12,500	55,900
2680	20 pile cluster	62,000	17,800	79,800
2700	1200K load, end bearing, 15 pile cluster	46,500	13,300	59,800
2720	1600K load, end bearing, 20 pile cluster	62,000	17,800	79,800
3740	75′ long, 50K load, end bearing, 1 pile	2,950	1,175	4,125
3760	Friction type, 2 pile cluster	5,675	2,350	8,025

A1020 Special Foundations

A1020 110	C.I.P. Concrete Piles	COST EACH		
		MAT.	INST.	TOTAL
3780	3 pile cluster	8,500	3,525	12,025
3800	100K load, end bearing, 2 pile cluster	5,900	2,350	8,250
3820	Friction type, 3 pile cluster	8,500	3,525	12,025
3840	5 pile cluster	14,200	5,850	20,050
3860	200K load, end bearing, 4 pile cluster	11,800	4,700	16,500
3880	6 pile cluster	17,700	7,025	24,725
3900	Friction type, 6 pile cluster	17,000	7,025	24,025
3910	7 pile cluster	19,800	8,175	27,975
3920	400K load, end bearing, 7 pile cluster	20,700	8,175	28,875
3930	11 pile cluster	32,500	12,900	45,400
3940	Friction type, 12 pile cluster	34,000	14,100	48,100
3950	14 pile cluster	39,700	16,400	56,100
3960	800K load, end bearing, 15 pile cluster	70,500	21,500	92,000
3970	20 pile cluster	94,000	28,800	122,800
3980	1200K load, end bearing, 17 pile cluster	80,000	24,500	104,500
3990				

A1020 Special Foundations

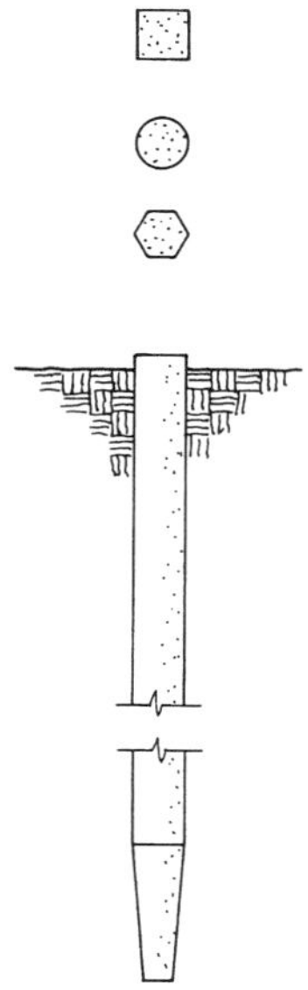

The Precast Concrete Pile System includes: pre-stressed concrete piles; standard steel driving point; and an allowance for cutoffs.

The Expanded System Listing shows costs per cluster of piles. Clusters range from one pile to twenty piles. Loads vary from 50 Kips to 1,600 Kips. Both end-bearing and friction type piles are listed.

Please see the reference section for cost of mobilization of the pile driving equipment and other design and cost information.

System Components	QUANTITY	UNIT	COST EACH MAT.	COST EACH INST.	COST EACH TOTAL
SYSTEM A1020 120 2220					
PRECAST CONCRETE PILE, 50′ LONG, 50K LOAD, END BEARING, 1 PILE					
Precast, prestressed conc. piles, 10″ square, no mobil.	53.000	V.L.F.	755.25	515.69	1,270.94
Steel pipe pile standard point, 8″ to 10″ diameter	1.000	Ea.	55.50	78	133.50
Piling special costs cutoffs concrete piles plain	1.000	Ea.		112	112
TOTAL			810.75	705.69	1,516.44

A1020 120	Precast Concrete Piles	COST EACH MAT.	COST EACH INST.	COST EACH TOTAL
2220	Precast conc pile, 50′ long, 50K load, end bearing, 1 pile	810	705	1,515
2240	Friction type, 2 pile cluster	2,400	1,475	3,875
2260	4 pile cluster	4,825	2,925	7,750
2280	100K load, end bearing, 2 pile cluster RA1020-100	1,625	1,400	3,025
2300	Friction type, 2 pile cluster	2,400	1,475	3,875
2320	4 pile cluster	4,825	2,925	7,750
2340	7 pile cluster	8,425	5,125	13,550
2360	200K load, end bearing, 3 pile cluster	2,425	2,125	4,550
2380	4 pile cluster	3,250	2,825	6,075
2400	Friction type, 8 pile cluster	9,650	5,850	15,500
2420	9 pile cluster	10,800	6,600	17,400
2440	14 pile cluster	16,900	10,300	27,200
2460	400K load, end bearing, 6 pile cluster	4,875	4,225	9,100
2480	8 pile cluster	6,475	5,625	12,100
2500	Friction type, 14 pile cluster	16,900	10,300	27,200
2520	16 pile cluster	19,300	11,700	31,000
2540	18 pile cluster	21,700	13,100	34,800
2560	800K load, end bearing, 12 pile cluster	10,500	9,175	19,675
2580	16 pile cluster	13,000	11,300	24,300
2600	1200K load, end bearing, 19 pile cluster	46,500	16,400	62,900
2620	20 pile cluster	48,900	17,300	66,200
2640	1600K load, end bearing, 19 pile cluster	46,500	16,400	62,900
4660	100′ long, 50K load, end bearing, 1 pile	1,550	1,200	2,750
4680	Friction type, 1 pile	2,275	1,250	3,525

A10 Foundations

A1020 Special Foundations

A1020 120	Precast Concrete Piles	COST EACH		
		MAT.	INST.	TOTAL
4700	2 pile cluster	4,550	2,500	7,050
4720	100K load, end bearing, 2 pile cluster	3,100	2,425	5,525
4740	Friction type, 2 pile cluster	4,550	2,500	7,050
4760	3 pile cluster	6,825	3,750	10,575
4780	4 pile cluster	9,075	5,000	14,075
4800	200K load, end bearing, 3 pile cluster	4,650	3,650	8,300
4820	4 pile cluster	6,200	4,850	11,050
4840	Friction type, 3 pile cluster	6,825	3,750	10,575
4860	5 pile cluster	11,400	6,250	17,650
4880	400K load, end bearing, 6 pile cluster	9,300	7,275	16,575
4900	8 pile cluster	12,400	9,700	22,100
4910	Friction type, 8 pile cluster	18,200	10,000	28,200
4920	10 pile cluster	22,700	12,500	35,200
4930	800K load, end bearing, 13 pile cluster	20,200	15,800	36,000
4940	16 pile cluster	24,800	19,400	44,200
4950	1200K load, end bearing, 19 pile cluster	89,000	28,400	117,400
4960	20 pile cluster	93,500	29,900	123,400
4970	1600K load, end bearing, 19 pile cluster	89,000	28,400	117,400

A1020 Special Foundations

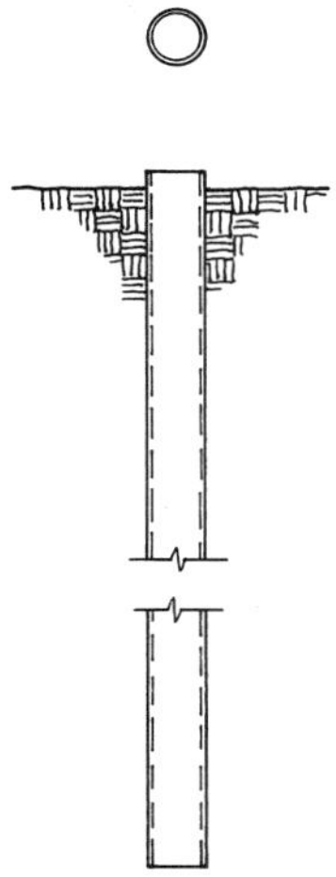

The Steel Pipe Pile System includes: steel pipe sections filled with 4,000 p.s.i. concrete; a standard steel driving point; splices when required and an allowance for cutoffs.

The Expanded System Listing shows costs per cluster of piles. Clusters range from one pile to twenty piles. Loads vary from 50 Kips to 1,600 Kips. Both end-bearing and friction-type piles are shown.

Please see the reference section for cost of mobilization of the pile driving equipment and other design and cost information.

System Components	QUANTITY	UNIT	COST EACH		
			MAT.	INST.	TOTAL
SYSTEM A1020 130 2220					
CONC. FILL STEEL PIPE PILE, 50′ LONG, 50K LOAD, END BEARING, 1 PILE					
Piles, steel, pipe, conc. filled, 12″ diameter	53.000	V.L.F.	1,855	867.61	2,722.61
Steel pipe pile, standard point, for 12″ or 14″ diameter pipe	1.000	Ea.	237	178	415
Pile cut off, concrete pile, thin steel shell	1.000	Ea.		16.15	16.15
TOTAL			2,092	1,061.76	3,153.76

A1020 130	Steel Pipe Piles	COST EACH		
		MAT.	INST.	TOTAL
2220	Conc. fill steel pipe pile, 50′ long, 50K load, end bearing, 1 pile	2,100	1,050	3,150
2240	Friction type, 2 pile cluster	4,175	2,125	6,300
2250	100K load, end bearing, 2 pile cluster (RA1020 -100)	4,175	2,125	6,300
2260	3 pile cluster	6,275	3,200	9,475
2300	Friction type, 4 pile cluster	8,375	4,250	12,625
2320	5 pile cluster	10,500	5,300	15,800
2340	10 pile cluster	20,900	10,600	31,500
2360	200K load, end bearing, 3 pile cluster	6,275	3,200	9,475
2380	4 pile cluster	8,375	4,250	12,625
2400	Friction type, 4 pile cluster	8,375	4,250	12,625
2420	8 pile cluster	16,700	8,500	25,200
2440	9 pile cluster	18,800	9,550	28,350
2460	400K load, end bearing, 6 pile cluster	12,600	6,350	18,950
2480	7 pile cluster	14,600	7,425	22,025
2500	Friction type, 9 pile cluster	18,800	9,550	28,350
2520	16 pile cluster	33,500	17,000	50,500
2540	19 pile cluster	39,700	20,200	59,900
2560	800K load, end bearing, 11 pile cluster	23,000	11,700	34,700
2580	14 pile cluster	29,300	14,900	44,200
2600	15 pile cluster	31,400	15,900	47,300
2620	Friction type, 17 pile cluster	35,600	18,100	53,700
2640	1200K load, end bearing, 16 pile cluster	33,500	17,000	50,500
2660	20 pile cluster	41,800	21,200	63,000
2680	1600K load, end bearing, 17 pile cluster	35,600	18,100	53,700
3700	100′ long, 50K load, end bearing, 1 pile	4,050	2,075	6,125
3720	Friction type, 1 pile	4,050	2,075	6,125

A1020 Special Foundations

A1020 130	Steel Pipe Piles	COST EACH		
		MAT.	INST.	TOTAL
3740	2 pile cluster	8,125	4,175	12,300
3760	100K load, end bearing, 2 pile cluster	8,125	4,175	12,300
3780	Friction type, 2 pile cluster	8,125	4,175	12,300
3800	3 pile cluster	12,200	6,250	18,450
3820	200K load, end bearing, 3 pile cluster	12,200	6,250	18,450
3840	4 pile cluster	16,200	8,325	24,525
3860	Friction type, 3 pile cluster	12,200	6,250	18,450
3880	4 pile cluster	16,200	8,325	24,525
3900	400K load, end bearing, 6 pile cluster	24,400	12,500	36,900
3910	7 pile cluster	28,400	14,600	43,000
3920	Friction type, 5 pile cluster	20,300	10,400	30,700
3930	8 pile cluster	32,500	16,700	49,200
3940	800K load, end bearing, 11 pile cluster	44,700	22,900	67,600
3950	14 pile cluster	57,000	29,200	86,200
3960	15 pile cluster	61,000	31,200	92,200
3970	1200K load, end bearing, 16 pile cluster	65,000	33,300	98,300
3980	20 pile cluster	81,000	41,700	122,700
3990	1600K load, end bearing, 17 pile cluster	69,000	35,400	104,400

A1020 Special Foundations

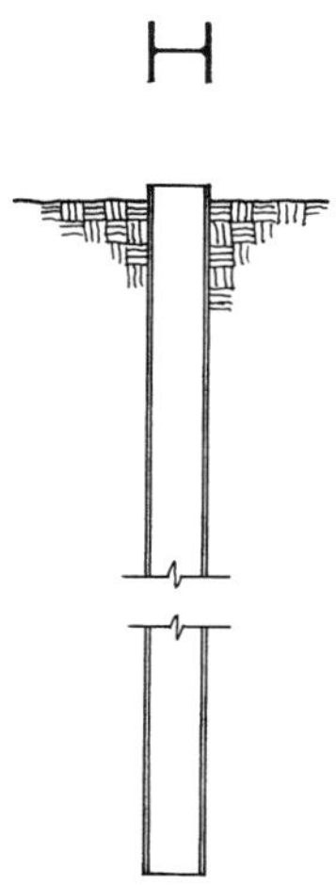

A Steel "H" Pile System includes: steel H sections; heavy duty driving point; splices where applicable and allowance for cutoffs.

The Expanded System Listing shows costs per cluster of piles. Clusters range from one pile to eighteen piles. Loads vary from 50 Kips to 2,000 Kips. All loads for Steel H Pile systems are given in terms of end bearing capacity.

Steel sections range from 10″ x 10″ to 14″ x 14″ in the Expanded System Listing. The 14″ x 14″ steel section is used for all H piles used in applications requiring a working load over 800 Kips.

Please see the reference section for cost of mobilization of the pile driving equipment and other design and cost information.

System Components	QUANTITY	UNIT	COST EACH MAT.	COST EACH INST.	COST EACH TOTAL
SYSTEM A1020 140 2220					
STEEL H PILES, 50′ LONG, 100K LOAD, END BEARING, 1 PILE					
Steel H piles 10″ x 10″, 42 #/L.F.	53.000	V.L.F.	1,139.50	589.89	1,729.39
Heavy duty point, 10″	1.000	Ea.	198	180	378
Pile cut off, steel pipe or H piles	1.000	Ea.		32.50	32.50
TOTAL			1,337.50	802.39	2,139.89

A1020 140	Steel H Piles	COST EACH MAT.	COST EACH INST.	COST EACH TOTAL
2220	Steel H piles, 50′ long, 100K load, end bearing, 1 pile	1,350	805	2,155
2260	2 pile cluster	2,675	1,600	4,275
2280	200K load, end bearing, 2 pile cluster (RA1020 -100)	2,675	1,600	4,275
2300	3 pile cluster	4,025	2,425	6,450
2320	400K load, end bearing, 3 pile cluster	4,025	2,425	6,450
2340	4 pile cluster	5,350	3,200	8,550
2360	6 pile cluster	8,025	4,800	12,825
2380	800K load, end bearing, 5 pile cluster	6,700	4,025	10,725
2400	7 pile cluster	9,375	5,625	15,000
2420	12 pile cluster	16,100	9,625	25,725
2440	1200K load, end bearing, 8 pile cluster	10,700	6,425	17,125
2460	11 pile cluster	14,700	8,825	23,525
2480	17 pile cluster	22,700	13,700	36,400
2500	1600K load, end bearing, 10 pile cluster	16,500	8,400	24,900
2520	14 pile cluster	23,100	11,800	34,900
2540	2000K load, end bearing, 12 pile cluster	19,800	10,100	29,900
2560	18 pile cluster	29,700	15,100	44,800
3580	100′ long, 50K load, end bearing, 1 pile	4,325	1,825	6,150
3600	100K load, end bearing, 1 pile	4,325	1,825	6,150
3620	2 pile cluster	8,650	3,675	12,325
3640	200K load, end bearing, 2 pile cluster	8,650	3,675	12,325
3660	3 pile cluster	13,000	5,500	18,500
3680	400K load, end bearing, 3 pile cluster	13,000	5,500	18,500
3700	4 pile cluster	17,300	7,350	24,650
3720	6 pile cluster	26,000	11,000	37,000
3740	800K load, end bearing, 5 pile cluster	21,600	9,175	30,775

A1020 Special Foundations

A1020 140	Steel H Piles	COST EACH		
		MAT.	INST.	TOTAL
3760	7 pile cluster	30,300	12,900	43,200
3780	12 pile cluster	52,000	22,000	74,000
3800	1200K load, end bearing, 8 pile cluster	34,600	14,700	49,300
3820	11 pile cluster	47,600	20,200	67,800
3840	17 pile cluster	73,500	31,200	104,700
3860	1600K load, end bearing, 10 pile cluster	43,300	18,300	61,600
3880	14 pile cluster	60,500	25,700	86,200
3900	2000K load, end bearing, 12 pile cluster	52,000	22,000	74,000
3920	18 pile cluster	78,000	33,000	111,000

A1020 Special Foundations

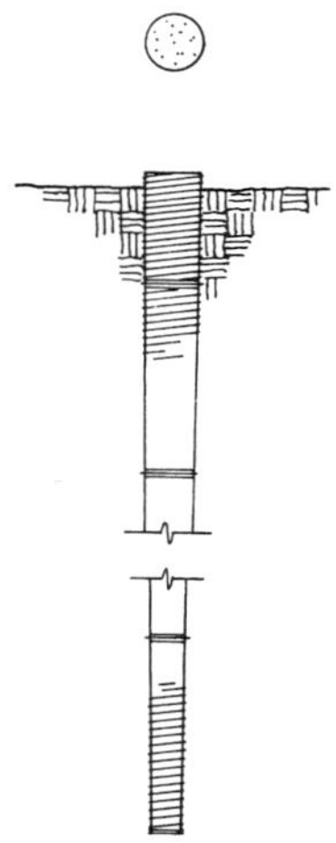

The Step Tapered Steel Pile System includes: step tapered piles filled with 4,000 p.s.i. concrete. The cost for splices and pile cutoffs is included.

The Expanded System Listing shows costs per cluster of piles. Clusters range from one pile to twenty-four piles. Both end bearing piles and friction piles are listed. Loads vary from 50 Kips to 1 600 Kips.

Please see the reference section for cost of mobilization of the pile driving equipment and other design and cost information.

System Components	QUANTITY	UNIT	COST EACH MAT.	COST EACH INST.	COST EACH TOTAL
SYSTEM A1020 150 1000					
STEEL PILE, STEP TAPERED, 50′ LONG, 50K LOAD, END BEARING, 1 PILE					
Steel shell step tapered conc. filled piles, 8″ tip 60 ton capacity to 60′	53.000	V.L.F.	2,093.50	487.07	2,580.57
Pile cutoff, steel pipe or H piles	1.000	Ea.		32.50	32.50
TOTAL			2,093.50	519.57	2,613.07

A1020 150	Step-Tapered Steel Piles	COST EACH MAT.	COST EACH INST.	COST EACH TOTAL
1000	Steel pile, step tapered, 50′ long, 50K load, end bearing, 1 pile	2,100	520	2,620
1200	Friction type, 3 pile cluster	6,275	1,550	7,825
1400	100K load, end bearing, 2 pile cluster RA1020 -100	4,175	1,050	5,225
1600	Friction type, 4 pile cluster	8,375	2,100	10,475
1800	200K load, end bearing, 4 pile cluster	8,375	2,100	10,475
2000	Friction type, 6 pile cluster	12,600	3,100	15,700
2200	400K load, end bearing, 7 pile cluster	14,700	3,650	18,350
2400	Friction type, 10 pile cluster	20,900	5,200	26,100
2600	800K load, end bearing, 14 pile cluster	29,300	7,275	36,575
2800	Friction type, 18 pile cluster	37,700	9,350	47,050
3000	1200K load, end bearing, 16 pile cluster	17,400	8,875	26,275
3200	Friction type, 21 pile cluster	22,800	11,700	34,500
3400	1600K load, end bearing, 18 pile cluster	19,600	10,000	29,600
3600	Friction type, 24 pile cluster	26,100	13,300	39,400
5000	100′ long, 50K load, end bearing, 1 pile	6,275	1,025	7,300
5200	Friction type, 2 pile cluster	12,600	2,075	14,675
5400	100K load, end bearing, 2 pile cluster	12,600	2,075	14,675
5600	Friction type, 3 pile cluster	18,800	3,100	21,900
5800	200K load, end bearing, 4 pile cluster	25,100	4,125	29,225
6000	Friction type, 5 pile cluster	31,400	5,175	36,575
6200	400K load, end bearing, 7 pile cluster	44,000	7,225	51,225
6400	Friction type, 8 pile cluster	50,500	8,275	58,775
6600	800K load, end bearing, 15 pile cluster	94,000	15,600	109,600
6800	Friction type, 16 pile cluster	100,500	16,500	117,000
7000	1200K load, end bearing, 17 pile cluster	57,000	27,000	84,000
7200	Friction type, 19 pile cluster	43,100	20,500	63,600
7400	1600K load, end bearing, 20 pile cluster	45,300	21,600	66,900
7600	Friction type, 22 pile cluster	49,900	23,800	73,700

A1020 Special Foundations

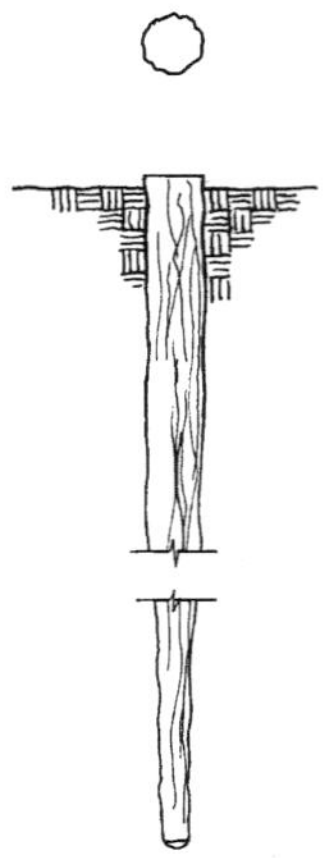

The Treated Wood Pile System includes: creosoted wood piles; a standard steel driving point; and an allowance for cutoffs.

The Expanded System Listing shows costs per cluster of piles. Clusters range from two piles to twenty piles. Loads vary from 50 Kips to 400 Kips. Both end-bearing and friction type piles are listed.

Please see the reference section for cost of mobilization of the pile driving equipment and other design and cost information.

System Components	QUANTITY	UNIT	COST EACH		
			MAT.	INST.	TOTAL
SYSTEM A1020 160 2220					
WOOD PILES, 25′ LONG, 50K LOAD, END BEARING, 3 PILE CLUSTER					
Wood piles, treated, 12″ butt, 8″ tip, up to 30′ long	81.000	V.L.F.	1,166.40	878.85	2,045.25
Point for driving wood piles	3.000	Ea.	99	91.50	190.50
Pile cutoff, wood piles	3.000	Ea.		48.45	48.45
TOTAL			1,265.40	1,018.80	2,284.20

A1020 160	Treated Wood Piles	COST EACH		
		MAT.	INST.	TOTAL
2220	Wood piles, 25′ long, 50K load, end bearing, 3 pile cluster	1,275	1,025	2,300
2240	Friction type, 3 pile cluster	1,175	930	2,105
2260	5 pile cluster	1,950	1,550	3,500
2280	100K load, end bearing, 4 pile cluster	1,675	1,350	3,025
2300	5 pile cluster	2,100	1,700	3,800
2320	6 pile cluster	2,525	2,025	4,550
2340	Friction type, 5 pile cluster	1,950	1,550	3,500
2360	6 pile cluster	2,325	1,850	4,175
2380	10 pile cluster	3,900	3,075	6,975
2400	200K load, end bearing, 8 pile cluster	3,375	2,725	6,100
2420	10 pile cluster	4,225	3,375	7,600
2440	12 pile cluster	5,050	4,075	9,125
2460	Friction type, 10 pile cluster	3,900	3,075	6,975
2480	400K load, end bearing, 16 pile cluster	6,750	5,450	12,200
2500	20 pile cluster	8,425	6,800	15,225
4520	50′ long, 50K load, end bearing, 3 pile cluster	3,125	1,525	4,650
4540	4 pile cluster	4,150	2,000	6,150
4560	Friction type, 2 pile cluster	1,975	915	2,890
4580	3 pile cluster	2,975	1,375	4,350
4600	100K load, end bearing, 5 pile cluster	5,200	2,500	7,700
4620	8 pile cluster	8,325	4,025	12,350
4640	Friction type, 3 pile cluster	2,975	1,375	4,350
4660	5 pile cluster	4,950	2,275	7,225
4680	200K load, end bearing, 9 pile cluster	9,350	4,525	13,875
4700	10 pile cluster	10,400	5,050	15,450
4720	15 pile cluster	15,600	7,550	23,150
4740	Friction type, 5 pile cluster	4,950	2,275	7,225
4760	6 pile cluster	5,950	2,750	8,700

RA1020 -100

A1020 Special Foundations

A1020 160	Treated Wood Piles	COST EACH		
		MAT.	INST.	TOTAL
4780	10 pile cluster	9,900	4,575	14,475
4800	400K load, end bearing, 18 pile cluster	18,700	9,075	27,775
4820	20 pile cluster	20,800	10,100	30,900
4840	Friction type, 9 pile cluster	8,925	4,125	13,050
4860	10 pile cluster	9,900	4,575	14,475
9000	Add for boot for driving tip, each pile	33	22.50	55.50

A1020 Special Foundations

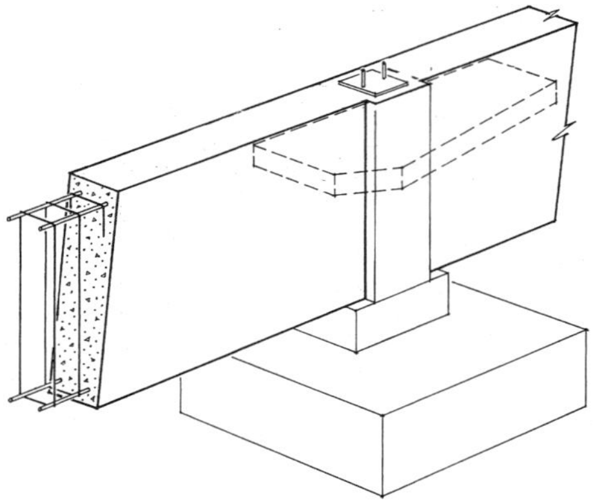

The Grade Beam System includes: excavation with a truck mounted backhoe; hand trim; backfill; forms (four uses); reinforcing steel; and 3,000 p.s.i. concrete placed from chute.

Superimposed loads vary in the listing from 1 Kip per linear foot (KLF) and above. In the Expanded System Listing, the span of the beams varies from 15′ to 40′. Depth varies from 28″ to 52″. Width varies from 12″ and wider.

Please see the reference section for further design and cost information.

System Components	QUANTITY	UNIT	COST PER L.F. MAT.	INST.	TOTAL
SYSTEM A1020 210 2220					
GRADE BEAM, 15′ SPAN, 28″ DEEP, 12″ WIDE, 8 KLF LOAD					
Excavation, trench, hydraulic backhoe, 3/8 CY bucket	.260	C.Y.		2.57	2.57
Trim sides and bottom of trench, regular soil	2.000	S.F.		1.96	1.96
Backfill, by hand, compaction in 6″ layers, using vibrating plate	.170	C.Y.		1.43	1.43
Forms in place, grade beam, 4 uses	4.700	SFCA	5.17	26.79	31.96
Reinforcing in place, beams & girders, #8 to #14	.019	Ton	19.95	18.53	38.48
Concrete ready mix, regular weight, 3000 psi	.090	C.Y.	10.62		10.62
Place and vibrate conc. for grade beam, direct chute	.090	C.Y.		1.78	1.78
TOTAL			35.74	53.06	88.80

A1020 210	Grade Beams	COST PER L.F. MAT.	INST.	TOTAL
2220	Grade beam, 15′ span, 28″ deep, 12″ wide, 8 KLF load	35.50	53	88.50
2240	14″ wide, 12 KLF load	37	53.50	90.50
2260	40″ deep, 12″ wide, 16 KLF load (RA1020 -230)	37.50	66.50	104
2280	20 KLF load	42.50	71	113.50
2300	52″ deep, 12″ wide, 30 KLF load	49.50	89.50	139
2320	40 KLF load	60	99	159
2340	50 KLF load	70.50	108	178.50
3360	20′ span, 28″ deep, 12″ wide, 2 KLF load	23	41.50	64.50
3380	16″ wide, 4 KLF load	30.50	46	76.50
3400	40″ deep, 12″ wide, 8 KLF load	37.50	66.50	104
3420	12 KLF load	46.50	74	120.50
3440	14″ wide, 16 KLF load	56.50	82.50	139
3460	52″ deep, 12″ wide, 20 KLF load	61	100	161
3480	14″ wide, 30 KLF load	83.50	119	202.50
3500	20″ wide, 40 KLF load	98	125	223
3520	24″ wide, 50 KLF load	121	142	263
4540	30′ span, 28″ deep, 12″ wide, 1 KLF load	24	42.50	66.50
4560	14″ wide, 2 KLF load	38	58	96
4580	40″ deep, 12″ wide, 4 KLF load	45	70	115
4600	18″ wide, 8 KLF load	63.50	85	148.50
4620	52″ deep, 14″ wide, 12 KLF load	80.50	116	196.50
4640	20″ wide, 16 KLF load	98	126	224
4660	24″ wide, 20 KLF load	121	143	264
4680	36″ wide, 30 KLF load	173	179	352
4700	48″ wide, 40 KLF load	228	217	445
5720	40′ span, 40″ deep, 12″ wide, 1 KLF load	33	62.50	95.50

A1020 Special Foundations

A1020 210	Grade Beams	COST PER L.F. MAT.	COST PER L.F. INST.	COST PER L.F. TOTAL
5740	2 KLF load	40.50	69.50	110
5760	52″ deep, 12″ wide, 4 KLF load	60	99	159
5780	20″ wide, 8 KLF load	95	123	218
5800	28″ wide, 12 KLF load	130	148	278
5820	38″ wide, 16 KLF load	178	180	358
5840	46″ wide, 20 KLF load	232	222	454

A1020 Special Foundations

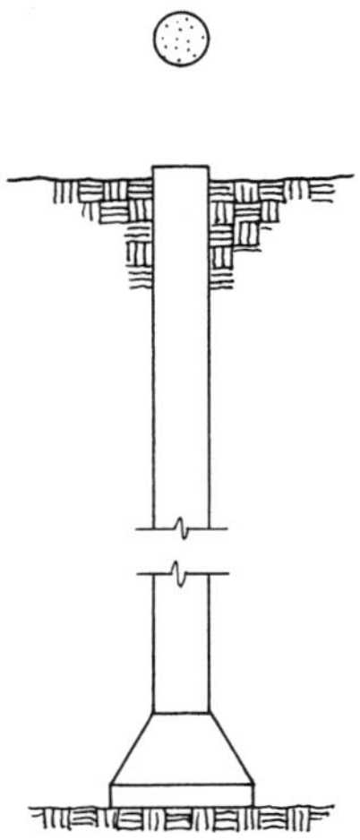

Caisson Systems are listed for three applications: stable ground, wet ground and soft rock. Concrete used is 3,000 p.s.i. placed from chute. Included are a bell at the bottom of the caisson shaft (if applicable) along with required excavation and disposal of excess excavated material up to two miles from job site.

The Expanded System lists cost per caisson. End-bearing loads vary from 200 Kips to 3,200 Kips. The dimensions of the caissons range from 2′ x 50′ to 7′ x 200′.

Please see the reference section for further design and cost information.

System Components	QUANTITY	UNIT	COST EACH		
			MAT.	INST.	TOTAL
SYSTEM A1020 310 2200					
CAISSON, STABLE GROUND, 3000 PSI CONC., 10 KSF BRNG, 200K LOAD, 2′X50′					
Caissons, drilled, to 50′, 24″ shaft diameter, .116 C.Y./L.F.	50.000	V.L.F.	837.50	1,550	2,387.50
Reinforcing in place, columns, #3 to #7	.060	Ton	63	105	168
4′ bell diameter, 24″ shaft, 0.444 C.Y.	1.000	Ea.	52.50	294	346.50
Load & haul excess excavation, 2 miles	6.240	C.Y.		44.92	44.92
TOTAL			953	1,993.92	2,946.92

A1020 310	Caissons	COST EACH		
		MAT.	INST.	TOTAL
2200	Caisson, stable ground, 3000 PSI conc, 10 KSF brng, 200K load, 2′-0″x50′-0	955	1,975	2,930
2400	400K load, 2′-6″x50′-0″	1,550	3,200	4,750
2600	800K load, 3′-0″x100′-0″ RA1020 -200	4,300	7,550	11,850
2800	1200K load, 4′-0″x100′-0″	7,450	9,575	17,025
3000	1600K load, 5′-0″x150′-0″	16,800	14,800	31,600
3200	2400K load, 6′-0″x150′-0″	24,600	19,400	44,000
3400	3200K load, 7′-0″x200′-0″	44,000	28,200	72,200
5000	Wet ground, 3000 PSI conc., 10 KSF brng, 200K load, 2′-0″x50′-0″	825	2,825	3,650
5200	400K load, 2′-6″x50′-0″	1,375	4,850	6,225
5400	800K load, 3′-0″x100′-0″	3,750	13,000	16,750
5600	1200K load, 4′-0″x100′-0″	6,450	19,500	25,950
5800	1600K load, 5′-0″x150′-0″	14,400	42,000	56,400
6000	2400K load, 6′-0″x150′-0″	21,100	52,000	73,100
6200	3200K load, 7′-0″x200′-0″	37,800	83,000	120,800
7800	Soft rock, 3000 PSI conc., 10 KSF brng, 200K load, 2′-0″x50′-0″	825	15,400	16,225
8000	400K load, 2′-6″x50′-0″	1,375	25,000	26,375
8200	800K load, 3′-0″x100′-0″	3,750	65,500	69,250
8400	1200K load, 4′-0″x100′-0″	6,450	96,500	102,950
8600	1600K load, 5′-0″x150′-0″	14,400	197,500	211,900
8800	2400K load, 6′-0″x150′-0″	21,100	236,000	257,100
9000	3200K load, 7′-0″x200′-0″	37,800	374,500	412,300

A1020 Special Foundations

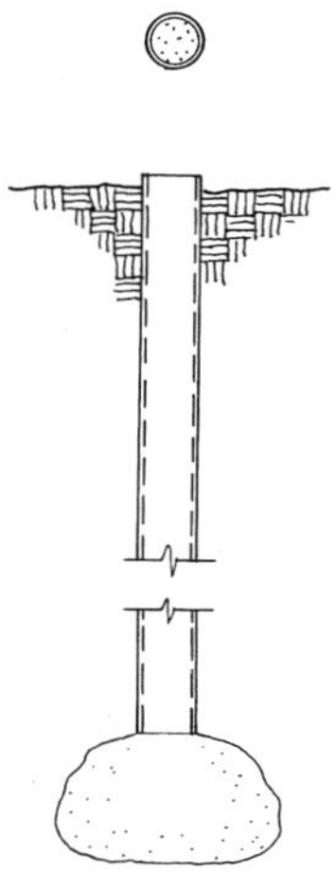

Pressure Injected Piles are usually uncased up to 25′ and cased over 25′ depending on soil conditions.

These costs include excavation and hauling of excess materials; steel casing over 25′; reinforcement; 3,000 p.s.i. concrete; plus mobilization and demobilization of equipment for a distance of up to fifty miles to and from the job site.

The Expanded System lists cost per cluster of piles. Clusters range from one pile to eight piles. End-bearing loads range from 50 Kips to 1,600 Kips.

Please see the reference section for further design and cost information.

System Components	QUANTITY	UNIT	COST EACH MAT.	COST EACH INST.	COST EACH TOTAL
SYSTEM A1020 710 4200					
PRESSURE INJECTED FOOTING, END BEARING, 50′ LONG, 50K LOAD, 1 PILE					
Pressure injected footings, cased, 30-60 ton cap., 12″ diameter	50.000	V.L.F.	907.50	1,497.50	2,405
Pile cutoff, concrete pile with thin steel shell	1.000	Ea.		16.15	16.15
TOTAL			907.50	1,513.65	2,421.15

A1020 710	Pressure Injected Footings	COST EACH MAT.	COST EACH INST.	COST EACH TOTAL
2200	Pressure injected footing, end bearing, 25′ long, 50K load, 1 pile	420	1,075	1,495
2400	100K load, 1 pile	420	1,075	1,495
2600	2 pile cluster RA1020 -100	845	2,125	2,970
2800	200K load, 2 pile cluster	845	2,125	2,970
3200	400K load, 4 pile cluster	1,675	4,275	5,950
3400	7 pile cluster	2,950	7,450	10,400
3800	1200K load, 6 pile cluster	3,525	8,025	11,550
4000	1600K load, 7 pile cluster	4,125	9,375	13,500
4200	50′ long, 50K load, 1 pile	910	1,525	2,435
4400	100K load, 1 pile	1,575	1,525	3,100
4600	2 pile cluster	3,150	3,025	6,175
4800	200K load, 2 pile cluster	3,150	3,025	6,175
5000	4 pile cluster	6,300	6,075	12,375
5200	400K load, 4 pile cluster	6,300	6,075	12,375
5400	8 pile cluster	12,600	12,100	24,700
5600	800K load, 7 pile cluster	11,000	10,600	21,600
5800	1200K load, 6 pile cluster	10,100	9,075	19,175
6000	1600K load, 7 pile cluster	11,700	10,600	22,300

A1030 Slab on Grade

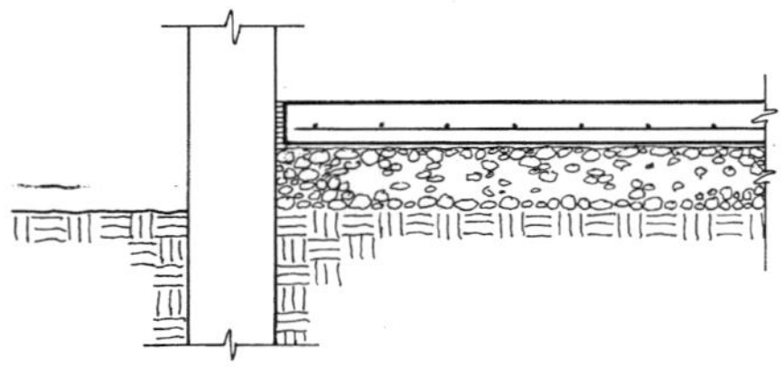

There are four types of Slab on Grade Systems listed: Non-industrial, Light industrial, Industrial and Heavy industrial. Each type is listed two ways: reinforced and non-reinforced. A Slab on Grade system includes three passes with a grader; 6″ of compacted gravel fill; polyethylene vapor barrier; 3500 p.s.i. concrete placed by chute; bituminous fibre expansion joint; all necessary edge forms (4 uses); steel trowel finish; and sprayed-on membrane curing compound.

The Expanded System Listing shows costs on a per square foot basis. Thicknesses of the slabs range from 4″ and above. Non-industrial applications are for foot traffic only with negligible abrasion. Light industrial applications are for pneumatic wheels and light abrasion. Industrial applications are for solid rubber wheels and moderate abrasion. Heavy industrial applications are for steel wheels and severe abrasion. All slabs are either shown unreinforced or reinforced with welded wire fabric.

System Components	QUANTITY	UNIT	COST PER S.F. MAT.	INST.	TOTAL
SYSTEM A1030 120 2220					
SLAB ON GRADE, 4″ THICK, NON INDUSTRIAL, NON REINFORCED					
Fine grade, 3 passes with grader and roller	.110	S.Y.		.52	.52
Gravel under floor slab, 6″ deep, compacted	1.000	S.F.	.48	.31	.79
Polyethylene vapor barrier, standard, 6 mil	1.000	S.F.	.04	.16	.20
Concrete ready mix, regular weight, 3500 psi	.012	C.Y.	1.45		1.45
Place and vibrate concrete for slab on grade, 4″ thick, direct chute	.012	C.Y.		.33	.33
Expansion joint, premolded bituminous fiber, 1/2″ x 6″	.100	L.F.	.10	.35	.45
Edge forms in place for slab on grade to 6″ high, 4 uses	.030	L.F.	.01	.11	.12
Cure with sprayed membrane curing compound	1.000	S.F.	.14	.10	.24
Finishing floor, monolithic steel trowel	1.000	S.F.		.93	.93
TOTAL			2.22	2.81	5.03

A1030 120	Plain & Reinforced	COST PER S.F. MAT.	INST.	TOTAL
2220	Slab on grade, 4″ thick, non industrial, non reinforced	2.22	2.81	5.03
2240	Reinforced	2.37	3.19	5.56
2260	Light industrial, non reinforced RA1030 -200	2.92	3.45	6.37
2280	Reinforced	3.07	3.83	6.90
2300	Industrial, non reinforced	3.59	7.15	10.74
2320	Reinforced	3.74	7.55	11.29
3340	5″ thick, non industrial, non reinforced	2.59	2.89	5.48
3360	Reinforced	2.74	3.27	6.01
3380	Light industrial, non reinforced	3.29	3.53	6.82
3400	Reinforced	3.44	3.91	7.35
3420	Heavy industrial, non reinforced	4.71	8.55	13.26
3440	Reinforced	4.81	9	13.81
4460	6″ thick, non industrial, non reinforced	3.07	2.82	5.89
4480	Reinforced	3.30	3.33	6.63
4500	Light industrial, non reinforced	3.79	3.46	7.25
4520	Reinforced	4.16	4.15	8.31
4540	Heavy industrial, non reinforced	5.20	8.65	13.85
4560	Reinforced	5.45	9.15	14.60
5580	7″ thick, non industrial, non reinforced	3.45	2.92	6.37
5600	Reinforced	3.75	3.47	7.22
5620	Light industrial, non reinforced	4.18	3.56	7.74
5640	Reinforced	4.48	4.11	8.59
5660	Heavy industrial, non reinforced	5.60	8.55	14.15
5680	Reinforced	5.80	9.05	14.85

A1030 Slab on Grade

A1030 120	Plain & Reinforced	COST PER S.F.		
		MAT.	INST.	TOTAL
6700	8" thick, non industrial, non reinforced	3.82	2.97	6.79
6720	Reinforced	4.07	3.43	7.50
6740	Light industrial, non reinforced	4.55	3.61	8.16
6760	Reinforced	4.80	4.07	8.87
6780	Heavy industrial, non reinforced	6	8.65	14.65
6800	Reinforced	6.35	9.10	15.45

A2010 Basement Excavation

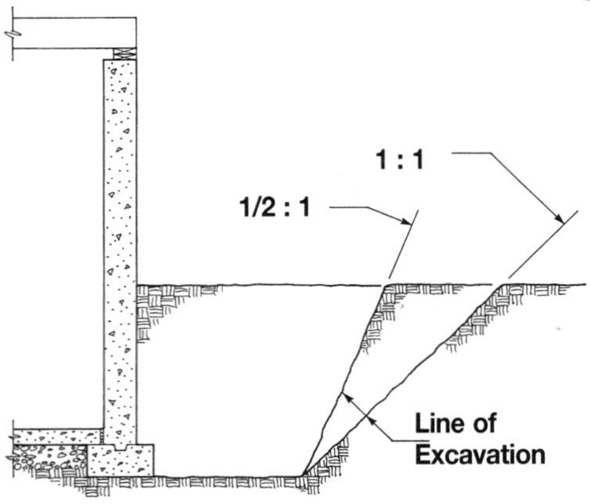

Pricing Assumptions: Two-thirds of excavation is by 2-1/2 C.Y. wheel mounted front end loader and one-third by 1-1/2 C.Y. hydraulic excavator.

Two-mile round trip haul by 12 C.Y. tandem trucks is included for excavation wasted and storage of suitable fill from excavated soil. For excavation in clay, all is wasted and the cost of suitable backfill with two-mile haul is included.

Sand and gravel assumes 15% swell and compaction; common earth assumes 25% swell and 15% compaction; clay assumes 40% swell and 15% compaction (non-clay).

In general, the following items are accounted for in the costs in the table below.

1. Excavation for building or other structure to depth and extent indicated.
2. Backfill compacted in place.
3. Haul of excavated waste.
4. Replacement of unsuitable material with bank run gravel.

Note: Additional excavation and fill beyond this line of general excavation for the building (as required for isolated spread footings, strip footings, etc.) are included in the cost of the appropriate component systems.

System Components	QUANTITY	UNIT	COST PER S.F. MAT.	COST PER S.F. INST.	COST PER S.F. TOTAL
SYSTEM A2010 110 2280					
EXCAVATE & FILL, 1000 S.F., 8′ DEEP, SAND, ON SITE STORAGE					
Excavating bulk shovel, 1.5 C.Y. bucket, 150 cy/hr	.262	C.Y.		.52	.52
Excavation, front end loader, 2-1/2 C.Y.	.523	C.Y.		1.32	1.32
Haul earth, 12 C.Y. dump truck	.341	L.C.Y.		2.43	2.43
Backfill, dozer bulk push 300′, including compaction	.562	C.Y.		2.28	2.28
TOTAL				6.55	6.55

A2010 110	Building Excavation & Backfill		COST PER S.F. MAT.	COST PER S.F. INST.	COST PER S.F. TOTAL
2220	Excav & fill, 1000 S.F., 4′ sand, gravel, or common earth, on site storage			1.08	1.08
2240	Off site storage			1.64	1.64
2260	Clay excavation, bank run gravel borrow for backfill		2.25	2.53	4.78
2280	8′ deep, sand, gravel, or common earth, on site storage	RG1010 -010		6.55	6.55
2300	Off site storage			14.55	14.55
2320	Clay excavation, bank run gravel borrow for backfill		9.15	11.80	20.95
2340	16′ deep, sand, gravel, or common earth, on site storage	RG1030 -400		17.85	17.85
2350	Off site storage			36	36
2360	Clay excavation, bank run gravel borrow for backfill		25.50	30.50	56
3380	4000 S.F., 4′ deep, sand, gravel, or common earth, on site storage			.57	.57
3400	Off site storage			1.14	1.14
3420	Clay excavation, bank run gravel borrow for backfill		1.15	1.27	2.42
3440	8′ deep, sand, gravel, or common earth, on site storage			4.63	4.63
3460	Off site storage			8.15	8.15
3480	Clay excavation, bank run gravel borrow for backfill		4.16	7.10	11.26
3500	16′ deep, sand, gravel, or common earth, on site storage			11.10	11.10
3520	Off site storage			22	22
3540	Clay, excavation, bank run gravel borrow for backfill		11.15	16.80	27.95
4560	10,000 S.F., 4′ deep, sand gravel, or common earth, on site storage			.32	.32
4580	Off site storage			.67	.67
4600	Clay excavation, bank run gravel borrow for backfill		.70	.78	1.48
4620	8′ deep, sand, gravel, or common earth, on site storage			4.03	4.03
4640	Off site storage			6.15	6.15
4660	Clay excavation, bank run gravel borrow for backfill		2.52	5.50	8.02
4680	16′ deep, sand, gravel, or common earth, on site storage			9.05	9.05
4700	Off site storage			15.55	15.55
4720	Clay excavation, bank run gravel borrow for backfill		6.70	12.60	19.30

A2010 Basement Excavation

A2010 110	Building Excavation & Backfill	COST PER S.F.		
		MAT.	INST.	TOTAL
5740	30,000 S.F., 4′ deep, sand, gravel, or common earth, on site storage		.18	.18
5760	Off site storage		.40	.40
5780	Clay excavation, bank run gravel borrow for backfill	.41	.46	.87
5800	8′ deep, sand & gravel, or common earth, on site storage		3.62	3.62
5820	Off site storage		4.78	4.78
5840	Clay excavation, bank run gravel borrow for backfill	1.42	4.47	5.89
5860	16′ deep, sand, gravel, or common earth, on site storage		7.80	7.80
5880	Off site storage		11.30	11.30
5900	Clay excavation, bank run gravel borrow for backfill	3.71	9.80	13.51
6910	100,000 S.F., 4′ deep, sand, gravel, or common earth, on site storage		.11	.11
6920	Off site storage		.23	.23
6930	Clay excavation, bank run gravel borrow for backfill	.20	.24	.44
6940	8′ deep, sand, gravel, or common earth, on site storage		3.38	3.38
6950	Off site storage		4	4
6960	Clay excavation, bank run gravel borrow for backfill	.77	3.84	4.61
6970	16′ deep, sand, gravel, or common earth, on site storage		7.10	7.10
6980	Off site storage		9	9
6990	Clay excavation, bank run gravel borrow for backfill	2	8.15	10.15

A2020 Basement Walls

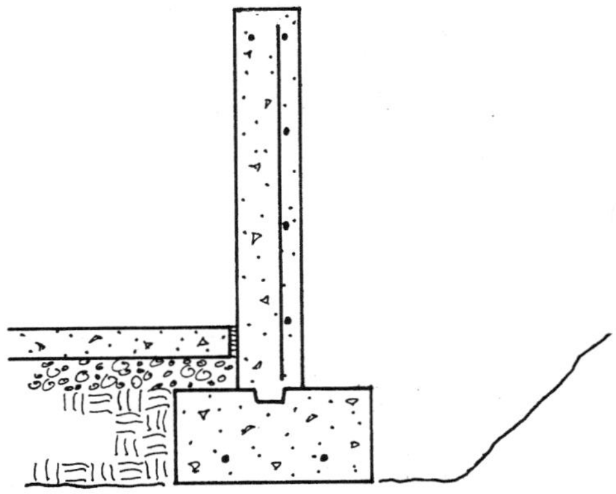

The Foundation Bearing Wall System includes: forms up to 16′ high (four uses); 3,000 p.s.i. concrete placed and vibrated; and form removal with breaking form ties and patching walls. The wall systems list walls from 6″ to 16″ thick and are designed with minimum reinforcement.

Excavation and backfill are not included.

Please see the reference section for further design and cost information.

System Components	QUANTITY	UNIT	COST PER L.F. MAT.	INST.	TOTAL
SYSTEM A2020 110 1500					
FOUNDATION WALL, CAST IN PLACE, DIRECT CHUTE, 4′ HIGH, 6″ THICK					
Formwork	8.000	SFCA	6.32	46	52.32
Reinforcing	3.300	Lb.	1.73	1.44	3.17
Unloading & sorting reinforcing	3.300	Lb.		.08	.08
Concrete, 3,000 psi	.074	C.Y.	8.73		8.73
Place concrete, direct chute	.074	C.Y.		2.43	2.43
Finish walls, break ties and patch voids, one side	4.000	S.F.	.20	4	4.20
TOTAL			16.98	53.95	70.93

A2020 110 Walls, Cast in Place

	WALL HEIGHT (FT.)	PLACING METHOD	CONCRETE (C.Y. per L.F.)	REINFORCING (LBS. per L.F.)	WALL THICKNESS (IN.)	COST PER L.F. MAT.	INST.	TOTAL
1500	4′	direct chute	.074	3.3	6	17	54	71
1520			.099	4.8	8	20.50	55.50	76
1540			.123	6.0	10	24	56.50	80.50
1560	RA2020 -210		.148	7.2	12	28	57.50	85.50
1580			.173	8.1	14	31	58.50	89.50
1600			.197	9.44	16	34.50	60	94.50
1700	4′	pumped	.074	3.3	6	17	55	72
1720			.099	4.8	8	20.50	57	77.50
1740			.123	6.0	10	24	58	82
1760			.148	7.2	12	28	59.50	87.50
1780			.173	8.1	14	31	61	92
1800			.197	9.44	16	34.50	62.50	97
3000	6′	direct chute	.111	4.95	6	25.50	81	106.50
3020			.149	7.20	8	31	83	114
3040			.184	9.00	10	36	84.50	120.50
3060			.222	10.8	12	41.50	86.50	128
3080			.260	12.15	14	47	87.50	134.50
3100			.300	14.39	16	52.50	90.50	143

A2020 Basement Walls

A2020 110 Walls, Cast in Place

	WALL HEIGHT (FT.)	PLACING METHOD	CONCRETE (C.Y. per L.F.)	REINFORCING (LBS. per L.F.)	WALL THICKNESS (IN.)	COST PER L.F.		
						MAT.	INST.	TOTAL
3200	6′	pumped	.111	4.95	6	25.50	82.50	108
3220			.149	7.20	8	31	86	117
3240			.184	9.00	10	36	87.50	123.50
3260			.222	10.8	12	41.50	90	131.50
3280			.260	12.15	14	47	91	138
3300			.300	14.39	16	52.50	93.50	146
5000	8′	direct chute	.148	6.6	6	34	108	142
5020			.199	9.6	8	41.50	114	155.50
5040			.250	12	10	49	113	162
5060			.296	14.39	12	55.50	115	170.50
5080			.347	16.19	14	58	116	174
5100			.394	19.19	16	69.50	120	189.50
5200	8′	pumped	.148	6.6	6	34	110	144
5220			.199	9.6	8	41.50	114	155.50
5240			.250	12	10	49	117	166
5260			.296	14.39	12	55.50	119	174.50
5280			.347	16.19	14	58	120	178
5300			.394	19.19	16	69.50	125	194.50
6020	10′	direct chute	.248	12	8	52	138	190
6040			.307	14.99	10	60.50	141	201.50
6060			.370	17.99	12	69.50	144	213.50
6080			.433	20.24	14	78	146	224
6100			.493	23.99	16	87	150	237
6220	10′	pumped	.248	12	8	52	142	194
6240			.307	14.99	10	60.50	145	205.50
6260			.370	17.99	12	69.50	150	219.50
6280			.433	20.24	14	78	152	230
6300			.493	23.99	16	87	156	243
7220	12′	pumped	.298	14.39	8	62.50	172	234.50
7240			.369	17.99	10	72.50	175	247.50
7260			.444	21.59	12	83.50	180	263.50
7280			.52	24.29	14	93.50	182	275.50
7300			.591	28.79	16	104	187	291
7420	12′	crane & bucket	.298	14.39	8	62.50	178	240.50
7440			.369	17.99	10	72.50	183	255.50
7460			.444	21.59	12	83.50	189	272.50
7480			.52	24.29	14	93.50	193	286.50
7500			.591	28.79	16	104	202	306
8220	14′	pumped	.347	16.79	8	72.50	200	272.50
8240			.43	20.99	10	84.50	203	287.50
8260			.519	25.19	12	97.50	209	306.50
8280			.607	28.33	14	109	213	322
8300			.69	33.59	16	122	218	340
8420	14′	crane & bucket	.347	16.79	8	72.50	208	280.50
8440			.43	20.99	10	84.50	212	296.50
8460			.519	25.19	12	97.50	221	318.50
8480			.607	28.33	14	109	226	335
8500			.69	33.59	16	122	233	355
9220	16′	pumped	.397	19.19	8	83	229	312
9240			.492	23.99	10	96.50	233	329.50
9260			.593	28.79	12	111	240	351
9280			.693	32.39	14	125	243	368
9300			.788	38.38	16	139	250	389

A2020 Basement Walls

A2020 110 Walls, Cast in Place

	WALL HEIGHT (FT.)	PLACING METHOD	CONCRETE (C.Y. per L.F.)	REINFORCING (LBS. per L.F.)	WALL THICKNESS (IN.)	COST PER L.F.		
						MAT.	INST.	TOTAL
9420	16′	crane & bucket	.397	19.19	8	83	238	321
9440			.492	23.99	10	96.50	243	339.50
9460			.593	28.79	12	111	252	363
9480			.693	32.39	14	125	258	383
9500			.788	38.38	16	139	266	405

A2020 Basement Walls

A2020 220 Subdrainage Piping

A2020 220	Subdrainage Piping	COST PER L.F. MAT.	COST PER L.F. INST.	COST PER L.F. TOTAL
2000	Piping, excavation & backfill excluded, PVC, perforated			
2110	3″ diameter	1.64	4.19	5.83
2130	4″ diameter	1.64	4.19	5.83
2140	5″ diameter	3.61	4.48	8.09
2150	6″ diameter	3.61	4.48	8.09
3000	Metal alum. or steel, perforated asphalt coated			
3150	6″ diameter	7.15	4.13	11.28
3160	8″ diameter	9.40	4.24	13.64
3170	10″ diameter	11.75	5.70	17.45
3180	12″ diameter	13.15	7.15	20.30
3220	18″ diameter	19.75	9.95	29.70
4000	Corrugated HDPE tubing			
4130	4″ diameter	.75	.78	1.53
4150	6″ diameter	1.74	1.03	2.77
4160	8″ diameter	3.47	1.33	4.80
4180	12″ diameter	7.05	4.62	11.67
4200	15″ diameter	8.55	5.25	13.80
4220	18″ diameter	13.30	7.40	20.70

B1010 Floor Construction

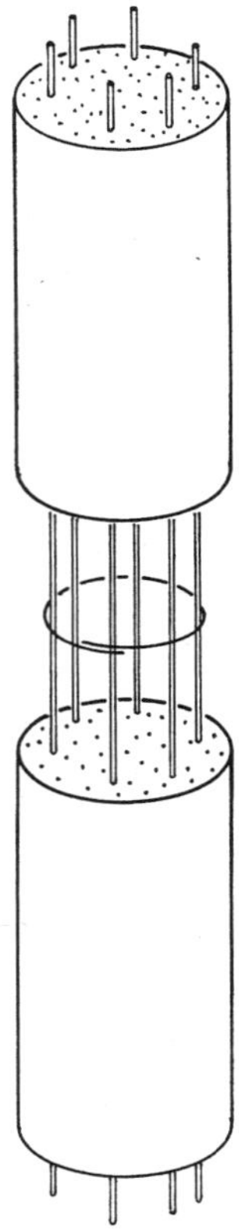

CONCRETE COLUMNS

General: It is desirable for purposes of consistency and simplicity to maintain constant column sizes throughout the building height. To do this, concrete strength may be varied (higher strength concrete at lower stories and lower strength concrete at upper stories), as well as varying the amount of reinforcing.

The first portion of the table provides probable minimum column sizes with related costs and weights per lineal foot of story height for bottom level columns.

The second portion of the table provides costs by column size for top level columns with minimum code reinforcement. Probable maximum loads for these columns are also given.

How to Use Table:

1. Enter the second portion (minimum reinforcing) of the table with the minimum allowable column size from the selected cast in place floor system.

   If the total load on the column does not exceed the allowable working load shown, use the cost per L.F. multiplied by the length of columns required to obtain the column cost.

2. If the total load on the column exceeds the allowable working load shown in the second portion of the table, enter the first portion of the table with the total load on the column and the minimum allowable column size from the selected cast in place floor system.

   Select a cost per L.F. for bottom level columns by total load or minimum allowable column size.

   Select a cost per L.F. for top level columns using the column size required for bottom level columns from the second portion of the table.

$$\frac{\text{Btm.} + \text{Top Col. Costs/L.F.}}{2} = \text{Avg. Col. Cost/L.F.}$$

Column Cost = Average Col. Cost/L.F. x Length of Cols. Required.

See reference section to determine total loads.

Design and Pricing Assumptions:
Normal wt. concrete, f'c = 4 or 6 KSI, placed by pump.
Steel, fy = 60 KSI, spliced every other level.
Minimum design eccentricity of 0.1t.
Assumed load level depth is 8″ (weights prorated to full story basis).
Gravity loads only (no frame or lateral loads included).

Please see the reference section for further design and cost information.

System Components	QUANTITY	UNIT	COST PER V.L.F. MAT.	INST.	TOTAL
SYSTEM B1010 201 1050					
ROUND TIED COLUMNS, 4 KSI CONCRETE, 100K MAX. LOAD, 10′ STORY, 12″ SIZE					
Forms in place, columns, round fiber tube, 12″ diam., 1 use	1.000	L.F.	3.12	15	18.12
Reinforcing in place, column ties	1.393	Lb.	4.23	6.77	11
Concrete ready mix, regular weight, 4000 psi	.029	C.Y.	3.60		3.60
Placing concrete, incl. vibrating, 12″ sq./round columns, pumped	.029	C.Y.		2.35	2.35
Finish, burlap rub w/grout	3.140	S.F.	.16	3.77	3.93
TOTAL			11.11	27.89	39

B1010 201 C.I.P. Column - Round Tied

	LOAD (KIPS)	STORY HEIGHT (FT.)	COLUMN SIZE (IN.)	COLUMN WEIGHT (P.L.F.)	CONCRETE STRENGTH (PSI)	COST PER V.L.F. MAT.	INST.	TOTAL
1050	100	10	12	110	4000	11.10	28	39.10
1060	RB1010 -112	12	12	111	4000	11.30	28.50	39.80
1070		14	12	112	4000	11.55	28.50	40.05
1080	150	10	12	110	4000	12.15	29.50	41.65
1090		12	12	111	4000	12.40	30	42.40
1100		14	14	153	4000	15.30	32.50	47.80
1120	200	10	14	150	4000	15.60	32.50	48.10
1140		12	14	152	4000	15.90	33	48.90

B1010 Floor Construction

B1010 201 — C.I.P. Column - Round Tied

	LOAD (KIPS)	STORY HEIGHT (FT.)	COLUMN SIZE (IN.)	COLUMN WEIGHT (P.L.F.)	CONCRETE STRENGTH (PSI)	COST PER V.L.F.		
						MAT.	INST.	TOTAL
1160	200	14	14	153	4000	16.20	33.50	49.70
1180	300	10	16	194	4000	19.50	36	55.50
1190		12	18	250	4000	26	41	67
1200		14	18	252	4000	26.50	41.50	68
1220	400	10	20	306	4000	30	46.50	76.50
1230		12	20	310	4000	30.50	47	77.50
1260		14	20	313	4000	31	48	79
1280	500	10	22	368	4000	36.50	52	88.50
1300		12	22	372	4000	37.50	53.50	91
1325		14	22	375	4000	38	54.50	92.50
1350	600	10	24	439	4000	41.50	58	99.50
1375		12	24	445	4000	42.50	59.50	102
1400		14	24	448	4000	43.50	60.50	104
1420	700	10	26	517	4000	52.50	65.50	118
1430		12	26	524	4000	53.50	67	120.50
1450		14	26	528	4000	54.50	68.50	123
1460	800	10	28	596	4000	58.50	72	130.50
1480		12	28	604	4000	59.50	74	133.50
1490		14	28	609	4000	60.50	75.50	136
1500	900	10	28	596	4000	62	77.50	139.50
1510		12	28	604	4000	63	79.50	142.50
1520		14	28	609	4000	64	81	145
1530	1000	10	30	687	4000	67.50	80	147.50
1540		12	30	695	4000	68.50	82	150.50
1620		14	30	701	4000	70	84.50	154.50
1640	100	10	12	110	6000	11.40	28	39.40
1660		12	12	111	6000	11.60	28.50	40.10
1680		14	12	112	6000	11.80	28.50	40.30
1700	150	10	12	110	6000	12.25	29.50	41.75
1710		12	12	111	6000	12.45	29.50	41.95
1720		14	12	112	6000	12.65	30	42.65
1730	200	10	12	110	6000	12.90	30.50	43.40
1740		12	12	111	6000	13.10	30.50	43.60
1760		14	12	112	6000	13.30	31	44.30
1780	300	10	14	150	6000	16.30	33	49.30
1790		12	14	152	6000	16.60	33.50	50.10
1800		14	16	153	6000	19.10	35.50	54.60
1810	400	10	16	194	6000	21	37	58
1820		12	16	196	6000	21.50	37.50	59
1830		14	18	252	6000	26.50	41	67.50
1850	500	10	18	247	6000	27.50	42.50	70
1870		12	18	250	6000	28	43	71
1880		14	20	252	6000	31	46.50	77.50
1890	600	10	20	306	6000	32	48	80
1900		12	20	310	6000	32.50	49	81.50
1905		14	20	313	6000	33	50	83
1910	700	10	22	368	6000	37.50	52	89.50
1915		12	22	372	6000	38	53.50	91.50
1920		14	22	375	6000	39	54.50	93.50
1925	800	10	24	439	6000	43	58	101
1930		12	24	445	6000	44	59.50	103.50
1935		14	24	448	6000	44.50	60.50	105

B1010 Floor Construction

B1010 201 C.I.P. Column - Round Tied

	LOAD (KIPS)	STORY HEIGHT (FT.)	COLUMN SIZE (IN.)	COLUMN WEIGHT (P.L.F.)	CONCRETE STRENGTH (PSI)	COST PER V.L.F.		
						MAT.	INST.	TOTAL
1940	900	10	24	439	6000	45.50	62	107.50
1945		12	24	445	6000	46.50	63.50	110
1970	1000	10	26	517	6000	55	67	122
1980		12	26	524	6000	56	68.50	124.50
1995		14	26	528	6000	57	70	127

B1010 202 C.I.P. Columns, Round Tied - Minimum Reinforcing

	LOAD (KIPS)	STORY HEIGHT (FT.)	COLUMN SIZE (IN.)	COLUMN WEIGHT (P.L.F.)	CONCRETE STRENGTH (PSI)	COST PER V.L.F.		
						MAT.	INST.	TOTAL
2500	100	10-14	12	107	4000	10.05	26.50	36.55
2510	200	10-14	16	190	4000	17.60	33	50.60
2520	400	10-14	20	295	4000	27	41.50	68.50
2530	600	10-14	24	425	4000	37.50	51	88.50
2540	800	10-14	28	580	4000	52.50	62.50	115
2550	1100	10-14	32	755	4000	67	73	140
2560	1400	10-14	36	960	4000	79.50	85.50	165
2570								

B1010 Floor Construction

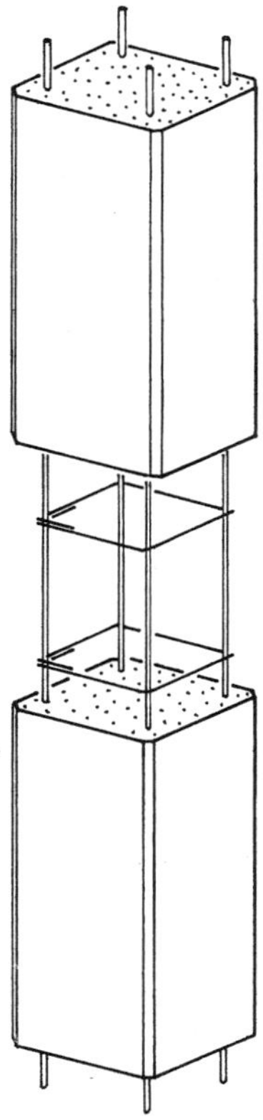

CONCRETE COLUMNS

General: It is desirable for purposes of consistency and simplicity to maintain constant column sizes throughout the building height. To do this, concrete strength may be varied (higher strength concrete at lower stories and lower strength concrete at upper stories), as well as varying the amount of reinforcing.

The first portion of the table provides probable minimum column sizes with related costs and weights per lineal foot of story height for bottom level columns.

The second portion of the table provides costs by column size for top level columns with minimum code reinforcement. Probable maximum loads for these columns are also given.

How to Use Table:

1. Enter the second portion (minimum reinforcing) of the table with the minimum allowable column size from the selected cast in place floor system.

   If the total load on the column does not exceed the allowable working load shown, use the cost per L.F. multiplied by the length of columns required to obtain the column cost.
2. If the total load on the column exceeds the allowable working load shown in the second portion of the table, enter the first portion of the table with the total load on the column and the minimum allowable column size from the selected cast in place floor system.

   Select a cost per L.F. for bottom level columns by total load or minimum allowable column size.

   Select a cost per L.F. for top level columns using the column size required for bottom level columns from the second portion of the table.

$$\frac{\text{Btm. + Top Col. Costs/L.F.}}{2} = \text{Avg. Col. Cost/L.F.}$$

Column Cost = Average Col. Cost/L.F. x Length of Cols. Required.

See reference section in back of book to determine total loads.

Design and Pricing Assumptions:
Normal wt. concrete, f'c = 4 or 6 KSI, placed by pump.
Steel, fy = 60 KSI, spliced every other level.
Minimum design eccentricity of 0.1t.
Assumed load level depth is 8" (weights prorated to full story basis).
Gravity loads only (no frame or lateral loads included).

Please see the reference section for further design and cost information.

System Components	QUANTITY	UNIT	COST PER V.L.F. MAT.	INST.	TOTAL
SYSTEM B1010 203 0640					
SQUARE COLUMNS, 100K LOAD, 10' STORY, 10" SQUARE					
Forms in place, columns, plywood, 10" x 10", 4 uses	3.323	SFCA	3.10	33.33	36.43
Chamfer strip, wood, 3/4" wide	4.000	L.F.	.72	4.52	5.24
Reinforcing in place, column ties	1.405	Lb.	3.74	5.98	9.72
Concrete ready mix, regular weight, 4000 psi	.026	C.Y.	3.22		3.22
Placing concrete, incl. vibrating, 12" sq./round columns, pumped	.026	C.Y.		2.11	2.11
Finish, break ties, patch voids, burlap rub w/grout	3.323	S.F.	.17	4	4.17
TOTAL			10.95	49.94	60.89

B1010 203 C.I.P. Column, Square Tied

	LOAD (KIPS)	STORY HEIGHT (FT.)	COLUMN SIZE (IN.)	COLUMN WEIGHT (P.L.F.)	CONCRETE STRENGTH (PSI)	COST PER V.L.F. MAT.	INST.	TOTAL
0640	100	10	10	96	4000	10.95	50	60.95
0680	RB1010 -112	12	10	97	4000	11.15	50.50	61.65
0700		14	12	142	4000	14.60	61	75.60
0710								

B1010 Floor Construction

B1010 203 C.I.P. Column, Square Tied

	LOAD (KIPS)	STORY HEIGHT (FT.)	COLUMN SIZE (IN.)	COLUMN WEIGHT (P.L.F.)	CONCRETE STRENGTH (PSI)	COST PER V.L.F.		
						MAT.	INST.	TOTAL
0740	150	10	10	96	4000	12.80	53	65.80
0780		12	12	142	4000	15.15	62	77.15
0800		14	12	143	4000	15.40	62	77.40
0840	200	10	12	140	4000	15.95	63	78.95
0860		12	12	142	4000	16.25	63.50	79.75
0900		14	14	196	4000	18.95	70.50	89.45
0920	300	10	14	192	4000	20	72	92
0960		12	14	194	4000	20.50	73	93.50
0980		14	16	253	4000	23.50	80.50	104
1020	400	10	16	248	4000	25	82.50	107.50
1060		12	16	251	4000	25.50	83.50	109
1080		14	16	253	4000	26	84.50	110.50
1200	500	10	18	315	4000	29.50	92.50	122
1250		12	20	394	4000	35.50	105	140.50
1300		14	20	397	4000	36.50	106	142.50
1350	600	10	20	388	4000	37	108	145
1400		12	20	394	4000	38	109	147
1600		14	20	397	4000	38.50	110	148.50
1900	700	10	20	388	4000	43	119	162
2100		12	22	474	4000	43	119	162
2300		14	22	478	4000	44	121	165
2600	800	10	22	388	4000	45	122	167
2900		12	22	474	4000	46	124	170
3200		14	22	478	4000	46.50	125	171.50
3400	900	10	24	560	4000	50.50	133	183.50
3800		12	24	567	4000	51.50	135	186.50
4000		14	24	571	4000	52.50	137	189.50
4250	1000	10	24	560	4000	54.50	140	194.50
4500		12	26	667	4000	58	147	205
4750		14	26	673	4000	59.50	149	208.50
5600	100	10	10	96	6000	11.20	50	61.20
5800		12	10	97	6000	11.40	50	61.40
6000		14	12	142	6000	15	61	76
6200	150	10	10	96	6000	13.05	53	66.05
6400		12	12	98	6000	15.50	62	77.50
6600		14	12	143	6000	15.80	62	77.80
6800	200	10	12	140	6000	16.35	63	79.35
7000		12	12	142	6000	16.60	63.50	80.10
7100		14	14	196	6000	19.45	70.50	89.95
7300	300	10	14	192	6000	20	71.50	91.50
7500		12	14	194	6000	20.50	72	92.50
7600		14	14	196	6000	21	73	94
7700	400	10	14	192	6000	21.50	74	95.50
7800		12	14	194	6000	22	74.50	96.50
7900		14	16	253	6000	25	80.50	105.50
8000	500	10	16	248	6000	26	82.50	108.50
8050		12	16	251	6000	26.50	83.50	110
8100		14	16	253	6000	27	84.50	111.50
8200	600	10	18	315	6000	30	93	123
8300		12	18	319	6000	30.50	94	124.50
8400		14	18	321	6000	31	95	126

B1010 Floor Construction

B1010 203 C.I.P. Column, Square Tied

	LOAD (KIPS)	STORY HEIGHT (FT.)	COLUMN SIZE (IN.)	COLUMN WEIGHT (P.L.F.)	CONCRETE STRENGTH (PSI)	COST PER V.L.F.		
						MAT.	INST.	TOTAL
8500	700	10	18	315	6000	31.50	96	127.50
8600		12	18	319	6000	32.50	97	129.50
8700		14	18	321	6000	33	98	131
8800	800	10	20	388	6000	37	105	142
8900		12	20	394	6000	37.50	106	143.50
9000		14	20	397	6000	38.50	108	146.50
9100	900	10	20	388	6000	40	110	150
9300		12	20	394	6000	40.50	111	151.50
9600		14	20	397	6000	41.50	112	153.50
9800	1000	10	22	469	6000	45	121	166
9840		12	22	474	6000	46	122	168
9900		14	22	478	6000	47	124	171

B1010 204 C.I.P. Column, Square Tied-Minimum Reinforcing

	LOAD (KIPS)	STORY HEIGHT (FT.)	COLUMN SIZE (IN.)	COLUMN WEIGHT (P.L.F.)	CONCRETE STRENGTH (PSI)	COST PER V.L.F.		
						MAT.	INST.	TOTAL
9913	150	10-14	12	135	4000	13.25	58.50	71.75
9918	300	10-14	16	240	4000	21.50	76.50	98
9924	500	10-14	20	375	4000	32	99	131
9930	700	10-14	24	540	4000	44	123	167
9936	1000	10-14	28	740	4000	58	150	208
9942	1400	10-14	32	965	4000	72	169	241
9948	1800	10-14	36	1220	4000	89.50	197	286.50
9954	2300	10-14	40	1505	4000	109	227	336

B1010 Floor Construction

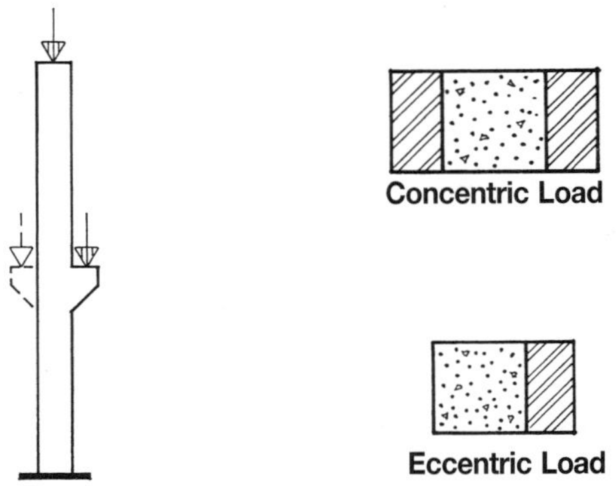

General: Data presented here is for plant produced members transported 50 miles to 100 miles to the site and erected.

Design and pricing assumptions:
Normal wt. concrete, f'c = 5 KSI

Main reinforcement, fy = 60 KSI
Ties, fy = 40 KSI

Minimum design eccentricity, 0.1t.

Concrete encased structural steel haunches are assumed where practical; otherwise galvanized rebar haunches are assumed.

Base plates are integral with columns.

Foundation anchor bolts, nuts and washers are included in price.

System Components	QUANTITY	UNIT	COST PER V.L.F. MAT.	INST.	TOTAL
SYSTEM B1010 206 0560					
PRECAST TIED COLUMN, CONCENTRIC LOADING, 100 K MAX.					
TWO STORY - 10′ PER STORY, 5 KSI CONCRETE, 12″X12″					
Precast column, two story-10′/story, 5 KSI conc., 12″x12″	1.000	Ea.	166		166
Anchor bolts	.050	Set	2	1.75	3.75
Steel bearing plates; top, bottom, haunches	3.250	Lb.	5.20		5.20
Erection crew	.013	Hr.		12.50	12.50
TOTAL			173.20	14.25	187.45

B1010 206 Tied, Concentric Loaded Precast Concrete Columns

	LOAD (KIPS)	STORY HEIGHT (FT.)	COLUMN SIZE (IN.)	COLUMN WEIGHT (P.L.F.)	LOAD LEVELS	COST PER V.L.F. MAT.	INST.	TOTAL
0560	100	10	12x12	164	2	173	14.25	187.25
0570		12	12x12	162	2	168	11.85	179.85
0580		14	12x12	161	2	168	11.65	179.65
0590	150	10	12x12	166	3	168	11.55	179.55
0600		12	12x12	169	3	168	10.35	178.35
0610		14	12x12	162	3	168	10.20	178.20
0620	200	10	12x12	168	4	167	12.35	179.35
0630		12	12x12	170	4	168	11.15	179.15
0640		14	12x12	220	4	167	11.05	178.05
0680	300	10	14x14	225	3	228	11.55	239.55
0690		12	14x14	225	3	228	10.35	238.35
0700		14	14x14	250	3	228	10.20	238.20
0710	400	10	16x16	255	4	199	12.35	211.35
0720		12	16x16	295	4	199	11.15	210.15
0750		14	16x16	305	4	160	11.05	171.05
0790	450	10	16x16	320	3	200	11.55	211.55
0800		12	16x16	315	3	199	10.35	209.35
0810		14	16x16	330	3	199	10.20	209.20
0820	600	10	18x18	405	4	251	12.35	263.35
0830		12	18x18	395	4	251	11.15	262.15
0840		14	18x18	410	4	251	11.05	262.05
0910	800	10	20x20	495	4	229	12.35	241.35
0920		12	20x20	505	4	228	11.15	239.15
0930		14	20x20	510	4	229	11.05	240.05

B1010 Floor Construction

B1010 206 Tied, Concentric Loaded Precast Concrete Columns

	LOAD (KIPS)	STORY HEIGHT (FT.)	COLUMN SIZE (IN.)	COLUMN WEIGHT (P.L.F.)	LOAD LEVELS	COST PER V.L.F.		
						MAT.	INST.	TOTAL
0970	900	10	22x22	625	3	277	11.55	288.55
0980		12	22x22	610	3	277	10.35	287.35
0990		14	22x22	605	3	277	10.20	287.20

B1010 207 Tied, Eccentric Loaded Precast Concrete Columns

	LOAD (KIPS)	STORY HEIGHT (FT.)	COLUMN SIZE (IN.)	COLUMN WEIGHT (P.L.F.)	LOAD LEVELS	COST PER V.L.F.		
						MAT.	INST.	TOTAL
1130	100	10	12x12	161	2	165	14.25	179.25
1140		12	12x12	159	2	165	11.85	176.85
1150		14	12x12	159	2	165	11.65	176.65
1160	150	10	12x12	161	3	163	11.55	174.55
1170		12	12x12	160	3	163	10.35	173.35
1180		14	12x12	159	3	162	10.20	172.20
1190	200	10	12x12	161	4	161	12.35	173.35
1200		12	12x12	160	4	161	11.15	172.15
1210		14	12x12	177	4	161	11.05	172.05
1250	300	10	12x12	185	3	221	11.55	232.55
1260		12	14x14	215	3	221	10.35	231.35
1270		14	14x14	215	3	221	10.20	231.20
1280	400	10	14x14	235	4	219	12.35	231.35
1290		12	16x16	285	4	218	11.15	229.15
1300		14	16x16	295	4	219	11.05	230.05
1360	450	10	14x14	245	3	194	11.55	205.55
1370		12	16x16	285	3	193	10.35	203.35
1380		14	16x16	290	3	193	10.20	203.20
1390	600	10	18x18	385	4	242	12.35	254.35
1400		12	18x18	380	4	243	11.15	254.15
1410		14	18x18	375	4	242	11.05	253.05
1480	800	10	20x20	490	4	220	12.35	232.35
1490		12	20x20	480	4	220	11.15	231.15
1500		14	20x20	475	4	220	11.05	231.05

B1010 Floor Construction

(A) Wide Flange

(B) Pipe

(C) Pipe, Concrete Filled

(D) Square Tube

(E) Square Tube Concrete Filled

(F) Rectangular Tube

(G) Rectangular Tube, Concrete Filled

General: The following pages provide data for seven types of steel columns: wide flange, round pipe, round pipe concrete filled, square tube, square tube concrete filled, rectangular tube and rectangular tube concrete filled.

Design Assumptions: Loads are concentric; wide flange and round pipe bearing capacity is for 36 KSI steel. Square and rectangular tubing bearing capacity is for 46 KSI steel.

The effective length factor K=1.1 is used for determining column values in the tables. K=1.1 is within a frequently used range for pinned connections with cross bracing.

How To Use Tables:

a. Steel columns usually extend through two or more stories to minimize splices. Determine floors with splices.

b. Enter Table No. below with load to column at the splice. Use the unsupported height.

c. Determine the column type desired by price or design.

Cost:

a. Multiply number of columns at the desired level by the total height of the column by the cost/VLF.

b. Repeat the above for all tiers.

Please see the reference section for further design and cost information.

B1010 208 Steel Columns

	LOAD (KIPS)	UNSUPPORTED HEIGHT (FT.)	WEIGHT (P.L.F.)	SIZE (IN.)	TYPE	COST PER V.L.F.		
						MAT.	INST.	TOTAL
1000	25	10	13	4	A	23	10.85	33.85
1020	RB1010 -130		7.58	3	B	13.40	10.85	24.25
1040			15	3-1/2	C	16.35	10.85	27.20
1060			6.87	3	D	12.15	10.85	23
1080			15	3	E	15.75	10.85	26.60
1100			8.15	4x3	F	14.40	10.85	25.25
1120			20	4x3	G	18.85	10.85	29.70
1200		16	16	5	A	26	8.15	34.15
1220			10.79	4	B	17.65	8.15	25.80
1240			36	5-1/2	C	24.50	8.15	32.65
1260			11.97	5	D	19.55	8.15	27.70
1280			36	5	E	26	8.15	34.15
1300			11.97	6x4	F	19.55	8.15	27.70
1320			64	8x6	G	38	8.15	46.15
1400		20	20	6	A	31	8.15	39.15
1420			14.62	5	B	22.50	8.15	30.65
1440			49	6-5/8	C	30.50	8.15	38.65
1460			11.97	5	D	18.50	8.15	26.65
1480			49	6	E	30.50	8.15	38.65
1500			14.53	7x5	F	22.50	8.15	30.65
1520			64	8x6	G	36	8.15	44.15
1600	50	10	16	5	A	28.50	10.85	39.35
1620			14.62	5	B	26	10.85	36.85
1640			24	4-1/2	C	19.50	10.85	30.35
1660			12.21	4	D	21.50	10.85	32.35
1680			25	4	E	22	10.85	32.85
1700			11.97	6x4	F	21	10.85	31.85
1720			28	6x3	G	25	10.85	35.85

B1010 Floor Construction

B1010 208 Steel Columns

	LOAD (KIPS)	UNSUPPORTED HEIGHT (FT.)	WEIGHT (P.L.F.)	SIZE (IN.)	TYPE	COST PER V.L.F.		
						MAT.	INST.	TOTAL
1800	50	16	24	8	A	39	8.15	47.15
1820			18.97	6	B	31	8.15	39.15
1840			36	5-1/2	C	24.50	8.15	32.65
1860			14.63	6	D	24	8.15	32.15
1880			36	5	E	26	8.15	34.15
1900			14.53	7x5	F	24	8.15	32.15
1920			64	8x6	G	38	8.15	46.15
1940								
2000	50	20	28	8	A	43.50	8.15	51.65
2020			18.97	6	B	29.50	8.15	37.65
2040			49	6-5/8	C	30.50	8.15	38.65
2060			19.02	6	D	29.50	8.15	37.65
2080			49	6	E	30.50	8.15	38.65
2100			22.42	8x6	F	34.50	8.15	42.65
2120			64	8x6	G	36	8.15	44.15
2200	75	10	20	6	A	35.50	10.85	46.35
2220			18.97	6	B	33.50	10.85	44.35
2240			36	4-1/2	C	49	10.85	59.85
2260			14.53	6	D	25.50	10.85	36.35
2280			28	4	E	22	10.85	32.85
2300			14.33	7x5	F	25.50	10.85	36.35
2320			35	6x4	G	28	10.85	38.85
2400		16	31	8	A	50.50	8.15	58.65
2420			28.55	8	B	46.50	8.15	54.65
2440			49	6-5/8	C	32	8.15	40.15
2460			17.08	7	D	28	8.15	36.15
2480			36	5	E	26	8.15	34.15
2500			23.34	7x5	F	38	8.15	46.15
2520			64	8x6	G	38	8.15	46.15
2600		20	31	8	A	48	8.15	56.15
2620			28.55	8	B	44	8.15	52.15
2640			81	8-5/8	C	46	8.15	54.15
2660			22.42	7	D	34.50	8.15	42.65
2680			49	6	E	30.50	8.15	38.65
2700			22.42	8x6	F	34.50	8.15	42.65
2720			64	8x6	G	36	8.15	44.15
2800	100	10	24	8	A	42.50	10.85	53.35
2820			28.57	6	B	50.50	10.85	61.35
2840			35	4-1/2	C	49	10.85	59.85
2860			17.08	7	D	30	10.85	40.85
2880			36	5	E	28.50	10.85	39.35
2900			19.02	7x5	F	33.50	10.85	44.35
2920			46	8x4	G	34.50	10.85	45.35
3000		16	31	8	A	50.50	8.15	58.65
3020			28.55	8	B	46.50	8.15	54.65
3040			56	6-5/8	C	47.50	8.15	55.65
3060			22.42	7	D	36.50	8.15	44.65
3080			49	6	E	32	8.15	40.15
3100			22.42	8x6	F	36.50	8.15	44.65
3120			64	8x6	G	38	8.15	46.15

B1010 Floor Construction

B1010 208 Steel Columns

	LOAD (KIPS)	UNSUPPORTED HEIGHT (FT.)	WEIGHT (P.L.F.)	SIZE (IN.)	TYPE	COST PER V.L.F. MAT.	INST.	TOTAL
3200	100	20	40	8	A	62	8.15	70.15
3220			28.55	8	B	44	8.15	52.15
3240			81	8-5/8	C	46	8.15	54.15
3260			25.82	8	D	40	8.15	48.15
3280			66	7	E	36	8.15	44.15
3300			27.59	8x6	F	42.50	8.15	50.65
3320			70	8x6	G	51.50	8.15	59.65
3400	125	10	31	8	A	55	10.85	65.85
3420			28.57	6	B	50.50	10.85	61.35
3440			81	8	C	52	10.85	62.85
3460			22.42	7	D	39.50	10.85	50.35
3480			49	6	E	34.50	10.85	45.35
3500			22.42	8x6	F	39.50	10.85	50.35
3520			64	8x6	G	41	10.85	51.85
3600		16	40	8	A	65.50	8.15	73.65
3620			28.55	8	B	46.50	8.15	54.65
3640			81	8	C	48.50	8.15	56.65
3660			25.82	8	D	42	8.15	50.15
3680			66	7	E	38	8.15	46.15
3700			27.59	8x6	F	45	8.15	53.15
3720			64	8x6	G	38	8.15	46.15
3800		20	48	8	A	74.50	8.15	82.65
3820			40.48	10	B	62.50	8.15	70.65
3840			81	8	C	46	8.15	54.15
3860			25.82	8	D	40	8.15	48.15
3880			66	7	E	36	8.15	44.15
3900			37.59	10x6	F	58	8.15	66.15
3920			60	8x6	G	51.50	8.15	59.65
4000	150	10	35	8	A	62	10.85	72.85
4020			40.48	10	B	71.50	10.85	82.35
4040			81	8-5/8	C	52	10.85	62.85
4060			25.82	8	D	45.50	10.85	56.35
4080			66	7	E	41	10.85	51.85
4100			27.48	7x5	F	48.50	10.85	59.35
4120			64	8x6	G	41	10.85	51.85
4200		16	45	10	A	73.50	8.15	81.65
4220			40.48	10	B	66	8.15	74.15
4240			81	8-5/8	C	48.50	8.15	56.65
4260			31.84	8	D	52	8.15	60.15
4280			66	7	E	38	8.15	46.15
4300			37.69	10x6	F	61.50	8.15	69.65
4320			70	8x6	G	54.50	8.15	62.65
4400		20	49	10	A	76	8.15	84.15
4420			40.48	10	B	62.50	8.15	70.65
4440			123	10-3/4	C	65.50	8.15	73.65
4460			31.84	8	D	49.50	8.15	57.65
4480			82	8	E	42	8.15	50.15
4500			37.69	10x6	F	58.50	8.15	66.65
4520			86	10x6	G	51	8.15	59.15

B1010 208 Steel Columns

	LOAD (KIPS)	UNSUPPORTED HEIGHT (FT.)	WEIGHT (P.L.F.)	SIZE (IN.)	TYPE	COST PER V.L.F.		
						MAT.	INST.	TOTAL
4600	200	10	45	10	A	79.50	10.85	90.35
4620			40.48	10	B	71.50	10.85	82.35
4640			81	8-5/8	C	52	10.85	62.85
4660			31.84	8	D	56.50	10.85	67.35
4680			82	8	E	47.50	10.85	58.35
4700			37.69	10x6	F	66.50	10.85	77.35
4720			70	8x6	G	59	10.85	69.85
4800		16	49	10	A	80	8.15	88.15
4820			49.56	12	B	81	8.15	89.15
4840			123	10-3/4	C	69	8.15	77.15
4860			37.60	8	D	61.50	8.15	69.65
4880			90	8	E	63.50	8.15	71.65
4900			42.79	12x6	F	70	8.15	78.15
4920			85	10x6	G	63.50	8.15	71.65
5000		20	58	12	A	90	8.15	98.15
5020			49.56	12	B	76.50	8.15	84.65
5040			123	10-3/4	C	65.50	8.15	73.65
5060			40.35	10	D	62.50	8.15	70.65
5080			90	8	E	60	8.15	68.15
5100			47.90	12x8	F	74	8.15	82.15
5120			93	10x6	G	77	8.15	85.15
5200	300	10	61	14	A	108	10.85	118.85
5220			65.42	12	B	116	10.85	126.85
5240			169	12-3/4	C	91.50	10.85	102.35
5260			47.90	10	D	84.50	10.85	95.35
5280			90	8	E	68	10.85	78.85
5300			47.90	12x8	F	84.50	10.85	95.35
5320			86	10x6	G	88	10.85	98.85
5400		16	72	12	A	118	8.15	126.15
5420			65.42	12	B	107	8.15	115.15
5440			169	12-3/4	C	85	8.15	93.15
5460			58.10	12	D	95	8.15	103.15
5480			135	10	E	81	8.15	89.15
5500			58.10	14x10	F	95	8.15	103.15
5600		20	79	12	A	122	8.15	130.15
5620			65.42	12	B	101	8.15	109.15
5640			169	12-3/4	C	80.50	8.15	88.65
5660			58.10	12	D	90	8.15	98.15
5680			135	10	E	77	8.15	85.15
5700			58.10	14x10	F	90	8.15	98.15
5800	400	10	79	12	A	140	10.85	150.85
5840			178	12-3/4	C	119	10.85	129.85
5860			68.31	14	D	121	10.85	131.85
5880			135	10	E	87.50	10.85	98.35
5900			62.46	14x10	F	110	10.85	120.85
6000		16	87	12	A	142	8.15	150.15
6040			178	12-3/4	C	111	8.15	119.15
6060			68.31	14	D	112	8.15	120.15
6080			145	10	E	105	8.15	113.15
6100			76.07	14x10	F	124	8.15	132.15

B1010 Floor Construction

B1010 208 Steel Columns

	LOAD (KIPS)	UNSUPPORTED HEIGHT (FT.)	WEIGHT (P.L.F.)	SIZE (IN.)	TYPE	COST PER V.L.F.		
						MAT.	INST.	TOTAL
6200	400	20	90	14	A	139	8.15	147.15
6240			178	12-3/4	C	105	8.15	113.15
6260			68.31	14	D	106	8.15	114.15
6280			145	10	E	99.50	8.15	107.65
6300			76.07	14x10	F	118	8.15	126.15
6400	500	10	99	14	A	175	10.85	185.85
6460			76.07	12	D	134	10.85	144.85
6480			145	10	E	113	10.85	123.85
6500			76.07	14x10	F	134	10.85	144.85
6600		16	109	14	A	178	8.15	186.15
6660			89.68	14	D	147	8.15	155.15
6700			89.68	16x12	F	147	8.15	155.15
6800		20	120	12	A	186	8.15	194.15
6860			89.68	14	D	139	8.15	147.15
6900			89.68	16x12	F	139	8.15	147.15
7000	600	10	120	12	A	212	10.85	222.85
7060			89.68	14	D	158	10.85	168.85
7100			89.68	16x12	F	158	10.85	168.85
7200		16	132	14	A	216	8.15	224.15
7260			103.30	16	D	169	8.15	177.15
7400		20	132	14	A	204	8.15	212.15
7460			103.30	16	D	160	8.15	168.15
7600	700	10	136	12	A	240	10.85	250.85
7660			103.30	16	D	182	10.85	192.85
7800		16	145	14	A	237	8.15	245.15
7860			103.30	16	D	169	8.15	177.15
8000		20	145	14	A	224	8.15	232.15
8060			103.30	16	D	160	8.15	168.15
8200	800	10	145	14	A	256	10.85	266.85
8300		16	159	14	A	260	8.15	268.15
8400		20	176	14	A	272	8.15	280.15
8800	900	10	159	14	A	281	10.85	291.85
8900		16	176	14	A	288	8.15	296.15
9000		20	193	14	A	299	8.15	307.15
9100	1000	10	176	14	A	310	10.85	320.85
9200		16	193	14	A	315	8.15	323.15
9300		20	211	14	A	325	8.15	333.15

B1010 Floor Construction

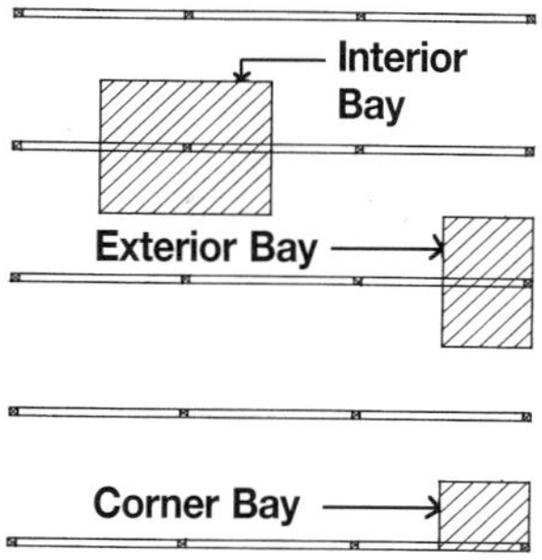

Description: Table below lists costs of columns per S.F. of bay for wood columns of various sizes and unsupported heights and the maximum allowable total load per S.F. by bay size.

Design Assumptions: Columns are concentrically loaded and are not subject to bending.

Fiber stress (f) is 1200 psi maximum.

Modulus of elasticity is 1,760,000. Use table to factor load capacity figures for modulus of elasticity other than 1,760,000.

The cost of columns per S.F. of exterior bay is proportional to the area supported. For exterior bays, multiply the costs below by two. For corner bays, multiply the cost by four.

Modulus of Elasticity	Factor
1,210,000 psi	0.69
1,320,000 psi	0.75
1,430,000 psi	0.81
1,540,000 psi	0.87
1,650,000 psi	0.94
1,760,000 psi	1.00

B1010 210 Wood Columns

	NOMINAL COLUMN SIZE (IN.)	BAY SIZE (FT.)	UNSUPPORTED HEIGHT (FT.)	MATERIAL (BF per M.S.F.)	TOTAL LOAD (P.S.F.)	COST PER S.F.		
						MAT.	INST.	TOTAL
1000	4 x 4	10 x 8	8	133	100	.18	.26	.44
1050			10	167	60	.23	.33	.56
1100			12	200	40	.28	.40	.68
1150			14	233	30	.32	.46	.78
1200		10 x 10	8	106	80	.15	.21	.36
1250			10	133	50	.18	.26	.44
1300			12	160	30	.22	.32	.54
1350			14	187	20	.26	.37	.63
1400		10 x 15	8	71	50	.10	.14	.24
1450			10	88	30	.12	.17	.29
1500			12	107	20	.15	.21	.36
1550			14	124	15	.17	.24	.41
1600		15 x 15	8	47	30	.06	.09	.15
1650			10	59	15	.08	.12	.20
1800		15 x 20	8	35	15	.05	.07	.12
1900								
2000	6 x 6	10 x 15	8	160	230	.25	.29	.54
2050			10	200	210	.32	.37	.69
2100			12	240	140	.38	.44	.82
2150			14	280	100	.44	.51	.95
2200		15 x 15	8	107	150	.17	.20	.37
2250			10	133	140	.21	.24	.45
2300			12	160	90	.25	.29	.54
2350			14	187	60	.29	.34	.63
2400		15 x 20	8	80	110	.13	.15	.28
2450			10	100	100	.16	.18	.34
2500			12	120	70	.19	.22	.41
2550			14	140	50	.22	.26	.48
2600		20 x 20	8	60	80	.09	.11	.20
2650			10	75	70	.12	.14	.26
2700			12	90	50	.14	.16	.30
2750			14	105	30	.17	.19	.36
2800		20 x 25	8	48	60	.08	.09	.17
2850			10	60	50	.09	.11	.20
2900			12	72	40	.11	.13	.24
2950			14	84	20	.13	.15	.28

B1010 Floor Construction

B1010 210 Wood Columns

	NOMINAL COLUMN SIZE (IN.)	BAY SIZE (FT.)	UNSUPPORTED HEIGHT (FT.)	MATERIAL (BF per M.S.F.)	TOTAL LOAD (P.S.F.)	COST PER S.F.		
						MAT.	INST.	TOTAL
3000	6 x 6	25 x 25	8	38	50	.06	.07	.13
3050			10	48	40	.08	.09	.17
3100			12	58	30	.09	.11	.20
3150			14	67	20	.11	.12	.23
3400	8 x 8	20 x 20	8	107	160	.15	.18	.33
3450			10	133	160	.19	.23	.42
3500			12	160	160	.23	.27	.50
3550			14	187	140	.26	.37	.63
3600		20 x 25	8	85	130	.12	.17	.29
3650			10	107	130	.15	.21	.36
3700			12	128	130	.21	.20	.41
3750			14	149	110	.25	.23	.48
3800		25 x 25	8	68	100	.11	.11	.22
3850			10	85	100	.14	.13	.27
3900			12	102	100	.17	.16	.33
3950			14	119	90	.20	.19	.39
4200	10 x 10	20 x 25	8	133	210	.22	.21	.43
4250			10	167	210	.28	.26	.54
4300			12	200	210	.33	.32	.65
4350			14	233	210	.38	.37	.75
4400		25 x 25	8	107	160	.18	.17	.35
4450			10	133	160	.22	.21	.43
4500			12	160	160	.26	.25	.51
4550			14	187	160	.31	.29	.60
4700	12 x 12	20 x 25	8	192	310	.31	.28	.59
4750			10	240	310	.38	.35	.73
4800			12	288	310	.46	.42	.88
4850			14	336	310	.54	.50	1.04
4900		25 x 25	8	154	240	.25	.23	.48
4950			10	192	240	.31	.28	.59
5000			12	230	240	.37	.34	.71
5050			14	269	240	.43	.40	.83

B1010 Floor Construction

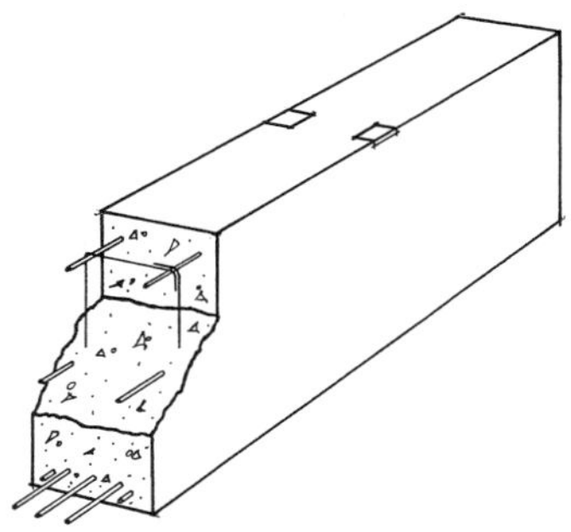

General: Beams priced in the following table are plant produced prestressed members transported to the site and erected.

Pricing Assumptions: Prices are based upon 10,000 S.F. to 20,000 S.F. projects and 50 mile to 100 mile transport.

Normal steel for connections is included in price. Deduct 20% from prices for Southern states. Add 10% to prices for Western states.

Design Assumptions: Normal weight concrete to 150 lbs./C.F. f'c = 5 KSI. Prestressing Steel: 250 KSI straight strand. Non-Prestressing Steel: fy = 60 KSI.

B1010 213 Rectangular Precast Beams

	SPAN (FT.)	SUPERIMPOSED LOAD (K.L.F.)	SIZE W X D (IN.)	BEAM WEIGHT (P.L.F.)	TOTAL LOAD (K.L.F.)	COST PER L.F.		
						MAT.	INST.	TOTAL
2200	15	2.32	12x16	200	2.52	162	19.80	181.80
2250		3.80	12x20	250	4.05	168	19.80	187.80
2300		5.60	12x24	300	5.90	174	20.50	194.50
2350		7.65	12x28	350	8.00	179	20.50	199.50
2400		5.85	18x20	375	6.73	179	20.50	199.50
2450		8.26	18x24	450	8.71	185	22	207
2500		11.39	18x28	525	11.91	197	22	219
2550		18.90	18x36	675	19.58	208	22	230
2600		10.78	24x24	600	11.38	203	22	225
2700		15.12	24x28	700	15.82	214	22	236
2750		25.23	24x36	900	26.13	232	23	255
2800	20	1.22	12x16	200	1.44	162	14.85	176.85
2850		2.03	12x20	250	2.28	168	15.40	183.40
2900		3.02	12x24	300	3.32	174	15.40	189.40
2950		4.15	12x28	350	4.50	179	15.40	194.40
3000		6.85	12x36	450	7.30	187	16.65	203.65
3050		3.13	18x20	375	3.50	179	15.40	194.40
3100		4.45	18x24	450	4.90	185	16.65	201.65
3150		6.18	18x28	525	6.70	197	16.65	213.65
3200		10.33	18x36	675	11.00	208	17.35	225.35
3400		15.48	18x44	825	16.30	226	18.90	244.90
3450		21.63	18x52	975	22.60	238	18.90	256.90
3500		5.80	24x24	600	6.40	203	16.65	219.65
3600		8.20	24x28	700	8.90	214	17.35	231.35
3700		13.80	24x36	900	14.70	232	18.90	250.90
3800		20.70	24x44	1100	21.80	249	21	270
3850		29.20	24x52	1300	30.50	272	26	298
3900	25	1.82	12x24	300	2.12	174	12.35	186.35
4000		2.53	12x28	350	2.88	179	13.30	192.30
4050		5.18	12x36	450	5.63	187	13.30	200.30
4100		6.55	12x44	550	7.10	197	13.30	210.30
4150		9.08	12x52	650	9.73	208	13.85	221.85
4200		1.86	18x20	375	2.24	179	13.30	192.30
4250		2.69	18x24	450	3.14	185	13.30	198.30
4300		3.76	18x28	525	4.29	197	13.30	210.30
4350		6.37	18x36	675	7.05	208	15.15	223.15
4400		9.60	18x44	872	10.43	226	16.65	242.65
4450		13.49	18x52	975	14.47	238	16.65	254.65

B1010 Floor Construction

B1010 213 Rectangular Precast Beams

	SPAN (FT.)	SUPERIMPOSED LOAD (K.L.F.)	SIZE W X D (IN.)	BEAM WEIGHT (P.L.F.)	TOTAL LOAD (K.L.F.)	COST PER L.F. MAT.	INST.	TOTAL
4500	25	18.40	18x60	1125	19.53	255	18.45	273.45
4600		3.50	24x24	600	4.10	203	15.15	218.15
4700		5.00	24x28	700	5.70	214	15.15	229.15
4800		8.50	24x36	900	9.40	232	16.65	248.65
4900		12.85	24x44	1100	13.95	249	18.45	267.45
5000		18.22	24x52	1300	19.52	272	21	293
5050		24.48	24x60	1500	26.00	290	24	314
5100	30	1.65	12x28	350	2.00	179	11.55	190.55
5150		2.79	12x36	450	3.24	187	11.55	198.55
5200		4.38	12x44	550	4.93	197	12.60	209.60
5250		6.10	12x52	650	6.75	208	12.60	220.60
5300		8.36	12x60	750	9.11	220	13.85	233.85
5400		1.72	18x24	450	2.17	185	11.55	196.55
5450		2.45	18x28	525	2.98	197	12.60	209.60
5750		4.21	18x36	675	4.89	208	12.60	220.60
6000		6.42	18x44	825	7.25	226	13.85	239.85
6250		9.07	18x52	975	10.05	238	15.40	253.40
6500		12.43	18x60	1125	13.56	255	17.35	272.35
6750		2.24	24x24	600	2.84	203	12.60	215.60
7000		3.26	24x28	700	3.96	214	13.85	227.85
7100		5.63	24x36	900	6.53	232	15.40	247.40
7200	35	8.59	24x44	1100	9.69	249	17	266
7300		12.25	24x52	1300	13.55	272	17	289
7400		16.54	24x60	1500	18.04	290	17	307
7500	40	2.23	12x44	550	2.78	197	10.40	207.40
7600		3.15	12x52	650	3.80	208	10.40	218.40
7700		4.38	12x60	750	5.13	220	11.55	231.55
7800		2.08	18x36	675	2.76	208	11.55	219.55
7900		3.25	18x44	825	4.08	226	11.55	237.55
8000		4.68	18x52	975	5.66	238	13	251
8100		6.50	18x60	1175	7.63	255	13	268
8200		2.78	24x36	900	3.60	232	16	248
8300		4.35	24x44	1100	5.45	249	16	265
8400		6.35	24x52	1300	7.65	272	17.35	289.35
8500		8.65	24x60	1500	10.15	290	17.35	307.35
8600	45	2.35	12x52	650	3.00	208	10.30	218.30
8800		3.29	12x60	750	4.04	220	10.30	230.30
9000		2.39	18x44	825	3.22	226	13.20	239.20
9020		3.49	18x52	975	4.47	238	13.20	251.20
9200		4.90	18x60	1125	6.03	255	15.40	270.40
9250		3.21	24x44	1100	4.31	249	16.80	265.80
9300		4.72	24x52	1300	6.02	272	16.80	288.80
9500		6.52	24x60	1500	8.02	290	16.80	306.80
9600	50	2.64	18x52	975	3.62	238	13.85	251.85
9700		3.75	18x60	1125	4.88	255	16.65	271.65
9750		3.58	24x52	1300	4.88	272	16.65	288.65
9800		5.00	24x60	1500	6.50	290	18.45	308.45
9950	55	3.86	24x60	1500	5.36	290	16.80	306.80

B1010 Floor Construction

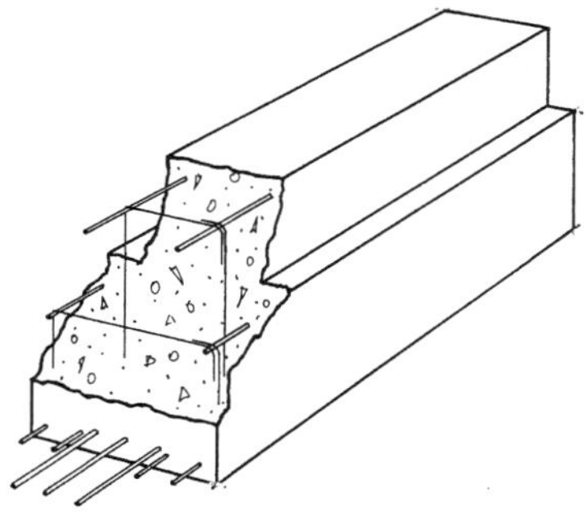

General: Beams priced in the following table are plant produced prestressed members transported to the site and erected.

Pricing Assumptions: Prices are based upon 10,000 S.F. to 20,000 S.F. projects and 50 mile to 100 mile transport.

Normal steel for connections is included in price. Deduct 20% from prices for Southern states. Add 10% to prices for Western states.

Design Assumptions: Normal weight concrete to 150 lbs./C.F. f'c = 5 KSI. Prestressing Steel: 250 KSI straight strand. Non-Prestressing Steel: fy = 60 KSI.

B1010 214 "T" Shaped Precast Beams

	SPAN (FT.)	SUPERIMPOSED LOAD (K.L.F.)	SIZE W X D (IN.)	BEAM WEIGHT (P.L.F.)	TOTAL LOAD (K.L.F.)	COST PER L.F.		
						MAT.	INST.	TOTAL
2300	15	2.8	12x16	260	3.06	185	20.50	205.50
2350		4.33	12x20	355	4.69	198	20.50	218.50
2400		6.17	12x24	445	6.62	208	22	230
2500		8.37	12x28	515	8.89	220	22	242
2550		13.54	12x36	680	14.22	243	22	265
2600		8.83	18x24	595	9.43	226	22	248
2650		12.11	18x28	690	12.8	238	22	260
2700		19.72	18x36	905	24.63	266	23	289
2800		11.52	24x24	745	12.27	238	22	260
2900		15.85	24x28	865	16.72	255	23	278
3000		26.07	24x36	1130	27.20	290	25.50	315.50
3100	20	1.46	12x16	260	1.72	185	15.40	200.40
3200		2.28	12x20	355	2.64	198	15.40	213.40
3300		3.28	12x24	445	3.73	208	16.65	224.65
3400		4.49	12x28	515	5.00	220	16.65	236.65
3500		7.32	12x36	680	8.00	243	17.35	260.35
3600		11.26	12x44	840	12.10	255	18.90	273.90
3700		4.70	18x24	595	5.30	226	16.65	242.65
3800		6.51	18x28	690	7.20	238	17.35	255.35
3900		10.7	18x36	905	11.61	266	18.90	284.90
4300		16.19	18x44	1115	17.31	295	21	316
4400		22.77	18x52	1330	24.10	320	21	341
4500		6.15	24x24	745	6.90	238	17.35	255.35
4600		8.54	24x28	865	9.41	255	17.35	272.35
4700		14.17	24x36	1130	15.30	290	21	311
4800		21.41	24x44	1390	22.80	320	26	346
4900		30.25	24x52	1655	31.91	355	26	381

B1010 Floor Construction

B1010 214 "T" Shaped Precast Beams

	SPAN (FT.)	SUPERIMPOSED LOAD (K.L.F.)	SIZE W X D (IN.)	BEAM WEIGHT (P.L.F.)	TOTAL LOAD (K.L.F.)	COST PER L.F.		
						MAT.	INST.	TOTAL
5000	25	2.68	12x28	515	3.2	220	13.30	233.30
5050		4.44	12x36	680	5.12	243	15.15	258.15
5100		6.90	12x44	840	7.74	255	15.15	270.15
5200		9.75	12x52	1005	10.76	300	15.85	315.85
5300		13.43	12x60	1165	14.60	355	16.65	371.65
5350		3.92	18x28	690	4.61	238	15.15	253.15
5400		6.52	18x36	905	7.43	266	15.85	281.85
5500		9.96	18x44	1115	11.08	295	17.55	312.55
5600		14.09	18x52	1330	15.42	320	18.45	338.45
5650		19.39	18x60	1540	20.93	345	21	366
5700		3.67	24x24	745	4.42	238	15.15	253.15
5750		5.15	24x28	865	6.02	255	16.65	271.65
5800		8.66	24x36	1130	9.79	290	18.45	308.45
5850		13.20	24x44	1390	14.59	320	21	341
5900		18.76	24x52	1655	20.42	355	22	377
5950		25.35	24x60	1916	27.27	380	24	404
6000	30	2.88	12x36	680	3.56	243	12.60	255.60
6100		4.54	12x44	840	5.38	255	13.85	268.85
6200		6.46	12x52	1005	7.47	300	13.85	313.85
6250		8.97	12x60	1165	10.14	355	15.40	370.40
6300		4.25	18x36	905	5.16	266	13.85	279.85
6350		6.57	18x44	1115	7.69	295	15.40	310.40
6400		9.38	18x52	1330	10.71	320	17.35	337.35
6500		13.00	18x60	1540	14.54	345	19.80	364.80
6700		3.31	24x28	865	4.18	255	15.40	270.40
6750		5.67	24x36	1130	6.80	290	17.35	307.35
6800		8.74	24x44	1390	10.13	320	18.45	338.45
6850		12.52	24x52	1655	14.18	355	19.80	374.80
6900		17.00	24x60	1215	18.92	380	21.50	401.50
7000	35	3.11	12x44	840	3.95	255	11.90	266.90
7100		4.48	12x52	1005	5.49	300	12.50	312.50
7200		6.28	12x60	1165	7.45	355	14	369
7300		4.53	18x44	1115	5.65	295	14	309
7500		6.54	18x52	1330	7.87	320	15.85	335.85
7600		9.14	18x60	1540	10.68	345	17	362
7700		3.87	24x36	1130	5.00	290	17	307
7800		6.05	24x44	1390	7.44	320	18.25	338.25
7900		8.76	24x52	1655	10.42	355	18.25	373.25
8000		12.00	24x60	1915	13.92	380	19.80	399.80
8100	40	3.19	12x52	1005	4.2	300	11.55	311.55
8200		4.53	12x60	1165	5.7	355	13	368
8300		4.70	18x52	1330	6.03	320	14.85	334.85
8400		6.64	18x60	1540	8.18	345	14.85	359.85
8600		4.31	24x44	1390	5.7	320	17.35	337.35
8700		6.32	24x52	1655	7.98	355	18.90	373.90
8800		8.74	24x60	1915	10.66	380	18.90	398.90
8900	45	3.34	12x60	1165	4.51	355	13.20	368.20
9000		4.92	18x60	1540	6.46	345	15.40	360.40
9250		4.64	24x52	1655	6.30	355	18.45	373.45
9900		6.5	24x60	1915	8.42	380	18.45	398.45
9950	50	4.9	24x60	1915	6.82	380	21	401

B1010 Floor Construction

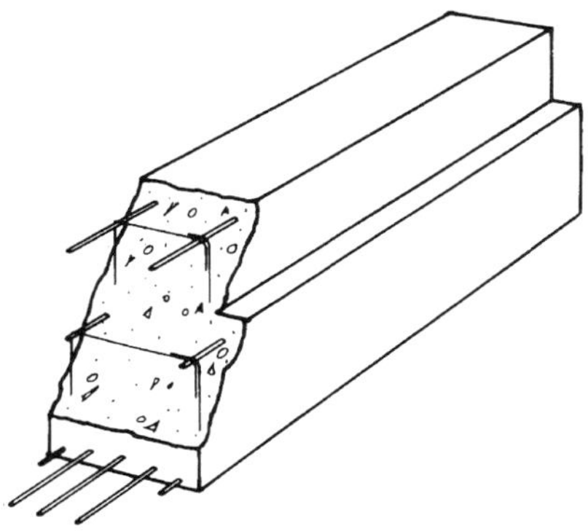

General: Beams priced in the following table are plant produced prestressed members transported to the site and erected.

Pricing Assumptions: Prices are based upon 10,000 S.F. to 20,000 S.F. projects and 50 mile to 100 mile transport.

Normal steel for connections is included in price. Deduct 20% from prices for Southern states. Add 10% to prices for Western states.

Design Assumptions: Normal weight concrete to 150 lbs./C.F. f'c = 5 KSI. Prestressing Steel: 250 KSI straight strand. Non-Prestressing Steel: fy = 60 KSI.

B1010 215 "L" Shaped Precast Beams

B1010 215	SPAN (FT.)	SUPERIMPOSED LOAD (K.L.F.)	SIZE W X D (IN.)	BEAM WEIGHT (P.L.F.)	TOTAL LOAD (K.L.F.)	COST PER L.F.		
						MAT.	INST.	TOTAL
2250	15	2.58	12x16	230	2.81	174	20.50	194.50
2300		4.10	12x20	300	4.40	174	20.50	194.50
2400		5.92	12x24	370	6.29	191	20.50	211.50
2450		8.09	12x28	435	8.53	192	20.50	212.50
2500		12.95	12x36	565	13.52	214	22	236
2600		8.55	18x24	520	9.07	208	22	230
2650		11.83	18x28	610	12.44	220	22	242
2700		19.30	18x36	790	20.09	238	22	260
2750		11.24	24x24	670	11.91	220	22	242
2800		15.40	24x28	780	16.18	232	23	255
2850		25.65	24x36	1015	26.67	261	24	285
2900	20	2.18	12x20	300	2.48	174	15.40	189.40
2950		3.17	12x24	370	3.54	191	16	207
3000		4.37	12x28	435	4.81	192	16	208
3100		7.04	12x36	565	7.60	214	17.35	231.35
3150		10.80	12x44	695	11.50	232	18.05	250.05
3200		15.08	12x52	825	15.91	249	18.90	267.90
3250		4.58	18x24	520	5.10	208	16	224
3300		6.39	18x28	610	7.00	220	16.65	236.65
3350		10.51	18x36	790	11.30	238	17.35	255.35
3400		15.73	18x44	970	16.70	261	18.90	279.90
3550		22.15	18x52	1150	23.30	278	19.80	297.80
3600		6.03	24x24	370	6.40	220	16.65	236.65
3650		8.32	24x28	780	9.10	232	18.90	250.90
3700		13.98	24x36	1015	15.00	261	19.80	280.80
3800		21.05	24x44	1245	22.30	284	23	307
3900		29.73	24x52	1475	31.21	315	26	341

B1010 Floor Construction

B1010 215 "L" Shaped Precast Beams

	SPAN (FT.)	SUPERIMPOSED LOAD (K.L.F.)	SIZE W X D (IN.)	BEAM WEIGHT (P.L.F.)	TOTAL LOAD (K.L.F.)	COST PER L.F.		
						MAT.	INST.	TOTAL
4000	25	2.64	12x28	435	3.08	192	13.30	205.30
4100		4.30	12x36	565	4.87	214	13.85	227.85
4200		6.65	12x44	695	7.35	232	14.45	246.45
4250		9.35	12x52	875	10.18	249	15.15	264.15
4300		12.81	12x60	950	13.76	266	15.85	281.85
4350		2.74	18x24	570	3.76	208	13.85	221.85
4400		3.67	18x28	610	4.28	220	14.45	234.45
4450		6.44	18x36	790	7.23	238	15.85	253.85
4500		9.72	18x44	970	10.69	261	17.55	278.55
4600		13.76	18x52	1150	14.91	278	17.55	295.55
4700		18.33	18x60	1330	20.16	300	19.60	319.60
4800		3.62	24x24	370	3.99	220	15.15	235.15
4900		5.04	24x28	780	5.82	232	15.85	247.85
5000		8.58	24x36	1015	9.60	261	16.65	277.65
5100		13.00	24x44	1245	14.25	284	19.60	303.60
5200		18.50	24x52	1475	19.98	315	22	337
5300	30	2.80	12x36	565	3.37	214	12.10	226.10
5400		4.42	12x44	695	5.12	232	13.20	245.20
5500		6.24	12x52	875	7.07	249	13.20	262.20
5600		8.60	12x60	950	9.55	266	14.60	280.60
5700		2.50	18x28	610	3.11	220	13.20	233.20
5800		4.23	18x36	790	5.02	238	13.20	251.20
5900		6.45	18x44	970	7.42	261	14.60	275.60
6000		9.20	18x52	1150	10.35	278	16.35	294.35
6100		12.67	18x60	1330	14.00	300	18.45	318.45
6200		3.26	24x28	780	4.04	232	14.60	246.60
6300		5.65	24x36	1015	6.67	261	16.35	277.35
6400		8.66	24x44	1245	9.90	284	17.35	301.35
6500	35	3.00	12x44	695	3.70	232	11.90	243.90
6600		4.27	12x52	825	5.20	249	11.90	260.90
6700		6.00	12x60	950	6.95	266	13.20	279.20
6800		2.90	18x36	790	3.69	238	11.90	249.90
6900		4.48	18x44	970	5.45	261	13.20	274.20
7000		6.45	18x52	1150	7.60	278	14	292
7100		8.95	18x60	1330	10.28	300	14.85	314.85
7200		3.88	24x36	1015	4.90	261	17	278
7300		6.03	24x44	1245	7.28	284	18.25	302.25
7400		8.71	24x52	1475	10.19	315	18.25	333.25
7500	40	3.15	12x52	825	3.98	249	10.95	259.95
7600		4.43	12x60	950	5.38	266	12.20	278.20
7700		3.20	18x44	970	4.17	261	13	274
7800		4.68	18x52	1150	5.83	278	13.85	291.85
7900		6.55	18x60	1330	7.88	300	13.85	313.85
8000		4.43	24x44	1245	5.68	284	18.90	302.90
8100		6.33	24x52	1475	7.81	315	18.90	333.90
8200	45	3.30	12x60	950	4.25	266	11.55	277.55
8300		3.45	18x52	1150	4.60	278	14.20	292.20
8500		4.44	18x60	1330	5.77	300	16.80	316.80
8750		3.16	24x44	1245	4.41	284	18.45	302.45
9000		4.68	24x52	1475	6.16	315	18.45	333.45
9900	50	3.71	18x60	1330	5.04	300	16.65	316.65
9950		3.82	24x52	1475	5.30	315	18.45	333.45

B1010 Floor Construction

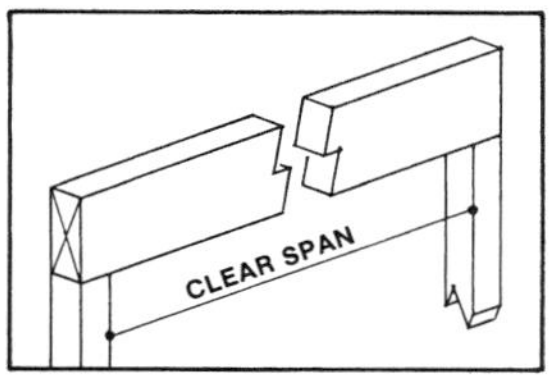

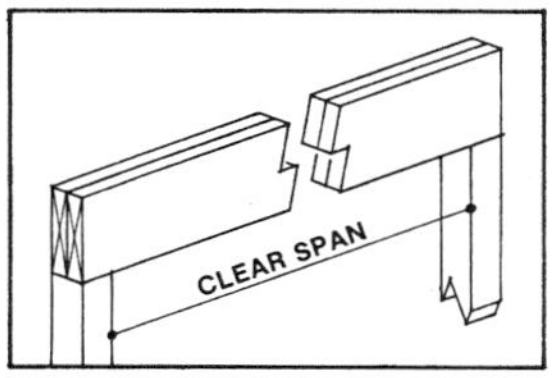

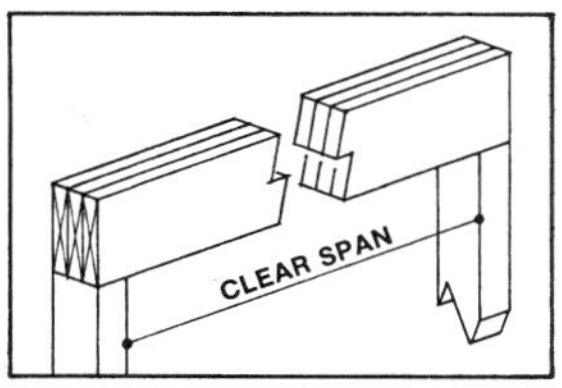

Description: Table below lists the cost per L.F. and total uniform load allowable for various size beams at various spans.

Design Assumptions: Fiber strength (f) is 1,000 PSI. Maximum deflection does not exceed 1/360 the span of the beam. Modulus of elasticity (E) is 1,100,000 psi. Anything less will result in excessive deflection in the longer members so that spans must be reduced or member size increased.

The total loads are in pounds per foot.

The span is in feet and is the unsupported clear span.

The costs are based on net quantities and do not include any allowance for waste or overlap.

The member sizes are from 2″ x 6″ to 4″ x 12″ and include one, two and three pieces of each member size to arrive at the costs, spans and loading. If more than a three-piece beam is required, its cost may be calculated directly. A four-piece 2″ x 6″ would cost two times as much as a two-piece 2″ x 6″ beam. The maximum span and total load cannot be doubled, however.

B1010 216 — Wood Beams

	BEAM SIZE (IN.)	TOTAL LOAD (P.L.F.) 6′ SPAN	TOTAL LOAD (P.L.F.) 10′ SPAN	TOTAL LOAD (P.L.F.) 14′ SPAN	TOTAL LOAD (P.L.F.) 18′ SPAN	COST PER L.F. MAT.	INST.	TOTAL
1000	2 - 2 x 6	238	RB1010 -240			1.34	4.55	5.89
1050	3 - 2 x 6	357				2.01	6.85	8.86
1100	2 - 2 x 8	314				1.87	5.25	7.12
1150	3 - 2 x 8	471	232			2.80	7.90	10.70
1200	2 - 2 x 10	400	240			3.43	5.40	8.83
1250	3 - 2 x 10	600	360			5.15	8.10	13.25
1300	2 - 2 x 12	487	292	208		3.74	5.90	9.64
1350	3 - 2 x 12	731	438	313		5.60	8.85	14.45
1400	2 - 2 x 14	574	344	246		4.67	6.20	10.87
1450	3 - 2 x 14	861	516	369	243	7	9.30	16.30
1550	2 - 3 x 6	397				4.13	5.10	9.23
1600	3 - 3 x 6	595				6.20	7.65	13.85
1650	1 - 3 x 8	261				2.95	2.95	5.90
1700	2 - 3 x 8	523	258			5.90	5.90	11.80
1750	3 - 3 x 8	785	388			8.85	8.85	17.70
1800	1 - 3 x 10	334	200			3.75	3.31	7.06
1850	2 - 3 x 10	668	400			7.50	6.65	14.15
1900	3 - 3 x 10	1,002	600	293		11.25	9.95	21.20
1950	1 - 3 x 12	406	243			4.50	3.60	8.10
2000	2 - 3 x 12	812	487	348		9	7.20	16.20
2050	3 - 3 x 12	1,218	731	522	248	13.50	10.80	24.30
2100	1 - 4 x 6	278				3.15	2.95	6.10
2150	2 - 4 x 6	556				6.30	5.90	12.20
2200	1 - 4 x 8	366				3.20	3.53	6.73
2250	2 - 4 x 8	733	362			6.40	7.05	13.45
2300	1 - 4 x 10	467	280			5.15	4	9.15
2350	2 - 4 x 10	935	560			10.35	8	18.35
2400	1 - 4 x 12	568	341			5.70	4.30	10
2450	2 - 4 x 12	1,137	682		232	11.40	8.60	20

B1010 Floor Construction

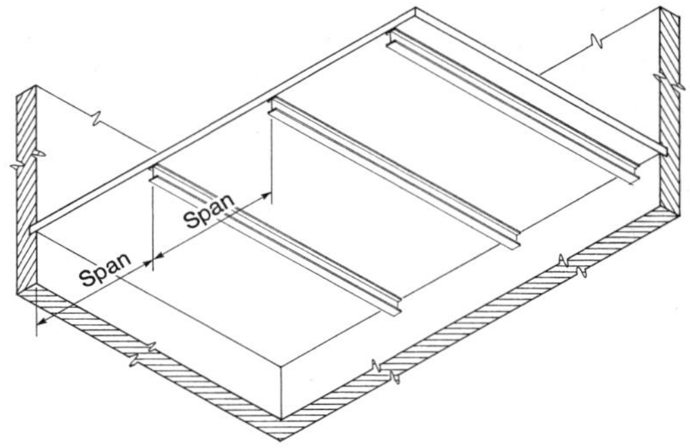

General: Solid concrete slabs of uniform depth reinforced for flexure in one direction and for temperature and shrinkage in the other direction.

Design and Pricing Assumptions:
Concrete f'c = 3 KSI normal weight placed by pump.
Reinforcement fy = 60 KSI.
Deflection≤ span/360.
Forms, four uses hung from steel beams plus edge forms.
Steel trowel finish for finish floor and cure.

System Components	QUANTITY	UNIT	COST PER S.F.		
			MAT.	INST.	TOTAL
SYSTEM B1010 217 2000					
6'-0" SINGLE SPAN, 4" SLAB DEPTH, 40 PSF SUPERIMPOSED LOAD					
Forms in place, floor slab forms hung from steel beams, 4 uses	1.000	S.F.	2.12	6.15	8.27
Edge forms to 6" high on elevated slab, 4 uses	.080	L.F.	.07	1.50	1.57
Reinforcing in place, elevated slabs #4 to #7	1.030	Lb.	.57	.46	1.03
Concrete, ready mix, regular weight, 3000 psi	.330	C.F.	1.46		1.46
Place and vibrate concrete, elevated slab less than 6", pumped	.330	C.F.		.55	.55
Finishing floor, monolithic steel trowel finish for finish floor	1.000	S.F.		.93	.93
Curing with sprayed membrane curing compound	1.000	S.F.	.14	.10	.24
TOTAL			4.36	9.69	14.05

B1010 217 Cast in Place Slabs, One Way

	SLAB DESIGN & SPAN (FT.)	SUPERIMPOSED LOAD (P.S.F.)	THICKNESS (IN.)	TOTAL LOAD (P.S.F.)		COST PER S.F.		
						MAT.	INST.	TOTAL
2000	Single 6	40	4	90	RB1010 -010	4.36	9.70	14.06
2100		75	4	125	RB1010 -100	4.36	9.70	14.06
2200		125	4	175		4.36	9.70	14.06
2300		200	4	250		4.55	9.85	14.40
2500	Single 8	40	4	90		4.31	8.55	12.86
2600		75	4	125		4.34	9.35	13.69
2700		125	4-1/2	181		4.52	9.40	13.92
2800		200	5	262		4.99	9.70	14.69
3000	Single 10	40	4	90		4.48	9.25	13.73
3100		75	4	125		4.48	9.25	13.73
3200		125	5	188		4.96	9.45	14.41
3300		200	7-1/2	293		6.50	10.60	17.10
3500	Single 15	40	5-1/2	90		5.30	9.35	14.65
3600		75	6-1/2	156		5.90	9.85	15.75
3700		125	7-1/2	219		6.40	10.10	16.50
3800		200	8-1/2	306		6.95	10.30	17.25
4000	Single 20	40	7-1/2	115		6.40	9.85	16.25
4100		75	9	200		6.95	9.80	16.75
4200		125	10	250		7.65	10.20	17.85
4300		200	10	324		8.20	10.65	18.85

B1010 Floor Construction

B1010 217 Cast in Place Slabs, One Way

	SLAB DESIGN & SPAN (FT.)	SUPERIMPOSED LOAD (P.S.F.)	THICKNESS (IN.)	TOTAL LOAD (P.S.F.)		COST PER S.F.		
						MAT.	INST.	TOTAL
4500	Multi 6	40	4	90		4.42	9.20	13.62
4600		75	4	125		4.42	9.20	13.62
4700		125	4	175		4.42	9.20	13.62
4800		200	4	250		4.42	9.20	13.62
5000	Multi 8	40	4	90		4.41	8.95	13.36
5100		75	4	125		4.41	8.95	13.36
5200		125	4	175		4.41	8.95	13.36
5300		200	4	250		4.41	8.95	13.36
5500	Multi 10	40	4	90		4.41	8.80	13.21
5600		75	4	125		4.41	8.90	13.31
5700		125	4	175		4.57	9	13.57
5800		200	5	263		5.05	9.40	14.45
6600	Multi 15	40	5	103		5.05	9.15	14.20
6800		75	5	138		5.25	9.30	14.55
7000		125	6-1/2	206		5.85	9.60	15.45
7100		200	6-1/2	281		6	9.70	15.70
7500	Multi 20	40	5-1/2	109		5.75	9.45	15.20
7600		75	6-1/2	156		6.25	9.75	16
7800		125	9	238		7.75	10.60	18.35
8000		200	9	313		7.75	10.60	18.35

B1010 Floor Construction

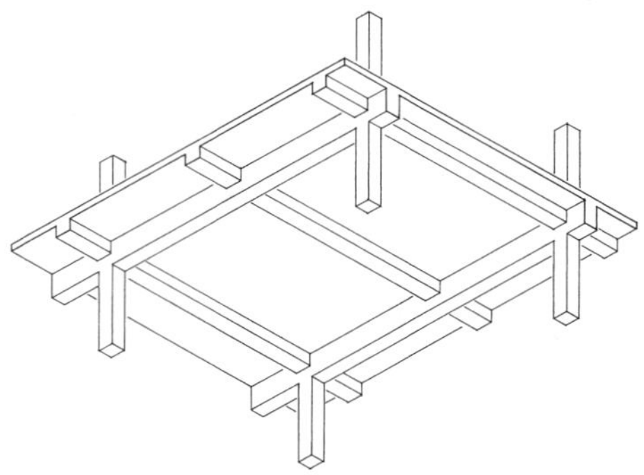

General: Solid concrete one-way slab cast monolithically with reinforced concrete support beams and girders.

Design and Pricing Assumptions:
Concrete f'c = 3 KSI, normal weight, placed by concrete pump.
Reinforcement, fy = 60 KSI.
Forms, four use.
Finish, steel trowel.
Curing, spray on membrane.
Based on 4 bay x 4 bay structure.

System Components	QUANTITY	UNIT	COST PER S.F. MAT.	COST PER S.F. INST.	COST PER S.F. TOTAL
SYSTEM B1010 219 3000					
BM. & SLAB ONE WAY 15′ X 15′ BAY, 40 PSF S.LOAD, 12″ MIN. COL.					
Forms in place, flat plate to 15′ high, 4 uses	.858	S.F.	1.15	5.32	6.47
Forms in place, exterior spandrel, 12″ wide, 4 uses	.142	SFCA	.16	1.58	1.74
Forms in place, interior beam. 12″ wide, 4 uses	.306	SFCA	.39	2.82	3.21
Reinforcing in place, elevated slabs #4 to #7	1.600	Lb.	.88	.72	1.60
Concrete, ready mix, regular weight, 3000 psi	.410	C.F.	1.79		1.79
Place and vibrate concrete, elevated slab less than 6″, pump	.410	C.F.		.67	.67
Finish floor, monolithic steel trowel finish for finish floor	1.000	S.F.		.93	.93
Cure with sprayed membrane curing compound	.010	C.S.F.	.14	.10	.24
TOTAL			4.51	12.14	16.65

B1010 219 Cast in Place Beam & Slab, One Way

	BAY SIZE (FT.)	SUPERIMPOSED LOAD (P.S.F.)	MINIMUM COL. SIZE (IN.)	SLAB THICKNESS (IN.)	TOTAL LOAD (P.S.F.)	COST PER S.F. MAT.	COST PER S.F. INST.	COST PER S.F. TOTAL
3000	15x15	40	12	4	120	4.51	12.15	16.66
3100	RB1010 -010	75	12	4	138	4.60	12.20	16.80
3200		125	12	4	188	4.73	12.30	17.03
3300		200	14	4	266	5	12.75	17.75
3600	15x20	40	12	4	102	4.63	12	16.63
3700	RB1010 -100	75	12	4	140	4.84	12.40	17.24
3800		125	14	4	192	5.10	12.80	17.90
3900		200	16	4	272	5.65	13.65	19.30
4200	20x20	40	12	5	115	5.10	11.70	16.80
4300		75	14	5	154	5.50	12.65	18.15
4400		125	16	5	206	5.75	13.30	19.05
4500		200	18	5	287	6.45	14.25	20.70
5000	20x25	40	12	5-1/2	121	5.30	11.75	17.05
5100		75	14	5-1/2	160	5.80	12.75	18.55
5200		125	16	5-1/2	215	6.25	13.45	19.70
5300		200	18	5-1/2	294	6.75	14.45	21.20
5500	25x25	40	12	6	129	5.55	11.55	17.10
5600		75	16	6	171	6.10	12.35	18.45
5700		125	18	6	227	7.10	14.15	21.25
5800		200	2	6	300	7.85	15.20	23.05
6500	25x30	40	14	6-1/2	132	5.70	11.75	17.45
6600		75	16	6-1/2	172	6.20	12.45	18.65
6700		125	18	6-1/2	231	7.25	14.10	21.35
6800		200	20	6-1/2	312	7.90	15.30	23.20

B1010 Floor Construction

B1010 219 Cast in Place Beam & Slab, One Way

	BAY SIZE (FT.)	SUPERIMPOSED LOAD (P.S.F.)	MINIMUM COL. SIZE (IN.)	SLAB THICKNESS (IN.)	TOTAL LOAD (P.S.F.)	COST PER S.F.		
						MAT.	INST.	TOTAL
7000	30x30	40	14	7-1/2	150	6.60	12.60	19.20
7100		75	18	7-1/2	191	7.30	13.30	20.60
7300		125	20	7-1/2	245	7.75	14.15	21.90
7400		200	24	7-1/2	328	8.65	15.65	24.30
7500	30x35	40	16	8	158	7	13.05	20.05
7600		75	18	8	196	7.45	13.45	20.90
7700		125	22	8	254	8.35	14.95	23.30
7800		200	26	8	332	9.05	15.40	24.45
8000	35x35	40	16	9	169	7.80	13.40	21.20
8200		75	20	9	213	8.45	14.65	23.10
8400		125	24	9	272	9.30	15.20	24.50
8600		200	26	9	355	10.20	16.20	26.40
9000	35x40	40	18	9	174	8	13.65	21.65
9300		75	22	9	214	8.65	14.75	23.40
9400		125	26	9	273	9.45	15.30	24.75
9600		200	30	9	355	10.35	16.30	26.65

B1010 Floor Construction

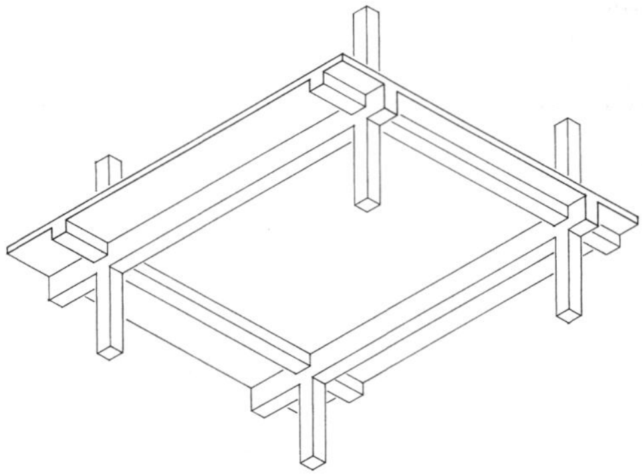

General: Solid concrete two-way slab cast monolithically with reinforced concrete support beams and girders.

Design and Pricing Assumptions:
Concrete f'c = 3 KSI, normal weight, placed by concrete pump.
Reinforcement, fy = 60 KSI.
Forms, four use.
Finish, steel trowel.
Curing, spray on membrane.
Based on 4 bay x 4 bay structure.

System Components	QUANTITY	UNIT	COST PER S.F. MAT.	COST PER S.F. INST.	COST PER S.F. TOTAL
SYSTEM B1010 220 2000					
15' X 15' BAY, 40 PSF S. LOAD, 12" MIN. COL					
Forms in place, flat plate to 15' high, 4 uses	.894	S.F.	1.20	5.54	6.74
Forms in place, exterior spandrel, 12" wide, 4 uses	.142	SFCA	.16	1.58	1.74
Forms in place, interior beam. 12" wide, 4 uses	.178	SFCA	.23	1.64	1.87
Reinforcing in place, elevated slabs #4 to #7	1.598	Lb.	.88	.72	1.60
Concrete, ready mix, regular weight, 3000 psi	.437	C.F.	1.91		1.91
Place and vibrate concrete, elevated slab less than 6", pump	.437	C.F.		.72	.72
Finish floor, monolithic steel trowel finish for finish floor	1.000	S.F.		.93	.93
Cure with sprayed membrane curing compound	.010	C.S.F.	.14	.10	.24
TOTAL			4.52	11.23	15.75

B1010 220 — Cast in Place Beam & Slab, Two Way

	BAY SIZE (FT.)	SUPERIMPOSED LOAD (P.S.F.)	MINIMUM COL. SIZE (IN.)	SLAB THICKNESS (IN.)	TOTAL LOAD (P.S.F.)	COST PER S.F. MAT.	COST PER S.F. INST.	COST PER S.F. TOTAL
2000	15x15	40	12	4-1/2	106	4.52	11.20	15.72
2200	RB1010 -010	75	12	4-1/2	143	4.65	11.35	16
2250		125	12	4-1/2	195	4.99	11.65	16.64
2300		200	14	4-1/2	274	5.40	12.35	17.75
2400	15 x 20	40	12	5-1/2	120	5.15	11.15	16.30
2600	RB1010 -100	75	12	5-1/2	159	5.60	11.95	17.55
2800		125	14	5-1/2	213	6.05	12.90	18.95
2900		200	16	5-1/2	294	6.60	13.75	20.35
3100	20 x 20	40	12	6	132	5.75	11.35	17.10
3300		75	14	6	167	6.05	12.40	18.45
3400		125	16	6	220	6.25	12.85	19.10
3600		200	18	6	301	6.95	13.70	20.65
4000	20 x 25	40	12	7	141	6.15	11.85	18
4300		75	14	7	181	7	12.95	19.95
4500		125	16	7	236	7.15	13.30	20.45
4700		200	18	7	317	7.90	14.35	22.25
5100	25 x 25	40	12	7-1/2	149	6.40	12	18.40
5200		75	16	7-1/2	185	7	12.85	19.85
5300		125	18	7-1/2	250	7.60	13.90	21.50
5400		200	20	7-1/2	332	9.05	15.05	24.10
6000	25 x 30	40	14	8-1/2	165	7.15	12	19.15
6200		75	16	8-1/2	201	7.60	13.05	20.65
6400		125	18	8-1/2	259	8.30	14.05	22.35

B1010 Floor Construction

B1010 220 Cast in Place Beam & Slab, Two Way

	BAY SIZE (FT.)	SUPERIMPOSED LOAD (P.S.F.)	MINIMUM COL. SIZE (IN.)	SLAB THICKNESS (IN.)	TOTAL LOAD (P.S.F.)	COST PER S.F.		
						MAT.	INST.	TOTAL
6600	25 x 30	200	20	8-1/2	341	9.40	14.85	24.25
7000	30 x 30	40	14	9	170	7.75	12.80	20.55
7100		75	18	9	212	8.50	13.80	22.30
7300		125	20	9	267	9.10	14.25	23.35
7400		200	24	9	351	9.90	15.05	24.95
7600	30 x 35	40	16	10	188	8.55	13.55	22.10
7700		75	18	10	225	9.05	14.05	23.10
8000		125	22	10	282	10.05	15.10	25.15
8200		200	26	10	371	10.85	15.55	26.40
8500	35 x 35	40	16	10-1/2	193	9.15	13.80	22.95
8600		75	20	10-1/2	233	9.55	14.30	23.85
9000		125	24	10-1/2	287	10.70	15.35	26.05
9100		200	26	10-1/2	370	11.05	16.30	27.35
9300	35 x 40	40	18	11-1/2	208	9.90	14.30	24.20
9500		75	22	11-1/2	247	10.35	14.65	25
9800		125	24	11-1/2	302	11.05	15.15	26.20
9900		200	26	11-1/2	383	11.85	16.15	28

B1010 Floor Construction

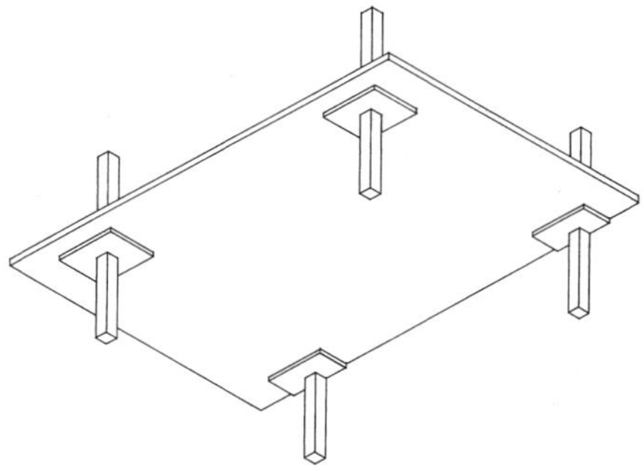

General: Flat Slab: Solid uniform depth concrete two-way slabs with drop panels at columns and no column capitals.

Design and Pricing Assumptions:
Concrete f'c = 3 KSI, placed by concrete pump.
Reinforcement, fy = 60 KSI.
Forms, four use.
Finish, steel trowel.
Curing, spray on membrane.
Based on 4 bay x 4 bay structure.

System Components	QUANTITY	UNIT	COST PER S.F. MAT.	COST PER S.F. INST.	COST PER S.F. TOTAL
SYSTEM B1010 222 1700					
15′ X 15′ BAY, 40 PSF S. LOAD, 12″ MIN. COL., 6″ SLAB, 1-1/2″ DROP, 117 PSF					
Forms in place, flat slab with drop panels, to 15′ high, 4 uses	.993	S.F.	1.53	6.31	7.84
Forms in place, exterior spandrel, 12″ wide, 4 uses	.034	SFCA	.04	.38	.42
Reinforcing in place, elevated slabs #4 to #7	1.588	Lb.	.87	.71	1.58
Concrete, ready mix, regular weight, 3000 psi	.513	C.F.	2.24		2.24
Place and vibrate concrete, elevated slab, 6″ to 10″ pump	.513	C.F.		.71	.71
Finish floor, monolithic steel trowel finish for finish floor	1.000	S.F.		.93	.93
Cure with sprayed membrane curing compound	.010	C.S.F.	.14	.10	.24
TOTAL			4.82	9.14	13.96

B1010 222 Cast in Place Flat Slab with Drop Panels

	BAY SIZE (FT.)	SUPERIMPOSED LOAD (P.S.F.)	MINIMUM COL. SIZE (IN.)	SLAB & DROP (IN.)	TOTAL LOAD (P.S.F.)	COST PER S.F. MAT.	COST PER S.F. INST.	COST PER S.F. TOTAL
1700	15 x 15	40	12	6 - 1-1/2	117	4.82	9.15	13.97
1720	RB1010 -010	75	12	6 - 2-1/2	153	4.94	9.20	14.14
1760		125	14	6 - 3-1/2	205	5.15	9.35	14.50
1780		200	16	6 - 4-1/2	281	5.45	9.60	15.05
1840	15 x 20	40	12	6-1/2 - 2	124	5.15	9.25	14.40
1860	RB1010 -100	75	14	6-1/2 - 4	162	5.40	9.50	14.90
1880		125	16	6-1/2 - 5	213	5.75	9.70	15.45
1900		200	18	6-1/2 - 6	293	5.90	9.85	15.75
1960	20 x 20	40	12	7 - 3	132	5.45	9.40	14.85
1980		75	16	7 - 4	168	5.80	9.65	15.45
2000		125	18	7 - 6	221	6.45	9.95	16.40
2100		200	20	8 - 6-1/2	309	6.55	10.10	16.65
2300	20 x 25	40	12	8 - 5	147	6.10	9.75	15.85
2400		75	18	8 - 6-1/2	184	6.60	10.10	16.70
2600		125	20	8 - 8	236	7.20	10.50	17.70
2800		200	22	8-1/2 - 8-1/2	323	7.45	10.75	18.20
3200	25 x 25	40	12	8-1/2 - 5-1/2	154	6.45	9.85	16.30
3400		75	18	8-1/2 - 7	191	6.80	10.15	16.95
4000		125	20	8-1/2 - 8-1/2	243	7.30	10.55	17.85
4400		200	24	9 - 8-1/2	329	7.65	10.80	18.45
5000	25 x 30	40	14	9-1/2 - 7	168	7	10.20	17.20
5200		75	18	9-1/2 - 7	203	7.45	10.60	18.05
5600		125	22	9-1/2 - 8	256	7.85	10.85	18.70
5800		200	24	10 - 10	342	8.35	11.20	19.55

B1010 Floor Construction

B1010 222 Cast in Place Flat Slab with Drop Panels

	BAY SIZE (FT.)	SUPERIMPOSED LOAD (P.S.F.)	MINIMUM COL. SIZE (IN.)	SLAB & DROP (IN.)	TOTAL LOAD (P.S.F.)	COST PER S.F.		
						MAT.	INST.	TOTAL
6400	30 x 30	40	14	10-1/2 - 7-1/2	182	7.60	10.50	18.10
6600		75	18	10-1/2 - 7-1/2	217	8.10	10.90	19
6800		125	22	10-1/2 - 9	269	8.45	11.15	19.60
7000		200	26	11 - 11	359	9.10	11.60	20.70
7400	30 x 35	40	16	11-1/2 - 9	196	8.30	10.90	19.20
7900		75	20	11-1/2 - 9	231	8.85	11.30	20.15
8000		125	24	11-1/2 - 11	284	9.20	11.60	20.80
9000	35 x 35	40	16	12 - 9	202	8.55	10.95	19.50
9400		75	20	12 - 11	240	9.20	11.50	20.70
9600		125	24	12 - 11	290	9.45	11.70	21.15

B1010 Floor Construction

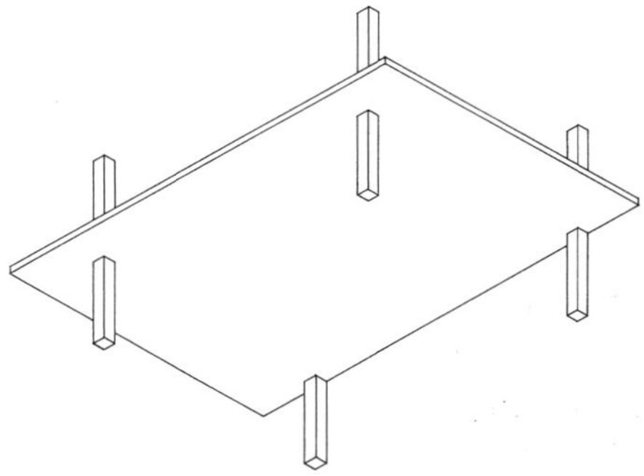

General: Flat Plates: Solid uniform depth concrete two-way slab without drops or interior beams. Primary design limit is shear at columns.

Design and Pricing Assumptions:
Concrete f'c to 4 KSI, placed by concrete pump.
Reinforcement, fy = 60 KSI.
Forms, four use.
Finish, steel trowel.
Curing, spray on membrane.
Based on 4 bay x 4 bay structure.

System Components	QUANTITY	UNIT	COST PER S.F. MAT.	INST.	TOTAL
SYSTEM B1010 223 2000					
15'X15' BAY, 40 PSF S. LOAD, 12" MIN. COL.					
Forms in place, flat plate to 15' high, 4 uses	.992	S.F.	1.33	6.15	7.48
Edge forms to 6" high on elevated slab, 4 uses	.065	L.F.	.01	.29	.30
Reinforcing in place, elevated slabs #4 to #7	1.706	Lb.	.94	.77	1.71
Concrete, ready mix, regular weight, 3000 psi	.459	C.F.	2.01		2.01
Place and vibrate concrete, elevated slab less than 6", pump	.459	C.F.		.75	.75
Finish floor, monolithic steel trowel finish for finish floor	1.000	S.F.		.93	.93
Cure with sprayed membrane curing compound	.010	C.S.F.	.14	.10	.24
TOTAL			4.43	8.99	13.42

B1010 223 — Cast in Place Flat Plate

	BAY SIZE (FT.)	SUPERIMPOSED LOAD (P.S.F.)	MINIMUM COL. SIZE (IN.)	SLAB THICKNESS (IN.)	TOTAL LOAD (P.S.F.)	COST PER S.F. MAT.	INST.	TOTAL
2000	15 x 15	40	12	5-1/2	109	4.43	9	13.43
2200	RB1010 -010	75	14	5-1/2	144	4.46	8.95	13.41
2400		125	20	5-1/2	194	4.64	9.05	13.69
2600		175	22	5-1/2	244	4.74	9.10	13.84
3000	15 x 20	40	14	7	127	5.15	9.10	14.25
3400	RB1010 -100	75	16	7-1/2	169	5.50	9.30	14.80
3600		125	22	8-1/2	231	6.05	9.55	15.60
3800		175	24	8-1/2	281	6.10	9.55	15.65
4200	20 x 20	40	16	7	127	5.15	9.10	14.25
4400		75	20	7-1/2	175	5.55	9.30	14.85
4600		125	24	8-1/2	231	6.05	9.50	15.55
5000		175	24	8-1/2	281	6.10	9.55	15.65
5600	20 x 25	40	18	8-1/2	146	6	9.50	15.50
6000		75	20	9	188	6.20	9.60	15.80
6400		125	26	9-1/2	244	6.75	9.90	16.65
6600		175	30	10	300	7	10	17
7000	25 x 25	40	20	9	152	6.25	9.55	15.80
7400		75	24	9-1/2	194	6.60	9.85	16.45
7600		125	30	10	250	7	10.05	17.05
8000								

B1010 Floor Construction

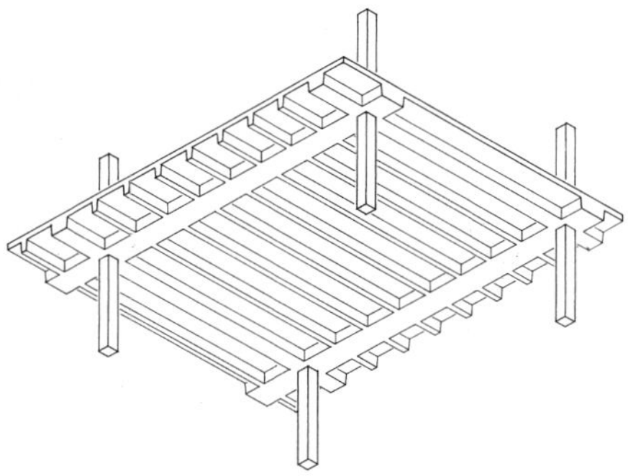

General: Combination of thin concrete slab and monolithic ribs at uniform spacing to reduce dead weight and increase rigidity. The ribs (or joists) are arranged parallel in one direction between supports.

Square end joists simplify forming. Tapered ends can increase span or provide for heavy load.

Costs for multiple span joists are provided in this section. Single span joist costs are not provided here.

Design and Pricing Assumptions:
Concrete f'c = 4 KSI, normal weight placed by concrete pump.
Reinforcement, fy = 60 KSI.
Forms, four use.
4-1/2″ slab.
30″ pans, sq. ends (except for shear req.).
6″ rib thickness.
Distribution ribs as required.
Finish, steel trowel.
Curing, spray on membrane.
Based on 4 bay x 4 bay structure.

System Components	QUANTITY	UNIT	COST PER S.F.		
			MAT.	INST.	TOTAL
SYSTEM B1010 226 2000					
15′ X 15′ BAY, 40 PSF S. LOAD, 12″ MIN. COLUMN					
Forms in place, floor slab, with 1-way joist pans, 4 use	.905	S.F.	3.51	6.24	9.75
Forms in place, exterior spandrel, 12″ wide, 4 uses	.170	SFCA	.19	1.90	2.09
Forms in place, interior beam. 12″ wide, 4 uses	.095	SFCA	.12	.87	.99
Edge forms, 7″-12″ high on elevated slab, 4 uses	.010	L.F.	.01	.07	.08
Reinforcing in place, elevated slabs #4 to #7	.628	Lb.	.35	.28	.63
Concrete ready mix, regular weight, 4000 psi	.555	C.F.	2.55		2.55
Place and vibrate concrete, elevated slab, 6″ to 10″ pump	.555	C.F.		.77	.77
Finish floor, monolithic steel trowel finish for finish floor	1.000	S.F.		.93	.93
Cure with sprayed membrane curing compound	.010	S.F.	.14	.10	.24
TOTAL			6.87	11.16	18.03

B1010 226 Cast in Place Multispan Joist Slab

	BAY SIZE (FT.)	SUPERIMPOSED LOAD (P.S.F.)	MINIMUM COL. SIZE (IN.)	RIB DEPTH (IN.)	TOTAL LOAD (P.S.F.)	COST PER S.F.		
						MAT.	INST.	TOTAL
2000	15 x 15	40	12	8	115	6.85	11.15	18
2100	RB1010 -010	75	12	8	150	6.90	11.15	18.05
2200		125	12	8	200	7.05	11.30	18.35
2300		200	14	8	275	7.20	11.70	18.90
2600	15 x 20	40	12	8	115	7	11.15	18.15
2800	RB1010 -100	75	12	8	150	7.10	11.80	18.90
3000		125	14	8	200	7.35	11.95	19.30
3300		200	16	8	275	7.70	12.15	19.85
3600	20 x 20	40	12	10	120	7.15	10.95	18.10
3900		75	14	10	155	7.45	11.65	19.10
4000		125	16	10	205	7.50	11.85	19.35
4100		200	18	10	280	7.80	12.35	20.15
4300	20 x 25	40	12	10	120	7.15	11.15	18.30
4400		75	14	10	155	7.40	11.70	19.10
4500		125	16	10	205	7.80	12.35	20.15
4600		200	18	12	280	8.15	12.95	21.10
4700	25 x 25	40	12	12	125	7.20	10.85	18.05
4800		75	16	12	160	7.60	11.45	19.05
4900		125	18	12	210	8.45	12.60	21.05
5000		200	20	14	291	8.80	12.85	21.65

B1010 Floor Construction

B1010 226 Cast in Place Multispan Joist Slab

	BAY SIZE (FT.)	SUPERIMPOSED LOAD (P.S.F.)	MINIMUM COL. SIZE (IN.)	RIB DEPTH (IN.)	TOTAL LOAD (P.S.F.)	COST PER S.F.		
						MAT.	INST.	TOTAL
5400	25 x 30	40	14	12	125	7.55	11.35	18.90
5600		75	16	12	160	7.75	11.75	19.50
5800		125	18	12	210	8.25	12.55	20.80
6000		200	20	14	291	8.85	13.05	21.90
6200	30 x 30	40	14	14	131	7.90	11.50	19.40
6400		75	18	14	166	8.10	11.85	19.95
6600		125	20	14	216	8.55	12.45	21
6700		200	24	16	297	9.15	12.95	22.10
6900	30 x 35	40	16	14	131	8.10	12	20.10
7000		75	18	14	166	8.25	12.05	20.30
7100		125	22	14	216	8.35	12.70	21.05
7200		200	26	16	297	9.20	13.40	22.60
7400	35 x 35	40	16	16	137	8.30	11.85	20.15
7500		75	20	16	172	8.70	12.30	21
7600		125	24	16	222	8.80	12.30	21.10
7700		200	26	20	309	9.50	13.10	22.60
8000	35 x 40	40	18	16	137	8.60	12.30	20.90
8100		75	22	16	172	8.95	12.80	21.75
8300		125	26	16	222	8.95	12.65	21.60
8400		200	30	20	309	9.80	13.25	23.05
8750	40 x 40	40	18	20	149	9.15	12.25	21.40
8800		75	24	20	184	9.40	12.50	21.90
8900		125	26	20	234	9.75	13	22.75
9100	40 x 45	40	20	20	149	9.50	12.65	22.15
9500		75	24	20	184	9.60	12.75	22.35
9800		125	28	20	234	9.85	13.25	23.10

B1010 Floor Construction

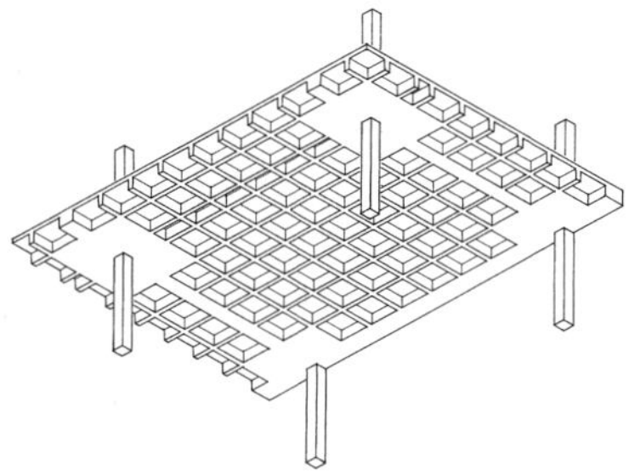

General: Waffle slabs are basically flat slabs with hollowed out domes on bottom side to reduce weight. Solid concrete heads at columns function as drops without increasing depth. The concrete ribs function as two-way right angle joist.

Joists are formed with standard sized domes. Thin slabs cover domes and are usually reinforced with welded wire fabric. Ribs have bottom steel and may have stirrups for shear.

Design and Pricing Assumptions:
Concrete f'c = 4 KSI, normal weight placed by concrete pump.
Reinforcement, fy = 60 KSI.
Forms, four use.
4-1/2" slab.
30" x 30" voids.
6" wide ribs.
(ribs @ 36" O.C.).
Rib depth filler beams as required.
Solid concrete heads at columns.
Finish, steel trowel.
Curing, spray on membrane.
Based on 4 bay x 4 bay structure.

System Components	QUANTITY	UNIT	COST PER S.F. MAT.	COST PER S.F. INST.	COST PER S.F. TOTAL
SYSTEM B1010 227 3900					
20′ X 20′ BAY, 40 PSF S. LOAD, 12″ MIN. COLUMN					
Forms in place, floor slab with 2-way waffle domes, 4 use	1.000	S.F.	4.05	7.35	11.40
Edge forms, 7″-12″ high on elevated slab, 4 uses	.052	SFCA	.03	.37	.40
Forms in place, bulkhead for slab with keyway, 1 use, 3 piece	.010	L.F.	.03	.08	.11
Reinforcing in place, elevated slabs #4 to #7	1.580	Lb.	.87	.71	1.58
Welded wire fabric rolls, 6 x 6 - W4 x W4 (4 x 4) 58 lb./c.s.f	1.000	S.F.	.35	.49	.84
Concrete ready mix, regular weight, 4000 psi	.690	C.F.	3.17		3.17
Place and vibrate concrete, elevated slab, over 10″, pump	.690	C.F.		1	1
Finish floor, monolithic steel trowel finish for finish floor	1.000	S.F.		.93	.93
Cure with sprayed membrane curing compound	.010	C.S.F.	.14	.10	.24
TOTAL			8.64	11.03	19.67

B1010 227 Cast in Place Waffle Slab

	BAY SIZE (FT.)	SUPERIMPOSED LOAD (P.S.F.)	MINIMUM COL. SIZE (IN.)	RIB DEPTH (IN.)	TOTAL LOAD (P.S.F.)	COST PER S.F. MAT.	COST PER S.F. INST.	COST PER S.F. TOTAL
3900	20 x 20	40	12	8	144	8.65	11	19.65
4000	RB1010 -010	75	12	8	179	8.80	11.15	19.95
4100		125	16	8	229	8.95	11.30	20.25
4200		200	18	8	304	9.40	11.65	21.05
4400	20 x 25	40	12	8	146	8.80	11.10	19.90
4500	RB1010 -100	75	14	8	181	9	11.25	20.25
4600		125	16	8	231	9.20	11.40	20.60
4700		200	18	8	306	9.60	11.70	21.30
4900	25 x 25	40	12	10	150	9.05	11.20	20.25
5000		75	16	10	185	9.30	11.40	20.70
5300		125	18	10	235	9.55	11.60	21.15
5500		200	20	10	310	9.80	11.85	21.65
5700	25 x 30	40	14	10	154	9.20	11.30	20.50
5800		75	16	10	189	9.45	11.50	20.95
5900		125	18	10	239	9.70	11.70	21.40
6000		200	20	12	329	10.65	12.20	22.85
6400	30 x 30	40	14	12	169	9.85	11.55	21.40
6500		75	18	12	204	10.05	11.70	21.75
6600		125	20	12	254	10.20	11.85	22.05
6700		200	24	12	329	11.05	12.50	23.55

B1010 Floor Construction

B1010 227 Cast in Place Waffle Slab

	BAY SIZE (FT.)	SUPERIMPOSED LOAD (P.S.F.)	MINIMUM COL. SIZE (IN.)	RIB DEPTH (IN.)	TOTAL LOAD (P.S.F.)	COST PER S.F.		
						MAT.	INST.	TOTAL
6900	30 x 35	40	16	12	169	10.05	11.65	21.70
7000		75	18	12	204	10.05	11.65	21.70
7100		125	22	12	254	10.50	12.05	22.55
7200		200	26	14	334	11.60	12.80	24.40
7400	35 x 35	40	16	14	174	10.60	11.95	22.55
7500		75	20	14	209	10.80	12.15	22.95
7600		125	24	14	259	11.05	12.30	23.35
7700		200	26	16	346	11.80	12.95	24.75
8000	35 x 40	40	18	14	176	10.85	12.15	23
8300		75	22	14	211	11.20	12.45	23.65
8500		125	26	16	271	11.75	12.70	24.45
8750		200	30	20	372	12.80	13.40	26.20
9200	40 x 40	40	18	14	176	11.20	12.45	23.65
9400		75	24	14	211	11.60	12.80	24.40
9500		125	26	16	271	11.95	12.85	24.80
9700	40 x 45	40	20	16	186	11.65	12.60	24.25
9800		75	24	16	221	12.05	12.95	25
9900		125	28	16	271	12.25	13.10	25.35

B1010 Floor Construction

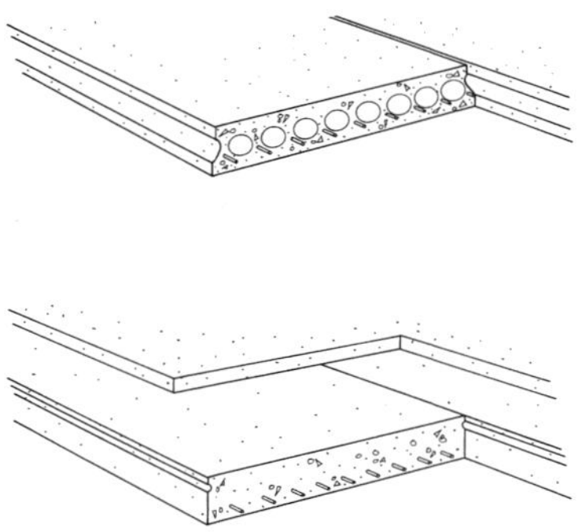

General: Units priced here are for plant produced prestressed members, transported to site and erected.

Normal weight concrete is most frequently used. Lightweight concrete may be used to reduce dead weight.

Structural topping is sometimes used on floors: insulating concrete or rigid insulation on roofs.

Camber and deflection may limit use by depth considerations.

Prices are based upon 10,000 S.F. to 20,000 S.F. projects, and 50 mile to 100 mile transport.

Concrete is f'c = 5 KSI and Steel is fy = 250 or 300 KSI

Note: Deduct from prices 20% for Southern states. Add to prices 10% for Western states.

Description of Table: Enter table at span and load. Most economical sections will generally consist of normal weight concrete without topping. If acceptable, note this price, depth and weight. For topping and/or lightweight concrete, note appropriate data.

Generally used on masonry and concrete bearing or reinforced concrete and steel framed structures.

The solid 4" slabs are used for light loads and short spans. The 6" to 12" thick hollow core units are used for longer spans and heavier loads. Cores may carry utilities.

Topping is used structurally for loads or rigidity and architecturally to level or slope surface.

Camber and deflection and change in direction of spans must be considered (door openings, etc.), especially untopped.

System Components	QUANTITY	UNIT	COST PER S.F. MAT.	INST.	TOTAL
SYSTEM B1010 230 2000					
10' SPAN, 40 LBS S.F. WORKING LOAD, 2" TOPPING					
Precast prestressed concrete roof/floor slabs 4" thick, grouted	1.000	S.F.	6.75	3.47	10.22
Edge forms to 6" high on elevated slab, 4 uses	.100	L.F.	.02	.45	.47
Welded wire fabric 6 x 6 - W1.4 x W1.4 (10 x 10), 21 lb/csf, 10% lap	.010	C.S.F.	.15	.38	.53
Concrete, ready mix, regular weight, 3000 psi	.170	C.F.	.74		.74
Place and vibrate concrete, elevated slab less than 6", pumped	.170	C.F.		.28	.28
Finishing floor, monolithic steel trowel finish for resilient tile	1.000	S.F.		1.22	1.22
Curing with sprayed membrane curing compound	.010	C.S.F.	.14	.10	.24
TOTAL			7.80	5.90	13.70

B1010 229 Precast Plank with No Topping

	SPAN (FT.)	SUPERIMPOSED LOAD (P.S.F.)	TOTAL DEPTH (IN.)	DEAD LOAD (P.S.F.)	TOTAL LOAD (P.S.F.)	COST PER S.F. MAT.	INST.	TOTAL
0720	10	40	4	50	90	6.75	3.47	10.22
0750	RB1010 -010	75	6	50	125	7.60	2.97	10.57
0770		100	6	50	150	7.60	2.97	10.57
0800	15	40	6	50	90	7.60	2.97	10.57
0820	RB1010 -100	75	6	50	125	7.60	2.97	10.57
0850		100	6	50	150	7.60	2.97	10.57
0875	20	40	6	50	90	7.60	2.97	10.57
0900		75	6	50	125	7.60	2.97	10.57
0920		100	6	50	150	7.60	2.97	10.57
0950	25	40	6	50	90	7.60	2.97	10.57
0970		75	8	55	130	8.45	2.60	11.05
1000		100	8	55	155	8.45	2.60	11.05
1200	30	40	8	55	95	8.45	2.60	11.05
1300		75	8	55	130	8.45	2.60	11.05
1400		100	10	70	170	8.80	2.31	11.11
1500	40	40	10	70	110	8.80	2.31	11.11
1600		75	12	70	145	9.10	2.08	11.18

B1010 Floor Construction

B1010 229 Precast Plank with No Topping

	SPAN (FT.)	SUPERIMPOSED LOAD (P.S.F.)	TOTAL DEPTH (IN.)	DEAD LOAD (P.S.F.)	TOTAL LOAD (P.S.F.)	COST PER S.F.		
						MAT.	INST.	TOTAL
1700	45	40	12	70	110	9.10	2.08	11.18

B1010 230 Precast Plank with 2″ Concrete Topping

	SPAN (FT.)	SUPERIMPOSED LOAD (P.S.F.)	TOTAL DEPTH (IN.)	DEAD LOAD (P.S.F.)	TOTAL LOAD (P.S.F.)	COST PER S.F.		
						MAT.	INST.	TOTAL
2000	10	40	6	75	115	7.80	5.90	13.70
2100		75	8	75	150	8.65	5.40	14.05
2200		100	8	75	175	8.65	5.40	14.05
2500	15	40	8	75	115	8.65	5.40	14.05
2600		75	8	75	150	8.65	5.40	14.05
2700		100	8	75	175	8.65	5.40	14.05
2800	20	40	8	75	115	8.65	5.40	14.05
2900		75	8	75	150	8.65	5.40	14.05
3000		100	8	75	175	8.65	5.40	14.05
3100	25	40	8	75	115	8.65	5.40	14.05
3200		75	8	75	150	8.65	5.40	14.05
3300		100	10	80	180	9.50	5.05	14.55
3400	30	40	10	80	120	9.50	5.05	14.55
3500		75	10	80	155	9.50	5.05	14.55
3600		100	10	80	180	9.50	5.05	14.55
3700	35	40	12	95	135	9.85	4.74	14.59
3800		75	12	95	170	9.85	4.74	14.59
3900		100	14	95	195	10.15	4.51	14.66
4000	40	40	12	95	135	9.85	4.74	14.59
4500		75	14	95	170	10.15	4.51	14.66
5000	45	40	14	95	135	10.15	4.51	14.66

B1010 Floor Construction

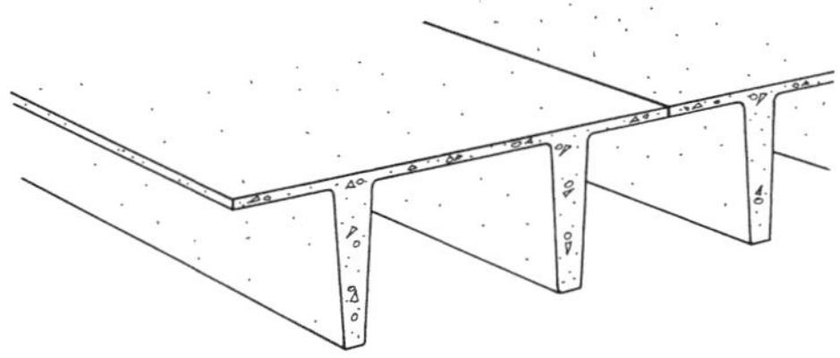

Most widely used for moderate span floors and roofs. At shorter spans, they tend to be competitive with hollow core slabs. They are also used as wall panels.

System Components	QUANTITY	UNIT	COST PER S.F. MAT.	COST PER S.F. INST.	COST PER S.F. TOTAL
SYSTEM B1010 235 6700					
PRECAST, DOUBLE "T", 2" TOPPING, 30' SPAN, 30 PSF SUP. LOAD, 18" X 8'					
Double "T" beams, reg. wt, 18" x 8' w, 30' span	1.000	S.F.	9.69	1.74	11.43
Edge forms to 6" high on elevated slab, 4 uses	.050	L.F.	.01	.23	.24
Concrete, ready mix, regular weight, 3000 psi	.250	C.F.	1.09		1.09
Place and vibrate concrete, elevated slab less than 6", pumped	.250	C.F.		.41	.41
Finishing floor, monolithic steel trowel finish for finish floor	1.000	S.F.		.93	.93
Curing with sprayed membrane curing compound	.010	C.S.F.	.14	.10	.24
TOTAL			10.93	3.41	14.34

B1010 234 Precast Double "T" Beams with No Topping

	SPAN (FT.)	SUPERIMPOSED LOAD (P.S.F.)	DBL. "T" SIZE D (IN.) W (FT.)	CONCRETE "T" TYPE	TOTAL LOAD (P.S.F.)	COST PER S.F. MAT.	COST PER S.F. INST.	COST PER S.F. TOTAL
1500	30	30	18x8	Reg. Wt.	92	9.70	1.74	11.44
1600	RB1010 -010	40	18x8	Reg. Wt.	102	9.85	2.25	12.10
1700		50	18x8	Reg. Wt	112	9.85	2.25	12.10
1800		75	18x8	Reg. Wt.	137	9.90	2.34	12.24
1900		100	18x8	Reg. Wt.	162	9.90	2.34	12.24
2000	40	30	20x8	Reg. Wt.	87	8.85	1.45	10.30
2100		40	20x8	Reg. Wt.	97	8.90	1.74	10.64
2200	RB1010 -100	50	20x8	Reg. Wt.	107	8.90	1.74	10.64
2300		75	20x8	Reg. Wt.	132	8.95	1.88	10.83
2400		100	20x8	Reg. Wt.	157	9.10	2.29	11.39
2500	50	30	24x8	Reg. Wt.	103	8.80	1.30	10.10
2600		40	24x8	Reg. Wt.	113	8.90	1.59	10.49
2700		50	24x8	Reg. Wt.	123	8.95	1.70	10.65
2800		75	24x8	Reg. Wt.	148	8.95	1.72	10.67
2900		100	24x8	Reg. Wt.	173	9.10	2.14	11.24
3000	60	30	24x8	Reg. Wt.	82	8.95	1.72	10.67
3100		40	32x10	Reg. Wt.	104	8.95	1.41	10.36
3150		50	32x10	Reg. Wt.	114	8.90	1.22	10.12
3200		75	32x10	Reg. Wt.	139	8.95	1.32	10.27
3250		100	32x10	Reg. Wt.	164	9.05	1.66	10.71
3300	70	30	32x10	Reg. Wt.	94	8.95	1.30	10.25
3350		40	32x10	Reg. Wt.	104	8.95	1.32	10.27
3400		50	32x10	Reg. Wt.	114	9.05	1.65	10.70
3450		75	32x10	Reg. Wt.	139	9.15	1.98	11.13
3500		100	32x10	Reg. Wt.	164	9.35	2.64	11.99
3550	80	30	32x10	Reg. Wt.	94	9.05	1.65	10.70
3600		40	32x10	Reg. Wt.	104	9.25	2.31	11.56
3900		50	32x10	Reg. Wt.	114	9.35	2.64	11.99

B1010 Floor Construction

B1010 234 Precast Double "T" Beams with No Topping

	SPAN (FT.)	SUPERIMPOSED LOAD (P.S.F.)	DBL. "T" SIZE D (IN.) W (FT.)	CONCRETE "T" TYPE	TOTAL LOAD (P.S.F.)	COST PER S.F. MAT.	COST PER S.F. INST.	COST PER S.F. TOTAL
4300	50	30	20x8	Lt. Wt.	66	9.80	1.59	11.39
4400		40	20x8	Lt. Wt.	76	9.80	1.46	11.26
4500		50	20x8	Lt. Wt.	86	9.90	1.75	11.65
4600		75	20x8	Lt. Wt.	111	9.95	2.01	11.96
4700		100	20x8	Lt. Wt.	136	10.10	2.43	12.53
4800	60	30	24x8	Lt. Wt.	70	9.80	1.59	11.39
4900		40	32x10	Lt. Wt.	88	9.80	1.09	10.89
5000		50	32x10	Lt. Wt.	98	9.85	1.33	11.18
5200		75	32x10	Lt. Wt.	123	9.90	1.53	11.43
5400		100	32x10	Lt. Wt.	148	10.05	1.85	11.90
5600	70	30	32x10	Lt. Wt.	78	9.80	1.09	10.89
5750		40	32x10	Lt. Wt.	88	9.85	1.32	11.17
5900		50	32x10	Lt. Wt.	98	9.90	1.52	11.42
6000		75	32x10	Lt. Wt.	123	10.05	1.85	11.90
6100		100	32x10	Lt. Wt.	148	10.25	2.51	12.76
6200	80	30	32x10	Lt. Wt.	78	9.90	1.52	11.42
6300		40	32x10	Lt. Wt.	88	10.05	1.85	11.90
6400		50	32x10	Lt. Wt.	98	10.15	2.18	12.33

B1010 235 Precast Double "T" Beams With 2″ Topping

	SPAN (FT.)	SUPERIMPOSED LOAD (P.S.F.)	DBL. "T" SIZE D (IN.) W (FT.)	CONCRETE "T" TYPE	TOTAL LOAD (P.S.F.)	COST PER S.F. MAT.	COST PER S.F. INST.	COST PER S.F. TOTAL
6700	30	30	18x8	Reg. Wt.	117	10.95	3.41	14.36
6750		40	18x8	Reg. Wt.	127	10.95	3.41	14.36
6800		50	18x8	Reg. Wt.	137	11.05	3.70	14.75
6900		75	18x8	Reg. Wt.	162	11.05	3.70	14.75
7000		100	18x8	Reg. Wt.	187	11.05	3.84	14.89
7100	40	30	18x8	Reg. Wt.	120	8.60	3.26	11.86
7200		40	20x8	Reg. Wt.	130	10.05	3.12	13.17
7300		50	20x8	Reg. Wt.	140	10.15	3.41	13.56
7400		75	20x8	Reg. Wt.	165	10.20	3.55	13.75
7500		100	20x8	Reg. Wt.	190	10.35	3.96	14.31
7550	50	30	24x8	Reg. Wt.	120	10.15	3.26	13.41
7600		40	24x8	Reg. Wt.	130	10.20	3.37	13.57
7750		50	24x8	Reg. Wt.	140	10.20	3.39	13.59
7800		75	24x8	Reg. Wt.	165	10.30	3.81	14.11
7900		100	32x10	Reg. Wt.	189	10.20	3.13	13.33
8000	60	30	32x10	Reg. Wt.	118	10.05	2.66	12.71
8100		40	32x10	Reg. Wt.	129	10.15	2.89	13.04
8200		50	32x10	Reg. Wt.	139	10.20	3.13	13.33
8300		75	32x10	Reg. Wt.	164	10.30	3.33	13.63
8350		100	32x10	Reg. Wt.	189	10.40	3.65	14.05
8400	70	30	32x10	Reg. Wt.	119	10.20	2.99	13.19
8450		40	32x10	Reg. Wt.	129	10.30	3.32	13.62
8500		50	32x10	Reg. Wt.	139	10.40	3.65	14.05
8550		75	32x10	Reg. Wt.	164	10.60	4.31	14.91
8600	80	30	32x10	Reg. Wt.	119	10.60	4.31	14.91
8800	50	30	20x8	Lt. Wt.	105	11.05	3.39	14.44
8850		40	24x8	Lt. Wt.	121	11.10	3.55	14.65
8900		50	24x8	Lt. Wt.	131	11.20	3.81	15.01
8950		75	24x8	Lt. Wt.	156	11.20	3.81	15.01
9000		100	24x8	Lt. Wt.	181	11.35	4.23	15.58

B1010 Floor Construction

B1010 235	Precast Double "T" Beams With 2" Topping							
	SPAN (FT.)	SUPERIMPOSED LOAD (P.S.F.)	DBL. "T" SIZE D (IN.) W (FT.)	CONCRETE "T" TYPE	TOTAL LOAD (P.S.F.)	COST PER S.F.		
						MAT.	INST.	TOTAL
9200	60	30	32x10	Lt. Wt.	103	11	2.76	13.76
9300		40	32x10	Lt. Wt.	113	11.10	3	14.10
9350		50	32x10	Lt. Wt.	123	11.10	3	14.10
9400		75	32x10	Lt. Wt.	148	11.15	3.20	14.35
9450		100	32x10	Lt. Wt.	173	11.25	3.52	14.77
9500	70	30	32x10	Lt. Wt.	103	11.15	3.19	14.34
9550		40	32x10	Lt. Wt.	113	11.15	3.19	14.34
9600		50	32x10	Lt. Wt.	123	11.25	3.52	14.77
9650		75	32x10	Lt. Wt.	148	11.40	3.86	15.26
9700	80	30	32x10	Lt. Wt.	103	11.25	3.52	14.77
9800		40	32x10	Lt. Wt.	113	11.40	3.85	15.25
9900		50	32x10	Lt. Wt.	123	11.50	4.18	15.68

B1010 Floor Construction

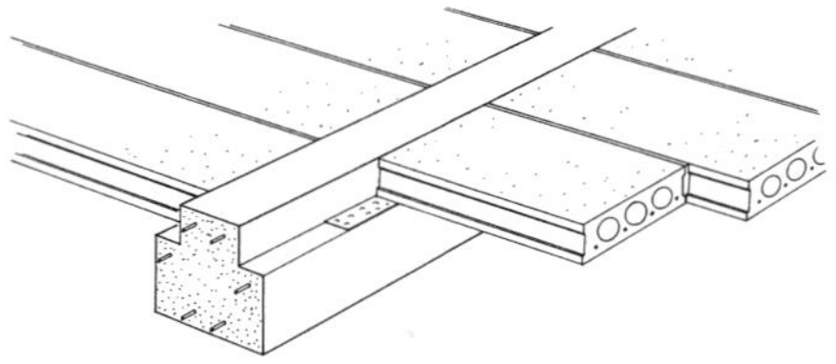

General: Units priced here are for plant produced prestressed members transported to the site and erected.

System has precast prestressed concrete beams and precast, prestressed hollow core slabs spanning the longer direction when applicable.

Camber and deflection must be considered when using untopped hollow core slabs.

Design and Pricing Assumptions:
Prices are based on 10,000 S.F. to 20,000 S.F. projects and 50 mile to 100 mile transport.

Concrete is f'c 5 KSI and prestressing steel is fy = 250 or 300 KSI.

System Components	QUANTITY	UNIT	COST PER S.F. MAT.	COST PER S.F. INST.	COST PER S.F. TOTAL
SYSTEM B1010 236 5200					
20′ X 20′ BAY, 6″ DEPTH, 40 PSF S. LOAD, 110 PSF TOTAL LOAD					
Precast concrete beam, T-shaped, 20′ span, 12″ x 20″	.038	L.F.	9.50	.74	10.24
Precast concrete beam, L-shaped, 20′ span, 12″ x 20″	.025	L.F.	5.57	.49	6.06
Precast prestressed concrete roof/floor slabs 6″ deep, grcuted	1.000	S.F.	7.60	2.97	10.57
TOTAL			22.67	4.20	26.87

B1010 236 Precast Beam & Plank with No Topping

	BAY SIZE (FT.)	SUPERIMPOSED LOAD (P.S.F.)	PLANK THICKNESS (IN.)	TOTAL DEPTH (IN.)	TOTAL LOAD (P.S.F.)	COST PER S.F. MAT.	COST PER S.F. INST.	COST PER S.F. TOTAL
5200	20x20	40	6	20	110	22.50	4.20	26.70
5400		75	6	24	148	17.60	3.77	21.37
5600		100	6	24	173	23	4.26	27.26
5800	20x25	40	8	24	113	21	3.65	24.65
6000		75	8	24	149	21	3.65	24.65
6100		100	8	36	183	21.50	3.59	25.09
6200	25x25	40	8	28	118	22	3.47	25.47
6400		75	8	36	158	22	3.46	25.46
6500		100	8	36	183	22	3.46	25.46
7000	25x30	40	8	36	110	20	3.35	23.35
7200		75	10	36	159	21	3.07	24.07
7400		100	12	36	188	21	2.80	23.80
7600	30x30	40	8	36	121	20.50	3.24	23.74
8000		75	10	44	140	20.50	3.25	23.75
8250		100	12	52	206	22.50	2.75	25.25
8500	30x35	40	12	44	135	19.60	2.65	22.25
8750		75	12	52	176	20.50	2.66	23.16
9000	35x35	40	12	52	141	20.50	2.60	23.10
9250		75	12	60	181	21.50	2.59	24.09
9500	35x40	40	12	52	137	20.50	2.60	23.10
9750	40x40	40	12	60	141	21.50	2.57	24.07

B1010 Floor Construction

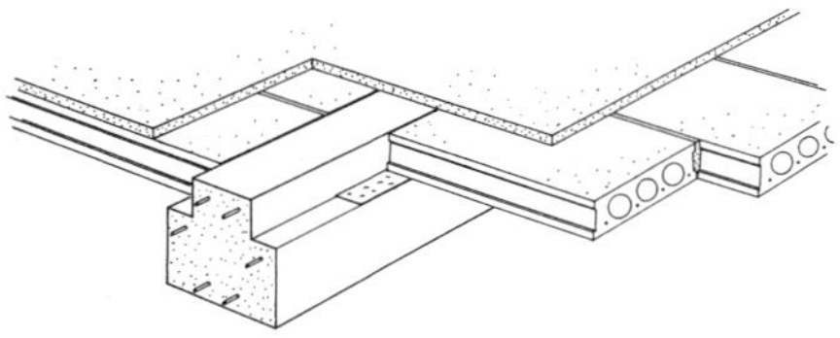

General: Beams and hollow core slabs priced here are for plant produced prestressed members transported to the site and erected.

The 2″ structural topping is applied after the beams and hollow core slabs are in place and is reinforced with W.W.F.

Design and Pricing Assumptions:

Prices are based on 10,000 S.F. to 20,000 S.F. projects and 50 mile to 100 mile transport.

Concrete for prestressed members is f'c 5 KSI.

Concrete for topping is f'c 3000 PSI and placed by pump.

Prestressing steel is fy = 250 or 300 KSI.

W.W.F. is 6 x 6 – W1.4 x W1.4 (10 x 10).

System Components	QUANTITY	UNIT	COST PER S.F. MAT.	INST.	TOTAL
SYSTEM B1010 238 4300					
20′ X 20′ BAY, 6″ PLANK, 40 PSF S. LOAD, 135 PSF TOTAL LOAD					
Precast concrete beam, T-shaped, 20′ span, 12″ x 20″	.038	L.F.	9.31	.73	10.04
Precast concrete beam, L-shaped, 20′ span, 12″ x 20″	.025	L.F.	5.57	.49	6.06
Precast prestressed concrete roof/floor slabs 6″ deep, grouted	1.000	S.F.	7.60	2.97	10.57
Edge forms to 6″ high on elevated slab, 4 uses	.050	L.F.	.01	.23	.24
Forms in place, bulkhead for slab with keyway, 1 use, 2 piece	.013	L.F.	.03	.09	.12
Welded wire fabric rolls, 6 x 6 - W1.4 x W1.4 (10 x 10), 21 lb/csf	.010	C.S.F.	.15	.38	.53
Concrete, ready mix, regular weight, 3000 psi	.170	C.F.	.74		.74
Place and vibrate concrete, elevated slab less than 6″, pump	.170	C.F.		.28	.28
Finish floor, monolithic steel trowel finish for finish floor	1.000	S.F.		.93	.93
Cure with sprayed membrane curing compound	.010	C.S.F.	.14	.10	.24
TOTAL			23.55	6.20	29.75

B1010 238 Precast Beam & Plank with 2″ Topping

	BAY SIZE (FT.)	SUPERIMPOSED LOAD (P.S.F.)	PLANK THICKNESS (IN.)	TOTAL DEPTH (IN.)	TOTAL LOAD (P.S.F.)	COST PER S.F. MAT.	INST.	TOTAL
4300	20x20	40	6	22	135	23.50	6.20	29.70
4400	RB1010 -010	75	6	24	173	24	6.30	30.30
4500		100	6	28	200	24.50	6.25	30.75
4600	20x25	40	6	26	134	22	6.05	28.05
5000	RB1010 -100	75	8	30	177	22.50	5.60	28.10
5200		100	8	30	202	22.50	5.60	28.10
5400	25x25	40	6	38	143	22.50	5.80	28.30
5600		75	8	38	183	17.90	5.50	23.40
6000		100	8	46	216	18.20	5.05	23.25
6200	25x30	40	8	38	144	21	5.25	26.25
6400		75	10	46	200	21.50	4.95	26.45
6600		100	10	46	225	22	4.96	26.96
7000	30x30	40	8	46	150	21.50	5.15	26.65
7200		75	10	54	181	23	5.15	28.15
7600		100	10	54	231	23	5.15	28.15
7800	30x35	40	10	54	166	21.50	4.76	26.26
8000		75	12	54	200	21.50	4.52	26.02
8200	35x35	40	10	62	170	22	4.70	26.70
9300		75	12	62	206	23.50	4.50	28
9500	35x40	40	12	62	167	22.50	4.44	26.94
9600	40x40	40	12	62	173	22	4.42	26.42

B1010 Floor Construction

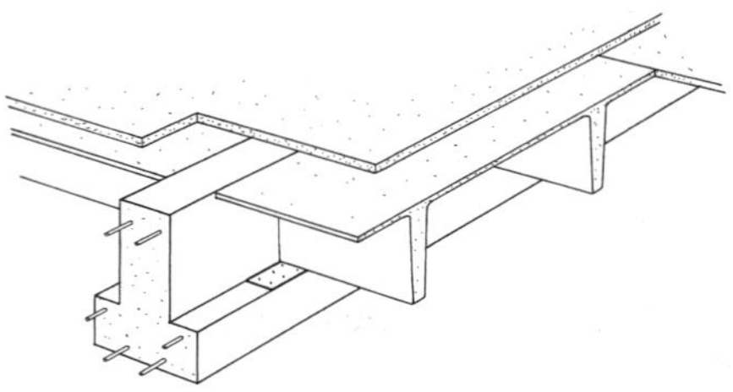

General: Beams and double tees priced here are for plant produced prestressed members transported to the site and erected.

The 2″ structural topping is applied after the beams and double tees are in place and is reinforced with W.W.F.

Design and Pricing Assumptions:
Prices are based on 10,000 S.F. to 20,000 S.F. projects and 50 mile to 100 mile transport.

Concrete for prestressed members is f'c 5 KSI.

Concrete for topping is f'c 3000 PSI and placed by pump.

Prestressing steel is fy = 250 or 300 KSI.

W.W.F. is 6 x 6 – W1.4 x W1.4 (10x10).

System Components	QUANTITY	UNIT	COST PER S.F. MAT.	INST.	TOTAL
SYSTEM B1010 239 3000					
25′ X 30′ BAY, 38″ DEPTH, 130 PSF T.L., 2″ TOPPING					
Precast concrete beam, T-shaped, 25′ span, 12″ x 36″	.025	L.F.	7.53	.47	8
Precast concrete beam, L-shaped, 25′ span, 12″ x 28″	.017	L.F.	4.03	.28	4.31
Double T, standard weight, 16″ x 8′ w, 25′ span	.989	S.F.	9.17	1.72	10.89
Edge forms to 6″ high on elevated slab, 4 uses	.037	L.F.	.01	.17	.18
Forms in place, bulkhead for slab with keyway, 1 use, 2 piece	.010	L.F.	.02	.07	.09
Welded wire fabric rolls, 6 x 6 - W1.4 x W1.4 (10 x 10), 21 lb/csf	1.000	S.F.	.15	.38	.53
Concrete, ready mix, regular weight, 3000 psi	.170	C.F.	.74		.74
Place and vibrate concrete, elevated slab less than 6″, pumped	.170	C.F.		.28	.28
Finishing floor, monolithic steel trowel finish for finish floor	1.000	S.F.		.93	.93
Curing with sprayed membrane curing compound	.010	S.F.	.14	.10	.24
TOTAL			21.79	4.40	26.19

B1010 239 Precast Double "T" & 2″ Topping on Precast Beams

	BAY SIZE (FT.)	SUPERIMPOSED LOAD (P.S.F.)	DEPTH (IN.)		TOTAL LOAD (P.S.F.)	COST PER S.F. MAT.	INST.	TOTAL
3000	25x30	40	38		130	22	4.40	26.40
3100	RB1010 -100	75	38		168	22	4.40	26.40
3300		100	46		196	21.50	4.37	25.87
3600	30x30	40	46		150	22	4.26	26.26
3750		75	46		174	22	4.26	26.26
4000		100	54		203	23	4.25	27.25
4100	30x40	40	46		136	19.60	3.85	23.45
4300		75	54		173	20	3.84	23.84
4400		100	62		204	21	3.85	24.85
4600	30x50	40	54		138	18.25	3.55	21.80
4800		75	54		181	18.10	3.58	21.68
5000		100	54		219	18.15	3.27	21.42
5200	30x60	40	62		151	17.85	3.22	21.07
5400		75	62		192	17.40	3.26	20.66
5600		100	62		215	17.50	3.26	20.76
5800	35x40	40	54		139	20.50	3.78	24.28
6000		75	62		179	20.50	3.62	24.12
6250		100	62		212	20.50	3.75	24.25
6500	35x50	40	62		142	18.95	3.52	22.47
6750		75	62		186	18.95	3.65	22.60
7300		100	62		231	19.05	3.34	22.39

B1010 Floor Construction

B1010 239 Precast Double "T" & 2″ Topping on Precast Beams

	BAY SIZE (FT.)	SUPERIMPOSED LOAD (P.S.F.)	DEPTH (IN.)		TOTAL LOAD (P.S.F.)	COST PER S.F.		
						MAT.	INST.	TOTAL
7600	35x60	40	54		154	18.30	3.08	21.38
7750		75	54		179	18.45	3.09	21.54
8000		100	62		224	18.85	3.14	21.99
8250	40x40	40	62		145	22.50	3.79	26.29
8400		75	62		187	21.50	3.81	25.31
8750		100	62		223	22	3.89	25.89
9000	40x50	40	62		151	19.70	3.55	23.25
9300		75	62		193	19.15	2.62	21.77
9800	40x60	40	62		164	22	2.52	24.52

B1010 Floor Construction

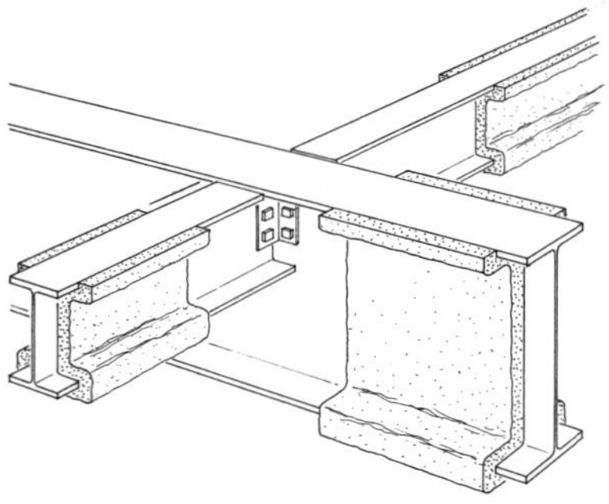

General: The following table is based upon structural W shape beam and girder framing. Non-composite action is assumed between beams and decking. Deck costs not included.

The deck spans the short direction. The steel beams and girders are fireproofed with sprayed fiber fireproofing.

Design and Pricing Assumptions:
Structural steel is A36, with high strength A325 bolts.

Fireproofing is sprayed fiber (non-asbestos).

Total load includes steel, deck & live load.

Spandrels are assumed the same as interior beams and girders to allow for exterior wall loads and bracing or moment connections. No columns included in price.

See Tables B1010 258 and B1020 128 for metal deck costs.

System Components	QUANTITY	UNIT	COST PER S.F. MAT.	INST.	TOTAL
SYSTEM B1010 241 1350 15′ X 20′ BAY, 40 P.S.F. L.L., 12″ DEPTH, .535 P.S.F. FIREPROOF, 50 PSF T.LOAD					
Structural steel	3.200	Lb.	4.77	1.41	6.18
Spray mineral fiber/cement for fire proof., 1″ thick on beams	.535	S.F.	.32	.57	.89
TOTAL			5.09	1.98	7.07

B1010 241 W Shape Beams & Girders

	BAY SIZE (FT.) BEAM X GIRD	SUPERIMPOSED LOAD (P.S.F.)	STEEL FRAMING DEPTH (IN.)	FIREPROOFING (S.F. PER S.F.)	TOTAL LOAD (P.S.F.)	COST PER S.F. MAT.	INST.	TOTAL
1350	15x20	40	12	.535	50	5.10	1.98	7.08
1400		40	16	.65	90	6.65	2.55	9.20
1450		75	18	.694	125	8.75	3.20	11.95
1500		125	24	.796	175	12.10	4.44	16.54
1550		200	24	.89	263	13.65	4.01	17.66
1600	20x15	40	14	.659	50	5.15	2.12	7.27
1650		40	14	.69	90	6.95	2.68	9.63
1700		75	14	.806	125	8.55	3.24	11.79
1800		125	16	.86	175	10.05	3.87	13.92
1900		200	18	1.00	250	11.90	3.56	15.46
2000	20x20	40	12	.55	50	5.70	2.17	7.87
2050		40	14	.579	90	7.80	2.82	10.62
2100		75	16	.672	125	9.35	3.36	12.71
2150		125	16	.714	175	11.15	4.07	15.22
2200		200	24	.841	263	13.95	4.02	17.97
2300	20x20	40	14	.67	50	5.75	2.30	8.05
2400		40	14	.718	90	7.85	2.97	10.82
2500		75	18	.751	125	9.10	3.36	12.46
2550		125	21	.879	175	12.45	4.62	17.07
2600		200	21	.976	250	14.90	4.39	19.29
2650	20x20	40	14	.746	50	5.80	2.37	8.17
2700		40	14	.839	90	7.95	3.09	11.04
2750		75	18	.894	125	10.05	3.78	13.83
2800		125	21	.959	175	13.40	4.98	18.38
2850		200	21	1.10	250	16.40	4.86	21.26
2900	20x25	40	16	.53	50	6.25	2.32	8.57
2950		40	18	.621	96	9.90	3.48	13.38
3000		75	18	.651	131	11.40	3.96	15.36
3050		125	24	.77	200	15.05	5.35	20.40

B1010 Floor Construction

B1010 241 W Shape Beams & Girders

	BAY SIZE (FT.) BEAM X GIRD	SUPERIMPOSED LOAD (P.S.F.)	STEEL FRAMING DEPTH (IN.)	FIREPROOFING (S.F. PER S.F.)	TOTAL LOAD (P.S.F.)	COST PER S.F.		
						MAT.	INST.	TOTAL
3100	20x25	200	27	.855	275	17.15	4.80	21.95
3300	20x25	40	14	.608	50	6.30	2.41	8.71
3350		40	21	.751	90	8.80	3.27	12.07
3400	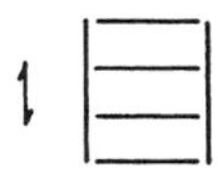	75	24	.793	125	10.90	3.93	14.83
3450		125	24	.846	175	13	4.77	17.77
3500		200	24	.947	256	16.35	4.69	21.04
3550	20x25	40	14	.72	50	7.25	2.79	10.04
3600		40	21	.802	90	8.80	3.32	12.12
3650	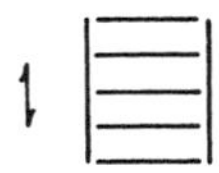	75	24	.924	125	11.30	4.16	15.46
3700		125	24	.964	175	13.70	5.10	18.80
3750		200	27	1.09	250	17.30	5.05	22.35
3800	25x20	40	12	.512	50	6.25	2.31	8.56
3850		40	16	.653	90	8.45	3.08	11.53
3900	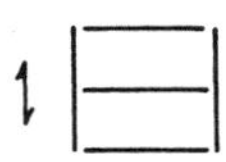	75	18	.726	125	10.85	3.85	14.70
4000		125	21	.827	175	13.60	4.93	18.53
4100		200	24	.928	250	16.60	4.73	21.33
4200	25x20	40	12	.65	50	6.35	2.46	8.81
4300		40	18	.702	90	9.35	3.39	12.74
4400	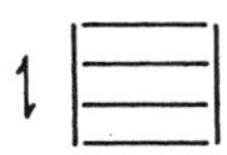	75	21	.829	125	10.90	3.96	14.86
4500		125	24	.914	175	13.35	4.93	18.28
4600		200	24	1.015	250	16.35	4.76	21.11
4700	25x20	40	14	.769	50	6.70	2.68	9.38
4800		40	16	.938	90	10.10	3.82	13.92
4900	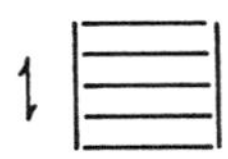	75	18	.969	125	12.20	4.47	16.67
5000		125	24	1.136	175	17.05	6.30	23.35
5100		200	24	1.239	250	22	6.30	28.30
5200	25x25	40	18	.486	50	6.85	2.46	9.31
5300		40	18	.592	96	10.20	3.53	13.73
5400	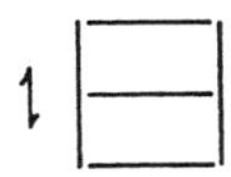	75	21	.668	131	12.30	4.24	16.54
5450		125	24	.738	191	16.25	5.65	21.90
5500		200	30	.861	272	18.90	5.20	24.10
5550	25x25	40	18	.597	50	6.60	2.49	9.09
5600		40	18	.704	90	10.55	3.74	14.29
5650	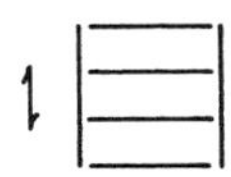	75	21	.777	125	12.10	4.26	16.36
5700		125	24	.865	175	15.40	5.55	20.95
5750		200	27	.96	250	18.65	5.25	23.90
5800	25x25	40	18	.71	50	7.25	2.78	10.03
5850		40	21	.767	90	10.60	3.81	14.41
5900	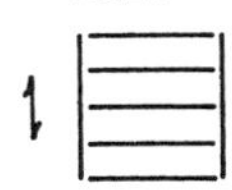	75	24	.887	125	12.75	4.56	17.31
5950		125	24	.972	175	16.05	5.85	21.90
6000		200	30	1.10	250	19.65	5.60	25.25
6050	25x30	40	24	.547	50	8.65	3.04	11.69
6100		40	24	.629	103	12.30	4.19	16.49
6150	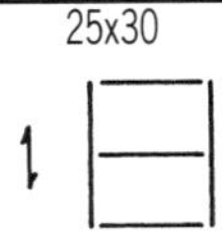	75	30	.726	138	14.75	4.99	19.74
6200		125	30	.751	206	17.45	6.10	23.55
6250		200	33	.868	281	21	5.70	26.70
6300	25x30	40	21	.568	50	7.80	2.81	10.61
6350		40	21	.694	90	10.55	3.73	14.28
6400	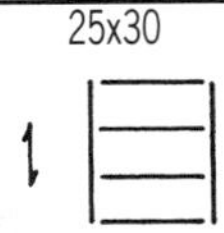	75	24	.776	125	13.55	4.70	18.25
6450		125	30	.904	175	16.30	5.85	22.15
6500		200	33	1.008	263	19.55	5.50	25.05

B1010 Floor Construction

B1010 241 W Shape Beams & Girders

	BAY SIZE (FT.) BEAM X GIRD	SUPERIMPOSED LOAD (P.S.F.)	STEEL FRAMING DEPTH (IN.)	FIREPROOFING (S.F. PER S.F.)	TOTAL LOAD (P.S.F.)	COST PER S.F.		
						MAT.	INST.	TOTAL
6550	25x30	40	16	.632	50	8.10	2.96	11.06
6600		40	21	.76	90	11.20	3.99	15.19
6650		75	24	.857	125	13.30	4.70	18
6700		125	30	.983	175	16.65	6	22.65
6750		200	33	1.11	250	21	5.95	26.95
6800	30x25	40	16	.532	50	7.45	2.68	10.13
6850		40	21	.672	96	11.45	3.98	15.43
6900		75	24	.702	131	13.50	4.62	18.12
6950		125	27	1.020	175	17.60	6.10	23.70
7000		200	30	1.160	250	22.50	7.70	30.20
7100	30x25	40	18	.569	50	7.80	2.81	10.61
7150		40	24	.740	90	10.85	3.87	14.72
7200		75	24	.787	125	13.55	4.71	18.26
7300		125	24	.874	175	16.90	6	22.90
7400		200	30	1.013	250	21	5.80	26.80
7450	30x25	40	16	.637	50	8.15	2.97	11.12
7500		40	24	.839	90	11.55	4.15	15.70
7550		75	24	.919	125	13.95	4.94	18.89
7600		125	27	1.02	175	17.60	6.35	23.95
7650		200	30	1.160	250	22	6.20	28.20
7700	30x30	40	21	.52	50	8.35	2.93	11.28
7750		40	24	.629	103	12.90	4.37	17.27
7800		75	30	.715	138	15.30	5.15	20.45
7850		125	36	.822	206	20	6.95	26.95
7900		200	36	.878	281	22.50	6.05	28.55
7950	30x30	40	24	.619	50	8.70	3.12	11.82
8000		40	24	.706	90	11.75	4.09	15.84
8020		75	27	.818	125	13.90	4.83	18.73
8040		125	30	.910	175	17.80	6.30	24.10
8060		200	33	.999	263	22	6.05	28.05
8080	30x30	40	18	.631	50	9.30	3.31	12.61
8100		40	24	.805	90	12.70	4.47	17.17
8120		75	27	.899	125	15.15	5.25	20.40
8150		125	30	1.010	175	18.80	6.70	25.50
8200		200	36	1.148	250	22.50	6.25	28.75
8250	30x35	40	21	.508	50	9.55	3.27	12.82
8300		40	24	.651	109	14.10	4.75	18.85
8350		75	33	.732	150	17.10	5.70	22.80
8400		125	36	.802	225	21.50	7.30	28.80
8450		200	36	.888	300	28	7.35	35.35
8500	30x35	40	24	.554	50	8.40	2.98	11.38
8520		40	24	.655	90	12.30	4.23	16.53
8540		75	30	.751	125	15.35	5.20	20.55
8600		125	33	.845	175	18.70	6.55	25.25
8650		200	36	.936	263	24.50	6.60	31.10
8700	30x35	40	21	.644	50	9	3.23	12.23
8720		40	24	.733	90	12.95	4.48	17.43
8740		75	30	.833	125	16.30	5.55	21.85
8760		125	36	.941	175	18.75	6.60	25.35
8780		200	36	1.03	250	25	6.75	31.75

B1010 Floor Construction

B1010 241	W Shape Beams & Girders							
	BAY SIZE (FT.) BEAM X GIRD	SUPERIMPOSED LOAD (P.S.F.)	STEEL FRAMING DEPTH (IN.)	FIREPROOFING (S.F. PER S.F.)	TOTAL LOAD (P.S.F.)	COST PER S.F.		
						MAT.	INST.	TOTAL
8800	35x30	40	24	.540	50	9.25	3.21	12.46
8850	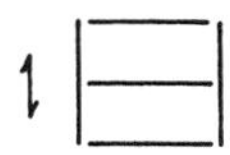	40	30	.670	103	13.80	4.68	18.48
8900		75	33	.748	138	16.85	5.65	22.50
8950		125	36	.824	206	21	7.15	28.15
8980		200	36	.874	281	25	6.65	31.65
9000	35x30	40	24	.619	50	9	3.21	12.21
9050	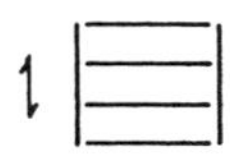	40	24	.754	90	13.25	4.59	17.84
9100		75	27	.844	125	15.70	5.40	21.10
9200		125	30	.856	175	19.30	6.70	26
9250		200	33	.953	263	21.50	5.85	27.35
9300	35x30	40	24	.705	50	9.65	3.48	13.13
9350	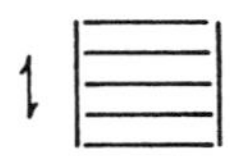	40	24	.833	90	13.60	4.76	18.36
9400		75	30	.963	125	16.65	5.80	22.45
9450		125	33	1.078	175	21	7.40	28.40
9500		200	36	1.172	250	25	7	32
9550	35x35	40	27	.560	50	9.85	3.42	13.27
9600	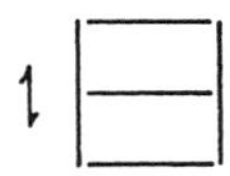	40	36	.706	109	18	5.95	23.95
9650		75	36	.750	150	18.60	6.20	24.80
9820		125	36	.797	225	25	8.40	33.40
9840		200	36	.914	300	29.50	7.75	37.25
9860	35x35	40	24	.580	50	9	3.17	12.17
9880	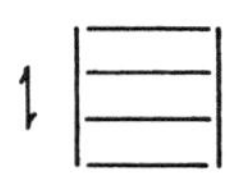	40	30	.705	90	13.85	4.71	18.56
9890		75	33	.794	125	16.85	5.70	22.55
9900		125	36	.878	175	21	7.20	28.20
9920		200	36	.950	263	25.50	6.85	32.35
9930	35x35	40	24	.689	50	9.65	3.47	13.12
9940	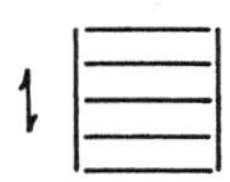	40	30	.787	90	14.15	4.89	19.04
9960		75	33	.871	125	18.10	6.10	24.20
9970		125	36	.949	175	19.95	7	26.95
9980		200	36	1.060	250	27.50	7.40	34.90

B1010 Floor Construction

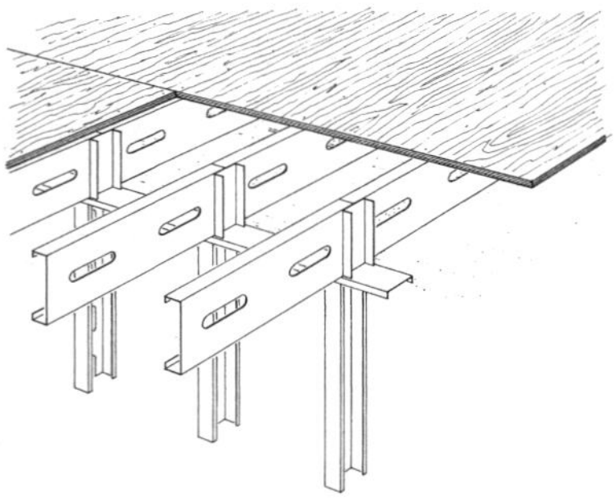

Description: Table below lists costs for light gauge CEE or PUNCHED DOUBLE joists to suit the span and loading with the minimum thickness subfloor required by the joist spacing.

Design Assumptions:
Maximum live load deflection is 1/360 of the clear span.

Maximum total load deflection is 1/240 of the clear span.

Bending strength is 20,000 psi.

8% allowance has been added to framing quantities for overlaps, double joists at openings under partitions, etc.; 5% added to glued & nailed subfloor for waste.

Maximum span is in feet and is the unsupported clear span.

System Components	QUANTITY	UNIT	COST PER S.F. MAT.	COST PER S.F. INST.	COST PER S.F. TOTAL
SYSTEM B1010 244 1500					
LIGHT GAUGE STL & PLYWOOD FLR SYS,40 PSF S.LOAD,15'SPAN,8"DEPTH,16" O.C.					
Light gauge steel, 8" deep, 16" O.C., 12 ga.	1.870	Lb.	1.98	1.36	3.34
Subfloor plywood CDX 5/8"	1.050	S.F.	.88	.92	1.80
TOTAL			2.86	2.28	5.14

B1010 244 Light Gauge Steel Floor Systems

	SPAN (FT.)	SUPERIMPOSED LOAD (P.S.F.)	FRAMING DEPTH (IN.)	FRAMING SPAC. (IN.)	TOTAL LOAD (P.S.F.)	COST PER S.F. MAT.	COST PER S.F. INST.	COST PER S.F. TOTAL
1500	15	40	8	16	54	2.86	2.28	5.14
1550			8	24	54	3	2.34	5.34
1600		65	10	16	80	3.73	2.74	6.47
1650			10	24	80	3.69	2.69	6.38
1700		75	10	16	90	3.73	2.74	6.47
1750			10	24	90	3.69	2.69	6.38
1800		100	10	16	116	4.86	3.47	8.33
1850			10	24	116	3.69	2.69	6.38
1900		125	10	16	141	4.86	3.47	8.33
1950			10	24	141	3.69	2.69	6.38
2500	20	40	8	16	55	3.29	2.57	5.86
2550			8	24	55	3	2.34	5.34
2600		65	8	16	80	3.81	2.94	6.75
2650			10	24	80	3.70	2.71	6.41
2700		75	10	16	90	3.52	2.61	6.13
2750			12	24	90	4.12	2.37	6.49
2800		100	10	16	115	3.52	2.61	6.13
2850			10	24	116	4.75	3.04	7.79
2900		125	12	16	142	5.50	2.97	8.47
2950			12	24	141	5.10	3.22	8.32
3500	25	40	10	16	55	3.73	2.74	6.47
3550			10	24	55	3.69	2.69	6.38
3600		65	10	16	81	4.86	3.47	8.33
3650			12	24	81	4.57	2.56	7.13
3700		75	12	16	92	5.50	2.96	8.46
3750			12	24	91	5.10	3.22	8.32
3800		100	12	16	117	6.20	3.27	9.47
3850		125	12	16	143	7.05	4.29	11.34
4500	30	40	12	16	57	5.50	2.96	8.46
4550			12	24	56	4.55	2.55	7.10
4600		65	12	16	82	6.15	3.26	9.41
4650		75	12	16	92	6.15	3.26	9.41

B1010 Floor Construction

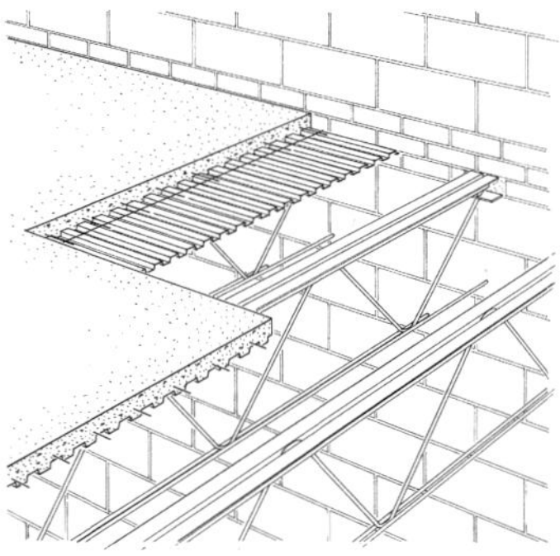

Description: Table below lists cost per S.F. for a floor system on bearing walls using open web steel joists, galvanized steel slab form and 2-1/2" concrete slab reinforced with welded wire fabric.

Design and Pricing Assumptions:
Concrete f'c = 3 KSI placed by pump.
WWF 6 x 6 – W1.4 x W1.4 (10 x 10)
Joists are spaced as shown.
Slab form is 28 gauge galvanized.
Joists costs include appropriate bridging. Deflection is limited to 1/360 of the span. Screeds and steel trowel finish.

Design Loads	Min.	Max.
Joists	3.0 PSF	7.6 PSF
Slab Form	1.0	1.0
2-1/2" Concrete	27.0	27.0
Ceiling	3.0	3.0
Misc.	9.0	9.4
	43.0 PSF	48.0 PSF

System Components	QUANTITY	UNIT	COST PER S.F. MAT.	INST.	TOTAL
SYSTEM B1010 246 1050					
SPAN 20' S.LOAD 40 PSF, JOIST SPACING 2'-0" O.C., 14-1/2" DEPTH					
Open web joists	2.800	Lb.	2.55	1.03	3.58
Slab form, galvanized steel 9/16" deep, 28 gauge	1.050	S.F.	2.28	1.16	3.44
Welded wire fabric rolls, 6 x 6 - W1.4 x W1.4 (10 x 10), 21 lb/csf	1.000	S.F.	.15	.38	.53
Concrete, ready mix, regular weight, 3000 psi	.210	C.F.	.92		.92
Place and vibrate concrete, elevated slab less than 6", pumped	.210	C.F.		.34	.34
Finishing floor, monolithic steel trowel finish for finish floor	1.000	S.F.		.93	.93
Curing with sprayed membrane curing compound	.010	C.S.F.	.14	.10	.24
TOTAL			6.04	3.94	9.98

B1010 246 Deck & Joists on Bearing Walls

	SPAN (FT.)	SUPERIMPOSED LOAD (P.S.F.)	JOIST SPACING FT. - IN.	DEPTH (IN.)	TOTAL LOAD (P.S.F.)	COST PER S.F. MAT.	INST.	TOTAL
1050	20	40	2-0	14-1/2	83	6.05	3.94	9.99
1070		65	2-0	16-1/2	109	6.50	4.17	10.67
1100		75	2-0	16-1/2	119	6.50	4.17	10.67
1120		100	2-0	18-1/2	145	6.55	4.18	10.73
1150		125	1-9	18-1/2	170	7.15	4.43	11.58
1170	25	40	2-0	18-1/2	84	6.55	4.18	10.73
1200		65	2-0	20-1/2	109	6.75	4.27	11.02
1220		75	2-0	20-1/2	119	7.20	4.45	11.65
1250		100	2-0	22-1/2	145	7.30	4.50	11.80
1270		125	1-9	22-1/2	170	8.30	4.90	13.20
1300	30	40	2-0	22-1/2	84	7.30	4.10	11.40
1320		65	2-0	24-1/2	110	7.90	4.28	12.18
1350		75	2-0	26-1/2	121	8.20	4.36	12.56
1370		100	2-0	26-1/2	146	8.70	4.52	13.22
1400		125	2-0	24-1/2	172	10.20	4.97	15.17
1420	35	40	2-0	26-1/2	85	9.65	4.80	14.45
1450		65	2-0	28-1/2	111	9.95	4.91	14.86
1470		75	2-0	28-1/2	121	9.95	4.91	14.86
1500		100	1-11	28-1/2	147	10.85	5.15	16
1520		125	1-8	28-1/2	172	11.90	5.50	17.40
1550	5/8" gyp. fireproof.							
1560	On metal furring, add					.86	3.27	4.13

B1010 Floor Construction

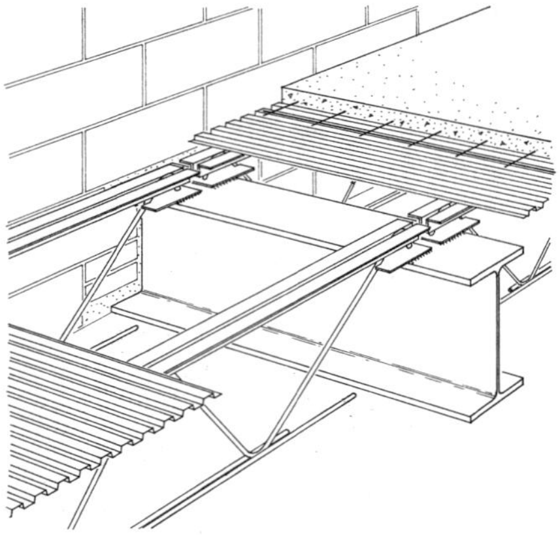

Table below lists costs for a floor system on exterior bearing walls and interior columns and beams using open web steel joists, galvanized steel slab form, 2-1/2″ concrete slab reinforced with welded wire fabric.

Design and Pricing Assumptions:
- Structural Steel is A36.
- Concrete f'c = 3 KSI placed by pump.
- WWF 6 x 6 – W1.4 x W1.4 (10 x 10)
- Columns are 12′ high.
- Building is 4 bays long by 4 bays wide.
- Joists are 2′ O.C. ± and span the long direction of the bay.

Joists at columns have bottom chords extended and are connected to columns.

Slab form is 28 gauge galvanized. Column costs in table are for columns to support 1 floor plus roof loading in a 2-story building; however, column costs are from ground floor to 2nd floor only. Joist costs include appropriate bridging. Deflection is limited to 1/360 of the span. Screeds and steel trowel finish.

Design Loads	Min.	Max.
S.S & Joists	4.4 PSF	11.5 PSF
Slab Form	1.0	1.0
2-1/2″ Concrete	27.0	27.0
Ceiling	3.0	3.0
Misc.	7.6	5.5
	43.0 PSF	48.0 PSF

System Components	QUANTITY	UNIT	COST PER S.F. MAT.	INST.	TOTAL
SYSTEM B1010 248 1200					
15′ X 20′ BAY, W.F. STEEL, STEEL JOISTS, SLAB FORM, CONCRETE SLAB					
Struct. steel, bolted-apts., nursing homes, etc., st. bearing, 1-2 stories	1.248	Lb.	1.37	.40	1.77
Open web joists	3.140	Lb.	2.86	1.16	4.02
Slab form, galvanized steel 9/16″ deep, 28 gauge	1.020	S.F.	1.55	.78	2.33
Welded wire fabric 6x6 - W1.4 x W1.4 (10 x 10), 21 lb/CSF roll, 10% lap	1.000	S.F.	.15	.38	.53
Concrete, ready mix, regular weight, 3000 psi	.210	C.F.	.92		.92
Place and vibrate concrete, elevated slab less than 6″, pumped	.210	C.F.		.34	.34
Finishing floor, monolithic steel trowel finish for finish floor	1.000	S.F.		.93	.93
Curing with sprayed membrane curing compound	.010	C.S.F.	.14	.10	.24
TOTAL			6.99	4.09	11.08

B1010 248 Steel Joists on Beam & Wall

	BAY SIZE (FT.)	SUPERIMPOSED LOAD (P.S.F.)	DEPTH (IN.)	TOTAL LOAD (P.S.F.)	COLUMN ADD	COST PER S.F. MAT.	INST.	TOTAL
1200	15x20 RB1010 -100	40	17	83		7	4.09	11.09
1210					columns	.45	.14	.59
1220	15x20	65	19	108		7.20	4.17	11.37
1230					columns	.45	.14	.59
1250	15x20	75	19	119		7.60	4.35	11.95
1260					columns	.54	.16	.70
1270	15x20	100	19	144		8	4.49	12.49
1280					columns	.54	.16	.70
1300	15x20	125	19	170		8.30	4.57	12.87
1310					columns	.72	.21	.93
1350	20x20	40	19	83		7.20	4.16	11.36
1360					columns	.41	.12	.53
1370	20x20	65	23	109		7.95	4.44	12.39
1380					columns	.54	.16	.70
1400	20x20	75	23	119		8.25	4.53	12.78
1410					columns	.54	.16	.70
1420	20x20	100	23	144		8.55	4.61	13.16
1430					columns	.54	.16	.70
1450	20x20	125	23	170		9.50	4.96	14.46
1460					columns	.65	.19	.84

B1010 Floor Construction

B1010 248 Steel Joists on Beam & Wall

	BAY SIZE (FT.)	SUPERIMPOSED LOAD (P.S.F.)	DEPTH (IN.)	TOTAL LOAD (P.S.F.)	COLUMN ADD	COST PER S.F.		
						MAT.	INST.	TOTAL
1500	20x25	40	23	84		7.55	4.31	11.86
1510					columns	.44	.13	.57
1520	20x25	65	26	110		8.20	4.52	12.72
1530					columns	.44	.13	.57
1550	20x25	75	26	120		8.65	4.71	13.36
1560					columns	.52	.15	.67
1570	20x25	100	26	145		9.10	4.88	13.98
1580					columns	.52	.15	.67
1670	20x25	125	29	170		10.35	5.35	15.70
1680					columns	.61	.18	.79
1720	25x25	40	23	84		8.20	4.49	12.69
1730					columns	.42	.13	.55
1750	25x25	65	29	110		8.60	4.64	13.24
1760					columns	.42	.13	.55
1770	25x25	75	26	120		9.55	4.98	14.53
1780					columns	.48	.14	.62
1800	25x25	100	29	145		10.75	5.45	16.20
1810					columns	.48	.14	.62
1820	25x25	125	29	170		11.30	5.60	16.90
1830					columns	.54	.16	.70
1870	25x30	40	29	84		9.50	4.57	14.07
1880					columns	.41	.12	.53
1900	25x30	65	29	110		9.55	4.59	14.14
1910					columns	.41	.12	.53
1920	25x30	75	29	120		10.05	4.75	14.80
1930					columns	.45	.13	.58
1950	25x30	100	29	145		10.90	5	15.90
1960					columns	.45	.13	.58
1970	25x30	125	32	170		12.45	5.45	17.90
1980					columns	.51	.15	.66
2020	30x30	40	29	84		9.35	4.54	13.89
2030					columns	.37	.11	.48
2050	30x30	65	29	110		10.60	4.90	15.50
2060					columns	.37	.11	.48
2070	30x30	75	32	120		10.90	4.99	15.89
2080					columns	.43	.13	.56
2100	30x30	100	35	145		12.05	5.35	17.40
2110					columns	.50	.15	.65
2120	30x30	125	35	172		13.90	5.90	19.80
2130					columns	.60	.18	.78
2170	30x35	40	29	85		10.55	4.90	15.45
2180					columns	.36	.10	.46
2200	30x35	65	29	111		12.55	5.50	18.05
2210					columns	.41	.13	.54
2220	30x35	75	32	121		12.55	5.50	18.05
2230					columns	.42	.13	.55
2250	30x35	100	35	148		12.75	5.55	18.30
2260					columns	.52	.15	.67
2270	30x35	125	38	173		14.30	6.05	20.35
2280					columns	.53	.15	.68
2320	35x35	40	32	85		10.85	4.99	15.84
2330					columns	.37	.11	.48

B1010 Floor Construction

B1010 248 Steel Joists on Beam & Wall

	BAY SIZE (FT.)	SUPERIMPOSED LOAD (P.S.F.)	DEPTH (IN.)	TOTAL LOAD (P.S.F.)	COLUMN ADD	COST PER S.F.		
						MAT.	INST.	TOTAL
2350	35x35	65	35	111		13.25	5.70	18.95
2360					columns	.44	.13	.57
2370	35x35	75	35	121		13.55	5.80	19.35
2380					columns	.44	.13	.57
2400	35x35	100	38	148		14.10	5.95	20.05
2410					columns	.55	.16	.71
2420	35x35	125	41	173		16.05	6.55	22.60
2430					columns	.55	.16	.71
2460	5/8 gyp. fireproof.							
2475	On metal furring, add					.86	3.27	4.13

B1010 Floor Construction

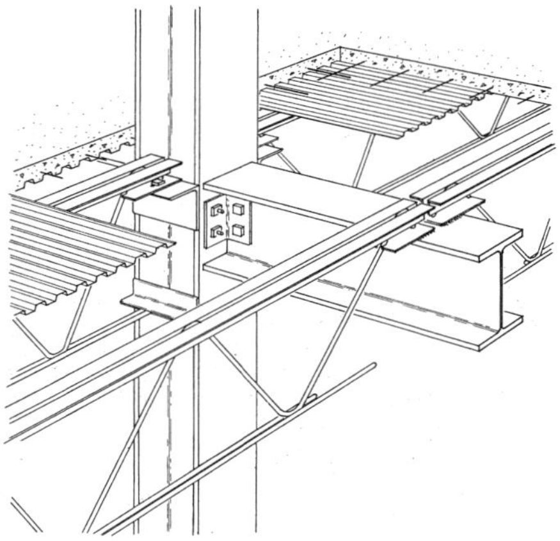

Table below lists costs for a floor system on steel columns and beams using open web steel joists, galvanized steel slab form, and 2-1/2″ concrete slab reinforced with welded wire fabric.

Design and Pricing Assumptions:
- Structural Steel is A36.
- Concrete f'c = 3 KSI placed by pump.
- WWF 6 x 6 – W1.4 x W1.4 (10 x 10)
- Columns are 12′ high.
- Building is 4 bays long by 4 bays wide.
- Joists are 2′ O.C. ± and span the long direction of the bay.

Joists at columns have bottom chords extended and are connected to columns.

Slab form is 28 gauge galvanized. Column costs in table are for columns to support 1 floor plus roof loading in a 2-story building; however, column costs are from ground floor to 2nd floor only. Joist costs include appropriate bridging. Deflection is limited to 1/360 of the span. Screeds and steel trowel finish.

Design Loads	Min.	Max.
S.S. & Joists	6.3 PSF	15.3 PSF
Slab Form	1.0	1.0
2-1/2″ Concrete	27.0	27.0
Ceiling	3.0	3.0
Misc.	5.7	1.7
	43.0 PSF	48.0 PSF

System Components	QUANTITY	UNIT	COST PER S.F. MAT.	INST.	TOTAL
SYSTEM B1010 250 2350					
15′ X 20′ BAY 40 PSF S. LOAD, 17″ DEPTH, 83 PSF TOTAL LOAD					
Struct. steel, bolted-apts., nursing homes, etc., st. bearing, 1-2 stories	1.974	Lb.	2.88	.85	3.73
Open web joists	3.140	Lb.	2.86	1.16	4.02
Slab form, galvanized steel 9/16″ deep, 28 gauge	1.020	S.F.	1.55	.78	2.33
Welded wire fabric rolls, 6 x 6 - W1.4 x W1.4 (10 x 10), 21 lb/csf	1.000	S.F.	.15	.38	.53
Concrete, ready mix, regular weight, 3000 psi	.210	C.F.	.92		.92
Place and vibrate concrete, elevated slab less than 6″, pumped	.210	C.F.		.34	.34
Finishing floor, monolithic steel trowel finish for finish floor	1.000	S.F.		.93	.93
Curing with sprayed membrane curing compound	.010	S.F.	.14	.10	.24
TOTAL			8.50	4.54	13.04

B1010 250 Steel Joists, Beams & Slab on Columns

	BAY SIZE (FT.)	SUPERIMPOSED LOAD (P.S.F.)	DEPTH (IN.)	TOTAL LOAD (P.S.F.)	COLUMN ADD	COST PER S.F. MAT.	INST.	TOTAL
2350	15x20 RB1010 -100	40	17	83		8.50	4.54	13.04
2400					column	1.38	.41	1.79
2450	15x20	65	19	108		9.45	4.81	14.26
2500					column	1.38	.41	1.79
2550	15x20	75	19	119		9.80	4.96	14.76
2600					column	1.51	.44	1.95
2650	15x20	100	19	144		10.40	5.15	15.55
2700					column	1.51	.44	1.95
2750	15x20	125	19	170		10.95	5.30	16.25
2800					column	2.01	.59	2.60
2850	20x20	40	19	83		9.25	4.75	14
2900					column	1.13	.33	1.46
2950	20x20	65	23	109		10.20	5.05	15.25
3000					column	1.51	.44	1.95
3100	20x20	75	26	119		10.75	5.25	16
3200					column	1.51	.44	1.95
3400	20x20	100	23	144		11.20	5.35	16.55
3450					column	1.51	.44	1.95
3500	20x20	125	23	170		12.40	5.80	18.20
3600					column	1.81	.53	2.34

B1010 Floor Construction

B1010 250 Steel Joists, Beams & Slab on Columns

	BAY SIZE (FT.)	SUPERIMPOSED LOAD (P.S.F.)	DEPTH (IN.)	TOTAL LOAD (P.S.F.)	COLUMN ADD	COST PER S.F.		
						MAT.	INST.	TOTAL
3700	20x25	40	44	83		9.90	4.98	14.88
3800					column	1.21	.35	1.56
3900	20x25	65	26	110		10.85	5.25	16.10
4000					column	1.21	.35	1.56
4100	20x25	75	26	120		11.30	5.45	16.75
4200					column	1.45	.43	1.88
4300	20x25	100	26	145		12	5.70	17.70
4400					column	1.45	.43	1.88
4500	20x25	125	29	170		13.30	6.15	19.45
4600					column	1.69	.49	2.18
4700	25x25	40	23	84		10.75	5.20	15.95
4800					column	1.16	.34	1.50
4900	25x25	65	29	110		11.35	5.40	16.75
5000					column	1.16	.34	1.50
5100	25x25	75	26	120		12.55	5.85	18.40
5200					column	1.35	.39	1.74
5300	25x25	100	29	145		13.90	6.35	20.25
5400					column	1.35	.39	1.74
5500	25x25	125	32	170		14.70	6.60	21.30
5600					column	1.50	.44	1.94
5700	25x30	40	29	84		12.25	5.35	17.60
5800					column	1.12	.33	1.45
5900	25x30	65	29	110		12.75	5.50	18.25
6000					column	1.12	.33	1.45
6050	25x30	75	29	120		13.10	5.65	18.75
6100					column	1.24	.37	1.61
6150	25x30	100	29	145		14.20	5.95	20.15
6200					column	1.24	.37	1.61
6250	25x30	125	32	170		16	6.50	22.50
6300					column	1.43	.42	1.85
6350	30x30	40	29	84		12.25	5.35	17.60
6400					column	1.04	.30	1.34
6500	30x30	65	29	110		13.95	5.85	19.80
6600					column	1.04	.30	1.34
6700	30x30	75	32	120		14.20	5.95	20.15
6800					column	1.19	.35	1.54
6900	30x30	100	35	145		15.75	6.40	22.15
7000					column	1.39	.41	1.80
7100	30x30	125	35	172		18	7.05	25.05
7200					column	1.55	.46	2.01
7300	30x35	40	29	85		13.80	5.85	19.65
7400					column	.89	.26	1.15
7500	30x35	65	29	111		16.05	6.50	22.55
7600					column	1.15	.34	1.49
7700	30x35	75	32	121		16.05	6.50	22.55
7800					column	1.17	.34	1.51
7900	30x35	100	35	148		16.55	6.65	23.20
8000					column	1.43	.42	1.85
8100	30x35	125	38	173		18.35	7.20	25.55
8200					column	1.46	.43	1.89
8300	35x35	40	32	85		14.20	5.90	20.10
8400					column	1.02	.30	1.32

B1010 Floor Construction

B1010 250 Steel Joists, Beams & Slab on Columns

	BAY SIZE (FT.)	SUPERIMPOSED LOAD (P.S.F.)	DEPTH (IN.)	TOTAL LOAD (P.S.F.)	COLUMN ADD	COST PER S.F.		
						MAT.	INST.	TOTAL
8500	35x35	65	35	111		16.90	6.75	23.65
8600					column	1.23	.37	1.60
9300	35x35	75	38	121		17.30	6.85	24.15
9400					column	1.23	.37	1.60
9500	35x35	100	38	148		18.65	7.30	25.95
9600					column	1.52	.44	1.96
9750	35x35	125	41	173		19.25	7.45	26.70
9800					column	1.54	.46	2
9810	5/8 gyp. fireproof.							
9815	On metal furring, add					.86	3.27	4.13

B1010 Floor Construction

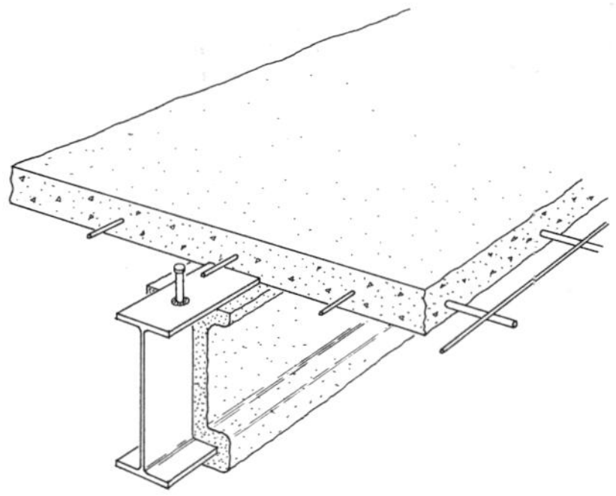

General: Composite construction of W shape flange beams and concrete slabs is most efficiently used when loads are heavy and spans are moderately long. It is stiffer with less deflection than non-composite construction of similar depth and spans.

In practice, composite construction is typically shallower in depth than non-composite would be.

Design Assumptions:
Steel, fy = 36 KSI
Beams unshored during construction
Deflection limited to span/360
Shear connectors, welded studs
Concrete, f'c = 3 KSI

System Components	QUANTITY	UNIT	COST PER S.F. MAT.	INST.	TOTAL
SYSTEM B1010 252 3800					
20′ X 25′ BAY, 40 PSF S. LOAD, 4″ THICK SLAB, 20″ TOTAL DEPTH					
Structural steel	3.820	Lb.	5.69	1.68	7.37
Welded shear connectors 3/4″ diameter 3-3/8″ long	.140	Ea.	.09	.29	.38
Forms in place, floor slab forms hung from steel beams, 4 uses	1.000	S.F.	2.12	6.15	8.27
Edge forms to 6″ high on elevated slab, 4 uses	.050	L.F.	.01	.23	.24
Reinforcing in place, elevated slabs #4 to #7	1.190	Lb.	.65	.54	1.19
Concrete, ready mix, regular weight, 3000 psi	.330	C.F.	1.44		1.44
Place and vibrate concrete, elevated slab less than 6″, pumped	.330	C.F.		.54	.54
Finishing floor, monolithic steel trowel finish for resilient tile	1.000	S.F.		1.22	1.22
Curing with sprayed membrane curing compound	.010	S.F.	.14	.10	.24
Spray mineral fiber/cement for fire proof, 1″ thick on beams	.520	S.F.	.31	.55	.86
TOTAL			10.45	11.30	21.75

B1010 252 Composite Beam & Cast in Place Slab

	BAY SIZE (FT.)	SUPERIMPOSED LOAD (P.S.F.)	SLAB THICKNESS (IN.)	TOTAL DEPTH (FT.-IN.)	TOTAL LOAD (P.S.F.)	COST PER S.F. MAT.	INST.	TOTAL
3800	20x25	40	4	1 - 8	94	10.45	11.30	21.75
3900	RB1010 -100	75	4	1 - 8	130	12.15	12	24.15
4000		125	4	1 - 10	181	14.15	12.80	26.95
4100		200	5	2 - 2	272	18.40	14.40	32.80
4200	25x25	40	4-1/2	1 - 8-1/2	99	10.90	11.45	22.35
4300		75	4-1/2	1 - 10-1/2	136	13.15	12.30	25.45
4400		125	5-1/2	2 - 0-1/2	200	15.55	13.30	28.85
4500		200	5-1/2	2 - 2-1/2	278	19.35	14.70	34.05
4600	25x30	40	4-1/2	1 - 8-1/2	100	12.20	11.90	24.10
4700		75	4-1/2	1 - 10-1/2	136	14.30	12.70	27
4800		125	5-1/2	2 - 2-1/2	202	17.55	13.95	31.50
4900		200	5-1/2	2 - 5-1/2	279	21	15.20	36.20
5000	30x30	40	4	1 - 8	95	12.20	11.90	24.10
5200		75	4	2 - 1	131	14.20	12.55	26.75
5400		125	4	2 - 4	183	17.55	13.85	31.40
5600		200	5	2 - 10	274	22.50	15.65	38.15

B1010 Floor Construction

B1010 252 Composite Beam & Cast in Place Slab

	BAY SIZE (FT.)	SUPERIMPOSED LOAD (P.S.F.)	SLAB THICKNESS (IN.)	TOTAL DEPTH (FT.-IN.)	TOTAL LOAD (P.S.F.)	COST PER S.F.		
						MAT.	INST.	TOTAL
5800	30x35	40	4	2 - 1	95	12.95	12.10	25.05
6000		75	4	2 - 4	139	15.65	13.05	28.70
6250		125	4	2 - 4	185	20.50	14.75	35.25
6500		200	5	2 - 11	276	25	16.40	41.40
7000	35x30	40	4	2 - 1	95	13.30	12.25	25.55
7200		75	4	2 - 4	139	15.85	13.20	29.05
7400		125	4	2 - 4	185	20.50	14.75	35.25
7600		200	5	2 - 11	276	25	16.40	41.40
8000	35x35	40	4	2 - 1	96	13.65	12.35	26
8250		75	4	2 - 4	133	16.25	13.35	29.60
8500		125	4	2 - 10	185	19.55	14.50	34.05
8750		200	5	3 - 5	276	25.50	16.60	42.10
9000	35x40	40	4	2 - 4	97	15.20	12.90	28.10
9250		75	4	2 - 4	134	17.80	13.70	31.50
9500		125	4	2 - 7	186	21.50	15.15	36.65
9750		200	5	3 - 5	278	28	17.35	45.35

B1010 Floor Construction

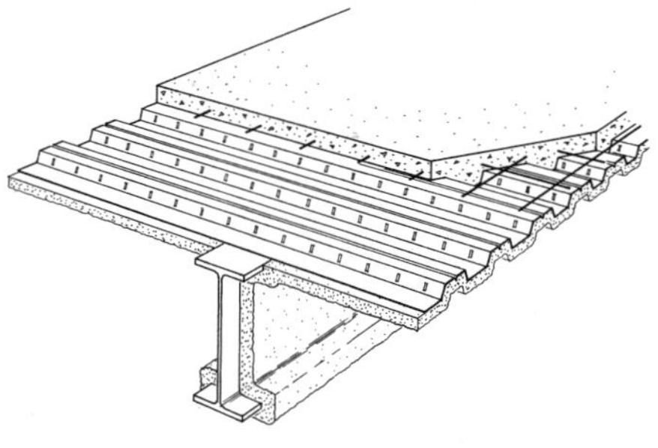

Description: Table below lists costs per S.F. for floors using steel beams and girders, composite steel deck, concrete slab reinforced with W.W.F. and sprayed fiber fireproofing (non-asbestos) on the steel beams and girders and on the steel deck.

Design and Pricing Assumptions:
Structural Steel is A36, high strength bolted.
Composite steel deck varies from 2″–20 gauge to 3″–16 gauge galvanized.
WWF 6 x 6 – W1.4 x W1.4 (10 x 10)
Concrete f'c = 3 KSI.
Steel trowel finish and cure.
Spandrels are assumed the same weight as interior beams and girders to allow for exterior wall loads and bracing or moment connections.

System Components	QUANTITY	UNIT	COST PER S.F. MAT.	INST.	TOTAL
SYSTEM B1010 254 0540					
W SHAPE BEAMS & DECK, FIREPROOFED, 15′ X 20′, 5″ SLAB, 40 PSF LOAD					
Structural steel	4.470	Lb.	6.13	1.81	7.94
Metal decking, non-cellular composite, galv 3″ deep, 20 gauge	1.050	S.F.	2.73	1.07	3.80
Sheet metal edge closure form, 12″, w/2 bends, 18 ga, galv	.058	L.F.	.27	.15	.42
Welded wire fabric 6 x 6 - W1.4 x W1.4 (10 x 10), 21 lb/CSF roll, 10% lap	1.000	S.F.	.15	.38	.53
Concrete ready mix, regular weight, 3000 psi	.011	C.Y.	1.30		1.30
Place and vibrate concrete, elevated slab less than 6″, pumped	.011	C.Y.		.38	.38
Finishing floor, monolithic steel trowel finish for finish floor	1.000	S.F.		.93	.93
Curing with sprayed membrane curing compound	.010	C.S.F.	.14	.10	.24
Sprayed mineral fiber/cement for fireproof, 1″ thick on decks	1.000	S.F.	.88	1.29	2.17
Sprayed mineral fiber/cement for fireproof, 1″ thick on beams	.615	S.F.	.36	.66	1.02
TOTAL			11.96	6.77	18.73

B1010 254 W Shape, Composite Deck, & Slab

	BAY SIZE (FT.) BEAM X GIRD	SUPERIMPOSED LOAD (P.S.F.)	SLAB THICKNESS (IN.)	TOTAL DEPTH (FT.-IN.)	TOTAL LOAD (P.S.F.)	COST PER S.F. MAT.	INST.	TOTAL
0540	15x20	40	5	1-7	89	11.95	6.80	18.75
0560	RB1010 -100	75	5	1-9	125	13.15	7.20	20.35
0580		125	5	1-11	176	15.40	7.85	23.25
0600		200	5	2-2	254	18.95	8.95	27.90
0620	20x15	40	4	1-4	89	12.05	6.95	19
0640		75	4	1-4	119	13.45	7.20	20.65
0660		125	4	1-6	170	15.05	7.70	22.75
0680		200	4	2-1	247	17.65	8.65	26.30
0700	20x20	40	5	1-9	90	13	7.05	20.05
0720		75	5	1-9	126	14.70	7.65	22.35
0740		125	5	1-11	177	16.70	8.30	25
0760		200	5	2-5	255	20	9.25	29.25
0780	20x25	40	5	1-9	90	13.25	7	20.25
0800		75	5	1-9	127	16.55	8.10	24.65
0820		125	5	1-11	180	19.80	8.80	28.60
0840		200	5	2-5	256	22	9.95	31.95
0860	25x20	40	5	1-11	90	13.45	7.15	20.60
0880		75	5	2-5	127	15.75	7.95	23.70
0900		125	5	2-5	178	17.75	8.65	26.40
0920		200	5	2-5	257	23	10.25	33.25
0940	25x25	40	5	1-11	91	15.05	7.55	22.60
0960		75	5	2-5	178	17.35	8.30	25.65
0980		125	5	2-5	181	21	9.20	30.20
1000		200	5-1/2	2-8-1/2	263	24	10.65	34.65

B1010 Floor Construction

B1010 254 W Shape, Composite Deck, & Slab

	BAY SIZE (FT.) BEAM X GIRD	SUPERIMPOSED LOAD (P.S.F.)	SLAB THICKNESS (IN.)	TOTAL DEPTH (FT.-IN.)	TOTAL LOAD (P.S.F.)	COST PER S.F.		
						MAT.	INST.	TOTAL
1400	25x30	40	5	2-5	91	14.90	7.70	22.60
1500		75	5	2-5	128	18.20	8.60	26.80
1600		125	5	2-8	180	20.50	9.50	30
1700		200	5	2-11	259	26	11	37
1800	30x25	40	5	2-5	92	15.90	8.05	23.95
1900		75	5	2-5	129	18.75	8.95	27.70
2000		125	5	2-8	181	21.50	9.85	31.35
2100		200	5-1/2	2-11	200	26	11.30	37.30
2200	30x30	40	5	2-2	92	16.55	8.20	24.75
2300		75	5	2-5	129	19.40	9.15	28.55
2400		125	5	2-11	182	23	10.25	33.25
2500		200	5	3-2	263	31.50	12.80	44.30
2600	30x35	40	5	2-5	94	18.05	8.70	26.75
2700		75	5	2-11	131	21	9.60	30.60
2800		125	5	3-2	183	24.50	10.75	35.25
2900		200	5-1/2	3-5-1/2	268	30.50	12.35	42.85
3400	35x30	40	5	2-5	93	17.45	8.55	26
3500		75	5	2-8	130	21	9.60	30.60
3600		125	5	2-11	183	25	10.85	35.85
3700		200	5	3-5	262	30.50	12.60	43.10
3800	35x35	40	5	2-8	94	18.60	8.65	27.25
3900		75	5	2-11	131	21.50	9.65	31.15
4000		125	5	3-5	184	26	11	37
4100		200	5-1/2	3-5-1/2	270	33.50	13	46.50
4200	35x40	40	5	2-11	94	19.40	9.15	28.55
4300		75	5	3-2	131	22.50	10.15	32.65
4400		125	5	3-5	184	26.50	11.45	37.95
4500		200	5	3-5-1/2	264	34	13.70	47.70

B1010 Floor Construction

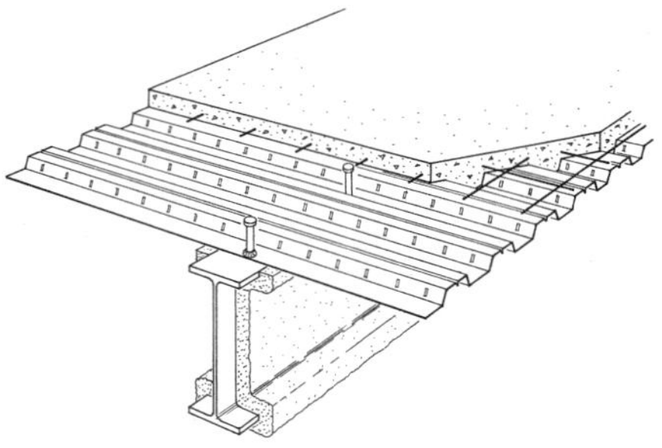

Description: Table below lists costs ($/S.F.) for a floor system using composite steel beams with welded shear studs, composite steel deck, and light weight concrete slab reinforced with W.W.F. Price includes sprayed fiber fireproofing on steel beams.

Design and Pricing Assumptions:
Structural steel is A36, high strength bolted.
Composite steel deck varies from 22 gauge to 16 gauge, galvanized.
Shear Studs are 3/4".
W.W.F., 6 x 6 – W1.4 x W1.4 (10 x 10)
Concrete f'c = 3 KSI, lightweight.
Steel trowel finish and cure.
Fireproofing is sprayed fiber (non-asbestos).

Spandrels are assumed the same as interior beams and girders to allow for exterior wall loads and bracing or moment connections.

System Components	QUANTITY	UNIT	COST PER S.F. MAT.	INST.	TOTAL
SYSTEM B1010 256 2400					
20' X 25' BAY, 40 PSF S. LOAD, 5-1/2" SLAB, 17-1/2" TOTAL THICKNESS					
Structural steel	4.320	Lb.	6.44	1.90	8.34
Welded shear connectors 3/4" diameter 4-7/8" long	.163	Ea.	.13	.35	.48
Metal decking, non-cellular composite, galv. 3" deep, 22 gauge	1.050	S.F.	2.46	1.01	3.47
Sheet metal edge closure form, 12", w/2 bends, 18 ga, galv	.045	L.F.	.21	.11	.32
Welded wire fabric rolls, 6 x 6 - W1.4 x W1.4 (10 x 10), 21 lb/csf	1.000	S.F.	.15	.38	.53
Concrete ready mix, light weight, 3,000 PSI	.333	C.F.	3.05		3.05
Place and vibrate concrete, elevated slab less than 6", pumped	.333	C.F.		.55	.55
Finishing floor, monolithic steel trowel finish for finish floor	1.000	S.F.		.93	.93
Curing with sprayed membrane curing compound	.010	C.S.F.	.14	.10	.24
Shores, erect and strip vertical to 10' high	.020	Ea.		.45	.45
Sprayed mineral fiber/cement for fireproof, 1" thick on beams	.483	S.F.	.28	.52	.80
TOTAL			12.86	6.30	19.16

B1010 256 Composite Beams, Deck & Slab

	BAY SIZE (FT.)	SUPERIMPOSED LOAD (P.S.F.)	SLAB THICKNESS (IN.)	TOTAL DEPTH (FT.-IN.)	TOTAL LOAD (P.S.F.)	COST PER S.F. MAT.	INST.	TOTAL
2400	20x25	40	5-1/2	1 - 5-1/2	80	12.85	6.30	19.15
2500	RB1010 -100	75	5-1/2	1 - 9-1/2	115	13.35	6.30	19.65
2750		125	5-1/2	1 - 9-1/2	167	16.25	7.35	23.60
2900		200	6-1/4	1 - 11-1/2	251	18.30	7.90	26.20
3000	25x25	40	5-1/2	1 - 9-1/2	82	12.65	6	18.65
3100		75	5-1/2	1 - 11-1/2	118	14.05	6.05	20.10
3200		125	5-1/2	2 - 2-1/2	169	14.65	6.55	21.20
3300		200	6-1/4	2 - 6-1/4	252	19.80	7.65	27.45
3400	25x30	40	5-1/2	1 - 11-1/2	83	12.90	5.95	18.85
3600		75	5-1/2	1 - 11-1/2	119	13.85	6.05	19.90
3900		125	5-1/2	1 - 11-1/2	170	16.10	6.80	22.90
4000		200	6-1/4	2 - 6-1/4	252	19.90	7.70	27.60
4200	30x30	40	5-1/2	1 - 11-1/2	81	13	6.15	19.15
4400		75	5-1/2	2 - 2-1/2	116	14	6.40	20.40
4500		125	5-1/2	2 - 5-1/2	168	16.90	7.15	24.05
4700		200	6-1/4	2 - 9-1/4	252	20	8.30	28.30
4900	30x35	40	5-1/2	2 - 2-1/2	82	13.55	6.35	19.90
5100		75	5-1/2	2 - 5-1/2	117	14.80	6.50	21.30
5300		125	5-1/2	2 - 5-1/2	169	17.40	7.30	24.70
5500		200	6-1/4	2 - 9-1/4	254	20.50	8.30	28.80
5750	35x35	40	5-1/2	2 - 5-1/2	84	14.45	6.35	20.80
6000		75	5-1/2	2 - 5-1/2	121	16.45	6.80	23.25

B1010 Floor Construction

B1010 256 Composite Beams, Deck & Slab

	BAY SIZE (FT.)	SUPERIMPOSED LOAD (P.S.F.)	SLAB THICKNESS (IN.)	TOTAL DEPTH (FT.-IN.)	TOTAL LOAD (P.S.F.)	COST PER S.F.		
						MAT.	INST.	TOTAL
7000	35x35	125	5-1/2	2 - 8-1/2	170	19.30	7.80	27.10
7200		200	5-1/2	2 - 11-1/2	254	22	8.60	30.60
7400	35x40	40	5-1/2	2 - 5-1/2	85	15.95	6.85	22.80
7600		75	5-1/2	2 - 5-1/2	121	17.30	7.05	24.35
8000		125	5-1/2	2 - 5-1/2	171	19.80	7.90	27.70
9000		200	5-1/2	2 - 11-1/2	255	24	8.95	32.95

B1010 Floor Construction

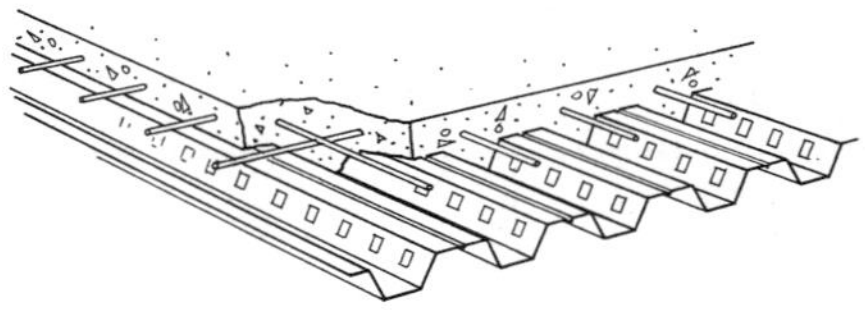

How to Use Tables: Enter any table at superimposed load and support spacing for deck.

Cellular decking tends to be stiffer and has the potential for utility integration with deck structure.

System Components	QUANTITY	UNIT	COST PER S.F. MAT.	COST PER S.F. INST.	COST PER S.F. TOTAL
SYSTEM B1010 258 0900					
SPAN 6′, LOAD 125 PSF, DECK 1-1/2″-22 GA., 4″ SLAB					
Metal decking, open, galv., 1-1/2″ deep, 22 ga., over 50 sq.	1.050	S.F.	1.72	.63	2.35
Welded wire fabric rolls, 6 x 6 - W1.4 x W1.4 (10 x 10), 21 lb/csf	1.000	S.F.	.15	.38	.53
Edge forms to 6″ high on elevated slab, 4 uses	.050	L.F.	.01	.23	.24
Concrete ready mix, regular weight, 3000 psi	.253	C.F.	1.06		1.06
Place and vibrate concrete, elevated slab less than 6″, pumped	.253	C.F.		.32	.32
Finishing floor, monolithic steel trowel finish for finish floor	1.000	S.F.		.93	.93
Curing with sprayed membrane curing compound	.010	C.S.F.	.14	.10	.24
TOTAL			3.08	2.59	5.67

B1010 258 Metal Deck/Concrete Fill

	SUPERIMPOSED LOAD (P.S.F.)	DECK SPAN (FT.)	DECK GAGE DEPTH	SLAB THICKNESS (IN.)	TOTAL LOAD (P.S.F.)	COST PER S.F. MAT.	COST PER S.F. INST.	COST PER S.F. TOTAL
0900	125	6	22 1-1/2	4	164	3.08	2.59	5.67
0920		7	20 1-1/2	4	164	3.39	2.74	6.13
0950		8	20 1-1/2	4	165	3.39	2.74	6.13
0970		9	18 1-1/2	4	165	3.96	2.74	6.70
1000		10	18 2	4	165	3.86	2.89	6.75
1020		11	18 3	5	169	4.75	3.12	7.87
1050	150	6	22 1-1/2	4	189	3.09	2.62	5.71
1070		7	22 1-1/2	4	189	3.09	2.62	5.71
1100		8	20 1-1/2	4	190	3.39	2.74	6.13
1120		9	18 1-1/2	4	190	3.96	2.74	6.70
1150		10	18 3	5	190	4.75	3.12	7.87
1170		11	18 3	5-1/2	194	4.75	3.12	7.87
1200	200	6	22 1-1/2	4	239	3.09	2.62	5.71
1220		7	22 1-1/2	4	239	3.09	2.62	5.71
1250		8	20 1-1/2	4	239	3.39	2.74	6.13
1270		9	18 2	4	241	4.55	2.94	7.49
1300		10	16 2	4	240	5.35	3	8.35
1320	250	6	22 1-1/2	4	289	3.09	2.62	5.71
1350		7	22 1-1/2	4	289	3.09	2.62	5.71
1370		8	18 2	4	290	4.55	2.94	7.49
1400		9	16 2	4	290	5.35	3	8.35

B1010 Floor Construction

B1010 260 Cellular Composite Deck

	SUPERIMPOSED LOAD (P.S.F.)	DECK SPAN (FT.)	DECK & PLATE GAUGE	SLAB THICKNESS (IN.)	TOTAL LOAD (P.S.F.)	COST PER S.F.		
						MAT.	INST.	TOTAL
1450	150	10	20 20	5	195	11.75	4.35	16.10
1470		11	20 20	5	195	11.75	4.35	16.10
1500		12	20 20	5	195	11.75	4.35	16.10
1520		13	20 20	5	195	11.75	4.35	16.10
1570	200	10	20 20	5	250	11.75	4.35	16.10
1650		11	20 20	5	250	11.75	4.35	16.10
1670		12	20 20	5	250	11.75	4.35	16.10
1700		13	18 20	5	250	13.90	4.39	18.29

B1010 Floor Construction

Description: Table below lists the S.F. costs for wood joists and a minimum thickness plywood subfloor.

Design Assumptions: 10% allowance has been added to framing quantities for overlaps, waste, double joists at openings or under partitions, etc. 5% added to subfloor for waste.

System Components	QUANTITY	UNIT	COST PER S.F. MAT.	COST PER S.F. INST.	COST PER S.F. TOTAL
SYSTEM B1010 261 2500					
WOOD JOISTS 2″ X 6″, 12″ O.C.					
Framing joists, fir, 2″ x 6″	1.100	B.F.	.74	1.05	1.79
Subfloor plywood CDX 1/2″	1.050	S.F.	.79	.83	1.62
TOTAL			1.53	1.88	3.41

B1010 261	Wood Joist	COST PER S.F. MAT.	COST PER S.F. INST.	COST PER S.F. TOTAL
2500	Wood joists, 2″x6″, 12″ O.C.	1.53	1.88	3.41
2550	16″ O.C.	1.35	1.62	2.97
2600	24″ O.C.	1.42	1.52	2.94
2900	2″x8″, 12″ O.C.	1.81	2.02	3.83
2950	16″ O.C.	1.56	1.73	3.29
3000	24″ O.C.	1.57	1.60	3.17
3300	2″x10″, 12″ O.C.	2.41	2.30	4.71
3350	16″ O.C.	2	1.93	3.93
3400	24″ O.C.	1.85	1.73	3.58
3700	2″x12″, 12″ O.C.	2.85	2.33	5.18
3750	16″ O.C.	2.33	1.95	4.28
3800	24″ O.C.	2.08	1.75	3.83
4100	2″x14″, 12″ O.C.	3.41	2.53	5.94
4150	16″ O.C.	2.77	2.11	4.88
4200	24″ O.C.	2.37	1.86	4.23
4500	3″x6″, 12″ O.C.	3.10	2.24	5.34
4550	16″ O.C.	2.53	1.89	4.42
4600	24″ O.C.	2.21	1.71	3.92
4900	3″x8″, 12″ O.C.	4.04	2.21	6.25
4950	16″ O.C.	3.22	1.86	5.08
5000	24″ O.C.	2.67	1.69	4.36
5300	3″x10″, 12″ O.C.	4.92	2.51	7.43
5350	16″ O.C.	3.90	2.09	5.99
5400	24″ O.C.	3.12	1.84	4.96
5700	3″x12″, 12″ O.C.	4.95	2.18	7.13
5750	16″ O.C.	4.51	2.47	6.98
5800	24″ O.C.	3.53	2.09	5.62
6100	4″x6″, 12″ O.C.	4.09	2.28	6.37
6150	16″ O.C.	3.27	1.92	5.19
6200	24″ O.C.	2.70	1.73	4.43

B1010 Floor Construction

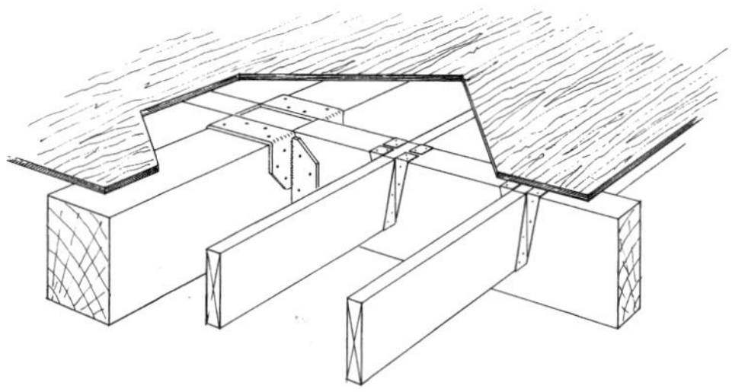

Description: Table lists the S.F. costs, total load, and member sizes, for various bay sizes and loading conditions.

Design Assumptions: Dead load = girder, beams, and joist weight plus 3/4" plywood floor.

Maximum deflection is 1/360 of the clear span.

Lumber is stress grade f(w) = 1,800 PSI

System Components	QUANTITY	UNIT	COST PER S.F.		
			MAT.	INST.	TOTAL
SYSTEM B1010 264 2000					
15' X 15' BAY, S. LOAD 40 P.S.F.					
Beams and girders, structural grade, 8" x 12"	.730	B.F.	1.30	.25	1.55
Framing joists, fir 4" x 12"	.660	B.F.	.94	.44	1.38
Framing joists, 2" x 6"	.840	B.F.	.56	.80	1.36
Beam to girder saddles	.510	Lb.	3.71	.78	4.49
Column caps	.510	Lb.	1.02	.10	1.12
Drilling, bolt holes	.510	Lb.		.38	.38
Machine bolts	.510	Lb.	.16	.17	.33
Joist hangers 18 ga.	.213	Ea.	.32	.77	1.09
Subfloor plywood CDX 3/4"	1.050	S.F.	1.05	1	2.05
TOTAL			9.06	4.69	13.75

B1010 264 Wood Beam & Joist

	BAY SIZE (FT.)	SUPERIMPOSED LOAD (P.S.F.)	GIRDER BEAM (IN.)	JOISTS (IN.)	TOTAL LOAD (P.S.F.)	COST PER S.F.		
						MAT.	INST.	TOTAL
2000	15x15	40	8 x 12 4 x 12	2 x 6 @ 16	53	9.05	4.69	13.74
2050	RB1010 -100	75	8 x 16 4 x 16	2 x 8 @ 16	90	12.60	5.10	17.70
2100		125	12 x 16 6 x 16	2 x 8 @ 12	144	19.50	6.55	26.05
2150		200	14 x 22 12 x 16	2 x 10 @ 12	227	39	10.05	49.05
2500	15x20	40	10 x 16 8 x 12	2 x 6 @ 16	58	11.40	4.78	16.18
2550		75	12 x 14 8 x 14	2 x 8 @ 16	96	15.45	5.55	21
2600		125	10 x 18 12 x 14	2 x 8 @ 12	152	24	7.45	31.45
2650		200	14 x 20 14 x 16	2 x 10 @ 12	234	34	9.15	43.15
3000	20x20	40	10 x 14 10 x 12	2 x 8 @ 16	63	11.90	4.71	16.61
3050		75	12 x 16 8 x 16	2 x 10 @ 16	102	16.85	5.25	22.10
3100		125	14 x 22 12 x 16	2 x 10 @ 12	163	33	8.40	41.40

B1010 Floor Construction

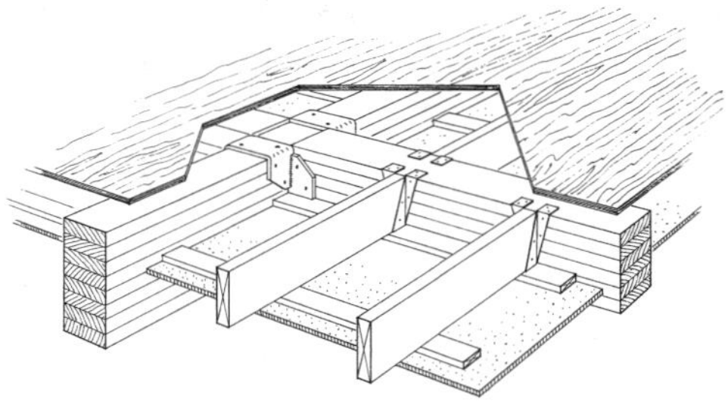

Description: Table below lists the S.F. costs, total load, and member sizes, for various bay sizes and loading conditions.

Design Assumptions: Dead load = girder, beams, and joist weight plus 3/4″ plywood floor.

Maximum deflection is 1/360 of the clear span.

Lumber is stress grade f(w) = 2,500 PSI

System Components	QUANTITY	UNIT	COST PER S.F. MAT.	INST.	TOTAL
SYSTEM B1010 265 2000					
15′ X 15′ BAY, 40 PSF L.L., JOISTS 2 X 6, @ 16″ O.C.					
Laminated framing, straight beams	1.000	B.F.	3.18	1.06	4.24
Framing joists, 2″ x 6″	.840	B.F.	.56	.80	1.36
Saddles	3.375	Lb.	3.71	.78	4.49
Column caps	3.375	Lb.	1.02	.10	1.12
Drilling, bolt holes	3.375	Lb.		.35	.35
Machine bolts	3.375	Lb.	.14	.14	.28
Joist & beam hangers 18 ga.	.213	Ea.	.32	.77	1.09
Subfloor plywood CDX 3/4″	1.050	S.F.	1.05	1	2.05
TOTAL			9.98	5	14.98

B1010 265 Laminated Wood Floor Beams

	BAY SIZE (FT.)	SUPERIMPOSED LOAD (P.S.F.)	GIRDER (IN.)	BEAM (IN.)	TOTAL LOAD (P.S.F.)	COST PER S.F. MAT.	INST.	TOTAL
2000	15x15	40	5 x 12-3/8	5 x 8-1/4	54	10	5	15
2050	RB1010 -100	75	5 x 16-1/2	5 x 11	92	12.80	5.75	18.55
2100		125	8-5/8 x 15-1/8	8-5/8 x 11	147	18.90	8.10	27
2150		200	8-5/8 x 22	8-5/8 x 16-1/2	128	35.50	11.80	47.30
2500	15x20	40	5 x 13-3/4	5 x 11	55	10	4.95	14.95
2550		75	5 x 17-3/8	5 x 15-1/8	92	14.45	6	20.45
2600		125	8-5/8 x 17-7/8	8-5/8 x 13-3/4	148	22.50	8.65	31.15
2650		200	8-5/8 x 22	8-5/8 x 17-7/8	226	31	10.75	41.75
3000	20x20	40	5 x 16-1/2	5 x 13-3/4	55	10.35	4.82	15.17
3050		75	5 x 20-5/8	5 x 16-1/2	92	14.50	5.85	20.35
3100		125	8-5/8 x 22	8-5/8 x 15-1/8	148	22	8.10	30.10
3150		200	8-5/8 x 28-7/8	8-5/8 x 20-5/8	229	33	10.90	43.90

B1010 Floor Construction

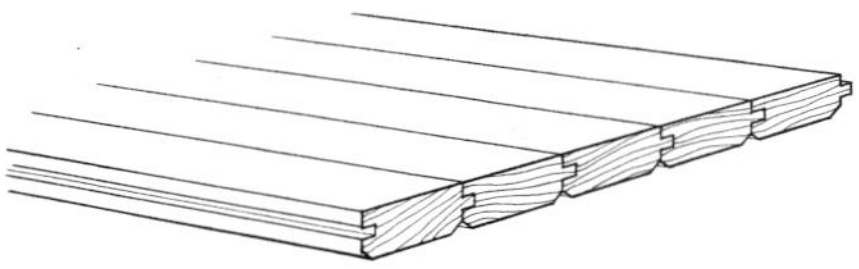

Description: Table below lists the S.F. costs and maximum spans for commonly used wood decking materials for various loading conditions.

Design Assumptions: Total applied load is the superimposed load plus dead load.

Maximum deflection is 1/180 of the clear span which is not suitable if plaster ceilings will be supported by the roof or floor.

Modulus of elasticity (E) and fiber strength (f) are as shown in the table to the right.

No allowance for waste has been included.

Supporting beams or purlins are not included in costs below.

Decking Material Characteristics

Deck Material	Modulus of Elasticity	Fiber Strength
Cedar	1,100,000	1000 psi
Douglas fir	1,760,000	1200 psi
Hemlock	1,600,000	1200 psi
White spruce	1,320,000	1200 psi

B1010 266 Wood Deck

	NOMINAL THK (IN.)	TYPE WOOD	MAXIMUM SPAN			COST PER S.F.		
			MAXIMUM SPAN 40 P.S.F. LOAD	MAXIMUM SPAN 100 P.S.F.	MAXIMUM SPAN 250 P.S.F. LOAD	MAT.	INST.	TOTAL
1000	2	cedar	6.5	5	3.5	7.10	3.40	10.50
1050		douglas fir	8	6	4	3	3.40	6.40
1100		hemlock	7.5	5.5	4	3.06	3.40	6.46
1150		white spruce	7	5	3.5	1.95	3.40	5.35
1200	3	cedar	11	8	6	10.65	3.72	14.37
1250		douglas fir	13	10	7	4.50	3.72	8.22
1300		hemlock	12	9	7	4.59	3.72	8.31
1350		white spruce	11	8	6	2.93	3.72	6.65
1400	4	cedar	15	11	8	14.20	4.76	18.96
1450		douglas fir	18	13	10	6	4.76	10.76
1500		hemlock	17	13	9	6.10	4.76	10.86
1550		white spruce	16	12	8	3.90	4.76	8.66
1600	6	cedar	23	17	13	21.50	5.95	27.45
1650		douglas fir	24+	21	15	9	5.95	14.95
1700		hemlock	24+	20	15	9.15	5.95	15.10
1750		white spruce	24	18	13	5.85	5.95	11.80

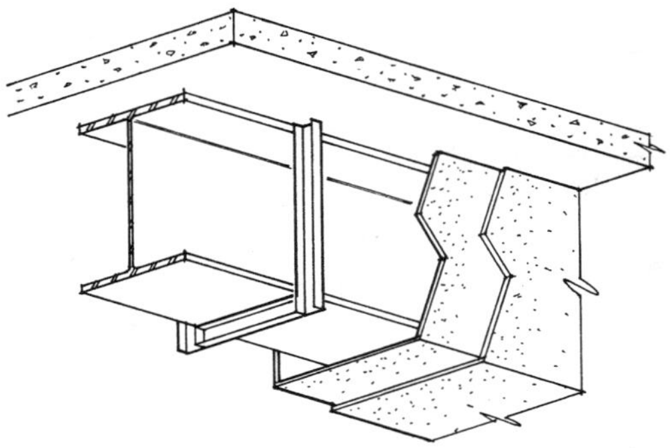

The table below lists fireproofing costs for steel beams by type, beam size, thickness and fire rating. Weights listed are for the fireproofing material only.

System Components	QUANTITY	UNIT	COST PER L.F. MAT.	INST.	TOTAL
SYSTEM B1010 710 1300					
FIREPROOFING, 5/8" F.R. GYP. BOARD, 12"X 4" BEAM, 2" THICK, 2 HR. F.R.					
1-1/4" x 1-1/4", galvanized	.020	C.L.F.	.37	3.40	3.77
L bead for drywall, galvanized	.020	C.L.F.	.42	3.96	4.38
Furring, beams & columns, 3/4" galv. channels, 24" O.C.	2.330	S.F.	.58	7.04	7.62
Drywall on beam, no finish, 2 layers at 5/8" thick	3.000	S.F.	2.61	11.88	14.49
Drywall, taping and finishing joints, add	3.000	S.F.	.15	1.77	1.92
TOTAL			4.13	28.05	32.18

B1010 710 Steel Beam Fireproofing

	ENCASEMENT SYSTEM	BEAM SIZE (IN.)	THICKNESS (IN.)	FIRE RATING (HRS.)	WEIGHT (P.L.F.)	COST PER L.F. MAT.	INST.	TOTAL
0400	Concrete	12x4	1	1	77	6.65	28.50	35.15
0450	3000 PSI		1-1/2	2	93	7.75	31	38.75
0500			2	3	121	8.65	34	42.65
0550		14x5	1	1	100	8.70	35	43.70
0600			1-1/2	2	122	10	39	49
0650			2	3	142	11.30	43	54.30
0700		16x7	1	1	147	10.15	37.50	47.65
0750			1-1/2	2	169	10.90	38.50	49.40
0800			2	3	195	12.35	42.50	54.85
0850		18x7-1/2	1	1	172	11.75	43.50	55.25
0900			1-1/2	2	196	13.20	47.50	60.70
0950			2	3	225	14.75	52	66.75
1000		24x9	1	1	264	15.80	52.50	68.30
1050			1-1/2	2	295	17.15	55.50	72.65
1100			2	3	328	18.50	59	77.50
1150		30x10-1/2	1	1	366	20.50	66	86.50
1200			1-1/2	2	404	22.50	70.50	93
1250			2	3	449	23.50	74.50	98
1300	5/8" fire rated	12x4	2	2	15	4.13	28	32.13
1350	Gypsum board		2-5/8	3	24	5.80	34.50	40.30
1400		14x5	2	2	17	4.46	29	33.46
1450			2-5/8	3	27	6.40	37	43.40
1500		16x7	2	2	20	5	32.50	37.50
1550			2-5/8	3	31	6.35	31.50	37.85
1600		18x7-1/2	2	2	22	5.40	35	40.40
1650			2-5/8	3	34	7.70	44	51.70

B1010 Floor Construction

B1010 710 Steel Beam Fireproofing

	ENCASEMENT SYSTEM	BEAM SIZE (IN.)	THICKNESS (IN.)	FIRE RATING (HRS.)	WEIGHT (P.L.F.)	COST PER L.F.		
						MAT.	INST.	TOTAL
1700	5/8" fire rated	24x9	2	2	27	6.70	43	49.70
1750	Gypsum board		2-5/8	3	42	9.50	54	63.50
1800		30x10-1/2	2	2	33	7.90	50.50	58.40
1850			2-5/8	3	51	11.20	63.50	74.70
1900	Gypsum	12x4	1-1/8	3	18	5.40	22	27.40
1950	Plaster on	14x5	1-1/8	3	21	6.10	24.50	30.60
2000	Metal lath	16x7	1-1/8	3	25	7.10	27.50	34.60
2050		18x7-1/2	1-1/8	3	32	8.10	31	39.10
2100		24x9	1-1/8	3	35	10	37.50	47.50
2150		30x10-1/2	1-1/8	3	44	12.05	45.50	57.55
2200	Perlite plaster	12x4	1-1/8	2	16	4.96	24	28.96
2250	On metal lath		1-1/4	3	20	5.30	25	30.30
2300			1-1/2	4	22	5.65	26.50	32.15
2350		14x5	1-1/8	2	18	6.50	29.50	36
2400			1-1/4	3	23	6.85	30.50	37.35
2450			1-1/2	4	25	7.60	33	40.60
2500		16x7	1-1/8	2	21	6.05	29.50	35.55
2550			1-1/4	3	26	6.35	31	37.35
2600			1-1/2	4	29	6.75	32	38.75
2650		18x7-1/2	1-1/8	2	26	7.50	34.50	42
2700			1-1/4	3	33	7.85	36	43.85
2750			1-1/2	4	36	8.60	38	46.60
2800		24x9	1-1/8	2	30	7.80	38.50	46.30
2850			1-1/4	3	38	8.10	40	48.10
2900			1-1/2	4	41	8.50	41	49.50
2950		30x10-1/2	1-1/8	2	37	10.10	48	58.10
3000			1-1/4	3	46	10.50	49	59.50
3050			1-1/2	4	51	11.20	51.50	62.70
3100	Sprayed Fiber	12x4	5/8	1	12	1.28	2.32	3.60
3150	(non asbestos)		1-1/8	2	21	2.30	4.18	6.48
3200			1-1/4	3	22	2.56	4.64	7.20
3250		14x5	5/8	1	14	1.28	2.32	3.60
3300			1-1/8	2	26	2.30	4.18	6.48
3350			1-1/4	3	29	2.56	4.64	7.20
3400		16x7	5/8	1	18	1.62	2.93	4.55
3450			1-1/8	2	32	2.91	5.30	8.21
3500			1-1/4	3	36	3.24	5.85	9.09
3550		18x7-1/2	5/8	1	20	1.79	3.24	5.03
3600			1-1/8	2	35	3.22	5.85	9.07
3650			1-1/4	3	39	3.55	6.45	10
3700		24x9	5/8	1	25	2.26	4.09	6.35
3750			1-1/8	2	45	4.09	7.40	11.49
3800			1-1/4	3	50	4.54	8.20	12.74
3850		30x10-1/2	5/8	1	31	2.74	4.96	7.70
3900			1-1/8	2	55	4.93	8.95	13.88
3950			1-1/4	3	61	5.45	9.95	15.40
4000	On Decking	Flat	1		6	.59	.67	1.26
4050	Per S.F.	Corrugated	1		7	.88	1.29	2.17
4100		Fluted	1		7	.88	1.29	2.17

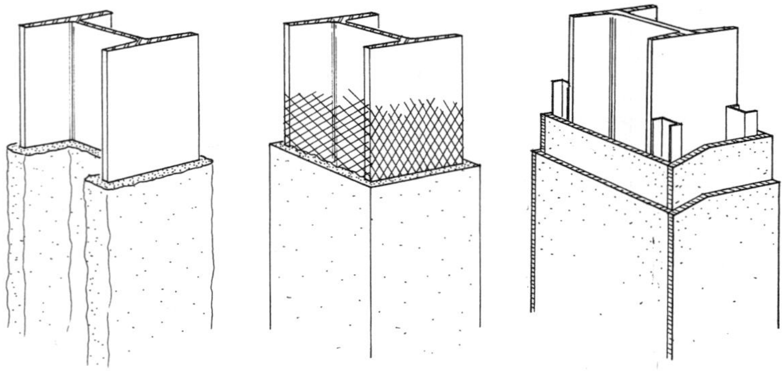

Listed below are costs per V.L.F. for fireproofing by material, column size, thickness and fire rating. Weights listed are for the fireproofing material only.

System Components	QUANTITY	UNIT	COST PER V.L.F. MAT.	INST.	TOTAL
SYSTEM B1010 720 3000					
CONCRETE FIREPROOFING, 8″ STEEL COLUMN, 1″ THICK, 1 HR. FIRE RATING					
Forms in place, columns, plywood, 4 uses	3.330	SFCA	2.93	29.97	32.90
Welded wire fabric, 2 x 2 #14 galv. 21 lb./C.S.F., column wrap	2.700	S.F.	1.24	5.45	6.69
Concrete ready mix, regular weight, 3000 psi	.621	C.F.	2.71		2.71
Place and vibrate concrete, 12″ sq./round columns, pumped	.621	C.F.		1.86	1.86
TOTAL			6.88	37.28	44.16

B1010 720	Steel Column Fireproofing							
	ENCASEMENT SYSTEM	COLUMN SIZE (IN.)	THICKNESS (IN.)	FIRE RATING (HRS.)	WEIGHT (P.L.F.)	COST PER V.L.F. MAT.	INST.	TOTAL
3000	Concrete	8	1	1	110	6.90	37.50	44.40
3050			1-1/2	2	133	7.90	41.50	49.40
3100			2	3	145	8.90	45	53.90
3150		10	1	1	145	9	46	55
3200			1-1/2	2	168	10	49.50	59.50
3250			2	3	196	11.10	53	64.10
3300		14	1	1	258	11.40	54	65.40
3350			1-1/2	2	294	12.40	57.50	69.90
3400			2	3	325	13.65	61.50	75.15
3450	Gypsum board	8	1/2	2	8	3.66	23	26.66
3500	1/2″ fire rated	10	1/2	2	11	3.85	24	27.85
3550	1 layer	14	1/2	2	18	3.96	24.50	28.46
3600	Gypsum board	8	1	3	14	5.10	29.50	34.60
3650	1/2″ fire rated	10	1	3	17	5.45	31	36.45
3700	2 layers	14	1	3	22	5.65	32	37.65
3750	Gypsum board	8	1-1/2	3	23	6.80	37	43.80
3800	1/2″ fire rated	10	1-1/2	3	27	7.65	41	48.65
3850	3 layers	14	1-1/2	3	35	8.45	45	53.45
3900	Sprayed fiber	8	1-1/2	2	6.3	4	7.30	11.30
3950	Direct application		2	3	8.3	5.50	10	15.50
4000			2-1/2	4	10.4	7.15	12.95	20.10
4050		10	1-1/2	2	7.9	4.83	8.75	13.58
4100			2	3	10.5	6.60	12	18.60
4150			2-1/2	4	13.1	8.50	15.45	23.95

B1010 Floor Construction

B1010 720 Steel Column Fireproofing

	ENCASEMENT SYSTEM	COLUMN SIZE (IN.)	THICKNESS (IN.)	FIRE RATING (HRS.)	WEIGHT (P.L.F.)	COST PER V.L.F.		
						MAT.	INST.	TOTAL
4200	Sprayed fiber	14	1-1/2	2	10.8	6	10.85	16.85
4250	Direct application		2	3	14.5	8.20	14.85	23.05
4300			2-1/2	4	18	10.50	19.05	29.55
4350	3/4" gypsum plaster	8	3/4	1	23	7.20	31	38.20
4400	On metal lath	10	3/4	1	28	8.20	35	43.20
4450		14	3/4	1	38	9.95	43	52.95
4500	Perlite plaster	8	1	2	18	7.85	34	41.85
4550	On metal lath		1-3/8	3	23	8.55	37.50	46.05
4600			1-3/4	4	35	10.10	44.50	54.60
4650	Perlite plaster	10	1	2	21	9	39	48
4700			1-3/8	3	27	10.15	44.50	54.65
4750			1-3/4	4	41	11.40	51	62.40
4800		14	1	2	29	11.25	49	60.25
4850			1-3/8	3	35	12.65	55.50	68.15
4900			1-3/4	4	53	13.15	58	71.15
4950	1/2 gypsum plaster	8	7/8	1	13	5.90	26	31.90
5000	On 3/8" gypsum lath	10	7/8	1	16	6.95	30	36.95
5050		14	7/8	1	21	8.40	37	45.40
5100	5/8" gypsum plaster	8	1	1-1/2	20	6.25	28	34.25
5150	On 3/8" gypsum lath	10	1	1-1/2	24	7.35	33	40.35
5200		14	1	1-1/2	33	8.90	40.50	49.40
5250	1"perlite plaster	8	1-3/8	2	23	7.90	29.50	37.40
5300	On 3/8" gypsum lath	10	1-3/8	2	28	9.05	33.50	42.55
5350		14	1-3/8	2	37	11.40	42	53.40
5400	1-3/8" perlite plaster	8	1-3/4	3	27	7.35	32.50	39.85
5450	On 3/8" gypsum lath	10	1-3/4	3	33	8.40	37.50	45.90
5500		14	1-3/4	3	43	10.45	47.50	57.95
5550	Concrete masonry	8	4-3/4	4	126	11.80	33.50	45.30
5600	Units 4" thick	10	4-3/4	4	166	14.80	42	56.80
5650	75% solid	14	4-3/4	4	262	17.70	50.50	68.20

B1020 Roof Construction

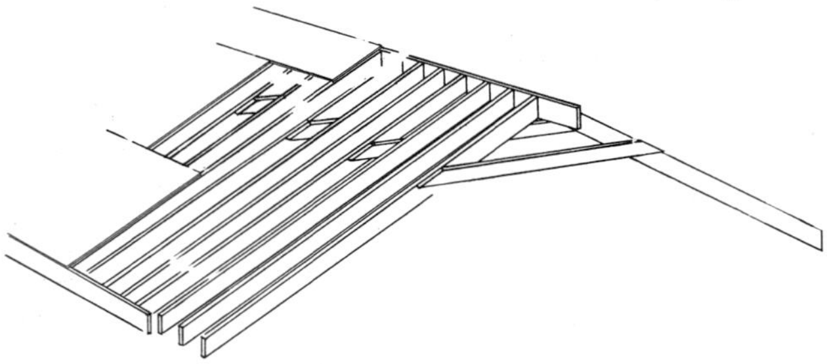

The table below lists prices per S.F. for roof rafters and sheathing by nominal size and spacing. Sheathing is 5/16″ CDX for 12″ and 16″ spacing and 3/8″ CDX for 24″ spacing.

Factors for Converting Inclined to Horizontal

Roof Slope	Approx. Angle	Factor	Roof Slope	Approx. Angle	Factor
Flat	0°	1.000	12 in 12	45.0°	1.414
1 in 12	4.8°	1.003	13 in 12	47.3°	1.474
2 in 12	9.5°	1.014	14 in 12	49.4°	1.537
3 in 12	14.0°	1.031	15 in 12	51.3°	1.601
4 in 12	18.4°	1.054	16 in 12	53.1°	1.667
5 in 12	22.6°	1.083	17 in 12	54.8°	1.734
6 in 12	26.6°	1.118	18 in 12	56.3°	1.803
7 in 12	30.3°	1.158	19 in 12	57.7°	1.873
8 in 12	33.7°	1.202	20 in 12	59.0°	1.943
9 in 12	36.9°	1.250	21 in 12	60.3°	2.015
10 in 12	39.8°	1.302	22 in 12	61.4°	2.088
11 in 12	42.5°	1.357	23 in 12	62.4°	2.162

System Components	QUANTITY	UNIT	COST PER S.F. MAT.	INST.	TOTAL
SYSTEM B1020 102 2500					
RAFTER 2″X4″, 12″ O.C.					
Framing joists, fir 2″x 4″	.730	B.F.	.48	1.04	1.52
Sheathing plywood on roof CDX 5/16″	1.050	S.F.	.72	.78	1.50
TOTAL			1.20	1.82	3.02

B1020 102	Wood/Flat or Pitched	COST PER S.F. MAT.	INST.	TOTAL
2500	Flat rafter, 2″x4″, 12″ O.C.	1.20	1.82	3.02
2550	16″ O.C.	1.08	1.56	2.64
2600	24″ O.C.	.95	1.35	2.30
2900	2″x6″, 12″ O.C.	1.46	1.83	3.29
2950	16″ O.C.	1.28	1.57	2.85
3000	24″ O.C.	1.08	1.34	2.42
3300	2″x8″, 12″ O.C.	1.74	1.97	3.71
3350	16″ O.C.	1.49	1.68	3.17
3400	24″ O.C.	1.23	1.42	2.65
3700	2″x10″, 12″ O.C.	2.34	2.25	4.59
3750	16″ O.C.	1.93	1.88	3.81
3800	24″ O.C.	1.51	1.55	3.06
4100	2″x12″, 12″ O.C.	2.78	2.28	5.06
4150	16″ O.C.	2.42	2.08	4.50
4200	24″ O.C.	1.74	1.57	3.31
4500	2″x14″, 12″ O.C.	3.34	3.31	6.65
4550	16″ O.C.	2.70	2.69	5.39
4600	24″ O.C.	2.03	2.10	4.13
4900	3″x6″, 12″ O.C.	2.99	1.94	4.93
4950	16″ O.C.	2.43	1.65	4.08
5000	24″ O.C.	1.85	1.40	3.25
5300	3″x8″, 12″ O.C.	3.97	2.16	6.13
5350	16″ O.C.	3.15	1.81	4.96
5400	24″ O.C.	2.33	1.51	3.84
5700	3″x10″, 12″ O.C.	4.85	2.46	7.31
5750	16″ O.C.	3.83	2.04	5.87
5800	24″ O.C.	2.78	1.66	4.44
6100	3″x12″, 12″ O.C.	5.65	2.96	8.61
6150	16″ O.C.	4.44	2.42	6.86
6200	24″ O.C.	3.19	1.91	5.10

B1020 Roof Construction

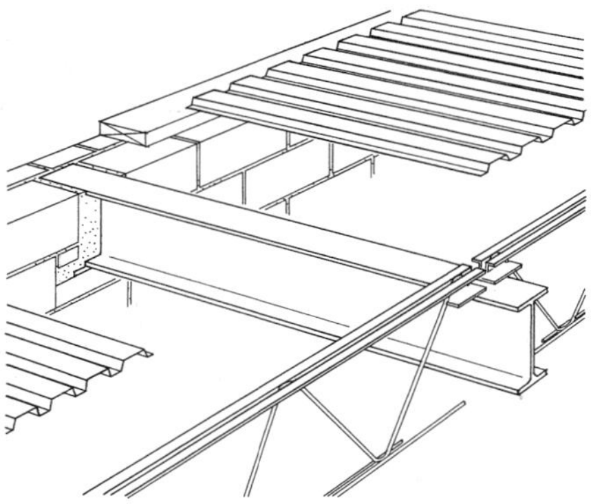

Table below lists the cost per S.F. for a roof system with steel columns, beams, and deck using open web steel joists and 1-1/2" galvanized metal deck. Perimeter of system is supported on bearing walls.

Design and Pricing Assumptions:
Columns are 18′ high.
Joists are 5′-0″ O.C. and span the long direction of the bay.

Joists at columns have bottom chords extended and are connected to columns. Column costs are not included but are listed separately per S.F. of floor.

Roof deck is 1-1/2″, 22 gauge galvanized steel. Joist cost includes appropriate bridging. Deflection is limited to 1/240 of the span. Fireproofing is not included.

Costs/S.F. are based on a building 4 bays long and 4 bays wide.

Design Loads	Min.	Max.
Joists & Beams	3 PSF	5 PSF
Deck	2	2
Insulation	3	3
Roofing	6	6
Misc.	6	6
Total Dead Load	20 PSF	22 PSF

System Components	QUANTITY	UNIT	COST PER S.F. MAT.	INST.	TOTAL
SYSTEM B1020 108 1200					
METAL DECK & STEEL JOISTS,15′X20′ BAY,20 PSF S. LOAD,16″ DP,40 PSF T.LOAD					
Structural steel	.488	Lb.	.71	.16	.87
Open web joists	1.022	Lb.	.93	.38	1.31
Metal decking, open, galvanized, 1-1/2″ deep, 22 gauge	1.050	S.F.	1.72	.63	2.35
TOTAL			3.36	1.17	4.53

B1020 108 Steel Joists, Beams & Deck on Columns & Walls

	BAY SIZE (FT.)	SUPERIMPOSED LOAD (P.S.F.)	DEPTH (IN.)	TOTAL LOAD (P.S.F.)	COLUMN ADD	COST PER S.F. MAT.	INST.	TOTAL
1200	15x20	20	16	40		3.36	1.17	4.53
1300					columns	.81	.19	1
1400		30	16	50		3.66	1.26	4.92
1500					columns	.81	.19	1
1600		40	18	60		3.97	1.35	5.32
1700					columns	.81	.19	1
1800	20x20	20	16	40		4	1.32	5.32
1900					columns	.61	.14	.75
2000		30	18	50		4	1.32	5.32
2100					columns	.61	.14	.75
2200		40	18	60		4.48	1.48	5.96
2300					columns	.61	.14	.75
2400	20x25	20	18	40		3.97	1.36	5.33
2500					columns	.49	.11	.60
2600		30	18	50		4.16	1.40	5.56
2700					columns	.49	.11	.60
2800		40	20	60		4.42	1.51	5.93
2900					columns	.65	.15	.80
3000	25x25	20	18	40		4.31	1.43	5.74
3100					columns	.39	.09	.48
3200		30	22	50		4.62	1.52	6.14
3300					columns	.52	.12	.64
3400		40	20	60		5.10	1.68	6.78
3500					columns	.52	.12	.64

B1020 Roof Construction

B1020 108 Steel Joists, Beams & Deck on Columns & Walls

	BAY SIZE (FT.)	SUPERIMPOSED LOAD (P.S.F.)	DEPTH (IN.)	TOTAL LOAD (P.S.F.)	COLUMN ADD	COST PER S.F.		
						MAT.	INST.	TOTAL
3600	25x30	20	22	40		4.54	1.40	5.94
3700					columns	.44	.10	.54
3800		30	20	50		5.15	1.56	6.71
3900					columns	.44	.10	.54
4000		40	25	60		5.35	1.60	6.95
4100					columns	.52	.12	.64
4200	30x30	20	25	42		5	1.50	6.50
4300					columns	.36	.09	.45
4400		30	22	52		5.45	1.63	7.08
4500					columns	.44	.10	.54
4600		40	28	62		5.65	1.68	7.33
4700					columns	.44	.10	.54
4800	30x35	20	22	42		5.20	1.56	6.76
4900					columns	.37	.09	.46
5000		30	28	52		5.45	1.63	7.08
5100					columns	.37	.09	.46
5200		40	25	62		5.90	1.75	7.65
5300					columns	.44	.10	.54
5400	35x35	20	28	42		5.20	1.57	6.77
5500					columns	.32	.08	.40
5600		30	25	52		6.30	1.85	8.15
5700					columns	.37	.09	.46
5800		40	28	62		6.40	1.88	8.28
5900					columns	.41	.10	.51

B1020 Roof Construction

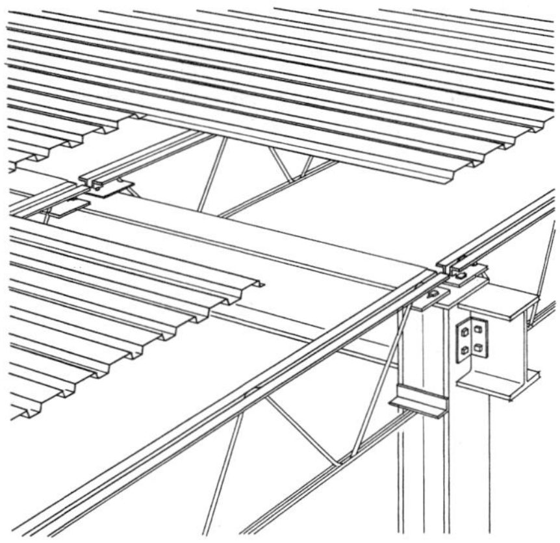

Description: Table below lists the cost per S.F. for a roof system with steel columns, beams, and deck, using open web steel joists and 1-1/2″ galvanized metal deck.

Design and Pricing Assumptions:
Columns are 18′ high.
Building is 4 bays long by 4 bays wide.
Joists are 5′-0″ O.C. and span the long direction of the bay.
Joists at columns have bottom chords extended and are connected to columns.
Column costs are not included but are listed separately per S.F. of floor.

Roof deck is 1-1/2″, 22 gauge galvanized steel. Joist cost includes appropriate bridging. Deflection is limited to 1/240 of the span. Fireproofing is not included.

Design Loads	Min.	Max.
Joists & Beams	3 PSF	5 PSF
Deck	2	2
Insulation	3	3
Roofing	6	6
Misc.	6	6
Total Dead Load	20 PSF	22 PSF

System Components	QUANTITY	UNIT	COST PER S.F. MAT.	INST.	TOTAL
SYSTEM B1020 112 1100					
METAL DECK AND JOISTS, 15′ X 20′ BAY, 20 PSF S. LOAD					
Structural steel	.954	Lb.	1.39	.33	1.72
Open web joists	1.260	Lb.	1.15	.47	1.62
Metal decking, open, galvanized, 1-1/2″ deep, 22 gauge	1.050	S.F.	1.72	.63	2.35
TOTAL			4.26	1.43	5.69

B1020 112 Steel Joists, Beams, & Deck on Columns

	BAY SIZE (FT.)	SUPERIMPOSED LOAD (P.S.F.)	DEPTH (IN.)	TOTAL LOAD (P.S.F.)	COLUMN ADD	COST PER S.F. MAT.	INST.	TOTAL
1100	15x20	20	16	40		4.26	1.43	5.69
1200					columns	2.26	.53	2.79
1300		30	16	50		4.68	1.52	6.20
1400					columns	2.26	.53	2.79
1500		40	18	60		4.75	1.58	6.33
1600					columns	2.26	.53	2.79
1700	20x20	20	16	40		4.61	1.48	6.09
1800					columns	1.69	.39	2.08
1900		30	18	50		5.10	1.61	6.71
2000					columns	1.69	.39	2.08
2100		40	18	60		5.65	1.78	7.43
2200					columns	1.69	.39	2.08
2300	20x25	20	18	40		4.84	1.59	6.43
2400					columns	1.35	.31	1.66
2500		30	18	50		5.15	1.67	6.82
2600					columns	1.35	.31	1.66
2700		40	20	60		5.45	1.77	7.22
2800					columns	1.81	.42	2.23
2900	25x25	20	18	40		5.70	1.80	7.50
3000					columns	1.08	.25	1.33
3100		30	22	50		5.90	1.84	7.74
3200					columns	1.45	.34	1.79
3300		40	20	60		6.55	2.04	8.59
3400					columns	1.45	.34	1.79

B1020 Roof Construction

B1020 112 Steel Joists, Beams, & Deck on Columns

	BAY SIZE (FT.)	SUPERIMPOSED LOAD (P.S.F.)	DEPTH (IN.)	TOTAL LOAD (P.S.F.)	COLUMN ADD	COST PER S.F. MAT.	COST PER S.F. INST.	COST PER S.F. TOTAL
3500	25x30	20	22	40		5.55	1.63	7.18
3600					columns	1.21	.28	1.49
3700		30	20	50		6.40	1.87	8.27
3800					columns	1.21	.28	1.49
3900		40	25	60		6.70	1.93	8.63
4000					columns	1.45	.34	1.79
4100	30x30	20	25	42		6.30	1.81	8.11
4200					columns	1	.24	1.24
4300		30	22	52		6.95	1.98	8.93
4400					columns	1.21	.28	1.49
4500		40	28	62		7.30	2.07	9.37
4600					columns	1.21	.28	1.49
4700	30x35	20	22	42		6.50	1.88	8.38
4800					columns	1.03	.24	1.27
4900		30	28	52		6.85	1.97	8.82
5000					columns	1.03	.24	1.27
5100		40	25	62		7.50	2.14	9.64
5200					columns	1.21	.28	1.49
5300	35x35	20	28	42		6.85	1.96	8.81
5400					columns	.88	.20	1.08
5500		30	25	52		7.70	2.17	9.87
5600					columns	1.03	.24	1.27
5700		40	28	62		8.30	2.32	10.62
5800					columns	1.14	.26	1.40

B1020 Roof Construction

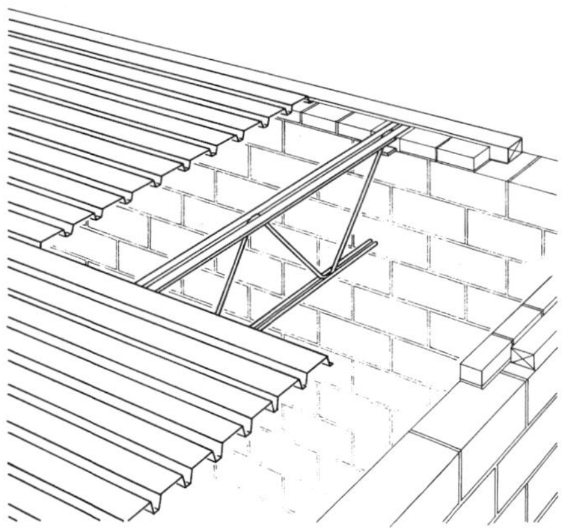

Description: Table below lists cost per S.F. for a roof system using open web steel joists and 1-1/2″ galvanized metal deck. The system is assumed supported on bearing walls or other suitable support. Costs for the supports are not included.

Design and Pricing Assumptions:
Joists are 5′-0″ O.C.
Roof deck is 1-1/2″, 22 gauge galvanized.

System Components	QUANTITY	UNIT	COST PER S.F.		
			MAT.	INST.	TOTAL
SYSTEM B1020 116 1100 METAL DECK AND JOISTS, 20′ SPAN, 20 PSF S. LOAD					
Open web joists, horiz. bridging T.L. lot, to 30′ span	1.114	Lb.	1.01	.41	1.42
Metal decking, open type, galv 1-1/2″ deep	1.050	S.F.	1.72	.63	2.35
TOTAL			2.73	1.04	3.77

B1020 116 Steel Joists & Deck on Bearing Walls

	BAY SIZE (FT.)	SUPERIMPOSED LOAD (P.S.F.)	DEPTH (IN.)	TOTAL LOAD (P.S.F.)		COST PER S.F.		
						MAT.	INST.	TOTAL
1100	20	20	13-1/2	40		2.73	1.04	3.77
1200		30	15-1/2	50		2.79	1.06	3.85
1300		40	15-1/2	60		2.99	1.14	4.13
1400	25	20	17-1/2	40		2.99	1.15	4.14
1500		30	17-1/2	50		3.22	1.24	4.46
1600		40	19-1/2	60		3.27	1.26	4.53
1700	30	20	19-1/2	40		3.26	1.26	4.52
1800		30	21-1/2	50		3.33	1.28	4.61
1900		40	23-1/2	60		3.58	1.38	4.96
2000	35	20	23-1/2	40		3.54	1.18	4.72
2100		30	25-1/2	50		3.64	1.21	4.85
2200		40	25-1/2	60		3.86	1.28	5.14
2300	40	20	25-1/2	41		3.92	1.30	5.22
2400		30	25-1/2	51		4.16	1.37	5.53
2500		40	25-1/2	61		4.30	1.41	5.71
2600	45	20	27-1/2	41		4.71	1.54	6.25
2700		30	31-1/2	51		4.98	1.62	6.60
2800		40	31-1/2	61		5.25	1.70	6.95
2900	50	20	29-1/2	42		5.25	1.71	6.96
3000		30	31-1/2	52		5.75	1.85	7.60
3100		40	31-1/2	62		6.10	1.97	8.07
3200	60	20	37-1/2	42		6.25	2.20	8.45
3300		30	37-1/2	52		6.90	2.43	9.33
3400		40	37-1/2	62		6.90	2.43	9.33
3500	70	20	41-1/2	42		6.90	2.43	9.33
3600		30	41-1/2	52		7.35	2.59	9.94
3700		40	41-1/2	64		8.95	3.13	12.08

B1020 Roof Construction

B1020 116 Steel Joists & Deck on Bearing Walls

	BAY SIZE (FT.)	SUPERIMPOSED LOAD (P.S.F.)	DEPTH (IN.)	TOTAL LOAD (P.S.F.)		COST PER S.F.		
						MAT.	INST.	TOTAL
3800	80	20	45-1/2	44		8.70	3.04	11.74
3900		30	45-1/2	54		8.70	3.04	11.74
4000		40	45-1/2	64		9.60	3.36	12.96
4100	90	20	53-1/2	44		8.10	2.73	10.83
4200		30	53-1/2	54		8.55	2.89	11.44
4300		40	53-1/2	65		10.20	3.43	13.63
4400	100	20	57-1/2	44		8.55	2.89	11.44
4500		30	57-1/2	54		10.20	3.43	13.63
4600		40	57-1/2	65		11.25	3.78	15.03
4700	125	20	69-1/2	44		10.20	3.42	13.62
4800		30	69-1/2	56		11.85	3.97	15.82
4900		40	69-1/2	67		13.45	4.51	17.96

B1020 Roof Construction

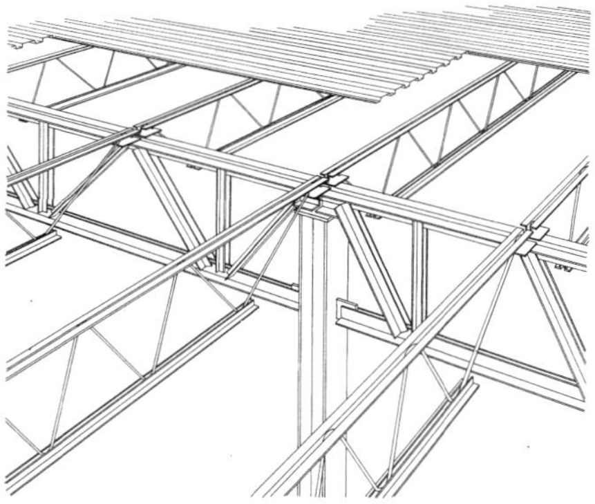

Description: Table below lists costs for a roof system supported on exterior bearing walls and interior columns. Costs include bracing, joist girders, open web steel joists and 1-1/2″ galvanized metal deck.

Design and Pricing Assumptions:
Columns are 18′ high.
Joists are 5′-0″ O.C.
Joist girders and joists have bottom chords connected to columns. Roof deck is 1-1/2″, 22 gauge galvanized steel. Costs include bridging and bracing. Deflection is limited to 1/240 of the span.

Fireproofing is not included.
Costs/S.F. are based on a building 4 bays long and 4 bays wide.
Costs for bearing walls are not included.

Column costs are not included but are listed separately per S.F. of floor.

System Components	QUANTITY	UNIT	COST PER S.F. MAT.	INST.	TOTAL
SYSTEM B1020 120 2050					
30′ X 30′ BAY SIZE, 20 PSF SUPERIMPOSED LOAD					
Joist girders, 40-ton job lots, shop primer	.475	Lb.	.39	.15	.54
Open web joists, K series, horiz bridging, 30′ to 50′ spans	.067	Lb.	1.28	.37	1.65
Cross bracing, rods, shop fabricated, 1″ diameter	.062	Lb.	.11	.15	.26
Metal decking, open, galv., 1-1/2″ deep, 22 ga., over 50 sq.	1.050	S.F.	1.72	.63	2.35
TOTAL			3.50	1.30	4.80

B1020 120 Steel Joists, Joist Girders & Deck on Columns & Walls

	BAY SIZE (FT.) GIRD X JOISTS	SUPERIMPOSED LOAD (P.S.F.)	DEPTH (IN.)	TOTAL LOAD (P.S.F.)	COLUMN ADD	COST PER S.F. MAT.	INST.	TOTAL
2050	30x30	20	17-1/2	40		3.50	1.30	4.80
2100					columns	.43	.10	.53
2150		30	17-1/2	50		3.78	1.38	5.16
2200					columns	.43	.10	.53
2250		40	21-1/2	60		3.98	1.47	5.45
2300					columns	.43	.10	.53
2350	30x35	20	32-1/2	40		3.98	1.45	5.43
2400					columns	.37	.09	.46
2450		30	36-1/2	50		4.15	1.51	5.66
2500					columns	.37	.09	.46
2550		40	36-1/2	60		4.46	1.60	6.06
2600					columns	.43	.10	.53
2650	35x30	20	36-1/2	40		3.68	1.37	5.05
2700					columns	.37	.09	.46
2750		30	36-1/2	50		4.06	1.50	5.56
2800					columns	.37	.09	.46
2850		40	36-1/2	60		5.40	1.89	7.29
2900					columns	.43	.10	.53
3000	35x35	20	36-1/2	40		4.34	1.79	6.13
3050					columns	.32	.08	.40
3100		30	36-1/2	50		4.48	1.84	6.32
3150					columns	.32	.08	.40
3200		40	36-1/2	60		4.86	1.97	6.83
3250					columns	.41	.10	.51

B1020 Roof Construction

B1020 120 Steel Joists, Joist Girders & Deck on Columns & Walls

	BAY SIZE (FT.) GIRD X JOISTS	SUPERIMPOSED LOAD (P.S.F.)	DEPTH (IN.)	TOTAL LOAD (P.S.F.)	COLUMN ADD	COST PER S.F. MAT.	INST.	TOTAL
3300	35x40	20	36-1/2	40		4.40	1.63	6.03
3350					columns	.33	.08	.41
3400		30	36-1/2	50		4.55	1.67	6.22
3450					columns	.36	.09	.45
3500		40	36-1/2	60		4.95	1.81	6.76
3550					columns	.36	.09	.45
3600	40x35	20	40-1/2	40		4.24	1.58	5.82
3650					columns	.33	.08	.41
3700		30	40-1/2	50		4.42	1.64	6.06
3750					columns	.36	.09	.45
3800		40	40-1/2	60		5.40	1.94	7.34
3850					columns	.36	.09	.45
3900	40x40	20	40-1/2	41		5.10	1.82	6.92
3950					columns	.32	.08	.40
4000		30	40-1/2	51		5.30	1.88	7.18
4050					columns	.32	.08	.40
4100		40	40-1/2	61		5.70	1.99	7.69
4150					columns	.32	.08	.40
4200	40x45	20	40-1/2	41		5.20	1.92	7.12
4250					columns	.28	.06	.34
4300		30	40-1/2	51		5.85	2.13	7.98
4350					columns	.28	.06	.34
4400		40	40-1/2	61		6.50	2.33	8.83
4450					columns	.32	.08	.40
4500	45x40	20	52-1/2	41		5.15	1.91	7.06
4550					columns	.28	.06	.34
4600		30	52-1/2	51		4.77	1.99	6.76
4650					columns	.32	.08	.40
4700		40	52-1/2	61		5.90	2.15	8.05
4750					columns	.32	.08	.40
4800	45x45	20	52-1/2	41		4.89	1.83	6.72
4850					columns	.25	.06	.31
4900		30	52-1/2	51		6.10	2.20	8.30
4950					columns	.28	.06	.34
5000		40	52-1/2	61		6.25	2.25	8.50
5050					columns	.32	.08	.40
5100	45x50	20	52-1/2	41		5.75	2.07	7.82
5150					columns	.23	.05	.28
5200		30	52-1/2	51		6.55	2.33	8.88
5250					columns	.25	.06	.31
5300		40	52-1/2	61		7	2.45	9.45
5350					columns	.33	.08	.41
5400	50x45	20	56-1/2	41		5.55	2.02	7.57
5450					columns	.23	.05	.28
5500		30	56-1/2	51		6.25	2.24	8.49
5550					columns	.25	.06	.31
5600		40	56-1/2	61		7.15	2.51	9.66
5650					columns	.31	.07	.38

B1020 Roof Construction

B1020 120 Steel Joists, Joist Girders & Deck on Columns & Walls

	BAY SIZE (FT.) GIRD X JOISTS	SUPERIMPOSED LOAD (P.S.F.)	DEPTH (IN.)	TOTAL LOAD (P.S.F.)	COLUMN ADD	COST PER S.F.		
						MAT.	INST.	TOTAL
5700	50x50	20	56-1/2	42		6.25	2.25	8.50
5750					columns	.23	.05	.28
5800		30	56-1/2	53		6.75	2.43	9.18
5850					columns	.29	.06	.35
5900		40	59	64		7.20	2.75	9.95
5950					columns	.32	.08	.40
6000	55x50	20	60-1/2	43		6.35	2.31	8.66
6050					columns	.39	.09	.48
6100		30	64-1/2	54		7.30	2.59	9.89
6150					columns	.48	.11	.59
6200		40	67	65		7.40	2.86	10.26
6250					columns	.48	.11	.59
6300	60x50	20	62-1/2	43		6.30	2.29	8.59
6350					columns	.36	.09	.45
6400		30	68-1/2	54		7.25	2.60	9.85
6450					columns	.44	.10	.54
6500		40	71	65		7.35	2.85	10.20
6550					columns	.48	.11	.59

B1020 Roof Construction

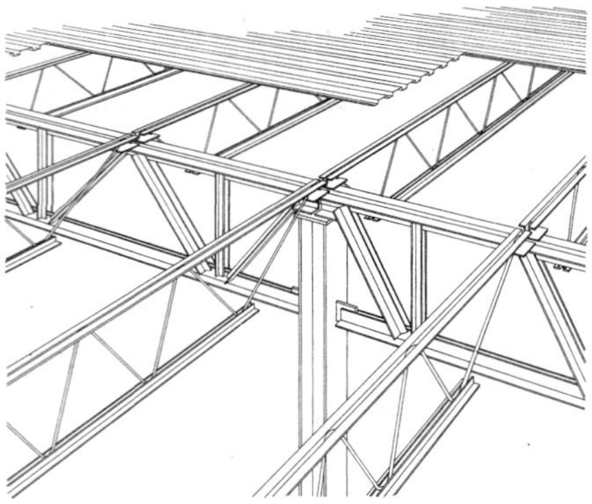

Description: Table below lists the cost per S.F. for a roof system supported on columns. Costs include joist girders, open web steel joists and 1-1/2" galvanized metal deck.

Design and Pricing Assumptions:
Columns are 18′ high.
Joists are 5′-0″ O.C.
Joist girders and joists have bottom chords connected to columns. Roof deck is 1-1/2", 22 gauge galvanized steel. Costs include bridging and bracing. Deflection is limited to 1/240 of the span. Fireproofing is not included.

Costs/S.F. are based on a building 4 bays long and 4 bays wide.

Costs for columns are not included, but are listed separately per S.F. of area.

System Components	QUANTITY	UNIT	COST PER S.F. MAT.	INST.	TOTAL
SYSTEM B1020 124 2000					
30′ X 30′ BAY SIZE, 20 PSF S. LOAD, 17-1/2″ DEPTH, 40 PSF T. LOAD					
Joist girders, 40-ton job lots, shop primer	.758	Lb.	.62	.23	.85
Open web joists, spans to 30′ average	1.580	Lb.	1.44	.59	2.03
Cross bracing, rods, shop fabricated, 1″ diameter	.120	Lb.	.21	.28	.49
Metal decking, open, galv., 1-1/2″ deep, 22 ga., over 50 sq.	1.050	S.F.	1.72	.63	2.35
TOTAL			3.99	1.73	5.72

B1020 124 Steel Joists & Joist Girders on Columns

	BAY SIZE (FT.) GIRD X JOISTS	SUPERIMPOSED LOAD (P.S.F.)	DEPTH (IN.)	TOTAL LOAD (P.S.F.)	COLUMN ADD	COST PER S.F. MAT.	INST.	TOTAL
2000	30x30	20	17-1/2	40		3.99	1.73	5.72
2050					columns	1.21	.28	1.49
2100		30	17-1/2	50		4.32	1.88	6.20
2150					columns	1.21	.28	1.49
2200		40	21-1/2	60		4.44	1.91	6.35
2250					columns	1.21	.28	1.49
2300	30x35	20	32-1/2	40		4.27	1.62	5.89
2350					columns	1.03	.24	1.27
2400		30	36-1/2	50		4.50	1.70	6.20
2450					columns	1.03	.24	1.27
2500		40	36-1/2	60		4.80	1.80	6.60
2550					columns	1.21	.28	1.49
2600	35x30	20	36-1/2	40		4.11	1.76	5.87
2650					columns	1.03	.24	1.27
2700		30	36-1/2	50		4.53	1.94	6.47
2750					columns	1.03	.24	1.27
2800		40	36-1/2	60		4.61	1.97	6.58
2850					columns	1.21	.28	1.49
2900	35x35	20	36-1/2	40		4.64	1.97	6.61
2950					columns	.89	.21	1.10
3000		30	36-1/2	50		4.80	2.01	6.81
3050					columns	1.03	.24	1.27
3100		40	36-1/2	60		5.20	2.17	7.37
3150					columns	1.14	.27	1.41

B1020 Roof Construction

B1020 124	Steel Joists & Joist Girders on Columns							
	BAY SIZE (FT.) GIRD X JOISTS	SUPERIMPOSED LOAD (P.S.F.)	DEPTH (IN.)	TOTAL LOAD (P.S.F.)	COLUMN ADD	COST PER S.F.		
						MAT.	INST.	TOTAL
3200	35x40	20	36-1/2	40		5.55	2.19	7.74
3250					columns	.90	.21	1.11
3300		30	36-1/2	50		5.70	2.25	7.95
3350					columns	1	.24	1.24
3400		40	36-1/2	60		6.25	2.42	8.67
3450					columns	1	.24	1.24
3500	40x35	20	40-1/2	40		4.89	2.02	6.91
3550					columns	.90	.21	1.11
3600		30	40-1/2	50		5	2.06	7.06
3650					columns	1	.24	1.24
3700		40	40-1/2	60		5.35	2.17	7.52
3750					columns	1	.24	1.24
3800	40x40	20	40-1/2	41		5.60	2.19	7.79
3850					columns	.88	.20	1.08
3900		30	40-1/2	51		5.80	2.26	8.06
3950					columns	.88	.20	1.08
4000		40	40-1/2	61		6.30	2.41	8.71
4050					columns	.88	.20	1.08
4100	40x45	20	40-1/2	41		6.05	2.50	8.55
4150					columns	.78	.19	.97
4200		30	40-1/2	51		6.55	2.65	9.20
4250					columns	.78	.19	.97
4300		40	40-1/2	61		7.30	2.88	10.18
4350					columns	.88	.20	1.08
4400	45x40	20	52-1/2	41		5.85	2.45	8.30
4450					columns	.78	.19	.97
4500		30	52-1/2	51		6.05	2.52	8.57
4550					columns	.78	.19	.97
4600		40	52-1/2	61		6.75	2.74	9.49
4650					columns	.88	.20	1.08
4700	45x45	20	52-1/2	41		6.15	2.49	8.64
4750					columns	.69	.16	.85
4800		30	52-1/2	51		6.90	2.73	9.63
4850					columns	.78	.19	.97
4900		40	52-1/2	61		7.10	2.79	9.89
4950					columns	.89	.21	1.10
5000	45x50	20	52-1/2	41		6.35	2.51	8.86
5050					columns	.62	.15	.77
5100		30	52-1/2	51		7.30	2.81	10.11
5150					columns	.70	.16	.86
5200		40	52-1/2	61		7.80	2.96	10.76
5250					columns	.80	.19	.99
5300	50x45	20	56-1/2	41		6.15	2.47	8.62
5350					columns	.62	.15	.77
5400		30	56-1/2	51		7	2.73	9.73
5450					columns	.70	.16	.86
5500		40	56-1/2	61		8	3.03	11.03
5550					columns	.80	.19	.99

B1020 Roof Construction

B1020 124 Steel Joists & Joist Girders on Columns

	BAY SIZE (FT.) GIRD X JOISTS	SUPERIMPOSED LOAD (P.S.F.)	DEPTH (IN.)	TOTAL LOAD (P.S.F.)	COLUMN ADD	COST PER S.F.		
						MAT.	INST.	TOTAL
5600	50x50	20	56-1/2	42		6.90	2.71	9.61
5650					columns	.63	.15	.78
5700		30	56-1/2	53		7.55	2.92	10.47
5750					columns	.81	.19	1
5800		40	59	64		7.90	3.23	11.13
5850					columns	.89	.21	1.10
5900	55x50	20	60-1/2	43		6.95	2.72	9.67
5950					columns	.66	.15	.81
6000		30	64-1/2	54		8	3.04	11.04
6050					columns	.81	.19	1
6100		40	67	65		8.05	3.29	11.34
6150					columns	.81	.19	1
6200	60x50	20	62-1/2	43		6.95	2.72	9.67
6250					columns	.60	.14	.74
6300		30	68-1/2	54		8.05	3.05	11.10
6350					columns	.74	.18	.92
6400		40	71	65		8.10	3.29	11.39
6450					columns	.80	.19	.99

B1020 Roof Construction

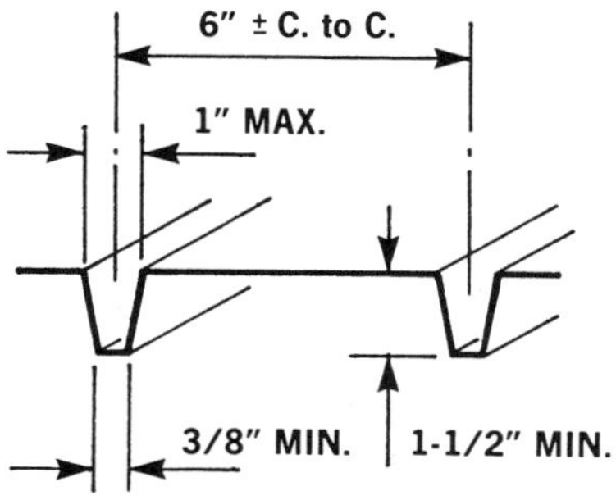

NARROW RIB DECK

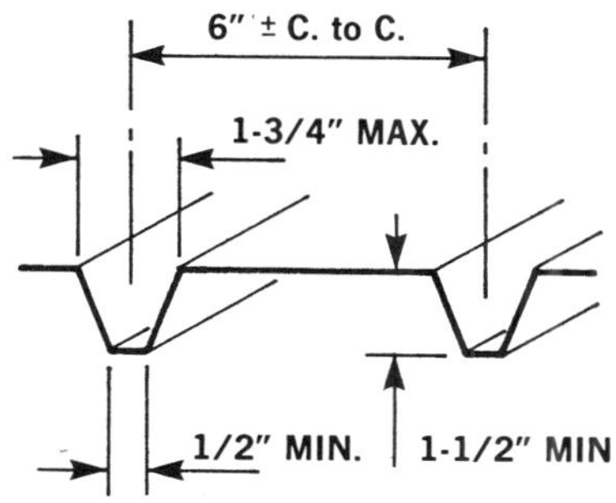

INTERMEDIATE RIB DECK

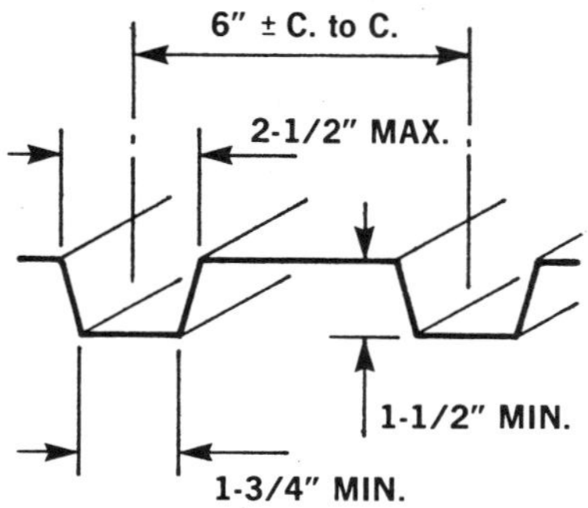

WIDE RIB DECK

General: Steel roof deck has the principal advantage of lightweight and speed of erection. It is covered by varying thicknesses of rigid insulation and roofing material.

Design Assumptions: Maximum spans shown in tables are based on construction and maintenance loads as specified by the Steel Deck Institute. Deflection is limited to 1/240 of the span.

Note: Costs shown in tables are based on projects over 5000 S.F. Under 5000 S.F. multiply costs by 1.16. For over four stories or congested site add 50% to the installation costs.

B1020 125 Narrow Rib, Single Span Steel Deck

	TOTAL LOAD (P.S.F.)	MAX. SPAN (FT. - IN.)	DECK GAUGE	WEIGHT (P.S.F.)	DEPTH (IN.)	COST PER S.F.		
						MAT.	INST.	TOTAL
1250	85	3-10	22	1.80	1-1/2	1.64	.60	2.24
1300	66	4-10	20	2.15	1-1/2	1.92	.72	2.64
1350	60	5-11	18	2.85	1-1/2	2.47	.72	3.19
1360								

B1020 126 Narrow Rib, Multi Span Steel Deck

	TOTAL LOAD (P.S.F.)	MAX. SPAN (FT.-IN.)	DECK GAUGE	WEIGHT (P.S.F.)	DEPTH (IN.)	COST PER S.F.		
						MAT.	INST.	TOTAL
1500	69	4-9	22	1.80	1-1//2	1.64	.60	2.24
1550	55	5-11	20	2.15	1-1/2	1.92	.72	2.64
1600	55	6-11	18	2.85	1-1/2	2.47	.72	3.19
1610								

B1020 127 Intermediate Rib, Single Span Steel Deck

	TOTAL LOAD (P.S.F.)	MAX. SPAN (FT.-IN.)	DECK GAUGE	WEIGHT (P.S.F.)	DEPTH (IN.)	COST PER S.F.		
						MAT.	INST.	TOTAL
1750	72	4-6	22	1.75	1-1/2	1.64	.60	2.24
1800	65	5-3	20	2.10	1-1/2	1.92	.72	2.64
1850	64	6-2	18	2.80	1-1/2	2.47	.72	3.19
1860								

B1020 Roof Construction

B1020 128 Intermediate Rib, Multi Span Steel Deck

	TOTAL LOAD (P.S.F.)	MAX. SPAN (FT.-IN.)	DECK GAUGE	WEIGHT (P.S.F.)	DEPTH (IN.)	COST PER S.F.		
						MAT.	INST.	TOTAL
2000	60	5-6	22	1.75	1-1/2	1.64	.60	2.24
2050	57	6-3	20	2.10	1-1/2	1.92	.72	2.64
2100	57	7-4	18	2.80	1-1/2	2.47	.72	3.19
2110								

B1020 129 Wide Rib, Single Span Steel Deck

	TOTAL LOAD (P.S.F.)	MAX. SPAN (FT.-IN.)	DECK GAUGE	WEIGHT (P.S.F.)	DEPTH (IN.)	COST PER S.F.		
						MAT.	INST.	TOTAL
2250	66	5-6	22	1.80	1-1/2	1.64	.60	2.24
2300	56	6-3	20	2.15	1-1/2	1.92	.72	2.64
2350	46	7-6	18	2.85	1-1/2	2.47	.72	3.19
2375								

B1020 130 Wide Rib, Multi Span Steel Deck

	TOTAL LOAD (P.S.F.)	MAX. SPAN (FT.-IN.)	DECK GAUGE	WEIGHT (P.S.F.)	DEPTH (IN.)	COST PER S.F.		
						MAT.	INST.	TOTAL
2500	76	6-6	22	1.80	1-1/2	1.64	.60	2.24
2550	64	7-5	20	2.15	1-1/2	1.92	.72	2.64
2600	55	8-10	18	2.80	1-1/2	2.47	.72	3.19
2610								

B1020 Roof Construction

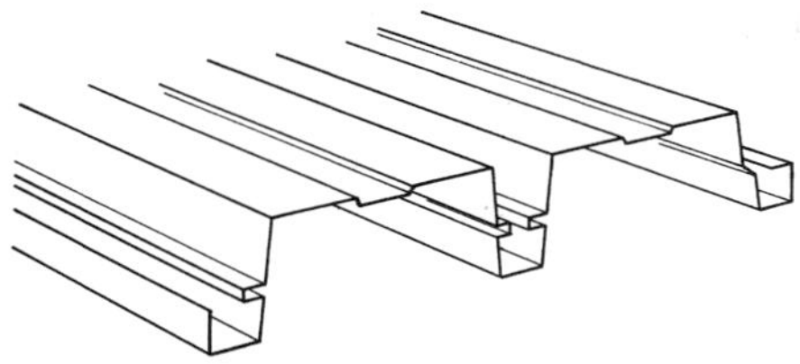

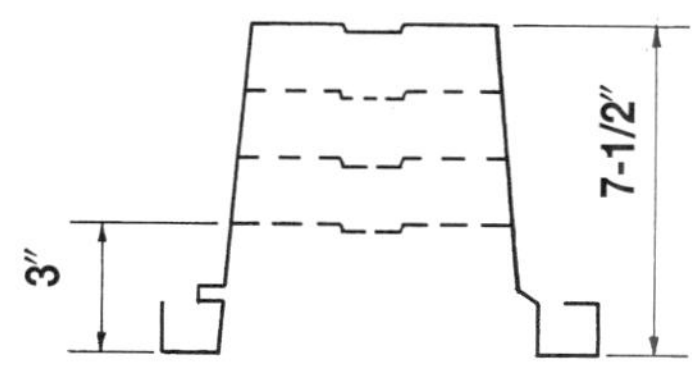

Design & Pricing Assumptions:
Costs in table below are based upon 5000 S.F. or more.
Deflection is limited to 1/240 of the span.

B1020 131 Single Span, Steel Deck

	TOTAL LOAD (P.S.F.)	DECK SUPPORT SPACING (FT.)	DEPTH (IN.)	DECK GAUGE	WEIGHT (P.S.F.)	COST PER S.F.		
						MAT.	INST.	TOTAL
0850	30	10	3	20	2.5	3.61	.91	4.52
0900		12	3	18	3.0	4.67	.96	5.63
0950		14	3	16	4.0	6.15	1.02	7.17
1000		16	4.5	20	3.0	4.05	1.14	5.19
1050		18	4.5	18	4.0	5.35	1.25	6.60
1100		20	4.5	16	5.0	6.95	1.31	8.26
1200		24	6.0	16	6.0	8	1.59	9.59
1300		28	7.5	16	7.0	9.45	1.93	11.38
1350	40	10	3.0	20	2.5	3.61	.91	4.52
1400		12	3.0	18	3.0	4.67	.96	5.63
1450		14	3.0	16	4.0	6.15	1.02	7.17
1500		16	4.5	20	3.0	4.05	1.14	5.19
1550		18	4.5	18	4.0	5.35	1.25	6.60
1600		20	6.0	16	6.0	8	1.59	9.59
1700		24	7.5	16	7.0	9.45	1.93	11.38
1750	50	10	3.0	20	2.5	3.61	.91	4.52
1800		12	3.0	18	3.0	4.67	.96	5.63
1850		14	3.0	16	4.0	6.15	1.02	7.17
1900		16	4.5	18	4.0	5.35	1.25	6.60
1950		18	4.5	16	5.0	6.95	1.31	8.26
2000		20	6.0	16	6.0	6.40	1.54	7.94
2100		24	7.5	16	7.0	9.45	1.93	11.38

B1020 132 Double Span, Steel Deck

	TOTAL LOAD (P.S.F.)	DECK SUPPORT SPACING (FT.)	DEPTH (IN.)	DECK GAUGE	WEIGHT (P.S.F.)	COST PER S.F.		
						MAT.	INST.	TOTAL
2250	30	10	3	20	2.5	3.61	.91	4.52
2300		12	3	20	2.5	3.61	.91	4.52
2350		14	3	18	3.0	4.67	.96	5.63
2400		16	3	16	4.0	6.15	1.02	7.17
2450	40	10	3	20	2.5	3.61	.91	4.52
2500		12	3	20	2.5	3.61	.91	4.52
2550		14	3	18	3.0	4.67	.96	5.63
2600		16	3	16	4.0	6.15	1.02	7.17
2650	50	10	3	20	2.5	3.61	.91	4.52
2700		12	3	18	3.0	4.67	.96	5.63
2750		14	3	16	4.0	6.15	1.02	7.17
2800		16	4.5	20	4.0	4.05	1.14	5.19

B1020 Roof Construction

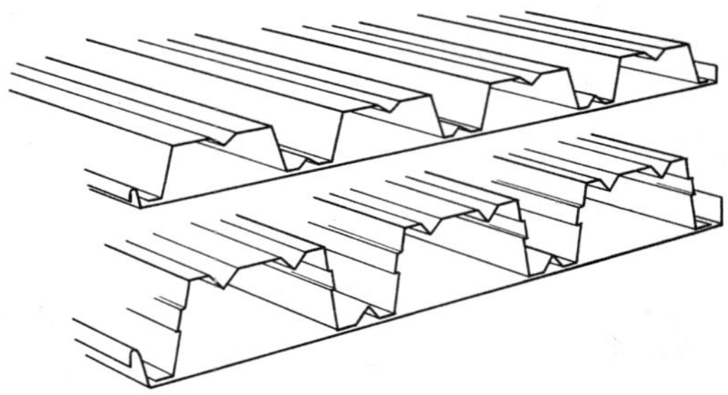

Design & Pricing Assumptions:
Costs in table below are based upon 15,000 S.F. or more.
No utility outlets or modifications included in cost.
Deflection is limited to 1/240 of the span.

B1020 134 — Cellular Steel Deck, Single Span

	TOTAL LOAD (P.S.F.)	DECK SUPPORT SPACING (FT.)	DEPTH (IN.)	WEIGHT (P.S.F.)		COST PER S.F.		
						MAT.	INST.	TOTAL
1150	30	10	1-5/8	4		10	2.11	12.11
1200		12	1-5/8	5		10.45	2.21	12.66
1250		14	3	5		11.95	2.28	14.23
1300		16	3	5		11.90	2.39	14.29
1350		18	3	5		11.90	2.39	14.29
1400		20	3	6		13.40	2.50	15.90
1450	40	8	1-5/8	4		10	2.11	12.11
1500		10	1-5/8	5		10.45	2.21	12.66
1550		12	3	5		11.90	2.39	14.29
1600		14	3	5		11.90	2.39	14.29
1650		16	3	5		11.90	2.39	14.29
1700		18	3	6		13.40	2.50	15.90
1750	50	8	1-5/8	4		10	2.11	12.11
1800		10	1-5/8	5		10.45	2.21	12.66
1850		12	3	4		11.95	2.28	14.23
1900		14	3	5		11.90	2.39	14.29
1950		16	3	6		13.40	2.50	15.90
1960								

B1020 135 — Cellular Steel Deck, Double Span

	TOTAL LOAD (P.S.F.)	DECK SUPPORT SPACING (FT.)	DEPTH (IN.)	WEIGHT (P.S.F.)		COST PER S.F.		
						MAT.	INST.	TOTAL
2150	30	10	1-5/8	4		10	2.11	12.11
2200		12	1-5/8	5		10.45	2.21	12.66
2250		14	1-5/8	5		10.45	2.21	12.66
2300		16	3	5		11.90	2.39	14.29
2350		18	3	5		11.90	2.39	14.29
2400		20	3	6		13.40	2.50	15.90
2450	40	8	1-5/8	4		10	2.11	12.11
2500		10	1-5/8	5		10.45	2.21	12.66
2550		12	1-5/8	5		10.45	2.21	12.66
2600		14	3	5		11.90	2.39	14.29
2650		16	3	5		11.90	2.39	14.29
2700		18	3	6		13.40	2.50	15.90

B1020 Roof Construction

B1020 135 Cellular Steel Deck, Double Span

	TOTAL LOAD (P.S.F.)	DECK SUPPORT SPACING (FT.)	DEPTH (IN.)	WEIGHT (P.S.F.)		COST PER S.F.		
						MAT.	INST.	TOTAL
2750	50	8	1-5/8	4		10	2.11	12.11
2800		10	1-5/8	5		10.45	2.21	12.66
2850		12	1-5/8	5		10.45	2.21	12.66
2900		14	3	5		11.90	2.39	14.29
2950		16	3	6		13.40	2.50	15.90
2960								

B1020 136 Cellular Steel Deck, Triple Span

	TOTAL LOAD (P.S.F.)	DECK SUPPORT SPACING (FT.)	DEPTH (IN.)	WEIGHT (P.S.F.)		COST PER S.F.		
						MAT.	INST.	TOTAL
3200	30	10	1-5/8	4		10	2.11	12.11
3210								
3250	40	8	1-5/8	4		10	2.11	12.11
3300		10	1-5/8	4		10	2.11	12.11
3350		12	1-5/8	5		10.45	2.21	12.66
3400	50	8	1-5/8	4		10	2.11	12.11
3450		10	1-5/8	5		10.45	2.21	12.66
3500		12	1-5/8	5		10.45	2.21	12.66

B1020 310 Canopies

	Canopies	COST PER S.F.		
		MAT.	INST.	TOTAL
0100	Canopies, wall hung, prefinished aluminum, 8′ x 10′	32.50	22.50	55
0110	12′ x 40′	26.50	8.10	34.60
0120	Canopies, wall hung, canvas awnings, with frame and lettering	98	13.20	111.20
0130	Freestanding (walkways), vinyl coated steel, 12′ wide	24.50	9.25	33.75

B2010 Exterior Walls

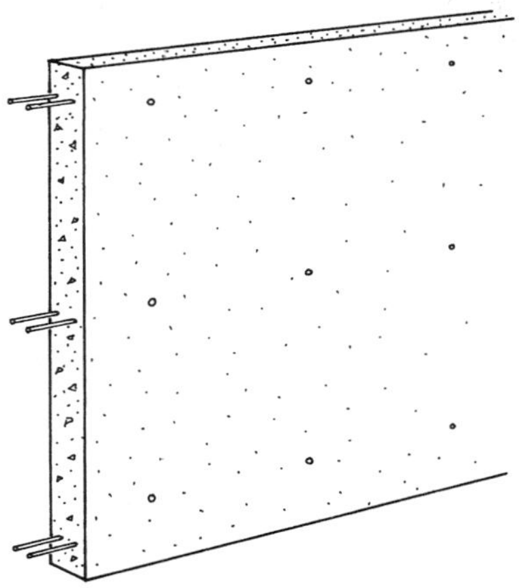

The table below describes a concrete wall system for exterior closure. There are several types of wall finishes priced from plain finish to a finish with 3/4″ rustication strip.

Design Assumptions:
Conc. f'c = 3000 to 5000 psi
Reinf. fy = 60,000 psi

System Components	QUANTITY	UNIT	COST PER S.F. MAT.	COST PER S.F. INST.	COST PER S.F. TOTAL
SYSTEM B2010 101 2100					
CONC. WALL, REINFORCED, 8′ HIGH, 6″ THICK, PLAIN FINISH, 3,000 PSI					
Forms in place, wall, job built plyform to 8′ high, 4 uses	2.000	SFCA	2.20	13.70	15.90
Reinforcing in place, walls, #3 to #7	.752	Lb.	.41	.33	.74
Concrete ready mix, regular weight, 3000 psi	.018	C.Y.	2.12		2.12
Place and vibrate concrete, walls 6″ thick, pump	.018	C.Y.		.88	.88
Finish wall, break ties, patch voids	2.000	S.F.	.10	2	2.10
TOTAL			4.83	16.91	21.74

B2010 101	Cast In Place Concrete	COST PER S.F. MAT.	COST PER S.F. INST.	COST PER S.F. TOTAL
2100	Conc wall reinforced, 8′ high, 6″ thick, plain finish, 3000 PSI	4.83	16.90	21.73
2200	4000 PSI	4.94	16.90	21.84
2300	5000 PSI	5.05	16.90	21.95
2400	Rub concrete 1 side, 3000 PSI	4.83	20	24.83
2500	4000 PSI	4.94	20	24.94
2600	5000 PSI	5.05	20	25.05
2700	Aged wood liner, 3000 PSI	6.05	19.20	25.25
2800	4000 PSI	6.20	19.20	25.40
2900	5000 PSI	6.30	19.20	25.50
3000	Sand blast light 1 side, 3000 PSI	5.40	19.15	24.55
3100	4000 PSI	5.50	19.15	24.65
3300	5000 PSI	5.65	19.15	24.80
3400	Sand blast heavy 1 side, 3000 PSI	6	23.50	29.50
3500	4000 PSI	6.10	23.50	29.60
3600	5000 PSI	6.20	23.50	29.70
3700	3/4″ bevel rustication strip, 3000 PSI	4.95	17.95	22.90
3800	4000 PSI	5.05	17.95	23
3900	5000 PSI	5.20	17.95	23.15
4000	8″ thick, plain finish, 3000 PSI	5.70	17.35	23.05
4100	4000 PSI	5.85	17.35	23.20
4200	5000 PSI	6.05	17.35	23.40
4300	Rub concrete 1 side, 3000 PSI	5.70	20	25.70
4400	4000 PSI	5.85	20	25.85
4500	5000 PSI	6.05	20	26.05
4550	8″ thick, aged wood liner, 3000 PSI	6.95	19.65	26.60
4600	4000 PSI	7.10	19.65	26.75

B2010 Exterior Walls

B2010 101	Cast In Place Concrete	COST PER S.F.		
		MAT.	INST.	TOTAL
4700	5000 PSI	7.25	19.65	26.90
4750	Sand blast light 1 side, 3000 PSI	6.30	19.60	25.90
4800	4000 PSI	6.45	19.60	26.05
4900	5000 PSI	6.60	19.60	26.20
5000	Sand blast heavy 1 side, 3000 PSI	6.85	24	30.85
5100	4000 PSI	7	24	31
5200	5000 PSI	7.20	24	31.20
5300	3/4" bevel rustication strip, 3000 PSI	5.85	18.35	24.20
5400	4000 PSI	6	18.35	24.35
5500	5000 PSI	6.15	18.35	24.50
5600	10" thick, plain finish, 3000 PSI	6.55	17.75	24.30
5700	4000 PSI	6.75	17.75	24.50
5800	5000 PSI	6.95	17.75	24.70
5900	Rub concrete 1 side, 3000 PSI	6.55	21	27.55
6000	4000 PSI	6.75	21	27.75
6100	5000 PSI	6.95	21	27.95
6200	Aged wood liner, 3000 PSI	7.80	20	27.80
6300	4000 PSI	8	20	28
6400	5000 PSI	8.20	20	28.20
6500	Sand blast light 1 side, 3000 PSI	7.15	20	27.15
6600	4000 PSI	7.30	20	27.30
6700	5000 PSI	7.55	20	27.55
6800	Sand blast heavy 1 side, 3000 PSI	7.70	24.50	32.20
6900	4000 PSI	7.90	24.50	32.40
7000	5000 PSI	8.10	24.50	32.60
7100	3/4" bevel rustication strip, 3000 PSI	6.70	18.75	25.45
7200	4000 PSI	6.85	18.75	25.60
7300	5000 PSI	7.10	18.75	25.85
7400	12" thick, plain finish, 3000 PSI	7.60	18.20	25.80
7500	4000 PSI	7.80	18.20	26
7600	5000 PSI	8.05	18.20	26.25
7700	Rub concrete 1 side, 3000 PSI	7.60	21.50	29.10
7800	4000 PSI	7.80	21.50	29.30
7900	5000 PSI	8.05	21.50	29.55
8000	Aged wood liner, 3000 PSI	8.80	20.50	29.30
8100	4000 PSI	9.05	20.50	29.55
8200	5000 PSI	9.30	20.50	29.80
8300	Sand blast light 1 side, 3000 PSI	8.15	20.50	28.65
8400	4000 PSI	8.35	20.50	28.85
8500	5000 PSI	8.65	20.50	29.15
8600	Sand blast heavy 1 side, 3000 PSI	8.75	25	33.75
8700	4000 PSI	8.95	25	33.95
8800	5000 PSI	9.20	25	34.20
8900	3/4" bevel rustication strip, 3000 PSI	7.70	19.25	26.95
9000	4000 PSI	7.90	19.25	27.15
9500	5000 PSI	8.20	19.25	27.45

B2010 Exterior Walls

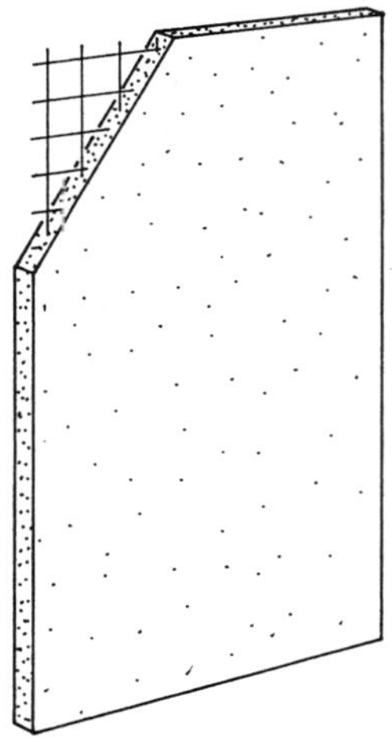

Precast concrete wall panels are either solid or insulated with plain, colored or textured finishes. Transportation is an important cost factor. Prices below are based on delivery within fifty miles of a plant. Engineering data is available from fabricators to assist with construction details. Usual minimum job size for economical use of panels is about 5000 S.F. Small jobs can double the prices below. For large, highly repetitive jobs, deduct up to 15% from the prices below.

B2010 102 Flat Precast Concrete

	THICKNESS (IN.)	PANEL SIZE (FT.)	FINISHES	RIGID INSULATION (IN)	TYPE	COST PER S.F.		
						MAT.	INST.	TOTAL
3000	4	5x18	smooth gray	none	low rise	12.65	7.90	20.55
3050		6x18				10.60	6.65	17.25
3100		8x20				21	3.20	24.20
3150		12x20				19.90	3.03	22.93
3200	6	5x18	smooth gray	2	low rise	13.70	8.55	22.25
3250		6x18				11.65	7.20	18.85
3300		8x20				22.50	3.91	26.41
3350		12x20				20.50	3.58	24.08
3400	8	5x18	smooth gray	2	low rise	29	4.92	33.92
3450		6x18				27.50	4.68	32.18
3500		8x20				25	4.31	29.31
3550		12x20				23	3.97	26.97
3600	4	4x8	white face	none	low rise	48.50	4.60	53.10
3650		8x8				36.50	5.15	41.65
3700		10x10				32	3.02	35.02
3750		20x10				29	2.74	31.74
3800	5	4x8	white face	none	low rise	49.50	4.69	54.19
3850		8x8				37.50	3.54	41.04
3900		10x10				33.50	3.14	36.64
3950		20x20				30.50	2.87	33.37
4000	6	4x8	white face	none	low rise	51.50	4.86	56.36
4050		8x8				39	3.69	42.69
4100		10x10				34.50	3.25	37.75
4150		20x10				31.50	2.99	34.49
4200	6	4x8	white face	2	low rise	52.50	5.55	58.05
4250		8x8				40.50	4.36	44.86
4300		10x10				35.50	3.92	39.42
4350		20x10				31.50	2.99	34.49
4400	7	4x8	white face	none	low rise	53	4.97	57.97
4450		8x8				40.50	3.83	44.33
4500		10x10				36.50	3.43	39.93
4550		20x10				33.50	3.14	36.64
4600	7	4x8	white face	2	low rise	54	5.65	59.65
4650		8x8				41.50	4.50	46
4700		10x10				37.50	4.10	41.60

B2010 Exterior Walls

B2010 102 Flat Precast Concrete

	THICKNESS (IN.)	PANEL SIZE (FT.)	FINISHES	RIGID INSULATION (IN)	TYPE	COST PER S.F.		
						MAT.	INST.	TOTAL
4750	7	20x10	white face	2	low rise	34.50	3.81	38.31
4800	8	4x8	white face	none	low rise	54	5.10	59.10
4850		8x8				41.50	3.92	45.42
4900		10x10				37.50	3.52	41.02
4950		20x10				34.50	3.24	37.74
5000	8	4x8	white face	2	low rise	55	5.75	60.75
5050		8x8				42.50	4.59	47.09
5100		10x10				43.50	4.84	48.34
5150		20x10				43.50	4.84	48.34

B2010 103 Fluted Window or Mullion Precast Concrete

	THICKNESS (IN.)	PANEL SIZE (FT.)	FINISHES	RIGID INSULATION (IN)	TYPE	COST PER S.F.		
						MAT.	INST.	TOTAL
5200	4	4x8	smooth gray	none	high rise	29	18.15	47.15
5250		8x8				20.50	12.95	33.45
5300		10x10				40	6.15	46.15
5350		20x10				35.50	5.35	40.85
5400	5	4x8	smooth gray	none	high rise	29.50	18.45	47.95
5450		8x8				21.50	13.45	34.95
5500		10x10				42	6.40	48.40
5550		20x10				37	5.65	42.65
5600	6	4x8	smooth gray	none	high rise	30.50	18.95	49.45
5650		8x8				22	13.90	35.90
5700		10x10				43	6.55	49.55
5750		20x10				38.50	5.85	44.35
5800	6	4x8	smooth gray	2	high rise	31.50	19.65	51.15
5850		8x8				23.50	14.55	38.05
5900		10x10				44.50	7.25	51.75
5950		20x10				39.50	6.50	46
6000	7	4x8	smooth gray	none	high rise	31	19.35	50.35
6050		8x8				23	14.30	37.30
6100		10x10				45	6.85	51.85
6150		20x10				39.50	6	45.50
6200	7	4x8	smooth gray	2	high rise	32	20	52
6250		8x8				24	14.95	38.95
6300		10x10				46	7.55	53.55
6350		20x10				40.50	6.70	47.20
6400	8	4x8	smooth gray	none	high rise	31.50	19.65	51.15
6450		8x8				23.50	14.60	38.10
6500		10x10				46.50	7.05	53.55
6550		20x10				41.50	6.30	47.80
6600	8	4x8	smooth gray	2	high rise	32.50	20.50	53
6650		8x8				24.50	15.30	39.80
6700		10x10				47.50	7.75	55.25
6750		20x10				42.50	6.95	49.45

B2010 104 Ribbed Precast Concrete

	THICKNESS (IN.)	PANEL SIZE (FT.)	FINISHES	RIGID INSULATION(IN.)	TYPE	COST PER S.F.		
						MAT.	INST.	TOTAL
6800	4	4x8	aggregate	none	high rise	34.50	18.15	52.65
6850		8x8				25.50	13.25	38.75
6900		10x10				42.50	4.81	47.31

B20 Exterior Enclosure

B2010 Exterior Walls

B2010 104 — Ribbed Precast Concrete

	THICKNESS (IN.)	PANEL SIZE (FT.)	FINISHES	RIGID INSULATION(IN.)	TYPE	COST PER S.F.		
						MAT.	INST.	TOTAL
6950	4	20x10	aggregate	none	high rise	37.50	4.25	41.75
7000	5	4x8	aggregate	none	high rise	35	18.45	53.45
7050		8x8				26	13.65	39.65
7100		10x10				44.50	4.99	49.49
7150		20x10				39.50	4.45	43.95
7200	6	4x8	aggregate	none	high rise	36	18.80	54.80
7250		8x8				26.50	14	40.50
7300		10x10				45.50	5.15	50.65
7350		20x10				41	4.62	45.62
7400	6	4x8	aggregate	2	high rise	37	19.50	56.50
7450		8x8				28	14.65	42.65
7500		10x10				47	5.80	52.80
7550		20x10				42	5.30	47.30
7600	7	4x8	aggregate	none	high rise	36.50	19.25	55.75
7650		8x8				27.50	14.35	41.85
7700		10x10				47	5.30	52.30
7750		20x10				42	4.75	46.75
7800	7	4x8	aggregate	2	high rise	38	19.90	57.90
7850		8x8				28.50	15	43.50
7900		10x10				48.50	6	54.50
7950		20x10				43.50	5.40	48.90
8000	8	4x8	aggregate	none	high rise	37.50	19.60	57.10
8050		8x8				28	14.75	42.75
8100		10x10				49	5.50	54.50
8150		20x10				44	4.97	48.97
8200	8	4x8	aggregate	2	high rise	38.50	20.50	59
8250		8x8				29.50	15.45	44.95
8300		10x10				50	7.90	57.90
8350		20x10				45	5.65	50.65

B2010 105 — Precast Concrete Specialties

	TYPE	SIZE				COST PER L.F.		
						MAT.	INST.	TOTAL
8400	Coping, precast	5" wide				19.15	13.70	32.85
8450	Stock units	10" wide				20.50	14.70	35.20
8460		12" wide				23.50	15.85	39.35
8480		14" wide				25.50	17.15	42.65
8500	Window sills	6" wide				16.10	14.70	30.80
8550	Precast	10" wide				26	17.15	43.15
8600		14" wide				26	20.50	46.50
8610								

B2010 Exterior Walls

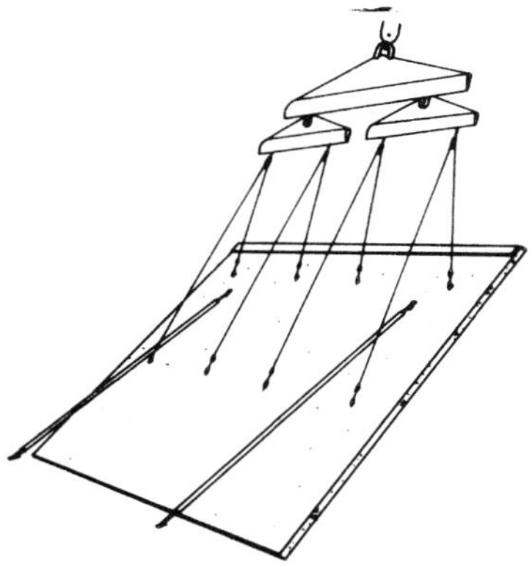

The advantage of tilt up construction is in the low cost of forms and placing of concrete and reinforcing. Tilt up has been used for several types of buildings, including warehouses, stores, offices, and schools. The panels are cast in forms on the ground, or floor slab. Most jobs use 5-1/2″ thick solid reinforced concrete panels.

Design Assumptions:
Conc. f'c = 3000 psi
Reinf. fy = 60,000

System Components	QUANTITY	UNIT	COST PER S.F. MAT.	COST PER S.F. INST.	COST PER S.F. TOTAL
SYSTEM B2010 106 3200					
TILT-UP PANELS, 20′ X 25′, BROOM FINISH, 5-1/2″ THICK, 3000 PSI					
Apply liquid bond release agent	500.000	S.F.	.02	.14	.16
Edge forms in place for slab on grade	120.000	L.F.	.09	.90	.99
Reinforcing in place	.350	Ton	.74	.81	1.55
Footings, form braces, steel	1.000	Set	1.02		1.02
Slab lifting inserts	1.000	Set	.17		.17
Framing, less than 4″ angles	1.000	Set	.17	1.69	1.86
Concrete, ready mix, regular weight, 3000 psi	8.550	C.Y.	2.02		2.02
Place and vibrate concrete for slab on grade, 4″ thick, direct chute	8.550	C.Y.		.46	.46
Finish floor, monolithic broom finish	500.000	S.F.		.84	.84
Cure with curing compound, sprayed	500.000	S.F.	.14	.10	.24
Erection crew	.058	Day		1.42	1.42
TOTAL			4.37	6.36	10.73

B2010 106	Tilt-Up Concrete Panel	COST PER S.F. MAT.	COST PER S.F. INST.	COST PER S.F. TOTAL
3200	Tilt-up conc panels, broom finish, 5-1/2″ thick, 3000 PSI	4.37	6.35	10.72
3250	5000 PSI	4.48	6.25	10.73
3300	6″ thick, 3000 PSI	4.79	6.50	11.29
3350	5000 PSI	4.94	6.40	11.34
3400	7-1/2″ thick, 3000 PSI	6.15	6.75	12.90
3450	5000 PSI	6.30	6.65	12.95
3500	8″ thick, 3000 PSI	6.55	6.90	13.45
3550	5000 PSI	6.80	6.80	13.60
3700	Steel trowel finish, 5-1/2″ thick, 3000 PSI	4.37	6.45	10.82
3750	5000 PSI	4.48	6.30	10.78
3800	6″ thick, 3000 PSI	4.79	6.60	11.39
3850	5000 PSI	4.94	6.45	11.39
3900	7-1/2″ thick, 3000 PSI	6.15	6.80	12.95
3950	5000 PSI	6.30	6.70	13
4000	8″ thick, 3000 PSI	6.55	7.05	13.60
4050	5000 PSI	6.80	6.90	13.70
4200	Exp. aggregate finish, 5-1/2″ thick, 3000 PSI	4.69	6.50	11.19
4250	5000 PSI	4.81	6.40	11.21
4300	6″ thick, 3000 PSI	5.10	6.65	11.75
4350	5000 PSI	5.25	6.55	11.80
4400	7-1/2″ thick, 3000 PSI	6.45	6.90	13.35
4450	5000 PSI	6.65	6.80	13.45

B2010 Exterior Walls

B2010 106	Tilt-Up Concrete Panel	COST PER S.F.		
		MAT.	INST.	TOTAL
4500	8" thick, 3000 PSI	6.90	7.05	13.95
4550	5000 PSI	7.10	6.95	14.05
4600	Exposed aggregate & vert. rustication 5-1/2" thick, 3000 PSI	7.05	8.05	15.10
4650	5000 PSI	7.15	7.95	15.10
4700	6" thick, 3000 PSI	7.50	8.20	15.70
4750	5000 PSI	7.60	8.10	15.70
4800	7-1/2" thick, 3000 PSI	8.80	8.45	17.25
4850	5000 PSI	9	8.35	17.35
4900	8" thick, 3000 PSI	9.25	8.60	17.85
4950	5000 PSI	9.45	8.50	17.95
5000	Vertical rib & light sandblast, 5-1/2" thick, 3000 PSI	6.85	10.50	17.35
5050	5000 PSI	6.95	10.40	17.35
5100	6" thick, 3000 PSI	7.25	10.65	17.90
5150	5000 PSI	7.40	10.55	17.95
5200	7-1/2" thick, 3000 PSI	8.60	10.90	19.50
5250	5000 PSI	8.80	10.75	19.55
5300	8" thick, 3000 PSI	9.05	11.10	20.15
5350	5000 PSI	9.25	11	20.25
6000	Broom finish w/2" polystyrene insulation, 6" thick, 3000 PSI	4.21	7.85	12.06
6050	5000 PSI	4.37	7.85	12.22
6100	Broom finish 2" fiberplank insulation, 6" thick, 3000 PSI	4.75	7.75	12.50
6150	5000 PSI	4.91	7.75	12.66
6200	Exposed aggregate w/2" polystyrene insulation, 6" thick, 3000 PSI	4.43	7.85	12.28
6250	5000 PSI	4.59	7.85	12.44
6300	Exposed aggregate 2" fiberplank insulation, 6" thick, 3000 PSI	4.97	7.75	12.72
6350	5000 PSI	5.15	7.75	12.90

B2010 Exterior Walls

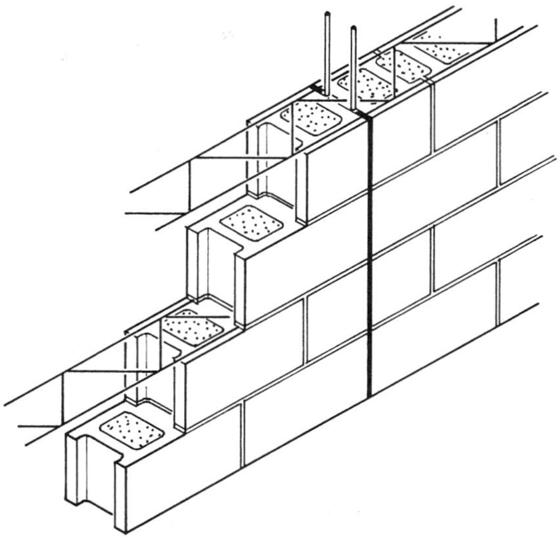

Exterior concrete block walls are defined in the following terms; structural reinforcement, weight, percent solid, size, strength and insulation. Within each of these categories, two to four variations are shown. No costs are included for brick shelf or relieving angles.

System Components	QUANTITY	UNIT	COST PER S.F. MAT.	COST PER S.F. INST.	COST PER S.F. TOTAL
SYSTEM B2010 109 1400					
UNREINFORCED CONCRETE BLOCK WALL, 8″ X 8″ X 16″, PERLITE CORE FILL					
Concrete block wall, 8″ thick	1.000	S.F.	2.71	7	9.71
Perlite insulation	1.000	S.F.	1.49	.43	1.92
Horizontal joint reinforcing, alternate courses	.800	S.F.	.18	.18	.36
Control joint	.050	L.F.	.08	.07	.15
TOTAL			4.46	7.68	12.14

B2010 109 Concrete Block Wall - Regular Weight

	TYPE	SIZE (IN.)	STRENGTH (P.S.I.)	CORE FILL		COST PER S.F. MAT.	COST PER S.F. INST.	COST PER S.F. TOTAL
1200	Hollow	4x8x16	2,000	none		2.18	6.30	8.48
1250			4,500	none		2.66	6.30	8.96
1300	RB2010 -200	6x8x16	2,000	perlite		3.78	7.10	10.88
1310				styrofoam		4.09	6.75	10.84
1340				none		2.77	6.75	9.52
1350			4,500	perlite		4.26	7.10	11.36
1360				styrofoam		4.57	6.75	11.32
1390				none		3.25	6.75	10
1400		8x8x16	2,000	perlite		4.46	7.70	12.16
1410				styrofoam		4.46	7.25	11.71
1440				none		2.97	7.25	10.22
1450			4,500	perlite		5.45	7.70	13.15
1460				styrofoam		5.45	7.25	12.70
1490				none		3.95	7.25	11.20
1500		12x8x16	2,000	perlite		7.20	10.25	17.45
1510				styrofoam		6.45	9.40	15.85
1540				none		4.76	9.40	14.16
1550			4,500	perlite		7.70	10.25	17.95
1560				styrofoam		7	9.40	16.40
1590				none		5.30	9.40	14.70
2000	75% solid	4x8x16	2,000	none		2.62	6.35	8.97
2050			4,500	none		3.26	6.35	9.61

B2010 Exterior Walls

B2010 109 Concrete Block Wall - Regular Weight

	TYPE	SIZE (IN.)	STRENGTH (P.S.I.)	CORE FILL		COST PER S.F.		
						MAT.	INST.	TOTAL
2100	75% solid	6x8x16	2,000	perlite		3.86	7.05	10.91
2140				none		3.35	6.85	10.20
2150			4,500	perlite		4.50	7.05	11.55
2190				none		3.99	6.85	10.84
2200		8x8x16	2,000	perlite		4.31	7.55	11.86
2240				none		3.56	7.35	10.91
2250			4,500	perlite		5.60	7.55	13.15
2290				none		4.85	7.35	12.20
2300		12x8x16	2,000	perlite		7	9.90	16.90
2340				none		5.80	9.50	15.30
2350			4,500	perlite		7.70	9.90	17.60
2390				none		6.50	9.50	16
2500	Solid	4x8x16	2,000	none		2.45	6.50	8.95
2550			4,500	none		3.66	6.45	10.11
2600		6x8x16	2,000	none		2.83	7	9.83
2650			4,500	none		4.48	6.95	11.43
2700		8x8x16	2,000	none		4.13	7.55	11.68
2750			4,500	none		5.45	7.45	12.90
2800		12x8x16	2,000	none		6.15	9.80	15.95
2850			4,500	none		7.30	9.65	16.95

B2010 110 Concrete Block Wall - Lightweight

	TYPE	SIZE (IN.)	WEIGHT (P.C.F.)	CORE FILL		COST PER S.F.		
						MAT.	INST.	TOTAL
3100	Hollow	8x4x16	105	perlite		4.40	7.25	11.65
3110				styrofoam		4.40	6.80	11.20
3140				none		2.91	6.80	9.71
3150			85	perlite		5.95	7.10	13.05
3160				styrofoam		5.95	6.65	12.60
3190				none		4.45	6.65	11.10
3200		4x8x16	105	none		2.30	6.15	8.45
3250			85	none		2.92	6.05	8.97
3300		6x8x16	105	perlite		4.14	6.95	11.09
3310				styrofoam		4.45	6.60	11.05
3340				none		3.13	6.60	9.73
3350			85	perlite		4.75	6.80	11.55
3360				styrofoam		5.05	6.45	11.50
3390				none		3.74	6.45	10.19
3400		8x8x16	105	perlite		5.25	7.50	12.75
3410				styrofoam		5.25	7.05	12.30
3440				none		3.78	7.05	10.83
3450			85	perlite		5.45	7.35	12.80
3460				styrofoam		5.45	6.90	12.35
3490				none		3.95	6.90	10.85
3500		12x8x16	105	perlite		7.20	9.95	17.15
3510				styrofoam		6.50	9.10	15.60
3540				none		4.78	9.10	13.88
3550			85	perlite		8.90	9.75	18.65
3560				styrofoam		8.20	8.90	17.10
3590				none		6.50	8.90	15.40
3600		4x8x24	105	none		1.75	6.90	8.65
3650			85	none		4.05	5.80	9.85
3690	For stacked bond add						.40	.40

B2010 Exterior Walls

B2010 110 Concrete Block Wall - Lightweight

	TYPE	SIZE (IN.)	WEIGHT (P.C.F.)	CORE FILL		COST PER S.F.		
						MAT.	INST.	TOTAL
3700	Hollow	6x8x24	105	perlite		3.33	7.55	10.88
3710				styrofoam		3.64	7.20	10.84
3740				none		2.32	7.20	9.52
3750			85	perlite		6.60	6.40	13
3760				styrofoam		6.90	6.05	12.95
3790				none		5.60	6.05	11.65
3800		8x8x24	105	perlite		4.33	8.15	12.48
3810				styrofoam		4.33	7.70	12.03
3840				none		2.84	7.70	10.54
3850			85	perlite		8.30	6.85	15.15
3860				styrofoam		8.30	6.40	14.70
3890				none		6.80	6.40	13.20
3900		12x8x24	105	perlite		6.05	9.60	15.65
3910				styrofoam		5.30	8.75	14.05
3940				none		3.61	8.75	12.36
3950			85	perlite		11	9.40	20.40
3960				styrofoam		10.30	8.55	18.85
3990				none		8.60	8.55	17.15
4000	75% solid	4x8x16	105	none		3.31	6.25	9.56
4050			85	none		3.59	6.10	9.69
4100		6x8x16	105	perlite		5.05	6.90	11.95
4140				none		4.56	6.70	11.26
4150			85	perlite		5.15	6.70	11.85
4190				none		4.62	6.50	11.12
4200		8x8x16	105	perlite		6.25	7.35	13.60
4240				none		5.50	7.15	12.65
4250			85	perlite		5.60	7.20	12.80
4290				none		4.85	7	11.85
4300		12x8x16	105	perlite		8.15	9.65	17.80
4340				none		6.95	9.25	16.20
4350			85	perlite		9.25	9.40	18.65
4390				none		8.05	9	17.05
4500	Solid	4x8x16	105	none		2.08	6.45	8.53
4550			85	none		4.65	6.15	10.80
4600		6x8x16	105	none		3.65	6.95	10.60
4650			85	none		5.35	6.60	11.95
4700		8x8x16	105	none		4.12	7.45	11.57
4750			85	none		8	7.05	15.05
4800		12x8x16	105	none		4.68	9.65	14.33
4850			85	none		12.15	9.10	21.25
4900	For stacked bond, add						.40	.40

B2010 111 Reinforced Concrete Block Wall - Regular Weight

	TYPE	SIZE (IN.)	STRENGTH (P.S.I.)	VERT. REINF & GROUT SPACING		COST PER S.F.		
						MAT.	INST.	TOTAL
5200	Hollow	4x8x16	2,000	#4 @ 48"		2.34	7	9.34
5250			4,500	#4 @ 48"		2.82	7	9.82
5300		6x8x16	2,000	#4 @ 48"		3.03	7.50	10.53
5330				#5 @ 32"		3.25	7.75	11
5340				#5 @ 16"		3.71	8.75	12.46
5350			4,500	#4 @ 28"		3.51	7.50	11.01
5380				#5 @ 32"		3.73	7.75	11.48
5390				#5 @ 16"		4.19	8.75	12.94

B2010 Exterior Walls

B2010 111 — Reinforced Concrete Block Wall - Regular Weight

	TYPE	SIZE (IN.)	STRENGTH (P.S.I.)	VERT. REINF & GROUT SPACING		COST PER S.F.		
						MAT.	INST.	TOTAL
5400	Hollow	8x8x16	2,000	#4 @ 48″		3.36	8	11.36
5430				#5 @ 32″		3.53	8.45	11.98
5440				#5 @ 16″		4.08	9.65	13.73
5450		8x8x16	4,500	#4 @ 48″		4.27	8.10	12.37
5480				#5 @ 32″		4.51	8.45	12.96
5490				#5 @ 16″		5.05	9.65	14.70
5500		12x8x16	2,000	#4 @ 48″		5.20	10.25	15.45
5530				#5 @ 32″		5.55	10.65	16.20
5540				#5 @ 16″		6.30	11.85	18.15
5550			4,500	#4 @ 48″		5.75	10.25	16
5580				#5 @ 32″		6.05	10.65	16.70
5590				#5 @ 16″		6.80	11.85	18.65
6100	75% solid	6x8x16	2,000	#4 @ 48″		3.49	7.45	10.94
6130				#5 @ 32″		3.64	7.70	11.34
6140				#5 @ 16″		3.93	8.50	12.43
6150			4,500	#4 @ 48″		4.13	7.45	11.58
6180				#5 @ 32″		4.28	7.70	11.98
6190				#5 @ 16″		4.57	8.50	13.07
6200		8x8x16	2,000	#4 @ 48″		3.72	8.05	11.77
6230				#5 @ 32″		3.88	8.30	12.18
6240				#5 @ 16″		4.18	9.30	13.48
6250			4,500	#4 @ 48″		5	8.05	13.05
6280				#5 @ 32″		5.15	8.30	13.45
6290				#5 @ 16″		5.45	9.30	14.75
6300		12x8x16	2,000	#4 @ 48″		6.05	10.20	16.25
6330				#5 @ 32″		6.25	10.50	16.75
6340				#5 @ 16″		6.70	11.50	18.20
6350			4,500	#4 @ 48″		6.75	10.20	16.95
6380				#4 @ 32″		6.95	10.50	17.45
6390				#5 @ 16″		7.40	11.50	18.90
6500	Solid-double Wythe	2-4x8x16	2,000	#4 @ 48″ E.W.		5.75	14.75	20.50
6530				#5 @ 16″ E.W.		6.45	15.65	22.10
6550			4,500	#4 @ 48″ E.W.		8.15	14.65	22.80
6580				#5 @ 16″ E.W.		8.85	15.55	24.40
6600		2-6x8x16	2,000	#4 @ 48″ E.W.		6.50	15.80	22.30
6630				#5 @ 16″ E.W.		7.20	16.70	23.90
6650			4,000	#4 @ 48″ E.W.		9.80	15.70	25.50
6680				#5 @ 16″ E.W.		10.50	16.60	27.10

B2010 112 — Reinforced Concrete Block Wall - Lightweight

	TYPE	SIZE (IN.)	WEIGHT (P.C.F.)	VERT REINF. & GROUT SPACING		COST PER S.F.		
						MAT.	INST.	TOTAL
7100	Hollow	8x4x16	105	#4 @ 48″		3.23	7.65	10.88
7130				#5 @ 32″		3.47	8	11.47
7140				#5 @ 16″		4.02	9.20	13.22
7150			85	#4 @ 48″		4.77	7.50	12.27
7180				#5 @ 32″		5	7.85	12.85
7190				#5 @ 16″		5.55	9.05	14.60
7200		4x8x16	105	#4 @ 48″		2.46	6.85	9.31
7250			85	#4 @ 48″		3.08	6.75	9.83

B2010 Exterior Walls

B2010 112 Reinforced Concrete Block Wall - Lightweight

	TYPE	SIZE (IN.)	WEIGHT (P.C.F.)	VERT REINF. & GROUT SPACING		COST PER S.F.		
						MAT.	INST.	TOTAL
7300	Hollow	6x8x16	105	#4 @ 48"		3.39	7.35	10.74
7330				#5 @ 32"		3.61	7.60	11.21
7340				#5 @ 16"		4.07	8.60	12.67
7350			85	#4 @ 48"		4	7.20	11.20
7380				#5 @ 32"		4.22	7.45	11.67
7390				#5 @ 16"		4.68	8.45	13.13
7400		8x8x16	105	#4 @ 48"		4.10	7.90	12
7430				#5 @ 32"		4.34	8.25	12.59
7440				#5 @ 16"		4.89	9.45	14.34
7450		8x8x16	85	#4 @ 48"		4.27	7.75	12.02
7480				#5 @ 32"		4.51	8.10	12.61
7490				#5 @ 16"		5.05	9.30	14.35
7510		12x8x16	105	#4 @ 48"		5.25	9.95	15.20
7530				#5 @ 32"		5.55	10.35	15.90
7540				#5 @ 16"		6.30	11.55	17.85
7550			85	#4 @ 48"		6.95	9.75	16.70
7580				#5 @ 32"		7.25	10.15	17.40
7590				#5 @ 16"		8	11.35	19.35
7600		4x8x24	105	#4 @ 48"		1.91	7.60	9.51
7650			85	#4 @ 48"		4.21	6.50	10.71
7700		6x8x24	105	#4 @ 48"		2.58	7.95	10.53
7730				#5 @ 32"		2.80	8.20	11
7740				#5 @ 16"		3.26	9.20	12.46
7750			85	#4 @ 48"		5.85	6.80	12.65
7780				#5 @ 32"		6.05	7.05	13.10
7790				#5 @ 16"		6.55	8.05	14.60
7800		8x8x24	105	#4 @ 48"		3.16	8.55	11.71
7840				#5 @ 16"		3.95	10.10	14.05
7850			85	#4 @ 48"		7.15	7.25	14.40
7880				#5 @ 32"		7.35	7.60	14.95
7890				#5 @ 16"		7.90	8.80	16.70
7900		12x8x24	105	#4 @ 48"		4.07	9.60	13.67
7930				#5 @ 32"		4.38	10	14.38
7940				#5 @ 16"		5.15	11.20	16.35
7950			85	#4 @ 48"		9.05	9.40	18.45
7980				#5 @ 32"		9.35	9.80	19.15
7990				#5 @ 16"		10.10	11	21.10
8100	75% solid	6x8x16	105	#4 @ 48"		4.70	7.30	12
8130				#5 @ 32"		4.85	7.55	12.40
8140				#5 @ 16"		5.15	8.35	13.50
8150			85	#4 @ 48"		4.76	7.10	11.86
8180				#5 @ 32"		4.91	7.35	12.26
8190				#5 @ 16"		5.20	8.15	13.35
8200		8x8x16	105	#4 @ 48"		5.65	7.85	13.50
8230				#5 @ 32"		5.85	8.10	13.95
8240				#5 @ 16"		6.15	9.10	15.25
8250			85	#4 @ 48"		5	7.70	12.70
8280				#5 @ 32"		5.15	7.95	13.10
8290				#5 @ 16"		5.45	8.95	14.40

B2010 Exterior Walls

B2010 112 Reinforced Concrete Block Wall - Lightweight

	TYPE	SIZE (IN.)	WEIGHT (P.C.F.)	VERT REINF. & GROUT SPACING		COST PER S.F.		
						MAT.	INST.	TOTAL
8300	75% solid	12x8x16	105	#4 @ 48"		7.20	9.95	17.15
8330				#5 @ 32"		7.40	10.25	17.65
8340				#5 @ 16"		7.85	11.25	19.10
8350			85	#4 @ 48"		8.30	9.70	18
8380				#5 @ 32"		8.50	10	18.50
8390				#5 @ 16"		8.95	11	19.95
8500	Solid-double	2-4x8x16	105	#4 @ 48"		5	14.65	19.65
8530	Wythe		105	#5 @ 16"		5.70	15.55	21.25
8550			85	#4 @ 48"		10.15	14.05	24.20
8580				#5 @ 16"		10.85	14.95	25.80
8600		2-6x8x16	105	#4 @ 48"		8.15	15.70	23.85
8630				#5 @ 16"		8.85	16.60	25.45
8650			85	#4 @ 48"		11.50	15	26.50
8680				#5 @ 16"		12.25	15.90	28.15
8900	For stacked bond add						.40	.40
8910								

B2010 Exterior Walls

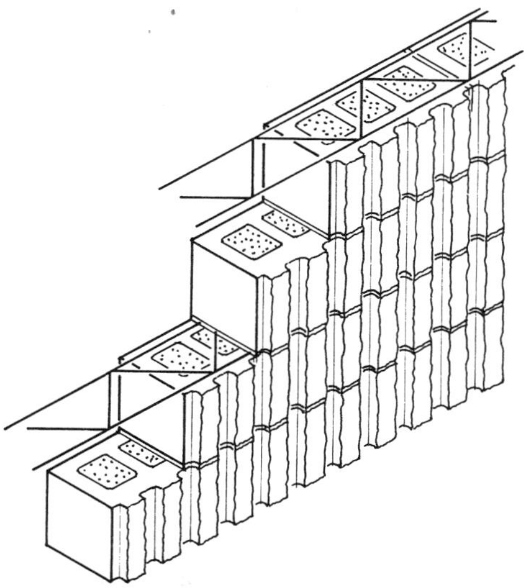

Exterior split ribbed block walls are defined in the following terms; structural reinforcement, weight, percent solid, size, number of ribs and insulation. Within each of these categories two to four variations are shown. No costs are included for brick shelf or relieving angles. Costs include control joints every 20′ and horizontal reinforcing.

System Components	QUANTITY	UNIT	COST PER S.F. MAT.	COST PER S.F. INST.	COST PER S.F. TOTAL
SYSTEM B2010 113 1430					
UNREINFORCED SPLIT RIB BLOCK WALL, 8″X8″X16″, 8 RIBS(HEX), PERLITE FILL					
Split ribbed block wall, 8″ thick	1.000	S.F.	5.85	8.60	14.45
Perlite insulation	1.000	S.F.	1.49	.43	1.92
Horizontal joint reinforcing, alternate courses	.800	L.F.	.18	.18	.36
Control joint	.050	L.F.	.08	.07	.15
TOTAL			7.60	9.28	16.88

B2010 113 Split Ribbed Block Wall - Regular Weight

	TYPE	SIZE (IN.)	RIBS	CORE FILL		COST PER S.F. MAT.	COST PER S.F. INST.	COST PER S.F. TOTAL
1220	Hollow	4x8x16	4	none		4.41	7.80	12.21
1250			8	none		4.78	7.80	12.58
1280			16	none		5.15	7.90	13.05
1330	RB2010 -200	6x8x16	4	none		6.45	8.60	15.05
1340				styrofoam		6.75	8.25	15
1350				none		5.45	8.25	13.70
1360			16	perlite		6.90	8.75	15.65
1370				styrofoam		7.20	8.40	15.60
1380				none		5.90	8.40	14.30
1430		8x8x16	8	perlite		7.60	9.30	16.90
1440				styrofoam		7.45	8.85	16.30
1450				none		6.10	8.85	14.95
1460			16	perlite		8.10	9.45	17.55
1470				styrofoam		7.95	9	16.95
1480				none		6.60	9	15.60
1530		12x8x16	8	perlite		9.60	12.40	22
1540				styrofoam		8.90	11.55	20.45
1550				none		7.20	11.55	18.75
1560			16	perlite		10.10	12.60	22.70
1570				styrofoam		9.40	11.75	21.15
1580				none		7.70	11.75	19.45

B2010 Exterior Walls

B2010 113 Split Ribbed Block Wall - Regular Weight

	TYPE	SIZE (IN.)	RIBS	CORE FILL		COST PER S.F.		
						MAT.	INST.	TOTAL
2120	75% solid	4x8x16	4	none		5.50	7.90	13.40
2150			8	none		6	7.90	13.90
2180			16	none		6.50	8.05	14.55
2230		6x8x16	8	perlite		7.35	8.60	15.95
2250				none		6.85	8.40	15.25
2260			16	perlite		7.90	8.75	16.65
2280				none		7.40	8.55	15.95
2330		8x8x16	8	perlite		8.40	9.20	17.60
2350				none		7.65	9	16.65
2360			16	perlite		9	9.35	18.35
2380				none		8.25	9.15	17.40
2430		12x8x16	8	perlite		10.15	12.15	22.30
2450				none		8.95	11.75	20.70
2460			16	perlite		10.85	12.35	23.20
2480				none		9.65	11.95	21.60
2520	Solid	4x8x16	4	none		6.25	8.05	14.30
2550			8	none		7.40	8.05	15.45
2580			16	none		7.40	8.15	15.55
2620		6x8x16	8	none		8.40	8.55	16.95
2650			16	none		8.40	8.65	17.05
2680		8x8x16	8	none		9.40	9.15	18.55
2720			16	none		9.40	9.30	18.70
2750		12x8x16	8	none		10.90	11.95	22.85
2780			16	none		10.90	12.20	23.10

B2010 114 Split Ribbed Block Wall - Lightweight

	TYPE	SIZE (IN.)	RIBS	CORE FILL		COST PER S.F.		
						MAT.	INST.	TOTAL
3250	Hollow	4x8x16	8	none		5.80	7.60	13.40
3280			16	none		6.25	7.70	13.95
3330		6x8x16	8	perlite		7.60	8.40	16
3340				styrofoam		7.90	8.05	15.95
3350				none		6.60	8.05	14.65
3360			16	perlite		8.15	8.50	16.65
3370				styrofoam		8.45	8.15	16.60
3380				none		7.15	8.15	15.30
3430		8x8x16	8	perlite		8.90	9.05	17.95
3440				styrofoam		8.90	8.60	17.50
3450				none		7.40	8.60	16
3460			16	perlite		9.50	9.15	18.65
3470				styrofoam		9.50	8.70	18.20
3480				none		8	8.70	16.70
3530		12x8x16	8	perlite		11.05	12	23.05
3540				styrofoam		10.35	11.15	21.50
3550				none		8.65	11.15	19.80
3560			16	perlite		11.70	12.20	23.90
3570				styrofoam		11	11.35	22.35
3580				none		9.30	11.35	20.65
4150	75% solid	4x8x16	8	none		7.35	7.70	15.05
4180			16	none		7.95	7.80	15.75

B2010 Exterior Walls

B2010 114 Split Ribbed Block Wall - Lightweight

	TYPE	SIZE (IN.)	RIBS	CORE FILL		COST PER S.F.		
						MAT.	INST.	TOTAL
4230	75% solid	6x8x16	8	perlite		8.85	8.35	17.20
4250				none		8.35	8.15	16.50
4260			16	perlite		9.55	8.45	18
4280				none		9.05	8.25	17.30
4330		8x8x16	8	perlite		10.10	8.90	19
4350				none		9.35	8.70	18.05
4360			16	perlite		10.85	9.05	19.90
4380		8x8x16	16	none		10.10	8.85	18.95
4430		12x8x16	8	perlite		12.05	11.75	23.80
4450				none		10.85	11.35	22.20
4460			16	perlite		12.90	11.95	24.85
4480				none		11.70	11.55	23.25
4550	Solid	4x8x16	8	none		8.35	7.80	16.15
4580			16	none		9.05	7.90	16.95
4650		6x8x16	8	none		9.50	8.25	17.75
4680			16	none		10.30	8.40	18.70
4750		8x8x16	8	none		10.60	8.85	19.45
4780			16	none		11.50	9	20.50
4850		12x8x16	8	none		12.30	11.55	23.85
4880			16	none		13.30	11.75	25.05

B2010 115 Reinforced Split Ribbed Block Wall - Regular Weight

	TYPE	SIZE (IN.)	RIBS	VERT. REINF. & GROUT SPACING		COST PER S.F.		
						MAT.	INST.	TOTAL
5200	Hollow	4x8x16	4	#4 @ 48"		4.57	8.50	13.07
5230			8	#4 @ 48"		4.94	8.50	13.44
5260			16	#4 @ 48"		5.30	8.60	13.90
5330		6x8x16	8	#4 @ 48"		5.70	9	14.70
5340				#5 @ 32"		5.90	9.25	15.15
5350				#5 @ 16"		6.40	10.25	16.65
5360			16	#4 @ 48"		6.15	9.15	15.30
5370				#5 @ 32"		6.35	9.40	15.75
5380				#5 @ 16"		6.85	10.40	17.25
5430		8x8x16	8	#4 @ 48"		6.45	9.70	16.15
5440				#5 @ 32"		6.65	10.05	16.70
5450				#5 @ 16"		7.20	11.25	18.45
5460			16	#4 @ 48"		6.95	9.85	16.80
5470				#5 @ 32"		7.15	10.20	17.35
5480				#5 @ 16"		7.70	11.40	19.10
5530		12x8x16	8	#4 @ 48"		7.65	12.40	20.05
5540				#5 @ 32"		7.95	12.80	20.75
5550				#5 @ 16"		8.70	14	22.70
5560			16	#4 @ 48"		8.15	12.60	20.75
5570				#5 @ 32"		8.45	13	21.45
5580				#5 @ 16"		9.20	14.20	23.40
6230	75% solid	6x8x16	8	#4 @ 48"		7	9	16
6240				#5 @ 32"		7.15	9.25	16.40
6250				#5 @ 16"		7.40	10.05	17.45
6260			16	#4 @ 48"		7.55	9.15	16.70
6270				#5 @ 32"		7.70	9.40	17.10
6280				#5 @ 16"		7.95	10.20	18.15

B2010 Exterior Walls

B2010 115 Reinforced Split Ribbed Block Wall - Regular Weight

	TYPE	SIZE (IN.)	RIBS	VERT. REINF. & GROUT SPACING		COST PER S.F.		
						MAT.	INST.	TOTAL
6330	75% solid	8x8x16	8	#4 @ 48"		7.80	9.70	17.50
6340				#5 @ 32"		8	9.95	17.95
6350				#5 @ 16"		8.30	10.95	19.25
6360			16	#4 @ 48"		8.40	9.85	18.25
6370				#5 @ 32"		8.60	10.10	18.70
6380				#5 @ 16"		8.90	11.10	20
6430		12x8x16	8	#4 @ 48"		9.20	12.45	21.65
6440				#5 @ 32"		9.40	12.75	22.15
6450				#5 @ 16"		9.85	13.75	23.60
6460			16	#4 @ 48"		9.90	12.65	22.55
6470				#5 @ 32"		10.10	12.95	23.05
6480				#5 @ 16"		10.55	13.95	24.50
6500	Solid-double Wythe	2-4x8x16	4	#4 @ 48"		7.25	9.70	16.95
6520				#5 @ 16"		7.60	10.15	17.75
6530			8	#4 @ 48"		8.40	9.70	18.10
6550				#5 @ 16"		8.75	10.15	18.90
6560			16	#4 @ 48"		8.40	9.80	18.20
6580				#5 @ 16"		8.75	10.25	19
6630		2-6x8x16	8	#4 @ 48"		9.40	10.25	19.65
6650				#5 @ 16"		9.75	10.70	20.45
6660			16	#4 @ 48"		9.40	10.35	19.75
6680				#5 @ 16"		9.75	10.80	20.55

B2010 116 Reinforced Split Ribbed Block Wall - Lightweight

	TYPE	SIZE (IN.)	RIBS	VERT. REINF. & GROUT SPACING		COST PER S.F.		
						MAT.	INST.	TOTAL
7230	Hollow	4x8x16	8	#4 @ 48"		5.95	8.30	14.25
7260			16	#4 @ 48"		6.40	8.40	14.80
7330		6x8x16	8	#4 @ 48"		6.85	8.80	15.65
7340				#5 @ 32"		7.05	9.05	16.10
7350				#5 @ 16"		7.55	10.05	17.60
7360			16	#4 @ 48"		7.40	8.90	16.30
7370				#5 @ 32"		7.60	9.15	16.75
7380				#5 @ 16"		8.10	10.15	18.25
7430		8x8x16	8	#4 @ 48"		7.75	9.45	17.20
7440				#5 @ 32"		7.95	9.80	17.75
7450				#5 @ 16"		8.50	11	19.50
7460			16	#4 @ 48"		8.35	9.55	17.90
7470				#5 @ 32"		8.55	9.90	18.45
7480				#5 @ 16"		9.10	11.10	20.20
7530		12x8x16	8	#4 @ 48"		9.10	12	21.10
7540				#5 @ 32"		9.40	12.40	21.80
7550				#5 @ 16"		10.15	13.60	23.75
7560			16	#4 @ 48"		9.75	12.20	21.95
7570				#5 @ 32"		10.05	12.60	22.65
7580				#5 @ 16"		10.80	13.80	24.60
8230	75% solid	6x8x16	8	#4 @ 48"		8.50	8.75	17.25
8240				#5 @ 32"		8.65	9	17.65
8250				#5 @ 16"		8.90	9.80	18.70
8260			16	#4 @ 48"		9.20	8.85	18.05
8270				#5 @ 32"		9.35	9.10	18.45
8280				#5 @ 16"		9.60	9.90	19.50

B2010 Exterior Walls

B2010 116 Reinforced Split Ribbed Block Wall - Lightweight

	TYPE	SIZE (IN.)	RIBS	VERT. REINF. & GROUT SPACING	COST PER S.F. MAT.	COST PER S.F. INST.	COST PER S.F. TOTAL
8330	75% solid	8x8x16	8	#4 @ 48"	9.50	9.40	18.90
8340				#5 @ 32"	9.70	9.65	19.35
8350				#5 @ 16"	10	10.65	20.65
8360			16	#4 @ 48"	10.25	9.55	19.80
8370				#5 @ 32"	10.45	9.80	20.25
8380				#5 @ 16"	10.75	10.80	21.55
8430		12x8x16	8	#4 @ 48"	11.10	12.05	23.15
8440				#5 @ 32"	11.30	12.35	23.65
8450				#5 @ 16"	11.75	13.35	25.10
8460			16	#4 @ 48"	11.95	12.25	24.20
8470				#5 @ 32"	12.15	12.55	24.70
8480				#5 @ 16"	12.60	13.55	26.15
8530	Solid double	2-4x8x16	8	#4 @ 48" E.W	9.45	9.75	19.20
8550	Wythe			#5 @ 16" E.W.	10.15	10.65	20.80
8560			16	#4 @ 48" E.W.	10.15	9.85	20
8580				#5 @ 16" E.W.	10.85	10.75	21.60
8630		2-6x8x16	8	#4 @ 48" E.W.	10.55	10.25	20.80
8650				#5 @ 16" E.W.	11.30	11.15	22.45
8660			16	#4 @ 48" E.W.	11.35	10.40	21.75
8680				#5 @ 16" E.W.	12.10	11.30	23.40
8900	For stacked bond, add					.40	.40

B2010 Exterior Walls

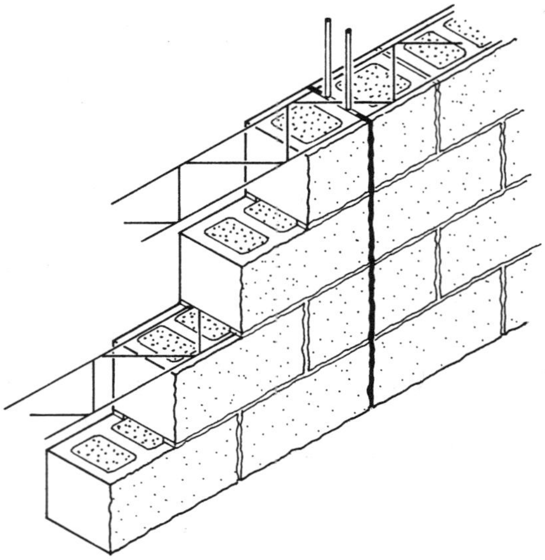

Exterior split face block walls are defined in the following terms; structural reinforcement, weight, percent solid, size, scores and insulation. Within each of these categories two to four variations are shown. No costs are included for brick shelf or relieving angles. Costs include control joints every 20′ and horizontal reinforcing.

System Components	QUANTITY	UNIT	COST PER S.F. MAT.	COST PER S.F. INST.	COST PER S.F. TOTAL
SYSTEM B2010 117 1600					
UNREINFORCED SPLIT FACE BLOCK WALL, 8″X8″X16″, 0 SCORES, PERLITE FILL					
Split face block wall, 8″ thick	1.000	S.F.	4.88	8.90	13.78
Perlite insulation	1.000	S.F.	1.49	.43	1.92
Horizontal joint reinforcing, alternate course	.800	L.F.	.18	.18	.36
Control joint	.050	L.F.	.08	.07	.15
TOTAL			6.63	9.58	16.21

B2010 117	Split Face Block Wall - Regular Weight						
	TYPE	SIZE (IN.)	SCORES	CORE FILL	COST PER S.F. MAT.	COST PER S.F. INST.	COST PER S.F. TOTAL
1200	Hollow	8x4x16	0	perlite	7.35	10.25	17.60
1210				styrofoam	7.35	9.80	17.15
1240				none	5.85	9.80	15.65
1250	RB2010 -200		1	perlite	7.75	10.25	18
1260				styrofoam	7.75	9.80	17.55
1290				none	6.25	9.80	16.05
1300		12x4x16	0	perlite	9.65	11.65	21.30
1310				styrofoam	8.95	10.80	19.75
1340				none	7.25	10.80	18.05
1350			1	perlite	10.10	11.65	21.75
1360				styrofoam	9.40	10.80	20.20
1390				none	7.70	10.80	18.50
1400		4x8x16	0	none	4.03	7.70	11.73
1450			1	none	4.36	7.80	12.16
1500		6x8x16	0	perlite	5.60	8.90	14.50
1510				styrofoam	5.90	8.55	14.45
1540				none	4.57	8.55	13.12
1550			1	perlite	5.95	9	14.95
1560				styrofoam	6.25	8.65	14.90
1590				none	4.94	8.65	13.59

B2010 Exterior Walls

B2010 117 Split Face Block Wall - Regular Weight

	TYPE	SIZE (IN.)	SCORES	CORE FILL		COST PER S.F.		
						MAT.	INST.	TOTAL
1600	Hollow	8x8x16	0	perlite		6.65	9.60	16.25
1610				styrofoam		6.65	9.15	15.80
1640				none		5.15	9.15	14.30
1650		8x8x16	1	perlite		7.05	9.75	16.80
1660				styrofoam		7.05	9.30	16.35
1690				none		5.55	9.30	14.85
1700		12x8x16	0	perlite		8.60	12.60	21.20
1710				styrofoam		7.90	11.75	19.65
1740				none		6.20	11.75	17.95
1750			1	perlite		9.10	12.80	21.90
1760				styrofoam		8.40	11.95	20.35
1790				none		6.70	11.95	18.65
1800	75% solid	8x4x16	0	perlite		7.85	10.15	18
1840				none		7.10	9.95	17.05
1850			1	perlite		8.40	10.15	18.55
1890				none		7.65	9.95	17.60
1900		12x4x16	0	perlite		9.90	11.40	21.30
1940				none		8.70	11	19.70
1950			1	perlite		10.55	11.40	21.95
1990				none		9.35	11	20.35
2000		4x8x16	0	none		5.05	7.80	12.85
2050			1	none		5.45	7.90	13.35
2100		6x8x16	0	perlite		6.20	10.35	16.55
2140				none		5.70	10.15	15.85
2150			1	perlite		6.70	9	15.70
2190				none		6.20	8.80	15
2200		8x8x16	0	perlite		7.15	9.50	16.65
2240				none		6.40	9.30	15.70
2250			1	perlite		7.65	9.65	17.30
2290				none		6.90	9.45	16.35
2300		12x8x16	0	perlite		8.85	12.35	21.20
2340				none		7.65	11.95	19.60
2350			1	perlite		9.50	12.60	22.10
2390				none		8.30	12.20	20.50
2400	Solid	8x4x16	0	none		7.95	10.15	18.10
2450			1	none		8.55	10.15	18.70
2500		12x4x16	0	none		9.65	11.25	20.90
2550			1	none		10.40	13.15	23.55
2600		4x8x16	0	none		5.70	7.90	13.60
2650			1	none		6.20	8.05	14.25
2700		6x8x16	0	none		6.45	8.80	15.25
2750			1	none		7	8.95	15.95
2800		8x8x16	0	none		7.20	9.45	16.65
2850			1	none		7.85	9.60	17.45
2900		12x8x16	0	none		8.65	12.20	20.85
2950			1	none		9.35	12.40	21.75

B2010 118 Split Face Block Wall - Lightweight

	TYPE	SIZE (IN.)	SCORES	CORE FILL		COST PER S.F.		
						MAT.	INST.	TOTAL
3200	Hollow	8x4x16	0	perlite		8.40	11.55	19.95
3210				styrofoam		8.40	11.10	19.50
3240				none		6.90	11.10	18

B2010 Exterior Walls

B2010 118 Split Face Block Wall - Lightweight

	TYPE	SIZE (IN.)	SCORES	CORE FILL		COST PER S.F.		
						MAT.	INST.	TOTAL
3250	Hollow	8x4x16	1	perlite		8.90	9.90	18.80
3260				styrofoam		8.90	9.45	18.35
3290				none		7.40	9.45	16.85
3300		12x4x16	0	perlite		10.85	11.25	22.10
3310				styrofoam		10.15	10.40	20.55
3340				none		8.45	10.40	18.85
3350		12x4x16	1	perlite		11.45	13.05	24.50
3360				styrofoam		10.75	12.20	22.95
3390				none		9.05	12.20	21.25
3400		4x8x16	0	none		4.86	7.50	12.36
3450			1	none		5.25	7.60	12.85
3500		6x8x16	0	perlite		6.50	8.60	15.10
3510				styrofoam		6.80	8.25	15.05
3540				none		5.50	8.25	13.75
3550			1	perlite		7	8.75	15.75
3560				styrofoam		7.30	8.40	15.70
3590				none		6	8.40	14.40
3600		8x8x16	0	perlite		7.65	9.30	16.95
3610				styrofoam		7.65	8.85	16.50
3640				none		6.15	8.85	15
3650			1	perlite		8.20	9.45	17.65
3660				styrofoam		8.20	9	17.20
3690				none		6.70	9	15.70
3700		12x8x16	0	perlite		9.80	12.20	22
3710				styrofoam		9.10	11.35	20.45
3740				none		7.40	11.35	18.75
3750			1	perlite		10.45	12.40	22.85
3760				styrofoam		9.75	11.55	21.30
3790				none		8.05	11.55	19.60
3800	75% solid	8x4x16	0	perlite		8.40	10.15	18.55
3840				none		7.65	9.95	17.60
3850			1	perlite		9.90	9.80	19.70
3890				none		9.15	9.60	18.75
3900		12x4x16	0	perlite		10.55	11.40	21.95
3940				none		9.35	11	20.35
3950			1	perlite		12.25	12.80	25.05
3990				none		11.05	12.40	23.45
4000		4x8x16	0	none		6.10	7.60	13.70
4050			1	none		6.65	7.70	14.35
4100		6x8x16	0	perlite		7.45	8.60	16.05
4140				none		6.95	8.40	15.35
4150			1	perlite		8.05	8.75	16.80
4190				none		7.55	8.55	16.10
4200		8x8x16	0	perlite		8.50	9.20	17.70
4240				none		7.75	9	16.75
4250			1	perlite		9.15	9.35	18.50
4290				none		8.40	9.15	17.55
4300		12x8x16	0	perlite		10.45	11.95	22.40
4340				none		9.25	11.55	20.80
4350			1	perlite		11.25	12.15	23.40
4390				none		10.05	11.75	21.80

B2010 Exterior Walls

B2010 118 Split Face Block Wall - Lightweight

	TYPE	SIZE (IN.)	SCORES	CORE FILL		COST PER S.F. MAT.	INST.	TOTAL
4400	Solid	8x4x16	0	none		9.50	9.80	19.30
4450			1	none		10.25	9.80	20.05
4500		12x4x16	0	none		11.50	10.80	22.30
4550			1	none		12.40	12.65	25.05
4600		4x8x16	0	none		6.95	7.70	14.65
4650			1	none		7.60	7.80	15.40
4700		6x8x16	0	none		7.85	8.55	16.40
4750			1	none		8.55	8.65	17.20
4800		8x8x16	0	none		8.75	9.15	17.90
4850			1	none		9.55	9.30	18.85
4900		12x8x16	0	none		10.45	11.75	22.20
4950			1	none		11.40	11.95	23.35

B2010 119 Reinforced Split Face Block Wall - Regular Weight

	TYPE	SIZE (IN.)	SCORES	VERT. REINF. & GROUT SPACING		COST PER S.F. MAT.	INST.	TOTAL
5200	Hollow	8x4x16	0	#4 @ 48″		6.20	10.65	16.85
5210				#5 @ 32″		6.40	11	17.40
5240				#5 @ 16″		6.95	12.20	19.15
5250			1	#4 @ 48″		6.60	10.65	17.25
5260				#5 @ 32″		6.80	11	17.80
5290				#5 @ 16″		7.35	12.20	19.55
5300		12x4x16	0	#4 @ 48″		7.70	11.65	19.35
5310				#5 @ 32″		8	12.05	20.05
5340				#5 @ 16″		8.75	13.25	22
5350			1	#4 @ 48″		8.15	11.65	19.80
5360				#5 @ 32″		8.45	12.05	20.50
5390				#5 @ 16″		9.20	13.25	22.45
5400		4x8x16	0	#4 @ 48″		4.19	8.40	12.59
5450			1	#4 @ 48″		4.52	8.50	13.02
5500		6x8x16	0	#4 @ 48″		4.83	9.30	14.13
5510				#5 @ 32″		5.05	9.55	14.60
5540				#5 @ 16″		5.50	10.55	16.05
5550			1	#4 @ 48″		5.20	9.40	14.60
5560				#5 @ 32″		5.40	9.65	15.05
5590				#5 @ 16″		5.90	10.65	16.55
5600		8x8x16	0	#4 @ 48″		5.45	10	15.45
5610				#5 @ 32″		5.70	10.35	16.05
5640				#5 @ 16″		5.90	10.95	16.85
5650			1	#4 @ 48″		9.55	13.75	23.30
5660				#5 @ 32″		6.10	10.50	16.60
5690				#5 @ 16″		6.65	11.70	18.35
5700		12x8x16	0	#4 @ 48″		6.65	12.60	19.25
5710				#5 @ 32″		6.95	13	19.95
5740				#5 @ 16″		7.70	14.20	21.90
5750			1	#4 @ 48″		7.15	12.80	19.95
5760				#5 @ 32″		7.45	13.20	20.65
5790				#5 @ 16″		8.20	14.40	22.60

B2010 Exterior Walls

B2010 119 Reinforced Split Face Block Wall - Regular Weight

	TYPE	SIZE (IN.)	SCORES	VERT. REINF. & GROUT SPACING		COST PER S.F. MAT.	COST PER S.F. INST.	COST PER S.F. TOTAL
6000	75% solid	8x4x16	0	#4 @ 48″		7.25	10.65	17.90
6010				#5 @ 32″		7.45	10.90	18.35
6040				#5 @ 16″		7.75	11.90	19.65
6050			1	#4 @ 48″		7.80	10.65	18.45
6060				#5 @ 32″		8	10.90	18.90
6090				#5 @ 16″		8.30	11.90	20.20
6100		12x4x16	0	#4 @ 48″		8.95	11.70	20.65
6110				#5 @ 32″		9.15	12	21.15
6140				#5 @ 16″		9.55	12.80	22.35
6150			1	#4 @ 48″		9.60	11.70	21.30
6160				#5 @ 32″		9.80	12	21.80
6190				#5 @ 16″		10.25	13	23.25
6200		6x8x16	0	#4 @ 48″		5.85	10.75	16.60
6210				#5 @ 32″		6	11	17
6240				#5 @ 16″		6.25	11.80	18.05
6250			1	#4 @ 48″		6.35	9.40	15.75
6260				#5 @ 32″		6.50	9.65	16.15
6290				#5 @ 16″		6.75	10.45	17.20
6300		8x8x16	0	#4 @ 48″		6.55	10	16.55
6310				#5 @ 32″		6.75	10.25	17
6340				#5 @ 16″		7.05	11.25	18.30
6350			1	#4 @ 48″		7.05	10.15	17.20
6360				#5 @ 32″		7.25	10.40	17.65
6390				#5 @ 16″		7.55	11.40	18.95
6400		12x8x16	0	#4 @ 48″		7.90	12.65	20.55
6410				#5 @ 32″		8.10	12.95	21.05
6440				#5 @ 16″		8.55	13.95	22.50
6450			1	#4 @ 48″		8.55	12.90	21.45
6460				#5 @ 32″		8.75	13.20	21.95
6490				#5 @ 16″		9.20	14.20	23.40
6700	Solid-double	2-4x8x16	0	#4 @ 48″ E.W.		12.25	17.55	29.80
6710	Wythe			#5 @ 32″ E.W.		12.50	17.70	30.20
6740				#5 @ 16″ E.W.		12.95	18.45	31.40
6750			1	#4 @ 48″ E.W.		13.25	17.85	31.10
6760				#5 @ 32″ E.W.		13.50	18	31.50
6790				#5 @ 16″ E.W.		13.95	18.75	32.70
6800		2-6x8x16	0	#4 @ 48″ E.W.		13.70	19.40	33.10
6810				#5 @ 32″ E.W.		14	19.55	33.55
6840				#5 @ 16″ E.W.		14.45	20	34.45
6850			1	#4 @ 48″ E.W.		14.80	19.70	34.50
6860				#5 @ 32″ E.W.		15.10	19.85	34.95
6890				#5 @ 16″ E.W.		15.55	20.50	36.05

B2010 120 Reinforced Split Face Block Wall - Lightweight

	TYPE	SIZE (IN.)	SCORES	VERT. REINF. & GROUT SPACING		COST PER S.F. MAT.	COST PER S.F. INST.	COST PER S.F. TOTAL
7200	Hollow	8x4x16	0	#4 @ 48″		7.25	11.95	19.20
7210				#5 @ 32″		7.45	12.30	19.75
7240				#5 @ 16″		8	13.50	21.50
7250			1	#4 @ 48″		7.75	10.30	18.05
7260				#5 @ 32″		7.95	10.65	18.60
7290				#5 @ 16″		8.50	11.85	20.35

B2010 Exterior Walls

B2010 120 Reinforced Split Face Block Wall - Lightweight

	TYPE	SIZE (IN.)	SCORES	VERT. REINF. & GROUT SPACING		COST PER S.F.		
						MAT.	INST.	TOTAL
7300	Hollow	12x4x16	0	#4 @ 48"		8.90	11.25	20.15
7340				#5 @ 16"		9.95	12.85	22.80
7350			1	#4 @ 48"		9.50	13.05	22.55
7360				#5 @ 32"		9.80	13.45	23.25
7390				#5 @ 16"		10.55	14.65	25.20
7400		4x8x16	0	#4 @ 48"		5	8.20	13.20
7450			1	#4 @ 48"		5.40	8.30	13.70
7500		6x8x16	0	#4 @ 48"		5.75	9	14.75
7510				#5 @ 32"		5.95	9.25	15.20
7540				#5 @ 16"		6.45	10.25	16.70
7550			1	#4 @ 48"		6.25	9.15	15.40
7560				#5 @ 32"		6.45	9.40	15.85
7590				#5 @ 16"		6.95	10.40	17.35
7600		8x8x16	0	#4 @ 48"		6.50	9.70	16.20
7610				#5 @ 32"		6.70	10.05	16.75
7640				#5 @ 16"		7.25	11.25	18.50
7650			1	#4 @ 48"		7.05	9.85	16.90
7660				#5 @ 32"		7.25	10.20	17.45
7690				#5 @ 16"		7.80	11.40	19.20
7700		12x8x16	0	#4 @ 48"		7.85	12.20	20.05
7710				#5 @ 32"		8.15	12.60	20.75
7740		12x8x16	0	#5 @ 16"		8.90	13.80	22.70
7750			1	#4 @ 48"		8.50	12.40	20.90
7760				#5 @ 16"		8.80	12.80	21.60
7790				#5 @ 16"		9.55	14	23.55
8000	75% solid	8x4x16	0	#4 @ 48"		7.80	10.65	18.45
8010				#5 @ 32"		8	10.90	18.90
8040				#5 @ 16"		8.30	11.90	20.20
8050			1	#4 @ 48"		9.30	10.30	19.60
8060				#5 @ 32"		9.50	10.55	20.05
8090				#5 @ 16"		9.80	11.55	21.35
8100		12x4x16	0	#4 @ 48"		9.60	11.70	21.30
8110				#5 @ 32"		9.80	12	21.80
8140				#5 @ 16"		10.25	13	23.25
8150			1	#4 @ 48"		11.30	13.10	24.40
8160				#5 @ 32"		11.50	13.40	24.90
8190				#5 @ 16"		11.95	14.40	26.35
8200		6x8x16	0	#4 @ 48"		7.10	9	16.10
8210				#5 @ 32"		7.25	9.25	16.50
8240				#5 @ 16"		7.50	10.05	17.55
8250			1	#4 @ 48"		7.70	9.15	16.85
8260				#5 @ 32"		7.85	9.40	17.25
8290				#5 @ 16"		8.10	10.20	18.30
8300		8x8x16	0	#4 @ 48"		7.90	9.70	17.60
8310				#5 @ 32"		8.10	9.95	18.05
8340				#5 @ 16"		8.40	10.95	19.35
8350			1	#4 @ 48"		8.55	9.85	18.40
8360				#5 @ 32"		8.75	10.10	18.85
8390				#5 @ 16"		9.05	11.10	20.15

B2010 Exterior Walls

B2010 120 Reinforced Split Face Block Wall - Lightweight

	TYPE	SIZE (IN.)	SCORES	VERT. REINF. & GROUT SPACING		COST PER S.F.		
						MAT.	INST.	TOTAL
8400	75% solid	12x8x16	0	#4 @ 48"		9.50	12.25	21.75
8410				#5 @ 32"		9.70	12.55	22.25
8440				#5 @ 16"		10.15	13.55	23.70
8450			1	#4 @ 48"		10.30	12.45	22.75
8460				#5 @ 32"		10.50	12.75	23.25
8490				#5 @ 16"		10.95	13.75	24.70
8700	Solid-double Wythe	2-4x8x16	0	#4 @ 48" E.W.		14.75	17.15	31.90
8710				#5 @ 32" E.W.		15	17.30	32.30
8740				#5 @ 16" E.W.		15.45	18.05	33.50
8750			1	#4 @ 48" E.W.		16.05	17.35	33.40
8760				#5 @ 32" E.W.		16.30	17.50	33.80
8790				#5 @ 16" E.W.		16.75	18.25	35
8800		2-6x8x16	0	#4 @ 48" E.W.		16.50	18.90	35.40
8810				#5 @ 32" E.W.		16.80	19.05	35.85
8840				#5 @ 16" E.W.		17.25	19.80	37.05
8850			1	#4 @ 48" E.W.		17.90	19.10	37
8860				#5 @ 32" E.W.		18.20	19.25	37.45
8890				#5 @ 16" E.W.		18.65	20	38.65

B2010 Exterior Walls

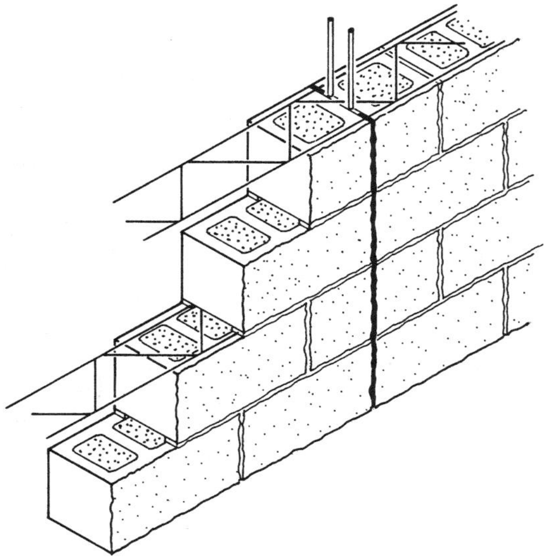

Exterior ground face block walls are defined in the following terms; structural reinforcement weight, percent solid, size, scores and insulation. Within each of these categories two to four variations are shown. No costs are included for brick shelf or relieving angles. Costs include control joints every 20′ and horizontal reinforcing.

System Components	QUANTITY	UNIT	COST PER S.F. MAT.	COST PER S.F. INST.	COST PER S.F. TOTAL
SYSTEM B2010 121 1600					
UNREINF. GROUND FACE BLOCK WALL, 8″X8″X16″, 0 SCORES, PERLITE FILL					
Ground face block wall, 8″ thick	1.000	S.F.	10.55	9.05	19.60
Perlite insulation	1.000	S.F.	1.49	.43	1.92
Horizontal joint reinforcing, alternate course	.800	L.F.	.18	.18	.36
Control joint	.050	L.F.	.08	.07	.15
TOTAL			12.30	9.73	22.03

B2010 121 Ground Face Block Wall

	TYPE	SIZE (IN.)	SCORES	CORE FILL		COST PER S.F. MAT.	COST PER S.F. INST.	COST PER S.F. TOTAL
1200	Hollow	4x8x16	0	none		8.60	7.80	16.40
1250			1	none		9.40	7.90	17.30
1300			2 to 5	none		10.20	8.05	18.25
1400	RB2010 -200	6x8x16	0	perlite		10.60	9	19.60
1410				styrofoam		10.90	8.65	19.55
1440				none		9.60	8.65	18.25
1450			1	perlite		11.50	9.15	20.65
1460				styrofoam		11.80	8.80	20.60
1490				none		10.50	8.80	19.30
1500			2 to 5	perlite		12.35	9.30	21.65
1510				styrofoam		12.65	8.95	21.60
1540				none		11.35	8.95	20.30
1600		8x8x16	0	perlite		12.30	9.75	22.05
1610				styrofoam		12.30	9.30	21.60
1640				none		10.80	9.30	20.10
1650			1	perlite		13.30	9.90	23.20
1660				styrofoam		13.30	9.45	22.75
1690				none		11.80	9.45	21.25
1700			2 to 5	perlite		14.30	10.05	24.35
1710				styrofoam		14.30	9.60	23.90
1740				none		12.80	9.60	22.40

B2010 Exterior Walls

B2010 121 Ground Face Block Wall

	TYPE	SIZE (IN.)	SCORES	CORE FILL		COST PER S.F.		
						MAT.	INST.	TOTAL
1800	Hollow	12x8x16	0	perlite		16.90	12.80	29.70
1810				styrofoam		16.20	11.95	28.15
1840				none		14.50	11.95	26.45
1850			1	perlite		18.25	13.05	31.30
1860				styrofoam		17.55	12.20	29.75
1890				none		15.85	12.20	28.05
1900		12x8x16	2 to 5	perlite		19.55	13.25	32.80
1910				styrofoam		18.85	12.40	31.25
1940				none		17.15	12.40	29.55
2200	75% solid	4x8x16	0	none		10.95	7.90	18.85
2250			1	none		12	8.05	20.05
2300			2 to 5	none		13.05	8.15	21.20
2400		6x8x16	0	perlite		12.75	9	21.75
2440				none		12.25	8.80	21.05
2450			1	perlite		13.90	9.15	23.05
2490				none		13.40	8.95	22.35
2500			2 to 5	perlite		15.05	9.30	24.35
2540				none		14.55	9.10	23.65
2600		8x8x16	0	perlite		14.55	9.65	24.20
2640				none		13.80	9.45	23.25
2650			1	perlite		15.80	9.80	25.60
2690				none		15.05	9.60	24.65
2700			2 to 5	perlite		17.10	10	27.10
2740				none		16.35	9.80	26.15
2800		12x8x16	0	perlite		19.65	12.60	32.25
2840				none		18.45	12.20	30.65
2850			1	perlite		21.50	12.80	34.30
2890				none		20	12.40	32.40
2900			2 to 5	perlite		23	13.05	36.05
2940				none		22	12.65	34.65
3200	Solid	4x8x16	0	none		12.55	8.05	20.60
3250			1	none		13.75	8.15	21.90
3300			2 to 5	none		14.95	8.25	23.20
3400		6x8x16	0	none		14	8.95	22.95
3450			1	none		15.35	9.10	24.45
3500			2 to 5	none		16.65	9.25	25.90
3600		8x8x16	0	none		15.75	9.60	25.35
3650			1	none		17.25	9.80	27.05
3700			2 to 5	none		18.70	9.95	28.65
3800		12x8x16	0	none		21.50	12.40	33.90
3850			1	none		23.50	12.65	36.15
3900			2 to 5	none		25.50	12.90	38.40

B2010 122 Reinforced Ground Face Block Wall

	TYPE	SIZE (IN.)	SCORES	VERT. REINF. & GROUT SPACING		COST PER S.F.		
						MAT.	INST.	TOTAL
5200	Hollow	4x8x16	0	#4 @ 48"		8.75	8.50	17.25
5250			1	#4 @ 48"		9.55	8.60	18.15
5300			2 to 5	#4 @ 48"		10.35	8.75	19.10
5400		6x8x16	0	#4 @ 48"		9.85	9.40	19.25
5420				#5 @ 32"		10.05	9.65	19.70
5430				#5 @ 16"		10.55	10.65	21.20

B2010 Exterior Walls

B2010 122 Reinforced Ground Face Block Wall

	TYPE	SIZE (IN.)	SCORES	VERT. REINF. & GROUT SPACING		COST PER S.F. MAT.	INST.	TOTAL
5450	Hollow	6x8x16	1	#4 @ 48"		10.75	9.55	20.30
5470				#5 @ 32"		10.95	9.80	20.75
5480				#5 @ 16"		11.45	10.80	22.25
5500			2 to 5	#4 @ 48"		11.60	9.70	21.30
5520				#5 @ 32"		11.80	9.95	21.75
5530				#5 @ 16"		12.30	10.95	23.25
5600		8x8x16	0	#4 @ 48"		11.15	10.15	21.30
5620				#5 @ 32"		11.35	10.50	21.85
5630				#5 @ 16"		11.90	11.70	23.60
5650			1	#4 @ 48"		12.15	10.30	22.45
5670				#5 @ 32"		12.35	10.65	23
5680				#5 @ 16"		12.90	11.85	24.75
5700			2 to 5	#4 @ 48"		13.15	10.45	23.60
5720				#5 @ 32"		13.35	10.80	24.15
5730				#5 @ 16"		13.90	12	25.90
5800		12x8x16	0	#4 @ 48"		14.95	12.80	27.75
5820				#5 @ 32"		15.25	13.20	28.45
5830				#5 @ 16"		16	14.40	30.40
5850			1	#4 @ 48"		16.30	13.05	29.35
5870				#5 @ 32"		16.60	13.45	30.05
5880				#5 @ 16"		17.35	14.65	32
5900			2 to 5	#4 @ 48"		17.60	13.25	30.85
5920				#5 @ 32"		17.90	13.65	31.55
5930				#5 @ 16"		18.65	14.85	33.50
6200	75% solid	6x8x16	0	#4 @ 48"		12.40	9.30	21.70
6220				#5 @ 32"		12.60	9.50	22.10
6230				#5 @ 16"		12.95	10.20	23.15
6250			1	#4 @ 48"		13.55	9.45	23
6270				#5 @ 32"		13.75	9.65	23.40
6280				#5 @ 16"		14.10	10.35	24.45
6300			2 to 5	#4 @ 48"		14.70	9.60	24.30
6330				#5 @ 16"		15.25	10.50	25.75
6400		8x8x16	0	#4 @ 48"		14	10	24
6420				#5 @ 32"		14.20	10.25	24.45
6430				#5 @ 16"		14.60	11	25.60
6450			1	#4 @ 48"		15.25	10.15	25.40
6470				#5 @ 32"		15.45	10.40	25.85
6480				#5 @ 16"		15.85	11.15	27
6500			2 to 5	#4 @ 48"		16.55	10.35	26.90
6520				#5 @ 32"		16.75	10.60	27.35
6530				#5 @ 16"		17.15	11.35	28.50
6600		12x8x16	0	#4 @ 48"		18.70	12.75	31.45
6620				#5 @ 32"		18.95	13	31.95
6630				#5 @ 16"		19.40	13.80	33.20
6650			1	#4 @ 48"		20.50	12.95	33.45
6670				#5 @ 32"		20.50	13.20	33.70
6680				#5 @ 16"		21	14	35
6700			2 to 5	#4 @ 48"		22	13.20	35.20
6720				#5 @ 32"		22.50	13.45	35.95
6730				#5 @ 16"		23	14.25	37.25

B2010 Exterior Walls

B2010 122 Reinforced Ground Face Block Wall

	TYPE	SIZE (IN.)	SCORES	VERT. REINF. & GROUT SPACING		COST PER S.F.		
						MAT.	INST.	TOTAL
6800	Solid-double Wythe	2-4x8x16	0	#4 @ 48" E.W.		26	17.70	43.70
6830				#5 @ 16" E.W.		26.50	18.45	44.95
6850			1	#4 @ 48" E.W.		28	17.90	45.90
6880				#5 @ 16" E.W.		29	18.65	47.65
6900			2 to 5	#4 @ 48" E.W.		30.50	18.10	48.60
6930				#5 @ 16" E.W.		31	18.85	49.85
7000		2-6x8x16	0	#4 @ 48" E.W.		28.50	19.50	48
7030				#5 @ 16" E.W.		29.50	20	49.50
7050			1	#4 @ 48" E.W.		31.50	19.80	51.30
7080				#5 @ 16" E.W.		32	20.50	52.50
7100			2 to 5	#4 @ 48" E.W.		34	20	54
7130				#5 @ 16" E.W.		34.50	20.50	55

B2010 Exterior Walls

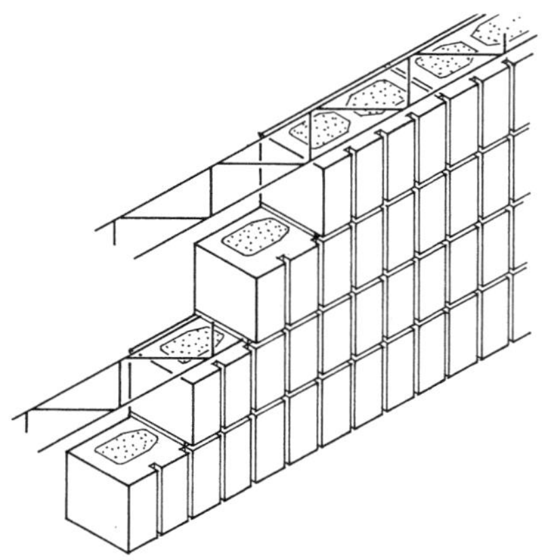

Exterior miscellaneous block walls are defined in the following terms: structural reinforcement finish texture, percent solid, size, weight and insulation. Within each of these categories two to four variations are shown. No costs are included for brick shelf or relieving angles. Costs include control joints every 20' and horizontal reinforcing.

System Components	QUANTITY	UNIT	COST PER S.F. MAT.	INST.	TOTAL
SYSTEM B2010 123 2700					
UNREINFORCED HEXAGONAL BLOCK WALL, 8"X8"X16", 125 PCF, PERLITE FILL					
Hexagonal block partition, 8" thick	1.000	S.F.	4.80	9.05	13.85
Perlite insulation	1.000	S.F.	1.49	.43	1.92
Horizontal joint reinf., alternate courses	.800	L.F.	.18	.18	.36
Control joint	.050	L.F.	.08	.07	.15
TOTAL			6.55	9.73	16.28

B2010 123 Miscellaneous Block Wall

	TYPE	SIZE (IN.)	WEIGHT (P.C.F.)	CORE FILL		COST PER S.F. MAT.	INST.	TOTAL
1100	Deep groove-hollow	4x8x16	125	none		2.79	7.80	10.59
1150			105	none		3.32	7.60	10.92
1200	RB2010 -200	6x8x16	125	perlite		4.99	8.60	13.59
1210				styrofoam		5.30	8.25	13.55
1240				none		3.98	8.25	12.23
1250			105	perlite		5.80	7.95	13.75
1260				styrofoam		6.10	7.60	13.70
1290				none		4.77	7.60	12.37
1300		8x8x16	125	perlite		5.60	9.45	15.05
1310				styrofoam		5.60	9	14.60
1340				none		4.13	9	13.13
1350			105	perlite		6.40	9.15	15.55
1360				styrofoam		6.40	8.70	15.10
1390				none		4.91	8.70	13.61
1400	Deep groove-75% solid	4x8x16	125	none		3.42	7.90	11.32
1450			105	none		4.11	7.70	11.81
1500		6x8x16	125	perlite		5.45	8.60	14.05
1540				none		4.93	8.40	13.33
1550			105	perlite		6.45	8.35	14.80
1590				none		5.95	8.15	14.10

B2010 Exterior Walls

B2010 123 Miscellaneous Block Wall

	TYPE	SIZE (IN.)	WEIGHT (P.C.F.)	CORE FILL		COST PER S.F. MAT.	INST.	TOTAL
1600	Deep groove 75% solid	8x8x16	125	perlite		5.80	9.35	15.15
1640				none		5.05	9.15	14.20
1650			105	perlite		6.85	9.05	15.90
1690				none		6.10	8.85	14.95
1700	Deep groove-solid	4x8x16	125	none		3.85	8.05	11.90
1750			105	none		4.64	7.80	12.44
1800		6x8x16	125	none		5.55	8.55	14.10
1850			105	none		6.75	8.25	15
1900		8x8x16	125	none		5.70	9.30	15
1950			105	none		8.05	9	17.05
2100	Fluted	4x8x16	125	none		4.32	7.80	12.12
2150			105	none		5.40	7.60	13
2200		6x8x16	125	perlite		7.30	8.60	15.90
2210				styrofoam		7.60	8.25	15.85
2240				none		6.30	8.25	14.55
2300		8x8x16	125	perlite		7.35	9.45	16.80
2310				styrofoam		7.35	9	16.35
2340				none		5.85	9	14.85
2500	Hex-hollow	4x8x16	125	none		4.10	7.80	11.90
2550			105	none		5.15	7.60	12.75
2600		6x8x16	125	perlite		5.45	9	14.45
2610				styrofoam		5.75	8.65	14.40
2640				none		4.43	8.65	13.08
2650			105	perlite		6.55	8.75	15.30
2660				styrofoam		6.85	8.40	15.25
2690				none		5.55	8.40	13.95
2700		8x8x16	125	perlite		6.55	9.75	16.30
2710				styrofoam		6.55	9.30	15.85
2740				none		5.05	9.30	14.35
2750			105	perlite		7.75	9.45	17.20
2760				styrofoam		7.75	9	16.75
2790				none		6.25	9	15.25
2800	Hex-solid	4x8x16	125	none		5.10	8.05	13.15
2850			105	none		6.40	7.80	14.20
3100	Slump block	4x4x16	125	none		4.81	6.45	11.26
3200		6x4x16	125	perlite		8.10	7	15.10
3210				styrofoam		8.40	6.65	15.05
3240				none		7.10	6.65	13.75
3300		8x4x16	125	perlite		8.35	7.35	15.70
3310				styrofoam		8.35	6.90	15.25
3340				none		6.85	6.90	13.75
3400		12x4x16	125	perlite		17.10	9.05	26.15
3410				styrofoam		16.20	8.20	24.40
3440				none		14.70	8.20	22.90
3600		6x6x16	125	perlite		7.60	7.20	14.80
3610				styrofoam		7.90	6.85	14.75
3640				none		6.60	6.85	13.45
3700		8x6x16	125	perlite		11.40	7.55	18.95
3710				styrofoam		11.40	7.10	18.50
3740				none		9.90	7.10	17
3800		12x6x16	125	perlite		18.15	9.75	27.90
3810				styrofoam		17.25	8.90	26.15
3840				none		15.75	8.90	24.65

B2010 Exterior Walls

B2010 124 Reinforced Misc. Block Walls

	TYPE	SIZE (IN.)	WEIGHT (P.C.F.)	VERT. REINF. & GROUT SPACING		COST PER S.F. MAT.	INST.	TOTAL
5100	Deep groove-hollow	4x8x16	125	#4 @ 48"		2.95	8.50	11.45
5150			105	#4 @ 48"		3.48	8.30	11.78
5200		6x8x16	125	#4 @ 48"		4.24	9	13.24
5220				#5 @ 32"		4.46	9.25	13.71
5230				#5 @ 16"		4.92	10.25	15.17
5250			105	#4 @ 48"		5.05	8.35	13.40
5270				#5 @ 32"		5.25	8.60	13.85
5280				#5 @ 16"		5.70	9.60	15.30
5300		8x8x16	125	#4 @ 48"		4.45	9.85	14.30
5320				#5 @ 32"		4.69	10.20	14.89
5330				#5 @ 16"		5.25	11.40	16.65
5350			105	#4 @ 48"		5.25	9.55	14.80
5370				#5 @ 32"		5.45	9.90	15.35
5380				#5 @ 16"		6	11.10	17.10
5500	Deep groove-75% solid	6x8x16	125	#4 @ 48"		5.10	8.90	14
5520				#5 @ 32"		5.30	9.10	14.40
5530				#5 @ 16"		5.65	9.80	15.45
5550			105	#4 @ 48"		6.10	8.65	14.75
5570				#5 @ 32"		6.35	8.95	15.30
5580				#5 @ 16"		6.65	9.55	16.20
5600		8x8x16	125	#4 @ 48"		5.30	9.70	15
5620				#5 @ 32"		5.45	9.95	15.40
5630				#5 @ 16"		5.85	10.70	16.55
5650			105	#4 @ 48"		6.30	9.40	15.70
5670				#5 @ 32"		6.50	9.65	16.15
5680				#5 @ 16"		6.90	10.40	17.30
5700	Deep groove-solid	2-4x8x16	125	#4 @ 48" E.W.		8.40	17.70	26.10
5730	Double wythe			#5 @ 16" E.W.		9.10	18.50	27.60
5750			105	#4 @ 48" E.W.		9.95	17.20	27.15
5780				#5 @ 16" E.W.		10.65	18	28.65
5800		2-6x8x16	125	#4 @ 48" E.W.		11.80	18.70	30.50
5830				#5 @ 16" E.W.		12.50	19.50	32
5850			105	#4 @ 48" E.W.		14.20	18.10	32.30
5880				#5 @ 16" E.W.		14.90	18.90	33.80
6100	Fluted	4x8x16	125	#4 @ 48"		4.48	8.50	12.98
6150			105	#4 @ 48"		5.55	8.30	13.85
6200		6x8x16	125	#4 @ 48"		6.55	9	15.55
6220				#5 @ 32"		6.75	9.25	16
6230				#5 @ 16"		7.25	10.25	17.50
6300		8x8x16	125	#4 @ 48"		6.20	9.85	16.05
6320				#5 @ 32"		6.40	10.20	16.60
6330				#5 @ 16"		6.95	11.40	18.35
6500	Hex-hollow	4x8x16	125	#4 @ 48"		4.26	8.50	12.76
6550			105	#4 @ 48"		5.30	8.30	13.60
6620				#5 @ 32"		4.91	9.65	14.56
6630				#5 @ 16"		5.35	10.65	16
6650		6x8x16	105	#4 @ 48"		5.80	9.15	14.95
6670				#5 @ 32"		6	9.40	15.40
6680				#5 @ 16"		6.50	10.40	16.90
6700		8x8x16	125	#4 @ 48"		5.40	10.15	15.55
6720				#5 @ 32"		5.60	10.50	16.10
6730				#5 @ 16"		6.15	11.70	17.85

B2010 Exterior Walls

B2010 124 Reinforced Misc. Block Walls

	TYPE	SIZE (IN.)	WEIGHT (P.C.F.)	VERT. REINF. & GROUT SPACING		COST PER S.F.		
						MAT.	INST.	TOTAL
6750	Hex hollow	8x8x16	105	#4 @ 48″		6.60	9.85	16.45
6770				#5 @ 32″		6.80	10.20	17
6800	Hex-solid	2-4x8x16	125	#4 @ 48″ E.W.		10.75	17.60	28.35
6830	Double wythe			#5 @ 16″ E.W.		11.55	18.50	30.05
6850			105	#4 @ 48″ E.W.		13.45	17.10	30.55
6880				#5 @ 16″ E.W.		14.20	18	32.20
7100	Slump block	4x4x16	125	#4 @ 48″		4.97	7.15	12.12
7200		6x4x16	125	#4 @ 48″		7.35	7.40	14.75
7220				#5 @ 32″		7.55	8	15.55
7230				#5 @ 16″		8.05	8.65	16.70
7300		8x4x16	125	#4 @ 48″		7.20	7.75	14.95
7320				#5 @ 32″		7.40	8.10	15.50
7330				#5 @ 16″		7.95	9.30	17.25
7400		12x4x16	125	#4 @ 48″		15.15	9.05	24.20
7420				#5 @ 32″		15.45	9.45	24.90
7430				#5 @ 16″		16.20	10.65	26.85
7600		6x6x16	125	#4 @ 48″		6.85	7.60	14.45
7620				#5 @ 32″		7.05	7.85	14.90
7630				#5 @ 16″		7.55	8.85	16.40
7700		8x6x16	125	#4 @ 48″		10.25	7.95	18.20
7720				#5 @ 32″		10.45	8.30	18.75
7730				#5 @ 16″		11	9.50	20.50
7800		12x6x16	125	#4 @ 48″		16.20	9.75	25.95
7820				#5 @ 32″		16.50	10.15	26.65
7830				#5 @ 16″		17.25	11.35	28.60

B2010 Exterior Walls

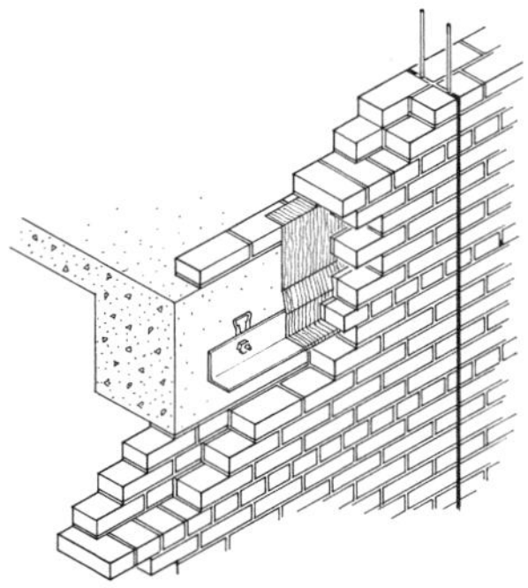

Exterior solid brick walls are defined in the following terms; structural reinforcement, type of brick, thickness, and bond. Sixteen different types of face bricks are presented, with single wythes shown in four different bonds. Six types of reinforced single wythe walls are also given, and twice that many for reinforced double wythe. Shelf angles are included in system components. These walls do not include ties and as such are not tied to a backup wall.

System Components	QUANTITY	UNIT	COST PER S.F. MAT.	COST PER S.F. INST.	COST PER S.F. TOTAL
SYSTEM B2010 125 1000					
BRICK WALL, COMMON, SINGLE WYTHE, 4″ THICK, RUNNING BOND					
Common brick wall, running bond	1.000	S.F.	4.42	11.40	15.82
Wash brick	1.000	S.F.	.05	1.01	1.06
Control joint, backer rod	.100	L.F.		.12	.12
Control joint, sealant	.100	L.F.	.03	.25	.28
Shelf angle	1.000	Lb.	1.12	1.03	2.15
Flashing	.100	S.F.	.14	.39	.53
TOTAL			5.76	14.20	19.96

B2010 125 Solid Brick Walls - Single Wythe

	TYPE	THICKNESS (IN.)	BOND			COST PER S.F. MAT.	COST PER S.F. INST.	COST PER S.F. TOTAL
1000	Common	4	running			5.75	14.20	19.95
1010			common			6.95	16.25	23.20
1050			Flemish			7.75	19.70	27.45
1100			English			8.70	21	29.70
1150	Standard	4	running			5.60	14.70	20.30
1160			common			6.60	17	23.60
1200			Flemish			7.35	20.50	27.85
1250			English			8.25	21.50	29.75
1300	Glazed	4	running			21	15.30	36.30
1310			common			25.50	17.80	43.30
1350			Flemish			28	21.50	49.50
1400			English			31.50	23	54.50
1450	Engineer	4	running			5.55	12.90	18.45
1460			common			6.55	14.70	21.25
1500			Flemish			7.25	17.80	25.05
1550			English			8.15	18.70	26.85
1600	Economy	4	running			8.35	11.25	19.60
1610			common			9.85	12.90	22.75
1650			Flemish			10.90	15.30	26.20
1700			English			23.50	16.25	39.75
1750	Double	4	running			12.15	9.20	21.35
1760			common			14.40	10.40	24.80
1800			Flemish			15.90	12.35	28.25
1850			English			17.90	13.10	31

B2010 Exterior Walls

B2010 125 Solid Brick Walls - Single Wythe

	TYPE	THICKNESS (IN.)	BOND			COST PER S.F. MAT.	INST.	TOTAL
1900	Fire	4-1/2	running			13.55	12.90	26.45
1910			common			16.15	14.70	30.85
1950			Flemish			17.90	17.80	35.70
2000			English			20	18.70	38.70
2050	King	3-1/2	running			4.58	11.55	16.13
2060			common			5.35	13.30	18.65
2100			Flemish			6.50	15.90	22.40
2150			English			7.20	16.60	23.80
2200	Roman	4	running			8.90	13.30	22.20
2210			common			10.65	15.30	25.95
2250			Flemish			11.80	18.25	30.05
2300			English			13.35	19.70	33.05
2350	Norman	4	running			6.90	11	17.90
2360			common			8.15	12.50	20.65
2400			Flemish			19.65	15	34.65
2450			English			10.15	15.90	26.05
2500	Norwegian	4	running			6.10	9.80	15.90
2510			common			7.25	11.15	18.40
2550			Flemish			8	13.30	21.30
2600			English			8.95	13.95	22.90
2650	Utility	4	running			6.50	8.65	15.15
2660			common			7.65	9.70	17.35
2700			Flemish			8.45	11.55	20
2750			English			9.45	12.15	21.60
2800	Triple	4	running			5.10	7.80	12.90
2810			common			5.90	8.75	14.65
2850			Flemish			6.45	10.30	16.75
2900			English			7.15	10.75	17.90
2950	SCR	6	running			8.25	11.25	19.50
2960			common			9.90	12.90	22.80
3000			Flemish			11	15.30	26.30
3050			English			12.50	16.25	28.75
3100	Norwegian	6	running			7.20	10	17.20
3110			common			8.50	11.40	19.90
3150			Flemish			9.40	13.50	22.90
3200			English			10.55	14.20	24.75
3250	Jumbo	6	running			6	8.85	14.85
3260			common			7.15	10	17.15
3300			Flemish			7.90	11.85	19.75
3350			English			8.95	12.50	21.45

B2010 126 Solid Brick Walls - Double Wythe

	TYPE	THICKNESS (IN.)	COLLAR JOINT THICKNESS (IN.)			COST PER S.F. MAT.	INST.	TOTAL
4100	Common	8	3/4			11.15	23.50	34.65
4150	Standard	8	1/2			10.60	24.50	35.10
4200	Glazed	8	1/2			41.50	25.50	67
4250	Engineer	8	1/2			10.55	22	32.55
4300	Economy	8	3/4			16.10	18.70	34.80
4350	Double	8	3/4			14.85	15.10	29.95
4400	Fire	8	3/4			26.50	22	48.50
4450	King	7	3/4			8.60	19.15	27.75

B2010 Exterior Walls

B2010 126 Solid Brick Walls - Double Wythe

	TYPE	THICKNESS (IN.)	COLLAR JOINT THICKNESS (IN.)			COST PER S.F. MAT.	COST PER S.F. INST.	COST PER S.F. TOTAL
4500	Roman	8	1			17.25	22.50	39.75
4550	Norman	8	3/4			13.15	18.25	31.40
4600	Norwegian	8	3/4			11.65	16.20	27.85
4650	Utility	8	3/4			12.45	14	26.45
4700	Triple	8	3/4			9.60	12.60	22.20
4750	SCR	12	3/4			15.90	18.75	34.65
4800	Norwegian	12	3/4			13.70	16.55	30.25

B2010 127 Solid Brick Walls, Reinforced

	TYPE	THICKNESS (IN.)	WYTHE	REINF. & SPACING		COST PER S.F. MAT.	COST PER S.F. INST.	COST PER S.F. TOTAL
7200	Common	4"	1	#4 @ 48"vert		5.95	14.75	20.70
7220				#5 @ 32"vert		6.10	14.95	21.05
7230				#5 @ 16"vert		6.40	15.60	22
7300	Utility	4"	1	#4 @ 48"vert		6.70	9.20	15.90
7320				#5 @ 32"vert		6.85	9.40	16.25
7330				#5 @ 16"vert		7.10	10.05	17.15
7400	SCR	6"	1	#4 @ 48"vert		8.50	11.80	20.30
7420				#5 @ 32"vert		8.65	12	20.65
7430				#5 @ 16"vert		9	12.70	21.70
7500	Jumbo	6"	1	#4 @ 48"vert		6.20	9.35	15.55
7520				#5 @ 32"vert		6.35	9.50	15.85
7530				#5 @ 16"vert		6.60	10.10	16.70
7600		8"	1	#4 @ 48"vert		8.25	9.55	17.80
7620				#5 @ 32"vert		8.40	9.80	18.20
7630				#5 @ 16"vert		8.80	10.60	19.40
7700	King	9"	2	#4 @ 48" E.W.		8.95	20	28.95
7720				#5 @ 32" E.W.		9.25	20	29.25
7730				#5 @ 16" E.W.		9.65	21	30.65
7800	Common	10"	2	#4 @ 48" E.W.		11.50	24.50	36
7820				#5 @ 32" E.W.		11.80	24.50	36.30
7830				#5 @ 16" E.W.		12.20	25.50	37.70
7900	Standard	10"	2	#4 @ 48" E.W.		10.95	25.50	36.45
7920				#5 @ 32" E.W.		11.25	25.50	36.75
7930				#5 @ 16" E.W.		11.65	26.50	38.15
8000	Engineer	10"	2	#4 @ 48" E.W.		10.90	22.50	33.40
8020				#5 @ 32" E.W.		11.20	23	34.20
8030				#5 @ 16" E.W.		11.60	23.50	35.10
8100	Economy	10"	2	#4 @ 48" E.W.		16.45	19.65	36.10
8120				#5 @ 32" E.W.		16.75	19.80	36.55
8130				#5 @ 16" E.W.		17.15	20.50	37.65
8200	Double	10"	2	#4 @ 48" E.W.		15.20	16.05	31.25
8220				#5 @ 32" E.W.		15.50	16.20	31.70
8230				#5 @ 16" E.W.		15.90	16.80	32.70
8300	Fire	10"	2	#4 @ 48" E.W.		27	22.50	49.50
8320				#5 @ 32" E.W.		27	23	50
8330				#5 @ 16" E.W.		27.50	23.50	51
8400	Roman	10"	2	#4 @ 48" E.W.		17.60	23.50	41.10
8420				#5 @ 32" E.W.		17.90	23.50	41.40
8430				#5 @ 16" E.W.		18.30	24	42.30

B2010 Exterior Walls

B2010 127 Solid Brick Walls, Reinforced

	TYPE	THICKNESS (IN.)	WYTHE	REINF. & SPACING		COST PER S.F.		
						MAT.	INST.	TOTAL
8500	Norman	10"	2	#4 @ 48" E.W.		13.50	19.20	32.70
8520				#5 @ 32" E.W.		13.80	19.35	33.15
8530				#5 @ 16" E.W.		14.20	19.95	34.15
8600	Norwegian	10"	2	#4 @ 48" E.W.		12	17.15	29.15
8620				#5 @ 32" E.W.		12.30	17.30	29.60
8630				#5 @ 16" E.W.		12.70	17.90	30.60
8700	Utility	10"	2	#4 @ 48" E.W.		12.80	14.95	27.75
8720				#5 @ 32" E.W.		13.10	15.10	28.20
8730				#5 @ 16" E.W.		13.50	15.70	29.20
8800	Triple	10"	2	#4 @ 48" E.W.		9.95	13.55	23.50
8820				#5 @ 32" E.W.		10.25	13.70	23.95
8830				#5 @ 16" E.W.		10.65	14.30	24.95
8900	SCR	14"	2	#4 @ 48" E.W.		16.30	19.65	35.95
8920				#5 @ 32" E.W.		16.60	19.80	36.40
8930				#5 @ 16" E.W.		17	20.50	37.50
9000	Jumbo	14"	2	#4 @ 48" E.W.		11.80	15.35	27.15
9030				#5 @ 16" E.W.		12.50	16.10	28.60
9100	Oversized, 6" x 16"	4"	1	#5@12" vert		13.15	9.65	22.80
9110		8"	1	#5@12" vert		16.70	8.80	25.50
9120	Oversized, 8" x 16"	4"	1	#5@12" vert		14.35	10.85	25.20
9130		8"	1	#5@12" vert		17.75	9.40	27.15

B2010 Exterior Walls

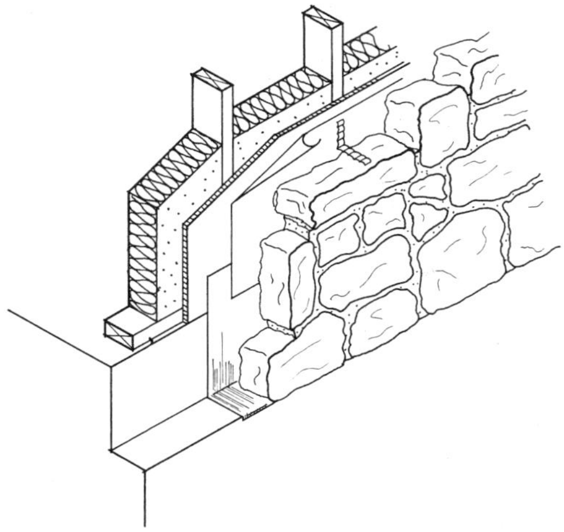

The table below lists costs per S.F. for stone veneer walls on various backup using different stone. Typical components for a system are shown in the component block below.

System Components	QUANTITY	UNIT	COST PER S.F. MAT.	COST PER S.F. INST.	COST PER S.F. TOTAL
SYSTEM B2010 128 2000					
ASHLAR STONE VENEER, 4", 2" X 4" STUD 16" O.C. BACK UP, 8' HIGH					
Sawn face, split joints, low priced stone	1.000	S.F.	13.15	18.75	31.90
Framing, 2" x 4" studs 8' high 16" O.C.	.920	B.F.	.49	1.49	1.98
Wall ties for stone veneer, galv. corrg 7/8" x 7", 22 gauge	.700	Ea.	.12	.38	.50
Asphalt felt sheathing paper, 15 lb.	1.000	S.F.	.06	.16	.22
Sheathing plywood on wall CDX 1/2"	1.000	S.F.	.75	1.06	1.81
Fiberglass insulation, 3-1/2", cr-11	1.000	S.F.	.52	.44	.96
Flashing, copper, paperback 1 side, 3 oz	.125	S.F.	.30	.22	.52
TOTAL			15.39	22.50	37.89

B2010 128	Stone Veneer	COST PER S.F. MAT.	COST PER S.F. INST.	COST PER S.F. TOTAL
2000	Ashlar veneer, 4", 2" x 4" stud backup, 16" O.C., 8' high, low priced stone	15.40	22.50	37.90
2050	2" x 6" stud backup, 16" O.C.	17.15	26.50	43.65
2100	Metal stud backup, 8' high, 16" O.C.	16.25	23	39.25
2150	24" O.C.	15.90	22.50	38.40
2200	Conc. block backup, 4" thick	16.90	26.50	43.40
2300	6" thick	17.55	27	44.55
2350	8" thick	17.70	27.50	45.20
2400	10" thick	18.25	29	47.25
2500	12" thick	19.50	31	50.50
3100	High priced stone, wood stud backup, 10' high, 16" O.C.	22	26	48
3200	Metal stud backup, 10' high, 16" O.C.	23	26	49
3250	24" O.C.	22.50	25.50	48
3300	Conc. block backup, 10' high, 4" thick	23.50	30	53.50
3350	6" thick	24	30	54
3400	8" thick	25	32	57
3450	10" thick	25	32	57
3500	12" thick	26	34	60
4000	Indiana limestone 2" thk., sawn finish, wood stud backup, 10' high, 16" O.C	34	13.85	47.85
4100	Metal stud backup, 10' high, 16" O.C.	36.50	17.55	54.05
4150	24" O.C.	36.50	17.55	54.05
4200	Conc. block backup, 4" thick	36	18.05	54.05
4250	6" thick	36.50	18.30	54.80

B2010 Exterior Walls

B2010 128	Stone Veneer	COST PER S.F. MAT.	COST PER S.F. INST.	COST PER S.F. TOTAL
4300	8" thick	36.50	18.90	55.40
4350	10" thick	37	20.50	57.50
4400	12" thick	38.50	22.50	61
4450	2" thick, smooth finish, wood stud backup, 8' high, 16" O.C.	34	13.85	47.85
4550	Metal stud backup, 8' high, 16" O.C.	35	13.95	48.95
4600	24" O.C.	34.50	13.45	47.95
4650	Conc. block backup, 4" thick	36	17.85	53.85
4700	6" thick	36.50	18.10	54.60
4750	8" thick	36.50	18.70	55.20
4800	10" thick	37	20.50	57.50
4850	12" thick	38.50	22.50	61
5350	4" thick, smooth finish, wood stud backup, 8' high, 16" O.C.	36.50	13.85	50.35
5450	Metal stud backup, 8' high, 16" O.C.	37.50	14.15	51.65
5500	24" O.C.	37	13.65	50.65
5550	Conc. block backup, 4" thick	38.50	18.05	56.55
5600	6" thick	39	18.30	57.30
5650	8" thick	43	23	66
5700	10" thick	39.50	20.50	60
5750	12" thick	41	22.50	63.50
6000	Granite, gray or pink, 2" thick, wood stud backup, 8' high, 16" O.C.	33.50	25	58.50
6100	Metal studs, 8' high, 16" O.C.	34.50	25.50	60
6150	24" O.C.	34	25	59
6200	Conc. block backup, 4" thick	35.50	29.50	65
6250	6" thick	36	29.50	65.50
6300	8" thick	36	30	66
6350	10" thick	36.50	31.50	68
6400	12" thick	38	33.50	71.50
6900	4" thick, wood stud backup, 8' high, 16" O.C.	44.50	29.50	74
7000	Metal studs, 8' high, 16" O.C.	45.50	29.50	75
7050	24" O.C.	45	29	74
7100	Conc. block backup, 4" thick	46.50	33.50	80
7150	6" thick	47	33.50	80.50
7200	8" thick	47	34.50	81.50
7250	10" thick	47.50	35.50	83
7300	12" thick	49	37.50	86.50

B2010 Exterior Walls

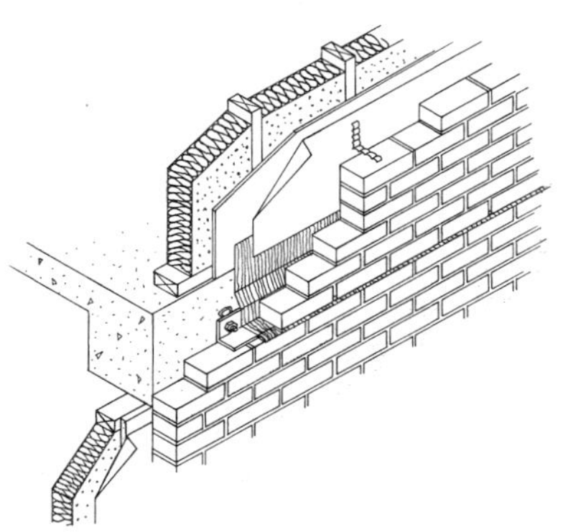

Exterior brick veneer/stud backup walls are defined in the following terms: type of brick and studs, stud spacing and bond. All systems include a back-up wall, a control joint every 20′, a brick shelf every 12′ of height, ties to the backup and the necessary dampproofing, flashing and insulation.

System Components	QUANTITY	UNIT	COST PER S.F. MAT.	COST PER S.F. INST.	COST PER S.F. TOTAL
SYSTEM B2010 129 1100					
STANDARD BRICK VENEER, 2″ X 4″ STUD BACKUP @ 16″ O.C., RUNNING BOND					
Standard brick wall, 4″ thick, running bond	1.000	S.F.	4.24	11.90	16.14
Wash smooth brick	1.000	S.F.	.05	1.01	1.06
Joint backer rod	.100	L.F.		.12	.12
Sealant	.100	L.F.	.03	.25	.28
Wall ties, corrugated, 7/8″ x 7″, 22 gauge	.003	Ea.	.05	.16	.21
Shelf angle	1.000	Lb.	1.12	1.03	2.15
Wood stud partition, backup, 2″ x 4″ @ 16″ O.C.	1.000	S.F.	.46	1.19	1.65
Sheathing, plywood, CDX, 1/2″	1.000	S.F.	.75	.85	1.60
Building paper, asphalt felt, 15 lb.	1.000	S.F.	.06	.16	.22
Fiberglass insulation, batts, 3-1/2″ thick paper backing	1.000	S.F.	.52	.44	.96
Flashing, copper, paperbacked	.100	S.F.	.14	.39	.53
TOTAL			7.42	17.50	24.92

B2010 129 Brick Veneer/Wood Stud Backup

	FACE BRICK	STUD BACKUP	STUD SPACING (IN.)	BOND	FACE	COST PER S.F. MAT.	COST PER S.F. INST.	COST PER S.F. TOTAL
1100	Standard	2x4-wood	16	running		7.40	17.50	24.90
1120				common		8.45	19.80	28.25
1140				Flemish		9.20	23	32.20
1160				English		10.10	24.50	34.60
1400		2x6-wood	16	running		7.90	17.45	25.35
1420				common		8.70	19.95	28.65
1440				Flemish		9.45	23	32.45
1460				English		10.35	24.50	34.85
1500			24	running		7.50	17.35	24.85
1520				common		8.50	19.65	28.15
1540				Flemish		9.25	23	32.25
1560				English		10.15	24	34.15
1700	Glazed	2x4-wood	16	running		23	18.10	41.10
1720				common		27	20.50	47.50
1740				Flemish		29.50	24.50	54
1760				English		33	25.50	58.50

B2010 Exterior Walls

B2010 129 Brick Veneer/Wood Stud Backup

	FACE BRICK	STUD BACKUP	STUD SPACING (IN.)	BOND	FACE	COST PER S.F.		
						MAT.	INST.	TOTAL
2000	Glazed	2x6-wood	16	running		23	18.25	41.25
2020				common		27.50	20.50	48
2040				Flemish		30	24.50	54.50
2060				English		33.50	25.50	59
2100			24	running		23	17.95	40.95
2120				common		27.50	20.50	48
2140				Flemish		30	24	54
2160				English		33.50	25.50	59
2300	Engineer	2x4-wood	16	running		7.40	15.70	23.10
2320				common		8.40	17.50	25.90
2340				Flemish		9.10	20.50	29.60
2360				English		10	21.50	31.50
2600		2x6-wood	16	running		7.65	15.85	23.50
2620				common		8.65	17.65	26.30
2640				Flemish		9.35	20.50	29.85
2660				English		10.25	21.50	31.75
2700			24	running		7.50	15.55	23.05
2720				common		8.45	17.35	25.80
2740				Flemish		9.15	20.50	29.65
2760				English		10.05	21.50	31.55
2900	Roman	2x4-wood	16	running		10.75	16.10	26.85
2920				common		12.50	18.10	30.60
2940				Flemish		13.65	21	34.65
2960				English		15.20	22.50	37.70
3200		2x6-wood	16	running		11	16.25	27.25
3220				common		12.75	18.25	31
3240				Flemish		13.90	21	34.90
3260				English		15.45	22.50	37.95
3300			24	running		10.80	15.95	26.75
3320				common		12.55	17.95	30.50
3340				Flemish		13.70	21	34.70
3360				English		15.25	22.50	37.75
3500	Norman	2x4-wood	16	running		8.75	13.80	22.55
3520				common		10	15.30	25.30
3540				Flemish		21.50	17.80	39.30
3560				English		12	18.70	30.70
3800		2x6-wood	16	running		9	13.95	22.95
3820				common		10.25	15.45	25.70
3840				Flemish		21	17.15	38.15
3860				English		12.25	18.85	31.10
3900			24	running		8.80	13.65	22.45
3920				common		10.05	15.15	25.20
3940				Flemish		21.50	17.65	39.15
3960				English		12.05	18.55	30.60
4100	Norwegian	2x4-wood	16	running		7.95	12.60	20.55
4120				common		9.10	13.95	23.05
4140				Flemish		9.85	16.10	25.95
4160				English		10.80	16.75	27.55
4400		2x6-wood	16	running		8.20	12.75	20.95
4420				common		9.35	14.10	23.45
4440				Flemish		10.10	16.25	26.35
4460				English		11.05	16.90	27.95

B2010 129 Brick Veneer/Wood Stud Backup

	FACE BRICK	STUD BACKUP	STUD SPACING (IN.)	BOND	FACE	COST PER S.F.		
						MAT.	INST.	TOTAL
4500	Norwegian	2x6-wood	24	running		8.05	12.45	20.50
4520				common		9.15	13.80	22.95
4540				Flemish		9.90	15.95	25.85
4560				English		10.85	16.60	27.45
4600	Oversized, 4″ x 2-1/4″ x 16″	2x4-wood	16	running	plain face	9.05	12.40	21.45
4610					1 to 3 slot face	9.45	12.90	22.35
4620					4 to 7 slot face	9.85	13.35	23.20
4630	Oversized, 4″ x 2-3/4″ x 16″	2x4-wood	16	running	plain face	9.30	11.95	21.25
4640					1 to 3 slot face	9.75	12.45	22.20
4650					4 to 7 slot face	10.10	12.80	22.90
4660	Oversized, 4″ x 4″ x 16″	2x4-wood	16	running	plain face	7.15	11.30	18.45
4670					1 to 3 slot face	7.45	11.75	19.20
4680					4 to 7 slot face	7.70	12.10	19.80
4690	Oversized, 4″ x 8″ x 16″	2x4-wood	16	running	plain face	7.85	10.50	18.35
4700					1 to 3 slot face	8.20	10.90	19.10
4710					4 to 7 slot face	8.50	11.20	19.70
4720	Oversized, 4″ x 2-1/4″ x 16″	2x6-wood	16	running	plain face	9.30	12.55	21.85
4730					1 to 3 slot face	9.70	13.05	22.75
4740					4 to 7 slot face	10.10	13.45	23.55
4750	Oversized, 4″ x 2-3/4″ x 16″	2x6-wood	16	running	plain face	9.55	12.10	21.65
4760					1 to 3 slot face	10	12.55	22.55
4770					4 to 7 slot face	10.35	12.95	23.30
4780	Oversized, 4″ x 4″ x 16″	2x6-wood	16	running	plain face	7.40	11.45	18.85
4790					1 to 3 slot face	7.70	11.85	19.55
4800					4 to 7 slot face	7.95	12.20	20.15
4810	Oversized, 4″ x 8″ x 16″	2x6-wood	16	running	plain face	8.10	10.65	18.75
4820					1 to 3 slot face	8.45	11	19.45
4830					4 to 7 slot face	8.75	11.30	20.05

B2010 130 Brick Veneer/Metal Stud Backup

	FACE BRICK	STUD BACKUP	STUD SPACING (IN.)	BOND	FACE	COST PER S.F.		
						MAT.	INST.	TOTAL
5050	Standard	16 ga x 6″LB	16	running		9	18.35	27.35
5100		25ga.x6″NLB	24	running		6.90	17.10	24
5120				common		7.90	19.40	27.30
5140				Flemish		8.30	22	30.30
5160				English		9.55	24	33.55
5200		20ga.x3-5/8″NLB	16	running		7	17.85	24.85
5220				common		8.05	20	28.05
5240				Flemish		8.80	23.50	32.30
5260				English		9.70	24.50	34.20
5300			24	running		6.90	17.30	24.20
5320				common		7.90	19.60	27.50
5340				Flemish		8.65	23	31.65
5360				English		9.55	24	33.55
5400		16ga.x3-5/8″LB	16	running		7.85	18.10	25.95
5420				common		8.85	20.50	29.35
5440				Flemish		9.60	23.50	33.10
5460				English		10.50	25	35.50
5500			24	running		7.50	17.60	25.10
5520				common		8.50	19.90	28.40
5540				Flemish		9.25	23	32.25
5560				English		10.15	24.50	34.65

B2010 130 Brick Veneer/Metal Stud Backup

	FACE BRICK	STUD BACKUP	STUD SPACING (IN.)	BOND	FACE	COST PER S.F.		
						MAT.	INST.	TOTAL
5700	Glazed	25ga.x6"NLB	24	running		22.50	17.70	40.20
5720				common		26.50	20	46.50
5740				Flemish		29	24	53
5760				English		32.50	25	57.50
5800		20ga.x3-5/8"NLB	24	running		22.50	17.90	40.40
5820				common		26.50	20.50	47
5840				Flemish		29	24	53
5860				English		32.50	25.50	58
6000		16ga.x3-5/8"LB	16	running		23.50	18.70	42.20
6020				common		27.50	21	48.50
6040				Flemish		30	25	55
6060				English		33.50	26	59.50
6100			24	running		23	18.20	41.20
6120				common		27.50	20.50	48
6140				Flemish		30	24.50	54.50
6160				English		33.50	25.50	59
6300	Engineer	25ga.x6"NLB	24	running		6.90	15.30	22.20
6320				common		7.85	17.10	24.95
6340				Flemish		8.55	20	28.55
6360				English		9.45	21	30.45
6400		20ga.x3-5/8"NLB	16	running		7	16.05	23.05
6420				common		8	17.85	25.85
6440				Flemish		8.70	21	29.70
6460				English		9.60	22	31.60
6500		20ga x3-5/8"NLB	24	running		6.55	14.60	21.15
6520				common		7.85	17.30	25.15
6540				Flemish		8.55	20.50	29.05
6560				English		9.45	21.50	30.95
6600		16ga.x3-5/8"LB	16	running		7.85	16.30	24.15
6620				common		8.80	18.10	26.90
6640				Flemish		9.50	21	30.50
6660				English		10.40	22	32.40
6700			24	running		7.50	15.80	23.30
6720				common		8.45	17.60	26.05
6740				Flemish		9.15	20.50	29.65
6760				English		10.05	21.50	31.55
6900	Roman	25ga.x6"NLB	24	running		10.20	15.70	25.90
6920				common		11.95	17.65	29.60
6940				Flemish		13.10	20.50	33.60
6960				English		14.65	22	36.65
7000		20ga.x3-5/8"NLB	16	running		10.35	16.45	26.80
7020				common		12.10	18.45	30.55
7040				Flemish		13.25	21.50	34.75
7060				English		14.80	23	37.80
7100			24	running		10.20	15.90	26.10
7120				common		11.95	17.90	29.85
7140				Flemish		13.10	21	34.10
7160				English		14.65	22.50	37.15
7200		16ga.x3-5/8"LB	16	running		11.15	16.70	27.85
7220				common		12.90	18.70	31.60
7240				Flemish		14.05	21.50	35.55
7260				English		15.60	23	38.60

B2010 Exterior Walls

B2010 130 Brick Veneer/Metal Stud Backup

	FACE BRICK	STUD BACKUP	STUD SPACING (IN.)	BOND	FACE	COST PER S.F.		
						MAT.	INST.	TOTAL
7300	Roman	16ga.x3-5/8"LB	24	running		10.80	16.20	27
7320				common		12.55	18.20	30.75
7340				Flemish		13.70	21	34.70
7360				English		15.25	22.50	37.75
7500	Norman	25ga.x6"NLB	24	running		8.20	13.40	21.60
7520				common		9.45	14.90	24.35
7540				Flemish		21	17.40	38.40
7560				English		11.45	18.30	29.75
7600		20ga.x3-5/8"NLB	24	running		8.20	13.60	21.80
7620				common		9.45	15.10	24.55
7640				Flemish		21	17.60	38.60
7660				English		11.45	18.50	29.95
7800		16ga.x3-5/8"LB	16	running		9.15	14.40	23.55
7820				common		10.40	15.90	26.30
7840				Flemish		22	18.40	40.40
7860				English		12.40	19.30	31.70
7900			24	running		8.80	13.90	22.70
7920				common		10.05	15.40	25.45
7940				Flemish		21.50	17.90	39.40
7960				English		12.05	18.80	30.85
8100	Norwegian	25ga.x6"NLB	24	running		7.45	12.20	19.65
8120				common		8.55	13.55	22.10
8140				Flemish		9.30	15.70	25
8160				English		10.25	16.35	26.60
8200		20ga.x3-5/8"NLB	16	running		7.55	12.95	20.50
8220				common		8.70	14.30	23
8240				Flemish		9.45	16.45	25.90
8260				English		10.40	17.10	27.50
8300			24	running		7.40	12.40	19.80
8320				common		8.55	13.75	22.30
8340				Flemish		9.30	15.90	25.20
8360				English		10.25	16.55	26.80
8400		16ga.x3-5/8"LB	16	running		8.40	13.20	21.60
8420				common		9.50	14.55	24.05
8440				Flemish		10.25	16.70	26.95
8460				English		11.20	17.35	28.55
8500			24	running		8.05	12.70	20.75
8520				common		9.15	14.05	23.20
8540				Flemish		9.90	16.20	26.10
8560				English		10.85	16.85	27.70
8600	Oversized, 4" x 2-1/4" x 16"	25 ga. x 6" NLB	24	running	plain face	8.95	12.05	21
8610					1 to 3 slot face	9.40	12.55	21.95
8620					4 to 7 slot face	9.75	12.95	22.70
8630	Oversized, 4" x 2-3/4" x 16"	25 ga. x 6" NLB	24	running	plain face	9.20	11.60	20.80
8640					1 to 3 slot face	9.70	12.05	21.75
8650					4 to 7 slot face	10.05	12.45	22.50
8660	Oversized, 4" x 4" x 16"	25 ga. x 6" NLB	24	running	plain face	7.10	10.95	18.05
8670					1 to 3 slot face	7.40	11.35	18.75
8680					4 to 7 slot face	7.65	11.70	19.35
8690	Oversized, 4" x 8" x 16"	25 ga x 6" NLB	24	running	plain face	7.80	10.15	17.95
8700					1 to 3 slot face	8.15	10.50	18.65
8710					4 to 7 slot face	8.45	10.80	19.25

B2010 Exterior Walls

B2010 130 — Brick Veneer/Metal Stud Backup

	FACE BRICK	STUD BACKUP	STUD SPACING (IN.)	BOND	FACE	COST PER S.F.		
						MAT.	INST.	TOTAL
8720	Oversized, 4" x 2-1/4" x 16"	20 ga. x 3-5/8" NLB	24	running	plain face	8.95	12.20	21.15
8730					1 to 3 slot face	9.40	12.75	22.15
8740					4 to 7 slot face	9.75	13.15	22.90
8750	Oversized, 4" x 2-3/4" x 16"	20 ga. x 3-5/8" NLB	24	running	plain face	9.20	11.75	20.95
8760					1 to 3 slot face	9.65	12.25	21.90
8770					4 to 7 slot face	10	12.65	22.65
8780	Oversized, 4" x 4" x 16"	20 ga. x 3-5/8" NLB	24	running	plain face	7.05	11.10	18.15
8790					1 to 3 slot face	7.35	11.55	18.90
8800					4 to 7 slot face	7.60	11.90	19.50
8810	Oversized, 4" x 8" x 16"	20 ga x 3-5/8" NLB	24	running	plain face	7.75	10.35	18.10
8820					1 to 3 slot face	8.10	10.70	18.80
8830					4 to 7 slot face	8.40	11	19.40

B2010 Exterior Walls

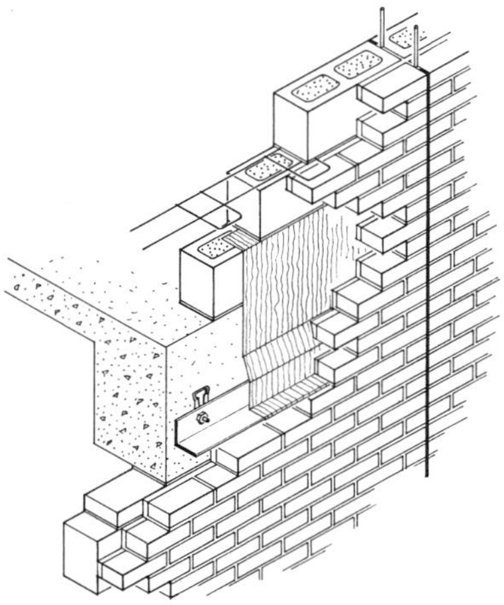

Exterior brick face composite walls are defined in the following terms: type of face brick and backup masonry, thickness of backup masonry and insulation. A special section is included on triple wythe construction at the back. Seven types of face brick are shown with various thicknesses of seven types of backup. All systems include a brick shelf, ties to the backup and necessary dampproofing, flashing, and control joints every 20′.

System Components	QUANTITY	UNIT	COST PER S.F. MAT.	INST.	TOTAL
SYSTEM B2010 132 1120					
COMPOSITE WALL, STANDARD BRICK FACE, 6″ C.M.U. BACKUP, PERLITE FILL					
Face brick veneer, standard, running bond	1.000	S.F.	4.24	11.90	16.14
Wash brick	1.000	S.F.	.05	1.01	1.06
Concrete block backup, 6″ thick	1.000	S.F.	2.53	6.55	9.08
Wall ties	.300	Ea.	.09	.16	.25
Perlite insulation, poured	1.000	S.F.	1.01	.34	1.35
Flashing, aluminum	.100	S.F.	.14	.39	.53
Shelf angle	1.000	Lb.	1.12	1.03	2.15
Control joint	.050	L.F.	.08	.07	.15
Backer rod	.100	L.F.		.12	.12
Sealant	.100	L.F.	.03	.25	.28
Collar joint	1.000	S.F.	.43	.67	1.10
TOTAL			9.72	22.49	32.21

B2010 132 Brick Face Composite Wall - Double Wythe

	FACE BRICK	BACKUP MASONRY	BACKUP THICKNESS (IN.)	BACKUP CORE FILL		COST PER S.F. MAT.	INST.	TOTAL
1000	Standard	common brick	4	none		10.60	27	37.60
1040		SCR brick	6	none		13.10	24	37.10
1080		conc. block	4	none		8.10	21.50	29.60
1120			6	perlite		9.70	22.50	32.20
1160				styrofoam		10.05	22	32.05
1200			8	perlite		10.40	23	33.40
1240				styrofoam		10.40	22.50	32.90
1280		L.W. block	4	none		8.20	21.50	29.70
1320			6	perlite		10.10	22.50	32.60
1360				styrofoam		10.40	22	32.40
1400			8	perlite		11.20	23	34.20
1440				styrofoam		11.20	22.50	33.70
1520		glazed block	4	none		17.35	23	40.35
1560			6	perlite		19.25	24	43.25
1600				styrofoam		19.55	23.50	43.05
1640			8	perlite		21	24.50	45.50
1680				styrofoam		21	24	45

B2010 Exterior Walls

B2010 132 Brick Face Composite Wall - Double Wythe

	FACE BRICK	BACKUP MASONRY	BACKUP THICKNESS (IN.)	BACKUP CORE FILL		COST PER S.F.		
						MAT.	INST.	TOTAL
1720	Standard	clay tile	4	none		13	21	34
1760			6	none		15.35	21.50	36.85
1800			8	none		17.60	22	39.60
1840		glazed tile	4	none		19	27.50	46.50
1880								
2000	Glazed	common brick	4	none		26	27.50	53.50
2040		SCR brick	6	none		28.50	24.50	53
2080		conc. block	4	none		23.50	22.50	46
2120			6	perlite		25	23	48
2160				styrofoam		25.50	22.50	48
2200			8	perlite		26	23.50	49.50
2240				styrofoam		26	23	49
2280		L.W. block	4	none		23.50	22	45.50
2320			6	perlite		25.50	23	48.50
2360				styrofoam		26	22.50	48.50
2400			8	perlite		26.50	23.50	50
2440				styrofoam		26.50	23	49.50
2520		glazed block	4	none		33	24	57
2560			6	perlite		35	24.50	59.50
2600				styrofoam		35	24	59
2640			8	perlite		36.50	25	61.50
2680				styrofoam		36.50	24.50	61
2720		clay tile	4	none		28.50	21.50	50
2760			6	none		31	22	53
2800			8	none		33	22.50	55.50
2840		glazed tile	4	none		34.50	28	62.50
2880								
3000	Engineer	common brick	4	none		10.60	25	35.60
3040		SCR brick	6	none		13.05	22	35.05
3080		conc. block	4	none		8.10	19.90	28
3120			6	perlite		9.70	20.50	30.20
3160				styrofoam		10	20.50	30.50
3200			8	perlite		10.35	21	31.35
3240				styrofoam		10.35	21	31.35
3280		L.W. block	4	none		6.25	14.10	20.35
3320			6	perlite		10.05	20.50	30.55
3360				styrofoam		10.40	20	30.40
3400			8	perlite		11.20	21	32.20
3440				styrofoam		11.20	20.50	31.70
3520		glazed block	4	none		17.30	21.50	38.80
3560			6	perlite		19.25	22	41.25
3600				styrofoam		19.55	21.50	41.05
3640			8	perlite		21	22.50	43.50
3680				styrofoam		21	22	43
3720		clay tile	4	none		12.95	19.05	32
3760			6	none		15.30	19.65	34.95
3800			8	none		17.55	20.50	38.05
3840		glazed tile	4	none		18.95	25.50	44.45
4000	Roman	common brick	4	none		13.90	25.50	39.40
4040		SCR brick	6	none		16.40	22.50	38.90

B2010 Exterior Walls

B2010 132 Brick Face Composite Wall - Double Wythe

	FACE BRICK	BACKUP MASONRY	BACKUP THICKNESS (IN.)	BACKUP CORE FILL		COST PER S.F.		
						MAT.	INST.	TOTAL
4080	Roman	conc. block	4	none		11.40	20.50	31.90
4120			6	perlite		13.05	21	34.05
4160				styrofoam		13.35	20.50	33.85
4200			8	perlite		13.70	21.50	35.20
4240				styrofoam		13.70	21	34.70
4280		L.W. block	4	none		11.55	20	31.55
4320			6	perlite		13.40	21	34.40
4360				styrofoam		13.70	20.50	34.20
4400			8	perlite		14.50	21.50	36
4440				styrofoam		14.50	21	35.50
4520		glazed block	4	none		20.50	22	42.50
4560			6	perlite		22.50	22.50	45
4600				styrofoam		23	22	45
4640			8	perlite		24	23	47
4680				styrofoam		24	22.50	46.50
4720		claytile	4	none		16.30	19.45	35.75
4760			6	none		18.65	20	38.65
4800			8	none		21	20.50	41.50
4840		glazed tile	4	none		22.50	26	48.50
5000	Norman	common brick	4	none		11.90	23.50	35.40
5040		SCR brick	6	none		14.40	20.50	34.90
5080		conc. block	4	none		9.40	18	27.40
5120			6	perlite		11.05	18.80	29.85
5160				styrofoam		11.35	18.45	29.80
5200			8	perlite		11.70	19.35	31.05
5240				styrofoam		11.70	18.90	30.60
5280		L.W. block	4	none		9.55	17.85	27.40
5320			6	perlite		11.40	18.75	30.15
5360				styrofoam		11.70	18.30	30
5400			8	perlite		12.50	19.15	31.65
5440				styrofoam		12.50	18.70	31.20
5520		glazed block	4	none		18.65	19.50	38.15
5560			6	perlite		20.50	20	40.50
5600				styrofoam		21	19.85	40.85
5640			8	perlite		22	20.50	42.50
5680				styrofoam		22	20.50	42.50
5720		clay tile	4	none		14.30	17.15	31.45
5760			6	none		16.65	17.75	34.40
5800			8	none		18.90	18.45	37.35
5840		glazed tile	4	none		20.50	24	44.50
6000	Norwegian	common brick	4	none		11.15	22	33.15
6040		SCR brick	6	none		13.60	19.15	32.75
6080		conc. block	4	none		8.65	16.80	25.45
6120			6	perlite		10.25	17.60	27.85
6160				styrofoam		10.55	17.25	27.80
6200			8	perlite		10.90	18.15	29.05
6240				styrofoam		10.90	17.70	28.60
6280		L.W. block	4	none		8.75	16.65	25.40
6320			6	perlite		10.60	17.45	28.05
6360				styrofoam		10.95	17.10	28.05
6400			8	perlite		11.75	17.95	29.70
6440				styrofoam		11.75	17.50	29.25

B2010 132 Brick Face Composite Wall - Double Wythe

	FACE BRICK	BACKUP MASONRY	BACKUP THICKNESS (IN.)	BACKUP CORE FILL		COST PER S.F.		
						MAT.	INST.	TOTAL
6520	Norwegian	glazed block	4	none		17.85	18.30	36.15
6560			6	perlite		19.80	19	38.80
6600				styrofoam		20	18.65	38.65
6640			8	perlite		21.50	19.60	41.10
6680				styrofoam		21.50	19.15	40.65
6720		clay tile	4	none		13.50	15.95	29.45
6760			6	none		15.85	16.55	32.40
6800			8	none		18.10	17.25	35.35
6840		glazed tile	4	none		19.50	22.50	42
6880								
7000	Utility	common brick	4	none		11.50	21	32.50
7040		SCR brick	6	none		14	18	32
7080		conc. block	4	none		9	15.65	24.65
7120			6	perlite		10.65	16.45	27.10
7160				styrofoam		10.95	16.10	27.05
7200			8	perlite		11.30	17	28.30
7240				styrofoam		11.30	16.55	27.85
7280		L.W. block	4	none		9.15	15.50	24.65
7320			6	perlite		11	16.30	27.30
7360				styrofoam		11.30	15.95	27.25
7400			8	perlite		12.10	16.80	28.90
7440				styrofoam		12.10	16.35	28.45
7520		glazed block	4	none		18.25	17.15	35.40
7560			6	perlite		20	17.85	37.85
7600				styrofoam		20.50	17.50	38
7640			8	perlite		21.50	18.45	39.95
7680				styrofoam		21.50	18	39.50
7720		clay tile	4	none		13.90	14.80	28.70
7760			6	none		16.25	15.40	31.65
7800			8	none		18.50	16.10	34.60
7840		glazed tile	4	none		19.90	21.50	41.40
7880								

B2010 133 Brick Face Composite Wall - Triple Wythe

	FACE BRICK	MIDDLE WYTHE	INSIDE MASONRY	TOTAL THICKNESS (IN.)		COST PER S.F.		
						MAT.	INST.	TOTAL
8000	Standard	common brick	standard brick	12		15.40	38.50	53.90
8100		4" conc. brick	standard brick	12		14.65	39.50	54.15
8120		4" conc. brick	common brick	12		14.85	39	53.85
8200	Glazed	common brick	standard brick	12		31	39	70
8300		4" conc. brick	standard brick	12		30	40.50	70.50
8320		4" conc. brick	glazed brick	12		30.50	40	70.50
8400	Engineer	common brick	standard brick	12		15.40	37	52.40
8500		4" conc. brick	standard brick	12		14.65	38	52.65
8520		4" conc. brick	engineer brick	12		14.85	37.50	52.35
8600	Roman	common brick	standard brick	12		18.70	37	55.70
8700		4" conc. brick	standard brick	12		18	38.50	56.50
8720		4" conc. brick	Roman brick	12		18.15	38	56.15
8800	Norman	common brick	standard brick	12		16.70	35	51.70
8900		4" conc. brick	standard brick	12		16	36	52
8920		4" conc. brick	Norman brick	12		16.15	35.50	51.65

B2010 Exterior Walls

B2010 133 — Brick Face Composite Wall - Triple Wythe

	FACE BRICK	MIDDLE WYTHE	INSIDE MASONRY	TOTAL THICKNESS (IN.)		COST PER S.F.		
						MAT.	INST.	TOTAL
9000	Norwegian	common brick	standard brick	12		15.95	33.50	49.45
9100		4″ conc. brick	standard brick	12		15.20	35	50.20
9120		4″ conc. brick	Norwegian brick	12		15.40	34.50	49.90
9200	Utility	common brick	standard brick	12		16.30	32.50	48.80
9300		4″ conc. brick	standard brick	12		15.40	33.50	48.90
9320		4″ conc. brick	utility brick	12		15.75	33	48.75

B2010 Exterior Walls

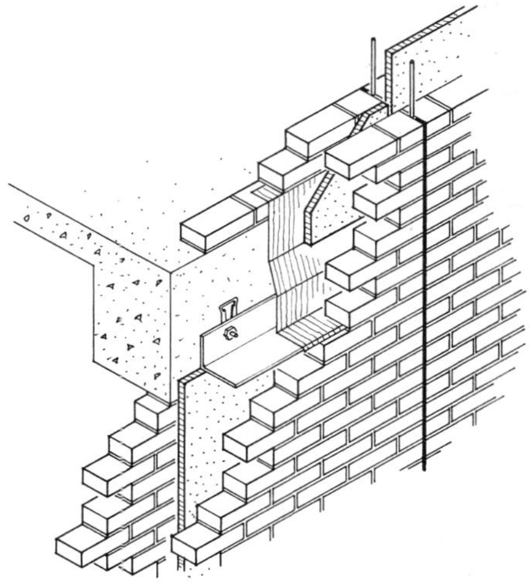

Exterior brick face cavity walls are defined in the following terms: cavity treatment, type of face brick, backup masonry, total thickness and insulation. Seven types of face brick are shown with various types of backup. All systems include a brick shelf, ties to the backups and necessary dampproofing, flashing, and control joints every 20′.

System Components	QUANTITY	UNIT	COST PER S.F. MAT.	COST PER S.F. INST.	COST PER S.F. TOTAL
SYSTEM B2010 134 1000					
CAVITY WALL, STANDARD BRICK FACE, COMMON BRICK BACKUP, POLYSTYRENE					
Face brick veneer, standard, running bond	1.000	S.F.	4.24	11.90	16.14
Wash brick	1.000	S.F.	.05	1.01	1.06
Common brick wall backup, 4″ thick	1.000	S.F.	4.42	11.40	15.82
Wall ties	.300	L.F.	.13	.16	.29
Polystyrene insulation board, 1″ thick	1.000	S.F.	.29	.74	1.03
Flashing, aluminum	.100	S.F.	.14	.39	.53
Shelf angle	1.000	Lb.	1.12	1.03	2.15
Control joint	.050	L.F.	.08	.07	.15
Backer rod	.100	L.F.		.12	.12
Sealant	.100	L.F.	.03	.25	.28
TOTAL			10.50	27.07	37.57

B2010 134 Brick Face Cavity Wall

	FACE BRICK	BACKUP MASONRY	TOTAL THICKNESS (IN.)	CAVITY INSULATION		COST PER S.F. MAT.	COST PER S.F. INST.	COST PER S.F. TOTAL
1000	Standard	4″ common brick	10	polystyrene		10.50	27	37.50
1020				none		10.20	26.50	36.70
1040		6″ SCR brick	12	polystyrene		13	24	37
1060				none		12.70	23.50	36.20
1080		4″ conc. block	10	polystyrene		8	22	30
1100				none		7.70	21	28.70
1120		6″ conc. block	12	polystyrene		8.65	22	30.65
1140				none		8.35	21.50	29.85
1160		4″ L.W. block	10	polystyrene		8.10	21.50	29.60
1180				none		7.85	21	28.85
1200		6″ L.W. block	12	polystyrene		9	22	31
1220				none		8.70	21.50	30.20
1240		4″ glazed block	10	polystyrene		17.25	23.50	40.75
1260				none		16.95	22.50	39.45
1280		6″ glazed block	12	polystyrene		17.25	23.50	40.75
1300				none		16.95	22.50	39.45

B2010 Exterior Walls

B2010 134 Brick Face Cavity Wall

	FACE BRICK	BACKUP MASONRY	TOTAL THICKNESS (IN.)	CAVITY INSULATION		COST PER S.F.		
						MAT.	INST.	TOTAL
1320	Standard	4" clay tile	10	polystyrene		12.90	21	33.90
1340				none		12.60	20	32.60
1360		4" glazed tile	10	polystyrene		18.90	27.50	46.40
1380				none		18.60	27	45.60
1500	Glazed	4" common brick	10	polystyrene		26	27.50	53.50
1520				none		25.50	27	52.50
1540		6" SCR brick	12	polystyrene		28.50	24.50	53
1560				none		28	24	52
1580		4" conc. block	10	polystyrene		23.50	22.50	46
1600				none		23	21.50	44.50
1620		6" conc. block	12	polystyrene		24	23	47
1640				none		24	22	46
1660		4" L.W. block	10	polystyrene		23.50	22	45.50
1680				none		23.50	21.50	45
1700		6" L.W. block	12	polystyrene		24.50	22.50	47
1720				none		24	22	46
1740		4" glazed block	10	polystyrene		32.50	24	56.50
1760				none		32.50	23	55.50
1780		6" glazed block	12	polystyrene		33.50	24	57.50
1800				none		33.50	23.50	57
1820		4" clay tile	10	polystyrene		28.50	21.50	50
1840				none		28	21	49
1860		4" glazed tile	10	polystyrene		30.50	22	52.50
1880				none		30.50	21.50	52
2000	Engineer	4" common brick	10	polystyrene		10.50	25.50	36
2020				none		10.20	24.50	34.70
2040		6" SCR brick	12	polystyrene		13	22.50	35.50
2060				none		12.70	21.50	34.20
2080		4" conc. block	10	polystyrene		8	19.95	27.95
2100				none		7.70	19.25	26.95
2120		6" conc. block	12	polystyrene		8.65	20.50	29.15
2140				none		8.35	19.70	28.05
2160		4" L.W. block	10	polystyrene		8.10	19.80	27.90
2180				none		7.80	19.10	26.90
2200		6" L.W. block	12	polystyrene		9	20.50	29.50
2220				none		8.70	19.55	28.25
2240		4" glazed block	10	polystyrene		17.20	21.50	38.70
2260				none		16.95	20.50	37.45
2280		6" glazed block	12	polystyrene		18.15	22	40.15
2300				none		17.85	21	38.85
2320		4" clay tile	10	polystyrene		12.85	19.10	31.95
2340				none		12.60	18.40	31
2360		4" glazed tile	10	polystyrene		18.85	26	44.85
2380				none		18.60	25	43.60
2500	Roman	4" common brick	10	polystyrene		13.80	25.50	39.30
2520				none		13.50	25	38.50
2540		6" SCR brick	12	polystyrene		16.30	22.50	38.80
2560				none		16.05	22	38.05
2580		4" conc. block	10	polystyrene		11.30	20.50	31.80
2600				none		11	19.65	30.65
2620		6" conc. block	12	polystyrene		11.95	21	32.95
2640				none		11.65	20	31.65

B2010 Exterior Walls

B2010 134 Brick Face Cavity Wall

	FACE BRICK	BACKUP MASONRY	TOTAL THICKNESS (IN.)	CAVITY INSULATION		COST PER S.F.		
						MAT.	INST.	TOTAL
2660	Roman	4" L.W. block	10	polystyrene		11.45	20	31.45
2680				none		11.15	19.50	30.65
2700		6" L.W. block	12	polystyrene		12.30	20.50	32.80
2720				none		12	19.95	31.95
2740		4" glazed block	10	polystyrene		20.50	22	42.50
2760				none		20.50	21	41.50
2780		6" glazed block	12	polystyrene		21.50	22	43.50
2800				none		21	21.50	42.50
2820		4" clay tile	10	polystyrene		16.20	19.50	35.70
2840				none		15.90	18.80	34.70
2860		4" glazed tile	10	polystyrene		22	26	48
2880				none		22	25.50	47.50
3000	Norman	4" common brick	10	polystyrene		11.80	23.50	35.30
3020				none		11.50	22.50	34
3040		6" SCR brick	12	polystyrene		14.30	20.50	34.80
3060				none		14.05	19.70	33.75
3080		4" conc. block	10	polystyrene		9.30	18.05	27.35
3100				none		9	17.35	26.35
3120		6" conc. block	12	polystyrene		9.95	18.50	28.45
3140				none		9.65	17.80	27.45
3160		4" L.W. block	10	polystyrene		9.45	17.90	27.35
3180				none		9.15	17.20	26.35
3200		6" L.W. block	12	polystyrene		10.30	18.35	28.65
3220				none		10	17.65	27.65
3240		4" glazed block	10	polystyrene		18.55	19.55	38.10
3260				none		18.25	18.85	37.10
3320		4" clay tile	10	polystyrene		14.20	17.20	31.40
3340				none		13.90	16.50	30.40
3360		4" glazed tile	10	polystyrene		20	24	44
3380				none		19.90	23	42.90
3500	Norwegian	4" common brick	10	polystyrene		11.05	22	33.05
3520				none		10.75	21.50	32.25
3540		6" SCR brick	12	polystyrene		13.55	19.20	32.75
3560				none		13.25	18.50	31.75
3580		4" conc. block	10	polystyrene		8.55	16.85	25.40
3600				none		8.25	16.15	24.40
3620		6" conc. block	12	polystyrene		9.20	17.30	26.50
3640				none		8.90	16.60	25.50
3660		4" L.W. block	10	polystyrene		8.65	16.70	25.35
3680				none		8.35	16	24.35
3700		6" L.W. block	12	polystyrene		9.55	17.15	26.70
3720				none		9.25	16.45	25.70
3740		4" glazed block	10	polystyrene		17.75	18.35	36.10
3760				none		17.50	17.65	35.15
3780		6" glazed block	12	polystyrene		18.70	18.70	37.40
3800				none		18.40	18	36.40
3820		4" clay tile	10	polystyrene		13.40	16	29.40
3840				none		13.15	15.30	28.45
3860		4" glazed tile	10	polystyrene		19.40	22.50	41.90
3880				none		19.15	22	41.15

B2010 Exterior Walls

B2010 134 Brick Face Cavity Wall

	FACE BRICK	BACKUP MASONRY	TOTAL THICKNESS (IN.)	CAVITY INSULATION		COST PER S.F.		
						MAT.	INST.	TOTAL
4000	Utility	4″ common brick	10	polystyrene		11.40	21	32.40
4020				none		11.10	20.50	31.60
4040		6″ SCR brick	12	polystyrene		13.90	18.05	31.95
4060				none		13.65	17.35	31
4080		4″ conc. block	10	polystyrene		8.90	15.70	24.60
4100				none		8.60	15	23.60
4120		6″ conc. block	12	polystyrene		9.55	16.15	25.70
4140				none		9.25	15.45	24.70
4160		4″ L.W. block	10	polystyrene		9.05	15.55	24.60
4180				none		8.75	14.85	23.60
4200		6″ L.W. block	12	polystyrene		9.90	16	25.90
4220				none		9.60	15.30	24.90
4240		4″ glazed block	10	polystyrene		18.15	17.20	35.35
4260				none		17.85	16.50	34.35
4280		6″ glazed block	12	polystyrene		19.05	17.55	36.60
4300				none		18.80	16.85	35.65
4320		4″ clay tile	10	polystyrene		13.80	14.85	28.65
4340				none		13.50	14.15	27.65
4360		4″ glazed tile	10	polystyrene		19.80	21.50	41.30
4380				none		19.50	21	40.50

B2010 135 Brick Face Cavity Wall - Insulated Backup

	FACE BRICK	BACKUP MASONRY	TOTAL THICKNESS (IN.)	BACKUP CORE FILL		COST PER S.F.		
						MAT.	INST.	TOTAL
5100	Standard	6″ conc. block	10	perlite		9.35	22	31.35
5120				styrofoam		9.65	21.50	31.15
5140		8″ conc. block	12	perlite		10	22.50	32.50
5160				styrofoam		10	22	32
5180		6″ L.W. block	10	perlite		9.70	21.50	31.20
5200				styrofoam		10	21.50	31.50
5220		8″ L.W. block	12	perlite		10.85	22	32.85
5240				styrofoam		10.85	21.50	32.35
5260		6″ glazed block	10	perlite		18.85	23	41.85
5280				styrofoam		19.15	23	42.15
5300		8″ glazed block	12	perlite		20.50	24	44.50
5320				styrofoam		20.50	23.50	44
5340		6″ clay tile	10	none		14.95	21	35.95
5360		8″ clay tile	12	none		17.20	21.50	38.70
5600	Glazed	6″ conc. block	10	perlite		25	22.50	47.50
5620				styrofoam		25	22	47
5640		8″ conc. block	12	perlite		25.50	23	48.50
5660				styrofoam		25.50	22.50	48
5680		6″ L.W. block	10	perlite		25	22.50	47.50
5700				styrofoam		25.50	22	47.50
5720		8″ L.W. block	12	perlite		26.50	23	49.50
5740				styrofoam		26.50	22.50	49
5760		6″ glazed block	10	perlite		34.50	24	58.50
5780				styrofoam		34.50	23.50	58
5800		8″ glazed block	12	perlite		36	24.50	60.50
5820				styrofoam		36	24	60
5840		6″ clay tile	10	none		30.50	21.50	52
5860		8″ clay tile	8	none		32.50	22	54.50

B2010 Exterior Walls

B2010 135 Brick Face Cavity Wall - Insulated Backup

	FACE BRICK	BACKUP MASONRY	TOTAL THICKNESS (IN.)	BACKUP CORE FILL		COST PER S.F.		
						MAT.	INST.	TOTAL
6100	Engineer	6" conc. block	10	perlite		9.30	20	29.30
6120				styrofoam		9.65	19.70	29.35
6140		8" conc. block	12	perlite		10	20.50	30.50
6160				styrofoam		10	20	30
6180		6" L.W. block	10	perlite		9.70	19.85	29.55
6200				styrofoam		10	19.55	29.55
6220		8" L.W. block	12	perlite		10.80	20.50	31.30
6240				styrofoam		10.80	19.95	30.75
6260		6" glazed block	10	perlite		18.85	21.50	40.35
6280				styrofoam		19.15	21	40.15
6300		8" glazed block	12	perlite		20.50	22	42.50
6320				styrofoam		20.50	21.50	42
6340		6" clay tile	10	none		14.95	19	33.95
6360		8" clay tile	12	none		17.20	19.70	36.90
6600	Roman	6" conc. block	10	perlite		12.65	20.50	33.15
6620				styrofoam		12.95	20	32.95
6640		8" conc. block	12	perlite		13.35	21	34.35
6660				styrofoam		13.35	20.50	33.85
6680		6" L.W. block	10	perlite		13	20.50	33.50
6700				styrofoam		13.30	19.95	33.25
6720		8" L.W. block	12	perlite		14.15	21	35.15
6740				styrofoam		14.15	20.50	34.65
6760		6" glazed block	10	perlite		22	22	44
6780				styrofoam		22.50	21.50	44
6800		8" glazed block	12	perlite		23.50	22.50	46
6820				styrofoam		23.50	22	45.50
6840		6" clay tile	10	none		18.25	19.40	37.65
6860		8" clay tile	12	none		20.50	20	40.50
7100	Norman	6" conc. block	10	perlite		10.65	18.10	28.75
7120				styrofoam		10.95	17.80	28.75
7140		8" conc. block	12	perlite		11.35	18.65	30
7160				styrofoam		11.35	18.25	29.60
7180		6" L.W. block	10	perlite		11	17.95	28.95
7200				styrofoam		11.30	17.65	28.95
7220		8" L.W. block	12	perlite		12.15	18.45	30.60
7240				styrofoam		12.15	18.05	30.20
7260		6" glazed block	10	perlite		20	19.50	39.50
7280				styrofoam		20.50	19.20	39.70
7300		8" glazed block	12	perlite		21.50	20	41.50
7320				styrofoam		21.50	19.70	41.20
7340		6" clay tile	10	none		16.25	17.10	33.35
7360		8" clay tile	12	none		18.55	17.80	36.35
7600	Norwegian	6" conc. block	10	perlite		9.85	16.90	26.75
7620				styrofoam		10.20	16.60	26.80
7640		8" conc. block	12	perlite		10.55	17.45	28
7660				styrofoam		10.55	17.05	27.60
7680		6" L.W. block	10	perlite		10.25	16.75	27
7700				styrofoam		10.55	16.45	27
7720		8" L.W. block	12	perlite		11.35	17.25	28.60
7740				styrofoam		11.35	16.85	28.20
7760		6" glazed block	10	perlite		19.40	18.30	37.70
7780				styrofoam		19.70	18	37.70

B2010 Exterior Walls

B2010 135 Brick Face Cavity Wall - Insulated Backup

	FACE BRICK	BACKUP MASONRY	TOTAL THICKNESS (IN.)	BACKUP CORE FILL		COST PER S.F.		
						MAT.	INST.	TOTAL
7800	Norwegian	8″ glazed block	12	perlite		21	18.90	39.90
7820				styrofoam		21	18.50	39.50
7840		6″ clay tile	10	none		15.50	15.90	31.40
7860		8″ clay tile	12	none		17.75	16.60	34.35
8100	Utility	6″ conc. block	10	perlite		10.25	15.75	26
8120				styrofoam		10.55	15.45	26
8140		8″ conc. block	12	perlite		10.95	16.30	27.25
8160				styrofoam		10.95	15.90	26.85
8180		6″ L.W. block	10	perlite		10.60	15.60	26.20
8200				styrofoam		10.90	15.30	26.20
8220		8″ L.W. block	12	perlite		11.75	16.10	27.85
8240				styrofoam		11.75	15.70	27.45
8260		6″ glazed block	10	perlite		19.75	17.15	36.90
8280				styrofoam		20	16.85	36.85
8300		8″ glazed block	12	perlite		21.50	17.75	39.25
8320				styrofoam		21.50	17.35	38.85
8340		6″ clay tile	10	none		15.85	14.75	30.60
8360		8″ clay tile	12	none		18.15	15.45	33.60

B2010 Exterior Walls

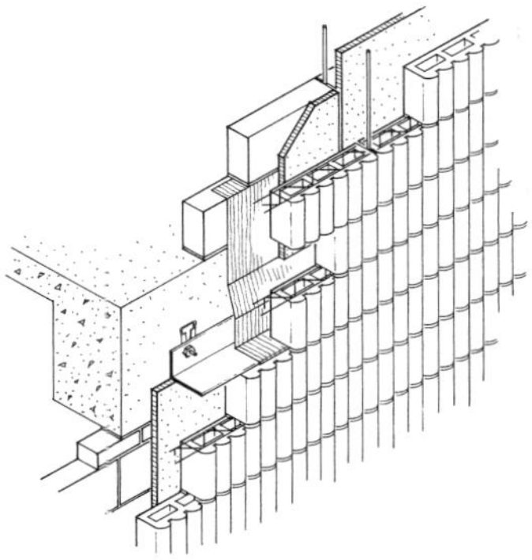

Exterior block face cavity walls are defined in the following terms: cavity treatment, type of face block, backup masonry, total thickness and insulation. Multiple types of face block are shown with various types of backup. All systems include a brick shelf and necessary dampproofing, flashing, and control joints every 20'.

System Components	QUANTITY	UNIT	COST PER S.F. MAT.	COST PER S.F. INST.	COST PER S.F. TOTAL
SYSTEM B2010 137 1600					
CAVITY WALL, FLUTED BLOCK FACE, 4" C.M.U. BACKUP, POLYSTYRENE BOARD					
Fluted block partition, 4" thick	1.000	S.F.	3.59	7.85	11.44
Conc. block wall backup 4" thick	1.000	S.F.	1.92	6.10	8.02
Horizontal joint reinforcing	.800	L.F.	.20	.22	.42
Polystyrene insulation board, 1" thick	1.000	S.F.	.29	.74	1.03
Flashing, aluminum	.100	S.F.	.14	.39	.53
Shelf angle	1.000	Lb.	1.12	1.03	2.15
Control joint	.050	L.F.	.08	.07	.15
Backer rod	.100	L.F.		.12	.12
Sealant	.100	L.F.	.03	.25	.28
TOTAL			7.37	16.77	24.14

B2010 137 Block Face Cavity Wall

	FACE BLOCK	BACKUP MASONRY	TOTAL THICKNESS (IN.)	CAVITY INSULATION		COST PER S.F. MAT.	COST PER S.F. INST.	COST PER S.F. TOTAL
1000	Deep groove	4" conc. block	10	polystyrene		7.35	16.75	24.10
1050	Reg. wt.			none		7.10	16.05	23.15
1060		6" conc. block	12	polystyrene		8	17.20	25.20
1110				none		7.70	16.50	24.20
1120		4" L.W. block	10	polystyrene		7.50	16.60	24.10
1170				none		7.20	15.90	23.10
1180		6" L.W. block	12	polystyrene		8.35	17.05	25.40
1230				none		8.05	16.35	24.40
1300	Deep groove	4" conc. block	10	polystyrene		8.15	16.50	24.65
1350	L.W.			none		7.85	15.80	23.65
1360		6" conc. block	12	polystyrene		8.75	16.95	25.70
1410				none		8.50	16.25	24.75
1420		4" L.W. block	10	polystyrene		8.30	16.35	24.65
1470				none		8	15.65	23.65
1480		6" L.W. block	12	polystyrene		9.15	16.80	25.95
1530				none		8.85	16.10	24.95

B2010 Exterior Walls

B2010 137 Block Face Cavity Wall

	FACE BLOCK	BACKUP MASONRY	TOTAL THICKNESS (IN.)	CAVITY INSULATION		COST PER S.F.		
						MAT.	INST.	TOTAL
1600	Fluted	4″ conc. block	10	polystyrene		7.35	16.75	24.10
1650	Reg. wt.			none		7.55	15.80	23.35
1660		6″ conc. block	12	polystyrene		8.45	16.95	25.40
1710				none		8.15	16.25	24.40
1720		4″ L.W. block	10	polystyrene		7.95	16.35	24.30
1770				none		7.65	15.65	23.30
1780		6″ L.W. block	12	polystyrene		8.80	16.80	25.60
1830				none		8.50	16.10	24.60
1900	Fluted	4″ conc. block	10	polystyrene		8.95	16.30	25.25
1950	L.W.			none		8.65	15.60	24.25
1960		6″ conc. block	12	polystyrene		9.55	16.75	26.30
2010				none		9.25	16.05	25.30
2020		4″ L.W. block	10	polystyrene		9.05	16.15	25.20
2070				none		8.75	15.45	24.20
2080		6″ L.W. block	12	polystyrene		9.90	16.60	26.50
2130				none		9.60	15.90	25.50
2200	Ground face	4″ conc. block	10	polystyrene		17.30	16.85	34.15
2250	1 Score			none		17	16.15	33.15
2260		6″ conc. block	12	polystyrene		17.90	17.30	35.20
2310				none		17.60	16.60	34.20
2320		4″ L.W. block	10	polystyrene		17.40	16.70	34.10
2370				none		17.10	16	33.10
2380		6″ L.W. block	12	polystyrene		18.25	17.15	35.40
2430				none		17.95	16.45	34.40
2500	Hexagonal	4″ conc. block	10	polystyrene		7.60	16.50	24.10
2550	Reg. wt.			none		7.35	15.80	23.15
2560		6″ conc. block	12	polystyrene		8.25	16.95	25.20
2610				none		7.95	16.25	24.20
2620		4″ L.W. block	10	polystyrene		7.75	16.35	24.10
2670				none		7.45	15.65	23.10
2680		6″ L.W. block	12	polystyrene		8.60	16.80	25.40
2730				none		8.30	16.10	24.40
2800	Hexagonal	4″ conc. block	10	polystyrene		8.65	16.30	24.95
2850	L.W.			none		8.35	15.60	23.95
2860		6″ conc. block	12	polystyrene		9.25	16.75	26
2910				none		8.95	16.05	25
2920		4″ L.W. block	10	polystyrene		8.75	16.15	24.90
2970				none		8.50	15.45	23.95
2980		6″ L.W. block	12	polystyrene		9.60	16.60	26.20
3030				none		9.35	15.90	25.25
3100	Slump block	4″ conc. block	10	polystyrene		8.35	15.15	23.50
3150	4x16			none		8.05	14.45	22.50
3160		6″ conc. block	12	polystyrene		8.95	15.60	24.55
3210				none		8.65	14.90	23.55
3220		4″ L.W. block	10	polystyrene		8.45	15	23.45
3270				none		8.15	14.30	22.45
3280		6″ L.W. block	12	polystyrene		9.30	15.45	24.75
3330				none		9	14.75	23.75

B2010 Exterior Walls

B2010 137 Block Face Cavity Wall

	FACE BLOCK	BACKUP MASONRY	TOTAL THICKNESS (IN.)	CAVITY INSULATION		COST PER S.F.		
						MAT.	INST.	TOTAL
3400	Split face	4" conc. block	10	polystyrene		9.75	16.75	26.50
3450	1 Score			none		9.45	16.05	25.50
3460	Reg. wt.	6" conc. block	12	polystyrene		10.35	17.20	27.55
3510				none		10.05	16.50	26.55
3520		4" L.W. block	10	polystyrene		9.85	16.60	26.45
3570				none		9.55	15.90	25.45
3580		6" L.W. block	12	polystyrene		10.70	17.05	27.75
3630				none		10.40	16.35	26.75
3700	Split face	4" conc. block	10	polystyrene		11.15	16.50	27.65
3750	1 Score			none		10.85	15.80	26.65
3760	L.W.	6" conc. block	12	polystyrene		11.75	16.95	28.70
3810				none		11.45	16.25	27.70
3820		4" L.W. block	10	polystyrene		11.25	16.35	27.60
3870				none		10.95	15.65	26.60
3880		6" L.W. block	12	polystyrene		12.10	16.80	28.90
3930				none		11.80	16.10	27.90
4000	Split rib	4" conc. block	10	polystyrene		10.95	16.75	27.70
4050	8 Rib			none		10.65	16.05	26.70
4060	Reg. wt.	6" conc. block	12	polystyrene		11.55	17.20	28.75
4110				none		11.25	16.50	27.75
4120		4" L.W. block	10	polystyrene		11.05	16.60	27.65
4170				none		10.75	15.90	26.65
4180		6" L.W. block	12	polystyrene		11.90	17.05	28.95
4230				none		11.60	16.35	27.95
4300	Split rib	4" conc. block	10	polystyrene		11.90	16.50	28.40
4350	8 Rib			none		11.60	15.80	27.40
4360	L.W.	6" conc. block	12	polystyrene		12.50	16.95	29.45
4410				none		12.20	16.25	28.45
4420		4" L.W. block	10	polystyrene		12	16.35	28.35
4470				none		11.70	15.65	27.35
4480		6" L.W. block	12	polystyrene		12.85	16.80	29.65
4530				none		12.55	16.10	28.65

B2010 138 Block Face Cavity Wall - Insulated Backup

	FACE BLOCK	BACKUP MASONRY	TOTAL THICKNESS (IN.)	BACKUP CORE INSULATION		COST PER S.F.		
						MAT.	INST.	TOTAL
5010	Deep groove	6" conc. block	10	perlite		8.70	16.80	25.50
5050	Reg. wt.			styrofoam		9	16.50	25.50
5060		8" conc. block	12	perlite		9.35	17.35	26.70
5110				styrofoam		9.20	16.95	26.15
5120		6" L.W. block	10	perlite		9.05	16.65	25.70
5170				styrofoam		9.35	16.35	25.70
5180		8" L.W. block	12	perlite		10.15	17.15	27.30
5230				styrofoam		10.15	16.75	26.90
5300	Deep groove	6" conc. block	10	perlite		9.50	16.55	26.05
5350	L.W.			styrofoam		9.80	16.25	26.05
5360		8" conc. block	12	perlite		10.15	17.10	27.25
5410				styrofoam		10.15	16.70	26.85
5420		6" L.W. block	10	perlite		9.85	16.40	26.25
5470				styrofoam		10.15	16.10	26.25
5480		8" L.W. block	12	perlite		10.95	16.90	27.85
5530				styrofoam		10.95	16.50	27.45

B2010 Exterior Walls

B2010 138 Block Face Cavity Wall - Insulated Backup

	FACE BLOCK	BACKUP MASONRY	TOTAL THICKNESS (IN.)	BACKUP CORE INSULATION		COST PER S.F.		
						MAT.	INST.	TOTAL
5600	Fluted	6" conc. block	10	perlite		9.15	16.55	25.70
5650	Reg.			styrofoam		9.50	16.25	25.75
5660		8" conc. block	12	perlite		9.85	17.10	26.95
5710				styrofoam		9.85	16.70	26.55
5720		6" L.W. block	10	perlite		9.55	16.40	25.95
5770				styrofoam		9.85	16.10	25.95
5780		8" L.W. block	12	perlite		10.65	16.90	27.55
5830				styrofoam		10.65	16.50	27.15
5900	Fluted	6 conc. block	10	perlite		10.25	16.35	26.60
5950	L.W.			styrofoam		10.55	16.05	26.60
5960		8" conc. block	12	perlite		10.90	16.90	27.80
6010				styrofoam		10.90	16.50	27.40
6020		6" L.W. block	10	perlite		10.60	16.20	26.80
6070				styrofoam		10.95	15.90	26.85
6080		8" L.W. block	12	perlite		11.75	16.70	28.45
6130				styrofoam		11.75	16.30	28.05
6200	Ground face	6" conc. block	10	perlite		18.60	16.90	35.50
6250	1 Score			styrofoam		18.90	16.60	35.50
6260		8" conc. block	12	perlite		19.25	17.45	36.70
6310				styrofoam		19.25	17.05	36.30
6320		6" L.W. block	10	perlite		18.95	16.75	35.70
6370				styrofoam		19.30	16.45	35.75
6380		8" L.W. block	12	perlite		20	17.25	37.25
6430				styrofoam		21.50	20.50	42
6500	Hexagonal	6" conc. block	10	perlite		8.95	16.55	25.50
6550	Reg. wt.			styrofoam		9.25	16.25	25.50
6560		8" conc. block	12	perlite		9.60	17.10	26.70
6610				styrofoam		9.60	16.70	26.30
6620		6" L.W. block	10	perlite		9.30	16.40	25.70
6670				styrofoam		9.60	16.10	25.70
6680		8" L.W. block	12	perlite		10.40	16.90	27.30
6730				styrofoam		10.40	16.50	26.90
6800	Hexagonal	6" conc. block	10	perlite		10	16.35	26.35
6850	L.W.			styrofoam		10.30	16.05	26.35
6860		8" conc. block	12	perlite		10.65	16.90	27.55
6910				styrofoam		10.65	16.50	27.15
6920		6" L.W. block	10	perlite		10.35	16.20	26.55
6970				styrofoam		10.65	15.90	26.55
6980		8" L.W. block	12	perlite		11.45	16.70	28.15
7030				styrofoam		11.45	16.30	27.75
7100	Slump block	6" conc. block	10	perlite		9.65	15.20	24.85
7150	4x16			styrofoam		9.95	14.90	24.85
7160		8" conc. block	12	perlite		10.30	15.75	26.05
7210				styrofoam		10.30	15.35	25.65
7220		6" L.W. block	10	perlite		10	15.05	25.05
7270				styrofoam		10.35	14.75	25.10
7280		8" L.W. block	12	perlite		11.15	15.55	26.70
7330				styrofoam		11.15	15.15	26.30

B2010 Exterior Walls

B2010 138 Block Face Cavity Wall - Insulated Backup

	FACE BLOCK	BACKUP MASONRY	TOTAL THICKNESS (IN.)	BACKUP CORE INSULATION		COST PER S.F.		
						MAT.	INST.	TOTAL
7400	Split face	6″ conc. block	10	perlite		11.05	16.80	27.85
7450	1 Score			styrofoam		11.35	16.50	27.85
7460	Reg. wt.	8″ conc. block	12	perlite		11.70	17.35	29.05
7510				styrofoam		11.70	16.95	28.65
7520		6″ L.W. block	10	perlite		11.40	16.65	28.05
7570				styrofoam		11.75	16.35	28.10
7580		8″ L.W. block	12	perlite		12.55	17.15	29.70
7630				styrofoam		12.55	16.75	29.30
7700	Split rib	6″ conc. block	10	perlite		12.25	16.80	29.05
7750	8 Rib			styrofoam		12.55	16.50	29.05
7760	Reg. wt.	8″ conc. block	12	perlite		12.90	17.35	30.25
7810				styrofoam		12.90	16.95	29.85
7820		6″ L.W. block	10	perlite		12.60	16.65	29.25
7870				styrofoam		12.95	16.35	29.30
7880		8″ L.W. block	12	perlite		13.75	17.15	30.90
7930				styrofoam		13.75	16.75	30.50
8000	Split rib	6″ conc. block	10	perlite		13.20	16.55	29.75
8050	8 Rib			styrofoam		13.60	16.30	29.90
8060	L.W.	8″ conc. block	12	perlite		13.85	17.10	30.95
8110				styrofoam		13.85	16.70	30.55
8120		6″ L.W. block	10	perlite		13.55	16.40	29.95
8170				styrofoam		13.90	16.10	30
8180		8″ L.W. block	12	perlite		14.70	16.90	31.60
8230				styrofoam		14.70	16.50	31.20

B2010 Exterior Walls

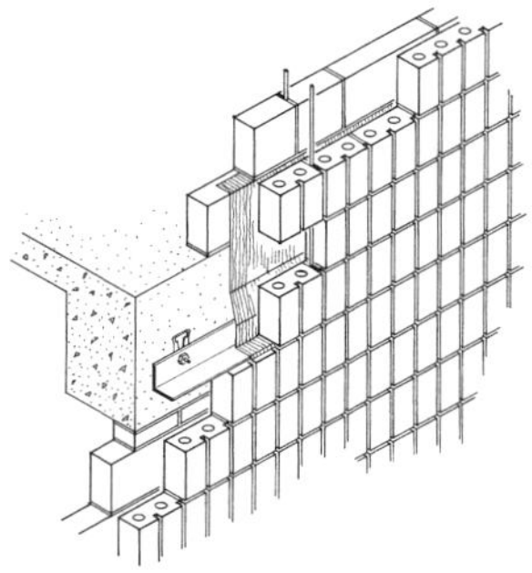

Exterior block face composite walls are defined in the following terms: type of face block and backup masonry, total thickness and insulation. All systems include shelf angles and necessary dampproofing flashing, and control joints every 20′.

System Components	QUANTITY	UNIT	COST PER S.F. MAT.	COST PER S.F. INST.	COST PER S.F. TOTAL
SYSTEM B2010 139 1000					
COMPOSITE WALL, GROOVED BLOCK FACE, 4″ C.M.U. BACKUP, PERLITE FILL					
Deep groove block veneer, 4″ thick	1.000	S.F.	3.59	7.85	11.44
Concrete block wall, backup, 4″ thick	1.000	S.F.	1.92	6.10	8.02
Horizontal joint reinforcing	.800	L.F.	.24	.18	.42
Perlite insulation, poured	1.000	S.F.	.66	.21	.87
Flashing, aluminum	.100	S.F.	.14	.39	.53
Shelf angle	1.000	Lb.	1.12	1.03	2.15
Control joint	.050	L.F.	.08	.07	.15
Backer rod	.100	L.F.		.12	.12
Sealant	.100	L.F.	.03	.25	.28
Collar joint	1.000	S.F.	.43	.67	1.10
TOTAL			8.21	16.87	25.08

B2010 139 Block Face Composite Wall

	FACE BLOCK	BACKUP MASONRY	TOTAL THICKNESS (IN.)	BACKUP CORE INSULATION		COST PER S.F. MAT.	COST PER S.F. INST.	COST PER S.F. TOTAL
1000	Deep groove	4″ conc. block	8	perlite		8.20	16.85	25.05
1050	Reg. wt.			none		7.55	16.65	24.20
1060		6″ conc. block	10	perlite		9.10	17.50	26.60
1110				styrofoam		9.45	17.15	26.60
1120		8″ conc. block	12	perlite		9.80	18.05	27.85
1170				styrofoam		9.80	17.60	27.40
1200		4″ L.W. block	8	perlite		8.35	16.70	25.05
1250				none		7.65	16.50	24.15
1260		6″ L.W. block	10	perlite		9.50	17.35	26.85
1310				styrofoam		9.80	17	26.80
1320		8″ L.W. block	12	perlite		10.60	17.85	28.45
1370				styrofoam		10.60	17.40	28

B2010 Exterior Walls

B2010 139 Block Face Composite Wall

	FACE BLOCK	BACKUP MASONRY	TOTAL THICKNESS (IN.)	BACKUP CORE INSULATION		COST PER S.F.		
						MAT.	INST.	TOTAL
1600	Deep groove	4″ conc. block	8	perlite		9	16.60	25.60
1650	L.W.			none		8.35	16.40	24.75
1660		6″ conc. block	10	perlite		9.90	17.25	27.15
1710				styrofoam		10.20	16.90	27.10
1720		8″ conc. block	12	perlite		10.60	17.80	28.40
1770				styrofoam		10.60	17.35	27.95
1800		4″ L.W. block	8	perlite		9.10	16.45	25.55
1850				none		8.45	16.25	24.70
1860		6″ L.W. block	10	perlite		10.25	17.10	27.35
1910				styrofoam		10.60	16.75	27.35
1920		8″ L.W. block	12	perlite		11.40	17.60	29
1970				styrofoam		11.40	17.15	28.55
2200	Fluted	4″ conc. block	8	perlite		8.70	16.60	25.30
2250	Reg. wt.			none		8	16.40	24.40
2260		6″ conc. block	10	perlite		9.60	17.25	26.85
2310				styrofoam		9.90	16.90	26.80
2320		8″ conc. block	12	perlite		10.25	17.80	28.05
2370				styrofoam		10.25	17.35	27.60
2400		4″ L.W. block	8	perlite		8.80	16.45	25.25
2450				styrofoam		8.15	16.25	24.40
2460		6″ L.W. block	10	perlite		9.95	17.10	27.05
2510				styrofoam		10.25	16.75	27
2520		8″ L.W. block	12	perlite		11.05	17.60	28.65
2570				styrofoam		11.05	17.15	28.20
2860	Fluted	6″ conc. block	10	perlite		10.80	17.45	28.25
2910	L.W.			styrofoam		11	16.70	27.70
2920		8″ conc. block	12	perlite		11.35	17.60	28.95
2970				styrofoam		11.35	17.15	28.50
3060		6″ L.W. block	10	perlite		11.05	16.90	27.95
3110				styrofoam		11.35	16.55	27.90
3120		8″ L.W. block	12	perlite		12.15	17.40	29.55
3170				styrofoam		12.15	16.95	29.10
3400	Ground face	4″ conc. block	8	perlite		18.10	16.95	35.05
3450	1 Score			none		17.45	16.75	34.20
3460		6″ conc. block	10	perlite		19.05	17.60	36.65
3510				styrofoam		19.35	17.25	36.60
3520		8″ conc. block	12	perlite		19.70	18.25	37.95
3570				styrofoam		19.70	17.70	37.40
3600		4″ L.W. block	8	perlite		18.25	16.80	35.05
3650				none		17.60	16.60	34.20
3660		6″ L.W. block	10	perlite		19.40	17.45	36.85
3710				styrofoam		19.70	17.10	36.80
3720		8″ L.W. block	12	perlite		20.50	17.95	38.45
3770				styrofoam		20.50	17.50	38

B2010 Exterior Walls

B2010 139 Block Face Composite Wall

	FACE BLOCK	BACKUP MASONRY	TOTAL THICKNESS (IN.)	BACKUP CORE INSULATION		COST PER S.F.		
						MAT.	INST.	TOTAL
4000	Hexagonal	4″ conc. block	8	perlite		8.45	16.60	25.05
4050	Reg. wt.			none		7.80	16.40	24.20
4060		6″ conc. block	10	perlite		9.35	17.25	26.60
4110				styrofoam		9.70	16.90	26.60
4120		8″ conc. block	12	perlite		10.10	17.85	27.95
4170				styrofoam		10.05	17.35	27.40
4200		4″ L.W. block	8	perlite		8.60	16.45	25.05
4250				none		7.90	16.25	24.15
4260		6″ L.W. block	10	perlite		9.75	17.10	26.85
4310				styrofoam		10.05	16.75	26.80
4320		8″ L.W. block	12	perlite		10.85	17.60	28.45
4370				styrofoam		10.85	17.15	28
4600	Hexagonal	4″ conc. block	8	perlite		9.50	16.40	25.90
4650	L.W.			none		8.85	16.20	25.05
4660		6″ conc. block	10	perlite		10.05	16.90	26.95
4710				styrofoam		10.70	16.70	27.40
4720		8″ conc. block	12	perlite		11.05	17.60	28.65
4770				styrofoam		11.05	17.15	28.20
4800		4″ L.W. block	8	perlite		9.60	16.25	25.85
4850				none		8.95	16.05	25
4860		6″ L.W. block	10	perlite		10.75	16.90	27.65
4910				styrofoam		11.05	16.55	27.60
4920		8″ L.W. block	12	perlite		11.90	17.40	29.30
4970				styrofoam		11.90	16.95	28.85
5200	Slump block	4″ conc. block	8	perlite		9.15	15.25	24.40
5250	4x16			none		8.50	15.05	23.55
5260		6″ conc. block	10	perlite		10.10	15.90	26
5310				styrofoam		10.40	15.55	25.95
5320		8″ conc. block	12	perlite		10.75	16.45	27.20
5370				styrofoam		10.75	16	26.75
5400		4″ L.W. block	8	perlite		9.30	15.10	24.40
5450				none		8.65	14.90	23.55
5460		6″ L.W. block	10	perlite		10.45	15.75	26.20
5510				styrofoam		10.75	15.40	26.15
5520		8″ L.W. block	12	perlite		11.55	16.25	27.80
5570				styrofoam		11.55	15.80	27.35
5800	Split face	4″ conc. block	8	perlite		10.55	16.85	27.40
5850	1 Score			none		9.90	16.65	26.55
5860	Reg. wt.	6″ conc. block	10	perlite		11.50	17.50	29
5910				styrofoam		11.80	17.15	28.95
5920		8″ conc. block	12	perlite		12.15	18.05	30.20
5970				styrofoam		12.15	17.60	29.75
6000		4″ L.W. block	8	perlite		10.70	16.70	27.40
6050				none		10.05	16.50	26.55
6060		6″ L.W. block	10	perlite		11.85	17.35	29.20
6110				styrofoam		12.15	17	29.15
6120		8″ L.W. block	12	perlite		12.95	17.85	30.80
6170				styrofoam		12.95	17.40	30.35

B2010 Exterior Walls

B2010 139 Block Face Composite Wall

	FACE BLOCK	BACKUP MASONRY	TOTAL THICKNESS (IN.)	BACKUP CORE INSULATION		COST PER S.F.		
						MAT.	INST.	TOTAL
6460	Split face	6″ conc. block	10	perlite		12.90	17.25	30.15
6510	1 Score			styrofoam		13.20	16.90	30.10
6520	L.W.	8″ conc. block	12	perlite		13.55	17.80	31.35
6570				styrofoam		13.55	17.35	30.90
6660		6″ L.W. block	10	perlite		13.25	17.10	30.35
6710				styrofoam		13.35	17	30.35
6720		8″ L.W. block	12	perlite		14.35	17.60	31.95
6770				styrofoam		14.15	17.40	31.55
7000	Split rib	4″ conc. block	8	perlite		11.75	16.85	28.60
7050	8 Rib			none		11.10	16.65	27.75
7060	Reg. wt.	6″ conc. block	10	perlite		12.70	17.50	30.20
7110				styrofoam		13	17.15	30.15
7120		8″ conc. block	12	perlite		13.35	18.05	31.40
7170				styrofoam		13.35	17.60	30.95
7200		4″ L.W. block	8	perlite		11.90	16.70	28.60
7250				none		11.25	16.50	27.75
7260		6″ L.W. block	10	perlite		13.05	17.35	30.40
7310				styrofoam		13.35	17	30.35
7320		8″ L.W. block	12	perlite		14.15	17.85	32
7370				styrofoam		14.15	17.40	31.55
7600	Split rib	4″ conc. block	8	perlite		12.70	16.60	29.30
7650	8 Rib			none		12.05	16.40	28.45
7660	L.W.	6″ conc. block	10	perlite		13.65	17.25	30.90
7710				styrofoam		13.95	16.90	30.85
7720		8″ conc. block	12	perlite		14.30	17.80	32.10
7770				styrofoam		14.30	17.35	31.65
7800		4″ conc. block	8	perlite		12.85	16.45	29.30
7850				none		12.20	16.25	28.45
7860		6″ conc. block	10	perlite		14	17.10	31.10
7910				styrofoam		14.30	16.75	31.05
7920		8″ conc. block	12	perlite		15.10	17.60	32.70
7970				styrofoam		15.10	17.15	32.25

B2010 Exterior Walls

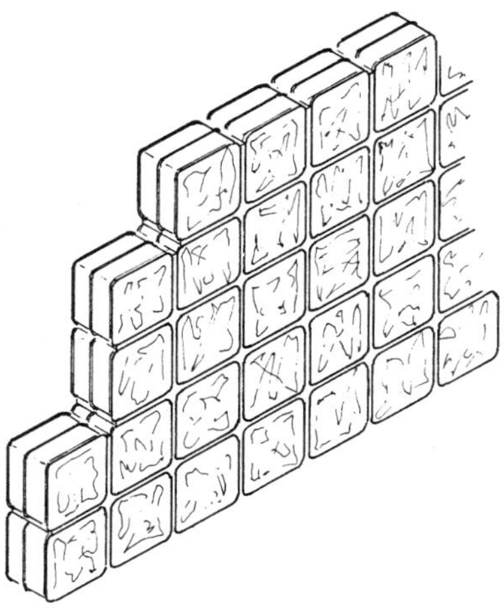

The table below lists costs per S.F. for glass block walls. Included in the costs are the following special accessories required for glass block walls.

Glass block accessories required for proper installation.

Wall ties: Galvanized double steel mesh full length of joint.

Fiberglass expansion joint at sides and top.

Silicone caulking: One gallon does 95 L.F.

Oakum: One lb. does 30 L.F.

Asphalt emulsion: One gallon does 600 L.F.

If block are not set in wall chase, use 2'-0" long wall anchors at 2'-0" O.C.

System Components	QUANTITY	UNIT	COST PER S.F. MAT.	INST.	TOTAL
SYSTEM B2010 140 2300					
GLASS BLOCK, 4" THICK, 6" X 6" PLAIN, UNDER 1,000 S.F.					
Glass block, 4" thick, 6" x 6" plain, under 1000 S.F.	4.100	Ea.	30.50	23	53.50
Glass block, cleaning blocks after installation, both sides add	2.000	S.F.	.18	2.62	2.80
TOTAL			30.68	25.62	56.30

B2010 140	Glass Block	COST PER S.F. MAT.	INST.	TOTAL
2300	Glass block 4" thick, 6"x6" plain, under 1,000 S.F.	30.50	25.50	56
2400	1,000 to 5,000 S.F.	30	22	52
2500	Over 5,000 S.F.	29	20.50	49.50
2600	Solar reflective, under 1,000 S.F.	43	35	78
2700	1,000 to 5,000 S.F.	42	30	72
2800	Over 5,000 S.F.	41	28	69
3500	8"x8" plain, under 1,000 S.F.	17.75	19	36.75
3600	1,000 to 5,000 S.F.	17.40	16.40	33.80
3700	Over 5,000 S.F.	16.90	14.80	31.70
3800	Solar reflective, under 1,000 S.F.	25	25.50	50.50
3900	1,000 to 5,000 S.F.	24.50	22	46.50
4000	Over 5,000 S.F.	23.50	19.70	43.20
5000	12"x12" plain, under 1,000 S.F.	26.50	17.60	44.10
5100	1,000 to 5,000 S.F.	26	14.80	40.80
5200	Over 5,000 S.F.	25	13.55	38.55
5300	Solar reflective, under 1,000 S.F.	37.50	23.50	61
5400	1,000 to 5,000 S.F.	36.50	19.70	56.20
5600	Over 5,000 S.F.	35	17.95	52.95
5800	3" thinline, 6"x6" plain, under 1,000 S.F.	24.50	25.50	50
5900	Over 5,000 S.F.	23.50	20.50	44
6000	Solar reflective, under 1,000 S.F.	34.50	35	69.50
6100	Over 5,000 S.F.	33	28	61
6200	8"x8" plain, under 1,000 S.F.	14	19	33
6300	Over 5,000 S.F.	13.40	14.80	28.20
6400	Solar reflective, under 1,000 S.F.	19.50	25.50	45
6500	Over 5,000 S.F.	18.65	19.70	38.35

B2010 Exterior Walls

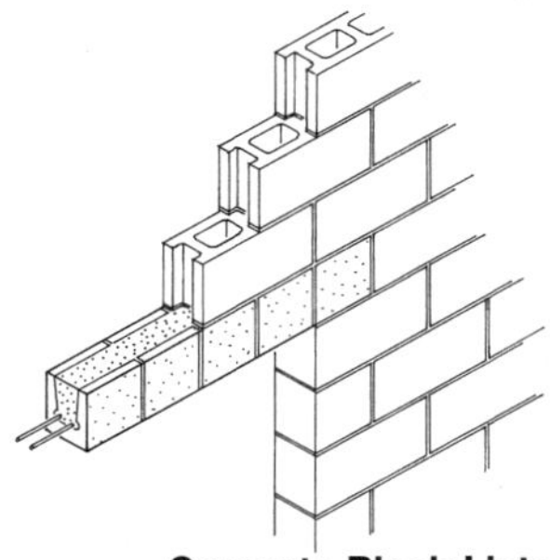

Concrete Block Lintel

Concrete block specialties are divided into lintels, pilasters, and bond beams. Lintels are defined by span, thickness, height and wall loading. Span refers to the clear opening but the cost includes 8″ of bearing at both ends.

Bond beams and pilasters are defined by height thickness and weight of the masonry unit itself. Components for bond beams also include grout and reinforcing. For pilasters, components include four #5 reinforcing bars and type N mortar.

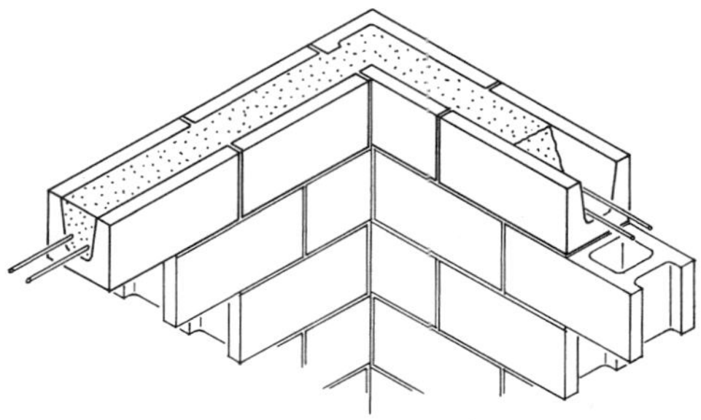

Bond Beam

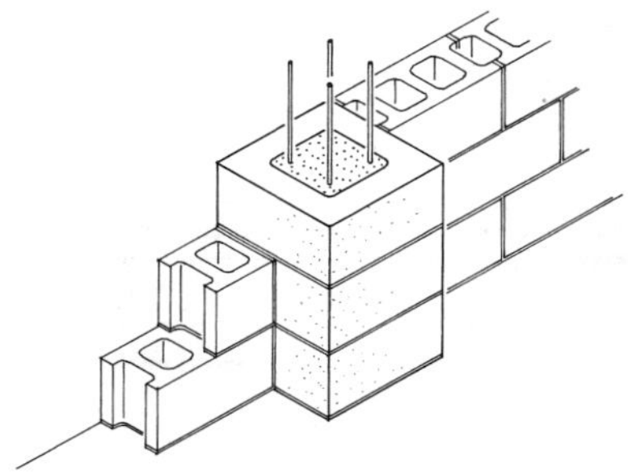

Pilaster

System Components	QUANTITY	UNIT	COST PER LINTEL		
			MAT.	INST.	TOTAL
SYSTEM B2010 144 3100					
CONCRETE BLOCK LINTEL, 6″ X 8″, LOAD 300 LB/L.F. 3′-4″ SPAN					
Lintel blocks, 6″ x 8″ x 8″	7.000	Ea.	10.15	24.01	34.16
Joint reinforcing, #3 & #4 steel bars, horizontal	3.120	Lb.	1.65	3.93	5.58
Grouting, 8″ deep, 6″ thick, .15 C.F./L.F.	.700	C.F.	4.27	5.61	9.88
Temporary shoring, lintel forms	1.000	Set	.94	13.48	14.42
Temporary shoring, wood joists	1.000	Set		24	24
Temporary shoring, vertical members	1.000	Set		43	43
Mortar, masonry cement, 1:3 mix, type N	.270	C.F.	1.63	.88	2.51
TOTAL			18.64	114.91	133.55

B2010 144	Concrete Block Lintel						
	SPAN	THICKNESS (IN.)	HEIGHT (IN.)	WALL LOADING P.L.F.	COST PER LINTEL		
					MAT.	INST.	TOTAL
3100	3′-4″	6	8	300	18.65	115	133.65
3150		6	16	1,000	30.50	124	154.50
3200		8	8	300	19.70	124	143.70
3300		8	8	1,000	21.50	128	149.50
3400		8	16	1,000	34	135	169
3450	4′-0″	6	8	300	22	127	149
3500		6	16	1,000	40	144	184
3600		8	8	300	23	134	157
3700		8	16	1,000	39	145	184

B2010 Exterior Walls

B2010 144 Concrete Block Lintel

	SPAN	THICKNESS (IN.)	HEIGHT (IN.)	WALL LOADING P.L.F.		COST PER LINTEL		
						MAT.	INST.	TOTAL
3800	4'-8"	6	8	300		24	130	154
3900		6	16	1,000		39.50	142	181.50
4000		8	8	300		26	143	169
4100		8	16	1,000		44	156	200
4200	5'-4"	6	8	300		29.50	144	173.50
4300		6	16	1,000		45.50	149	194.50
4400		8	8	300		31	157	188
4500		8	16	1,000		50	170	220
4600	6'-0"	6	8	300		35.50	174	209.50
4700		6	16	300		48.50	183	231.50
4800		6	16	1,000		51	189	240
4900		8	8	300		34.50	191	225.50
5000		8	16	1,000		56.50	208	264.50
5100	6'-8"	6	16	300		47	184	231
5200		6	16	1,000		50	191	241
5300		8	8	300		40	198	238
5400		8	16	1,000		62	219	281
5500	7'-4"	6	16	300		52.50	192	244.50
5600		6	16	1,000		65.50	206	271.50
5700		8	8	300		50	215	265
5800		8	16	300		74.50	231	305.50
5900		8	16	1,000		74.50	231	305.50
6000	8'-0"	6	16	300		56.50	200	256.50
6100		6	16	1,000		71	216	287
6200		8	16	300		83	243	326
6500		8	16	1,000		83	243	326

B2010 145 Concrete Block Specialties

	TYPE	HEIGHT	THICKNESS	WEIGHT (P.C.F.)		COST PER L.F.		
						MAT.	INST.	TOTAL
7100	Bond beam	8	8	125		5.45	7.70	13.15
7200				105		6.10	7.65	13.75
7300			12	125		6.85	9.70	16.55
7400				105		7.75	9.60	17.35
7500	Pilaster	8	16	125		12.90	25.50	38.40
7600			20	125		15.55	30.50	46.05

B2010 Exterior Walls

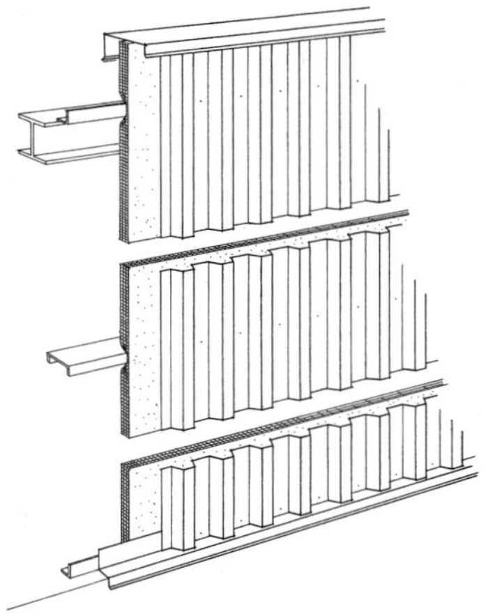

The table below lists costs for metal siding of various descriptions, not including the steel frame, or the structural steel, of a building. Costs are per S.F. including all accessories and insulation.

For steel frame support see System B2010 154.

System Components	QUANTITY	UNIT	COST PER S.F. MAT.	COST PER S.F. INST.	COST PER S.F. TOTAL
SYSTEM B2010 146 1400					
METAL SIDING ALUMINUM PANEL, CORRUGATED, .024″ THICK, NATURAL					
Corrugated aluminum siding, industrial type, .024″ thick	1.000	S.F.	2.53	3.01	5.54
Flashing alum mill finish .032″ thick	.111	S.F.	.17	.44	.61
Tapes, sealant , p.v.c. foam adheasive, 1/16″x1″	.004	C.L.F.	.07		.07
Closure strips for aluminum siding, corrugated, .032″ thick	.111	L.F.	.12	.32	.44
Pre-engineered steel building insulation, vinyl faced, 1-1/2″ thick, R5	1.000	S.F.	.37	.52	.89
TOTAL			3.26	4.29	7.55

B2010 146	Metal Siding Panel	COST PER S.F. MAT.	COST PER S.F. INST.	COST PER S.F. TOTAL
1400	Metal siding aluminum panel, corrugatec, .024″ thick, natural	3.26	4.29	7.55
1450	Painted	3.45	4.29	7.74
1500	.032″ thick, natural	3.48	4.29	7.77
1550	Painted	4.03	4.29	8.32
1600	Ribbed 4″ pitch, .032″ thick, natural	3.55	4.29	7.84
1650	Painted	4.09	4.29	8.38
1700	.040″ thick, natural	4.07	4.29	8.36
1750	Painted	4.64	4.29	8.93
1800	.050″ thick, natural	4.59	4.29	8.88
1850	Painted	5.20	4.29	9.49
1900	Ribbed 8″ pitch, .032″ thick, natural	3.39	4.11	7.50
1950	Painted	3.92	4.13	8.05
2000	.040″ thick, natural	3.91	4.14	8.05
2050	Painted	4.46	4.18	8.64
2100	.050″ thick, natural	4.42	4.16	8.58
2150	Painted	5.05	4.19	9.24
3000	Steel, corrugated or ribbed, 29 Ga., .0135″ thick, galvanized	2.14	3.90	6.04
3050	Colored	3.07	3.94	7.01
3100	26 Ga., .0179″ thick, galvanized	2.25	3.92	6.17
3150	Colored	3.25	3.96	7.21
3200	24 Ga., .0239″ thick, galvanized	2.80	3.94	6.74
3250	Colored	3.54	3.98	7.52
3300	22 Ga., .0299″ thick, galvanized	3.02	3.96	6.98
3350	Colored	4.24	4	8.24
3400	20 Ga., .0359″ thick, galvanized	3.02	3.96	6.98
3450	Colored	4.50	4.23	8.73

B2010 Exterior Walls

B2010 146	Metal Siding Panel	COST PER S.F.		
		MAT.	INST.	TOTAL
4100	Sandwich panels, factory fab., 1" polystyrene, steel core, 26 Ga., galv.	5.90	6.15	12.05
4200	Colored, 1 side	7.25	6.15	13.40
4300	2 sides	9.10	6.15	15.25
4400	2" polystyrene core, 26 Ga., galvanized	10.05	6.15	16.20
4500	Colored, 1 side	8.35	6.15	14.50
4600	2 sides	10.20	6.15	16.35
4700	22 Ga., baked enamel exterior	13.35	6.45	19.80
4800	Polyvinyl chloride exterior	14.10	6.45	20.55
5100	Textured aluminum, 4' x 8' x 5/16" plywood backing, single face	4.50	3.73	8.23
5200	Double face	5.85	3.73	9.58
5300	4' x 10' x 5/16" plywood backing, single face	4.73	3.73	8.46

B2010 Exterior Walls

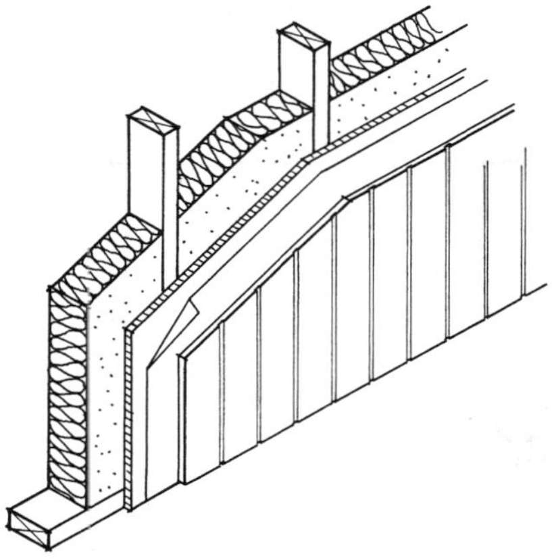

The table below lists costs per S.F. for exterior walls with wood siding. A variety of systems are presented using both wood and metal studs at 16″ and 24″ O.C.

System Components	QUANTITY	UNIT	COST PER S.F. MAT.	COST PER S.F. INST.	COST PER S.F. TOTAL
SYSTEM B2010 148 1400					
2″ X 4″ STUDS, 16″ O.C., INSUL. WALL, W/ 5/8″ TEXTURE 1-11 FIR PLYWOOD					
Partitions, 2″ x 4″ studs 8′ high 16″ O.C.	1.000	B.F.	.49	1.49	1.98
Sheathing plywood on wall CDX 1/2″	1.000	S.F.	.75	1.06	1.81
Building paper, asphalt felt sheathing paper 15 lb	1.000	S.F.	.06	.16	.22
Fiberglass insulation batts, paper or foil back, 3-1/2″, R11	1.000	S.F.	.52	.44	.96
Siding plywood texture 1-11 fir 5/8″, natural	1.000	S.F.	1.61	1.76	3.37
Exterior wood stain on other than shingles, 2 coats & sealer	1.000	S.F.	.25	1.14	1.39
TOTAL			3.68	6.05	9.73

B2010 148	Panel, Shingle & Lap Siding	COST PER S.F. MAT.	COST PER S.F. INST.	COST PER S.F. TOTAL
1400	Wood siding w/2″ x 4″ studs, 16″ O.C., insul. wall, 5/8″ text 1-11 fir ply.	3.68	6.05	9.73
1450	5/8″ text 1-11 cedar plywood	5.05	5.85	10.90
1500	1″ x 4″ vert T.&G. redwood	5.65	7	12.65
1600	1″ x 8″ vert T.&G. redwood	7.75	6.10	13.85
1650	1″ x 5″ rabbetted cedar bev. siding	6.75	6.30	13.05
1700	1″ x 6″ cedar drop siding	6.85	6.35	13.20
1750	1″ x 12″ rough sawn cedar	3.49	6.10	9.59
1800	1″ x 12″ sawn cedar, 1″ x 4″ battens	7	5.70	12.70
1850	1″ x 10″ redwood shiplap siding	7	5.85	12.85
1900	18″ no. 1 red cedar shingles, 5-1/2″ exposed	4.88	8	12.88
1950	6″ exposed	4.68	7.70	12.38
2000	6-1/2″ exposed	4.48	7.45	11.93
2100	7″ exposed	4.28	7.20	11.48
2150	7-1/2″ exposed	4.08	6.95	11.03
3000	8″ wide aluminum siding	4.54	5.45	9.99
3150	8″ plain vinyl siding	3.31	5.55	8.86
3250	8″ insulated vinyl siding	3.67	6.15	9.82
3400	2″ x 6″ studs, 16″ O.C., insul. wall, w/ 5/8″ text 1-11 fir plywood	4.11	6.20	10.31
3500	5/8″ text 1-11 cedar plywood	5.45	6.20	11.65
3600	1″ x 4″ vert T.&G. redwood	6.10	7.15	13.25
3700	1″ x 8″ vert T.&G. redwood	8.20	6.25	14.45
3800	1″ x 5″ rabbetted cedar bev siding	7.20	6.45	13.65
3900	1″ x 6″ cedar drop siding	7.25	6.50	13.75
4000	1″ x 12″ rough sawn cedar	3.92	6.25	10.17
4200	1″ x 12″ sawn cedar, 1″ x 4″ battens	7.45	5.85	13.30
4500	1″ x 10″ redwood shiplap siding	7.45	6.05	13.50

B2010 Exterior Walls

B2010 148	Panel, Shingle & Lap Siding	COST PER S.F.		
		MAT.	INST.	TOTAL
4550	18" no. 1 red cedar shingles, 5-1/2" exposed	5.30	8.15	13.45
4600	6" exposed	5.10	7.90	13
4650	6-1/2" exposed	4.91	7.60	12.51
4700	7" exposed	4.71	7.35	12.06
4750	7-1/2" exposed	4.51	7.10	11.61
4800	8" wide aluminum siding	4.97	5.60	10.57
4850	8" plain vinyl siding	4.82	9.75	14.57
4900	8" insulated vinyl siding	5.20	10.35	15.55
4950	8" fiber cement siding	4.70	10.55	15.25
5000	2" x 6" studs, 24" O.C., insul. wall, 5/8" text 1-11, fir plywood	3.94	5.85	9.79
5050	5/8" text 1-11 cedar plywood	5.30	5.85	11.15
5100	1" x 4" vert T.&G. redwood	7	10.85	17.85
5150	1" x 8" vert T.&G. redwood	8.05	5.90	13.95
5200	1" x 5" rabbetted cedar bev siding	7	6.10	13.10
5250	1" x 6" cedar drop siding	7.10	6.15	13.25
5300	1" x 12" rough sawn cedar	3.75	5.90	9.65
5400	1" x 12" sawn cedar, 1" x 4" battens	7.25	5.50	12.75
5450	1" x 10" redwood shiplap siding	7.30	5.65	12.95
5500	18" no. 1 red cedar shingles, 5-1/2" exposed	5.15	7.80	12.95
5550	6" exposed	4.94	7.50	12.44
5650	7" exposed	4.54	7	11.54
5700	7-1/2" exposed	4.34	6.75	11.09
5750	8" wide aluminum siding	4.80	5.25	10.05
5800	8" plain vinyl siding	3.57	5.35	8.92
5850	8" insulated vinyl siding	3.93	5.95	9.88
5875	8" fiber cement siding	4.53	10.15	14.68
5900	3-5/8" metal studs, 16 Ga. 16" O.C. insul. wall, 5/8" text 1-11 fir plywood	4.54	6.35	10.89
5950	5/8" text 1-11 cedar plywood	5.90	6.35	12.25
6000	1" x 4" vert T.&G. redwood	6.50	7.30	13.80
6050	1" x 8" vert T.&G. redwood	8.65	6.40	15.05
6100	1" x 5" rabbetted cedar bev siding	7.60	6.60	14.20
6150	1" x 6" cedar drop siding	7.70	6.65	14.35
6200	1" x 12" rough sawn cedar	4.35	6.40	10.75
6250	1" x 12" sawn cedar, 1" x 4" battens	7.85	6	13.85
6300	1" x 10" redwood shiplap siding	7.90	6.20	14.10
6350	18" no. 1 red cedar shingles, 5-1/2" exposed	5.75	8.30	14.05
6500	6" exposed	5.55	8.05	13.60
6550	6-1/2" exposed	5.35	7.75	13.10
6600	7" exposed	5.15	7.50	12.65
6650	7-1/2" exposed	5.15	7.30	12.45
6700	8" wide aluminum siding	5.40	5.75	11.15
6750	8" plain vinyl siding	4.17	5.65	9.82
6800	8" insulated vinyl siding	4.53	6.25	10.78
7000	3-5/8" metal studs, 16 Ga. 24" O.C. insul wall, 5/8" text 1-11 fir plywood	4.17	5.65	9.82
7050	5/8" text 1-11 cedar plywood	5.55	5.65	11.20
7100	1" x 4" vert T.&G. redwood	6.15	6.80	12.95
7150	1" x 8" vert T.&G. redwood	8.25	5.90	14.15
7200	1" x 5" rabbetted cedar bev siding	7.25	6.10	13.35
7250	1" x 6" cedar drop siding	7.35	6.15	13.50
7300	1" x 12" rough sawn cedar	3.98	5.90	9.88
7350	1" x 12" sawn cedar 1" x 4" battens	7.50	5.50	13
7400	1" x 10" redwood shiplap siding	7.50	5.65	13.15
7450	18" no. 1 red cedar shingles, 5-1/2" exposed	5.60	8.05	13.65
7500	6" exposed	5.35	7.80	13.15
7550	6-1/2" exposed	5.15	7.50	12.65
7600	7" exposed	4.97	7.25	12.22
7650	7-1/2" exposed	4.57	6.75	11.32
7700	8" wide aluminum siding	5.05	5.25	10.30

B2010 148	Panel, Shingle & Lap Siding	COST PER S.F.		
		MAT.	INST.	TOTAL
7750	8″ plain vinyl siding	3.80	5.35	9.15
7800	8″ insul. vinyl siding	4.16	5.95	10.11

B2010 Exterior Walls

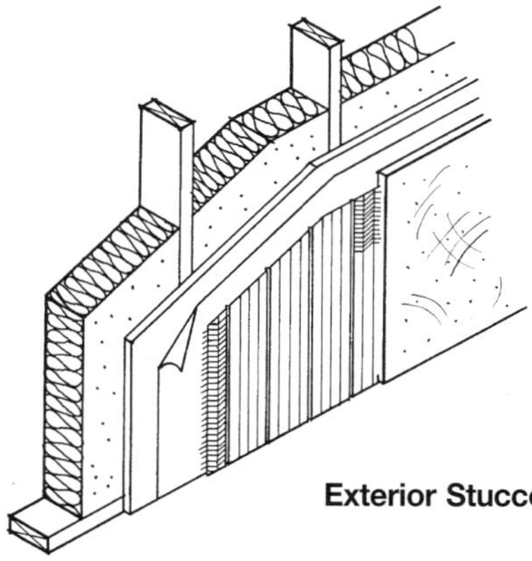

Exterior Stucco Wall

The table below lists costs for some typical stucco walls including all the components as demonstrated in the component block below. Prices are presented for backup walls using wood studs, metal studs and CMU.

System Components	QUANTITY	UNIT	COST PER S.F. MAT.	COST PER S.F. INST.	COST PER S.F. TOTAL
SYSTEM B2010 151 2100					
7/8″ CEMENT STUCCO, PLYWOOD SHEATHING, STUD WALL, 2″ X 4″, 16″ O.C.					
Framing 2x4 studs 8′ high 16″ O.C.	1.000	S.F.	.49	1.49	1.98
Plywood sheathing on ext. stud wall, 5/8″ thick	1.000	S.F.	.84	1.13	1.97
Building paper, asphalt felt sheathing paper 15 lb	1.000	S.F.	.06	.16	.22
Stucco, 3 coats 7/8″ thk, float finish, on frame construction	1.000	S.F.	.78	5.71	6.49
Fiberglass insulation batts, paper or foil back, 3-1/2″, R-11	1.000	S.F.	.52	.44	.96
Paint exterior stucco, brushwork, primer & 2 coats	1.000	S.F.	.24	.97	1.21
TOTAL			2.93	9.90	12.83

B2010 151	Stucco Wall	COST PER S.F. MAT.	COST PER S.F. INST.	COST PER S.F. TOTAL
2100	Cement stucco, 7/8″ th., plywood sheathing, stud wall, 2″ x 4″, 16″ O.C.	2.93	9.90	12.83
2200	24″ O.C.	2.82	9.60	12.42
2300	2″ x 6″, 16″ O.C.	3.36	10.05	13.41
2400	24″ O.C.	3.19	9.70	12.89
2500	No sheathing, metal lath on stud wall, 2″ x 4″, 16″ O.C.	2.03	8.60	10.63
2600	24″ O.C.	1.92	8.30	10.22
2700	2″ x 6″, 16″ O.C.	2.46	8.75	11.21
2800	24″ O.C.	2.29	8.40	10.69
2900	1/2″ gypsum sheathing, 3-5/8″ metal studs, 16″ O.C.	3.16	9	12.16
2950	24″ O.C.	2.79	8.50	11.29
3000	Cement stucco, 5/8″ th., 2 coats on std. CMU block, 8″x 16″, 8″ thick	3.43	11.80	15.23
3100	10″ thick	5	12.10	17.10
3200	12″ thick	5.15	13.90	19.05
3300	Std. light Wt. block 8″ x 16″, 8″ thick	4.19	11.60	15.79
3400	10″ thick	4.95	11.90	16.85
3500	12″ thick	5.20	13.65	18.85
3600	3 coat stucco, self furring metal lath 3.4 Lb/SY, on 8″ x 16″, 8″ thick	4.30	13.90	18.20
3700	10″ thick	5.05	12.80	17.85
3800	12″ thick	8.25	17.25	25.50
3900	Lt. Wt. block, 8″ thick	4.24	12.30	16.54
4000	10″ thick	5	12.60	17.60
4100	12″ thick	5.25	14.35	19.60

B2010 Exterior Walls

B2010 152	E.I.F.S.	COST PER S.F.		
		MAT.	INST.	TOTAL
5100	E.I.F.S., plywood sheathing, stud wall, 2" x 4", 16" O.C., 1" EPS	3.97	9.90	13.87
5110	2" EPS	4.26	9.90	14.16
5120	3" EPS	4.54	9.90	14.44
5130	4" EPS	4.83	9.90	14.73
5140	2" x 6", 16" O.C., 1" EPS	4.40	10.05	14.45
5150	2" EPS	4.87	10.20	15.07
5160	3" EPS	4.97	10.05	15.02
5170	4" EPS	5.25	10.05	15.30
5180	Cement board sheathing, 3-5/8" metal studs, 16" O.C., 1" EPS	4.85	12.25	17.10
5190	2" EPS	5.15	12.25	17.40
5200	3" EPS	5.45	12.25	17.70
5210	4" EPS	6.60	12.25	18.85
5220	6" metal studs, 16" O.C., 1" EPS	5.45	12.30	17.75
5230	2" EPS	5.90	12.45	18.35
5240	3" EPS	6	12.30	18.30
5250	4" EPS	7.15	12.30	19.45
5260	CMU block, 8" x 8" x 16", 1" EPS	5.05	13.95	19
5270	2" EPS	6.80	18.35	25.15
5280	3" EPS	5.60	13.95	19.55
5290	4" EPS	6.75	15.40	22.15
5300	8" x 10" x 16", 1" EPS	6.65	14.25	20.90
5310	2" EPS	6.90	14.25	21.15
5320	3" EPS	7.20	14.25	21.45
5330	4" EPS	7.50	14.25	21.75
5340	8" x 12" x 16", 1" EPS	6.80	16.05	22.85
5350	2" EPS	7.10	16.05	23.15
5360	3" EPS	7.35	16.05	23.40
5370	4" EPS	7.65	16.05	23.70

B2010 153	Oversized Brick Curtain Wall	COST PER S.F.		
		MAT.	INST.	TOTAL
1000	Oversized brk curtain wall, 6"x4"x16", incl rebar, insul, bond beam, plain	22	10.40	32.40
1010	1 to 3 slots in face	27.50	12.65	40.15
1020	4 to 7 slots in face	31	14.15	45.15
1030	8" x 4" x 16", incl rebar, insul & bond beam, plain face	26	11.85	37.85
1040	1 to 3 slots in face	33	14.50	47.50
1050	4 to 7 slots in face	37.50	16.25	53.75
1060	6" x 8" x 16", incl rebar, insul & bond beam, plain face	27	9.40	36.40
1070	1 to 3 slots in face	34	11.35	45.35
1080	4 to 7 slots in face	38.50	12.65	51.15
1090	8" x 8" x 16", incl rebar, insul & bond beam, plain face	36.50	10.10	46.60
1100	1 to 3 slots in face	46	12.25	58.25
1110	4 to 7 slots in face	52.50	13.70	66.20

B2010 Exterior Walls

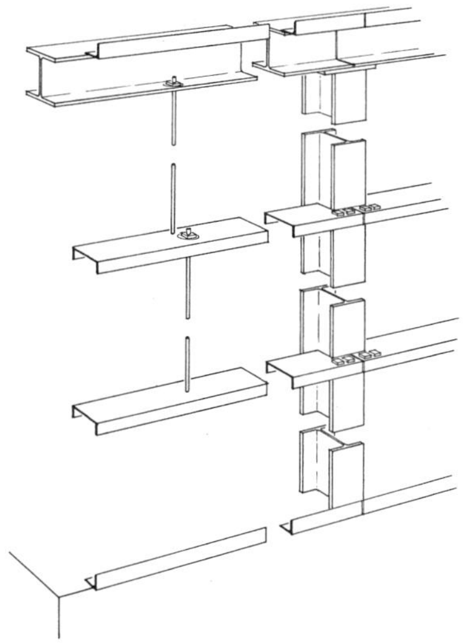

Description: The table below lists costs, $/S.F., for channel girts with sag rods and connector angles top and bottom for various column spacings, building heights and wind loads. Additive costs are shown for wind columns.

How to Use this Table: Add the cost of girts, sag rods, and angles to the framing costs of steel buildings clad in metal or composition siding. If the column spacing is in excess of the column spacing shown, use intermediate wind columns. Additive costs are shown under "Wind Columns".

System Components	QUANTITY	UNIT	COST PER S.F. MAT.	INST.	TOTAL
SYSTEM B2010 154 3000					
METAL SIDING SUPPORT, 18′ HIGH, 20 PSF WIND LOAD, 20′ SPACING					
Structural steel	1.417	Lb.	2.07	.48	2.55
Lightgage framing, angles less than 4″	.400	Lb.	.35	3.54	3.89
TOTAL			2.42	4.02	6.44
SYSTEM B2010 154 3100					
WIND COLUMNS					
Structural steel	.856	S.F.	1.25	.29	1.54

B2010 154	Metal Siding Support						
	BLDG. HEIGHT (FT.)	WIND LOAD (P.S.F.)	COL. SPACING (FT.)	INTERMEDIATE COLUMNS	COST PER S.F. MAT.	INST.	TOTAL
3000	18	20	20		2.42	4.02	6.44
3100				wind cols.	1.25	.29	1.54
3200	RB2010 -430	20	25		2.65	4.07	6.72
3300				wind cols.	1	.23	1.23
3400		20	30		2.92	4.14	7.06
3500				wind cols.	.83	.19	1.02
3600		20	35		3.21	4.21	7.42
3700				wind cols.	.71	.16	.87
3800		30	20		2.68	4.08	6.76
3900				wind cols.	1.25	.29	1.54
4000		30	25		2.92	4.14	7.06
4100				wind cols.	1	.23	1.23
4200		30	30		3.23	4.21	7.44
4300				wind cols.	1.13	.26	1.39
4400		30	35		4.29	4.46	8.75
4500				wind cols.	1.12	.26	1.38
4600	30	20	20		2.35	2.83	5.18
4700				wind cols.	1.77	.41	2.18
4800		20	25		2.64	2.90	5.54
4900				wind cols.	1.41	.33	1.74

B2010 Exterior Walls

B2010 154 Metal Siding Support

	BLDG. HEIGHT (FT.)	WIND LOAD (P.S.F.)	COL. SPACING (FT.)		INTERMEDIATE COLUMNS	COST PER S.F.		
						MAT.	INST.	TOTAL
5000	30	20	30			2.97	2.98	5.95
5100					wind cols.	1.39	.33	1.72
5200		20	35			3.32	3.06	6.38
5300					wind cols.	1.38	.32	1.70
5400		30	20			2.67	2.91	5.58
5500					wind cols.	2.09	.49	2.58
5600		30	25			2.96	2.97	5.93
5700					wind cols.	1.99	.47	2.46
5800		30	30			3.34	3.06	6.40
5900					wind cols.	1.87	.44	2.31
6000		30	35			4.61	3.36	7.97
6100					wind cols.	1.84	.43	2.27

B2020 Exterior Windows

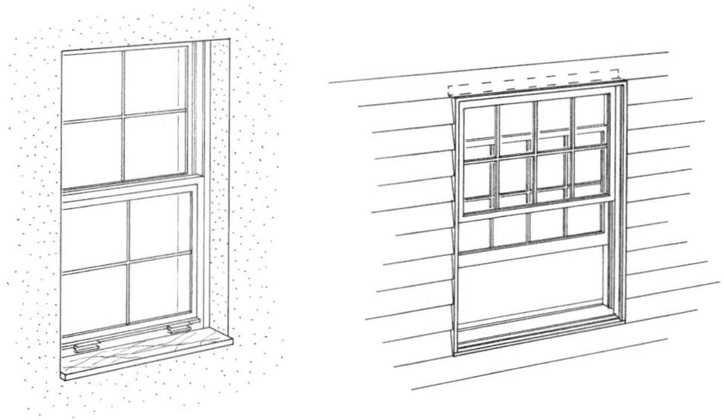

The table below lists window systems by material, type and size. Prices between sizes listed can be interpolated with reasonable accuracy. Prices include frame, hardware, and casing as illustrated in the component block below.

System Components	QUANTITY	UNIT	COST PER UNIT MAT.	INST.	TOTAL
SYSTEM B2020 102 3000					
WOOD, DOUBLE HUNG, STD. GLASS, 2'-8" X 4'-6"					
Framing rough opening, header w/jacks	14.680	B.F.	10.28	30.09	40.37
Casing, stock pine 11/16" x 2-1/2"	16.000	L.F.	24.80	38.08	62.88
Stool cap, pine, 11/16" x 3-1/2"	5.000	L.F.	11.40	14.85	26.25
Residential wood window, double hung, 2'-8" x 4'-6" standard glazed	1.000	Ea.	201	175	376
TOTAL			247.48	258.02	505.50

B2020 102			Wood Windows					
	MATERIAL	TYPE	GLAZING	SIZE	DETAIL	COST PER UNIT MAT.	INST.	TOTAL
3000	Wood	double hung	std. glass	2'-8" x 4'-6"		247	258	505
3050				3'-0" x 5'-6"		325	296	621
3100			insul. glass	2'-8" x 4'-6"		263	258	521
3150				3'-0" x 5'-6"		350	296	646
3200		sliding	std. glass	3'-4" x 2'-7"		330	215	545
3250				4'-4" x 3'-3"		345	235	580
3300				5'-4" x 6'-0"		435	283	718
3350			insul. glass	3'-4" x 2'-7"		405	251	656
3400				4'-4" x 3'-3"		415	273	688
3450				5'-4" x 6'-0"		530	320	850
3500		awning	std. glass	2'-10" x 1'-9"		230	129	359
3600				4'-4" x 2'-8"		375	125	500
3700			insul. glass	2'-10" x 1'-9"		281	149	430
3800				4'-4" x 2'-8"		460	137	597
3900		casement	std. glass	1'-10" x 3'-2"	1 lite	340	170	510
3950				4'-2" x 4'-2"	2 lite	600	225	825
4000				5'-11" x 5'-2"	3 lite	900	290	1,190
4050				7'-11" x 6'-3"	4 lite	1,275	350	1,625
4100				9'-11" x 6'-3"	5 lite	1,675	405	2,080
4150			insul. glass	1'-10" x 3'-2"	1 lite	340	170	510
4200				4'-2" x 4'-2"	2 lite	615	225	840
4250				5'-11" x 5'-2"	3 lite	950	290	1,240
4300				7'-11" x 6'-3"	4 lite	1,350	350	1,700
4350				9'-11" x 6'-3"	5 lite	1,675	405	2,080

B2020 Exterior Windows

B2020 102 Wood Windows

	MATERIAL	TYPE	GLAZING	SIZE	DETAIL	COST PER UNIT MAT.	COST PER UNIT INST.	COST PER UNIT TOTAL
4400	Wood	picture	std. glass	4'-6" x 4'-6"		475	291	766
4450				5'-8" x 4'-6"		535	325	860
4500		picture	insul. glass	4'-6" x 4'-6"		580	340	920
4550				5'-8" x 4'-6"		650	375	1,025
4600		fixed bay	std. glass	8' x 5'		1,650	535	2,185
4650				9'-9" x 5'-4"		1,175	750	1,925
4700			insul. glass	8' x 5'		2,175	535	2,710
4750				9'-9" x 5'-4"		1,275	750	2,025
4800		casement bay	std. glass	8' x 5'		1,725	615	2,340
4850			insul. glass	8' x 5'		1,875	615	2,490
4900		vert. bay	std. glass	8' x 5'		1,875	615	2,490
4950			insul. glass	8' x 5'		1,950	615	2,565

B2020 104 Steel Windows

	MATERIAL	TYPE	GLAZING	SIZE	DETAIL	COST PER UNIT MAT.	COST PER UNIT INST.	COST PER UNIT TOTAL
5000	Steel	double hung	1/4" tempered	2'-8" x 4'-6"		910	200	1,110
5050				3'-4" x 5'-6"		1,400	305	1,705
5100			insul. glass	2'-8" x 4'-6"		930	230	1,160
5150				3'-4" x 5'-6"		1,425	350	1,775
5200		horz. pivoted	std. glass	2' x 2'		276	66.50	342.50
5250				3' x 3'		620	150	770
5300				4' x 4'		1,100	266	1,366
5350				6' x 4'		1,650	400	2,050
5400			insul. glass	2' x 2'		281	76.50	357.50
5450				3' x 3'		635	172	807
5500				4' x 4'		1,125	305	1,430
5550				6' x 4'		1,675	460	2,135
5600		picture window	std. glass	3' x 3'		390	150	540
5650				6' x 4'		1,050	400	1,450
5700			insul. glass	3' x 3'		405	172	577
5750				6' x 4'		1,075	460	1,535
5800		industrial security	std. glass	2'-9" x 4'-1"		830	187	1,017
5850				4'-1" x 5'-5"		1,625	370	1,995
5900			insul. glass	2'-9" x 4'-1"		845	215	1,060
5950				4'-1" x 5'-5"		1,675	425	2,100
6000		comm. projected	std. glass	3'-9" x 5'-5"		1,350	340	1,690
6050				6'-9" x 4'-1"		1,825	460	2,285
6100			insul. glass	3'-9" x 5'-5"		1,375	390	1,765
6150				6'-9" x 4'-1"		1,875	525	2,400
6200		casement	std. glass	4'-2" x 4'-2"	2 lite	1,225	289	1,514
6250			insul. glass	4'-2" x 4'-2"		1,250	330	1,580
6300			std. glass	5'-11" x 5'-2"	3 lite	2,550	510	3,060
6350			insul. glass	5'-11" x 5'-2"		2,600	585	3,185

B2020 106 Aluminum Windows

	MATERIAL	TYPE	GLAZING	SIZE	DETAIL	COST PER UNIT MAT.	COST PER UNIT INST.	COST PER UNIT TOTAL
6400	Aluminum	double hung	insul. glass	3'-0" x 4'-0"		530	144	674
6450				4'-5" x 5'-3"		440	180	620
6500			insul. glass	3'-1" x 3'-2"		470	173	643
6550				4'-5" x 5'-3"		530	216	746

B2020 Exterior Windows

B2020 106 Aluminum Windows

	MATERIAL	TYPE	GLAZING	SIZE	DETAIL	COST PER UNIT		
						MAT.	INST.	TOTAL
6600	Aluminum	sliding	std. glass	3′ x 2′		239	144	383
6650				5′ x 3′		365	160	525
6700				8′ x 4′		385	240	625
6750				9′ x 5′		585	360	945
6800			insul. glass	3′ x 2′		255	144	399
6850				5′ x 3′		425	160	585
6900				8′ x 4′		620	240	860
6950				9′ x 5′		935	360	1,295
7000		single hung	std. glass	2′ x 3′		229	144	373
7050				2′-8″ x 6′-8″		400	180	580
7100				3′-4″ x 5′-0″		330	160	490
7150			insul. glass	2′ x 3′		277	144	421
7200				2′-8″ x 6′-8″		520	180	700
7250				3′-4″ x 5′		365	160	525
7300		double hung	std. glass	2′ x 3′		320	100	420
7350				2′-8″ x 6′-8″		950	296	1,246
7400				3′-4″ x 5′		890	277	1,167
7450			insul. glass	2′ x 3′		330	115	445
7500				2′-8″ x 6′-8″		975	340	1,315
7550				3′-4″ x 5′-0″		910	320	1,230
7600		casement	std. glass	3′-1″ x 3′-2″		283	162	445
7650				4′-5″ x 5′-3″		685	395	1,080
7700			insul. glass	3′-1″ x 3′-2″		296	187	483
7750				4′-5″ x 5′-3″		720	450	1,170
7800		hinged swing	std. glass	3′ x 4′		585	200	785
7850				4′ x 5′		980	335	1,315
7900			insul. glass	3′ x 4′		605	230	835
7950				4′ x 5′		1,000	385	1,385
8200		picture unit	std. glass	2′-0″ x 3′-0″		182	100	282
8250				2′-8″ x 6′-8″		540	296	836
8300				3′-4″ x 5′-0″		505	277	782
8350			insul. glass	2′-0″ x 3′-0″		191	115	306
8400				2′-8″ x 6′-8″		565	340	905
8450				3′-4″ x 5′-0″		530	320	850
8500		awning type	std. glass	3′-0″ x 3′-0″	2 lite	470	103	573
8550				3′-0″ x 4′-0″	3 lite	540	144	684
8600				3′-0″ x 5′-4″	4 lite	650	144	794
8650				4′-0″ x 5′-4″	4 lite	715	160	875
8700			insul. glass	3′-0″ x 3′-0″	2 lite	495	103	598
8750				3′-0″ x 4′-0″	3 lite	620	144	764
8800				3′-0″ x 5′-4″	4 lite	765	144	909
8850				4′-0″ x 5′-4″	4 lite	855	160	1,015

B2020 108 Vinyl Clad Windows

	MATERIAL	TYPE	GLAZING	SIZE	DETAIL	COST PER UNIT		
						MAT.	INST.	TOTAL
1000	Vinyl clad	casement	insul. glass	1′-4″ x 4′-0″		425	137	562
1010				2′-0″ x 3′-0″		355	137	492
1020				2′-0″ x 4′-0″		410	144	554
1030				2′-0″ x 5′-0″		460	152	612
1040				2′-0″ x 6′-0″		480	152	632
1050				2′-4″ x 4′-0″		460	152	612

B2020 Exterior Windows

B2020 108 Vinyl Clad Windows

	MATERIAL	TYPE	GLAZING	SIZE	DETAIL	COST PER UNIT		
						MAT.	INST.	TOTAL
1060	Vinyl clad	casement	insul. glass	2'-6" x 5'-0"		565	148	713
1070				3'-0" x 5'-0"		790	152	942
1080				4'-0" x 3'-0"		865	152	1,017
1090				4'-0" x 4'-0"		745	152	897
1100				4'-8" x 4'-0"		820	152	972
1110				4'-8" x 5'-0"		925	177	1,102
1120				4'-8" x 6'-0"		1,025	177	1,202
1130				6'-0" x 4'-0"		945	177	1,122
1140				6'-0" x 5'-0"		1,050	177	1,227
3000		double-hung	insul. glass	2'-0" x 4'-0"		440	132	572
3050				2'-0" x 5'-0"		500	132	632
3100				2'-4" x 4'-0"		455	137	592
3150				2'-4" x 4'-8"		455	137	592
3200				2'-4" x 6'-0"		515	137	652
3250				2'-6" x 4'-0"		455	137	592
3300				2'-8" x 4'-0"		680	177	857
3350				2'-8" x 5'-0"		535	152	687
3375				2'-8" x 6'-0"		545	152	697
3400				3'-0 x 3'-6"		430	137	567
3450				3'-0 x 4'-0"		495	144	639
3500				3'-0 x 4'-8"		530	144	674
3550				3'-0 x 5'-0"		545	152	697
3600				3'-0 x 6'-0"		490	152	642
3700				4'-0 x 5'-0"		675	163	838
3800				4'-0 x 6'-0"		825	163	988

B2020 Exterior Windows

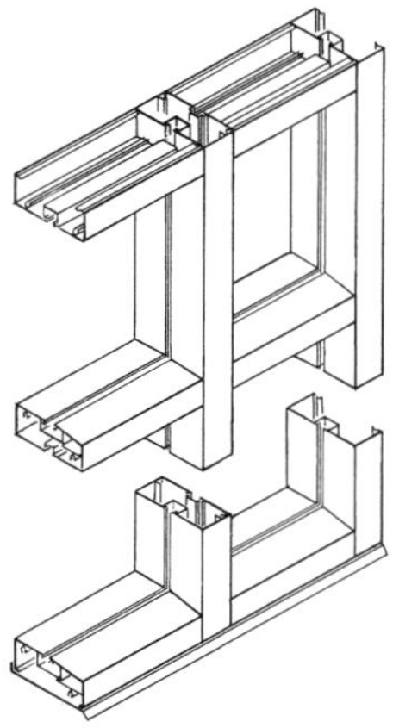

The table below lists costs per S.F of opening for framing with 1-3/4" x 4-1/2" clear anodized tubular aluminum framing. This is the type often used for 1/4" plate glass flush glazing.

For bronze finish, add 18% to material cost. For black finish, add 27% to material cost. For stainless steel, add 75% to material cost. For monumental grade, add 50% to material cost. This tube framing is usually installed by a glazing contractor.

Note: The costs below do not include the glass. For glazing, use Assembly B2020 220.

System Components	QUANTITY	UNIT	COST/S.F. OPNG. MAT.	INST.	TOTAL
SYSTEM B2020 210 1250					
ALUM FLUSH TUBE, FOR 1/4" GLASS, 5'X20' OPENING, 3 INTER. HORIZONTALS					
Flush tube frame, alum mill fin, 1-3/4"x4" open headr for 1/4" glass	.450	L.F.	6.71	6.39	13.10
Flush tube frame, alum mill fin, 1-3/4"x4" open sill for 1/4" glass	.050	L.F.	.61	.69	1.30
Flush tube frame, alum mill fin, 1-3/4"x4" closed back sill, 1/4" glass	.150	L.F.	3.08	2	5.08
Aluminum structural shapes, 1" to 10" members, under 1 ton	.040	Lb.	.17	.07	.24
Joints for tube frame, 90° clip type	.100	Ea.	2.90		2.90
Caulking/sealants, polysulfide, 1 or 2 part,1/2x1/4"bead 154 lf/gal	.500	L.F.	.30	.98	1.28
TOTAL			13.77	10.13	23.90

B2020 210	Tubular Aluminum Framing	COST/S.F. OPNG. MAT.	INST.	TOTAL
1100	Alum flush tube frame, for 1/4" glass, 1-3/4"x4", 5'x6'opng, no inter horiz	14.95	11.85	26.80
1150	One intermediate horizontal	20.50	14.15	34.65
1200	Two intermediate horizontals	26	16.40	42.40
1250	5' x 20' opening, three intermediate horizontals	13.75	10.10	23.85
1400	1-3/4" x 4-1/2", 5' x 6' opening, no intermediate horizontals	17.20	11.85	29.05
1450	One intermediate horizontal	23	14.15	37.15
1500	Two intermediate horizontals	28.50	16.40	44.90
1550	5' x 20' opening, three intermediate horizontals	15.45	10.10	25.55
1700	For insulating glass, 2"x4-1/2", 5'x6' opening, no intermediate horizontals	17.50	12.50	30
1750	One intermediate horizontal	22.50	14.95	37.45
1800	Two intermediate horizontals	27	17.35	44.35
1850	5' x 20' opening, three intermediate horizontals	14.95	10.70	25.65
2000	Thermal break frame, 2-1/4"x4-1/2", 5'x6'opng, no intermediate horizontals	18.30	12.70	31
2050	One intermediate horizontal	24.50	15.60	40.10
2100	Two intermediate horizontals	30.50	18.45	48.95
2150	5' x 20' opening, three intermediate horizontals	16.55	11.25	27.80

B2020 Exterior Windows

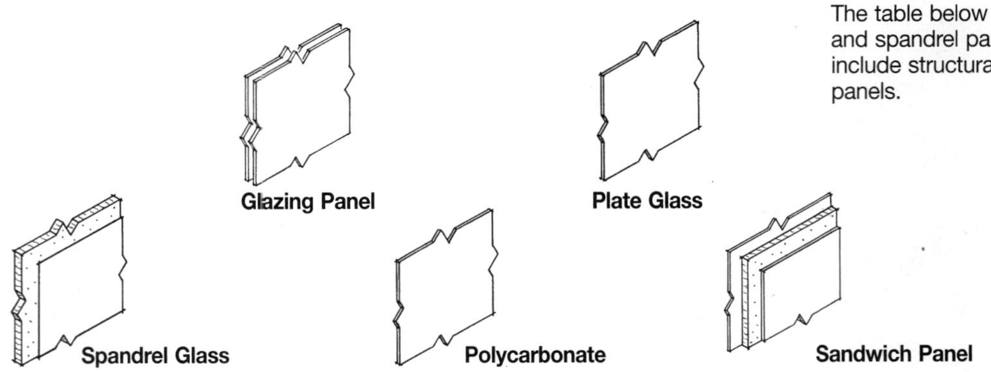

The table below lists costs of curtain wall and spandrel panels per S.F. Costs do not include structural framing used to hang the panels.

B2020 220	Curtain Wall Panels	COST PER S.F.		
		MAT.	INST.	TOTAL
1000	Glazing panel, insulating, 1/2" thick, 2 lites 1/8" float, clear	11.30	11.95	23.25
1100	Tinted	15.65	11.95	27.60
1200	5/8" thick units, 2 lites 3/16" float, clear	15.05	12.65	27.70
1400	1" thick units, 2 lites, 1/4" float, clear	18.85	15.15	34
1700	Light and heat reflective glass, tinted	34.50	13.35	47.85
2000	Plate glass, 1/4" thick, clear	7.05	9.45	16.50
2050	Tempered	9.90	9.45	19.35
2100	Tinted	10	9.45	19.45
2200	3/8" thick, clear	11.45	15.15	26.60
2250	Tempered	18.45	15.15	33.60
2300	Tinted	17.25	15.15	32.40
2400	1/2" thick, clear	19.40	20.50	39.90
2450	Tempered	27.50	20.50	48
2500	Tinted	30	20.50	50.50
2600	3/4" thick, clear	39	32.50	71.50
2650	Tempered	45.50	32.50	78
3000	Spandrel glass, panels, 1/4" plate glass insul w/fiberglass, 1" thick	18.55	9.45	28
3100	2" thick	22.50	9.45	31.95
3200	Galvanized steel backing, add	6.45		6.45
3300	3/8" plate glass, 1" thick	31.50	9.45	40.95
3400	2" thick	35.50	9.45	44.95
4000	Polycarbonate, masked, clear or colored, 1/8" thick	15.75	6.70	22.45
4100	3/16" thick	19.10	6.90	26
4200	1/4" thick	19.10	7.35	26.45
4300	3/8" thick	29.50	7.60	37.10
5000	Facing panel, textured alum., 4' x 8' x 5/16" plywood backing, sgl face	4.50	3.73	8.23
5100	Double face	5.85	3.73	9.58
5200	4' x 10' x 5/16" plywood backing, single face	4.73	3.73	8.46
5300	Double face	6.40	3.73	10.13
5400	4' x 12' x 5/16" plywood backing, single face	4.73	3.73	8.46
5500	Sandwich panel, 22 Ga. galv., both sides 2" insulation, enamel exterior	13.35	6.45	19.80
5600	Polyvinylidene fluoride exterior finish	14.10	6.45	20.55
5700	26 Ga., galv. both sides, 1" insulation, colored 1 side	7.25	6.15	13.40
5800	Colored 2 sides	9.10	6.15	15.25

B2030 Exterior Doors

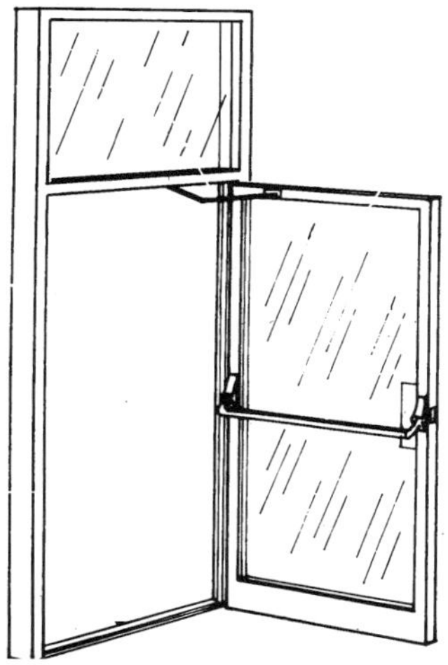

Costs are listed for exterior door systems by material, type and size. Prices between sizes listed can be interpolated with reasonable accuracy. Prices are per opening for a complete door system including frame.

B2030 110 Glazed Doors, Steel or Aluminum

	MATERIAL	TYPE	DOORS	SPECIFICATION	OPENING	COST PER OPNG.		
						MAT.	INST.	TOTAL
5600	St. Stl. & glass	revolving	stock unit	manual oper.	6'-0" x 7'-0"	52,000	9,550	61,550
5650				auto Cntrls.	6'-10" x 7'-0"	68,500	10,200	78,700
5700	Bronze	revolving	stock unit	manual oper.	6'-10" x 7'-0"	53,000	19,200	72,200
5750				auto Cntrls.	6'-10" x 7'-0"	69,500	19,900	89,400
5800	St. Stl. & glass	balanced	standard	economy	3'-0" x 7'-0"	10,300	1,600	11,900
5850				premium	3'-0" x 7'-0"	17,500	2,100	19,600
6300	Alum. & glass	w/o transom	narrow stile	w/panic Hrdwre.	3'-0" x 7'-0"	1,900	1,075	2,975
6350				dbl. door, Hrdwre.	6'-0" x 7'-0"	3,100	1,750	4,850
6400			wide stile	hdwre.	3'-0" x 7'-0"	2,100	1,050	3,150
6450				dbl. door, Hdwre.	6'-0" x 7'-0"	4,075	2,100	6,175
6500			full vision	hdwre.	3'-0" x 7'-0"	2,800	1,675	4,475
6550				dbl. door, Hdwre.	6'-0" x 7'-0"	3,750	2,375	6,125
6600			non-standard	hdwre.	3'-0" x 7'-0"	2,300	1,050	3,350
6650				dbl. door, Hdwre.	6'-0" x 7'-0"	4,600	2,100	6,700
6700			bronze fin.	hdwre.	3'-0" x 7'-0"	1,675	1,050	2,725
6750				dbl. door, Hrdwre.	6'-0" x 7'-0"	3,350	2,100	5,450
6800			black fin.	hdwre.	3'-0" x 7'-0"	2,350	1,050	3,400
6850				dbl. door, Hdwre.	6'-0" x 7'-0"	4,700	2,100	6,800
6900		w/transom	narrow stile	hdwre.	3'-0" x 10'-0"	2,400	1,225	3,625
6950				dbl. door, Hdwre.	6'-0" x 10'-0"	3,750	2,125	5,875
7000			wide stile	hdwre.	3'-0" x 10'-0"	2,725	1,450	4,175
7050				dbl. door, Hdwre.	6'-0" x 10'-0"	4,150	2,525	6,675
7100			full vision	hdwre.	3'-0" x 10'-0"	3,025	1,625	4,650
7150				dbl. door, Hdwre.	6'-0" x 10'-0"	4,525	2,775	7,300
7200			non-standard	hdwre.	3'-0" x 10'-0"	2,400	1,125	3,525
7250				dbl. door, Hdwre.	6'-0" x 10'-0"	4,775	2,275	7,050
7300			bronze fin.	hdwre.	3'-0" x 10'-0"	1,775	1,125	2,900
7350				dbl. door, Hdwre.	6'-0" x 10'-0"	3,525	2,275	5,800
7400			black fin.	hdwre.	3'-0" x 10'-0"	2,450	1,125	3,575
7450				dbl. door, Hdwre.	6'-0" x 10'-0"	4,875	2,275	7,150
7500		revolving	stock design	minimum	6'-10" x 7'-0"	24,900	3,850	28,750
7550				average	6'-0" x 7'-0"	29,300	4,800	34,100
7600				maximum	6'-10" x 7'-0"	34,600	6,400	41,000
7650				min., automatic	6'-10" x 7'-0"	41,500	4,500	46,000

B2030 Exterior Doors

B2030 110 Glazed Doors, Steel or Aluminum

	MATERIAL	TYPE	DOORS	SPECIFICATION	OPENING	COST PER OPNG. MAT.	COST PER OPNG. INST.	COST PER OPNG. TOTAL
7700	Alum. & glass	revolving	stock design	avg., automatic	6'-10" x 7'-0"	45,900	5,450	51,350
7750				max., automatic	6'-10" x 7'-0"	51,000	7,050	58,050
7800		balanced	standard	economy	3'-0" x 7'-0"	7,375	1,600	8,975
7850				premium	3'-0" x 7'-0"	8,825	2,050	10,875
7900		mall front	sliding panels	alum. fin.	16'-0" x 9'-0"	3,900	875	4,775
7950					24'-0" x 9'-0"	5,525	1,625	7,150
8000				bronze fin.	16'-0" x 9'-0"	4,550	1,025	5,575
8050					24'-0" x 9'-0"	6,425	1,900	8,325
8100			fixed panels	alum. fin.	48'-0" x 9'-0"	10,200	1,275	11,475
8150				bronze fin.	48'-0" x 9'-0"	11,900	1,475	13,375
8200		sliding entrance	5' x 7' door	electric oper.	12'-0" x 7'-6"	9,575	1,625	11,200
8250		sliding patio	temp. glass	economy	6'-0" x 7'-0"	1,825	297	2,122
8300			temp. glass	economy	12'-0" x 7'-0"	4,275	395	4,670
8350				premium	6'-0" x 7'-0"	2,750	445	3,195
8400					12'-0" x 7'-0"	6,425	595	7,020

B2030 Exterior Doors

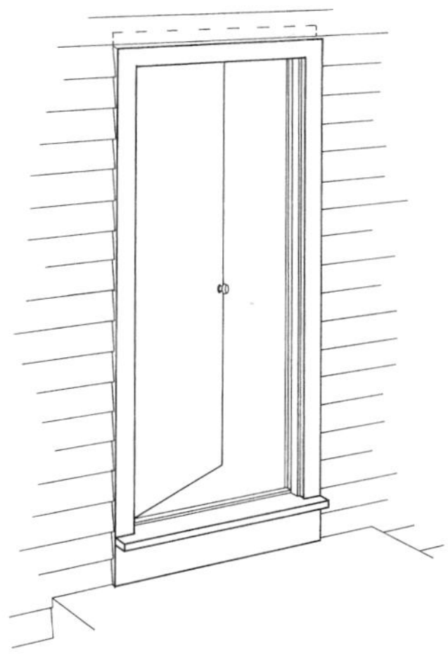

Costs are listed for exterior door systems by material, type and size. Prices between sizes listed can be interpolated with reasonable accuracy. Prices are per opening for a complete door system including frame as illustrated in the component block.

System Components	QUANTITY	UNIT	COST PER OPNG. MAT.	INST.	TOTAL
SYSTEM B2030 210 2500					
WOOD DOOR, SOLID CORE, SINGLE, HINGED, 3′-0″ X 7′-0″					
Exterior, flush, solid core, birch, 1-3/4″ x 2′-6″ x 7′-0″	1.000	Ea.	190	79.50	269.50
Wood door exterior frame, pine, w/trim, 5/4 x 5-3/16″ deep	1.000	Set	147.05	53.89	200.94
Hinges, full mortise-high freq, brass base, 4-1/2″ x 4-1/2″, US10	1.500	Pr.	124.50		124.50
Panic device for mortise locks, single door exit only	1.000	Ea.	660	149	809
Paint exterior door & frame one side 3′ x 7′, primer & 2 coats	1.000	Ea.	2.46	28.50	30.96
Sill, 8/4 x 8″ deep, oak, 2″ horns	1.000	Ea.	29	13.20	42.20
TOTAL			1,153.01	324.09	1,477.10

B2030 210 Wood Doors

	MATERIAL	TYPE	DOORS	SPECIFICATION	OPENING	COST PER OPNG. MAT.	INST.	TOTAL
2350	Birch	solid core	single door	hinged	2′-6″ x 6′-8″	1,150	325	1,475
2400					2′-6″ x 7′-0″	1,150	325	1,475
2450					2′-8″ x 7′-0″	1,150	325	1,475
2500					3′-0″ x 7′-0″	1,150	325	1,475
2550			double door		2′-6″ x 6′-8″	2,150	590	2,740
2600					2′-6″ x 7′-0″	2,150	590	2,740
2650					2′-8″ x 7′-0″	2,125	590	2,715
2700					3′-0″ x 7′-0″	2,000	515	2,515
2750	Wood	combination	storm & screen		3′-0″ x 6′-8″	335	79.50	414.50
2800					3′-0″ x 7′-0″	370	87.50	457.50
2850		overhead	panels, H.D.	manual oper.	8′-0″ x 8′-0″	1,250	595	1,845
2900					10′-0″ x 10′-0″	1,775	660	2,435
2950					12′-0″ x 12′-0″	2,450	795	3,245
3000					14′-0″ x 14′-0″	3,700	915	4,615
3050					20′-0″ x 16′-0″	6,200	1,825	8,025
3100				electric oper.	8′-0″ x 8′-0″	2,425	890	3,315
3150					10′-0″ x 10′-0″	2,950	955	3,905
3200					12′-0″ x 12′-0″	3,625	1,100	4,725
3250					14′-0″ x 14′-0″	4,875	1,200	6,075
3300					20′-0″ x 16′-0″	7,450	2,425	9,875

B2030 Exterior Doors

B2030 220	Steel Doors							
	MATERIAL	TYPE	DOORS	SPECIFICATION	OPENING	COST PER OPNG.		
						MAT.	INST.	TOTAL
3350	Steel 18 Ga.	hollow metal	1 door w/frame	no label	2′-6″ x 7′-0″	1,550	335	1,885
3400					2′-8″ x 7′-0″	1,550	335	1,885
3450					3′-0″ x 7′-0″	1,575	360	1,935
3500					3′-6″ x 7′-0″	1,675	350	2,025
3550		hollow metal	1 door w/frame	no label	4′-0″ x 8′-0″	1,975	355	2,330
3600			2 doors w/frame	no label	5′-0″ x 7′-0″	3,000	620	3,620
3650					5′-4″ x 7′-0″	3,000	620	3,620
3700					6′-0″ x 7′-0″	3,000	620	3,620
3750					7′-0″ x 7′-0″	3,250	655	3,905
3800					8′-0″ x 8′-0″	3,825	655	4,480
3850			1 door w/frame	"A" label	2′-6″ x 7′-0″	1,950	405	2,355
3900					2′-8″ x 7′-0″	1,950	410	2,360
3950					3′-0″ x 7′-0″	1,950	410	2,360
4000					3′-6″ x 7′-0″	2,100	415	2,515
4050					4′-0″ x 8′-0″	2,325	435	2,760
4100			2 doors w/frame	"A" label	5′-0″ x 7′-0″	3,750	745	4,495
4150					5′-4″ x 7′-0″	3,750	755	4,505
4200					6′-0″ x 7′-0″	3,750	755	4,505
4250					7′-0″ x 7′-0″	4,075	775	4,850
4300					8′-0″ x 8′-0″	4,075	775	4,850
4350	Steel 24 Ga.	overhead	sectional	manual oper.	8′-0″ x 8′-0″	990	595	1,585
4400					10′-0″ x 10′-0″	1,350	660	2,010
4450					12′-0″ x 12′-0″	1,625	795	2,420
4500					20′-0″ x 14′-0″	3,775	1,700	5,475
4550				electric oper.	8′-0″ x 8′-0″	2,175	890	3,065
4600					10′-0″ x 10′-0″	2,525	955	3,480
4650					12′-0″ x 12′-0″	2,800	1,100	3,900
4700					20′-0″ x 14′-0″	5,025	2,300	7,325
4750	Steel	overhead	rolling	manual oper.	8′-0″ x 8′-0″	1,300	900	2,200
4800					10′-0″ x 10′-0″	2,150	1,025	3,175
4850					12′-0″ x 12′-0″	2,175	1,200	3,375
4900					14′-0″ x 14′-0″	3,475	1,800	5,275
4950					20′-0″ x 12′-0″	2,425	1,600	4,025
5000					20′-0″ x 16′-0″	4,175	2,400	6,575
5050				electric oper.	8′-0″ x 8′-0″	2,600	1,200	3,800
5100					10′-0″ x 10′-0″	3,450	1,325	4,775
5150					12′-0″ x 12′-0″	3,475	1,500	4,975
5200					14′-0″ x 14′-0″	4,775	2,100	6,875
5250					20′-0″ x 12′-0″	3,625	1,900	5,525
5300					20′-0″ x 16′-0″	5,375	2,700	8,075
5350				fire rated	10′-0″ x 10′-0″	2,450	1,300	3,750
5400			rolling grille	manual oper.	10′-0″ x 10′-0″	2,850	1,450	4,300
5450					15′-0″ x 8′-0″	3,200	1,800	5,000
5500		vertical lift	1 door w/frame	motor operator	16″-0″ x 16″-0″	23,700	5,650	29,350
5550					32′-0″ x 24′-0″	53,500	3,775	57,275

B2030 Exterior Doors

B2030 230 Aluminum Doors

	MATERIAL	TYPE	DOORS	SPECIFICATION	OPENING	COST PER OPNG.		
						MAT.	INST.	TOTAL
6000	Aluminum	combination	storm & screen	hinged	3'-0" x 6'-8"	340	85	425
6050					3'-0" x 7'-0"	375	93.50	468.50
6100		overhead	rolling grille	manual oper.	12'-0" x 12'-0"	4,525	2,525	7,050
6150				motor oper.	12'-0" x 12'-0"	5,900	2,825	8,725
6200	Alum. & Fbrgls.	overhead	heavy duty	manual oper.	12'-0" x 12'-0"	3,350	795	4,145
6250				electric oper.	12'-0" x 12'-0"	4,525	1,100	5,625

B3010 Roof Coverings

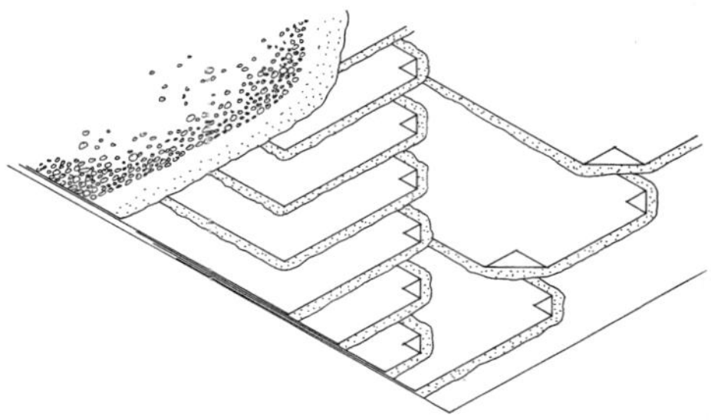

Multiple ply roofing is the most popular covering for minimum pitch roofs. Lines 1200 through 6300 list the costs of the various types, plies and weights per S.F.

System Components	QUANTITY	UNIT	COST PER S.F. MAT.	COST PER S.F. INST.	COST PER S.F. TOTAL
SYSTEM B3010 105 2500					
ASPHALT FLOOD COAT, W/GRAVEL, 3 PLY ORGANIC FELT					
Organic #30 base felt	1.000	S.F.	.11	.10	.21
Organic #15 felt, 3 plies	3.000	S.F.	.18	.29	.47
Asphalt mopping of felts	4.000	S.F.	.49	.91	1.40
Asphalt flood coat	1.000	S.F.	.31	.73	1.04
Gravel aggregate, washed river stone	4.000	Lb.	.07	.16	.23
TOTAL			1.16	2.19	3.35

B3010 105	Built-Up	COST PER S.F. MAT.	COST PER S.F. INST.	COST PER S.F. TOTAL
1200	Asphalt flood coat w/gravel; not incl. insul, flash., nailers			
1300				
1400	Asphalt base sheets & 3 plies #15 asphalt felt, mopped	1.09	1.98	3.07
1500	On nailable deck	1.13	2.08	3.21
1600	4 plies #15 asphalt felt, mopped	1.50	2.18	3.68
1700	On nailable deck	1.33	2.30	3.63
1800	Coated glass base sheet, 2 plies glass (type IV), mopped	1.21	1.98	3.19
1900	For 3 plies	1.47	2.18	3.65
2000	On nailable deck	1.39	2.30	3.69
2300	4 plies glass fiber felt (type IV), mopped	1.82	2.18	4
2400	On nailable deck	1.65	2.30	3.95
2500	Organic base sheet & 3 plies #15 organic felt, mopped	1.16	2.19	3.35
2600	On nailable deck	1.14	2.30	3.44
2700	4 plies #15 organic felt, mopped	1.43	1.98	3.41
2750				
2800	Asphalt flood coat, smooth surface, not incl. insul, flash., nailers			
2850				
2900	Asphalt base sheet & 3 plies #15 asphalt felt, mopped	1.16	1.82	2.98
3000	On nailable deck	1.08	1.90	2.98
3100	Coated glass fiber base sheet & 2 plies glass fiber felt, mopped	1.16	1.75	2.91
3200	On nailable deck	1.10	1.82	2.92
3300	For 3 plies, mopped	1.42	1.90	3.32
3400	On nailable deck	1.34	1.98	3.32
3700	4 plies glass fiber felt (type IV), mopped	1.68	1.90	3.58
3800	On nailable deck	1.60	1.98	3.58
3900	Organic base sheet & 3 plies #15 organic felt, mopped	1.18	1.82	3
4000	On nailable decks	1.09	1.90	2.99
4100	4 plies #15 organic felt, mopped	1.38	1.98	3.36
4200	Coal tar pitch with gravel surfacing			
4300	4 plies #15 tarred felt, mopped	2.20	2.08	4.28
4400	3 plies glass fiber felt (type IV), mopped	1.81	2.30	4.11
4500	Coated glass fiber base sheets 2 plies glass fiber felt, mopped	1.86	2.30	4.16

B3010 Roof Coverings

B3010 105	Built-Up	COST PER S.F.		
		MAT.	INST.	TOTAL
4600	On nailable decks	1.62	2.43	4.05
4800	3 plies glass fiber felt (type IV), mopped	2.52	2.08	4.60
4900	On nailable decks	2.28	2.18	4.46
5300	Asphalt mineral surface, roll roofing			
5400	1 ply #15 organic felt, 1 ply mineral surfaced			
5500	Selvage roofing, lap 19", nailed and mopped	.85	1.62	2.47
5600	3 plies glass fiber felt (type IV), 1 ply mineral surfaced			
5700	Selvage roofing, lapped 19", mopped	1.45	1.75	3.20
5800	Coated glass fiber base sheet			
5900	2 plies glass fiber, felt (type IV), 1 ply mineral surfaced			
6000	Selvage, roofing, lapped 19", mopped	1.53	1.75	3.28
6100	On nailable deck	1.40	1.82	3.22
6200	3 plies glass fiber felt (type IV), 1 ply mineral surfaced			
6300	Selvage roofing, lapped 19", mopped	1.45	1.75	3.20

B3010 Roof Coverings

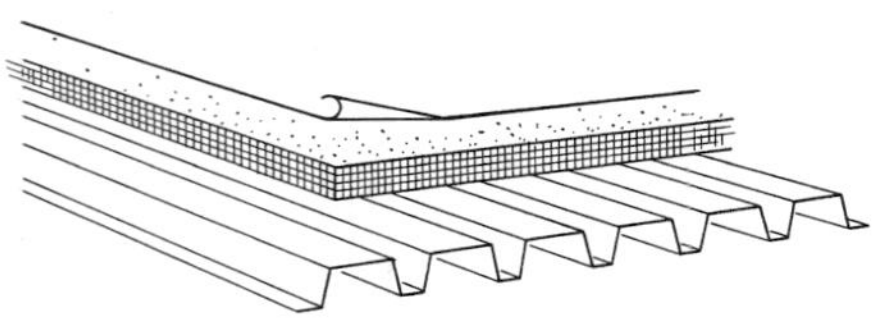

Fully Adhered

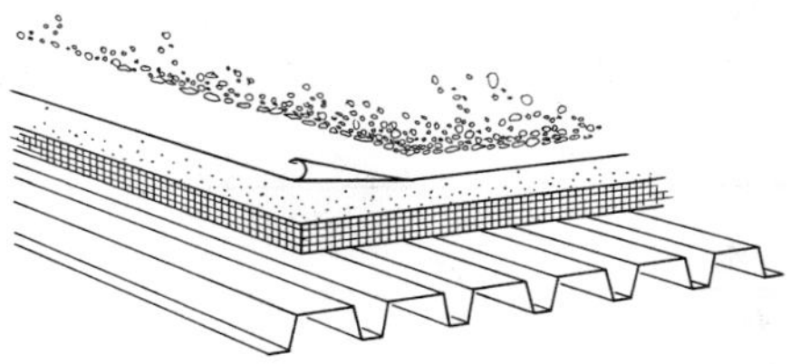

Ballasted

The systems listed below reflect only the cost for the single ply membrane.

B3010 120	Single Ply Membrane	COST PER S.F.		
		MAT.	INST.	TOTAL
1000	CSPE (Chlorosulfonated polyethylene), 45 mils, plate attached	2.76	.80	3.56
1100	Loosely laid with stone ballast	2.76	.80	3.56
2000	EPDM (Ethylene propylene diene monomer), 45 mils, fully adhered	1.22	1.08	2.30
2100	Loosely laid with stone ballast	.92	.55	1.47
2200	Mechanically fastened with batten strips	.86	.80	1.66
3300	60 mils, fully adhered	1.34	1.08	2.42
3400	Loosely laid with stone ballast	1.05	.55	1.60
3500	Mechanically fastened with batten strips	.98	.80	1.78
4000	Modified bit., SBS modified, granule surface cap sheet, mopped, 150 mils	1.40	2.18	3.58
4100	Smooth surface cap sheet, mopped, 145 mils	.88	2.07	2.95
4500	APP modified, granule surface cap sheet, torched, 180 mils	1.02	1.40	2.42
4600	Smooth surface cap sheet, torched, 170 mils	.83	1.33	2.16
5000	PIB (Polyisobutylene), 100 mils, fully adhered with contact cement	2.75	1.08	3.83
5100	Loosely laid with stone ballast	2.20	.55	2.75
5200	Partially adhered with adhesive	2.64	.80	3.44
5300	Hot asphalt attachment	2.53	.80	3.33
6000	Reinforced PVC, 48 mils, loose laid and ballasted with stone	1.29	.55	1.84
6100	Partially adhered with mechanical fasteners	1.23	.80	2.03
6200	Fully adhered with adhesive	1.70	1.08	2.78
6300	Reinforced PVC, 60 mils, loose laid and ballasted with stone	1.30	.55	1.85
6400	Partially adhered with mechanical fasteners	1.24	.80	2.04
6500	Fully adhered with adhesive	1.71	1.08	2.79

B3010 Roof Coverings

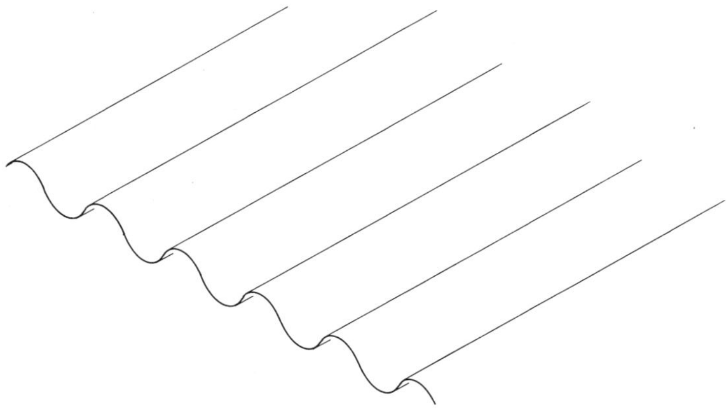

Listed below are installed prices for preformed roofing materials with applicable specifications for each material.

B3010 130	Preformed Metal Roofing	COST PER S.F.		
		MAT.	INST.	TOTAL
0200	Corrugated roofing, aluminum, mill finish, .0175" thick, .272 P.S.F.	1.10	1.94	3.04
0250	.0215" thick, .334 P.S.F.	1.38	1.94	3.32
0300	.0240" thick, .412 P.S.F.	1.98	1.94	3.92
0350	.0320" thick, .552 P.S.F.	2.83	1.94	4.77
0400	Painted, .0175" thick, .280 P.S.F.	1.60	1.94	3.54
0450	.0215" thick, .344 P.S.F.	1.71	1.94	3.65
0500	.0240" thick, .426 P.S.F.	2.42	1.94	4.36
0550	.0320" thick, .569 P.S.F.	3.50	1.94	5.44
0700	Fiberglass, 6 oz., .375 P.S.F.	2.37	2.33	4.70
0750	8 oz., .5 P.S.F.	4.62	2.33	6.95
0900	Steel, galvanized, 29 ga., .72 P.S.F.	1.80	2.12	3.92
0930	26 ga., .91 P.S.F.	1.86	2.22	4.08
0950	24 ga., 1.26 P.S.F.	2.15	2.33	4.48
0970	22 ga., 1.45 P.S.F.	3.19	2.45	5.64
1000	Colored, 28 ga., 1.08 P.S.F.	1.89	2.22	4.11
1050	26 ga., 1.43 P.S.F.	2.24	2.33	4.57

B3010 Roof Coverings

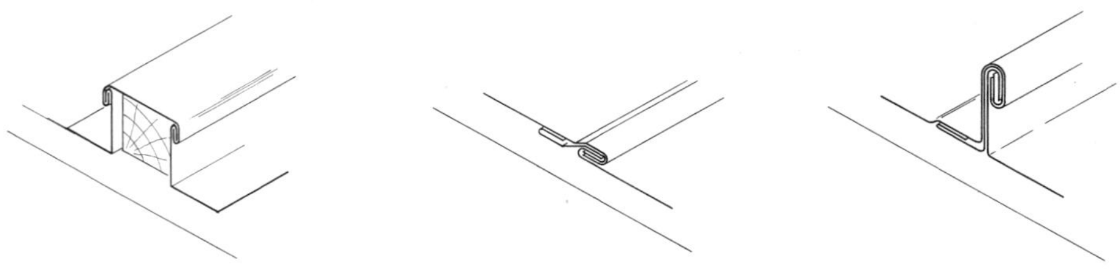

Formed metal roofing is practical on all sloped roofs from 1/4″ per foot of rise to vertical. Its use is more aesthetic than economical. Table below lists the various materials used and the weight per S.F.

System Components	QUANTITY	UNIT	COST PER S.F. MAT.	INST.	TOTAL
SYSTEM B3010 135 1000					
BATTEN SEAM, FORMED COPPER ROOFING 3″ MIN. SLOPE, 16 OZ., 1.2 PSF.					
Copper batten seam roofing, over 10 sq, 16 oz., 130 lb per sq	1.000	S.F.	13.75	6.35	20.10
Asphalt impregnated felt, 30 lb, 2 sq. per roll, not mopped	1.000	S.F.	.11	.10	.21
TOTAL			13.86	6.45	20.31

B3010 135	Formed Metal	COST PER S.F. MAT.	INST.	TOTAL
1000	Batten seam, formed copper roofing, 3″ min slope, 16 oz., 1.2 P.S.F.	13.85	6.45	20.30
1100	18 oz., 1.35 P.S.F.	15.35	7.10	22.45
1200	20 oz., 1.50 P.S.F.	18.10	7.10	25.20
1500	Lead, 3″ min slope, 3 Lb., 3.6 P.S.F.	15.10	5.95	21.05
2000	Zinc copper alloy, 3″ min slope, .020″ thick, .88 P.S.F.	14.10	5.95	20.05
2050	.027″ thick, 1.18 P.S.F.	17.85	6.20	24.05
2100	.032″ thick, 1.41 P.S.F.	20.50	6.45	26.95
2150	.040″ thick, 1.77 P.S.F.	25.50	6.75	32.25
3000	Flat seam, copper, 1/4″ min. slope, 16 oz., 1.2 P.S.F.	10.10	5.95	16.05
3100	18 oz., 1.35 P.S.F.	11.35	6.20	17.55
3200	20 oz., 1.50 P.S.F.	13.35	6.45	19.80
3500	Lead, 1/4″ min. slope, 3 Lb. 3.6 P.S.F.	12.85	5.50	18.35
4000	Zinc copper alloy, 1/4″ min. slope, .020″ thick, .84 P.S.F.	12.10	5.95	18.05
4050	.027″ thick, 1.13 P.S.F.	15.10	6.20	21.30
4100	.032″ thick, 1.33 P.S.F.	17.35	6.35	23.70
4150	.040″ thick, 1.67 P.S.F.	22	6.75	28.75
5000	Standing seam, copper, 2-1/2″ min. slope, 16 oz., 1.25 P.S.F.	10.85	5.50	16.35
5100	18 oz., 1.40 P.S.F.	12.10	5.95	18.05
5200	20 oz., 1.56 P.S.F.	14.35	6.45	20.80
5500	Lead, 2-1/2″ min. slope, 3 Lb., 3.6 P.S.F.	14.35	5.50	19.85
6000	Zinc copper alloy, 2-1/2″ min. slope, .020″ thick, .87 P.S.F.	13.60	5.95	19.55
6050	.027″ thick, 1.18 P.S.F.	17.10	6.20	23.30
6100	.032″ thick, 1.39 P.S.F.	19.60	6.45	26.05
6150	.040″ thick, 1.74 P.S.F.	24.50	6.75	31.25

B3010 Roof Coverings

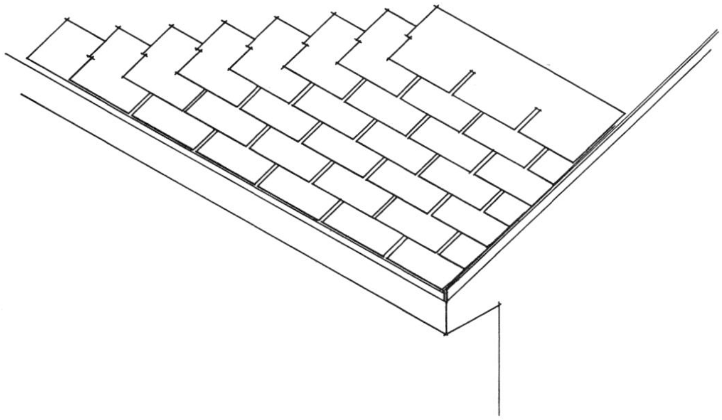

Shingles and tiles are practical in applications where the roof slope is more than 3-1/2″ per foot of rise. Table below lists the various materials and the weight per S.F.

System Components	QUANTITY	UNIT	COST PER S.F. MAT.	COST PER S.F. INST.	COST PER S.F. TOTAL
SYSTEM B3010 140 1200					
STRIP SHINGLES, 4″ SLOPE, MULTI-LAYERED, CLASS A					
Class A, 260-300 lb./sq., 4 bundles/sq.	1.000	S.F.	1.74	1.62	3.36
Building paper, asphalt felt sheathing paper 15 lb	1.000	S.F.	.06	.16	.22
TOTAL			1.80	1.78	3.58

B3010 140	Shingle & Tile	COST PER S.F. MAT.	COST PER S.F. INST.	COST PER S.F. TOTAL
1095	Asphalt roofing			
1100	Strip shingles, 4″ slope, inorganic class A 210-235 lb./sq.	.95	1.19	2.14
1150	Organic, class C, 235-240 lb./sq.	1.12	1.30	2.42
1200	Premium laminated multi-layered, class A, 260-300 lb./sq.	1.80	1.78	3.58
1250	Class C, 300-385 lb./sq.	2.71	2.05	4.76
1545	Metal roofing			
1550	Alum., shingles, colors, 3″ min slope, .019″ thick, 0.4 PSF	2.77	1.29	4.06
1600	.020″ thick, 0.6 PSF	3.05	1.29	4.34
1850	Steel, colors, 3″ min slope, 26 gauge, 1.0 PSF	4.53	2.69	7.22
2000	24 gauge, 1.7 PSF	4.43	2.69	7.12
2795	Slate roofing			
2800	Slate roofing, 4″ min. slope, shingles, 3/16″ thick, 8.0 PSF	6.05	3.35	9.40
2900	1/4″ thick, 9.0 PSF	6.05	3.35	9.40
3495	Wood roofing			
3500	4″ min slope, cedar shingles, 16″ x 5″, 5″ exposure 1.6 PSF	3.22	2.57	5.79
4000	Shakes, 18″, 8-1/2″ exposure, 2.8 PSF	2.09	3.16	5.25
4100	24″, 10″ exposure, 3.5 PSF	3.27	2.57	5.84
5095	Tile roofing			
5100	Aluminum, mission, 3″ min slope, .019″ thick, 0.65 PSF	9.20	2.48	11.68
5200	S type, 0.48 PSF	6.05	2.08	8.13
6000	Clay tile, flat shingle, interlocking, 15″, 166 pcs/sq, fireflashed blend	5.20	2.95	8.15
6500	Mission, 4″ minimum slope, 12.2 PSF	4.61	3.20	7.81

B3010 Roof Coverings

B3010 320	Roof Deck Rigid Insulation	COST PER S.F.		
		MAT.	INST.	TOTAL
0100	Fiberboard low density RB3010 -010			
0150	1" thick R2.78	.73	.73	1.46
0300	1 1/2" thick R4.17	1.04	.73	1.77
0350	2" thick R5.56	1.33	.73	2.06
0370	Fiberboard high density, 1/2" thick R1.3	.45	.62	1.07
0380	1" thick R2.5	.74	.73	1.47
0390	1 1/2" thick R3.8	1.07	.73	1.80
0410	Fiberglass, 3/4" thick R2.78	.79	.62	1.41
0450	15/16" thick R3.70	1.01	.62	1.63
0500	1-1/16" thick R4.17	1.23	.62	1.85
0550	1-5/16" thick R5.26	1.63	.62	2.25
0600	2-1/16" thick R8.33	1.72	.73	2.45
0650	2 7/16" thick R10	1.97	.73	2.70
1260	Perlite, 1/2" thick R1.32	.41	.60	1.01
1300	3/4" thick R2.08	.53	.73	1.26
1350	1" thick R2.78	.70	.73	1.43
1400	1 1/2" thick R4.17	.98	.73	1.71
1450	2" thick R5.56	1.29	.80	2.09
1510	Polyisocyanurate 2#/CF density, 1" thick	.64	.49	1.13
1550	1 1/2" thick	.82	.53	1.35
1600	2" thick	1.02	.58	1.60
1650	2 1/2" thick	1.33	.60	1.93
1700	3" thick	1.46	.62	2.08
1750	3 1/2" thick	2.21	.62	2.83
1800	Tapered for drainage	.77	.49	1.26
1810	Expanded polystyrene, 1#/CF density, 3/4" thick R2.89	.33	.47	.80
1820	2" thick R7.69	.69	.53	1.22
1825	Extruded Polystyrene			
1830	Extruded polystyrene, 15 PSI compressive strength, 1" thick, R5	.78	.47	1.25
1835	2" thick R10	.98	.53	1.51
1840	3" thick R15	1.84	.62	2.46
1900	25 PSI compressive strength, 1" thick R5	.88	.47	1.35
1950	2" thick R10	1.56	.53	2.09
2000	3" thick R15	2.32	.62	2.94
2050	4" thick R20	3.16	.62	3.78
2150	Tapered for drainage	.79	.47	1.26
2550	40 PSI compressive strength, 1" thick R5	.70	.47	1.17
2600	2" thick R10	1.23	.53	1.76
2650	3" thick R15	1.72	.62	2.34
2700	4" thick R20	2.22	.62	2.84
2750	Tapered for drainage	.96	.49	1.45
2810	60 PSI compressive strength, 1" thick R5	.93	.48	1.41
2850	2" thick R10	1.67	.54	2.21
2900	Tapered for drainage	1.18	.49	1.67
4000	Composites with 1-1/2" polyisocyanurate			
4010	1" fiberboard	1.43	.73	2.16
4020	1" perlite	1.29	.69	1.98
4030	7/16" oriented strand board	1.14	.73	1.87

B3010 Roof Coverings

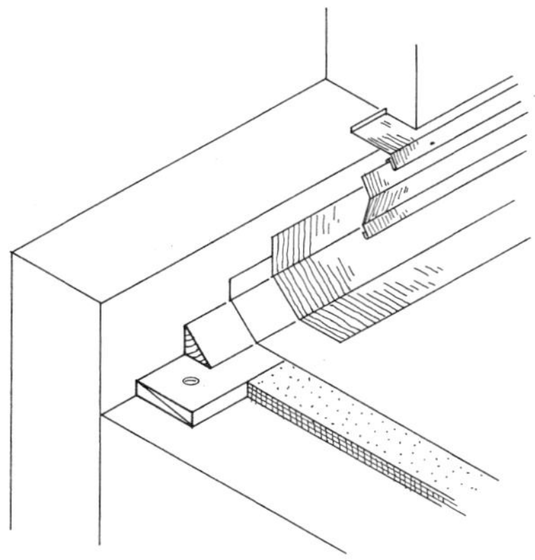

The table below lists the costs for various types of base flashing per L.F.

All the systems below are also based on using a 4" x 4" diagonally cut treated wood cant and a 2" x 8" treated wood nailer fastened at 4'-0" O.C.

System Components	QUANTITY	UNIT	COST PER L.F. MAT.	COST PER L.F. INST.	COST PER L.F. TOTAL
SYSTEM B3010 410 1000					
ALUMINUM BASE FLASHING MILL FINISH, .019" THICK					
Treated 2" x 8" wood blocking	1.333	B.F.	1.17	1.83	3
Anchor bolts, 4' O.C.	.250	Ea.	1.47	1.64	3.11
Treated wood cant, 4" x 4" split	1.000	L.F.	1.98	1.75	3.73
Single ply membrane, CSPE, 45 mils, heat welded seams, plate attachment	1.000	S.F.	2.76	.80	3.56
Base flashing, .019" thick, aluminum	1.000	L.F.	1.38	3.92	5.30
Reglet .025" thick counter flashing .032" thick	1.000	L.F.	2.18	2.64	4.82
Counter flashing, aluminum	1.000	L.F.	2.42	4.67	7.09
TOTAL			13.36	17.25	30.61

B3010 410	Base Flashing							
	TYPE	DESCRIPTION	SPECIFICATION	REGLET	COUNTER FLASHING	COST PER L.F. MAT.	COST PER L.F. INST.	COST PER L.F. TOTAL
1000	Aluminum	mill finish	.019" thick	alum. .025	alum. .032	13.35	17.25	30.60
1100			.050" thick	.025	.032	14.90	17.25	32.15
1200		fabric, 2 sides	.004" thick	.025	.032	13.75	15.05	28.80
1300			.016" thick	.025	.032	14	15.05	29.05
1400		mastic, 2 sides	.004" thick	.025	.032	13.75	15.05	28.80
1500			.016" thick	.025	.032	14.25	15.05	29.30
1700	Copper	sheets, plain	16 oz.	copper 16 oz.	copper 16 oz.	30	18.25	48.25
1800			20 oz.	16 oz.	16 oz.	33	18.45	51.45
1900			24 oz.	16 oz.	16 oz.	37.50	18.70	56.20
2000			32 oz.	16 oz.	16 oz.	42	19	61
2200	Sheets, Metal	galvanized	20 Ga.	galv. 24 Ga.	galv. 24 Ga.	12.35	39.50	51.85
2300			24 Ga.	24 Ga.	24 Ga	11.75	32	43.75
2400			30 Ga.	24 Ga.	24 Ga	11.10	24	35.10
2600	Rubber	butyl	1/32" thick	galv. 24 Ga.	galv. 24 Ga.	12.45	15.30	27.75
2700			1/16" thick	24 Ga.	24 Ga.	13.25	15.30	28.55
2800		neoprene	1/16" thick	24 Ga.	24 Ga.	13.10	15.30	28.40
2900			1/8" thick	24 Ga.	24 Ga.	16.95	15.30	32.25

B3010 Roof Coverings

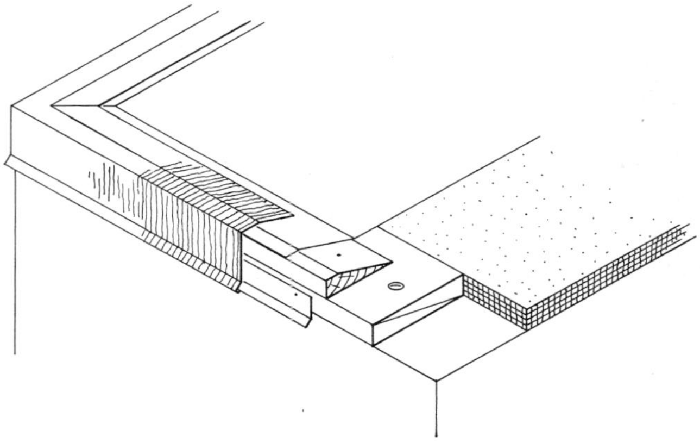

The table below lists the costs for various types of roof perimeter edge treatments.

All systems below include the cost per L.F. for a 2″ x 8″ treated wood nailer fastened at 4′-0″ O.C. and a diagonally cut 4″ x 6″ treated wood cant.

Roof edge and base flashing are assumed to be made from the same material.

System Components	QUANTITY	UNIT	COST PER L.F. MAT.	COST PER L.F. INST.	COST PER L.F. TOTAL
SYSTEM B3010 420 1000					
ALUMINUM ROOF EDGE, MILL FINISH, .050″ THICK, 4″ HIGH					
Treated 2″ x 8″ wood blocking	1.333	B.F.	1.17	1.83	3
Anchor bolts, 4′ O.C.	.250	Ea.	1.47	1.64	3.11
Treated wood cant	1.000	L.F.	1.98	1.75	3.73
Single ply membrane, CSFE, 45 mils, heat welded seams, plate attachment	1.000	S.F.	2.76	.80	3.56
Aluminum roof edge, .050′ thick, 4″ high	1.000	L.F.	7.05	4.83	11.88
TOTAL			14.43	10.85	25.28

B3010 420 Roof Edges

	EDGE TYPE	DESCRIPTION	SPECIFICATION	FACE HEIGHT		COST PER L.F. MAT.	COST PER L.F. INST.	COST PER L.F. TOTAL
1000	Aluminum	mill finish	.050″ thick	4″		14.45	10.85	25.30
1100				6″		14.65	11.20	25.85
1200				8″		15.70	11.60	27.30
1300		duranodic	.050″ thick	4″		15.50	10.85	26.35
1400				6″		15.95	11.20	27.15
1500				8″		17.40	11.60	29
1600		painted	.050″ thick	4″		15.60	10.85	26.45
1700				6″		16.95	11.20	28.15
1800				8″		17.90	11.60	29.50
2000	Copper	plain	16 oz.	4″		34	10.85	44.85
2100				6″		44	11.20	55.20
2200				8″		56	12.95	68.95
2300			20 oz.	4″		42.50	12.45	54.95
2400				6″		49.50	12	61.50
2500				8″		63.50	14	77.50
2700	Sheet Metal	galvanized	20 Ga.	4″		17.25	12.95	30.20
2800				6″		17.65	12.95	30.60
2900				8″		21	15.25	36.25
3000			24 Ga.	4″		14.30	10.85	25.15
3100				6″		14.55	10.85	25.40
3200				8″		16.90	12.45	29.35

B3010 Roof Coverings

B3010 430 — Flashing

	MATERIAL	BACKING	SIDES	SPECIFICATION	QUANTITY	COST PER S.F. MAT.	COST PER S.F. INST.	COST PER S.F. TOTAL
0040	Aluminum	none		.019″		1.38	3.92	5.30
0050				.032″		1.49	3.92	5.41
0100				.040″		2.54	3.92	6.46
0150				.050″		2.90	3.92	6.82
0300		fabric	2	.004″		1.76	1.72	3.48
0350				.016″		2.04	1.72	3.76
0400		mastic		.004″		1.76	1.72	3.48
0450				.005″		2.05	1.72	3.77
0500				.016″		2.26	1.72	3.98
0510								
0700	Copper	none		16 oz.	<500 lbs.	8.65	4.94	13.59
0710					>2000 lbs.	8.65	3.67	12.32
0750				20 oz.	<500 lbs.	11.35	5.15	16.50
0760					>2000 lbs.	10.75	3.92	14.67
0800				24 oz.	<500 lbs.	16.05	5.40	21.45
0810					>2000 lbs.	15.25	4.21	19.46
0850				32 oz.	<500 lbs.	20.50	5.70	26.20
0860					>2000 lbs.	19.35	4.37	23.72
1000		paper backed	1	2 oz.		2.02	1.72	3.74
1100				3 oz.		2.42	1.72	4.14
1200			2	2 oz.		2.26	1.72	3.98
1300				5 oz.		3.69	1.72	5.41
3000	Lead	none		2.5 lb.	to 12″ wide	5.95	4.21	10.16
3100					over 12″	4.39	4.21	8.60
3500	PVC black	none		.010″		.23	1.99	2.22
3600				.020″		.29	1.99	2.28
3700				.030″		.37	1.99	2.36
3800				.056″		.90	1.99	2.89
4000	Rubber butyl	none		1/32″		2.16	1.99	4.15
4100				1/16″		2.92	1.99	4.91
4200	Neoprene			1/16″		2.81	1.99	4.80
4300				1/8″		6.65	1.99	8.64
4500	Stainless steel	none		.015″	<500 lbs.	7.05	4.94	11.99
4600	Copper clad				>2000 lbs.	7.35	3.67	11.02
4700				.018″	<500 lbs.	8.65	5.70	14.35
4800					>2000 lbs.	8.25	3.92	12.17
5000	Plain			32 ga.		3.58	3.67	7.25
5100				28 ga.		4.62	3.67	8.29
5200				26 ga.		4.90	3.67	8.57
5300				24 ga.		5.45	3.67	9.12
5400	Terne coated			28 ga.		8.85	3.67	12.52
5500				26 ga.		9.90	3.67	13.57

B3010 Roof Coverings

B3010 610 Gutters

	SECTION	MATERIAL	THICKNESS	SIZE	FINISH	COST PER L.F.		
						MAT.	INST.	TOTAL
0050	Box	aluminum	.027"	5"	enameled	3.17	5.80	8.97
0100					mill	3.19	5.80	8.99
0200			.032"	5"	enameled	3.95	5.60	9.55
0300					mill	3.81	5.60	9.41
0500		copper	16 Oz.	5"	lead coated	20	5.60	25.60
0600					mill	8.85	5.60	14.45
0700				6"	lead coated	21.50	5.60	27.10
0800					mill	9.35	5.60	14.95
1000		steel galv.	28 Ga.	5"	enameled	2.40	5.60	8
1100					mill	2.31	5.60	7.91
1200			26 Ga.	5"	mill	2.64	5.60	8.24
1300				6"	mill	3.06	5.60	8.66
1500		copper clad ss		5"	mill	8.25	5.60	13.85
1600		stainless		5"	mill	10.40	5.60	16
1800		vinyl		4"	colors	1.43	5.15	6.58
1900				5"	colors	1.76	5.15	6.91
2200		wood clear,		3"x4"	treated	11	5.95	16.95
2300		hemlock or fir		4"x5"	treated	18.45	5.95	24.40
3000	Half round	copper	16 Oz.	4"	lead coated	16.15	5.60	21.75
3100					mill	10.45	5.60	16.05
3200				5"	lead coated	18.55	6.45	25
3300					mill	7.95	5.60	13.55
3600		steel galv.	28 Ga.	5"	enameled	2.40	5.60	8
3700					mill	2.31	5.60	7.91
3800			26 Ga.	5"	mill	2.64	5.60	8.24
4000		copper clad ss		5"	mill	8.25	5.60	13.85
4100				5"	mill	10.40	5.60	16
5000		vinyl		4"	white	1.49	5.15	6.64

B3010 620 Downspouts

	MATERIALS	SECTION	SIZE	FINISH	THICKNESS	COST PER V.L.F.		
						MAT.	INST.	TOTAL
0100	Aluminum	rectangular	2"x3"	embossed mill	.020"	1.16	3.68	4.84
0150				enameled	.020"	1.53	3.68	5.21
0200				enameled	.024"	2.26	3.89	6.15
0250			3"x4"	enameled	.024"	2.61	5	7.61
0300		round corrugated	3"	enameled	.020"	2.20	3.68	5.88
0350			4"	enameled	.025"	3.25	5	8.25
0500	Copper	rectangular corr.	2"x3"	mill	16 Oz.	9.60	3.68	13.28
0550				lead coated	16 Oz.	17.25	3.68	20.93
0600		smooth		mill	16 Oz.	12.15	3.68	15.83
0650				lead coated	16 Oz.	26.50	3.68	30.18
0700		rectangular corr.	3"x4"	mill	16 Oz.	11.45	4.83	16.28
0750				lead coated	16 Oz.	31.50	4.18	35.68
0800		smooth		mill	16 Oz.	15.35	4.83	20.18
0850				lead coated	16 Oz.	36.50	4.83	41.33
0900		round corrugated	2"	mill	16 Oz.	8.60	3.79	12.39
0950		smooth		lead coated	16 Oz.	24.50	3.68	28.18
1000		corrugated	3"	mill	16 Oz.	9.10	3.81	12.91
1050		smooth		lead coated	16 Oz.	24.50	3.68	28.18

B3010 Roof Coverings

B3010 620 Downspouts

	MATERIALS	SECTION	SIZE	FINISH	THICKNESS	COST PER V.L.F.		
						MAT.	INST.	TOTAL
1100	Copper	corrugated	4″	mill	16 Oz.	11.10	5	16.10
1150		smooth		lead coated	16 Oz.	26	4.83	30.83
1200		corrugated	5″	mill	16 Oz.	18.10	5.65	23.75
1250				lead coated	16 Oz.	25	5.40	30.40
1300	Steel	rectangular corr.	2″x3″	galvanized	28 Ga.	2.13	3.68	5.81
1350				epoxy coated	24 Ga.	2.56	3.68	6.24
1400		smooth		galvanized	28 Ga.	4.02	3.68	7.70
1450		rectangular corr.	3″x4″	galvanized	28 Ga.	2	4.83	6.83
1500				epoxy coated	24 Ga.	3.11	4.83	7.94
1550		smooth		galvanized	28 Ga.	4.47	4.83	9.30
1600		round corrugated	2″	galvanized	28 Ga.	2.31	3.68	5.99
1650			3″	galvanized	28 Ga.	2.31	3.68	5.99
1700			4″	galvanized	28 Ga.	2.77	4.83	7.60
1750			5″	galvanized	28 Ga.	4.10	5.40	9.50
1800				galvanized	26 Ga.	4.10	5.40	9.50
1850			6″	galvanized	28 Ga.	4.57	6.65	11.22
1900				galvanized	26 Ga.	3.85	6.65	10.50
2500	Stainless steel	rectangular	2″x3″	mill		42.50	3.68	46.18
2550	Tubing sch.5		3″x4″	mill		54	4.83	58.83
2600			4″x5″	mill		111	5.20	116.20
2650		round	3″	mill		42.50	3.68	46.18
2700			4″	mill		54	4.83	58.83
2750			5″	mill		111	5.20	116.20

B3010 630 Gravel Stop

	MATERIALS	SECTION	SIZE	FINISH	THICKNESS	COST PER L.F.		
						MAT.	INST.	TOTAL
5100	Aluminum	extruded	4″	mill	.050″	7.05	4.83	11.88
5200			4″	duranodic	.050″	8.10	4.83	12.93
5300			8″	mill	.050″	8.30	5.60	13.90
5400			8″	duranodic	.050″	10	5.60	15.60
5500			12″-2 pc.	mill	.050″	11.70	7	18.70
6000			12″-2 pc.	duranodic	.050″	11.90	7	18.90
6100	Stainless	formed	6″	mill	24 Ga.	16.85	5.20	22.05
6200			12″	mill	24 Ga.	25	7	32

B3020 Roof Openings

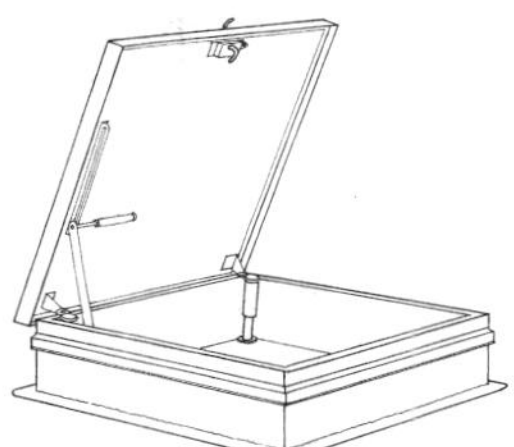

Roof Hatch

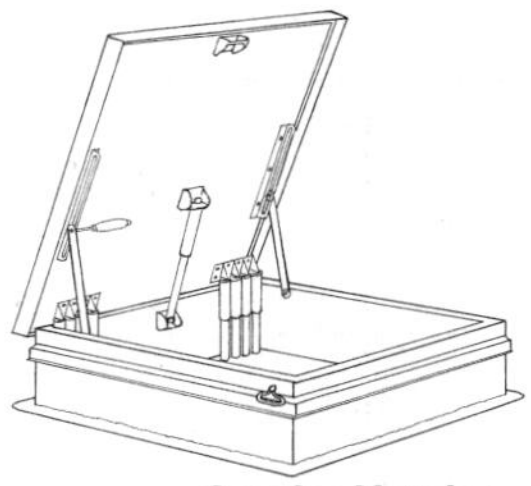

Smoke Hatch

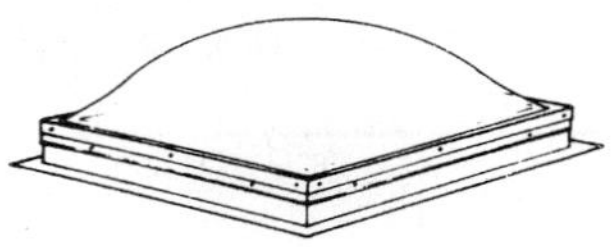

Skylight

B3020 110	Skylights	COST PER S.F.		
		MAT.	INST.	TOTAL
5100	Skylights, plastic domes, insul curbs, nom. size to 10 S.F., single glaze	32	14.55	46.55
5200	Double glazing	36.50	17.95	54.45
5300	10 S.F. to 20 S.F., single glazing	32.50	5.90	38.40
5400	Double glazing	31	7.40	38.40
5500	20 S.F. to 30 S.F., single glazing	24.50	5	29.50
5600	Double glazing	29.50	5.90	35.40
5700	30 S.F. to 65 S.F., single glazing	24	3.82	27.82
5800	Double glazing	30	5	35
6000	Sandwich panels fiberglass, 1-9/16" thick, 2 S.F. to 10 S.F.	20.50	11.65	32.15
6100	10 S.F. to 18 S.F.	18.25	8.80	27.05
6200	2-3/4" thick, 25 S.F. to 40 S.F.	29.50	7.90	37.40
6300	40 S.F. to 70 S.F.	24	7.05	31.05

B3020 210	Hatches	COST PER OPNG.		
		MAT.	INST.	TOTAL
0200	Roof hatches with curb and 1" fiberglass insulation, 2'-6"x3'-0", aluminum	1,075	233	1,308
0300	Galvanized steel 165 lbs.	700	233	933
0400	Primed steel 164 lbs.	775	233	1,008
0500	2'-6"x4'-6" aluminum curb and cover, 150 lbs.	1,400	259	1,659
0600	Galvanized steel 220 lbs.	975	259	1,234
0650	Primed steel 218 lbs.	1,050	259	1,309
0800	2'x6"x8'-0" aluminum curb and cover, 260 lbs.	2,075	355	2,430
0900	Galvanized steel, 360 lbs.	1,975	355	2,330
0950	Primed steel 358 lbs.	1,575	355	1,930
1200	For plexiglass panels, add to the above	495		495
2100	Smoke hatches, unlabeled not incl. hand winch operator, 2'-6"x3', galv	845	282	1,127
2200	Plain steel, 160 lbs.	940	282	1,222
2400	2'-6"x8'-0",galvanized steel, 360 lbs.	2,150	390	2,540
2500	Plain steel, 350 lbs.	1,725	390	2,115
3000	4'-0"x8'-0", double leaf low profile, aluminum cover, 359 lb.	2,925	291	3,216
3100	Galvanized steel 475 lbs.	2,425	291	2,716
3200	High profile, aluminum cover, galvanized curb, 361 lbs.	2,850	291	3,141

C Interiors

C1010 Partitions

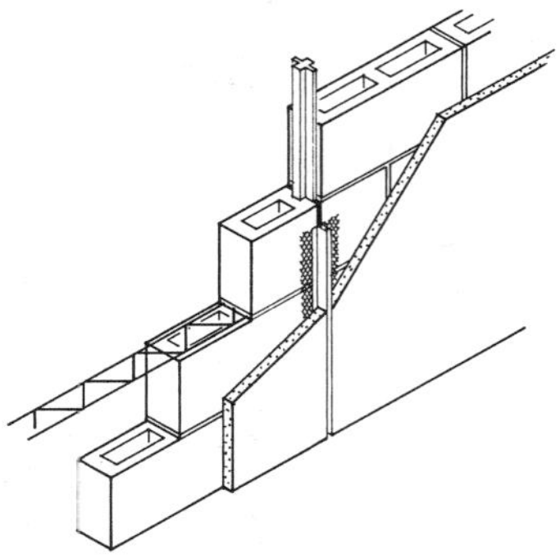

The Concrete Block Partition Systems are defined by weight and type of block, thickness, type of finish and number of sides finished. System components include joint reinforcing on alternate courses and vertical control joints.

System Components	QUANTITY	UNIT	COST PER S.F. MAT.	COST PER S.F. INST.	COST PER S.F. TOTAL
SYSTEM C1010 102 1020					
CONC. BLOCK PARTITION, 8″ X 16″, 4″ TK., 2 CT. GYP. PLASTER 2 SIDES					
Conc. block partition, 4″ thick	1.000	S.F.	1.92	6.10	8.02
Control joint	.050	L.F.	.18	.15	.33
Horizontal joint reinforcing	.800	L.F.	.08	.07	.15
Gypsum plaster, 2 coat, on masonry	2.000	S.F.	.77	5.89	6.66
TOTAL			2.95	12.21	15.16

C1010 102 Concrete Block Partitions - Regular Weight

	TYPE	THICKNESS (IN.)	TYPE FINISH	SIDES FINISHED		COST PER S.F. MAT.	COST PER S.F. INST.	COST PER S.F. TOTAL
1000	Hollow	4	none	0		2.18	6.30	8.48
1010			gyp. plaster 2 coat	1		2.57	9.25	11.82
1020				2		2.95	12.20	15.15
1100			lime plaster - 2 coat	1		2.42	9.25	11.67
1150			lime portland - 2 coat	1		2.43	9.25	11.68
1200			portland - 3 coat	1		2.45	9.75	12.20
1400			5/8″ drywall	1		2.82	8.50	11.32
1410				2		3.46	10.70	14.16
1500		6	none	0		2.77	6.75	9.52
1510			gyp. plaster 2 coat	1		3.16	9.70	12.86
1520				2		3.54	12.65	16.19
1600			lime plaster - 2 coat	1		3.01	9.70	12.71
1650			lime portland - 2 coat	1		3.02	9.70	12.72
1700			portland - 3 coat	1		3.04	10.20	13.24
1900			5/8″ drywall	1		3.41	8.95	12.36
1910				2		4.05	11.15	15.20
2000		8	none	0		2.97	7.25	10.22
2010			gyp. plaster 2 coat	1		3.36	10.20	13.56
2020			gyp. plaster 2 coat	2		3.74	13.15	16.89
2100			lime plaster - 2 coat	1		3.21	10.20	13.41
2150			lime portland - 2 coat	1		3.22	10.20	13.42
2200			portland - 3 coat	1		3.24	10.65	13.89
2400			5/8″ drywall	1		3.61	9.45	13.06
2410				2		4.25	11.65	15.90

C1010 Partitions

C1010 102 Concrete Block Partitions - Regular Weight

	TYPE	THICKNESS (IN.)	TYPE FINISH	SIDES FINISHED		COST PER S.F.		
						MAT.	INST.	TOTAL
2500	Hollow	10	none	0		3.51	7.60	11.11
2510			gyp. plaster 2 coat	1		3.90	10.50	14.40
2520				2		4.28	13.50	17.78
2600			lime plaster - 2 coat	1		3.75	10.50	14.25
2650			lime portland - 2 coat	1		3.76	10.50	14.26
2700			portland - 3 coat	1		3.78	11	14.78
2900			5/8" drywall	1		4.15	9.80	13.95
2910				2		4.79	11.95	16.74
3000	Solid	2	none	0		1.88	6.25	8.13
3010			gyp. plaster	1		2.27	9.20	11.47
3020				2		2.65	12.15	14.80
3100			lime plaster - 2 coat	1		2.12	9.20	11.32
3150			lime portland - 2 coat	1		2.13	9.20	11.33
3200			portland - 3 coat	1		2.15	9.70	11.85
3400			5/8" drywall	1		2.52	8.45	10.97
3410				2		3.16	10.65	13.81
3500		4	none	0		2.45	6.50	8.95
3510			gyp. plaster	1		2.95	9.50	12.45
3520				2		3.22	12.40	15.62
3600			lime plaster - 2 coat	1		2.69	9.45	12.14
3650			lime portland - 2 coat	1		2.70	9.45	12.15
3700			portland - 3 coat	1		2.72	9.95	12.67
3900			5/8" drywall	1		3.09	8.70	11.79
3910				2		3.73	10.90	14.63
4000		6	none	0		2.83	7	9.83
4010			gyp. plaster	1		3.22	9.95	13.17
4020				2		3.60	12.90	16.50
4100			lime plaster - 2 coat	1		3.07	9.95	13.02
4150			lime portland - 2 coat	1		3.08	9.95	13.03
4200			portland - 3 coat	1		3.10	10.45	13.55
4400			5/8" drywall	1		3.47	9.20	12.67
4410				2		4.11	11.40	15.51

C1010 104 Concrete Block Partitions - Lightweight

	TYPE	THICKNESS (IN.)	TYPE FINISH	SIDES FINISHED		COST PER S.F.		
						MAT.	INST.	TOTAL
5000	Hollow	4	none	0		2.30	6.15	8.45
5010			gyp. plaster	1		2.69	9.10	11.79
5020				2		3.07	12.05	15.12
5100			lime plaster - 2 coat	1		2.54	9.10	11.64
5150			lime portland - 2 coat	1		2.55	9.10	11.65
5200			portland - 3 coat	1		2.57	9.60	12.17
5400			5/8" drywall	1		2.94	8.35	11.29
5410				2		3.58	10.55	14.13
5500		6	none	0		3.13	6.60	9.73
5510			gyp. plaster	1		3.52	9.55	13.07
5520			gyp. plaster	2		3.90	12.50	16.40
5600			lime plaster - 2 coat	1		3.37	9.55	12.92
5650			lime portland - 2 coat	1		3.38	9.55	12.93
5700			portland - 3 coat	1		3.40	10.05	13.45
5900			5/8" drywall	1		3.77	8.80	12.57
5910				2		4.41	11	15.41

C1010 Partitions

C1010 104 Concrete Block Partitions - Lightweight

	TYPE	THICKNESS (IN.)	TYPE FINISH	SIDES FINISHED		COST PER S.F.		
						MAT.	INST.	TOTAL
6000	Hollow	8	none	0		3.78	7.05	10.83
6010			gyp. plaster	1		4.17	10	14.17
6020				2		4.55	12.95	17.50
6100			lime plaster - 2 coat	1		4.02	10	14.02
6150			lime portland - 2 coat	1		4.03	10	14.03
6200			portland - 3 coat	1		4.05	10.45	14.50
6400			5/8" drywall	1		4.42	9.25	13.67
6410				2		5.05	11.45	16.50
6500		10	none	0		4.55	7.40	11.95
6510			gyp. plaster	1		4.94	10.30	15.24
6520				2		5.30	13.30	18.60
6700			portland - 3 coat	1		4.82	10.80	15.62
6900			5/8" drywall	1		5.20	9.60	14.80
6910				2		5.85	11.75	17.60
7000	Solid	4	none	0		2.08	6.45	8.53
7010			gyp. plaster	1		2.47	9.40	11.87
7020				2		2.85	12.35	15.20
7100			lime plaster - 2 coat	1		2.32	9.40	11.72
7150			lime portland - 2 coat	1		2.33	9.40	11.73
7200			portland - 3 coat	1		2.35	9.90	12.25
7400			5/8" drywall	1		2.72	8.65	11.37
7410				2		3.36	10.85	14.21
7500		6	none	0		3.65	6.95	10.60
7510			gyp. plaster	1		4.18	9.95	14.13
7520				2		4.42	12.85	17.27
7600			lime plaster - 2 coat	1		3.89	9.90	13.79
7650			lime portland - 2 coat	1		3.90	9.90	13.80
7700			portland - 3 coat	1		3.92	10.40	14.32
7900			5/8" drywall	1		4.29	9.15	13.44
7910				2		4.93	11.35	16.28
8000		8	none	0		4.12	7.45	11.57
8010			gyp. plaster	1		4.51	10.40	14.91
8020				2		4.89	13.35	18.24
8100			lime plaster - 2 coat	1		4.36	10.40	14.76
8150			lime portland - 2 coat	1		4.37	10.40	14.77
8200			portland - 3 coat	1		4.39	10.85	15.24
8400			5/8" drywall	1		4.76	9.65	14.41
8410				2		5.40	11.85	17.25

C1010 Partitions

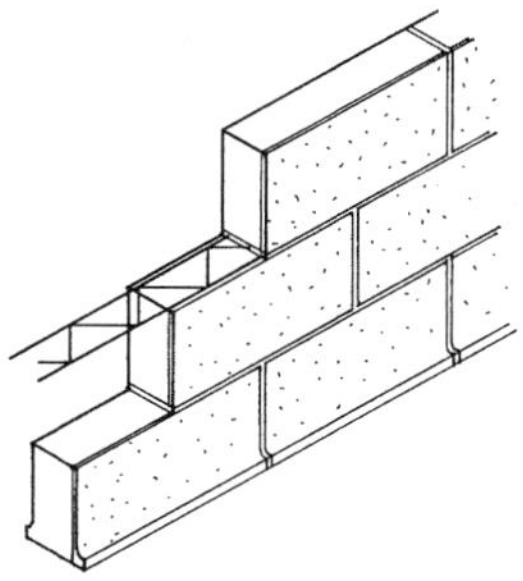

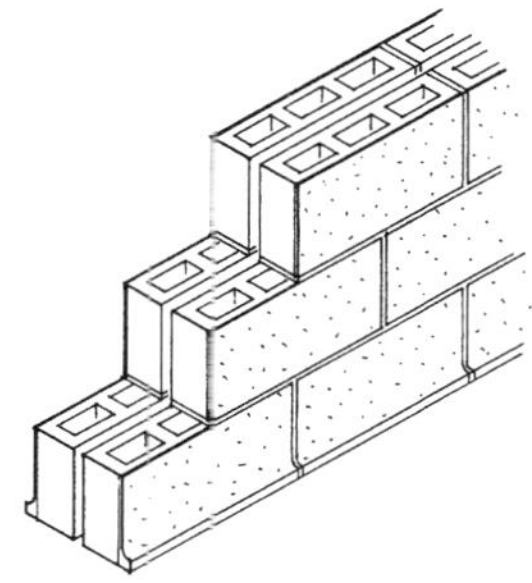

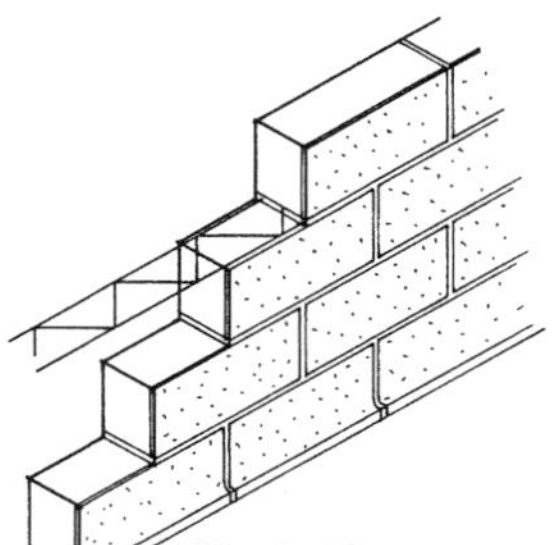

Single Wythe

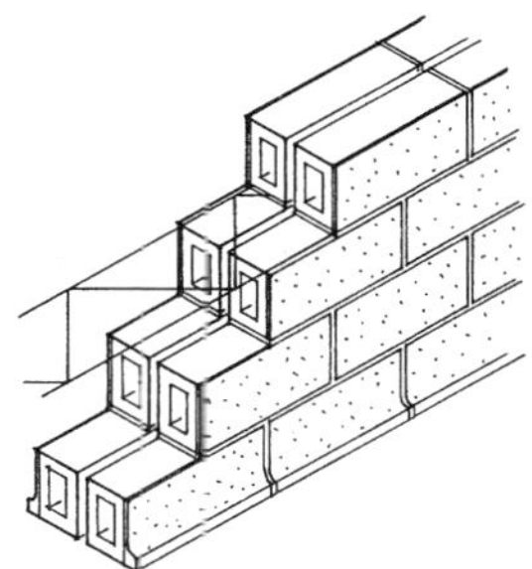

Double Wythe

C1010 120	Tile Partitions	COST PER S.F.		
		MAT.	INST.	TOTAL
1000	8W series 8"x16", 4" thick wall, reinf every 2 courses, glazed 1 side	15.05	7.60	22.65
1100	Glazed 2 sides	17.90	8.05	25.95
1200	Glazed 2 sides, using 2 wythes of 2" thick tile	22	14.60	36.60
1300	6" thick wall, horizontal reinf every 2 courses, glazed 1 side	26	7.95	33.95
1400	Glazed 2 sides, each face different color, 2" and 4" tile	26	14.90	40.90
1500	8" thick wall, glazed 2 sides using 2 wythes of 4" thick tile	30	15.20	45.20
1600	10" thick wall, glazed 2 sides using 1 wythe of 4" & 1 wythe of 6" tile	41	15.55	56.55
1700	Glazed 2 sides cavity wall, using 2 wythes of 4" thick tile	30	15.20	45.20
1800	12" thick wall, glazed 2 sides using 2 wythes of 6" thick tile	52	15.90	67.90
1900	Glazed 2 sides cavity wall, using 2 wythes of 4" thick tile	30	15.20	45.20
2100	6T series 5-1/3"x12" tile, 4" thick, non load bearing glazed one side	12.80	11.90	24.70
2200	Glazed two sides	17.05	13.45	30.50
2300	Glazed two sides, using two wythes of 2" thick tile	18.80	23.50	42.30
2400	6" thick, glazed one side	19.20	12.50	31.70
2500	Glazed two sides	22.50	14.20	36.70
2600	Glazed two sides using 2" thick tile and 4" thick tile	22	23.50	45.50
2700	8" thick, glazed one side	25.50	14.55	40.05
2800	Glazed two sides using two wythes of 4" thick tile	25.50	24	49.50
2900	Glazed two sides using 6" thick tile and 2" thick tile	28.50	24	52.50
3000	10" thick cavity wall, glazed two sides using two wythes of 4" tile	25.50	24	49.50
3100	12" thick, glazed two sides using 4" thick tile and 8" thick tile	38.50	26.50	65
3200	2" thick facing tile, glazed one side, on 6" concrete block	13.60	13.85	27.45
3300	On 8" concrete block	13.75	14.30	28.05
3400	On 10" concrete block	14.30	14.60	28.90

C1010 Partitions

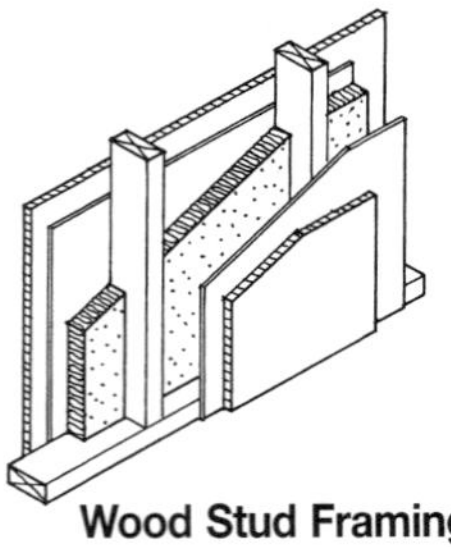

Wood Stud Framing

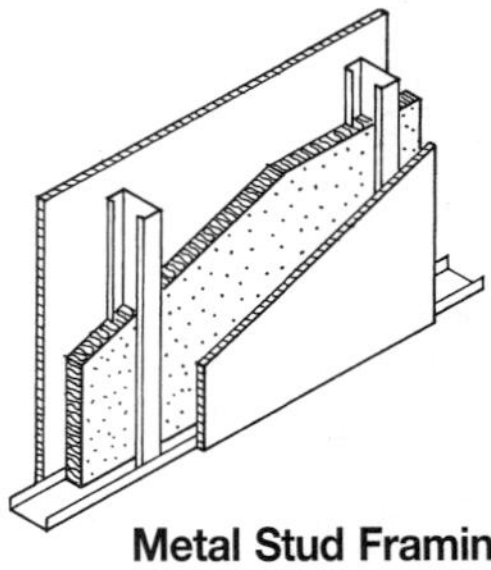

Metal Stud Framing

The Drywall Partitions/Stud Framing Systems are defined by type of drywall and number of layers, type and spacing of stud framing, and treatment on the opposite face. Components include taping and finishing.

Cost differences between regular and fire resistant drywall are negligible, and terminology is interchangeable. In some cases fiberglass insulation is included for additional sound deadening.

System Components	QUANTITY	UNIT	COST PER S.F. MAT.	INST.	TOTAL
SYSTEM C1010 124 1250					
DRYWALL PARTITION,5/8″ F.R.1 SIDE,5/8″ REG.1 SIDE,2″X4″STUDS,16″ O.C.					
Gypsum plasterboard, nailed/screwed to studs, 5/8″ fire resistant	1.000	S.F.	.39	.59	.98
Gypsum plasterboard, nailed/screwed to studs, 5/8″ regular	1.000	S.F.	.37	.59	.96
Taping and finishing joints	2.000	S.F.	.10	1.18	1.28
Framing, 2 x 4 studs @ 16″ O.C., 10′ high	1.000	S.F.	.46	1.19	1.65
TOTAL			1.32	3.55	4.87

C1010 124 Drywall Partitions/Wood Stud Framing

	FACE LAYER	BASE LAYER	FRAMING	OPPOSITE FACE	INSULATION	COST PER S.F. MAT.	INST.	TOTAL
1200	5/8″ FR drywall	none	2 x 4, @ 16″ O.C.	same	0	1.34	3.55	4.89
1250				5/8″ reg. drywall	0	1.32	3.55	4.87
1300				nothing	0	.90	2.37	3.27
1400		1/4″ SD gypsum	2 x 4 @ 16″ O.C.	same	1-1/2″ fiberglass	2.64	5.45	8.09
1450				5/8″ FR drywall	1-1/2″ fiberglass	2.20	4.80	7
1500				nothing	1-1/2″ fiberglass	1.76	3.62	5.38
1600		resil. channels	2 x 4 @ 16″, O.C.	same	1-1/2″ fiberglass	2.09	6.95	9.04
1650				5/8″ FR drywall	1-1/2″ fiberglass	1.93	5.55	7.48
1700				nothing	1-1/2″ fiberglass	1.49	4.36	5.85
1800		5/8″ FR drywall	2 x 4 @ 24″ O.C.	same	0	2.01	4.49	6.50
1850				5/8″ FR drywall	0	1.62	3.90	5.52
1900				nothing	0	1.18	2.72	3.90
1950		5/8′ FR drywall	2 x 4, 16″ O.C.	same	0	2.12	4.73	6.85
1955				5/8″ FR drywall	0	1.73	4.14	5.87
2000				nothing	0	1.29	2.96	4.25
2010		5/8″ FR drywall	staggered, 6″ plate	same	0	2.59	5.95	8.54
2015				5/8″ FR drywall	0	2.20	5.35	7.55
2020				nothing	0	1.76	4.17	5.93
2200		5/8″ FR drywall	2 rows-2 x 4 16″O.C.	same	2″ fiberglass	3.12	6.50	9.62
2250				5/8″ FR drywall	2″ fiberglass	2.73	5.90	8.63
2300				nothing	2″ fiberglass	2.29	4.74	7.03
2400	5/8″ WR drywall	none	2 x 4, @ 16″ O.C.	same	0	1.50	3.55	5.05
2450				5/8″ FR drywall	0	1.42	3.55	4.97
2500				nothing	0	.98	2.37	3.35
2600		5/8″ FR drywall	2 x 4, @ 24″ O.C.	same	0	2.17	4.49	6.66
2650				5/8″ FR drywall	0	1.70	3.90	5.60
2700				nothing	0	1.26	2.72	3.98

C1010 Partitions

C1010 124 Drywall Partitions/Wood Stud Framing

	FACE LAYER	BASE LAYER	FRAMING	OPPOSITE FACE	INSULATION	COST PER S.F.		
						MAT.	INST.	TOTAL
2800	5/8" VF drywall	none	2 x 4, @ 16" O.C.	same	0	2.14	3.83	5.97
2850				5/8" FR drywall	0	1.74	3.69	5.43
2900				nothing	0	1.30	2.51	3.81
3000		5/8" FR drywall	2 x 4 , 24" O.C.	same	0	2.81	4.77	7.58
3050				5/8" FR drywall	0	2.02	4.04	6.06
3100				nothing	0	1.58	2.86	4.44
3200	1/2" reg drywall	3/8" reg drywall	2 x 4, @ 16" O.C.	same	0	2.06	4.73	6.79
3250				5/8" FR drywall	0	1.70	4.14	5.84
3300				nothing	0	1.26	2.96	4.22

C1010 126 Drywall Partitions/Metal Stud Framing

	FACE LAYER	BASE LAYER	FRAMING	OPPOSITE FACE	INSULATION	COST PER S.F.		
						MAT.	INST.	TOTAL
5000	5/8" FR drywall	none	1-5/8" @ 16" O.C.	same	0	1.15	3.56	4.71
5010				5/8" reg. drywall	0	1.13	3.56	4.69
5020				nothing	0	.71	2.38	3.09
5030			2-1/2"" @ 16" O.C.	same	0	1.22	3.57	4.79
5040				5/8" reg. drywall	0	1.20	3.57	4.77
5050				nothing	0	.78	2.39	3.17
5060			3-5/8" @ 16" O.C.	same	0	1.29	3.60	4.89
5070				5/8" reg. drywall	0	1.27	3.60	4.87
5080				nothing	0	.85	2.42	3.27
5200			1-5/8" @ 24" O.C.	same	0	1.08	3.14	4.22
5250				5/8" reg. drywall	0	1.06	3.14	4.20
5300				nothing	0	.64	1.96	2.60
5310			2-1/2" @ 24" O.C.	same	0	1.13	3.15	4.28
5320				5/8" reg. drywall	0	1.11	3.15	4.26
5330				nothing	0	.69	1.97	2.66
5400			3-5/8" @ 24" O.C.	same	0	1.18	3.16	4.34
5450				5/8" reg. drywall	0	1.16	3.16	4.32
5500				nothing	0	.74	1.98	2.72
5530		1/4" SD gypsum	1-5/8" @ 16" O.C.	same	0	2.03	4.88	6.91
5535				5/8" FR drywall	0	1.15	3.56	4.71
5540				nothing	0	1.15	3.04	4.19
5545			2-1/2" @ 16" O.C.	same	0	2.10	4.89	6.99
5550				5/8" FR drywall	0	1.22	3.57	4.79
5555				nothing	0	1.22	3.05	4.27
5560			3-5/8" @ 16" O.C.	same	0	2.17	4.92	7.09
5565				5/8" FR drywall	0	1.29	3.60	4.89
5570				nothing	0	1.29	3.08	4.37
5600			1-5/8" @ 24" O.C.	same	0	1.96	4.46	6.42
5650				5/8" FR drywall	0	1.52	3.80	5.32
5700				nothing	0	1.08	2.62	3.70
5800			2-1/2" @ 24" O.C.	same	0	2.01	4.47	6.48
5850				5/8" FR drywall	0	1.57	3.81	5.38
5900				nothing	0	1.13	2.63	3.76
5910			3-5/8" @ 24" O.C.	same	0	2.06	4.48	6.54
5920				5/8" FR drywall	0	1.62	3.82	5.44
5930				nothing	0	1.18	2.64	3.82

C1010 Partitions

C1010 126 Drywall Partitions/Metal Stud Framing

	FACE LAYER	BASE LAYER	FRAMING	OPPOSITE FACE	INSULATION	COST PER S.F.		
						MAT.	INST.	TOTAL
6000	5/8" FR drywall	5/8" FR drywall	2-1/2" @ 16" O.C.	same	0	2.10	5.05	7.15
6050				5/8" FR drywall	0	1.71	4.47	6.18
6100				nothing	0	1.27	3.29	4.56
6110			3-5/8" @ 16" O.C.	same	0	2.07	4.78	6.85
6120				5/8" FR drywall	0	1.68	4.19	5.87
6130				nothing	0	1.24	3.01	4.25
6170			2-1/2" @ 24" O.C.	same	0	1.91	4.33	6.24
6180				5/8" FR drywall	0	1.52	3.74	5.26
6190				nothing	0	1.08	2.56	3.64
6200			3-5/8" @ 24" O.C.	same	0	1.96	4.34	6.30
6250				5/8"FR drywall	3-1/2" fiberglass	2.09	4.19	6.28
6300				nothing	0	1.13	2.57	3.70
6310	5/8" WR drywall	none	1-5/8" @ 16" O.C.	same	0	1.31	3.56	4.87
6320				5/8" WR drywall	0	1.78	4.15	5.93
6330				nothing	0	1.26	2.97	4.23
6340			2-1/2" @ 16" O.C.	same	0	2.32	4.75	7.07
6350				5/8" WR drywall	0	1.85	4.16	6.01
6360				nothing	0	1.33	2.98	4.31
6370			3-5/8" @ 16" O.C.	same	0	2.39	4.78	7.17
6380				5/8" WR drywall	0	1.92	4.19	6.11
6390				nothing	0	1.40	3.01	4.41
6400		none	1-5/8" @ 24" O.C.	same	0	1.24	3.14	4.38
6450				5/8" WR drywall	0	1.16	3.14	4.30
6500				nothing	0	.72	1.96	2.68
6510			2-1/2" @ 24" O.C.	same	0	2.23	4.33	6.56
6520				5/8" WR drywall	0	1.21	3.15	4.36
6530				nothing	0	.77	1.97	2.74
6600			3-5/8" @ 24" O.C.	same	0	1.34	3.16	4.50
6650				5/8" WR drywall	0	1.26	3.16	4.42
6700				nothing	0	.82	1.98	2.80
6800		5/8" FR drywall	2-1/2" @ 16" O.C.	same	0	2.26	5.05	7.31
6850				5/8" FR drywall	0	1.79	4.47	6.26
6900				nothing	0	1.35	3.29	4.64
6910			3-5/8" @ 16" O.C.	same	0	2.23	4.78	7.01
6920				5/8" FR drywall	0	1.76	4.19	5.95
6930				nothing	0	1.32	3.01	4.33
6940			2-1/2" @ 24" O.C.	same	0	1.91	4.33	6.24
6950				5/8" FR drywall	0	1.60	3.74	5.34
6960				nothing	0	1.16	2.56	3.72
7000			3-5/8" @ 24" O.C.	same	0	2.12	4.34	6.46
7050				5/8"FR drywall	3-1/2" fiberglass	2.17	4.19	6.36
7100				nothing	0	1.21	2.57	3.78
7110	5/8" VF drywall	none	1-5/8" @ 16" O.C.	same	0	1.95	3.84	5.79
7120				5/8" FR drywall	0	1.62	4.15	5.77
7130				nothing	0	1.18	2.97	4.15
7140			3-5/8" @ 16" O.C.	same	0	2.23	4.78	7.01
7150				5/8" FR drywall	0	1.76	4.19	5.95
7160				nothing	0	1.32	3.01	4.33
7200		none	1-5/8" @ 24" O.C.	same	0	1.88	3.42	5.30
7250				5/8" FR drywall	0	1.48	3.28	4.76
7300				nothing	0	1.04	2.10	3.14
7400			3-5/8" @ 24" O.C.	same	0	1.98	3.44	5.42
7450				5/8" FR drywall	0	1.58	3.30	4.88

C1010 Partitions

C1010 126 Drywall Partitions/Metal Stud Framing

	FACE LAYER	BASE LAYER	FRAMING	OPPOSITE FACE	INSULATION	COST PER S.F.		
						MAT.	INST.	TOTAL
7500	5/8" VF drywall	none	3-5/8" @ 24" O.C.	nothing	0	1.14	2.12	3.26
7600		5/8" FR drywall	2-1/2" @ 16" O.C.	same	0	2.90	5.35	8.25
7650				5/8" FR drywall	0	2.11	4.61	6.72
7700				nothing	0	1.67	3.43	5.10
7710			3-5/8" @ 16" O.C.	same	0	2.87	5.05	7.92
7720				5/8" FR drywall	0	2.08	4.33	6.41
7730				nothing	0	1.64	3.15	4.79
7740			2-1/2" @ 24" O.C.	same	0	2.71	4.61	7.32
7750				5/8" FR drywall	0	1.92	3.88	5.80
7760				nothing	0	1.48	2.70	4.18
7800			3-5/8" @ 24" O.C.	same	0	2.76	4.62	7.38
7850				5/8"FR drywall	3-1/2" fiberglass	2.49	4.33	6.82
7900				nothing	0	1.53	2.71	4.24

C1010 Partitions

C1010 128	Drywall Components	COST PER S.F.		
		MAT.	INST.	TOTAL
0140	Metal studs, 24" O.C. including track, load bearing, 18 gage, 2-1/2"	.70	1.11	1.81
0160	3-5/8"	.84	1.13	1.97
0180	4"	.75	1.16	1.91
0200	6"	1.12	1.18	2.30
0220	16 gage, 2-1/2"	.82	1.27	2.09
0240	3-5/8"	.98	1.29	2.27
0260	4"	1.03	1.32	2.35
0280	6"	1.30	1.35	2.65
0300	Non load bearing, 25 gage, 1-5/8"	.20	.78	.98
0340	3-5/8"	.30	.80	1.10
0360	4"	.33	.80	1.13
0380	6"	.40	.82	1.22
0400	20 gage, 2-1/2"	.33	.99	1.32
0420	3-5/8"	.37	1.01	1.38
0440	4"	.44	1.01	1.45
0460	6"	.52	1.02	1.54
0540	Wood studs including blocking, shoe and double top plate, 2"x4", 12" O.C.	.57	1.49	2.06
0560	16" O.C.	.46	1.19	1.65
0580	24" O.C.	.35	.95	1.30
0600	2"x6", 12" O.C.	.88	1.70	2.58
0620	16" O.C.	.71	1.32	2.03
0640	24" O.C.	.54	1.04	1.58
0642	Furring one side only, steel channels, 3/4", 12" O.C.	.42	2.38	2.80
0644	16" O.C.	.38	2.11	2.49
0646	24" O.C.	.25	1.60	1.85
0647	1-1/2" , 12" O.C.	.57	2.66	3.23
0648	16" O.C.	.51	2.33	2.84
0649	24" O.C.	.34	1.83	2.17
0650	Wood strips, 1" x 3", on wood, 12" O.C.	.43	1.08	1.51
0651	16" O.C.	.32	.81	1.13
0652	On masonry, 12" O.C.	.46	1.20	1.66
0653	16" O.C.	.35	.90	1.25
0654	On concrete, 12" O.C.	.46	2.29	2.75
0655	16" O.C.	.35	1.72	2.07
0665	Gypsum board, one face only, exterior sheathing, 1/2"	.51	1.06	1.57
0680	Interior, fire resistant, 1/2"	.40	.59	.99
0700	5/8"	.39	.59	.98
0720	Sound deadening board 1/4"	.44	.66	1.10
0740	Standard drywall 3/8"	.39	.59	.98
0760	1/2"	.36	.59	.95
0780	5/8"	.37	.59	.96
0800	Tongue & groove coreboard 1"	.83	2.48	3.31
0820	Water resistant, 1/2"	.45	.59	1.04
0840	5/8"	.47	.59	1.06
0860	Add for the following:, foil backing	.17		.17
0880	Fiberglass insulation, 3-1/2"	.52	.44	.96
0900	6"	.68	.44	1.12
0920	Rigid insulation 1"	.57	.59	1.16
0940	Resilient furring @ 16" O.C.	.22	1.86	2.08
0960	Taping and finishing	.05	.59	.64
0980	Texture spray	.04	.70	.74
1000	Thin coat plaster	.12	.74	.86
1050	Sound wall framing, 2x6 plates, 2x4 staggered studs, 12" O.C.	.69	1.46	2.15

C1010 Partitions

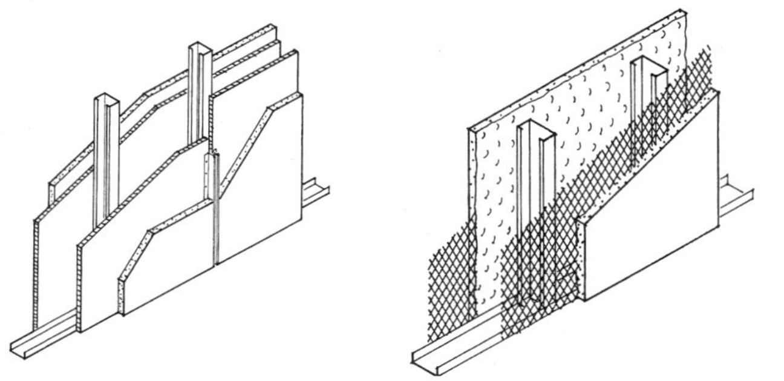

Plaster Partitions are defined as follows: type of plaster, type and spacing of framing, type of lath and treatment of opposite face.

Included in the system components are expansion joints. Metal studs are assumed to be non-loadbearing.

System Components	QUANTITY	UNIT	COST PER S.F. MAT.	COST PER S.F. INST.	COST PER S.F. TOTAL
SYSTEM C1010 140 1000					
GYP PLASTER PART'N, 2 COATS,2-1/2″ MTL. STUD,16″O.C.,3/8″GYP. LATH					
Gypsum plaster, 2 coats, on walls	2.000	S.F.	1.19	5.66	6.85
Gypsum lath, 3/8″ thick, on metal studs	2.000	S.F.	.75	1.65	2.40
Metal studs, 25 ga., 2-1/2″ @ 16″ O.C.	1.000	S.F.	.34	1.21	1.55
Expansion joint	.100	L.F.	.08	.21	.29
TOTAL			2.36	8.73	11.09

C1010 140 Plaster Partitions/Metal Stud Framing

	TYPE	FRAMING	LATH	OPPOSITE FACE		COST PER S.F. MAT.	COST PER S.F. INST.	COST PER S.F. TOTAL
1000	2 coat gypsum	2-1/2″ @ 16″O.C.	3/8″ gypsum	same		2.36	8.75	11.11
1010				nothing		1.38	5.05	6.43
1100		3-1/4″ @ 24″O.C.	1/2″ gypsum	same		2.16	8.45	10.61
1110				nothing		1.27	4.72	5.99
1500	2 coat vermiculite	2-1/2″ @ 16″O.C.	3/8″ gypsum	same		2.60	9.55	12.15
1510				nothing		1.51	5.50	7.01
1600		3-1/4″ @ 24″O.C.	1/2″ gypsum	same		2.40	9.25	11.65
1610				nothing		1.40	5.15	6.55
2000	3 coat gypsum	2-1/2″ @ 16″O.C.	3/8″ gypsum	same		2.40	9.90	12.30
2010				nothing		1.41	5.65	7.06
2020			3.4lb. diamond	same		2.70	9.90	12.60
2030				nothing		1.57	5.65	7.22
2040			2.75lb. ribbed	same		2.50	9.90	12.40
2050				nothing		1.47	5.65	7.12
2100		3-1/4″ @ 24″O.C.	1/2″ gypsum	same		2.20	9.65	11.85
2110				nothing		1.30	5.30	6.60
2120			3.4lb. ribbed	same		2.71	9.65	12.36
2130				nothing		1.55	5.30	6.85
2500	3 coat lime	2-1/2″ @ 16″O.C.	3.4lb. diamond	same		2.06	9.90	11.96
2510				nothing		1.25	5.65	6.90
2520			2.75lb. ribbed	same		1.86	9.90	11.76
2530				nothing		1.15	5.65	6.80
2600		3-1/4″ @ 24″O.C.	3.4lb. ribbed	same		2.07	9.65	11.72
2610				nothing		1.23	5.30	6.53

C1010 Partitions

C1010 140 Plaster Partitions/Metal Stud Framing

	TYPE	FRAMING	LATH	OPPOSITE FACE		COST PER S.F.		
						MAT.	INST.	TOTAL
3000	3 coat Portland	2-1/2" @ 16"O.C.	3.4lb. diamond	same		2.02	9.90	11.92
3010				nothing		1.23	5.65	6.88
3020			2.75lb. ribbed	same		1.82	9.90	11.72
3030				nothing		1.13	5.65	6.78
3100		3-1/4" @ 24"O.C.	3.4lb. ribbed	same		2.03	9.65	11.68
3110				nothing		1.21	5.30	6.51
3500	3 coat gypsum W/med. Keenes	2-1/2" @ 16"O.C.	3/8" gypsum	same		2.88	12.65	15.53
3510				nothing		1.65	7.05	8.70
3520			3.4lb. diamond	same		3.18	12.65	15.83
3530				nothing		1.81	7.05	8.86
3540			2.75lb. ribbed	same		2.98	12.65	15.63
3550				nothing		1.71	7.05	8.76
3600		3-1/4" @ 24"O.C.	1/2" gypsum	same		2.68	12.40	15.08
3610				nothing		1.54	6.70	8.24
3620			3.4lb. ribbed	same		3.19	12.40	15.59
3630				nothing		1.79	6.70	8.49
4000	3 coat gypsum W/hard Keenes	2-1/2" @ 16"O.C.	3/8" gypsum	same		2.88	14.05	16.93
4010				nothing		1.65	7.70	9.35
4020			3.4lb. diamond	same		3.18	14.05	17.23
4030				nothing		1.81	7.70	9.51
4040			2.75lb. ribbed	same		2.98	14.05	17.03
4050				nothing		1.71	7.70	9.41
4100		3-1/4" @ 24"O.C.	1/2" gypsum	same		2.68	13.75	16.43
4110				nothing		1.54	7.35	8.89
4120			3.4lb. ribbed	same		3.19	13.75	16.94
4130				nothing		1.79	7.35	9.14
4500	3 coat lime Portland	2-1/2" @ 16"O.C.	3.4lb. diamond	same		2.03	9.90	11.93
4510				nothing		1.23	5.65	6.88
4520			2.75lb. ribbed	same		1.83	9.90	11.73
4530				nothing		1.13	5.65	6.78
4600		3-1/4" @ 24"O.C.	3.4lb. ribbed	same		2.04	9.65	11.69
4610				nothing		1.21	5.30	6.51

C1010 142 Plaster Partitions/Wood Stud Framing

	TYPE	FRAMING	LATH	OPPOSITE FACE		COST PER S.F.		
						MAT.	INST.	TOTAL
5000	2 coat gypsum	2"x4" @ 16"O.C.	3/8" gypsum	same		2.48	8.55	11.03
5010				nothing		1.50	4.95	6.45
5100		2"x4" @ 24"O.C.	1/2" gypsum	same		2.21	8.40	10.61
5110				nothing		1.32	4.76	6.08
5500	2 coat vermiculite	2"x4" @ 16"O.C.	3/8" gypsum	same		2.72	9.30	12.02
5510				nothing		1.63	5.40	7.03
5600		2"x4" @ 24"O.C.	1/2" gypsum	same		2.45	9.15	11.60
5610				nothing		1.45	5.20	6.65

C1010 Partitions

C1010 142 Plaster Partitions/Wood Stud Framing

	TYPE	FRAMING	LATH	OPPOSITE FACE		COST PER S.F.		
						MAT.	INST.	TOTAL
6000	3 coat gypsum	2"x4" @ 16"O.C.	3/8" gypsum	same		2.52	9.70	12.22
6010				nothing		1.53	5.55	7.08
6020			3.4lb. diamond	same		2.82	9.80	12.62
6030				nothing		1.69	5.60	7.29
6040			2.75lb. ribbed	same		2.59	9.85	12.44
6050				nothing		1.57	5.60	7.17
6100		2"x4" @ 24"O.C.	1/2" gypsum	same		2.25	9.55	11.80
6110				nothing		1.35	5.35	6.70
6120			3.4lb. ribbed	same		2.51	9.65	12.16
6130				nothing		1.48	5.40	6.88
6500	3 coat lime	2"x4" @ 16"O.C.	3.4lb. diamond	same		2.18	9.80	11.98
6510				nothing		1.37	5.60	6.97
6520			2.75lb. ribbed	same		1.95	9.85	11.80
6530				nothing		1.25	5.60	6.85
6600		2"x4" @ 24"O.C.	3.4lb. ribbed	same		1.87	9.65	11.52
6610				nothing		1.16	5.40	6.56
7000	3 coat Portland	2"x4" @ 16"O.C.	3.4lb. diamond	same		2.14	9.80	11.94
7010				nothing		1.35	5.60	6.95
7020			2.75lb. ribbed	same		1.91	9.85	11.76
7030				nothing		1.23	5.60	6.83
7100		2"x4" @ 24"O.C.	3.4lb. ribbed	same		2.08	9.80	11.88
7110				nothing		1.26	5.45	6.71
7500	3 coat gypsum W/med Keenes	2"x4" @ 16"O.C.	3/8" gypsum	same		3	12.45	15.45
7510				nothing		1.77	6.95	8.72
7520			3.4lb. diamond	same		3.30	12.55	15.85
7530				nothing		1.93	7	8.93
7540			2.75lb. ribbed	same		3.07	12.60	15.67
7550				nothing		1.81	7	8.81
7600		2"x4" @ 24"O.C.	1/2" gypsum	same		2.73	12.30	15.03
7610				nothing		1.59	6.75	8.34
7620			3.4lb. ribbed	same		3.24	12.55	15.79
7630				nothing		1.84	6.85	8.69
8000	3 coat gypsum W/hard Keenes	2"x4" @ 16"O.C.	3/8" gypsum	same		3	13.85	16.85
8010				nothing		1.77	7.60	9.37
8020			3.4lb. diamond	same		3.30	13.95	17.25
8030				nothing		1.93	7.65	9.58
8040			2.75lb. ribbed	same		3.07	14	17.07
8050				nothing		1.81	7.65	9.46
8100		2"x4" @ 24"O.C.	1/2" gypsum	same		2.73	13.70	16.43
8110				nothing		1.59	7.40	8.99
8120			3.4lb. ribbed	same		3.24	13.90	17.14
8130				nothing		1.84	7.50	9.34
8500	3 coat lime Portland	2"x4" @ 16"O.C.	3.4lb. diamond	same		2.15	9.80	11.95
8510				nothing		1.35	5.60	6.95
8520			2.75lb. ribbed	same		1.92	9.85	11.77
8530				nothing		1.23	5.60	6.83
8600		2"x4" @ 24"O.C.	3.4lb. ribbed	same		2.09	9.80	11.89
8610				nothing		1.26	5.45	6.71

C1010 Partitions

C1010 144	Plaster Partition Components	COST PER S.F.		
		MAT.	INST.	TOTAL
0060	Metal studs, 16" O.C., including track, non load bearing, 25 gage, 1-5/8"	.34	1.21	1.55
0080	2-1/2"	.34	1.21	1.55
0100	3-1/4"	.41	1.24	1.65
0120	3-5/8"	.41	1.24	1.65
0140	4"	.45	1.25	1.70
0160	6"	.54	1.26	1.80
0180	Load bearing, 20 gage, 2-1/2"	.95	1.55	2.50
0200	3-5/8"	1.14	1.57	2.71
0220	4"	1.18	1.61	2.79
0240	6"	1.51	1.63	3.14
0260	16 gage 2-1/2"	1.13	1.75	2.88
0280	3-5/8"	1.35	1.80	3.15
0300	4"	1.41	1.83	3.24
0320	6"	1.78	1.86	3.64
0340	Wood studs, including blocking, shoe and double plate, 2"x4" , 12" O.C.	.57	1.49	2.06
0360	16" C.C.	.46	1.19	1.65
0380	24" C.C.	.35	.95	1.30
0400	2"x6" , 12" O.C.	.88	1.70	2.58
0420	16" C.C.	.71	1.32	2.03
0440	24" C.C.	.54	1.04	1.58
0460	Furring one face only, steel channels, 3/4" , 12" O.C.	.42	2.38	2.80
0480	16" C.C.	.38	2.11	2.49
0500	24" C.C.	.25	1.60	1.85
0520	1-1/2" , 12" O.C.	.57	2.66	3.23
0540	16" C.C.	.51	2.33	2.84
0560	24"O.C.	.34	1.83	2.17
0580	Wood strips 1"x3", on wood., 12" O.C.	.43	1.08	1.51
0600	16"O.C.	.32	.81	1.13
0620	On masonry, 12" O.C.	.46	1.20	1.66
0640	16" C.C.	.35	.90	1.25
0660	On concrete, 12" O.C.	.46	2.29	2.75
0680	16" C.C.	.35	1.72	2.07
0700	Gypsum lath. plain or perforated, nailed to studs, 3/8" thick	.37	.73	1.10
0720	1/2" thick	.30	.78	1.08
0740	Clipped to studs, 3/8" thick	.37	.83	1.20
0760	1/2" thick	.30	.89	1.19
0780	Metal lath, diamond painted, nailed to wood studs, 2.5 lb.	.50	.73	1.23
0800	3.4 lb.	.53	.78	1.31
0820	Screwed to steel studs, 2.5 lb.	.50	.78	1.28
0840	3.4 lb.	.51	.83	1.34
0860	Rib painted, wired to steel, 2.75 lb	.43	.83	1.26
0880	3.4 lb	.55	.89	1.44
0900	4.0 lb	.69	.95	1.64
0910				
0920	Gypsum plaster, 2 coats	.45	2.82	3.27
0940	3 coats	.65	3.42	4.07
0960	Perlite or vermiculite plaster, 2 coats	.72	3.24	3.96
0980	3 coats	.78	4	4.78
1000	Stucco, 3 coats, 1" thick, on wood framing	.78	5.70	6.48
1020	On masonry	.31	4.42	4.73
1100	Metal base galvanized and painted 2-1/2" high	.72	2.33	3.05

C1010 Partitions

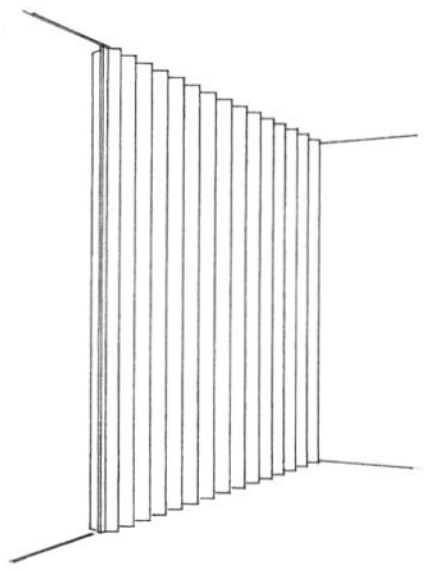

Folding Accordion

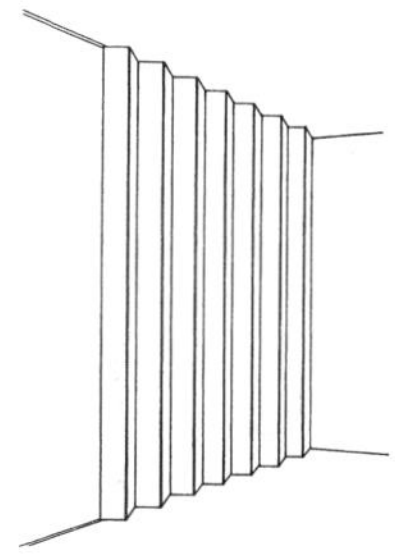

Folding Leaf

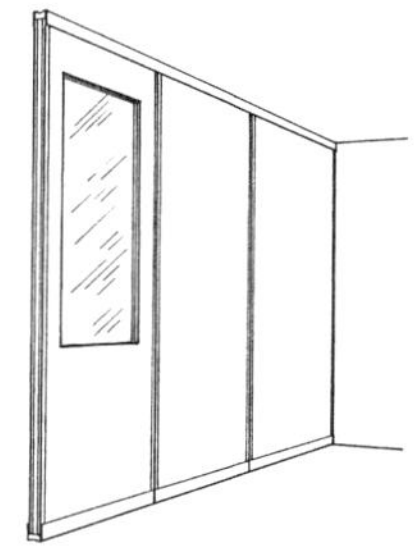

Movable and Borrow Lites

C1010 205	Partitions	COST PER S.F.		
		MAT.	INST.	TOTAL
0360	Folding accordion, vinyl covered, acoustical, 3 lb. S.F., 17 ft max. hgt	35	11.90	46.90
0380	5 lb. per S.F. 27 ft max height	48.50	12.50	61
0400	5.5 lb. per S.F., 17 ft. max height	56.50	13.20	69.70
0420	Commercial, 1.75 lb per S.F., 8 ft. max height	31.50	5.30	36.80
0440	2.0 Lb per S.F., 17 ft. max height	32.50	7.95	40.45
0460	Industrial, 4.0 lb. per S.F. 27 ft. max height	49	15.85	64.85
0480	Vinyl clad wood or steel, electric operation 6 psf	69	7.45	76.45
0500	Wood, non acoustic, birch or mahogany	37	3.96	40.96
0560	Folding leaf, alum framed acoustical 12′ high., 5.5 lb/S.F. standard trim	45	19.80	64.80
0580	Premium trim	54	39.50	93.50
0600	6.5 lb. per S.F., standard trim	47	19.80	66.80
0620	Premium trim	58.50	39.50	98
0640	Steel acoustical, 7.5 per S.F., vinyl faced, standard trim	66.50	19.80	86.30
0660	Premium trim	81.50	39.50	121
0665	Steel with vinyl face, economy	66.50	19.80	86.30
0670	Deluxe	81.50	39.50	121
0680	Wood acoustic type, vinyl faced to 18′ high 6 psf, economy trim	62.50	19.80	82.30
0700	Standard trim	74.50	26.50	101
0720	Premium trim	96.50	39.50	136
0740	Plastic lam. or hardwood faced, standard trim	64.50	19.80	84.30
0760	Premium trim	68.50	39.50	108
0780	Wood, low acoustical type to 12 ft. high 4.5 psf	47	24	71
0840	Demountable, trackless wall, cork finish, semi acous, 1-5/8″ th, unsealed	42.50	3.66	46.16
0860	Sealed	47	6.25	53.25
0880	Acoustic, 2″ thick, unsealed	40	3.90	43.90
0900	Sealed	61.50	5.30	66.80
0920	In-plant modular office system, w/prehung steel door			
0940	3″ thick honeycomb core panels			
0960	12′ x 12′, 2 wall	13.60	.74	14.34
0970	4 wall	13.80	.98	14.78
0980	16′ x 16′, 2 wall	13.75	.52	14.27
0990	4 wall	9.35	.52	9.87
1000	Gypsum, demountable, 3″ to 3-3/4″ thick x 9′ high, vinyl clad	7.35	2.78	10.13
1020	Fabric clad	18.30	3	21.30
1040	1.75 system, vinyl clad hardboard, paper honeycomb core panel			
1060	1-3/4″ to 2-1/2″ thick x 9′ high	12.30	2.78	15.08
1080	Unitized gypsum panel system, 2″ to 2-1/2″ thick x 9′ high			
1100	Vinyl clad gypsum	15.85	2.78	18.63
1120	Fabric clad gypsum	26	3	29
1140	Movable steel walls, modular system			
1160	Unitized panels, 48″ wide x 9′ high			
1180	Baked enamel, pre-finished	17.85	2.20	20.05
1200	Fabric clad	26	2.33	28.33

C1010 Partitions

C1010 210	Partitions	MAT.	INST.	TOTAL
		COST PER S.F.		
0100	Partitions, folding accordion, 1 25 lb. vinyl (residential), to 8′ high	27.50	3.96	31.46
0110	4 lb. vinyl (industrial), to 27′ high	49	15.85	64.85
0120	Acoustical, 3 lb. vinyl, to 17′ high	35	11.90	46.90
0130	4.5 lb. vinyl, fire rated 40, 20′ high	56.50	7.45	63.95
0485	Operable acoustic, no track, 1-5/8″ thick, economy	36.50	3.17	39.67
0490	With track, 3′ thick, premium trim	110	3.96	113.96

C1010 230	Demountable Partitions, L.F.	MAT.	INST.	TOTAL
		COST PER L.F.		
0100	Office-demountable gypsum, 3″ to 3-3/4″ thick x 9′ high, vinyl clad	66	25	91
0110	Fabric clad	165	27	192
0120	1.75 system, vinyl clad hardboard, paper honeycomb			
0130	core panel, 1-3/4″ to 2-1/2″ thick x 9′ high	111	25	136
0140	Steel, modular, unitized panels, 48″ wide x 9′ high			
0150	Baked enamel, pre-finished	161	19.80	180.80
0160	Portable, fiber core, 4′ high, fabric face	74.50	6.80	81.30

C1010 Partitions

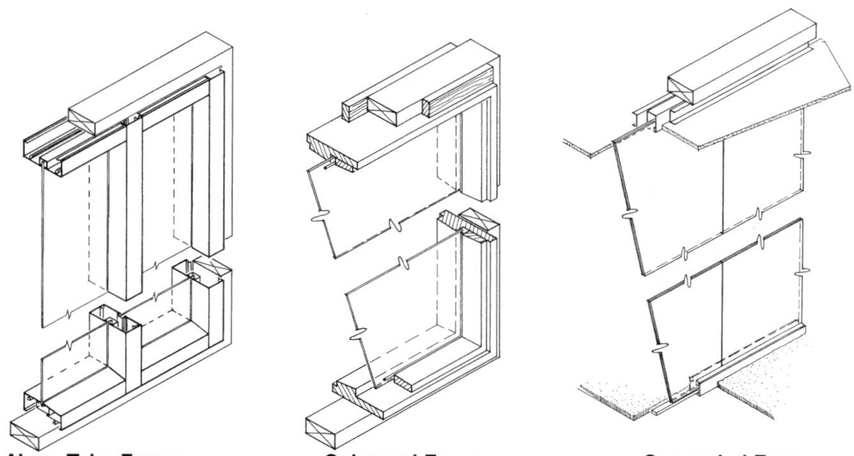

Interior Glazed Openings are defined as follows: framing material, glass, size and intermediate framing members. Components for each system include gasket setting or glazing bead and typical wood blocking.

System Components	QUANTITY	UNIT	COST PER OPNG. MAT.	INST.	TOTAL
SYSTEM C1010 710 1000					
GLAZED OPENING, ALUMINUM TUBE FRAME, FLUSH, 1/4″ FLOAT GLASS, 6′ X 4′					
Aluminum tube frame, flush, anodized bronze, head & jamb	16.500	L.F.	245.85	234.30	480.15
Aluminum tube frame, flush, anodized bronze, open sill	7.000	L.F.	85.05	96.95	182
Joints for tube frame, clip type	4.000	Ea.	116		116
Gasket setting, add	20.000	L.F.	127		127
Wood blocking	8.000	B.F.	5.28	26.40	31.68
Float glass, 1/4″ thick, clear, plain	24.000	S.F.	169.20	226.80	396
TOTAL			748.38	584.45	1,332.83

C1010 710 Interior Glazed Opening

	FRAME	GLASS	OPENING-SIZE W X H	INTERMEDIATE MULLION	INTERMEDIATE HORIZONTAL	COST PER OPNG. MAT.	INST.	TOTAL
1000	Aluminum flush	1/4″ float	6′x4′	0	0	750	585	1,335
1040	Tube		12′x4′	3	0	1,850	1,250	3,100
1080			4′x5′	0	0	680	505	1,185
1120			8′x5′	1	0	1,250	935	2,185
1160			12′x5′	2	0	1,825	1,375	3,200
1240		3/8″ float	9′x6′	2	0	2,025	1,575	3,600
1280			4′x8′-6″	0	1	1,300	1,000	2,300
1320			16′x10′	3	1	5,250	4,150	9,400
1400		1/4″ tempered	6′x4′	0	0	815	585	1,400
1440			12′x4′	3	0	1,975	1,250	3,225
1480			4′x5′	0	0	735	505	1,240
1520			8′x5′	1	0	1,375	935	2,310
1560			12′x5′	2	0	2,000	1,375	3,375
1640		3/8″ tempered	9′x6′	2	0	2,400	1,575	3,975
1680			4′x8′-6″	0	0	1,325	950	2,275
1720			16′x10′	3	0	5,625	3,900	9,525
1800		1/4″ one way mirror	6′x4′	0	0	1,100	575	1,675
1840			12′x4′	3	0	2,525	1,225	3,750

C1010 710	Interior Glazed Opening							
	FRAME	GLASS	OPENING-SIZE W X H	INTERMEDIATE MULLION	INTERMEDIATE HORIZONTAL	COST PER OPNG.		
						MAT.	INST.	TOTAL
1880	Aluminum flush	1/4" one way mirror	4'x5'	0	0	965	500	1,465
1920	Tube		8'x5'	1	0	1,825	920	2,745
1960			12'x5'	2	0	2,700	1,350	4,050
2040		3/8" laminated safety	9'x6'	2	0	2,725	1,550	4,275
2080		3/8" laminated safety	4'x8'-6"	0	0	1,550	930	2,480
2120			16'x10'	3	0	6,600	3,800	10,400
2200		1-3/16" bullet proof	6'x4'	0	0	4,000	2,075	6,075
2240			12'x4'	3	0	8,300	4,250	12,550
2280			4'x5'	0	0	3,400	1,750	5,150
2320			8'x5'	1	0	6,650	3,425	10,075
2360			12'x5'	2	0	9,900	5,100	15,000
2440		1' acoustical	9'x6'	2	0	3,550	1,400	4,950
2480			4'x8'-6"	0	0	2,075	845	2,920
2520			16'x10'	3	0	8,875	3,350	12,225
3000	Oakwood	1/4" float	6'x4'	0	0	425	420	845
3040			12'x4'	3	0	1,050	1,025	2,075
3080			4'x5'	0	0	370	345	715
3120			8'x5'	1	0	745	725	1,470
3160			12'x5'	2	0	1,125	1,100	2,225
3240		3/8" float	9'x6'	2	0	1,300	1,350	2,650
3280			4'x8'-6"	0	1	805	790	1,595
3320			16'x10'	3	1	3,675	3,850	7,525
3400		1/4" tempered	6'x4'	0	0	495	420	915
3440			12'x4'	3	0	1,200	1,025	2,225
3480			4'x5'	0	0	425	345	770
3520			8'x5'	1	0	855	725	1,580
3560			12'x5'	2	0	1,300	1,100	2,400
3640		3/8" tempered	9'x6'	2	0	1,675	1,350	3,025
3680			4'x8'-6"	0	0	940	715	1,655
3720			16'x10'	3	0	4,375	3,500	7,875
3800		1/4" one way mirror	6'x4'	0	0	770	410	1,180
3840			12'x4'	3	0	1,750	1,000	2,750
3880			4'x5'	0	0	655	335	990
3920			8'x5'	1	0	1,325	710	2,035
3960			12'x5'	2	0	1,975	1,075	3,050
4040		3/8" laminated safety	9'x6'	2	0	2,025	1,300	3,325
4080			4'x8'-6"	0	0	1,150	695	1,845
4120			16'x10'	3	0	5,350	3,425	8,775
4200		1-3/16" bullet proof	6'x4'	0	0	3,600	1,900	5,500
4240			12'x4'	3	0	7,400	3,975	11,375
4280			4'x5'	0	0	3,000	1,575	4,575
4360			12'x5'	2	0	9,025	4,800	13,825
4440		1" acoustical	9'x6'	2	0	2,725	1,125	3,850
4520			16'x10'	3	0	7,425	2,900	10,325

C1010 Partitions

C1010 710 Interior Glazed Opening

	FRAME	GLASS	OPENING-SIZE W X H	INTERMEDIATE MULLION	INTERMEDIATE HORIZONTAL	COST PER OPNG.		
						MAT.	INST.	TOTAL
5000	Concealed frame	1/2" float	6'x4'	1		740	900	1,640
5040	Butt glazed		9'x4'	2		1,050	1,275	2,325
5080			8'x5'	1		1,125	1,350	2,475
5120			16'x5'	3		2,125	2,550	4,675
5200		3/4" float	6'x8'	1		2,250	2,150	4,400
5240			9'x8'	2		3,275	3,100	6,375
5280			8'x10'	1		3,625	3,350	6,975
5320			16'x10'	3		6,975	6,375	13,350
5400		1/4" tempered	6'x4'	1		510	635	1,145
5440			9'x4'	2		715	885	1,600
5480			8'x5'	1		750	905	1,655
5520			16'x5'	3		1,375	1,650	3,025
5600		1/2" tempered	6'x8'	1		1,700	1,575	3,275
5640			9'x8'	2		2,450	2,225	4,675
5680			8'x10'	1		2,700	2,400	5,100
5720			16'x10'	3		5,125	4,450	9,575
5800		1/4" laminated safety	6'x4'	1		600	710	1,310
5840			9'x4'	2		855	1,000	1,855
5880			8'x5'	1		905	1,025	1,930
5920			16'x5'	3		1,675	1,900	3,575
6000		1/2" laminated safety	6'x8'	1		1,725	1,425	3,150
6040			9'x8'	2		2,500	2,000	4,500
6080			8'x10'	1		2,725	2,150	4,875
6120			16'x10'	3		5,225	3,975	9,200
6200		1-3/16" bullet proof	6'x4'	1		3,600	2,100	5,700
6240			9'x4'	2		5,375	3,100	8,475
6280			8'x5'	1		5,925	3,375	9,300
6320			16'x5'	3		11,700	6,575	18,275
6400		2" bullet proof	6'x8'	1		9,650	5,125	14,775
6440			9'x8'	2		14,400	7,550	21,950
6480			8'x10'	1		15,900	8,300	24,200
6520			16'x10'	3		31,600	16,300	47,900

C1020 Interior Doors

Sliding Panel-Mall Front

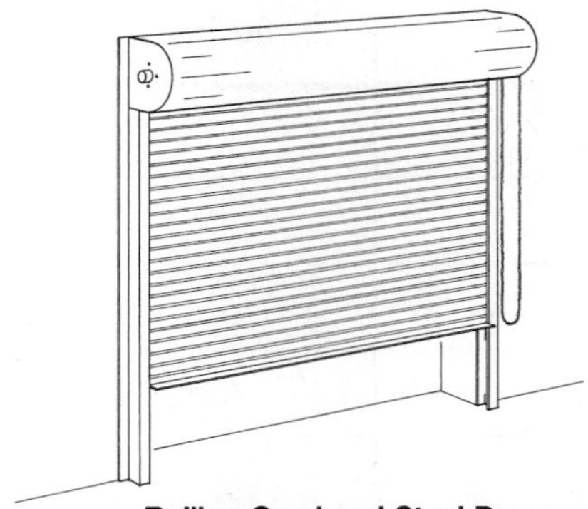

Rolling Overhead Steel Door

C1020 102	Special Doors	COST PER OPNG.		
		MAT.	INST.	TOTAL
3800	Sliding entrance door and system mill finish	11,000	2,875	13,875
3900	Bronze finish	12,000	3,100	15,100
4000	Black finish	12,500	3,225	15,725
4100	Sliding panel mall front, 16′x9′ opening, mill finish	3,900	875	4,775
4200	Bronze finish	5,075	1,150	6,225
4300	Black finish	6,250	1,400	7,650
4400	24′x9′ opening mill finish	5,525	1,625	7,150
4500	Bronze finish	7,175	2,125	9,300
4600	Black finish	8,850	2,600	11,450
4700	48′x9′ opening mill finish	10,200	1,275	11,475
4800	Bronze finish	13,300	1,650	14,950
4900	Black finish	16,300	2,050	18,350
5000	Rolling overhead steel door, manual, 8′ x 8′ high	1,300	900	2,200
5100	10′ x 10′ high	2,150	1,025	3,175
5200	20′ x 10′ high	3,500	1,450	4,950
5300	12′ x 12′ high	2,175	1,200	3,375
5400	Motor operated, 8′ x 8′ high	2,600	1,200	3,800
5500	10′ x 10′ high	3,450	1,325	4,775
5600	20′ x 10′ high	4,800	1,750	6,550
5700	12′ x 12′ high	3,475	1,500	4,975
5800	Roll up grille, aluminum, manual, 10′ x 10′ high, mill finish	3,150	1,750	4,900
5900	Bronze anodized	4,950	1,750	6,700
6000	Motor operated, 10′ x 10′ high, mill finish	4,525	2,050	6,575
6100	Bronze anodized	6,325	2,050	8,375
6200	Steel, manual, 10′ x 10′ high	2,850	1,450	4,300
6300	15′ x 8′ high	3,200	1,800	5,000
6400	Motor operated, 10′ x 10′ high	4,225	1,750	5,975
6500	15′ x 8′ high	4,575	2,100	6,675
8970	Counter door, rolling, 6′ high, 14′ wide, aluminum	3,025	850	3,875

C1020 Interior Doors

Steel Door, Half Glass
Steel Frame

Steel Door, Flush
Steel Frame

Wood Door, Flush
Wood Frame

The Metal Door/Metal Frame Systems are defined as follows: door type, design and size, frame type and depth. Included in the components for each system is painting the door and frame. No hardware has been included in the systems.

System Components

System Components	QUANTITY	UNIT	COST EACH MAT.	COST EACH INST.	COST EACH TOTAL
SYSTEM C1020 114 1200					
STEEL DOOR, HOLLOW, 20 GA., HALF GLASS, 3′-0″X7′-0″, D.W.FRAME, 4-7/8″ DP					
Steel door, flush, hollow core, 1-3/4″ thk, half gl, 20 Ga., 3′-0″ x 7′-0″	1.000	Ea.	650	66	716
Steel frame, 16 ga., up to 4-7/8″ deep, 3′-0″ x 7′-0″ single	1.000	Ea.	183	74.50	257.50
Hinges full mortise, avg. freq., steel base, USP, 4-1/2″ x 4-1/2″	1.500	Pr.	60.75		60.75
Float glass, 3/16″ thick, clear, tempered	5.000	S.F.	38.50	43.75	82.25
Paint door and frame each side, primer	1.000	Ea.	8.14	97	105.14
Paint door and frame each side, 2 coats	1.000	Ea.	9.98	162	171.98
TOTAL			950.37	443.25	1,393.62

C1020 114 Metal Door/Metal Frame

	TYPE	DESIGN	SIZE	FRAME	DEPTH	COST EACH MAT.	COST EACH INST.	COST EACH TOTAL
1000	Flush-hollow	20 ga. full panel	3′-0″ x 7′-0″	drywall K.D.	4-7/8″	740	405	1,145
1020				butt welded	8-3/4″	850	455	1,305
1160			6′-0″ x 7′-0″	drywall K.D.	4-7/8″	1,325	640	1,965
1180				butt welded	8-3/4″	1,450	690	2,140
1200		20 ga. half glass	3′-0″ x 7′-0″	drywall K.D.	4-7/8″	950	445	1,395
1220				butt welded	8-3/4″	1,050	495	1,545
1360			6′-0″ x 7′-0″	drywall K.D.	4-7/8″	1,725	720	2,445
1380				butt welded	8-3/4″	1,850	770	2,620
1800		18 ga. full panel	3′-0″ x 7′-0″	drywall K.D.	4-7/8″	830	405	1,235
1820				butt welded	8-3/4″	940	455	1,395
1960			6′-0″ x 7′-0″	drywall K.D.	4-7/8″	1,500	640	2,140
1980				butt welded	8-3/4″	1,625	690	2,315
2000		18 ga. half glass	3′-0″ x 7′-0″	drywall K.D.	4-7/8″	1,000	445	1,445
2020				butt welded	8-3/4″	1,125	500	1,625
2160			6′-0″ x 7′-0″	drywall K.D.	4-7/8″	1,850	725	2,575
2180				butt welded	8-3/4″	1,975	780	2,755

C1020 Interior Doors

C1020 114 Metal Door/Metal Frame

	TYPE	DESIGN	SIZE	FRAME	DEPTH	COST EACH MAT.	COST EACH INST.	COST EACH TOTAL
5000	Flush-hollow	16 ga. full panel	3'-0" x 7'-0"	drywall K.D.	4-7/8"	875	400	1,275
5020				butt welded	8-3/4"	985	450	1,435
5160			6'-0" x 7'-0"	drywall K.D.	4-7/8"	1,575	630	2,205
5180				butt welded	8-3/4"	1,700	685	2,385

C1020 116 Labeled Metal Door/Metal Frames

	TYPE	DESIGN	SIZE	FRAME	DEPTH	COST EACH MAT.	COST EACH INST.	COST EACH TOTAL
6000	Hollow-1-1/2 hour	20 ga. full panel	3'-0" x 7'-0"	drywall K.D.	4-7/8"	810	410	1,220
6020				butt welded	8-3/4"	975	460	1,435
6160			6'-0" x 7'-0"	drywall K.D.	4-7/8"	1,450	655	2,105
6180				butt welded	8-3/4"	1,650	705	2,355
6200		20 ga. vision	3'-0" x 7'-0"	drywall K.D.	4-7/8"	990	430	1,420
6220				butt welded	8-3/4"	1,150	480	1,630
6360			6'-0" x 7'-0"	drywall K.D.	4-7/8"	1,775	695	2,470
6380				butt welded	8-3/4"	1,975	745	2,720
6600		18 ga. full panel	3'-0" x 7'-0"	drywall K.D.	4-7/8"	875	415	1,290
6620				butt welded	8-3/4"	1,025	465	1,490
6760			6'-0" x 7'-0"	drywall K.D.	4-7/8"	1,625	660	2,285
6780				butt welded	8-3/4"	1,625	715	2,340
6800		18 ga. vision	3'-0" x 7'-0"	drywall K.D.	4-7/8"	1,225	435	1,660
6820				butt welded	8-3/4"	1,375	485	1,860
6960			6'-0" x 7'-0"	drywall K.D.	4-7/8"	2,250	700	2,950
6980				butt welded	8-3/4"	2,425	755	3,180
7185		16 ga. full panel	3'-0" x 7'-0"	drywall K.D.	4-7/8"	910	415	1,325
7190				butt welded	8-3/4"	1,075	465	1,540
7195			6'-0" x 7'-0"	drywall K.D.	4-7/8"	1,650	660	2,310
7197				butt welded	8-3/4"	1,850	715	2,565
7200	Hollow-3 hour	18 ga. full panel	3'-0" x 7'-0"	butt K.D.	5-3/4"	875	415	1,290
7220				butt welded	8-3/4"	1,025	465	1,490
7360			6'-0" x 7'-0"	butt K.D.	5-3/4"	1,550	660	2,210
7380				butt welded	8-3/4"	1,750	715	2,465
8000	Composite-1-1/2 hour	20 ga. full panel	3'-0" x 7'-0"	drywall K.D.	4-3/8"	1,075	415	1,490
8020				butt welded	8-3/4"	1,225	465	1,690
8160			6'-0" x 7'-0"	drywall K.D.	4-7/8"	1,950	660	2,610
8180				butt welded	8-3/4"	2,150	715	2,865
8200		2C ga. vision	3'-0" x 7'-0"	drywall K.D.	4-7/8"	1,400	435	1,835
8220				butt welded	8-3/4"	1,575	485	2,060
8360			6'-0" x 7'-0"	drywall K.D.	4-7/8"	2,625	700	3,325
8400	Mineral core-3 hour	18 ga. full panel	3'-0" x 7'-0"	drywall K.D.	5-3/4"	1,375	500	1,875
8420				butt welded	8-3/4"	1,200	470	1,670
8560			6'-0" x 7'-0"	drywall K.D.	5-3/4"	2,075	670	2,745

C1020 120 Wood Door/Wood Frame

	TYPE	FACE	SIZE	FRAME	DEPTH	COST EACH MAT.	COST EACH INST.	COST EACH TOTAL
1600	Hollow core/flush	lauan	3'-0" x 7'-0"	pine	3-5/8"	335	261	596
1620					5-3/16"	345	274	619
1760			6'-0" x 7'-0"	pine	3-5/8"	590	430	1,020
1780					5-3/16"	605	445	1,050

C1020 Interior Doors

C1020 120 Wood Door/Wood Frame

	TYPE	FACE	SIZE	FRAME	DEPTH	COST EACH		
						MAT.	INST.	TOTAL
1800	Hollow core/flush	birch	3'-0" x 7'-0"	pine	3-5/8"	325	261	586
1820					5-3/16"	345	280	625
1960			6'-0" x 7'-0"	pine	3-5/8"	565	425	990
1980					5-3/16"	580	440	1,020
2000		oak	3'-0" x 7'-0"	oak	3-5/8"	340	211	551
2020					5-3/16"	575	274	849
2160			6'-0" x 7'-0"	oak	3-5/8"	845	430	1,275
2180					5-3/16"	940	445	1,385
3000	Particle core/flush	lauan	3'-0" x 7'-0"	pine	3-5/8"	330	273	603
3020					5-3/16"	340	288	628
3160			6'-0" x 7'-0"	pine	3-5/8"	580	455	1,035
3180					5-3/16"	595	475	1,070
3200		birch	3'-0" x 7'-0"	pine	3-5/8"	305	273	578
3220					5-3/16"	315	288	603
3360			6'-0" x 7'-0"	pine	3-5/8"	530	455	985
3380					5-3/16"	545	475	1,020
3398								
3400		oak	3'-0" x 7'-0"	oak	3-5/8"	475	277	752
3420					5-3/16"	550	291	841
3560			6'-0" x 7'-0"	oak	3-5/8"	795	460	1,255
3580					5-3/16"	895	480	1,375
5000	Solid core/flush	birch	3'-0" x 7'-0"	pine	3-5/8"	400	267	667
5020					5-3/16"	410	281	691
5160			6'-0" x 7'-0"	pine	3-5/8"	720	445	1,165
5180					5-3/16"	735	460	1,195
5200		oak	3'-0" x 7'-0"	oak	3-5/8"	520	271	791
5220					5-3/16"	595	285	880
5360			6'-0" x 7'-0"	oak	3-5/8"	885	450	1,335
5380					5-3/16"	980	465	1,445
5400		M.D. overlay	3'-0" x 7'-0"	pine	3-5/8"	375	435	810

C1020 122 Wood Door/Metal Frame

	TYPE	FACE	SIZE	FRAME	DEPTH	COST EACH		
						MAT.	INST.	TOTAL
1600	Hollow core/flush	lauan	3'-0" x 7'-0"	drywall K.D.	4-7/8"	430	236	666
1620				butt welded	8-3/4"	575	246	821
1760			6'-0" x 7'-0"	drywall K.D.	4-7/8"	725	400	1,125
1780				butt welded	8-3/4"	870	410	1,280
1800		birch	3'-0" x 7'-0"	drywall K.D.	4-7/8"	425	236	661
1820				butt welded	8-3/4"	565	246	811
1960			6'-0" x 7'-0"	drywall K.D.	4-7/8"	710	400	1,110
1980				butt welded	8-3/4"	855	410	1,265
2000		oak	3'-0" x 7'-0"	drywall K.D.	4-7/8"	505	236	741
2020				butt welded	8-3/4"	645	246	891
2160			6'-0" x 7'-0"	drywall K.D.	4-7/8"	870	400	1,270
2180				butt welded	8-3/4"	1,025	410	1,435
3000	Particle core/flush	lauan	3'-0" x 7'-0"	drywall K.D.	4-7/8"	430	253	683
3020				butt welded	8-3/4"	575	263	838
3160			6'-0" x 7'-0"	drywall K.D.	4-7/8"	725	435	1,160
3180				butt welded	8-3/4"	870	445	1,315

C1020 Interior Doors

C1020 122 Wood Door/Metal Frame

	TYPE	FACE	SIZE	FRAME	DEPTH	COST EACH MAT.	COST EACH INST.	COST EACH TOTAL
3200	Particle core/flush	birch	3′-0″ x 7′-0″	drywall K.D.	4-7/8″	405	253	658
3220				butt welded	8-3/4″	550	263	813
3360			6′-0″ x 7′-0″	drywall K.D.	4-7/8″	675	435	1,110
3380				butt welded	8-3/4″	820	445	1,265
3400		oak	3′-0″ x 7′-0″	drywall K.D.	4-7/8″	480	253	733
3420				butt welded	8-3/4″	625	263	888
3560			6′-0″ x 7′-0″	drywall K.D.	4-7/8″	820	435	1,255
3580				butt welded	8-3/4″	965	445	1,410
5000	Solid core/flush	birch	2′-8″ x 6′-8″	drywall K.D.	4-7/8″	365	241	606
5020				butt welded	8-3/4″	510	251	761
5160			6′-0″ x 7′-0″	drywall K.D.	4-7/8″	640	435	1,075
5180				butt welded	8-3/4″	785	445	1,230
5200		oak	2′-8″ x 6′-8″	drywall K.D.	4-7/8″	380	241	621
5220				butt welded	8-3/4″	525	251	776
5360			6′-0″ x 7′-0″	drywall K.D.	4-7/8″	665	435	1,100
5380				butt welded	8-3/4″	810	445	1,255
6000	1 hr/flush	birch	3′-0″ x 7′-0″	drywall K.D.	4-7/8″	780	260	1,040
6020				butt welded	8-3/4″	945	270	1,215
6160			6′-0″ x 7′-0″	drywall K.D.	4-7/8″	1,400	450	1,850
6180				butt welded	8-3/4″	1,575	460	2,035
6200		oak	3′-0″ x 7′-0″	drywall K.D.	4-7/8″	805	260	1,065
6220				butt welded	8-3/4″	970	270	1,240
6360			6′-0″ x 7′-0″	drywall K.D.	4-7/8″	1,450	450	1,900
6380				butt welded	8-3/4″	1,625	460	2,085
6400		walnut	3′-0″ x 7′-0″	drywall K.D.	4-7/8″	860	260	1,120
6420				butt welded	8-3/4″	1,025	270	1,295
6560			6′-0″ x 7′-0″	drywall K.D.	4-7/8″	1,550	450	2,000
6580				butt welded	8-3/4″	1,750	460	2,210
7600	1-1/2 hr/flush	birch	3′-0″ x 7′-0″	drywall K.D.	4-7/8″	730	260	990
7620				butt welded	8-3/4″	895	270	1,165
7760			6′-0″ x 7′-0″	drywall K.D.	4-7/8″	1,300	450	1,750
7780				butt welded	8-3/4″	1,475	460	1,935
7800		oak	3′-0″ x 7′-0″	drywall K.D.	4-7/8″	800	260	1,060
7820				butt welded	8-3/4″	965	270	1,235
7960			6′-0″ x 7′-0″	drywall K.D.	4-7/8″	1,425	450	1,875
7980				butt welded	8-3/4″	1,625	460	2,085
8000		walnut	3′-0″ x 7′-0″	drywall K.D.	4-7/8″	815	260	1,075
8020				butt welded	8-3/4″	980	270	1,250
8160			6′-0″ x 7′-0″	drywall K. D.	4 7/8″	1,475	450	1,925
8180				butt welded	8-3/4″	1,650	460	2,110

C1020 124 Wood Fire Doors/Metal Frames

	FIRE RATING (HOURS)	CORE MATERIAL	THICKNESS (IN.)	FACE MATERIAL	SIZE W(FT)XH(FT)	COST EACH MAT.	COST EACH INST.	COST EACH TOTAL
0840	1	mineral	1-3/4″	birch	2′-8″ x 6′-8″	700	246	946
0860		3 ply stile			3′-0″ x 7′-0″	720	266	986
0880				oak	2′-8″ x 6′-8″	735	246	981
0900					3′-0″ x 7′-0″	745	266	1,011
0920				walnut	2′-8″ x 6′-8″	755	246	1,001
0940					3′-0″ x 7′-0″	800	266	1,066
0960				m.d. overlay	2′-8″ x 6′-8″	595	420	1,015
0980					3′-0″ x 7′-0″	685	430	1,115

C1020 Interior Doors

C1020 124 Wood Fire Doors/Metal Frames

	FIRE RATING (HOURS)	CORE MATERIAL	THICKNESS (IN.)	FACE MATERIAL	SIZE W(FT)XH(FT)	COST EACH		
						MAT.	INST.	TOTAL
1000	1-1/2	mineral	1-3/4"	birch	2'-8" x 6'-8"	595	246	841
1020		3 ply stile			3'-0" x 7'-0"	670	266	936
1040				oak	2'-8" x 6'-8"	715	246	961
1060					3'-0" x 7'-0"	740	266	1,006
1080				walnut	2'-8" x 6'-8"	680	246	926
1100					3'-0" x 7'-0"	755	266	1,021
1120				m.d. overlay	2'-8" x 6'-8"	640	420	1,060
1140					3'-0" x 7'-0"	695	430	1,125

C1020 Interior Doors

C1020 310	Hardware	MAT.	INST.	TOTAL
		COST EACH		
0060	**HINGES**			
0080				
0100	Full mortise, low frequency, steel base, 4-1/2" x 4-1/2", USP	12.50		12.50
0120	5" x 5" USP	26.50		26.50
0140	6" x 6" USP	51.50		51.50
0160	Average frequency, steel base, 4-1/2" x 4-1/2" USP	20.50		20.50
0180	5" x 5" USP	31.50		31.50
0200	6" x 6" USP	65.50		65.50
0220	High frequency, steel base, 4-1/2" x 4-1/2", USP	36.50		36.50
0240	5" x 5" USP	29		29
0260	6" x 6" USP	76		76
0280				
0300	**LOCKSETS**			
0320				
0340	Heavy duty cylindrical, passage door			
0360	Non-keyed, passage	77	49.50	126.50
0380	Privacy	82.50	49.50	132
0400	Keyed, single cylinder function	118	59.50	177.50
0420	Hotel	220	74.50	294.50
0440				
0460	For re-core cylinder, add	55		55
0480				
0500	**CLOSERS**			
0520				
0540	Rack & pinion			
0560	Adjustable backcheck, 3 way mount, all sizes, regular arm	211	99	310
0580	Hold open arm	211	99	310
0600	Fusible link	182	91.50	273.50
0620	Non-sized, regular arm	211	99	310
0640	4 way mount, non-sized, regular arm	211	99	310
0660	Hold open arm	220	99	319
0680				
0700	**PUSH, PULL**			
0720				
0740	Push plate, aluminum	14.20	49.50	63.70
0760	Bronze	28	49.50	77.50
0780	Pull handle, push bar, aluminum	134	54	188
0800	Bronze	180	59.50	239.50
0810				
0820	**PANIC DEVICES**			
0840				
0860	Narrow stile, rim mounted, bar, exit only	670	99	769
0880	Outside key & pull	980	119	1,099
0900	Bar and vertical rod, exit only	1,000	119	1,119
0920	Outside key & pull	1,000	149	1,149
0940	Bar and concealed rod, exit only	1,175	198	1,373
0960	Mortise, bar, exit only	660	149	809
0980	Touch bar, exit only	845	149	994
1000				
1020	**WEATHERSTRIPPING**			
1040				
1060	Interlocking, 3' x 7', zinc	48.50	198	246.50
1080	Spring type, 3' x 7', bronze	52	198	250

C1030 Fittings

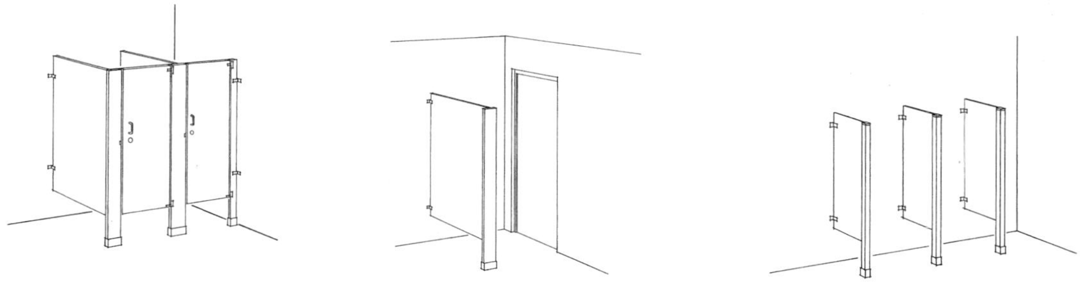

C1030 110	Toilet Partitions	COST PER UNIT		
		MAT.	INST.	TOTAL
0380	Toilet partitions, cubicles, ceiling hung, marble	1,975	560	2,535
0400	Painted metal	575	297	872
0420	Plastic laminate	580	297	877
0460	Stainless steel	1,225	297	1,522
0480	Handicap addition	495		495
0520	Floor and ceiling anchored, marble	2,175	450	2,625
0540	Painted metal	630	238	868
0560	Plastic laminate	905	238	1,143
0600	Stainless steel	1,350	238	1,588
0620	Handicap addition	405		405
0660	Floor mounted marble	1,350	375	1,725
0680	Painted metal	660	170	830
0700	Plastic laminate	585	170	755
0740	Stainless steel	1,575	170	1,745
0760	Handicap addition	370		370
0780	Juvenile deduction	45.50		45.50
0820	Floor mounted with headrail marble	1,275	375	1,650
0840	Painted metal	420	198	618
0860	Plastic laminate	845	198	1,043
0900	Stainless steel	1,125	198	1,323
0920	Handicap addition	410		410
0960	Wall hung, painted metal	700	170	870
1020	Stainless steel	1,825	170	1,995
1040	Handicap addition	410		410
1080	Entrance screens, floor mounted, 54″ high, marble	870	125	995
1100	Painted metal	266	79.50	345.50
1140	Stainless steel	1,000	79.50	1,079.50
1300	Urinal screens, floor mounted, 24″ wide, laminated plastic	224	149	373
1320	Marble	710	172	882
1340	Painted metal	253	149	402
1380	Stainless steel	640	149	789
1428	Wall mounted wedge type, painted metal	148	119	267
1430	Urinal screens, wall hung, plastic laminate/particle board	230	149	379
1460	Stainless steel	635	119	754
1500	Partitions, shower stall, single wall, painted steel, 2′-8″ x 2′-8″	1,050	280	1,330
1510	Fiberglass, 2′-8″ x 2′-8″	875	310	1,185
1520	Double wall, enameled steel, 2′-8″ x 2′-8″	1,125	280	1,405
1530	Stainless steel, 2′-8″ x 2′-8″	2,525	280	2,805
1540	Doors, plastic, 2′ wide	158	78	236
1550	Tempered glass, chrome/brass frame-deluxe	740	700	1,440
1560	Tub enclosure, sliding panels, tempered glass, aluminum frame	450	350	800
1570	Chrome/brass frame, clear glass	1,325	465	1,790

C1030 Fittings

C1030 210	Fabricated Compact Units and Cubicles, EACH	COST EACH		
		MAT.	INST.	TOTAL
0500	Telephone enclosure, shelf type, wall hung, recessed	820	238	1,058
0510	Surface mount	1,775	238	2,013
0520	Booth type, painted steel, flat recessed	2,500	795	3,295
0530	Stainless steel, curved surface	6,475	795	7,270

C1030 310	Storage Specialties, EACH	COST EACH		
		MAT.	INST.	TOTAL
0200	Lockers, steel, single tier, 5′ to 6′ high, per opening, 1 wide	181	50	231
0210	3 wide	320	58.50	378.50
0220	Two tier, unassembled	93.50	27	120.50
0230	Set up	135	35	170
0440	Parts bins, 3′ wide x 7′ high, 14 bins, 18″ x 12″ x 12″	390	93	483
0450	84 bins, 6″ x 6″ x 12″	1,050	116	1,166
0600	Shelving, metal industrial, braced, 3′ wide, 1′ deep	23.50	12.35	35.85
0610	2′ deep	34	13.10	47.10
0620	Enclosed, 3′ wide, 1′ deep	53.50	14.45	67.95
0630	2′ deep	71	17	88

C1030 510	Identifying/Visual Aid Specialties, EACH	COST EACH		
		MAT.	INST.	TOTAL
0100	Control boards, magnetic, porcelain finish, framed, 24″ x 18″	213	149	362
0110	96″ x 48″	1,175	238	1,413
0120	Directory boards, outdoor, black plastic, 36″ x 24″	840	595	1,435
0130	36″ x 36″	970	795	1,765
0140	Indoor, economy, open faced, 18″ x 24″	183	170	353
0150	36″ x 48″	276	198	474
0160	Building, aluminum, black felt panel, 24″ x 18″	345	297	642
0170	48″ x 72″	1,075	1,200	2,275
0500	Signs, interior electric exit sign, wall mounted, 6″	102	82.50	184.50
0510	Street, reflective alum., dbl. face, 4 way, w/bracket	269	39.50	308.50
0520	Letters, cast aluminum, 1/2″ deep, 4″ high	27.50	33	60.50
0530	1″ deep, 10″ high	58	33	91
0540	Plaques, cast aluminum, 20″ x 30″	2,025	297	2,322
0550	Cast bronze, 36″ x 48″	5,125	595	5,720

C1030 520	Identifying/Visual Aid Specialties, S.F.	COST PER S.F.		
		MAT.	INST.	TOTAL
0100	Bulletin board, cork sheets, no frame, 1/4″ thick	1.52	4.10	5.62
0110	1/2″ thick	4.55	4.10	8.65
0120	Aluminum frame, 1/4″ thick, 3′ x 5′	9.90	4.99	14.89
0130	4′ x 8′	9.45	2.64	12.09
0200	Chalkboards, wall hung, alum, frame & chalktrough	14.90	2.64	17.54
0210	Wood frame & chalktrough	10.85	2.84	13.69
0220	Sliding board, one board with back panel	57	2.54	59.54
0230	Two boards with back panel	89	2.54	91.54
0240	Liquid chalk type, alum. frame & chalktrough	13.65	2.64	16.29
0250	Wood frame & chalktrough	33.50	2.64	36.14

C1030 710	Bath and Toilet Accessories, EACH	COST EACH		
		MAT.	INST.	TOTAL
0100	Specialties, bathroom accessories, st. steel, curtain rod, 5′ long, 1″ diam	31	45.50	76.50
0110	1-1/2″ diam.	34	45.50	79.50
0120	Dispenser, towel, surface mounted	52	37	89
0130	Flush mounted with waste receptacle	340	59.50	399.50
0140	Grab bar, 1-1/4″ diam., 12″ long	33.50	25	58.50
0150	1-1/2″ diam. 36″ long	39	29.50	68.50

C1030 Fittings

C1030 710	Bath and Toilet Accessories, EACH	COST EACH		
		MAT.	INST.	TOTAL
0160	Mirror, framed with shelf, 18" x 24"	218	29.50	247.50
0170	72" x 24"	285	99	384
0180	Toilet tissue dispenser, surface mounted, single roll	21	19.80	40.80
0190	Double roll	27	25	52
0200	Towel bar, 18" long	45.50	26	71.50
0210	30" long	73	28.50	101.50
0300	Medicine cabinets, sliding mirror doors, 20" x 16" x 4-3/4", unlighted	146	85	231
0310	24" x 19" x 8-1/2", lighted	202	119	321
0320	Triple door, 30" x 32", unlighted, plywood body	360	85	445
0330	Steel body	410	85	495
0340	Oak door, wood body, beveled mirror, single door	219	85	304
0350	Double door	455	99	554

C1030 730	Bath and Toilet Accessories, L.F.	COST PER L.F.		
		MAT.	INST.	TOTAL
0100	Partitions, hospital curtain, ceiling hung, polyester oxford cloth	24.50	5.80	30.30
0110	Designer oxford cloth	17.20	7.35	24.55

C1030 830	Fabricated Cabinets, EACH	COST EACH		
		MAT.	INST.	TOTAL
0110	Household, base, hardwood, one top drawer & one door below x 12" wide	298	48	346
0115	24" wide	410	53.50	465
0120	Four drawer x 24" wide	385	53.50	438.50
0130	Wall, hardwood, 30" high with one door x 12" wide	258	54	312
0140	Two doors x 48" wide	565	64.50	629.50

C1030 830	Fabricated Counters, L.F.	COST PER L.F.		
		MAT.	INST.	TOTAL
0150	Counter top-laminated plastic, stock, economy	19	19.80	38.80
0160	Custom-square edge, 7/8" thick	16.85	44.50	61.35
0170	School, counter, wood, 32" high	264	59.50	323.50
0180	Metal, 84" high	470	79.50	549.50

C1030 910	Other Fittings, EACH	COST EACH		
		MAT.	INST.	TOTAL
0500	Mail boxes, horizontal, rear loaded, aluminum, 5" x 6" x 15" deep	47	17.50	64.50
0510	Front loaded, aluminum, 10" x 12" x 15" deep	102	29.50	131.50
0520	Vertical, front loaded, aluminum, 15" x 5" x 6" deep	47	17.50	64.50
0530	Bronze, duranodic finish	53	17.50	70.50
0540	Letter slot, post office	131	74.50	205.50
0550	Mail counter, window, post office, with grille	605	297	902
0600	Partition, woven wire, wall panel, 4' x 7' high	158	47.50	205.50
0610	Wall panel with window & shelf, 5' x 8' high	495	79.50	574.50
0620	Doors, swinging w/ no transom, 3' x 7' high	325	198	523
0630	Sliding, 6' x 10' high	880	297	1,177
0700	Turnstiles, one way, 4' arm, 46" diam., manual	2,300	238	2,538
0710	Electric	2,525	990	3,515
0720	3 arm, 5'-5" diam. & 7' high, manual	8,425	1,200	9,625
0730	Electric	10,200	1,975	12,175

C2010 Stair Construction

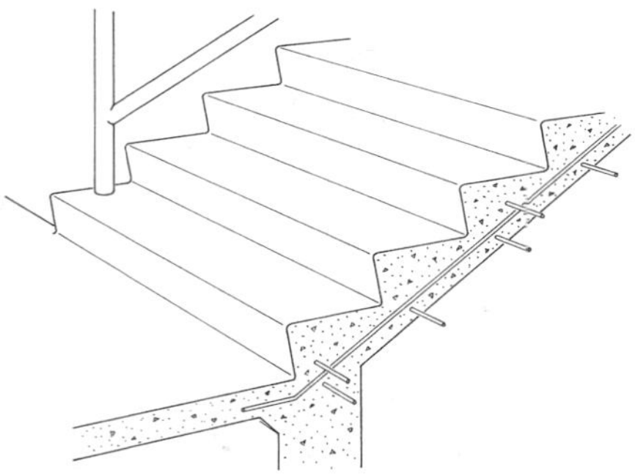

General Design: See reference section for code requirements. Maximum height between landings is 12′; usual stair angle is 20° to 50° with 30° to 35° best. Usual relation of riser to treads is:

Riser + tread = 17.5.
2x (Riser) + tread = 25.
Riser x tread = 70 or 75.

Maximum riser height is 7″ for commercial, 8-1/4″ for residential.
Usual riser height is 6-1/2″ to 7-1/4″.

Minimum tread width is 11″ for commercial and 9″ for residential.

For additional information please see reference section.

Cost Per Flight: Table below lists the cost per flight for 4′-0″ wide stairs.
Side walls are not included.
Railings are included.

System Components	QUANTITY	UNIT	COST PER FLIGHT MAT.	COST PER FLIGHT INST.	COST PER FLIGHT TOTAL
SYSTEM C2010 110 0560					
STAIRS, C.I.P. CONCRETE WITH LANDING, 12 RISERS					
Concrete in place, free standing stairs not incl. safety treads	48.000	L.F.	302.40	2,036.16	2,338.56
Concrete in place, free standing stair landing	32.000	S.F.	161.60	560.64	722.24
Cast alum nosing insert, abr surface, pre-drilled, 3″ wide x 4′ long	12.000	Ea.	966	209.40	1,175.40
Industrial railing, welded, 2 rail 3′-6″ high 1-1/2″ pipe	18.000	L.F.	540	216.72	756.72
Wall railing with returns, steel pipe	17.000	L.F.	289.85	204.68	494.53
TOTAL			2,259.85	3,227.60	5,487.45

C2010 110	Stairs		COST PER FLIGHT MAT.	COST PER FLIGHT INST.	COST PER FLIGHT TOTAL
0470	Stairs, C.I.P. concrete, w/o landing, 12 risers, w/o nosing	RC2010 -100	1,125	2,475	3,600
0480	With nosing		2,100	2,675	4,775
0550	W/landing, 12 risers, w/o nosing		1,300	3,025	4,325
0560	With nosing		2,250	3,225	5,475
0570	16 risers, w/o nosing		1,575	3,775	5,350
0580	With nosing		2,875	4,075	6,950
0590	20 risers, w/o nosing		1,875	4,575	6,450
0600	With nosing		3,475	4,925	8,400
0610	24 risers, w/o nosing		2,150	5,325	7,475
0620	With nosing		4,100	5,750	9,850
0630	Steel, grate type w/nosing & rails, 12 risers, w/o landing		5,700	1,250	6,950
0640	With landing		7,975	1,700	9,675
0660	16 risers, with landing		9,875	2,100	11,975
0680	20 risers, with landing		11,800	2,500	14,300
0700	24 risers, with landing		13,700	2,925	16,625
0710	Concrete fill metal pan & picket rail, 12 risers, w/o landing		7,375	1,250	8,625
0720	With landing		9,725	1,850	11,575
0740	16 risers, with landing		12,200	2,250	14,450
0760	20 risers, with landing		14,600	2,675	17,275
0780	24 risers, with landing		17,100	3,050	20,150
0790	Cast iron tread & pipe rail, 12 risers, w/o landing		7,375	1,250	8,625
0800	With landing		9,725	1,850	11,575
0820	16 risers, with landing		12,200	2,250	14,450
0840	20 risers, with landing		14,600	2,675	17,275
0860	24 risers, with landing		17,100	3,050	20,150
0870	Pan tread & flat bar rail, pre-assembled, 12 risers, w/o landing		6,250	895	7,145
0880	With landing		11,300	1,400	12,700
0900	16 risers, with landing		12,500	1,575	14,075
0920	20 risers, with landing		14,600	1,875	16,475
0940	24 risers, with landing		16,600	2,175	18,775

C2010 Stair Construction

C2010 110	Stairs	COST PER FLIGHT		
		MAT.	INST.	TOTAL
0950	Spiral steel, industrial checkered plate 4'-6' dia., 12 risers	5,825	820	6,645
0960	16 risers	7,750	1,075	8,825
0970	20 risers	9,700	1,375	11,075
0980	24 risers	11,600	1,625	13,225
0990	Spiral steel, industrial checkered plate 6'-0" dia., 12 risers	8,150	925	9,075
1000	16 risers	10,900	1,250	12,150
1010	20 risers	13,600	1,525	15,125
1020	24 risers	16,300	1,850	18,150
1030	Aluminum, spiral, stock units, 5'-0" dia., 12 risers	7,025	820	7,845
1040	16 risers	9,350	1,075	10,425
1050	20 risers	11,700	1,375	13,075
1060	24 risers	14,000	1,625	15,625
1070	Custom 5'-0" dia., 12 risers	12,300	820	13,120
1080	16 risers	16,400	1,075	17,475
1090	20 risers	20,500	1,375	21,875
1100	24 risers	24,600	1,625	26,225
1120	Wood, prefab box type, oak treads, wood rails 3'-6" wide, 14 risers	2,250	495	2,745
1150	Prefab basement type, oak treads, wood rails 3'-0" wide, 14 risers	1,100	122	1,222

C3010 Wall Finishes

C3010 230	Paint & Covering	COST PER S.F.		
		MAT.	INST.	TOTAL
0060	Painting, interior on plaster and drywall, brushwork, primer & 1 coat	.14	.75	.89
0080	Primer & 2 coats	.21	.99	1.20
0100	Primer & 3 coats	.29	1.22	1.51
0120	Walls & ceilings, roller work, primer & 1 coat	.14	.50	.64
0140	Primer & 2 coats	.21	.64	.85
0160	Woodwork incl. puttying, brushwork, primer & 1 coat	.14	1.08	1.22
0180	Primer & 2 coats	.21	1.43	1.64
0200	Primer & 3 coats	.29	1.95	2.24
0260	Cabinets and casework, enamel, primer & 1 coat	.14	1.22	1.36
0280	Primer & 2 coats	.22	1.50	1.72
0300	Masonry or concrete, latex, brushwork, primer & 1 coat	.31	1.01	1.32
0320	Primer & 2 coats	.40	1.45	1.85
0340	Addition for block filler	.23	1.25	1.48
0380	Fireproof paints, intumescent, 1/8" thick 3/4 hour	2.43	.99	3.42
0420	7/16" thick 2 hour	7	3.47	10.47
0440	1-1/16" thick 3 hour	11.40	6.95	18.35
0480	Miscellaneous metal brushwork, exposed metal, primer & 1 coat	.13	1.22	1.35
0500	Gratings, primer & 1 coat	.33	1.52	1.85
0600	Pipes over 12" diameter	.84	4.86	5.70
0700	Structural steel, brushwork, light framing 300-500 S.F./Ton	.11	1.77	1.88
0720	Heavy framing 50-100 S.F./Ton	.11	.89	1
0740	Spraywork, light framing 300-500 S.F./Ton	.12	.39	.51
0800	Varnish, interior wood trim, no sanding sealer & 1 coat	.08	1.22	1.30
0820	Hardwood floor, no sanding 2 coats	.17	.26	.43
0840	Wall coatings, acrylic glazed coatings, minimum	.34	.93	1.27
0860	Maximum	.72	1.59	2.31
0880	Epoxy coatings, solvent based	.44	.93	1.37
0900	Water based	.30	2.86	3.16
1100	High build epoxy 50 mil, solvent based	.75	1.25	2
1120	Water based	1.27	5.10	6.37
1140	Laminated epoxy with fiberglass solvent based	.80	1.65	2.45
1160	Water based	1.45	3.35	4.80
1180	Sprayed perlite or vermiculite 1/16" thick, solvent based	.30	.17	.47
1200	Water based	.81	.76	1.57
1260	Wall coatings, vinyl plastic, solvent based	.36	.66	1.02
1280	Water based	.90	2.03	2.93
1300	Urethane on smooth surface, 2 coats, solvent based	.34	.43	.77
1320	Water based	.65	.73	1.38
1340	3 coats, solvent based	.48	.58	1.06
1360	Water based	.87	1.03	1.90
1380	Ceramic-like glazed coating, cementitious, solvent based	.53	1.11	1.64
1400	Water based	.89	1.41	2.30
1420	Resin base, solvent based	.42	.76	1.18
1440	Water based	.64	1.47	2.11
1460	Wall coverings, aluminum foil	1.14	1.78	2.92
1500	Vinyl backing	6.10	2.04	8.14
1520	Cork tiles, 12"x12", light or dark, 3/16" thick	5.15	2.04	7.19
1540	5/16" thick	3.44	2.08	5.52
1580	Natural, non-directional, 1/2" thick	7.20	2.04	9.24
1600	12"x36", granular, 3/16" thick	1.44	1.27	2.71
1620	1" thick	1.85	1.32	3.17
1660	5/16" thick	7.25	2.08	9.33
1661	Paneling, prefinished plywood, birch	1.32	2.83	4.15
1662	Mahogany, African	3.08	2.97	6.05
1664	Oak or cherry	2.31	2.97	5.28
1665	Rosewood	3.25	3.72	6.97

C3010 Wall Finishes

C3010 230	Paint & Covering	COST PER S.F.		
		MAT.	INST.	TOTAL
1666	Teak	3.25	2.97	6.22
1680	Cork wallpaper, paper backed, natural	2.20	1.02	3.22
1700	Color	3.16	1.02	4.18
1720	Gypsum based, fabric backed, minimum	.85	.61	1.46
1740	Average	1.40	.68	2.08
1760	Small quantities	.77	.77	1.54
1780	Vinyl wall covering, fabric back, light weight	1.07	.77	1.84
1800	Medium weight	.99	1.02	2.01
1820	Heavy weight	1.51	1.13	2.64
1840	Wall paper, double roll, solid pattern, avg. workmanship	.65	.77	1.42
1860	Basic pattern, avg. workmanship	1.24	.91	2.15
1880	Basic pattern, quality workmanship	1.89	1.13	3.02
1900	Grass cloths with lining paper, minimum	1.06	1.22	2.28
1920	Maximum	3.14	1.40	4.54
1940	Ceramic tile, thin set, 4-1/4" x 4-1/4"	2.74	4.93	7.67
1960	12" x 12"	5	5.85	10.85

C3010 235	Paint Trim	COST PER L.F.		
		MAT.	INST.	TOTAL
2040	Painting, wood trim, to 6" wide, enamel, primer & 1 coat	.14	.61	.75
2060	Primer & 2 coats	.22	.77	.99
2080	Misc. metal brushwork, ladders	.66	6.10	6.76
2120	6" to 8" dia.	.22	2.56	2.78
2140	10" to 12" dia.	.66	3.83	4.49
2160	Railings, 2" pipe	.23	3.04	3.27
2180	Handrail, single	.18	1.22	1.40
2185	Caulking & Sealants, Polyurethane, In place, 1 or 2 component, 1/2" X 1/4"	.38	1.96	2.34

C3020 Floor Finishes

C3020 410	Tile & Covering	COST PER S.F.		
		MAT.	INST.	TOTAL
0060	Carpet tile, nylon, fusion bonded, 18" x 18" or 24" x 24", 24 oz.	3.72	.73	4.45
0080	35 oz.	4.33	.73	5.06
0100	42 oz.	5.65	.73	6.38
0140	Carpet, tufted, nylon, roll goods, 12' wide, 26 oz.	2.61	.78	3.39
0160	36 oz.	4.27	.78	5.05
0180	Woven, wool, 36 oz.	11.10	.83	11.93
0200	42 oz.	12.20	.83	13.03
0220	Padding, add to above, 2.7 density	.73	.39	1.12
0240	13.0 density	.98	.39	1.37
0260	Composition flooring, acrylic, 1/4" thick	1.83	5.70	7.53
0280	3/8" thick	2.31	6.60	8.91
0300	Epoxy, 3/8" thick	3.09	4.38	7.47
0320	1/2" thick	4.46	6.05	10.51
0340	Epoxy terrazzo, granite chips	6.60	6.55	13.15
0360	Recycled porcelain	10.05	8.70	18.75
0380	Mastic, hot laid, 1-1/2" thick, minimum	4.57	4.29	8.86
0400	Maximum	5.85	5.70	11.55
0420	Neoprene 1/4" thick, minimum	4.50	5.45	9.95
0440	Maximum	6.15	6.90	13.05
0460	Polyacrylate with ground granite 1/4", granite chips	3.87	4.02	7.89
0480	Recycled porcelain	7.10	6.20	13.30
0500	Polyester with colored quartz chips 1/16", minimum	3.50	2.79	6.29
0520	Maximum	5.25	4.38	9.63
0540	Polyurethane with vinyl chips, clear	8.35	2.79	11.14
0560	Pigmented	12.15	3.44	15.59
0600	Concrete topping, granolithic concrete, 1/2" thick	.36	4.71	5.07
0620	1" thick	.71	4.83	5.54
0640	2" thick	1.43	5.55	6.98
0660	Heavy duty 3/4" thick, minimum	.48	7.30	7.78
0680	Maximum	.89	8.70	9.59
0700	For colors, add to above, minimum	.52	1.68	2.20
0720	Maximum	.86	1.85	2.71
0740	Exposed aggregate finish, minimum	.22	.87	1.09
0760	Maximum	.37	1.16	1.53
0780	Abrasives, .25 P.S.F. add to above, minimum	.58	.64	1.22
0800	Maximum	.87	.64	1.51
0820	Dust on coloring, add, minimum	.52	.42	.94
0840	Maximum	.86	.87	1.73
0860	Floor coloring using 0.6 psf powdered color, 1/2" integral, minimum	5.75	4.71	10.46
0880	Maximum	6.10	4.71	10.81
1020	Integral topping and finish, 1:1:2 mix, 3/16" thick	.12	2.78	2.90
1040	1/2" thick	.32	2.93	3.25
1060	3/4" thick	.48	3.27	3.75
1080	1" thick	.64	3.70	4.34
1100	Terrazzo, minimum	3.69	17.50	21.19
1120	Maximum	6.80	22	28.80
1340	Cork tile, minimum	7.65	1.67	9.32
1360	Maximum	13.35	1.67	15.02
1380	Polyethylene, in rolls, minimum	4.58	1.91	6.49
1400	Maximum	7.50	1.91	9.41
1420	Polyurethane, thermoset, minimum	5.95	5.25	11.20
1440	Maximum	7.45	10.50	17.95
1460	Rubber, sheet goods, minimum	7.80	4.38	12.18
1480	Maximum	13.55	5.85	19.40
1500	Tile, minimum	6.30	1.32	7.62
1520	Maximum	11.20	1.91	13.11
1540	Synthetic turf, minimum	5.15	2.51	7.66
1560	1/2" thick	5.55	2.77	8.32

C3020 Floor Finishes

C3020 410	Tile & Covering	COST PER S.F.		
		MAT.	INST.	TOTAL
1580	Vinyl, composition tile, minimum	1.31	1.05	2.36
1600	Maximum	1.71	1.05	2.76
1620	Vinyl tile, 3/32"	4.08	1.05	5.13
1640	Maximum	3.58	1.05	4.63
1660	Sheet goods, plain pattern/colors	4.06	2.10	6.16
1680	Intricate pattern/colors	8.25	2.63	10.88
1720	Tile, ceramic natural clay	6.35	5.10	11.45
1730	Marble, synthetic 12"x12"x5/8"	11.35	15.65	27
1740	Porcelain type, minimum	5.70	5.10	10.80
1760	Maximum	7.60	6.05	13.65
1800	Quarry tile, mud set, minimum	8.75	6.70	15.45
1820	Maximum	9.55	8.50	18.05
1840	Thin set, deduct		1.34	1.34
1850	Tile, natural stone, marble, in mortar bed, 12" x 12" x 3/8" thick	17.60	25	42.60
1860	Terrazzo precast, minimum	5.30	7	12.30
1880	Maximum	10.25	7	17.25
1900	Non-slip, minimum	22.50	17.15	39.65
1920	Maximum	22.50	23	45.50
1960	Stone flooring, polished marble in mortar bed	19.25	25	44.25
2020	Wood block, end grain factory type, natural finish, 2" thick	4.94	4.76	9.70
2040	Fir, vertical grain, 1"x4", no finish, minimum	2.89	2.33	5.22
2060	Maximum	3.07	2.33	5.40
2080	Prefinished white oak, prime grade, 2-1/4" wide	5.30	3.50	8.80
2100	3-1/4" wide	5.75	3.21	8.96
2120	Maple strip, sanded and finished, minimum	5.85	5.10	10.95
2140	Maximum	6	5.10	11.10
2160	Oak strip, sanded and finished, minimum	4.65	5.10	9.75
2180	Maximum	5.50	5.10	10.60
2200	Parquetry, sanded and finished, plain pattern	6.95	5.30	12.25
2220	Intricate pattern	11.65	7.55	19.20
2260	Add for sleepers on concrete, treated, 24" O.C., 1"x2"	1.80	3.05	4.85
2280	1"x3"	1.95	2.38	4.33
2300	2"x4"	1	1.20	2.20
2320	2"x6"	.99	.92	1.91
2340	Underlayment, plywood, 3/8" thick	1.09	.79	1.88
2350	1/2" thick	1.29	.82	2.11
2360	5/8" thick	1.42	.85	2.27
2370	3/4" thick	1.63	.91	2.54
2380	Particle board, 3/8" thick	.45	.79	1.24
2390	1/2" thick	.50	.82	1.32
2400	5/8" thick	.59	.85	1.44
2410	3/4" thick	.75	.91	1.66
2420	Hardboard, 4' x 4', .215" thick	.69	.79	1.48
9200	Vinyl, composition tile, 12" x 12" x 1/8" thick, recycled content	2.43	1.05	3.48

C3030 Ceiling Finishes

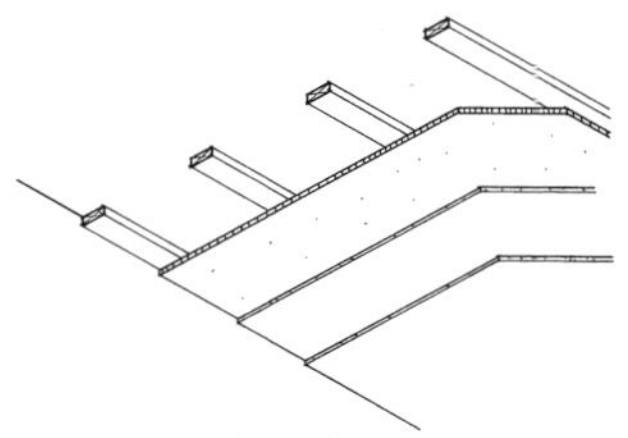

2 Coats of Plaster on Gypsum Lath on Wood Furring

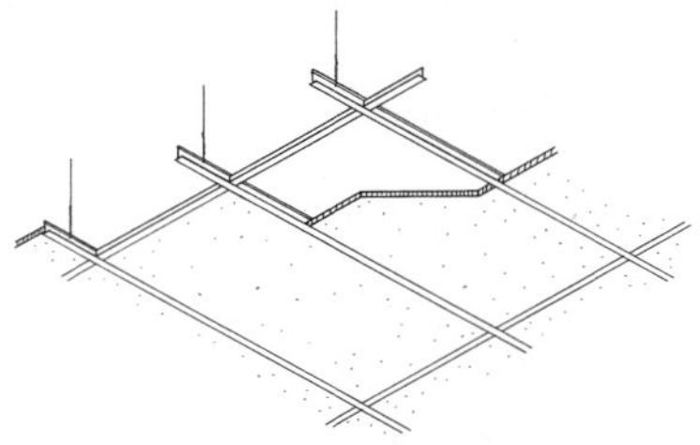

Fiberglass Board on Exposed Suspended Grid System

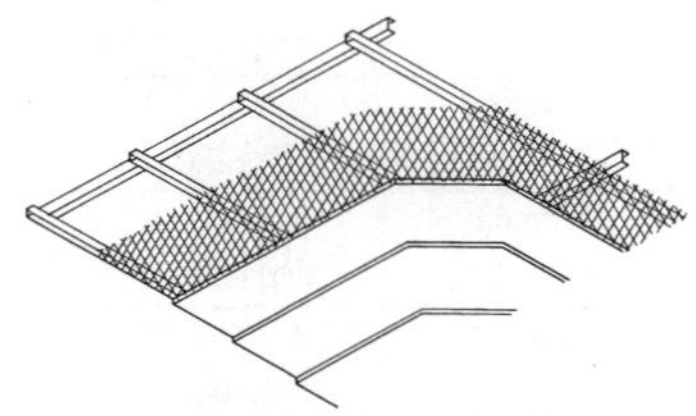

Plaster and Metal Lath on Metal Furring

System Components	QUANTITY	UNIT	COST PER S.F. MAT.	COST PER S.F. INST.	COST PER S.F. TOTAL
SYSTEM C3030 105 2400					
GYPSUM PLASTER, 2 COATS, 3/8″GYP. LATH, WOOD FURRING, FRAMING					
Gypsum plaster 2 coats no lath, on ceilings	.110	S.Y.	.45	3.24	3.69
Gypsum lath plain/perforated nailed, 3/8″ thick	.110	S.Y.	.37	.73	1.10
Add for ceiling installation	.110	S.Y.		.29	.29
Furring, 1″ x 3″ wood strips on wood joists	.750	L.F.	.32	1.28	1.60
Paint, primer and one coat	1.000	S.F.	.14	.50	.64
TOTAL			1.28	6.04	7.32

C3030 105 Plaster Ceilings

	TYPE	LATH	FURRING	SUPPORT		COST PER S.F. MAT.	COST PER S.F. INST.	COST PER S.F. TOTAL
2400	2 coat gypsum	3/8″ gypsum	1″x3″ wood, 16″ O.C.	wood		1.28	6.05	7.33
2500	Painted			masonry		1.31	6.15	7.46
2600				concrete		1.31	6.90	8.21
2700	3 coat gypsum	3.4# metal	1″x3″ wood, 16″ O.C.	wood		1.64	6.50	8.14
2800	Painted			masonry		1.67	6.60	8.27
2900				concrete		1.67	7.30	8.97
3000	2 coat perlite	3/8″ gypsum	1″x3″ wood, 16″ O.C.	wood		1.55	6.25	7.80
3100	Painted			masonry		1.58	6.40	7.98
3200				concrete		1.58	7.10	8.68
3300	3 coat perlite	3.4# metal	1″x3″ wood, 16″ O.C.	wood		1.71	6.40	8.11
3400	Painted			masonry		1.74	6.55	8.29
3500				concrete		1.74	7.25	8.99
3600	2 coat gypsum	3/8″ gypsum	3/4″ CRC, 12″ O.C.	1-1/2″ CRC, 48″O.C.		1.38	7.45	8.83
3700	Painted		3/4″ CRC, 16″ O.C.	1-1/2″ CRC, 48″O.C.		1.34	6.70	8.04
3800			3/4″ CRC, 24″ O.C.	1-1/2″ CRC, 48″O.C.		1.21	6.10	7.31
3900	2 coat perlite	3/8″ gypsum	3/4″ CRC, 12″ O.C.	1-1/2″ CRC, 48″O.C		1.65	7.95	9.60
4000	Painted		3/4″ CRC, 16″ O.C.	1-1/2″ CRC, 48″O.C.		1.61	7.20	8.81
4100			3/4″ CRC, 24″ O.C.	1-1/2″ CRC, 48″O.C.		1.48	6.60	8.08
4200	3 coat gypsum	3.4# metal	3/4″ CRC, 12″ O.C.	1-1/2″ CRC, 36″ O.C.		1.91	9.65	11.56
4300	Painted		3/4″ CRC, 16″ O.C.	1-1/2″ CRC, 36″ O.C.		1.87	8.90	10.77
4400			3/4″ CRC, 24″ O.C.	1-1/2″ CRC, 36″ O.C.		1.74	8.30	10.04
4500	3 coat perlite	3.4# metal	3/4″ CRC, 12″ O.C.	1-1/2″ CRC,36″ O.C.		2.04	10.50	12.54
4600	Painted		3/4″ CRC, 16″ O.C.	1-1/2″ CRC, 36″ O.C.		2	9.80	11.80
4700			3/4″ CRC, 24″ O.C.	1-1/2″ CRC, 36″ O.C.		1.87	9.20	11.07

C3030 Ceiling Finishes

C3030 110 — Drywall Ceilings

	TYPE	FINISH	FURRING	SUPPORT		COST PER S.F. MAT.	COST PER S.F. INST.	COST PER S.F. TOTAL
4800	1/2" F.R. drywall	painted and textured	1"x3" wood, 16" O.C.	wood		.95	3.71	4.66
4900				masonry		.98	3.83	4.81
5000				concrete		.98	4.55	5.53
5100	5/8" F.R. drywall	painted and textured	1"x3" wood, 16" O.C.	wood		.94	3.71	4.65
5200				masonry		.97	3.83	4.80
5300				concrete		.97	4.55	5.52
5400	1/2" F.R. drywall	painted and textured	7/8" resil. channels	24" O.C.		.77	3.60	4.37
5500			1"x2" wood	stud clips		.95	3.48	4.43
5600			1-5/8"metal studs	24" O.C.		.83	3.21	4.04
5700	5/8" F.R. drywall	painted and textured	1-5/8"metal studs	24" O.C.		.82	3.21	4.03
5702								

C3030 Ceiling Finishes

C3030 140	Plaster Ceiling Components	COST PER S.F. MAT.	INST.	TOTAL
0060	Plaster, gypsum incl. finish			
0080	3 coats	.65	3.83	4.48
0100	Perlite, incl. finish, 2 coats	.72	3.77	4.49
0120	3 coats	.78	4.71	5.49
0140	Thin coat on drywall	.12	.74	.86
0200	Lath, gypsum, 3/8" thick	.37	1.02	1.39
0220	1/2" thick	.30	1.07	1.37
0240	5/8" thick	.36	1.24	1.60
0260	Metal, diamond, 2.5 lb.	.50	.83	1.33
0280	3.4 lb.	.53	.89	1.42
0300	Flat rib, 2.75 lb.	.43	.83	1.26
0320	3.4 lb.	.55	.89	1.44
0440	Furring, steel channels, 3/4" galvanized 12" O.C.	.42	2.66	3.08
0460	16" O.C.	.38	1.93	2.31
0480	24" O.C.	.25	1.33	1.58
0500	1-1/2" galvanized , 12" O.C.	.57	2.94	3.51
0520	16" O.C.	.51	2.15	2.66
0540	24" O.C.	.34	1.43	1.77
0560	Wood strips, 1"x3", on wood, 12" O.C.	.43	1.70	2.13
0580	16" O.C.	.32	1.28	1.60
0600	24" O.C.	.22	.85	1.07
0620	On masonry, 12" O.C.	.46	1.86	2.32
0640	16" O.C.	.35	1.40	1.75
0660	24" O.C.	.23	.93	1.16
0680	On concrete, 12" O.C.	.46	2.83	3.29
0700	16" O.C.	.35	2.12	2.47
0720	24" O.C.	.23	1.42	1.65

C3030 210 Acoustical Ceilings

	TYPE	TILE	GRID	SUPPORT		COST PER S.F. MAT.	INST.	TOTAL
5800	5/8" fiberglass board	24" x 48"	tee	suspended		2.45	1.78	4.23
5900		24" x 24"	tee	suspended		2.70	1.95	4.65
6000	3/4" fiberglass board	24" x 48"	tee	suspended		3.96	1.82	5.78
6100		24" x 24"	tee	suspended		4.21	1.99	6.20
6500	5/8" mineral fiber	12" x 12"	1"x3" wood, 12" O.C.	wood		2.88	3.68	6.56
6600				masonry		2.91	3.84	6.75
6700				concrete		2.91	4.81	7.72
6800	3/4" mineral fiber	12" x 12"	1"x3" wood, 12" O.C.	wood		3.65	3.68	7.33
6900				masonry		3.65	3.68	7.33
7000				concrete		3.65	3.68	7.33
7100	3/4"mineral fiber on	12" x 12"	25 ga. channels	runners		4.10	4.59	8.69
7102	5/8" F.R. drywall							
7200	5/8" plastic coated	12" x 12"		adhesive backed		2.89	1.98	4.87
7201	Mineral fiber							
7202								
7300	3/4" plastic coated	12" x 12"		adhesive backed		3.66	1.98	5.64
7301	Mineral fiber							
7302								
7400	3/4" mineral fiber	12" x 12"	conceal 2" bar &	suspended		2.75	4.47	7.22
7401			channels					
7402								

C3030 Ceiling Finishes

C3030 240	Acoustical Ceiling Components	COST PER S.F.		
		MAT.	INST.	TOTAL
2480	Ceiling boards, eggcrate, acrylic, 1/2" x 1/2" x 1/2" cubes	1.97	1.19	3.16
2500	Polystyrene, 3/8" x 3/8" x 1/2" cubes	1.73	1.17	2.90
2520	1/2" x 1/2" x 1/2" cubes	2	1.19	3.19
2540	Fiberglass boards, plain, 5/8" thick	1.38	.95	2.33
2560	3/4" thick	2.89	.99	3.88
2580	Grass cloth faced, 3/4" thick	2.99	1.19	4.18
2600	1" thick	3.64	1.23	4.87
2620	Luminous panels, prismatic, acrylic	2.90	1.49	4.39
2640	Polystyrene	1.78	1.49	3.27
2660	Flat or ribbed, acrylic	4.37	1.49	5.86
2680	Polystyrene	2.52	1.49	4.01
2700	Drop pan, white, acrylic	6.25	1.49	7.74
2720	Polystyrene	4.90	1.49	6.39
2740	Mineral fiber boards, 5/8" thick, standard	.75	.88	1.63
2760	Plastic faced	2.76	1.49	4.25
2780	2 hour rating	1.25	.88	2.13
2800	Perforated aluminum sheets, .024 thick, corrugated painted	2.71	1.21	3.92
2820	Plain	5.05	1.19	6.24
3080	Mineral fiber, plastic coated, 12" x 12" or 12" x 24", 5/8" thick	2.45	1.98	4.43
3100	3/4" thick	3.22	1.98	5.20
3120	Fire rated, 3/4" thick, plain faced	1.44	1.98	3.42
3140	Mylar faced	2.12	1.98	4.10
3160	Add for ceiling primer	.14		.14
3180	Add for ceiling cement	.44		.44
3240	Suspension system, furring, 1" x 3" wood 12" O.C.	.43	1.70	2.13
3260	T bar suspension system, 2' x 4' grid	.84	.74	1.58
3280	2' x 2' grid	1.09	.91	2
3300	Concealed Z bar suspension system 12" module	.95	1.14	2.09
3320	Add to above for 1-1/2" carrier channels 4' O.C.	.13	1.26	1.39
3340	Add to above for carrier channels for recessed lighting	.24	1.29	1.53

D1010 Elevators and Lifts

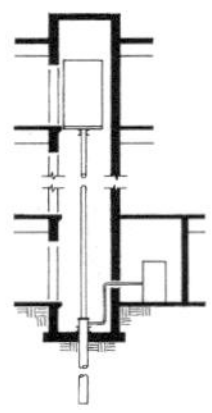

The hydraulic elevator obtains its motion from the movement of liquid under pressure in the piston connected to the car bottom. These pistons can provide travel to a maximum rise of 70′ and are sized for the intended load. As the rise reaches the upper limits the cost tends to exceed that of a geared electric unit.

System Components	QUANTITY	UNIT	COST EACH MAT.	COST EACH INST.	COST EACH TOTAL
SYSTEM D1010 110 2000					
PASS. ELEV., HYDRAULIC, 2500 LB., 5 FLOORS, 100 FPM					
Passenger elevator, hydraulic, 1500 lb capacity, 2 stop, 100 FPM	1.000	Ea.	42,500	18,900	61,400
Over 10′ travel height, passenger elevator, hydraulic, add	40.000	V.L.F.	31,200	10,400	41,600
Passenger elevator, hydraulic, 2500 lb. capacity over standard, add	1.000	Ea.	3,175		3,175
Over 2 stops, passenger elevator, hydraulic, add	3.000	Stop	3,225	21,000	24,225
Hall lantern	5.000	Ea.	2,850	1,180	4,030
Maintenance agreement for pass. elev. 9 months	1.000	Ea.	4,400		4,400
Position indicator at lobby	1.000	Ea.	460	59	519
TOTAL			87,810	51,539	139,349

D1010 110	Hydraulic	COST EACH MAT.	COST EACH INST.	COST EACH TOTAL
1300	Pass. elev., 1500 lb., 2 Floors, 100 FPM (RD1010 -010)	48,000	19,400	67,400
1400	5 Floors, 100 FPM	84,500	51,500	136,000
1600	2000 lb., 2 Floors, 100 FPM	49,000	19,400	68,400
1700	5 floors, 100 FPM	85,500	51,500	137,000
1900	2500 lb., 2 Floors, 100 FPM	51,000	19,400	70,400
2000	5 floors, 100 FPM	88,000	51,500	139,500
2200	3000 lb., 2 Floors, 100 FPM	52,500	19,400	71,900
2300	5 floors, 100 FPM	89,000	51,500	140,500
2500	3500 lb., 2 Floors, 100 FPM	56,000	19,400	75,400
2600	5 floors, 100 FPM	92,500	51,500	144,000
2800	4000 lb., 2 Floors, 100 FPM	57,500	19,400	76,900
2900	5 floors, 100 FPM	94,000	51,500	145,500
3100	4500 lb., 2 Floors, 100 FPM	60,500	19,400	79,900
3200	5 floors, 100 FPM	97,000	51,500	148,500
4000	Hospital elevators, 3500 lb., 2 Floors, 100 FPM	82,000	19,400	101,400
4100	5 floors, 100 FPM	131,500	51,500	183,000
4300	4000 lb., 2 Floors, 100 FPM	82,000	19,400	101,400
4400	5 floors, 100 FPM	131,500	51,500	183,000
4600	4500 lb., 2 Floors, 100 FPM	90,000	19,400	109,400
4800	5 floors, 100 FPM	139,500	51,500	191,000
4900	5000 lb., 2 Floors, 100 FPM	93,500	19,400	112,900
5000	5 floors, 100 FPM	143,000	51,500	194,500
6700	Freight elevators (Class "B"), 3000 lb., 2 Floors, 50 FPM	120,500	24,600	145,100
6800	5 floors, 100 FPM	193,000	64,500	257,500
7000	4000 lb., 2 Floors, 50 FPM	125,500	24,600	150,100
7100	5 floors, 100 FPM	198,000	64,500	262,500
7500	10,000 lb., 2 Floors, 50 FPM	157,000	24,600	181,600
7600	5 floors, 100 FPM	229,500	64,500	294,000
8100	20,000 lb., 2 Floors, 50 FPM	204,500	24,600	229,100
8200	5 Floors, 100 FPM	277,000	64,500	341,500

D1010 Elevators and Lifts

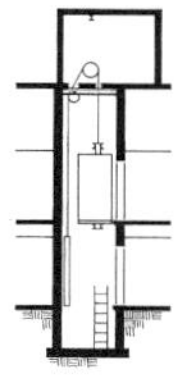

Geared traction elevators are the intermediate group both in cost and in operating areas. These are in buildings of four to fifteen floors and speed ranges from 200′ to 500′ per minute.

System Components	QUANTITY	UNIT	COST EACH MAT.	COST EACH INST.	COST EACH TOTAL
SYSTEM D1010 140 1600					
PASSENGER, 2500 LB., 5 FLOORS, 200 FPM					
Passenger elevator, geared, 2000 lb. capacity, 4 stop, 200 FPM	1.000	Ea.	103,500	37,800	141,300
Over 40′ travel height, passenger elevator electric, add	10.000	V.L.F.	7,850	2,600	10,450
Passenger elevator, electric, 2500 lb. cap., add	1.000	Ea.	4,350		4,350
Over 4 stops, passenger elevator, electric, add	1.000	Stop	3,450	7,000	10,450
Hall lantern	5.000	Ea.	2,850	1,180	4,030
Maintenance agreement for passenger elevator, 9 months	1.000	Ea.	4,400		4,400
Position indicator at lobby	1.000	Ea.	460	59	519
TOTAL			126,860	48,639	175,499

D1010 140	Traction Geared Elevators	COST EACH MAT.	COST EACH INST.	COST EACH TOTAL
1300	Passenger, 2000 Lb., 5 floors, 200 FPM	122,500	48,600	171,100
1500	15 floors, 350 FPM	251,000	147,000	398,000
1600	2500 Lb., 5 floors, 200 FPM (RD1010 -010)	127,000	48,600	175,600
1800	15 floors, 350 FPM	255,500	147,000	402,500
2200	3500 Lb., 5 floors, 200 FPM	129,000	48,600	177,600
2400	15 floors, 350 FPM	257,500	147,000	404,500
2500	4000 Lb., 5 floors, 200 FPM	130,000	48,600	178,600
2700	15 floors, 350 FPM	258,500	147,000	405,500
2800	4500 Lb., 5 floors, 200 FPM	133,000	48,600	181,600
3000	15 floors, 350 FPM	261,500	147,000	408,500
3100	5000 Lb., 5 floors, 200 FPM	135,500	48,600	184,100
3300	15 floors, 350 FPM	264,000	147,000	411,000
4000	Hospital, 3500 Lb., 5 floors, 200 FPM	130,500	48,600	179,100
4200	15 floors, 350 FPM	306,500	147,000	453,500
4300	4000 Lb., 5 floors, 200 FPM	130,500	48,600	179,100
4500	15 floors, 350 FPM	306,500	147,000	453,500
4600	4500 Lb., 5 floors, 200 FPM	137,000	48,600	185,600
4800	15 floors, 350 FPM	313,000	147,000	460,000
4900	5000 Lb., 5 floors, 200 FPM	139,000	48,600	187,600
5100	15 floors, 350 FPM	315,000	147,000	462,000
6000	Freight, 4000 Lb., 5 floors, 50 FPM class 'B'	156,000	50,500	206,500
6200	15 floors, 200 FPM class 'B'	360,000	175,000	535,000
6300	8000 Lb., 5 floors, 50 FPM class 'B'	181,500	50,500	232,000
6500	15 floors, 200 FPM class 'B'	385,500	175,000	560,500
7000	10,000 Lb., 5 floors, 50 FPM class 'B'	209,000	50,500	259,500
7200	15 floors, 200 FPM class 'B'	597,000	175,000	772,000
8000	20,000 Lb., 5 floors, 50 FPM class 'B'	231,000	50,500	281,500
8200	15 floors, 200 FPM class 'B'	619,000	175,000	794,000

D1010 Elevators and Lifts

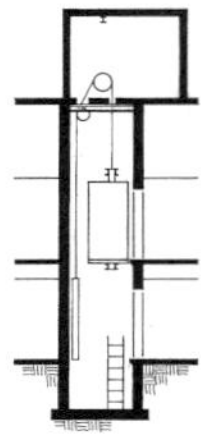

Gearless traction elevators are used in high rise situations where speeds to over 1000 FPM are required to move passengers efficiently. This type of installation is also the most costly.

System Components	QUANTITY	UNIT	COST EACH MAT.	COST EACH INST.	COST EACH TOTAL
SYSTEM D1010 150 1700					
PASSENGER, 2500 LB., 10 FLOORS, 200 FPM					
Passenger elevator, geared, 2000 lb. capacity, 4 stop 100 FPM	1.000	Ea.	103,500	37,800	141,300
Over 10' travel height, passenger elevator electric, add	60.000	V.L.F.	47,100	15,600	62,700
Passenger elevator, electric, 2500 lb. cap., add	1.000	Ea.	4,350		4,350
Over 2 stops, passenger elevator, electric, add	6.000	Stop	20,700	42,000	62,700
Gearless variable voltage machinery, overhead	1.000	Ea.	89,000	27,000	116,000
Hall lantern	9.000	Ea.	5,700	2,360	8,060
Position indicator at lobby	1.000	Ea.	460	59	519
Fireman services control	1.000	Ea.	8,550	1,575	10,125
Emergency power manual switching, add	1.000	Ea.	565	236	801
Maintenance agreement for passenger elevator, 9 months	1.000	Ea.	4,400		4,400
TOTAL			284,325	126,630	410,955

D1010 150	Traction Gearless Elevators	COST EACH MAT.	COST EACH INST.	COST EACH TOTAL
1700	Passenger, 2500 Lb., 10 floors, 200 FPM	284,500	126,500	411,000
1900	30 floors, 600 FPM	571,000	323,500	894,500
2000	3000 Lb., 10 floors, 200 FPM (RD1010 -010)	285,000	126,500	411,500
2200	30 floors, 600 FPM	571,500	323,500	895,000
2300	3500 Lb., 10 floors, 200 FPM	286,500	126,500	413,000
2500	30 floors, 600 FPM	573,000	323,500	896,500
2700	50 floors, 800 FPM	825,500	520,000	1,345,500
2800	4000 Lb., 10 floors, 200 FPM	287,500	126,500	414,000
3000	30 floors, 600 FPM	574,500	323,500	898,000
3200	50 floors, 800 FPM	826,500	520,000	1,346,500
3300	4500 Lb., 10 floors, 200 FPM	290,500	126,500	417,000
3500	30 floors, 600 FPM	577,000	323,500	900,500
3700	50 floors, 800 FPM	829,500	520,000	1,349,500
3800	5000 Lb., 10 floors, 200 FPM	293,000	126,500	419,500
4000	30 floors, 600 FPM	579,500	323,500	903,000
4200	50 floors, 800 FPM	831,500	520,000	1,351,500
6000	Hospital, 3500 Lb., 10 floors, 200 FPM	311,500	126,500	438,000
6200	30 floors, 600 FPM	698,500	323,500	1,022,000
6400	4000 Lb., 10 floors, 200 FPM	311,500	126,500	438,000
6600	30 floors, 600 FPM	698,500	323,500	1,022,000
6800	4500 Lb., 10 floors, 200 FPM	318,000	126,500	444,500
7000	30 floors, 600 FPM	705,000	323,500	1,028,500
7200	5000 Lb., 10 floors, 200 FPM	320,000	126,500	446,500
7400	30 floors, 600 FPM	707,000	323,500	1,030,500

D1020 Escalators and Moving Walks

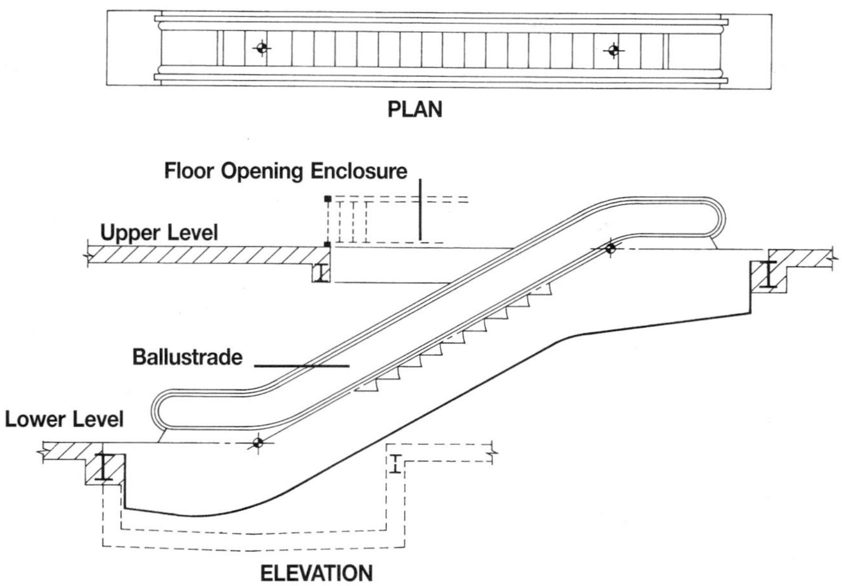

Moving stairs can be used for buildings where 600 or more people are to be carried to the second floor or beyond. Freight cannot be carried on escalators and at least one elevator must be available for this function.

Carrying capacity is 5000 to 8000 people per hour. Power requirement is 2 KW to 3 KW per hour and incline angle is 30°.

D1020 110 Moving Stairs

	TYPE	HEIGHT	WIDTH	BALUSTRADE MATERIAL		COST EACH		
						MAT.	INST.	TOTAL
0100	Escalator	10ft	32″	glass		95,000	52,500	147,500
0150				metal		103,000	52,500	155,500
0200			48″	glass		102,500	52,500	155,000
0250				metal		111,500	52,500	164,000
0300		15ft	32″	glass		100,000	61,000	161,000
0350				metal		108,500	61,000	169,500
0400			48″	glass		106,000	61,000	167,000
0450				metal		115,000	61,000	176,000
0500		20ft	32″	glass		106,500	74,000	180,500
0550				metal		115,500	74,000	189,500
0600			48″	glass		115,500	74,000	189,500
0650				metal		125,000	74,000	199,000
0700		25ft	32″	glass		116,000	91,000	207,000
0750				metal		125,500	91,000	216,500
0800			48″	glass		134,000	91,000	225,000
0850				metal		144,500	91,000	235,500

D1020 210 Moving Walks

	TYPE	STORY HEIGHT	DEGREE SLOPE	WIDTH	COST RANGE	COST PER L.F.		
						MAT.	INST.	TOTAL
1500	Ramp	10′-23′	12	3′-4″	minimum	2,650	1,025	3,675
1600					maximum	3,350	1,250	4,600
2000	Walk	0′	0	2′-0″	minimum	980	560	1,540
2500					maximum	1,350	820	2,170
3000				3′-4″	minimum	2,200	820	3,020
3500					maximum	2,600	955	3,555

D1090 Other Conveying Systems

D1090 410	Miscellaneous Types	COST EACH		
		MAT.	INST.	TOTAL
1050	Conveyors, Horiz. belt, ctr drive, 60 FPM, cap 40 lb./L.F., belt x 26.5'L	3,700	2,400	6,100
1100	24" belt, 42' length	5,625	3,000	8,625
1200	24" belt, 62' length	7,925	4,000	11,925
1500	Inclined belt, horiz. loading and end idler assembly, 16" belt x 27.5'L	8,025	4,000	12,025
1600	24" belt, 27.5' length	9,825	8,000	17,825
3500	Pneumatic tube system, single, 2 station-100', stock economy model, 3" dia.	3,725	12,200	15,925
3600	4" diameter	4,700	16,300	21,000
4000	Twin tube system base cost, minimum	6,325	1,950	8,275
4100	Maximum	12,700	5,850	18,550
4200	Add per L.F., of system, minimum	9.10	15.70	24.80
4300	Maximum	27.50	39	66.50
5000	Completely automatic system, 15 to 50 stations, 4" dia., cost/station	22,100	5,050	27,150
5100	50 to 144 stations, 4" diameter, cost per station	17,200	4,575	21,775

D2010 Plumbing Fixtures

Systems are complete with trim seat and rough-in (supply, waste and vent) for connection to supply branches and waste mains.

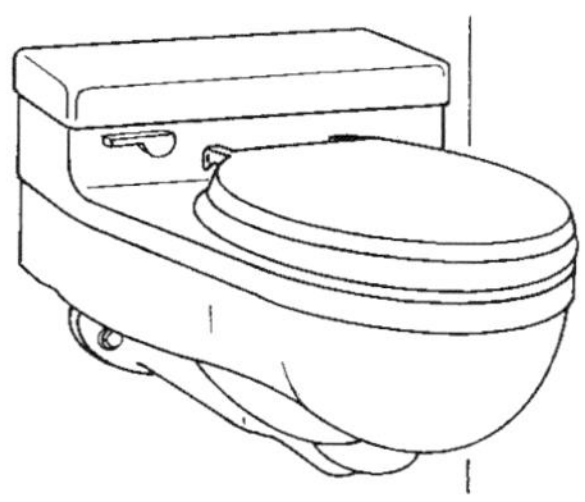

One Piece Wall Hung

Supply

Waste/Vent

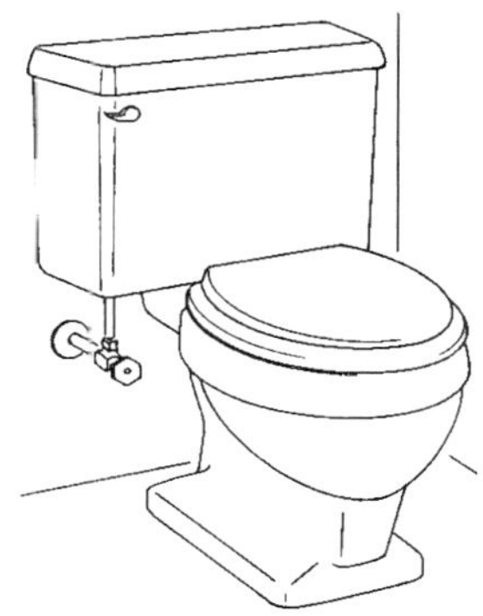

Floor Mount

System Components	QUANTITY	UNIT	COST EACH MAT.	COST EACH INST.	COST EACH TOTAL
SYSTEM D2010 110 1880					
WATER CLOSET, VITREOUS CHINA					
TANK TYPE, WALL HUNG, TWO PIECE					
Water closet, tank type vit china wall hung 2 pc. w/seat supply & stop	1.000	Ea.	580	243	823
Pipe Steel galvanized, schedule 40, threaded, 2″ diam.	4.000	L.F.	37.20	80	117.20
Pipe, CI soil, no hub, cplg 10′ OC, hanger 5′ OC, 4″ diam.	2.000	L.F.	43	44	87
Pipe, coupling, standard coupling, CI soil, no hub, 4″ diam.	2.000	Ea.	44	78	122
Copper tubing type L solder joint, hangar 10′ O.C., 1/2″ diam.	6.000	L.F.	24.30	53.10	77.40
Wrought copper 90° elbow for solder joints 1/2″ diam.	2.000	Ea.	2.48	71	73.48
Wrought copper Tee for solder joints 1/2″ diam.	1.000	Ea.	2.11	55	57.11
Supports/carrier, water closet, siphon jet, horiz, single, 4″ waste	1.000	Ea.	890	134	1,024
TOTAL			1,623.09	758.10	2,381.19

D2010 110	Water Closet Systems		COST EACH MAT.	COST EACH INST.	COST EACH TOTAL
1800	Water closet, vitreous china				
1840	Tank type, wall hung				
1880	Close coupled two piece	RD2010 -031	1,625	760	2,385
1920	Floor mount, one piece		1,225	805	2,030
1960	One piece low profile	RD2010 -400	1,200	805	2,005
2000	Two piece close coupled		660	805	1,465
2040	Bowl only with flush valve				
2080	Wall hung		2,225	860	3,085
2120	Floor mount		790	820	1,610
2160	Floor mount, ADA compliant with 18″ high bowl		785	840	1,625

D2010 Plumbing Fixtures

Systems are complete with trim, seat, flush valve and rough-in (supply, waste and vent) for connection to supply branches and waste mains.

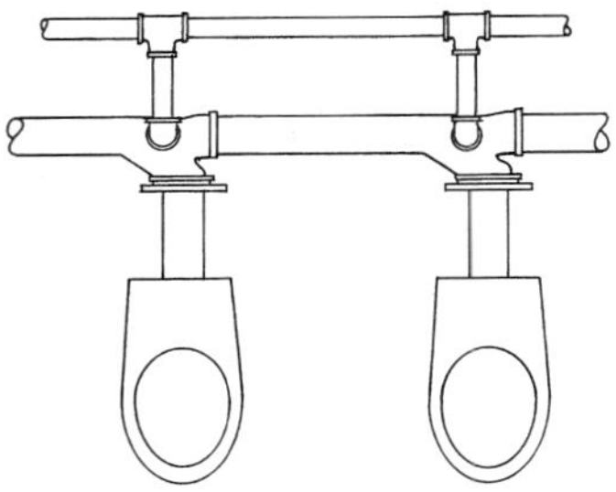

Side by Side

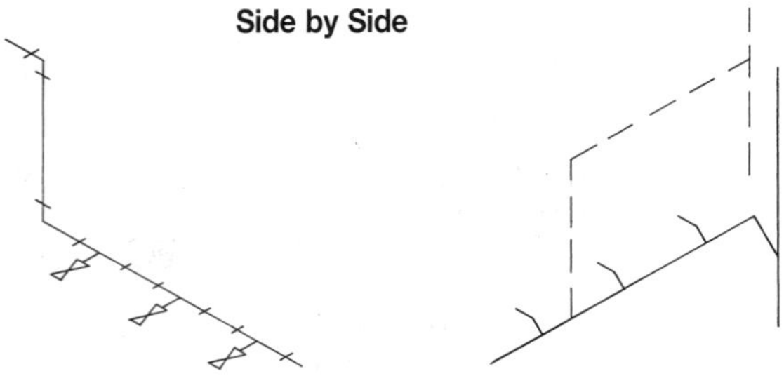

Supply

Waste/Vent

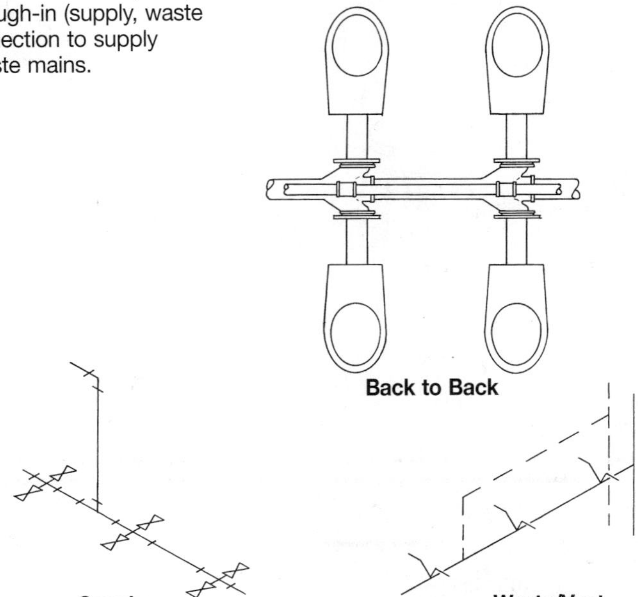

Back to Back

Supply

Waste/Vent

System Components	QUANTITY	UNIT	COST EACH MAT.	COST EACH INST.	COST EACH TOTAL
SYSTEM D2010 120 1760					
WATER CLOSETS, BATTERY MOUNT, WALL HUNG, SIDE BY SIDE, FIRST CLOSET					
Water closet, bowl only w/flush valve, seat, wall hung	1.000	Ea.	1,050	222	1,272
Pipe, CI soil, no hub, cplg 10′ OC, hanger 5′ OC, 4″ diam.	3.000	L.F.	64.50	66	130.50
Coupling, standard, CI, soil, no hub, 4″ diam.	2.000	Ea.	44	78	122
Copper tubing, type L, solder joints, hangers 10′ OC, 1″ diam.	6.000	L.F.	48.90	63	111.90
Copper tubing, type DWV solder joints, hangers 10′OC, 2″ diam.	6.000	L.F.	97.50	97.50	195
Wrought copper 90° elbow for solder joints 1″ diam.	1.000	Ea.	6.85	44.50	51.35
Wrought copper Tee for solder joints 1″ diam.	1.000	Ea.	15.30	71.50	86.80
Support/carrier, siphon jet, horiz, adjustable single, 4″ pipe	1.000	Ea.	890	134	1,024
Valve, gate, bronze, 125 lb, NRS, soldered 1″ diam.	1.000	Ea.	90	37.50	127.50
Wrought copper, DWV, 90° elbow, 2″ diam.	1.000	Ea.	28	71.50	99.50
TOTAL			2,335.05	885.50	3,220.55

D2010 120	Water Closets, Group		COST EACH MAT.	COST EACH INST.	COST EACH TOTAL
1760	Water closets, battery mount, wall hung, side by side, first closet	RD2010-031	2,325	885	3,210
1800	Each additional water closet, add		2,225	835	3,060
3000	Back to back, first pair of closets	RD2010-400	4,350	1,175	5,525
3100	Each additional pair of closets, back to back		4,275	1,150	5,425
9000	Back to back, first pair of closets, auto sensor flush valve, 1.28 gpf		4,875	1,275	6,150
9100	Ea additional pair of cls, back to back, auto sensor flush valve, 1.28 gpf		4,725	1,200	5,925

D2010 Plumbing Fixtures

Systems are complete with trim, flush valve and rough-in (supply, waste and vent) for connection to supply branches and waste mains.

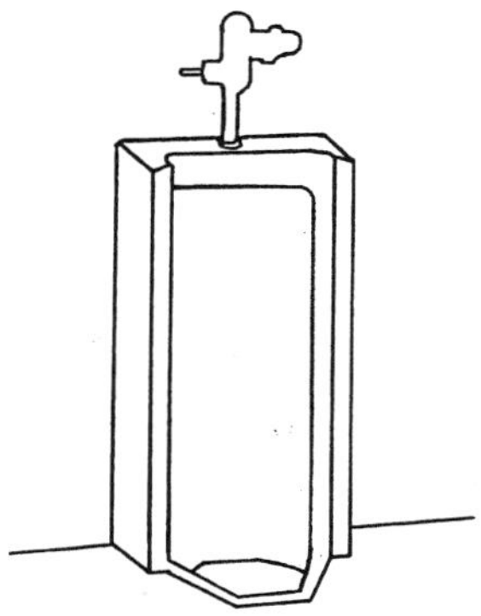

Stall Type

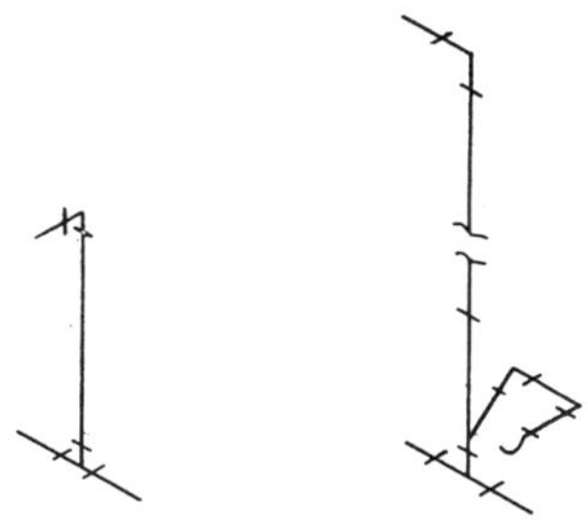

Supply Waste/Vent

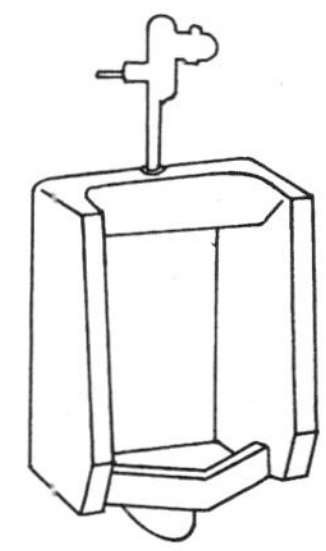

Wall Hung

System Components	QUANTITY	UNIT	COST EACH MAT.	INST.	TOTAL
SYSTEM D2010 210 2000					
URINAL, VITREOUS CHINA, WALL HUNG					
Urinal, wall hung, vitreous china, incl. hanger	1.000	Ea.	305	430	735
Pipe, steel, galvanized, schedule 40, threaded, 1-1/2" diam.	5.000	L.F.	36.25	80.50	116.75
Copper tubing type DWV, solder joint, hangers 10' OC, 2" diam.	3.000	L.F.	48.75	48.75	97.50
Combination Y & 1/8 bend for CI soil pipe, no hub, 3" diam.	1.000	Ea.	21.50		21.50
Pipe, CI, no hub, cplg. 10' OC, hanger 5' OC, 3" diam.	4.000	L.F.	68	80	148
Pipe coupling standard, CI soil, no hub, 3" diam.	3.000	Ea.	37.60	68	105.60
Copper tubing type L, solder joint, hanger 10' OC 3/4" diam.	5.000	L.F.	28.75	47	75.75
Wrought copper 90° elbow for solder joints 3/4" diam.	1.000	Ea.	2.78	37.50	40.28
Wrought copper Tee for solder joints, 3/4" diam.	1.000	Ea.	5.10	59.50	64.60
TOTAL			553.73	851.25	1,404.98

D2010 210	Urinal Systems		COST EACH MAT.	INST.	TOTAL
2000	Urinal, vitreous china, wall hung	RD2010 -031	555	850	1,405
2040	Stall type		1,225	1,025	2,250

D2010 Plumbing Fixtures

Systems are complete with trim and rough-in (supply, waste and vent) to connect to supply branches and waste mains.

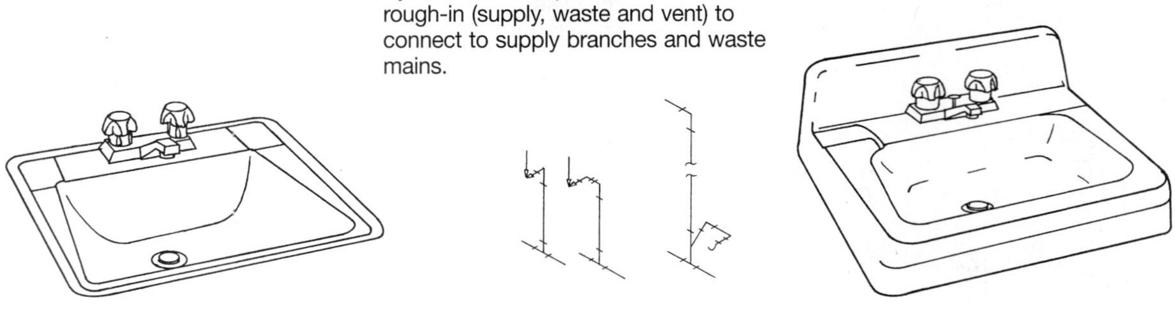

Vanity Top Supply Waste/Vent Wall Hung

System Components	QUANTITY	UNIT	COST EACH MAT.	COST EACH INST.	COST EACH TOTAL
SYSTEM D2010 310 1560					
LAVATORY W/TRIM, VANITY TOP, P.E. ON C.I., 20″ X 18″					
Lavatory w/trim, PE on CI, white, vanity top, 20″ x 18″ oval	1.000	Ea.	345	201	546
Pipe, steel, galvanized, schedule 40, threaded, 1-1/4″ diam.	4.000	L.F.	25.60	57.80	83.40
Copper tubing type DWV, solder joint, hanger 10′ OC 1-1/4″ diam.	4.000	L.F.	40.40	47.60	88
Wrought copper DWV, Tee, sanitary, 1-1/4″ diam.	1.000	Ea.	23	79.50	102.50
P trap w/cleanout, 20 ga., 1-1/4″ diam.	1.000	Ea.	86	39.50	125.50
Copper tubing type L, solder joint, hanger 10′ OC 1/2″ diam.	10.000	L.F.	40.50	88.50	129
Wrought copper 90° elbow for solder joints 1/2″ diam.	2.000	Ea.	2.48	71	73.48
Wrought copper Tee for solder joints, 1/2″ diam.	2.000	Ea.	4.22	110	114.22
Stop, chrome, angle supply, 1/2″ diam.	2.000	Ea.	18.90	65	83.90
TOTAL			586.10	759.90	1,346

D2010 310	Lavatory Systems	COST EACH MAT.	COST EACH INST.	COST EACH TOTAL
1560	Lavatory w/trim, vanity top, PE on CI, 20″ x 18″, Vanity top by others.	585	760	1,345
1600	19″ x 16″ oval	395	760	1,155
1640	18″ round	700	760	1,460
1680	Cultured marble, 19″ x 17″ (RD2010-400)	405	760	1,165
1720	25″ x 19″	435	760	1,195
1760	Stainless, self-rimming, 25″ x 22″	615	760	1,375
1800	17″ x 22″	605	760	1,365
1840	Steel enameled, 20″ x 17″	405	780	1,185
1880	19″ round	405	780	1,185
1920	Vitreous china, 20″ x 16″	500	795	1,295
1960	19″ x 16″	510	795	1,305
2000	22″ x 13″	510	795	1,305
2040	Wall hung, PE on CI, 18″ x 15″	905	840	1,745
2080	19″ x 17″	935	840	1,775
2120	20″ x 18″	770	840	1,610
2160	Vitreous china, 18″ x 15″	685	860	1,545
2200	19″ x 17″	640	860	1,500
2240	24″ x 20″	770	860	1,630
2300	20″ x 27″, handicap	1,425	930	2,355

D2010 Plumbing Fixtures

Systems are complete with trim and rough-in (supply, waste and vent) to connect to supply branches and waste mains.

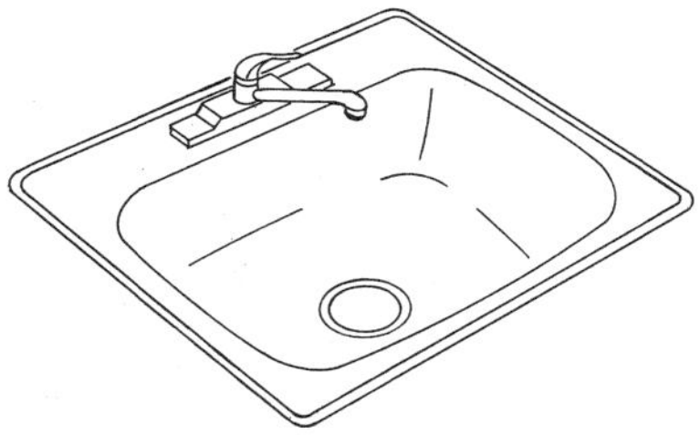

Countertop Single Bowl

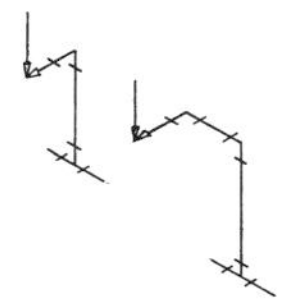

Supply

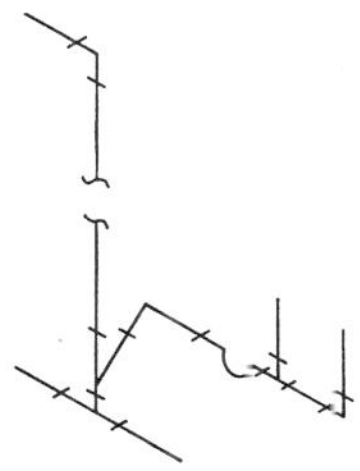

Waste/Vent

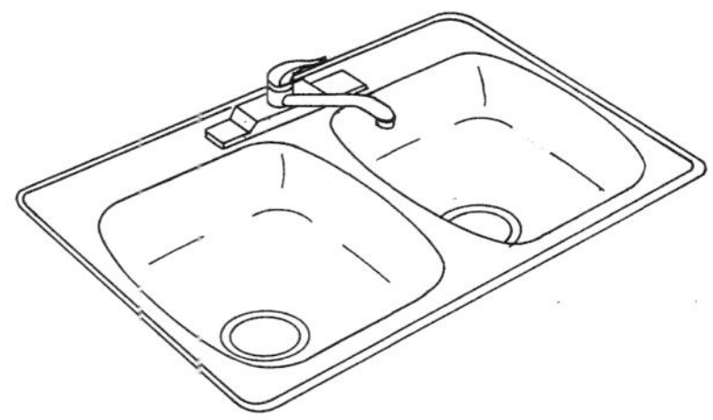

Countertop Double Bowl

System Components	QUANTITY	UNIT	COST EACH		
			MAT.	INST.	TOTAL
SYSTEM D2010 410 1720					
KITCHEN SINK W/TRIM, COUNTERTOP, P.E. ON C.I., 24″ X 21″, SINGLE BOWL					
Kitchen sink, counter top, PE on CI, 1 bowl, 24″ x 21″ OD	1.000	Ea.	325	230	555
Pipe, steel, galvanized, schedule 40, threaded, 1-1/4″ diam.	4.000	L.F.	25.60	57.80	83.40
Copper tubing, type DWV, solder, hangers 10′ OC 1-1/2″ diam.	6.000	L.F.	74.40	79.50	153.90
Wrought copper, DWV, Tee, sanitary, 1-1/2″ diam.	1.000	Ea.	29	89.50	118.50
P trap, standard, copper, 1-1/2″ diam.	1.000	Ea.	79	42	121
Copper tubing, type L, solder joints, hangers 10′ OC 1/2″ diam.	10.000	L.F.	40.50	88.50	129
Wrought copper 90° elbow for solder joints 1/2″ diam.	2.000	Ea.	2.48	71	73.48
Wrought copper Tee for solder joints, 1/2″ diam.	2.000	Ea.	4.22	110	114.22
Stop, angle supply, chrome, 1/2″ CTS	2.000	Ea.	18.90	65	83.90
TOTAL			599.10	833.30	1,432.40

D2010 410	Kitchen Sink Systems	COST EACH		
		MAT.	INST.	TOTAL
1720	Kitchen sink w/trim, countertop, PE on CI, 24″x21″, single bowl (RD2010 -031)	600	835	1,435
1760	30″ x 21″ single bowl	875	835	1,710
1800	32″ x 21″ double bowl	705	900	1,605
1880	Stainless steel, 19″ x 18″ single bowl	940	835	1,775
1920	25″ x 22″ single bowl	1,025	835	1,860
1960	33″ x 22″ double bowl	1,375	900	2,275
2000	43″ x 22″ double bowl	1,550	910	2,460
2040	44″ x 22″ triple bowl	1,575	950	2,525
2080	44″ x 24″ corner double bowl	1,100	910	2,010
2120	Steel, enameled, 24″ x 21″ single bowl	855	835	1,690
2160	32″ x 21″ double bowl	870	900	1,770
2240	Raised deck, PE on CI, 32″ x 21″, dual level, double bowl	795	1,150	1,945
2280	42″ x 21″ dual level, triple bowl	1,650	1,250	2,900

D2010 Plumbing Fixtures

Systems are complete with trim and rough-in (supply, waste and vent) to connect to supply branches and waste mains.

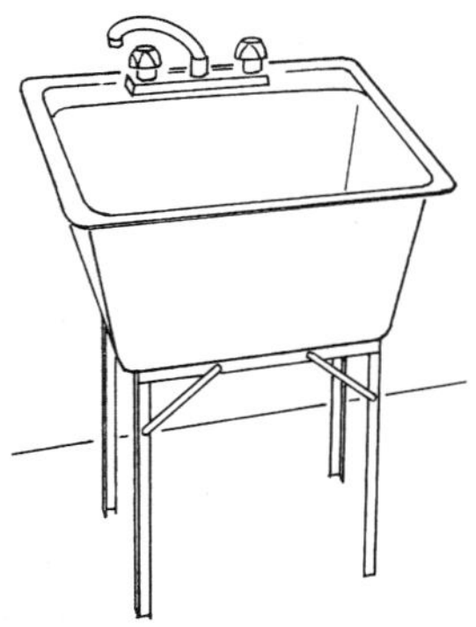

Single Compartment Sink

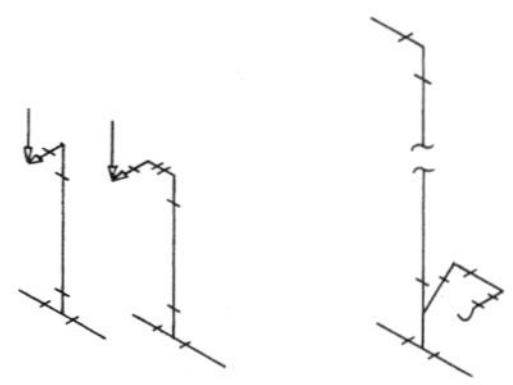

Supply Waste/Vent

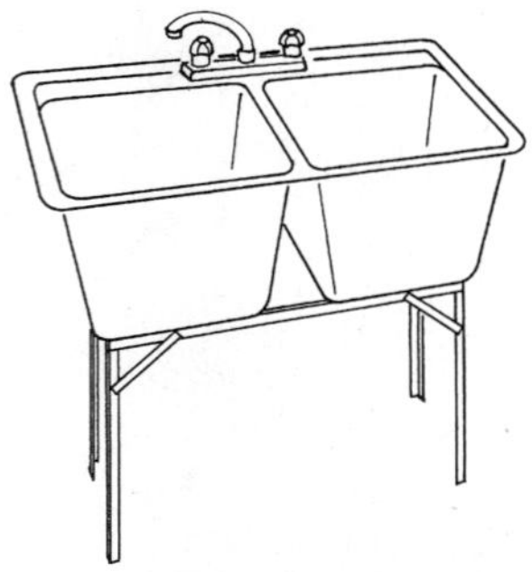

Double Compartment Sink

System Components	QUANTITY	UNIT	COST EACH MAT.	COST EACH INST.	COST EACH TOTAL
SYSTEM D2010 420 1760					
LAUNDRY SINK W/TRIM, PE ON CI, BLACK IRON FRAME					
24″ X 20″ OD, SINGLE COMPARTMENT					
Laundry sink PE on CI w/trim & frame, 24″ x 21″ OD, 1 compartment	1.000	Ea.	645	214	859
Pipe, steel, galvanized, schedule 40, threaded, 1-1/4″ diam	4.000	L.F.	25.60	57.80	83.40
Copper tubing, type DWV, solder joint, hanger 10′ OC 1-1/2″diam	6.000	L.F.	74.40	79.50	153.90
Wrought copper, DWV, Tee, sanitary, 1-1/2″ diam	1.000	Ea.	29	89.50	118.50
P trap, standard, copper, 1-1/2″ diam	1.000	Ea.	79	42	121
Copper tubing type L, solder joints, hangers 10′ OC, 1/2″ diam	10.000	L.F.	40.50	88.50	129
Wrought copper 90° elbow for solder joints 1/2″ diam	2.000	Ea.	2.48	71	73.48
Wrought copper Tee for solder joints, 1/2″ diam	2.000	Ea.	4.22	110	114.22
Stop, angle supply, 1/2″ diam	2.000	Ea.	18.90	65	83.90
TOTAL			919.10	817.30	1,736.40

D2010 420	Laundry Sink Systems		COST EACH MAT.	COST EACH INST.	COST EACH TOTAL
1740	Laundry sink w/trim, PE on CI, black iron frame	RD2010 -031			
1760	24″ x 20″, single compartment		920	815	1,735
1800	24″ x 23″ single compartment	RD2010 -400	950	815	1,765
1840	48″ x 21″ double compartment		1,175	885	2,060
1920	Molded stone, on wall, 22″ x 21″ single compartment		470	815	1,285
1960	45″x 21″ double compartment		675	885	1,560
2040	Plastic, on wall or legs, 18″ x 23″ single compartment		445	800	1,245
2080	20″ x 24″ single compartment		460	800	1,260
2120	36″ x 23″ double compartment		520	865	1,385
2160	40″ x 24″ double compartment		625	865	1,490

D2010 Plumbing Fixtures

Corrosion resistant laboratory sink systems are complete with trim and rough-in (supply, waste and vent) to connect to supply branches and waste mains.

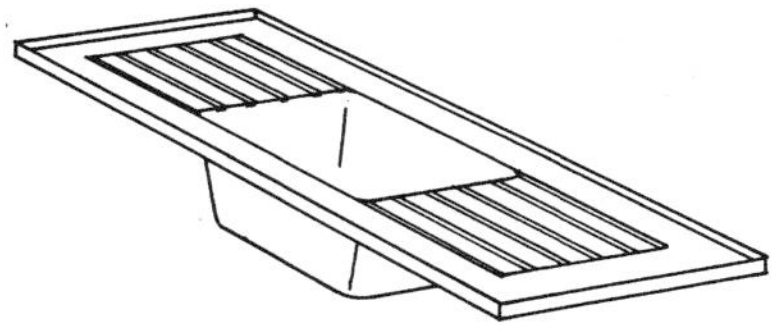

Laboratory Sink

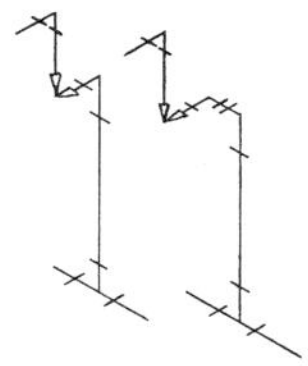

Supply

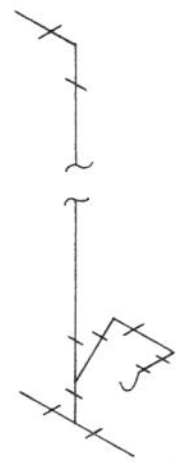

Waste/Vent

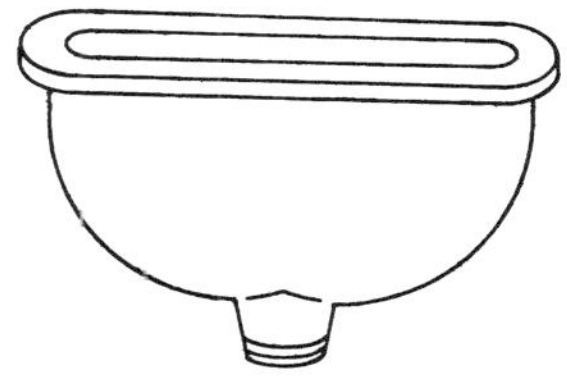

Polypropylene Cup Sink

System Components	QUANTITY	UNIT	COST EACH MAT.	COST EACH INST.	COST EACH TOTAL
SYSTEM D2010 430 1600					
LABORATORY SINK W/TRIM, STAINLESS STEEL, SINGLE BOWL					
DOUBLE DRAINBOARD, 54″ X 24″ O.D.					
Sink w/trim, stainless steel, 1 bowl, 2 drainboards 54″ x 24″ OD	1.000	Ea.	1,200	430	1,630
Pipe, polypropylene, schedule 40, acid resistant 1-1/2″ diam.	10.000	L.F.	90	189	279
Tee, sanitary, polypropylene, acid resistant, 1-1/2″ diam.	1.000	Ea.	20.50	71.50	92
P trap, polypropylene, acid resistant, 1-1/2″ diam.	1.000	Ea.	65	42	107
Copper tubing type L, solder joint, hanger 10′ O.C. 1/2″ diam.	10.000	L.F.	40.50	88.50	129
Wrought copper 90° elbow for solder joints 1/2″ diam.	2.000	Ea.	2.48	71	73.48
Wrought copper Tee for solder joints, 1/2″ diam.	2.000	Ea.	4.22	110	114.22
Stop, angle supply, chrome, 1/2″ diam.	2.000	Ea.	18.90	65	83.90
TOTAL			1,441.60	1,067	2,508.60

D2010 430	Laboratory Sink Systems		COST EACH MAT.	COST EACH INST.	COST EACH TOTAL
1580	Laboratory sink w/trim, stainless steel, single bowl,	RD2010 -031			
1590	Stainless steel, single bowl,				
1600	Double drainboard, 54″ x 24″ O.D.	RD2010 -400	1,450	1,075	2,525
1640	Single drainboard, 47″ x 24″O.D.		1,175	1,075	2,250
1670	Stainless steel, double bowl,				
1680	70″ x 24″ O.D.		1,575	1,075	2,650
1750	Polyethylene, single bowl,				
1760	Flanged, 14-1/2″ x 14-1/2″ O.D.		520	955	1,475
1800	18-1/2″ x 18-1/2″ O.D.		630	955	1,585
1840	23-1/2″ x 20-1/2″ O.D.		650	955	1,605
1920	Polypropylene, cup sink, oval, 7″ x 4″ O.D.		410	845	1,255
1960	10″ x 4-1/2″ O.D.		435	845	1,280

D2010 Plumbing Fixtures

Corrosion resistant laboratory sink systems are complete with trim and rough–in (supply, waste and vent) to connect to supply branches and waste mains.

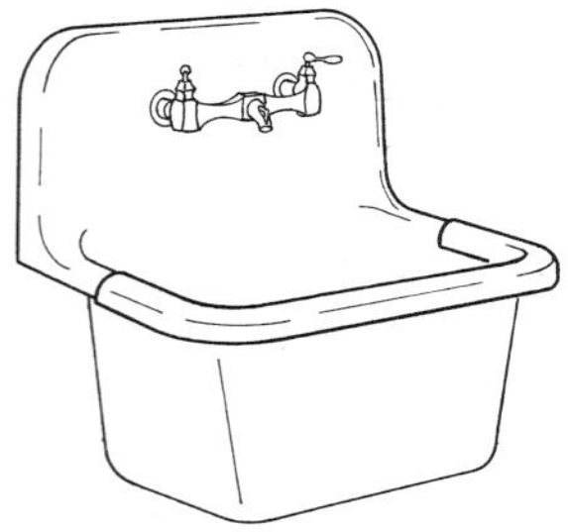

Wall Hung

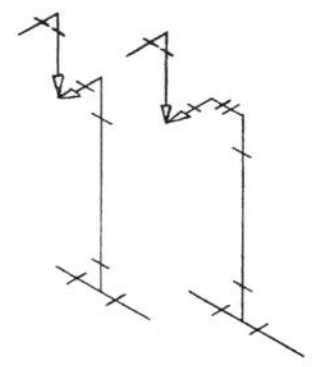

Supply

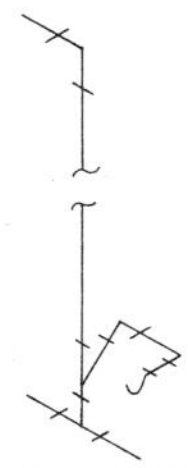

Waste/Vent

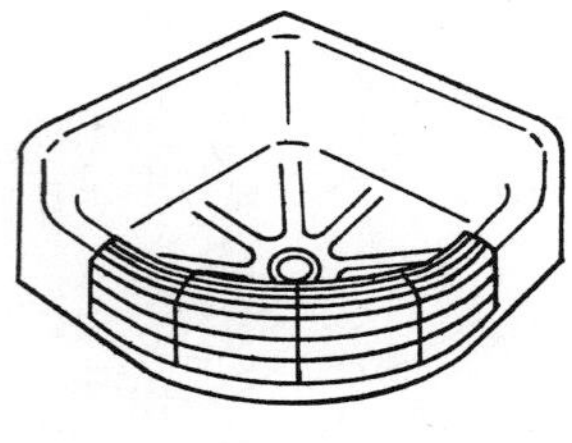

Corner, Floor

System Components	QUANTITY	UNIT	COST EACH		
			MAT.	INST.	TOTAL
SYSTEM D2010 440 4260					
SERVICE SINK, PE ON CI, CORNER FLOOR, 28"X28", W/RIM GUARD & TRIM					
Service sink, corner floor, PE on CI, 28" x 28", w/rim guard & trim	1.000	Ea.	1,125	292	1,417
Copper tubing type DWV, solder joint, hanger 10'OC 3" diam.	6.000	L.F.	174	132	306
Copper tubing type DWV, solder joint, hanger 10'OC 2" diam	4.000	L.F.	65	65	130
Wrought copper DWV, Tee, sanitary, 3" diam.	1.000	Ea.	151	184	335
P trap with cleanout & slip joint, copper 3" diam	1.000	Ea.	310	65	375
Copper tubing, type L, solder joints, hangers 10' OC, 1/2" diam	10.000	L.F.	40.50	88.50	129
Wrought copper 90° elbow for solder joints 1/2" diam	2.000	Ea.	2.48	71	73.48
Wrought copper Tee for solder joints, 1/2" diam	2.000	Ea.	4.22	110	114.22
Stop, angle supply, chrome, 1/2" diam	2.000	Ea.	18.90	65	83.90
TOTAL			1,891.10	1,072.50	2,963.60

D2010 440	Service Sink Systems	COST EACH		
		MAT.	INST.	TOTAL
4260	Service sink w/trim, PE on CI, corner floor, 28" x 28", w/rim guard (RD2010 -031)	1,900	1,075	2,975
4300	Wall hung w/rim guard, 22" x 18"	2,125	1,250	3,375
4340	24" x 20"	2,200	1,250	3,450
4380	Vitreous china, wall hung 22" x 20"	1,925	1,250	3,175

D2010 Plumbing Fixtures

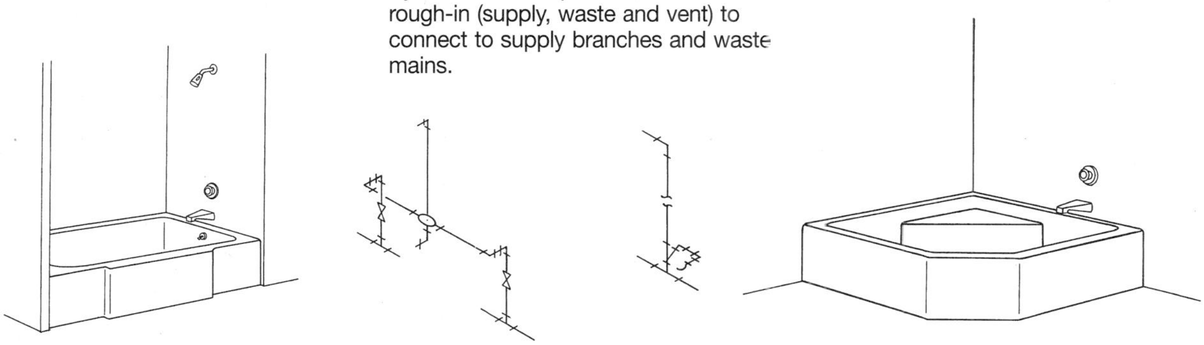

System Components	QUANTITY	UNIT	COST EACH MAT.	COST EACH INST.	COST EACH TOTAL
SYSTEM D2010 510 2000					
BATHTUB, RECESSED, PORCELAIN ENAMEL ON CAST IRON,, 48″ x 42″					
Bath tub, porcelain enamel on cast iron, w/fittings, 48″ x 42″	1.000	Ea.	2,925	320	3,245
Pipe, steel, galvanized, schedule 40, threaded, 1-1/4″ diam.	4.000	L.F.	25.60	57.80	83.40
Pipe, CI no hub soil w/couplings 10′ OC, hangers 5′ OC, 4″ diam.	3.000	L.F.	64.50	66	130.50
Combination Y and 1/8 bend for C.I. soil pipe, no hub, 4″ pipe size	1.000	Ea.	56		56
Drum trap, 3″ x 5″, copper, 1-1/2″ diam.	1.000	Ea.	117	44.50	161.50
Copper tubing type L, solder joints, hangers 10′ OC 1/2″ diam.	10.000	L.F.	40.50	88.50	129
Wrought copper 90° elbow, solder joints, 1/2″ diam.	2.000	Ea.	2.48	71	73.48
Wrought copper Tee, solder joints, 1/2″ diam.	2.000	Ea.	4.22	110	114.22
Stop, angle supply, 1/2″ diameter	2.000	Ea.	18.90	65	83.90
Copper tubing type DWV, solder joints, hanger 10′ OC 1-1/2″ diam.	3.000	L.F.	37.20	39.75	76.95
Pipe coupling, standard, C.I. soil no hub, 4″ pipe size	2.000	Ea.	44	78	122
TOTAL			3,335.40	940.55	4,275.95

D2010 510	Bathtub Systems		COST EACH MAT.	COST EACH INST.	COST EACH TOTAL
2000	Bathtub, recessed, P.E. on CI., 48″ x 42″	RD2010 -031	3,325	940	4,265
2040	72″ x 36″		3,425	1,050	4,475
2080	Mat bottom, 5′ long	RD2010 -400	1,650	915	2,565
2120	5′-6″ long		2,425	940	3,365
2160	Corner, 48″ x 42″		3,325	915	4,240
2200	Formed steel, enameled, 4′-6″ long		945	845	1,790

D2010 Plumbing Fixtures

Systems are complete with trim, flush valve and rough-in (supply, waste and vent) for connection to supply branches and waste mains.

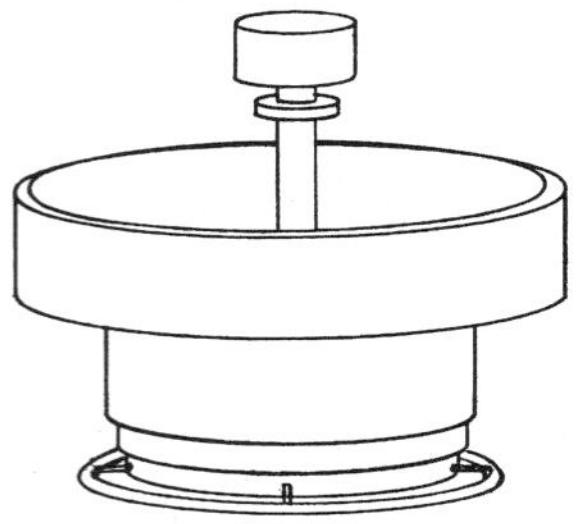

Circular Fountain

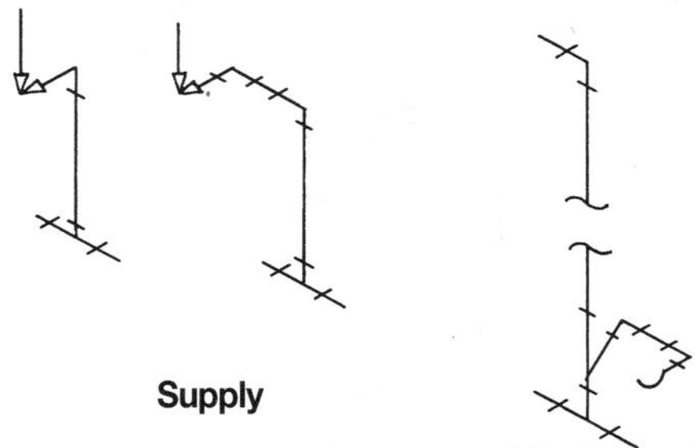

Supply

Waste/Vent

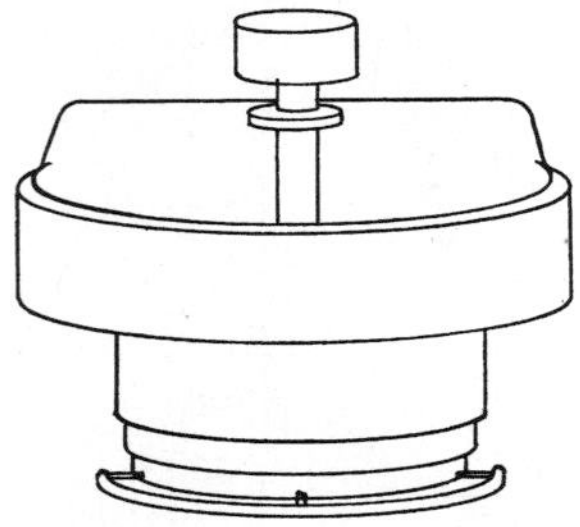

Semi-Circular Fountain

System Components	QUANTITY	UNIT	COST EACH MAT.	COST EACH INST.	COST EACH TOTAL
SYSTEM D2010 610 1760					
GROUP WASH FOUNTAIN, PRECAST TERRAZZO					
CIRCULAR, 36″ DIAMETER					
Wash fountain, group, precast terrazzo, foot control 36″ diam.	1.000	Ea.	8,075	665	8,740
Copper tubing type DWV, solder joint, hanger 10′ OC, 2″ diam.	10.000	L.F.	162.50	162.50	325
P trap, standard, copper, 2″ diam.	1.000	Ea.	122	47.50	169.50
Wrought copper, Tee, sanitary, 2″ diam.	1.000	Ea.	39.50	102	141.50
Copper tubing type L, solder joint, hanger 10′ OC 1/2″ diam.	20.000	L.F.	81	177	258
Wrought copper 90° elbow for solder joints 1/2″ diam.	3.000	Ea.	3.72	106.50	110.22
Wrought copper Tee for solder joints, 1/2″ diam.	2.000	Ea.	4.22	110	114.22
TOTAL			8,487.94	1,370.50	9,858.44

D2010 610	Group Wash Fountain Systems	COST EACH MAT.	COST EACH INST.	COST EACH TOTAL
1740	Group wash fountain, precast terrazzo (RD2010 -031)			
1760	Circular, 36″ diameter	8,500	1,375	9,875
1800	54″ diameter	10,500	1,500	12,000
1840	Semi-circular, 36″ diameter	7,500	1,375	8,875
1880	54″ diameter	9,950	1,500	11,450
1960	Stainless steel, circular, 36″ diameter	7,700	1,275	8,975
2000	54″ diameter	9,375	1,425	10,800
2040	Semi-circular, 36″ diameter	6,025	1,275	7,300
2080	54″ diameter	8,225	1,425	9,650
2160	Thermoplastic, circular, 36″ diameter	5,375	1,050	6,425
2200	54″ diameter	6,175	1,200	7,375
2240	Semi-circular, 36″ diameter	4,975	1,050	6,025
2280	54″ diameter	5,975	1,200	7,175

D2010 Plumbing Fixtures

Systems are complete with trim and rough-in (supply, waste and vent) for connection to supply branches and waste mains.

Three Wall

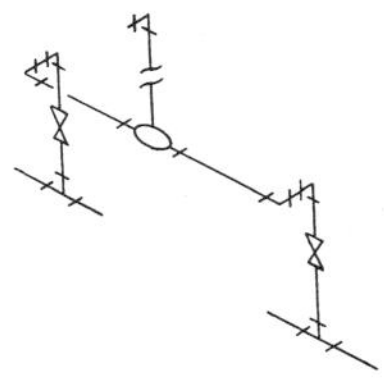

Supply

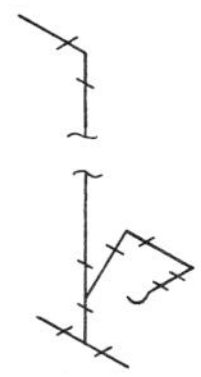

Waste/Vent

Corner Angle

System Components	QUANTITY	UNIT	COST EACH MAT.	COST EACH INST.	COST EACH TOTAL
SYSTEM D2010 710 1560					
SHOWER, STALL, BAKED ENAMEL, MOLDED STONE RECEPTOR, 30″ SQUARE					
Shower stall, enameled steel, molded stone receptor, 30″ square	1.000	Ea.	1,275	247	1,522
Copper tubing type DWV, solder joints, hangers 10′ OC, 2″ diam.	6.000	L.F.	74.40	79.50	153.90
Wrought copper DWV, Tee, sanitary, 2″ diam.	1.000	Ea.	29	89.50	118.50
Trap, standard, copper, 2″ diam.	1.000	Ea.	79	42	121
Copper tubing type L, solder joint, hanger 10′ OC 1/2″ diam.	16.000	L.F.	64.80	141.60	206.40
Wrought copper 90° elbow for solder joints 1/2″ diam.	3.000	Ea.	3.72	106.50	110.22
Wrought copper Tee for solder joints, 1/2″ diam.	2.000	Ea.	4.22	110	114.22
Stop and waste, straightway, bronze, solder joint 1/2″ diam.	2.000	Ea.	38.50	60	98.50
TOTAL			1,568.64	876.10	2,444.74

D2010 710	Shower Systems		COST EACH MAT.	COST EACH INST.	COST EACH TOTAL
1560	Shower, stall, baked enamel, molded stone receptor, 30″ square	RD2010 -031	1,575	875	2,450
1600	32″ square		1,575	885	2,460
1640	Terrazzo receptor, 32″ square	RD2010 -400	1,775	885	2,660
1680	36″ square		1,925	895	2,820
1720	36″ corner angle		2,200	895	3,095
1800	Fiberglass one piece, three walls, 32″ square		665	865	1,530
1840	36″ square		715	865	1,580
1880	Polypropylene, molded stone receptor, 30″ square		995	1,275	2,270
1920	32″ square		1,000	1,275	2,275
1960	Built-in head, arm, bypass, stops and handles		113	330	443
2050	Shower, stainless steel panels, handicap				
2100	w/fixed and handheld head, control valves, grab bar, and seat		4,000	3,950	7,950
2500	Shower, group with six heads, thermostatic mix valves & balancing valve		12,500	965	13,465
2520	Five heads		8,950	880	9,830

D2010 Plumbing Fixtures

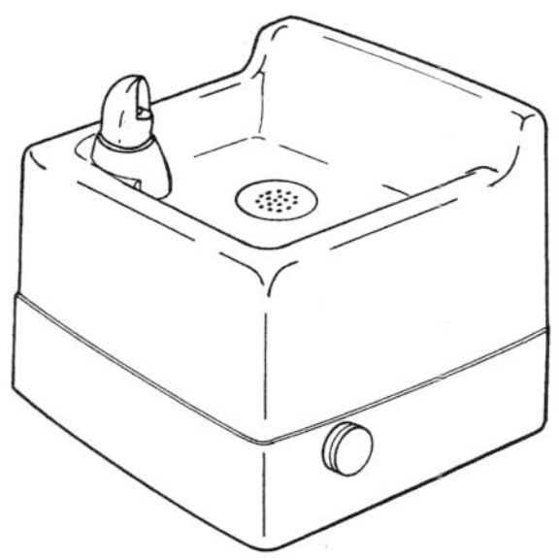

Wall Mounted, No Back

Systems are complete with trim and rough-in (supply, waste and vent) to connect to supply branches and waste mains.

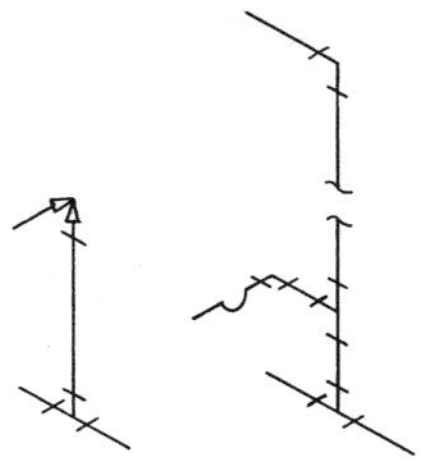

Supply Waste/Vent

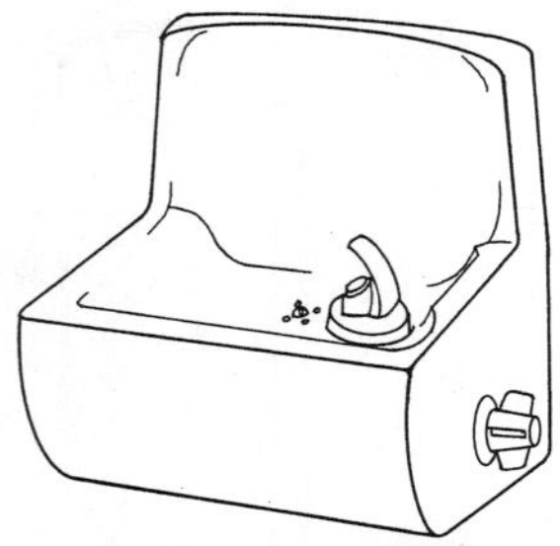

Wall Mounted, Low Back

System Components	QUANTITY	UNIT	COST EACH MAT.	COST EACH INST.	COST EACH TOTAL
SYSTEM D2010 810 1800					
DRINKING FOUNTAIN, ONE BUBBLER, WALL MOUNTED					
NON RECESSED, BRONZE, NO BACK					
Drinking fountain, wall mount, bronze, 1 bubbler	1.000	Ea.	1,100	179	1,279
Copper tubing, type L, solder joint, hanger 10′ OC 3/8″ diam.	5.000	L.F.	19.25	42.50	61.75
Stop, supply, straight, chrome, 3/8″ diam.	1.000	Ea.	19.25	30	49.25
Wrought copper 90° elbow for solder joints 3/8″ diam.	1.000	Ea.	3.42	32.50	35.92
Wrought copper Tee for solder joints, 3/8″ diam.	1.000	Ea.	5.80	51	56.80
Copper tubing, type DWV, solder joint, hanger 10′ OC 1-1/4″ diam.	4.000	L.F.	40.40	47.60	88
P trap, standard, copper drainage, 1-1/4″ diam.	1.000	Ea.	86	39.50	125.50
Wrought copper, DWV, Tee, sanitary, 1-1/4″ diam.	1.000	Ea.	23	79.50	102.50
TOTAL			1,297.12	501.60	1,798.72

D2010 810	Drinking Fountain Systems		COST EACH MAT.	COST EACH INST.	COST EACH TOTAL
1740	Drinking fountain, one bubbler, wall mounted	RD2010 -031			
1760	Non recessed				
1800	Bronze, no back	RD2010 -400	1,300	500	1,800
1840	Cast iron, enameled, low back		1,250	500	1,750
1880	Fiberglass, 12″ back		2,325	500	2,825
1920	Stainless steel, no back		1,375	500	1,875
1960	Semi-recessed, poly marble		1,300	500	1,800
2040	Stainless steel		1,675	500	2,175
2080	Vitreous china		1,225	500	1,725
2120	Full recessed, poly marble		2,025	500	2,525
2200	Stainless steel		1,875	500	2,375
2240	Floor mounted, pedestal type, aluminum		2,675	680	3,355
2320	Bronze		2,325	680	3,005
2360	Stainless steel		2,150	680	2,830

D2010 Plumbing Fixtures

Systems are complete with trim and rough-in (supply, waste and vent) for connection to supply branches and waste mains.

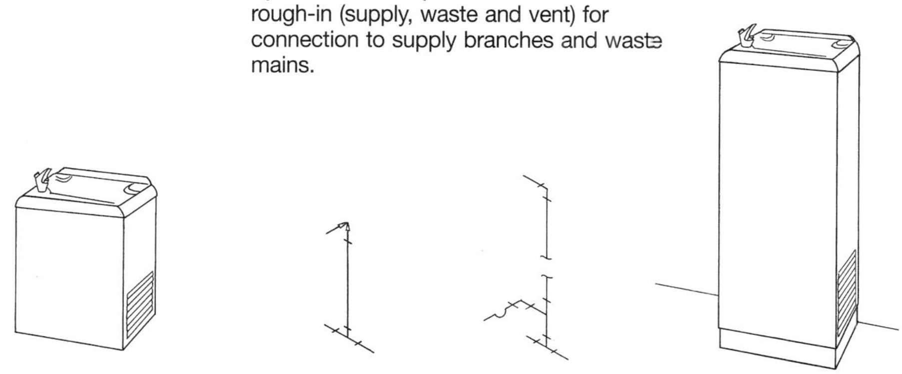

System Components	QUANTITY	UNIT	COST EACH MAT.	COST EACH INST.	COST EACH TOTAL
SYSTEM D2010 820 1840					
WATER COOLER, ELECTRIC, SELF CONTAINED, WALL HUNG, 8.2 G.P.H.					
Water cooler, wall mounted, 8.2 GPH	1.000	Ea.	890	320	1,210
Copper tubing type DWV, solder joint, hanger 10′ OC 1-1/4″ diam.	4.000	L.F.	40.40	47.60	88
Wrought copper DWV, Tee, sanitary 1-1/4″ diam.	1.000	Ea.	23	79.50	102.50
P trap, copper drainage, 1-1/4″ diam.	1.000	Ea.	86	39.50	125.50
Copper tubing type L, solder joint, hanger 10′ OC 3/8″ diam.	5.000	L.F.	19.25	42.50	61.75
Wrought copper 90° elbow for solder joints 3/8″ diam.	1.000	Ea.	3.42	32.50	35.92
Wrought copper Tee for solder joints, 3/8″ diam.	1.000	Ea.	5.80	51	56.80
Stop and waste, straightway, bronze, solder, 3/8″ diam.	1.000	Ea.	19.25	30	49.25
TOTAL			1,087.12	642.60	1,729.72

D2010 820	Water Cooler Systems		COST EACH MAT.	COST EACH INST.	COST EACH TOTAL
1840	Water cooler, electric, wall hung, 8.2 G.P.H.	RD2010 -031	1,075	645	1,720
1880	Dual height, 14.3 G.P.H.		1,375	665	2,040
1920	Wheelchair type, 7.5 G.P.H.	RD2010 -400	1,225	645	1,870
1960	Semi recessed, 8.1 G.P.H.		1,050	645	1,695
2000	Full recessed, 8 G.P.H.		2,550	695	3,245
2040	Floor mounted, 14.3 G.P.H.		1,225	560	1,785
2080	Dual height, 14.3 G.P.H.		1,575	680	2,255
2120	Refrigerated compartment type, 1.5 G.P.H.		1,875	560	2,435

D2010 Plumbing Fixtures

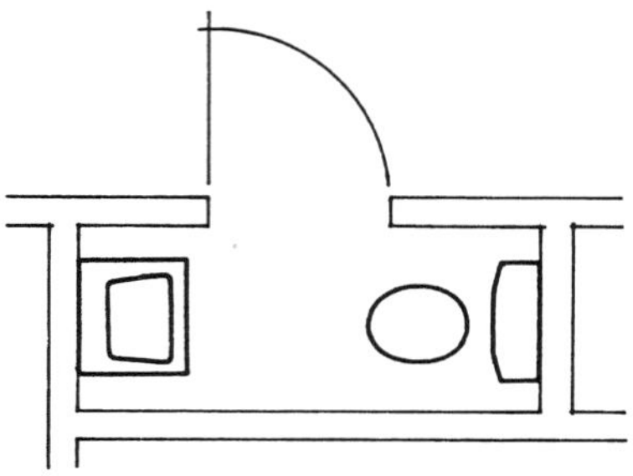

Two Fixture Bathroom Systems consisting of a lavatory, water closet, and rough-in service piping.

- Prices for plumbing and fixtures only.

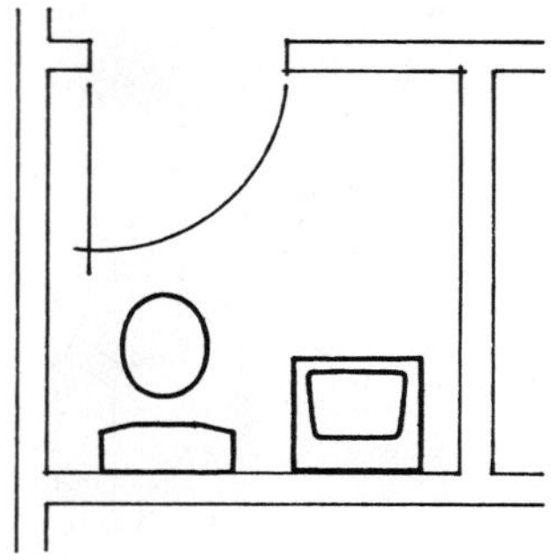

*Common wall is with an adjacent bathroom.

System Components	QUANTITY	UNIT	COST EACH MAT.	COST EACH INST.	COST EACH TOTAL
SYSTEM D2010 920 1180					
BATHROOM, LAVATORY & WATER CLOSET, 2 WALL PLUMBING, STAND ALONE					
Water closet, two piece, close coupled	1.000	Ea.	253	243	496
Water closet, rough-in waste & vent	1.000	Set	360	420	780
Lavatory w/ftngs., wall hung, white, PE on CI, 20″ x 18″	1.000	Ea.	291	161	452
Lavatory, rough-in waste & vent	1.000	Set	480	775	1,255
Copper tubing type L, solder joint, hanger 10′ OC 1/2″ diam.	10.000	L.F.	40.50	88.50	129
Pipe, steel, galvanized, schedule 40, threaded, 2″ diam.	12.000	L.F.	111.60	240	351.60
Pipe, CI soil, no hub, coupling 10′ OC, hanger 5′ OC, 4″ diam.	7.000	L.F.	150.50	164.50	315
TOTAL			1,686.60	2,092	3,778.60

D2010 920	Two Fixture Bathroom, Two Wall Plumbing		COST EACH MAT.	COST EACH INST.	COST EACH TOTAL
1180	Bathroom, lavatory & water closet, 2 wall plumbing, stand alone	RD2010 -031	1,675	2,100	3,775
1200	Share common plumbing wall*		1,550	1,800	3,350

D2010 922	Two Fixture Bathroom, One Wall Plumbing		COST EACH MAT.	COST EACH INST.	COST EACH TOTAL
2220	Bathroom, lavatory & water closet, one wall plumbing, stand alone	RD2010 -400	1,600	1,875	3,475
2240	Share common plumbing wall*		1,375	1,600	2,975

D2010 Plumbing Fixtures

Three Fixture Bathroom Systems consisting of a lavatory, water closet, bathtub or shower and rough-in service piping.

- Prices for plumbing and fixtures only.

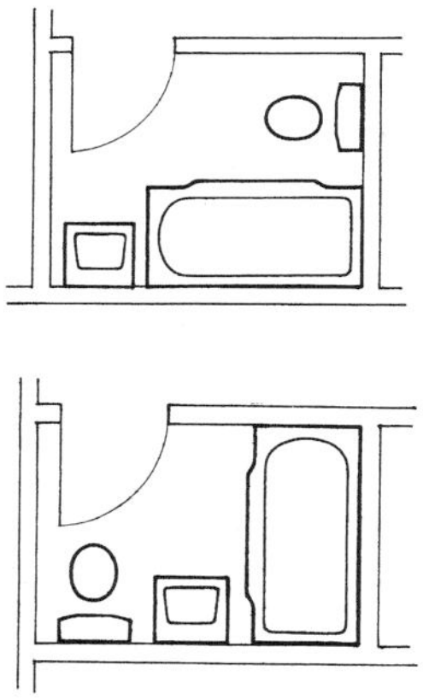

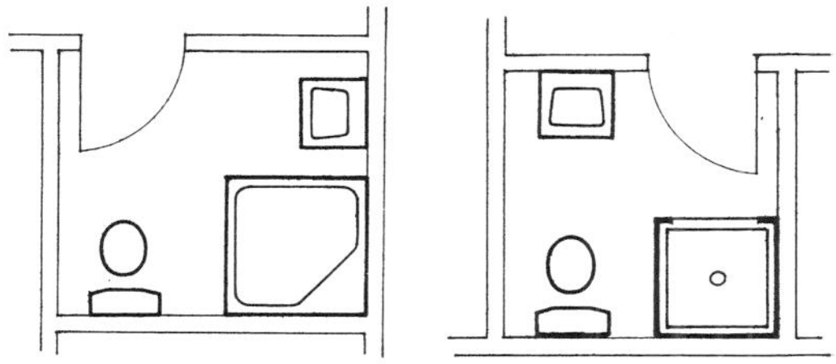

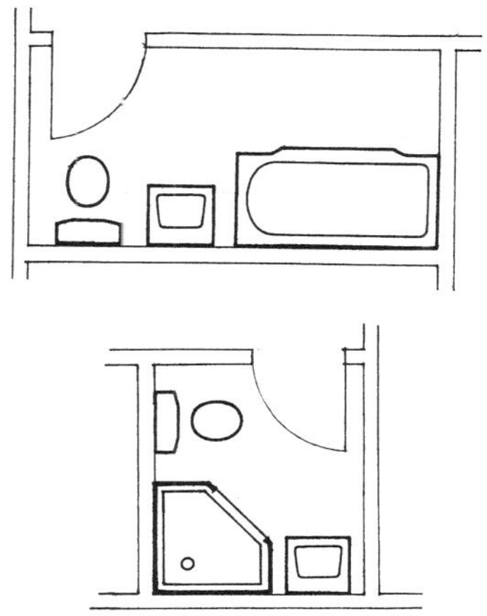

*Common wall is with an adjacent bathroom.

System Components	QUANTITY	UNIT	COST EACH MAT.	COST EACH INST.	COST EACH TOTAL
SYSTEM D2010 924 1170					
BATHROOM, LAVATORY, WATER CLOSET & BATHTUB					
ONE WALL PLUMBING, STAND ALONE					
Wtr closet, rough-in, supply, waste and vent	1.000	Set	360	420	780
Wtr closet, 2 pc close cpld vit china flr mntd w/seat supply & stop	1.000	Ea.	253	243	496
Lavatory w/ftngs, wall hung, white, PE on CI, 20" x 18"	1.000	Ea.	291	161	452
Lavatory, rough-in waste & vent	1.000	Set	480	775	1,255
Bathtub, white PE on CI, w/ftgs, mat bottom, recessed, 5' long	1.000	Ea.	1,250	292	1,542
Baths, rough-in waste and vent	1.000	Set	351	558	909
TOTAL			2,985	2,449	5,434

D2010 924	Three Fixture Bathroom, One Wall Plumbing	COST EACH MAT.	COST EACH INST.	COST EACH TOTAL
1150	Bathroom, three fixture, one wall plumbing (RD2010 -031)			
1160	Lavatory, water closet & bathtub			
1170	Stand alone	2,975	2,450	5,425
1180	Share common plumbing wall *	2,575	1,775	4,350

D2010 926	Three Fixture Bathroom, Two Wall Plumbing	COST EACH MAT.	COST EACH INST.	COST EACH TOTAL
2130	Bathroom, three fixture, two wall plumbing			
2140	Lavatory, water closet & bathtub			
2160	Stand alone	3,000	2,475	5,475
2180	Long plumbing wall common *	2,700	1,975	4,675
3610	Lavatory, bathtub & water closet			
3620	Stand alone	3,200	2,825	6,025
3640	Long plumbing wall common *	3,050	2,550	5,600
4660	Water closet, corner bathtub & lavatory			
4680	Stand alone	4,700	2,500	7,200
4700	Long plumbing wall common *	4,300	1,900	6,200
6100	Water closet, stall shower & lavatory			
6120	Stand alone	3,275	2,825	6,100
6140	Long plumbing wall common *	3,150	2,600	5,750
7060	Lavatory, corner stall shower & water closet			
7080	Stand alone	3,675	2,500	6,175
7100	Short plumbing wall common *	3,175	1,675	4,850

D2010 Plumbing Fixtures

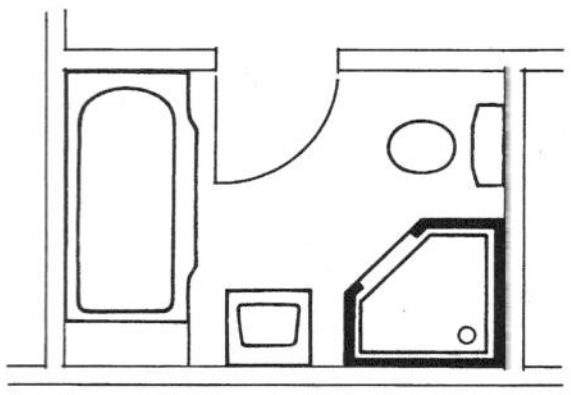

Four Fixture Bathroom Systems consisting of a lavatory, water closet, bathtub, shower and rough-in service piping.

- Prices for plumbing and fixtures only.

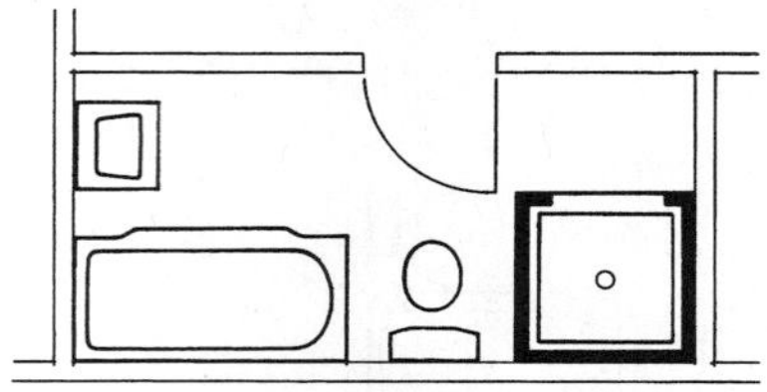

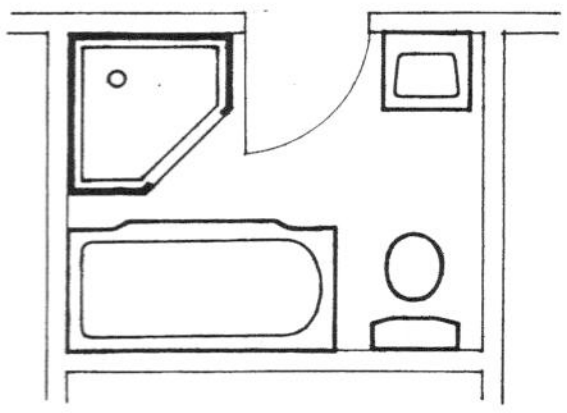

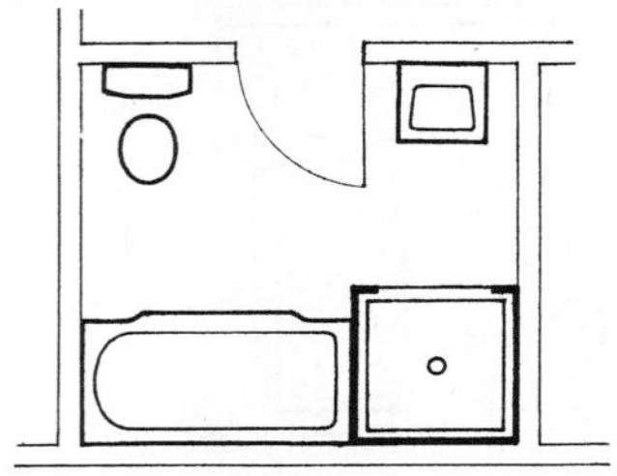

*Common wall is with an adjacent bathroom.

System Components	QUANTITY	UNIT	COST EACH MAT.	COST EACH INST.	COST EACH TOTAL
SYSTEM D2010 928 1160					
BATHROOM, BATHTUB, WATER CLOSET, STALL SHOWER & LAVATORY					
TWO WALL PLUMBING, STAND ALONE					
Wtr closet, 2 pc close cpld vit china flr mntd w/seat supply & stop	1.000	Ea.	253	243	496
Water closet, rough-in waste & vent	1.000	Set	360	420	780
Lavatory w/ftngs, wall hung, white PE on CI, 20" x 18"	1.000	Ea.	291	161	452
Lavatory, rough-in waste & vent	1.000	Set	48	77.50	125.50
Bathtub, white PE on CI, w/ftgs, mat bottom, recessed, 5' long	1.000	Ea.	1,250	292	1,542
Baths, rough-in waste and vent	1.000	Set	390	620	1,010
Shower stall, bkd enam, molded stone receptor, door & trim 32" sq.	1.000	Ea.	1,275	257	1,532
Shower stall, rough-in supply, waste & vent	1.000	Set	380	630	1,010
TOTAL			4,247	2,700.50	6,947.50

D2010 928	Four Fixture Bathroom, Two Wall Plumbing	COST EACH MAT.	COST EACH INST.	COST EACH TOTAL
1140	Bathroom, four fixture, two wall plumbing (RD2010 -031)			
1150	Bathtub, water closet, stall shower & lavatory			
1160	Stand alone	4,250	2,700	6,950
1180	Long plumbing wall common *	3,850	2,075	5,925
2260	Bathtub, lavatory, corner stall shower & water closet			
2280	Stand alone	4,875	2,700	7,575
2320	Long plumbing wall common *	4,475	2,100	6,575
3620	Bathtub, stall shower, lavatory & water closet			
3640	Stand alone	4,675	3,400	8,075
3660	Long plumbing wall (opp. door) common *	4,300	2,775	7,075

D2010 930	Four Fixture Bathroom, Three Wall Plumbing	COST EACH MAT.	COST EACH INST.	COST EACH TOTAL
4680	Bathroom, four fixture, three wall plumbing (RD2010 -400)			
4700	Bathtub, stall shower, lavatory & water closet			
4720	Stand alone	5,450	3,750	9,200
4760	Long plumbing wall (opposite door) common *	5,325	3,450	8,775

D2010 Plumbing Fixtures

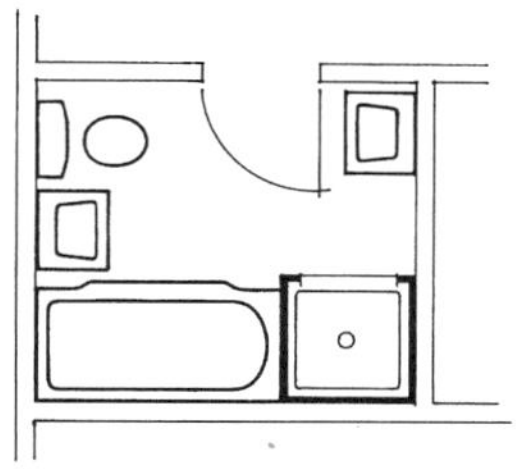

Five Fixture Bathroom Systems consisting of two lavatories, a water closet, bathtub, shower and rough-in service piping.

- Prices for plumbing and fixtures only.

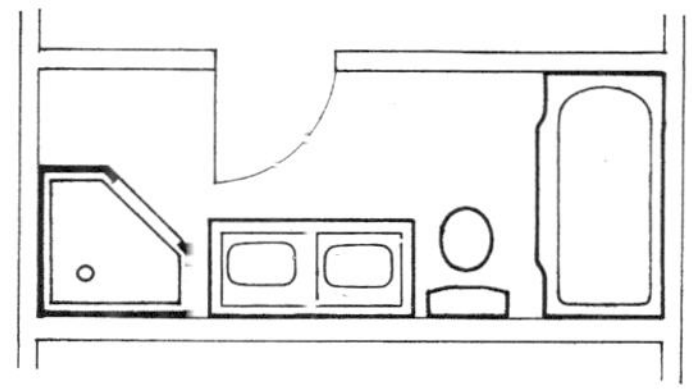

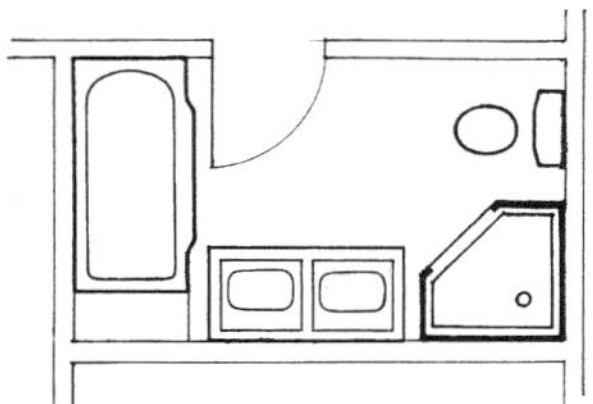

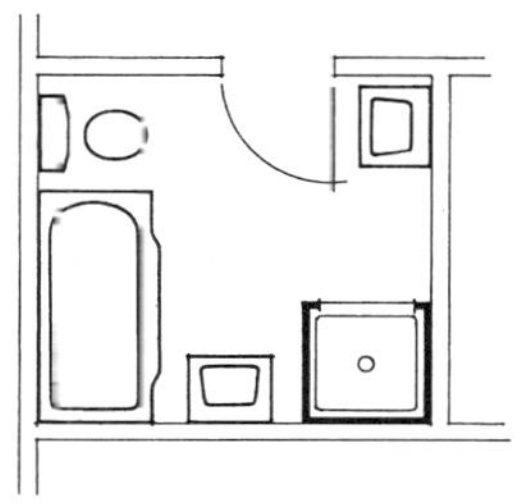

*Common wall is with an adjacent bathroom.

System Components	QUANTITY	UNIT	COST EACH MAT.	COST EACH INST.	COST EACH TOTAL
SYSTEM D2010 932 1360					
BATHROOM, BATHTUB, WATER CLOSET, STALL SHOWER & TWO LAVATORIES					
TWO WALL PLUMBING, STAND ALONE					
Wtr closet, 2 pc close cpld vit china flr mntd incl seat,supply & stop	1.000	Ea.	253	243	496
Water closet, rough-in waste & vent	1.000	Set	360	420	780
Lavatory w/ftngs, wall hung, white PE on CI, 20″ x 18″	2.000	Ea.	582	322	904
Lavatory, rough-in waste & vent	2.000	Set	960	1,550	2,510
Bathtub, white PE on CI, w/ftgs, mat bottom, recessed, 5′ long	1.000	Ea.	1,250	292	1,542
Baths, rough-in waste and vent	1.000	Set	390	620	1,010
Shower stall, bkd enam molded stone receptor, door & ftng, 32″ sq.	1.000	Ea.	1,275	257	1,532
Shower stall, rough-in supply, waste & vent	1.000	Set	380	630	1,010
TOTAL			5,450	4,334	9,784

D2010 932	Five Fixture Bathroom, Two Wall Plumbing		COST EACH MAT.	COST EACH INST.	COST EACH TOTAL
1320	Bathroom, five fixture, two wall plumbing	RD2010 -031			
1340	Bathtub, water closet, stall shower & two lavatories				
1360	Stand alone		5,450	4,325	9,775
1400	One short plumbing wall common *		5,050	3,725	8,775
1500	Bathtub, two lavatories, corner stall shower & water closet				
1520	Stand alone		6,075	4,350	10,425
1540	Long plumbing wall common*		5,475	3,375	8,850

D2010 934	Five Fixture Bathroom, Three Wall Plumbing		COST EACH MAT.	COST EACH INST.	COST EACH TOTAL
2360	Bathroom, five fixture, three wall plumbing	RD2010 -031			
2380	Water closet, bathtub, two lavatories & stall shower				
2400	Stand alone		6,075	4,350	10,425
2440	One short plumbing wall common *		5,675	3,725	9,400

D2010 936	Five Fixture Bathroom, One Wall Plumbing		COST EACH MAT.	COST EACH INST.	COST EACH TOTAL
4080	Bathroom, five fixture, one wall plumbing	RD2010 -400			
4100	Bathtub, two lavatories, corner stall shower & water closet				
4120	Stand alone		5,775	3,875	9,650
4160	Share common wall *		4,875	2,525	7,400

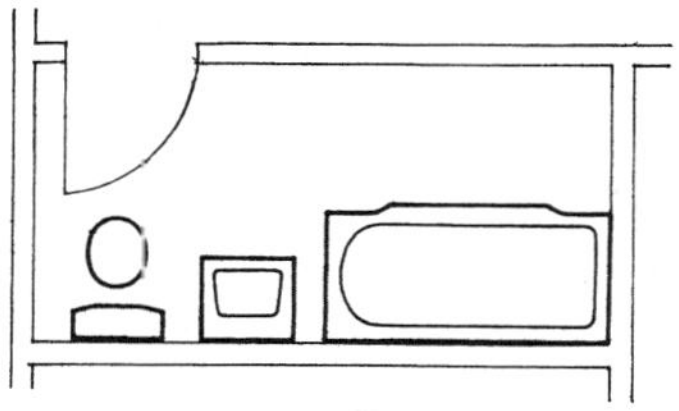

Example of Plumbing Cost Calculations: The bathroom system includes the individual fixtures such as bathtub, lavatory, shower and water closet. These fixtures are listed below as separate items merely as a checklist.

D2010 951 Plumbing Systems 20 Unit, 2 Story Apartment Building

	FIXTURE	SYSTEM	LINE	QUANTITY	UNIT	COST EACH		
						MAT.	INST.	TOTAL
0440	Bathroom	D2010 926	3640	20	Ea.	61,000	51,000	112,000
0480	Bathtub							
0520	Booster pump[1]	not req'd.						
0560	Drinking fountain							
0600	Garbage disposal[1]	not incl.						
0660								
0680	Grease interceptor							
0720	Water heater	D2020 250	2140	1	Ea.	13,500	3,475	16,975
0760	Kitchen sink	D2010 410	1960	20	Ea.	27,500	18,000	45,500
0800	Laundry sink	D2010 420	1840	4	Ea.	4,700	3,550	8,250
0840	Lavatory							
0900								
0920	Roof drain, 1 floor	D2040 210	4200	2	Ea.	2,125	1,850	3,975
0960	Roof drain, add'l floor	D2040 210	4240	20	L.F.	430	470	900
1000	Service sink	D2010 440	4300	1	Ea.	2,125	1,250	3,375
1040	Sewage ejector[1]	not req'd.						
1080	Shower							
1100								
1160	Sump pump							
1200	Urinal							
1240	Water closet							
1320								
1360	**SUB TOTAL**					111,000	79,500	190,500
1480	Water controls	RD2010-031		10%[2]		11,100	7,950	19,050
1520	Pipe & fittings[3]	RD2010-031		30%[2]		33,400	23,900	57,300
1560	Other							
1600	Quality/complexity	RD2010-031		15%[2]		16,700	11,900	28,600
1680								
1720	**TOTAL**					172,500	123,500	296,000
1740								

[1]**Note:** Cost for items such as booster pumps, backflow preventers, sewage ejectors, water meters, etc., may be obtained from RSMeans "Plumbing Cost Data" book, or be added as an increase in the Water Controls percentage. Water Controls, Pipe and Fittings, and the Quality/Complexity factors come from Table RD2010-031.

[2]Percentage of subtotal.

[3]Long easily discernable runs of pipe would be more accurately priced from Table D2090 810. If this is done, reduce the miscellaneous percentage in proportion.

D2020 Domestic Water Distribution

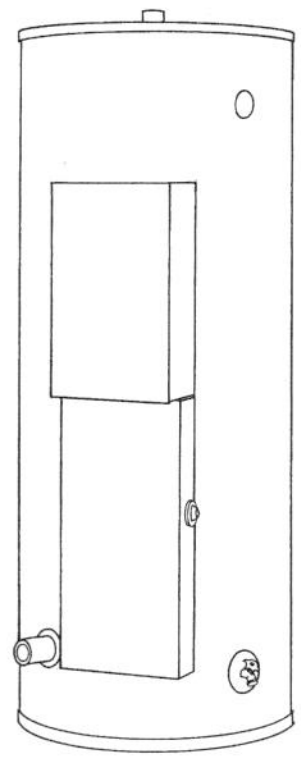

Systems below include piping and fittings within 10′ of heater. Electric water heaters do not require venting.

System Components	QUANTITY	UNIT	COST EACH MAT.	COST EACH INST.	COST EACH TOTAL
SYSTEM D2020 240 1820					
ELECTRIC WATER HEATER, COMMERCIAL, 100°F RISE					
50 GALLON TANK, 9 KW, 37 GPH					
Water heater, commercial, electric, 50 Gal, 9 KW, 37 GPH	1.000	Ea.	7,175	395	7,570
Copper tubing, type L, solder joint, hanger 10′ OC, 3/4″ diam	34.000	L.F.	195.50	319.60	515.10
Wrought copper 90° elbow for solder joints 3/4″ diam	5.000	Ea.	13.90	187.50	201.40
Wrought copper Tee for solder joints, 3/4″ diam	2.000	Ea.	10.20	119	129.20
Wrought copper union for soldered joints, 3/4″ diam.	2.000	Ea.	38.70	79	117.70
Valve, gate, bronze, 125 lb, NRS, soldered 3/4″ diam	2.000	Ea.	145	71	216
Relief valve, bronze, press & temp, self-close, 3/4″ IPS	1.000	Ea.	227	25.50	252.50
Wrought copper adapter, copper tubing to male, 3/4″ IPS	1.000	Ea.	4.36	42	46.36
TOTAL			7,809.66	1,238.60	9,048.26

D2020 240	Electric Water Heaters - Commercial Systems	COST EACH MAT.	COST EACH INST.	COST EACH TOTAL
1800	Electric water heater, commercial, 100°F rise (RD2020 -100)			
1820	50 gallon tank, 9 KW 37 GPH	7,800	1,250	9,050
1860	80 gal, 12 KW 49 GPH	9,900	1,525	11,425
1900	36 KW 147 GPH	13,500	1,650	15,150
1940	120 gal, 36 KW 147 GPH	14,600	1,800	16,400
1980	150 gal, 120 KW 490 GPH	45,100	1,925	47,025
2020	200 gal, 120 KW 490 GPH	47,500	1,975	49,475
2060	250 gal, 150 KW 615 GPH	53,000	2,300	55,300
2100	300 gal, 180 KW 738 GPH	57,000	2,425	59,425
2140	350 gal, 30 KW 123 GPH	42,500	2,625	45,125
2180	180 KW 738 GPH	59,000	2,625	61,625
2220	500 gal, 30 KW 123 GPH	55,500	3,075	58,575
2260	240 KW 984 GPH	89,500	3,075	92,575
2300	700 gal, 30 KW 123 GPH	67,000	3,500	70,500
2340	300 KW 1230 GPH	99,500	3,500	103,000
2380	1000 gal, 60 KW 245 GPH	81,000	4,900	85,900
2420	480 KW 1970 GPH	131,500	4,900	136,400
2460	1500 gal, 60 KW 245 GPH	118,500	6,050	124,550
2500	480 KW 1970 GPH	162,500	6,050	168,550

D2020 Domestic Water Distribution

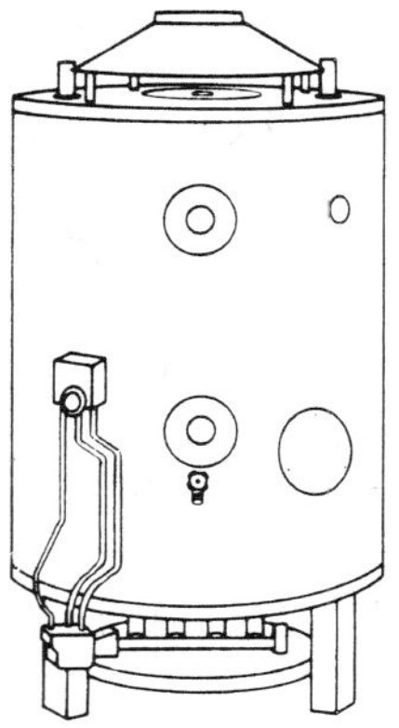

Units may be installed in multiples for increased capacity.

Included below is the heater with self-energizing gas controls, safety pilots, insulated jacket, hi-limit aquastat and pressure relief valve.

Installation includes piping and fittings within 10′ of heater. Gas heaters require vent piping (not included in these prices).

System Components	QUANTITY	UNIT	COST EACH MAT.	COST EACH INST.	COST EACH TOTAL
SYSTEM D2020 250 1780					
GAS FIRED WATER HEATER, COMMERCIAL, 100°F RISE					
75.5 MBH INPUT, 63 GPH					
Water heater, commercial, gas, 75.5 MBH, 63 GPH	1.000	Ea.	3,775	510	4,285
Copper tubing, type L, solder joint, hanger 10′ OC, 1-1/4″ diam	30.000	L.F.	331.50	369	700.50
Wrought copper 90° elbow for solder joints 1-1/4″ diam	4.000	Ea.	40.40	190	230.40
Wrought copper tee for solder joints, 1-1/4″ diam	2.000	Ea.	42	159	201
Wrought copper union for soldered joints, 1-1/4″ diam	2.000	Ea.	109	102	211
Valve, gate, bronze, 125 lb, NRS, soldered 1-1/4″ diam	2.000	Ea.	296	95	391
Relief valve, bronze, press & temp, self-close, 3/4″ IPS	1.000	Ea.	227	25.50	252.50
Copper tubing, type L, solder joints, 3/4″ diam	8.000	L.F.	46	75.20	121.20
Wrought copper 90° elbow for solder joints 3/4″ diam	1.000	Ea.	2.78	37.50	40.28
Wrought copper, adapter, CTS to MPT, 3/4″ IPS	1.000	Ea.	4.36	42	46.36
Pipe steel black, schedule 40, threaded, 3/4″ diam	10.000	L.F.	36.80	117	153.80
Pipe, 90° elbow, malleable iron black, 150 lb threaded, 3/4″ diam	2.000	Ea.	9.18	102	111.18
Pipe, union with brass seat, malleable iron black, 3/4″ diam	1.000	Ea.	20.50	55	75.50
Valve, gas stop w/o check, brass, 3/4″ IPS	1.000	Ea.	17.05	32.50	49.55
TOTAL			4,957.57	1,911.70	6,869.27

D2020 250	Gas Fired Water Heaters - Commercial Systems	COST EACH MAT.	COST EACH INST.	COST EACH TOTAL
1760	Gas fired water heater, commercial, 100°F rise			
1780	75.5 MBH input, 63 GPH	4,950	1,900	6,850
1820	95 MBH input, 86 GPH	8,550	1,900	10,450
1860	100 MBH input, 91 GPH (RD2020 -100)	8,800	2,000	10,800
1900	115 MBH input, 110 GPH	8,850	2,050	10,900
1980	155 MBH input, 150 GPH	11,100	2,325	13,425
2020	175 MBH input, 168 GPH	11,400	2,475	13,875
2060	200 MBH input, 192 GPH	11,700	2,800	14,500
2100	240 MBH input, 230 GPH	12,400	3,025	15,425
2140	300 MBH input, 278 GPH	13,500	3,475	16,975
2180	390 MBH input, 374 GPH	16,000	3,500	19,500
2220	500 MBH input, 480 GPH	21,600	3,800	25,400
2260	600 MBH input, 576 GPH	24,900	4,100	29,000

D2020 Domestic Water Distribution

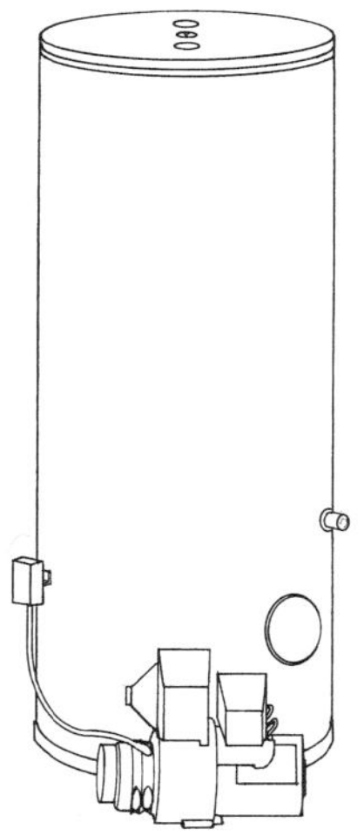

Units may be installed in multiples for increased capacity.

Included below is the heater, wired-in flame retention burners, cadmium cell primary controls, hi-limit controls, ASME pressure relief valves, draft controls, and insulated jacket.

Oil fired water heater systems include piping and fittings within 10′ of heater. Oil fired heaters require vent piping (not included in these systems).

System Components	QUANTITY	UNIT	COST EACH MAT.	INST.	TOTAL
SYSTEM D2020 260 1820					
OIL FIRED WATER HEATER, COMMERCIAL, 100°F RISE					
140 GAL., 140 MBH INPUT, 134 GPH					
Water heater, commercial, oil, 140 gal., 140 MBH input, 134 GPH	1.000	Ea.	25,700	605	26,305
Copper tubing, type L, solder joint, hanger 10′ OC, 3/4″ diam.	34.000	L.F.	195.50	319.60	515.10
Wrought copper 90° elbow for solder joints 3/4″ diam.	5.000	Ea.	13.90	187.50	201.40
Wrought copper Tee for solder joints, 3/4″ diam.	2.000	Ea.	10.20	119	129.20
Wrought copper union for soldered joints, 3/4″ diam.	2.000	Ea.	38.70	79	117.70
Valve, bronze, 125 lb, NRS, soldered 3/4″ diam.	2.000	Ea.	145	71	216
Relief valve, bronze, press & temp, self-close, 3/4″ IPS	1.000	Ea.	227	25.50	252.50
Wrought copper adapter, copper tubing to male, 3/4″ IPS	1.000	Ea.	4.36	42	46.36
Copper tubing, type L, solder joint, hanger 10′ OC, 3/8″ diam.	10.000	L.F.	38.50	85	123.50
Wrought copper 90° elbow for solder joints 3/8″ diam.	2.000	Ea.	6.84	65	71.84
Valve, globe, fusible, 3/8″ IPS	1.000	Ea.	14.40	30.50	44.90
TOTAL			26,394.40	1,629.10	28,023.50

D2020 260	Oil Fired Water Heaters - Commercial Systems	COST EACH MAT.	INST.	TOTAL
1800	Oil fired water heater, commercial, 100°F rise			
1820	140 gal., 140 MBH input, 134 GPH	26,400	1,625	28,025
1900	140 gal., 255 MBH input, 247 GPH	28,400	2,050	30,450
1940	140 gal., 270 MBH input, 259 GPH	35,000	2,325	37,325
1980	140 gal., 400 MBH input, 384 GPH	36,100	2,675	38,775
2060	140 gal., 720 MBH input, 691 GPH	38,400	2,800	41,200
2100	221 gal., 300 MBH input, 288 GPH	51,000	3,075	54,075
2140	221 gal., 600 MBH input, 576 GPH	56,500	3,125	59,625
2180	221 gal., 800 MBH input, 768 GPH	57,500	3,250	60,750
2220	201 gal., 1000 MBH input, 960 GPH	58,500	3,275	61,775
2260	201 gal., 1250 MBH input, 1200 GPH	59,500	3,350	62,850
2300	201 gal., 1500 MBH input, 1441 GPH	64,500	3,425	67,925
2340	411 gal., 600 MBH input, 576 GPH	65,000	3,500	68,500
2380	411 gal., 800 MBH input, 768 GPH	68,000	3,600	71,600
2420	411 gal., 1000 MBH input, 960 GPH	71,000	4,150	75,150
2460	411 gal., 1250 MBH input, 1200 GPH	72,500	4,275	76,775
2500	397 gal., 1500 MBH input, 1441 GPH	77,000	4,400	81,400
2540	397 gal., 1750 MBH input, 1681 GPH	78,500	4,550	83,050

RD2020 -100

D2020 Domestic Water Distribution

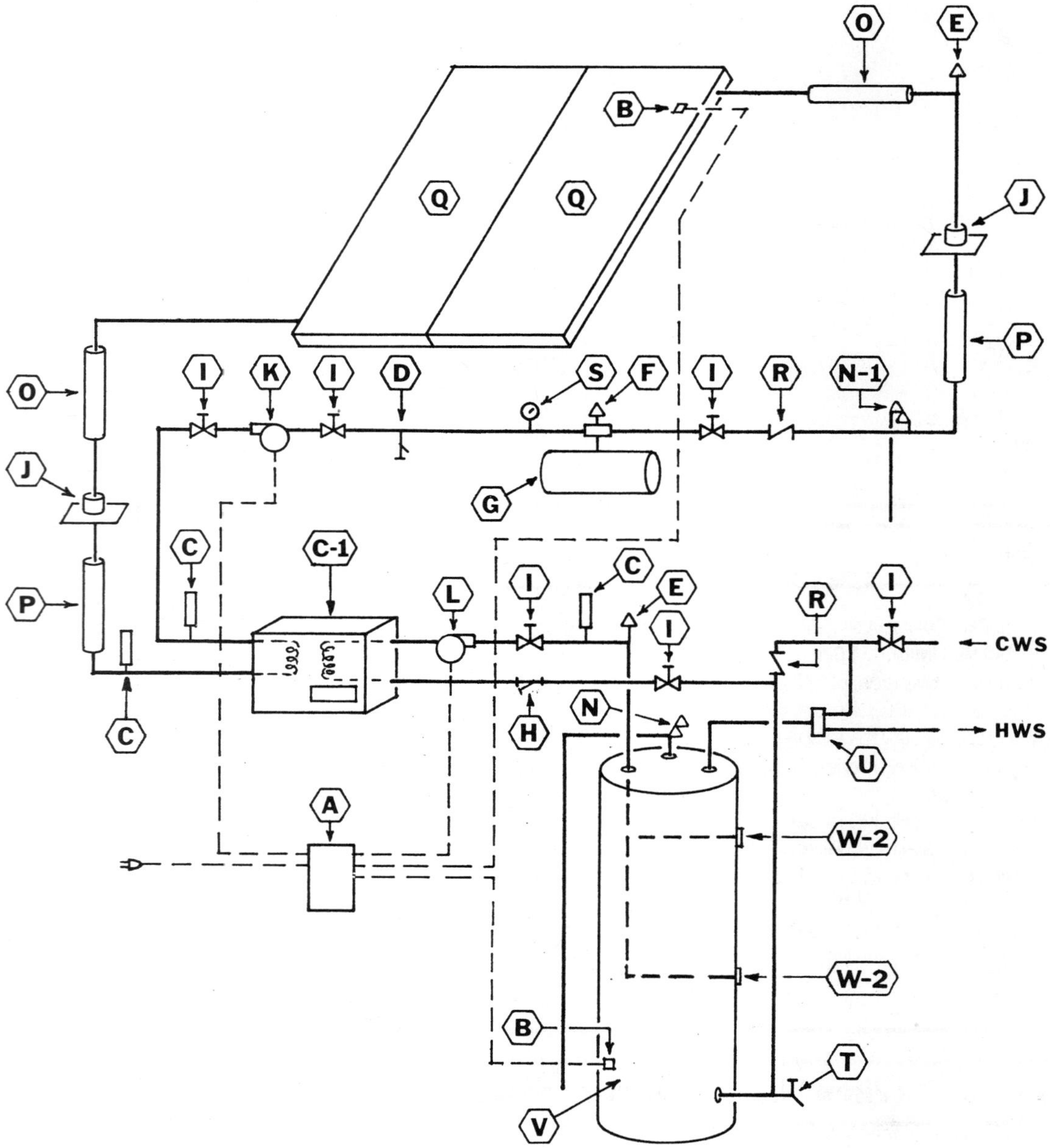

In this closed-loop indirect collection system, fluid with a low freezing temperature, such as propylene glycol, transports heat from the collectors to water storage. The transfer fluid is contained in a closed-loop consisting of collectors, supply and return piping, and a remote heat exchanger. The heat exchanger transfers heat energy from the fluid in the collector loop to potable water circulated in a storage loop. A typical two-or-three panel system contains 5 to 6 gallons of heat transfer fluid.

When the collectors become approximately 20°F warmer than the storage temperature, a controller activates the circulator on the collector and storage loops. The circulators will move the fluid and potable water through the heat exchanger until heat collection no longer occurs. At that point, the system shuts down. Since the heat transfer medium is a fluid with a very low freezing temperature, there is no need for it to be drained from the system between periods of collection.

D2020 Domestic Water Distribution

System Components		QUANTITY	UNIT	COST EACH MAT.	INST.	TOTAL
	SYSTEM D2020 265 2760					
	SOLAR, CLOSED LOOP, ADD-ON HOT WATER SYS., EXTERNAL HEAT EXCHANGER					
	3/4" TUBING, TWO 3'X7' BLACK CHROME COLLECTORS					
A,B,G,L,K,M	Heat exchanger fluid-fluid pkg incl 2 circulators, expansion tank, Check valve, relief valve, controller, hi temp cutoff, & 2 sensors	1.000	Ea.	880	515	1,395
C	Thermometer, 2" dial	3.000	Ea.	72	133.50	205.50
D, T	Fill & drain valve, brass, 3/4" connection	1.000	Ea.	11.25	30	41.25
E	Air vent, manual, 1/8" fitting	2.000	Ea.	5.36	45	50.36
F	Air purger	1.000	Ea.	51.50	59.50	111
H	Strainer, Y type, bronze body, 3/4" IPS	1.000	Ea.	45.50	38.50	84
I	Valve, gate, bronze, NRS, soldered 3/4" diam	6.000	Ea.	435	213	648
J	Neoprene vent flashing	2.000	Ea.	11	71	82
N-1, N	Relief valve temp & press, 150 psi 210°F self-closing 3/4" IPS	1.000	Ea.	24.50	24	48.50
O	Pipe covering, urethane, ultraviolet cover, 1" wall 3/4" diam	20.000	L.F.	62.20	130	192.20
P	Pipe covering, fiberglass, all service jacket, 1" wall, 3/4" diam	50.000	L.F.	49.50	260	309.50
Q	Collector panel solar energy blk chrome on copper, 1/8" temp glass 3'x7'	2.000	Ea.	2,200	270	2,470
	Roof clamps for solar energy collector panels	2.000	Set	6.22	36.80	43.02
R	Valve, swing check, bronze, regrinding disc, 3/4" diam	2.000	Ea.	232	71	303
S	Pressure gauge, 60 psi, 2" dial	1.000	Ea.	26.50	22.50	49
U	Valve, water tempering, bronze, sweat connections, 3/4" diam	1.000	Ea.	150	35.50	185.50
W-2, V	Tank water storage w/heating element, drain, relief valve, existing	1.000	Ea.			
	Copper tubing type L, solder joint, hanger 10' OC 3/4" diam	20.000	L.F.	115	188	303
	Copper tubing, type M, solder joint, hanger 10' OC 3/4" diam	70.000	L.F.	337.40	640.50	977.90
	Sensor wire, #22-2 conductor multistranded	.500	C.L.F.	7.83	33	40.83
	Solar energy heat transfer fluid, propylene glycol anti-freeze	6.000	Gal.	99	153	252
	Wrought copper fittings & solder, 3/4" diam	76.000	Ea.	211.28	2,850	3,061.28
	TOTAL			5,033.04	5,819.80	10,852.84

D2020 265	Solar, Closed Loop, Add-On Hot Water Systems	COST EACH MAT.	INST.	TOTAL
2550	Solar, closed loop, add-on hot water system, external heat exchanger			
2700	3/4" tubing, 3 ea. 3'x7' black chrome collectors	6,125	5,975	12,100
2720	3 ea. 3'x7' flat black absorber plate collectors RD3010 -600	5,725	6,000	11,725
2740	2 ea. 4'x9' flat black w/plastic glazing collectors	5,125	6,025	11,150
2760	2 ea. 3'x7' black chrome collectors	5,025	5,825	10,850
2780	1" tubing, 4 ea 2'x9' plastic absorber & glazing collectors	6,150	6,775	12,925
2800	4 ea. 3'x7' black chrome absorber collectors	8,000	6,800	14,800
2820	4 ea. 3'x7' flat black absorber collectors	7,450	6,825	14,275

D2020 Domestic Water Distribution

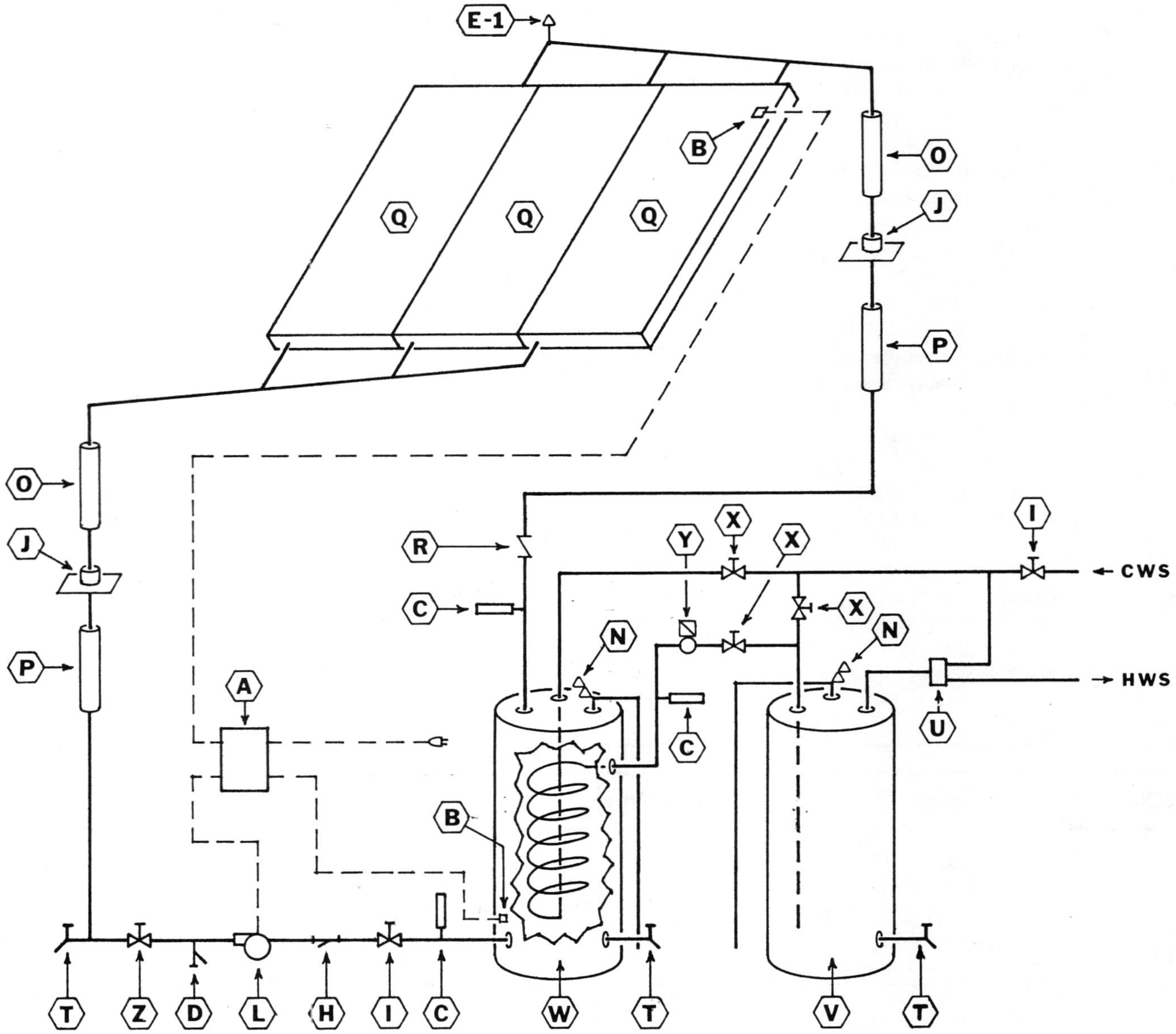

In the drainback indirect-collection system, the heat transfer fluid is distilled water contained in a loop consisting of collectors, supply and return piping, and an unpressurized holding tank. A large heat exchanger containing incoming potable water is immersed in the holding tank. When a controller activates solar collection, the distilled water is pumped through the collectors and heated and pumped back down to the holding tank. When the temperature differential between the water in the collectors and water in storage is such that collection no longer occurs, the pump turns off and gravity causes the distilled water in the collector loop to drain back to the holding tank. All the loop piping is pitched so that the water can drain out of the collectors and piping and not freeze there. As hot water is needed in the home, incoming water first flows through the holding tank with the immersed heat exchanger and is warmed and then flows through a conventional heater for any supplemental heating that is necessary.

D2020 Domestic Water Distribution

System Components

System Components	QUANTITY	UNIT	COST EACH MAT.	COST EACH INST.	COST EACH TOTAL
SYSTEM D2020 270 2760					
SOLAR, DRAINBACK, ADD ON, HOT WATER, IMMERSED HEAT EXCHANGER					
3/4" TUBING, THREE EA 3'X7' BLACK CHROME COLLECTOR					
A, B Differential controller 2 sensors, thermostat, solar energy system	1.000	Ea.	143	59.50	202.50
C Thermometer 2" dial	3.000	Ea.	72	133.50	205.50
D, T Fill & drain valve, brass, 3/4" connection	1.000	Ea.	11.25	30	41.25
E-1 Automatic air vent 1/8" fitting	1.000	Ea.	18.95	22.50	41.45
H Strainer, Y type, bronze body, 3/4" IPS	1.000	Ea.	45.50	38.50	84
I Valve, gate, bronze, NRS, soldered 3/4" diam	2.000	Ea.	145	71	216
J Neoprene vent flashing	2.000	Ea.	11	71	82
L Circulator, solar heated liquid, 1/20 HP	1.000	Ea.	289	107	396
N Relief valve temp. & press. 150 psi 210°F self-closing 3/4" IPS	1.000	Ea.	24.50	24	48.50
O Pipe covering, urethane, ultraviolet cover, 1" wall, 3/4" diam	20.000	L.F.	62.20	130	192.20
P Pipe covering, fiberglass, all service jacket, 1" wall, 3/4" diam	50.000	L.F.	49.50	260	309.50
Q Collector panel solar energy blk chrome on copper, 1/8" temp glas 3'x7'	3.000	Ea.	3,300	405	3,705
Roof clamps for solar energy collector panels	3.000	Set	9.33	55.20	64.53
R Valve, swing check, bronze, regrinding disc, 3/4" diam	1.000	Ea.	116	35.50	151.50
U Valve, water tempering, bronze sweat connections, 3/4" diam	1.000	Ea.	150	35.50	185.50
V Tank, water storage w/heating element, drain, relief valve, existing	1.000	Ea.			
W Tank, water storage immersed heat exchr elec 2"x1/2# insul 120 gal	1.000	Ea.	2,025	510	2,535
X Valve, globe, bronze, rising stem, 3/4" diam, soldered	3.000	Ea.	342	106.50	448.50
Y Flow control valve	1.000	Ea.	143	32.50	175.50
Z Valve, ball, bronze, solder 3/4" diam, solar loop flow control	1.000	Ea.	25.50	35.50	61
Copper tubing, type L, solder joint, hanger 10' OC 3/4" diam	20.000	L.F.	115	188	303
Copper tubing, type M, solder joint, hanger 10' OC 3/4" diam	70.000	L.F.	337.40	640.50	977.90
Sensor wire, #22-2 conductor, multistranded	.500	C.L.F.	7.83	33	40.83
Wrought copper fittings & solder, 3/4" diam	76.000	Ea.	211.28	2,850	3,061.28
TOTAL			7,654.24	5,874.20	13,528.44

D2020 270	Solar, Drainback, Hot Water Systems		COST EACH MAT.	COST EACH INST.	COST EACH TOTAL
2550	Solar, drainback, hot water, immersed heat exchanger				
2560	3/8" tubing, 3 ea. 4' x 4'-4" vacuum tube collectors		6,875	5,300	12,175
2760	3/4" tubing, 3 ea 3'x7' black chrome collectors, 120 gal tank	RD3010 -600	7,650	5,875	13,525
2780	3 ea. 3'x7' flat black absorber collectors, 120 gal tank		7,250	5,900	13,150
2800	2 ea. 4'x9' flat blk w/plastic glazing collectors 120 gal tank		6,650	5,925	12,575
2840	1" tubing, 4 ea. 2'x9' plastic absorber & glazing collectors, 120 gal tank		7,750	6,675	14,425
2860	4 ea. 3'x7' black chrome absorber collectors, 120 gal tank		9,600	6,700	16,300
2880	4 ea. 3'x7' flat black absorber collectors, 120 gal tank		9,050	6,725	15,775

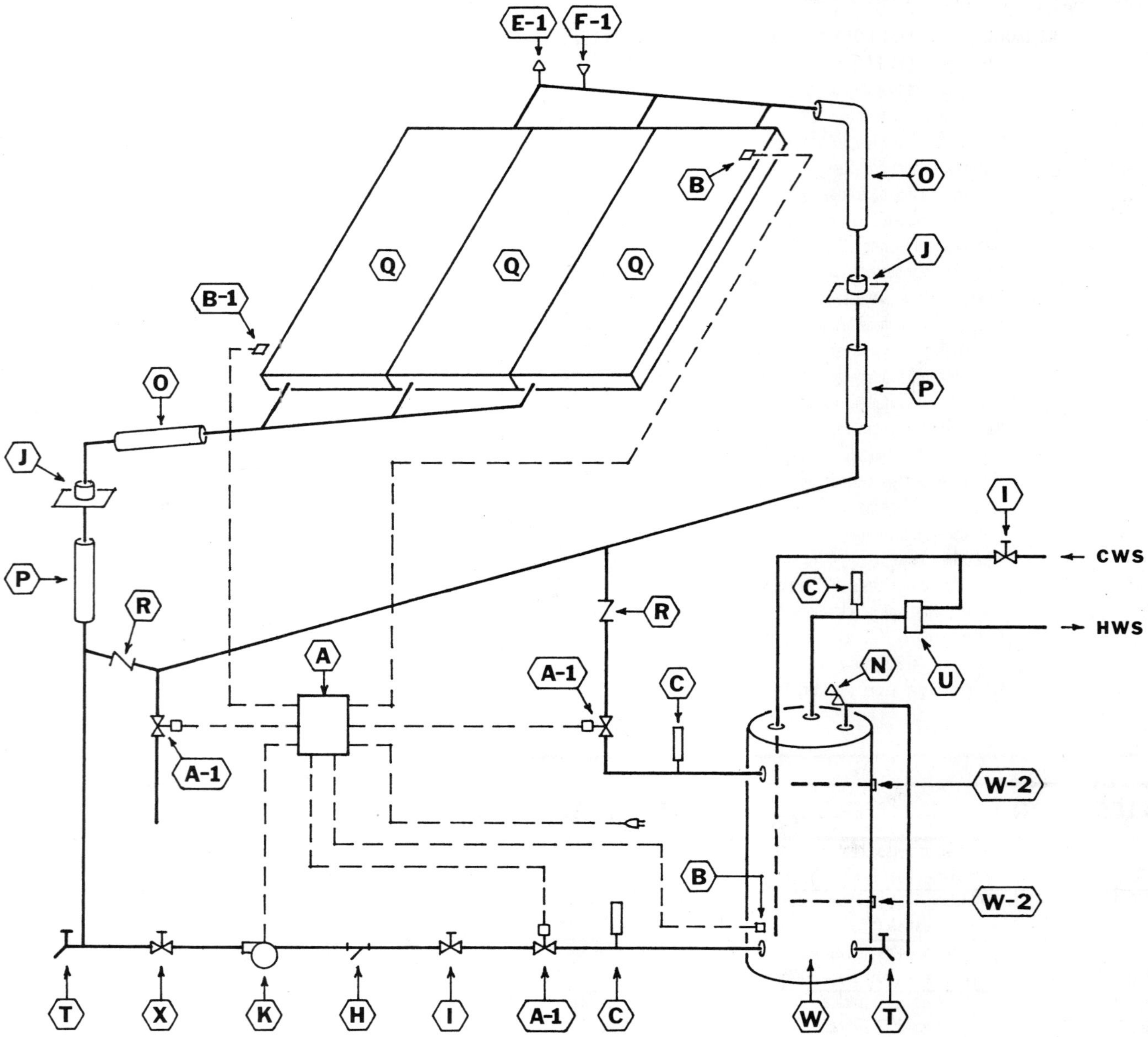

In the draindown direct-collection system, incoming domestic water is heated in the collectors. When the controller activates solar collection, domestic water is first heated as it flows through the collectors and is then pumped to storage. When conditions are no longer suitable for heat collection, the pump shuts off and the water in the loop drains down and out of the system by means of solenoid valves and properly pitched piping.

D2020 Domestic Water Distribution

System Components

System Components		QUANTITY	UNIT	COST EACH MAT.	COST EACH INST.	COST EACH TOTAL
	SYSTEM D2020 275 2760					
	SOLAR, DRAINDOWN, HOT WATER, DIRECT COLLECTION					
	3/4" TUBING, THREE 3'X7' BLACK CHROME COLLECTORS					
A, B	Differential controller, 2 sensors, thermostat, solar energy system	1.000	Ea.	143	59.50	202.50
A-1	Solenoid valve, solar heating loop, brass, 3/4" diam, 24 volts	3.000	Ea.	423	238.50	661.50
B-1	Solar energy sensor, freeze prevention	1.000	Ea.	25	22.50	47.50
C	Thermometer, 2" dial	3.000	Ea.	72	133.50	205.50
E-1	Vacuum relief valve, 3/4" diam	1.000	Ea.	32.50	22.50	55
F-1	Air vent, automatic, 1/8" fitting	1.000	Ea.	18.95	22.50	41.45
H	Strainer, Y type, bronze body, 3/4" IPS	1.000	Ea.	45.50	38.50	84
I	Valve, gate, bronze, NRS, soldered, 3/4" diam	2.000	Ea.	145	71	216
J	Vent flashing neoprene	2.000	Ea.	11	71	82
K	Circulator, solar heated liquid, 1/25 HP	1.000	Ea.	199	92	291
N	Relief valve temp & press 150 psi 210°F self-closing 3/4" IPS	1.000	Ea.	24.50	24	48.50
O	Pipe covering, urethane, ultraviolet cover, 1" wall, 3/4" diam	20.000	L.F.	62.20	130	192.20
P	Pipe covering, fiberglass, all service jacket, 1" wall, 3/4" diam	50.000	L.F.	49.50	260	309.50
	Roof clamps for solar energy collector panels	3.000	Set	9.33	55.20	64.53
Q	Collector panel solar energy blk chrome on copper, 1/8" temp glass 3'x7'	3.000	Ea.	3,300	405	3,705
R	Valve, swing check, bronze, regrinding disc, 3/4" diam, soldered	2.000	Ea.	232	71	303
T	Drain valve, brass, 3/4" connection	2.000	Ea.	22.50	60	82.50
U	Valve, water tempering, bronze, sweat connections, 3/4" diam	1.000	Ea.	150	35.50	185.50
W-2, W	Tank, water storage elec elem 2"x1/2# insul 120 gal	1.000	Ea.	2,025	510	2,535
X	Valve, globe, bronze, rising stem, 3/4" diam, soldered	1.000	Ea.	114	35.50	149.50
	Copper tubing, type L, solder joints, hangers 10' OC 3/4" diam	20.000	L.F.	115	188	303
	Copper tubing, type M, solder joints, hangers 10' OC 3/4" diam	70.000	L.F.	337.40	640.50	977.90
	Sensor wire, #22-2 conductor, multistranded	.500	C.L.F.	7.83	33	40.83
	Wrought copper fittings & solder, 3/4" diam	76.000	Ea.	211.28	2,850	3,061.28
	TOTAL			7,775.49	6,069.20	13,844.69

D2020 275	Solar, Draindown, Hot Water Systems		COST EACH MAT.	COST EACH INST.	COST EACH TOTAL
2550	Solar, draindown, hot water				
2580	1/2" tubing, 4 ea. 4' x 4'-4" vacuum tube collectors, 80 gal tank		7,825	5,925	13,750
2760	3/4" tubing, 3 ea. 3'x7' black chrome collectors, 120 gal tank	RD3010 -600	7,775	6,075	13,850
2780	3 ea. 3'x7' flat collectors, 120 gal tank		7,375	6,100	13,475
2800	2 ea. 4'x9' flat black & plastic glazing collectors, 120 gal tank		6,775	6,125	12,900
2840	1" tubing, 4 ea. 2'x9' plastic absorber & glazing collectors, 120 gal tank		11,500	8,075	19,575
2860	4 ea. 3'x7' black chrome absorber collectors, 120 gal tank		11,000	7,625	18,625
2880	4 ea. 3'x7' flat black absorber collectors, 120 gal tank		9,350	6,900	16,250

D2020 Domestic Water Distribution

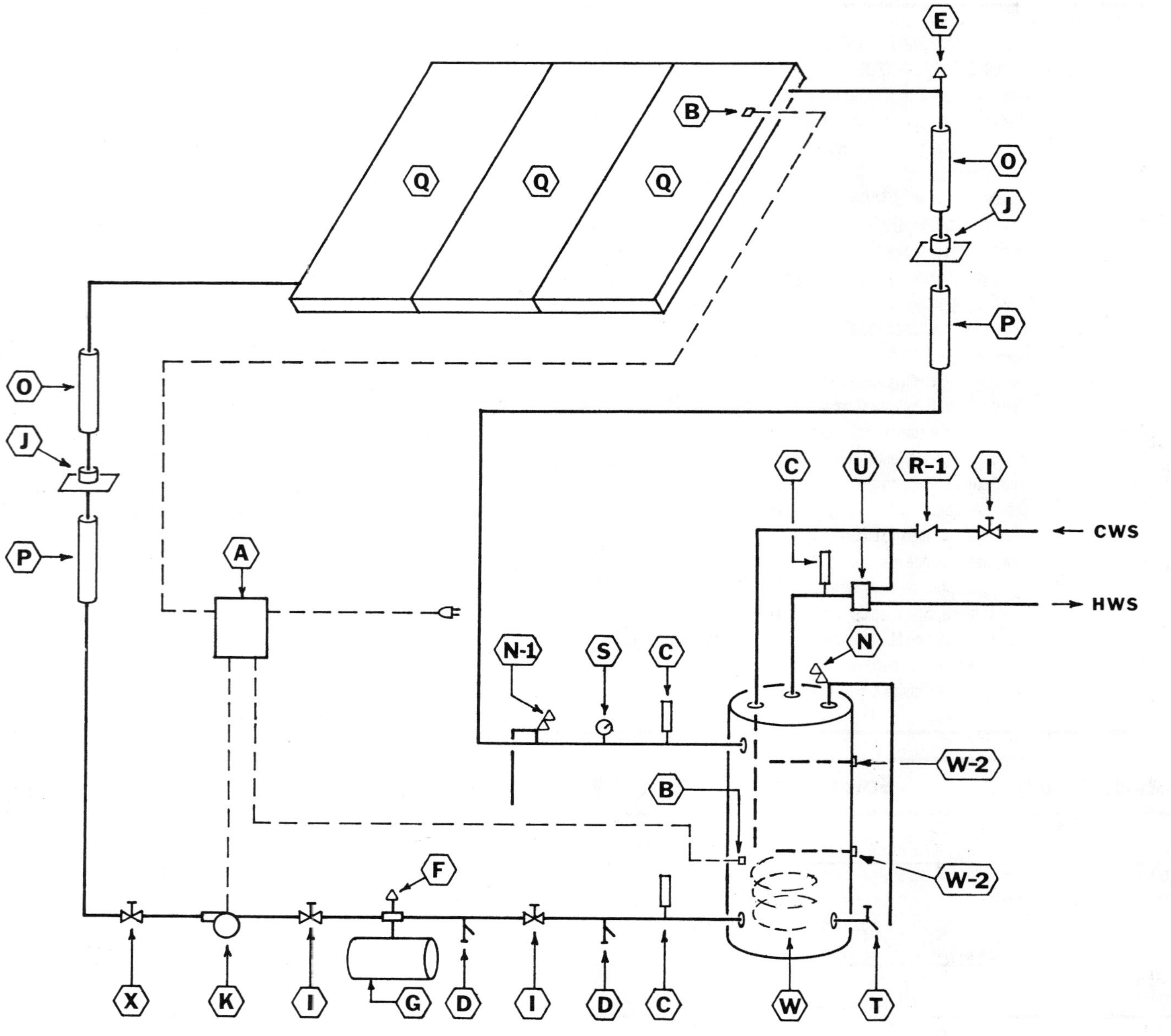

In this closed-loop indirect collection system, fluid with a low freezing temperature, such as propylene glycol, transports heat from the collectors to water storage. The transfer fluid is contained in a closed-loop consisting of collectors, supply and return piping, and a heat exchanger immersed in the storage tank. A typical two-or-three panel system contains 5 to 6 gallons of heat transfer fluid.

When the collectors become approximately 20°F warmer than the storage temperature, a controller activates the circulator. The circulator moves the fluid continuously through the collectors until the temperature difference between the collectors and storage is such that heat collection no longer occurs; at that point, the circulator shuts off. Since the heat transfer fluid has a very low freezing temperature, there is no need for it to be drained from the collectors between periods of collection.

D2020 Domestic Water Distribution

System Components	QUANTITY	UNIT	COST EACH MAT.	COST EACH INST.	COST EACH TOTAL
SYSTEM D2020 295 2760					
SOLAR, CLOSED LOOP, HOT WATER SYSTEM, IMMERSED HEAT EXCHANGER					
3/4" TUBING, THREE 3' X 7' BLACK CHROME COLLECTORS					
A, B Differential controller, 2 sensors, thermostat, solar energy system	1.000	Ea.	143	59.50	202.50
C Thermometer 2" dial	3.000	Ea.	72	133.50	205.50
D, T Fill & drain valves, brass, 3/4" connection	3.000	Ea.	33.75	90	123.75
E Air vent, manual, 1/8" fitting	1.000	Ea.	2.68	22.50	25.18
F Air purger	1.000	Ea.	51.50	59.50	111
G Expansion tank	1.000	Ea.	75	22.50	97.50
I Valve, gate, bronze, NRS, soldered 3/4" diam	3.000	Ea.	217.50	106.50	324
J Neoprene vent flashing	2.000	Ea.	11	71	82
K Circulator, solar heated liquid, 1/25 HP	1.000	Ea.	199	92	291
N-1, N Relief valve, temp & press 150 psi 210°F self-closing 3/4" IPS	2.000	Ea.	49	48	97
O Pipe covering, urethane ultraviolet cover, 1" wall, 3/4" diam	20.000	L.F.	62.20	130	192.20
P Pipe covering, fiberglass, all service jacket, 1" wall, 3/4" diam	50.000	L.F.	49.50	260	309.50
Roof clamps for solar energy collector panel	3.000	Set	9.33	55.20	64.53
Q Collector panel solar blk chrome on copper, 1/8" temp glass, 3'x7'	3.000	Ea.	3,300	405	3,705
R-1 Valve, swing check, bronze, regrinding disc, 3/4" diam, soldered	1.000	Ea.	116	35.50	151.50
S Pressure gauge, 60 psi, 2-1/2" dial	1.000	Ea.	26.50	22.50	49
U Valve, water tempering, bronze, sweat connections, 3/4" diam	1.000	Ea.	150	35.50	185.50
W-2, W Tank, water storage immersed heat exchr elec elem 2"x2# insul 120 Gal	1.000	Ea.	2,025	510	2,535
X Valve, globe, bronze, rising stem, 3/4" diam, soldered	1.000	Ea.	114	35.50	149.50
Copper tubing type L, solder joint, hanger 10' OC 3/4" diam	20.000	L.F.	115	188	303
Copper tubing, type M, solder joint, hanger 10' OC 3/4" diam	70.000	L.F.	337.40	640.50	977.90
Sensor wire, #22-2 conductor multistranded	.500	C.L.F.	7.83	33	40.83
Solar energy heat transfer fluid, propylene glycol, anti-freeze	6.000	Gal.	99	153	252
Wrought copper fittings & solder, 3/4" diam	76.000	Ea.	211.28	2,850	3,061.28
TOTAL			7,477.47	6,058.70	13,536.17

D2020 295	Solar, Closed Loop, Hot Water Systems	COST EACH MAT.	COST EACH INST.	COST EACH TOTAL
2550	Solar, closed loop, hot water system, immersed heat exchanger			
2560	3/8" tubing, 3 ea. 4' x 4'-4" vacuum tube collectors, 80 gal. tank	6,725	5,500	12,225
2580	1/2" tubing, 4 ea. 4' x 4'-4" vacuum tube collectors, 80 gal. tank RD3010 -600	7,550	5,925	13,475
2600	120 gal. tank	7,800	6,000	13,800
2640	2 ea. 3'x7' black chrome collectors, 80 gal. tank	5,825	5,650	11,475
2660	120 gal. tank	6,025	5,650	11,675
2760	3/4" tubing, 3 ea. 3'x7' black chrome collectors, 120 gal. tank	7,475	6,050	13,525
2780	3 ea. 3'x7' flat black collectors, 120 gal. tank	7,075	6,075	13,150
2840	1" tubing, 4 ea. 2'x9' plastic absorber & glazing collectors 120 gal. tank	7,400	6,825	14,225
2860	4 ea. 3'x7' black chrome collectors, 120 gal. tank	9,250	6,850	16,100

D2040 Rain Water Drainage

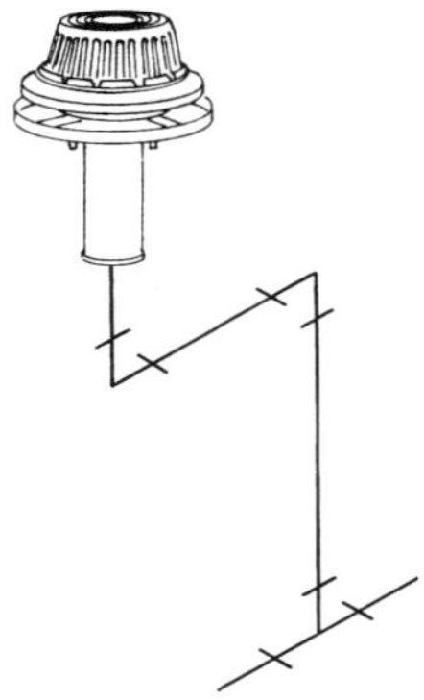

Design Assumptions: Vertical conductor size is based on a maximum rate of rainfall of 4″ per hour. To convert roof area to other rates multiply "Max. S.F. Roof Area" shown by four and divide the result by desired local rate. The answer is the local roof area that may be handled by the indicated pipe diameter.

Basic cost is for roof drain, 10′ of vertical leader and 10′ of horizontal, plus connection to the main.

Pipe Dia.	Max. S.F. Roof Area	Gallons per Min.
2″	544	23
3″	1610	67
4″	3460	144
5″	6280	261
6″	10,200	424
8″	22,000	913

System Components	QUANTITY	UNIT	COST EACH MAT.	COST EACH INST.	COST EACH TOTAL
SYSTEM D2040 210 1880					
ROOF DRAIN, DWV PVC PIPE, 2″ DIAM., 10′ HIGH					
Drain, roof, main, ABS, dome type 2″ pipe size	1.000	Ea.	137	92	229
Clamp, roof drain, underdeck	1.000	Ea.	29	53.50	82.50
Pipe, Tee, PVC DWV, schedule 40, 2″ pipe size	1.000	Ea.	5.15	64.50	69.65
Pipe, PVC, DWV, schedule 40, 2″ diam.	20.000	L.F.	126	440	566
Pipe, elbow, PVC schedule 40, 2″ diam.	2.000	Ea.	6.66	71	77.66
TOTAL			303.81	721	1,024.81

D2040 210	Roof Drain Systems	COST EACH MAT.	COST EACH INST.	COST EACH TOTAL
1880	Roof drain, DWV PVC, 2″ diam., piping, 10′ high	305	720	1,025
1920	For each additional foot add	6.30	22	28.30
1960	3″ diam., 10′ high	385	840	1,225
2000	For each additional foot add	9	24.50	33.50
2040	4″ diam., 10′ high	460	945	1,405
2080	For each additional foot add	11.25	27	38.25
2120	5″ diam., 10′ high	1,475	1,075	2,550
2160	For each additional foot add	30	30	60
2200	6″ diam., 10′ high	1,475	1,200	2,675
2240	For each additional foot add	18.60	33	51.60
2280	8″ diam., 10′ high	3,175	2,050	5,225
2320	For each additional foot add	39	41.50	80.50
3940	C.I., soil, single hub, service wt., 2″ diam. piping, 10′ high	715	785	1,500
3980	For each additional foot add	12.85	20.50	33.35
4120	3″ diam., 10′ high	920	850	1,770
4160	For each additional foot add	17	21.50	38.50
4200	4″ diam., 10′ high	1,050	930	1,980
4240	For each additional foot add	21.50	23.50	45
4280	5″ diam., 10′ high	1,475	1,025	2,500
4320	For each additional foot add	29	26.50	55.50
4360	6″ diam., 10′ high	1,850	1,100	2,950
4400	For each additional foot add	35.50	27.50	63
4440	8″ diam., 10′ high	3,625	2,225	5,850
4480	For each additional foot add	53	46	99
6040	Steel galv. sch 40 threaded, 2″ diam. piping, 10′ high	750	760	1,510
6080	For each additional foot add	9.30	20	29.30
6120	3″ diam., 10′ high	1,375	1,100	2,475
6160	For each additional foot add	18.55	30	48.55

D2040 Rain Water Drainage

D2040 210	Roof Drain Systems	COST EACH		
		MAT.	INST.	TOTAL
6200	4″ diam., 10′ high	1,950	1,400	3,350
6240	For each additional foot add	27	35.50	62.50
6280	5″ diam., 10′ high	1,875	1,175	3,050
6320	For each additional foot add	30	35	65
6360	6″ diam, 10′ high	2,375	1,500	3,875
6400	For each additional foot add	37.50	47.50	85
6440	8″ diam., 10′ high	4,400	2,175	6,575
6480	For each additional foot add	56.50	54	110.50

D2090 Other Plumbing Systems

Pipe Material Considerations:

1. Malleable iron fittings should be used for gas service.
2. Malleable fittings are used where there are stresses/strains due to expansion and vibration.
3. Cast iron fittings may be broken as an aid to disassembling of piping joints frozen by long use, temperature and mineral deposits.
4. Cast iron pipe is extensively used for underground and submerged service.
5. Type M (light wall) copper tubing is available in hard temper only and is used for nonpressure and less severe applications than K and L.
6. Type L (medium wall) copper tubing, available hard or soft for interior service.
7. Type K (heavy wall) copper tubing, available in hard or soft temper for use where conditions are severe. For underground and interior service.
8. Hard drawn tubing requires fewer hangers or supports but should not be bent. Silver brazed fittings should be used.

Piping costs include hangers spaced for the material type and size, and the couplings required.

D2090 810	Piping - Installed - Unit Costs	COST PER L.F.		
		MAT.	INST.	TOTAL
0840	Cast iron, soil, B & S, service weight, 2" diameter RD2010 -030	12.85	20.50	33.35
0860	3" diameter	17	21.50	38.50
0880	4" diameter	21.50	23.50	45
0900	5" diameter	29	26.50	55.50
0920	6" diameter	35.50	27.50	63
0940	8" diameter	53	46	99
0960	10" diameter	85	50.50	135.50
0980	12" diameter	120	56.50	176.50
1040	No hub, 1-1/2" diameter	12.65	18.10	30.75
1060	2" diameter	13.25	19.20	32.45
1080	3" diameter	17	20	37
1100	4" diameter	21.50	22	43.50
1120	5" diameter	29.50	24	53.50
1140	6" diameter	36	25.50	61.50
1160	8" diameter	60.50	39.50	100
1180	10" diameter	99.50	44.50	144
1220	Copper tubing, hard temper, solder, type K, 1/2" diameter	5.50	9.15	14.65
1260	3/4" diameter	8.95	9.65	18.60
1280	1" diameter	11.65	10.85	22.50
1300	1-1/4" diameter	14.30	12.75	27.05
1320	1-1/2" diameter	17.95	14.30	32.25
1340	2" diameter	27	17.85	44.85
1360	2-1/2" diameter	40	21.50	61.50
1380	3" diameter	54.50	24	78.50
1400	4" diameter	85	34	119
1480	5" diameter	161	40	201
1500	6" diameter	235	52.50	287.50
1520	8" diameter	410	59	469
1560	Type L, 1/2" diameter	4.05	8.85	12.90
1600	3/4" diameter	5.75	9.40	15.15
1620	1" diameter	8.15	10.50	18.65
1640	1-1/4" diameter	11.05	12.30	23.35
1660	1-1/2" diameter	13.85	13.75	27.60
1680	2" diameter	20	17	37
1700	2-1/2" diameter	30.50	21	51.50
1720	3" diameter	40.50	23	63.50
1740	4" diameter	63.50	33	96.50
1760	5" diameter	117	38	155
1780	6" diameter	160	50	210
1800	8" diameter	273	55.50	328.50
1840	Type M, 1/2" diameter	3.53	8.50	12.03
1880	3/4" diameter	4.82	9.15	13.97

D2090 Other Plumbing Systems

D2090 810	Piping - Installed - Unit Costs	COST PER L.F. MAT.	COST PER L.F. INST.	COST PER L.F. TOTAL
1900	1" diameter	7.25	10.20	17.45
1920	1-1/4" diameter	10.10	11.90	22
1940	1-1/2" diameter	13.30	13.25	26.55
1960	2" diameter	19.95	16.25	36.20
1980	2-1/2" diameter	29.50	20	49.50
2000	3" diameter	39	22	61
2020	4" diameter	63.50	32	95.50
2040	5" diameter	127	35.50	162.50
2060	6" diameter	175	47.50	222.50
2080	8" diameter	300	52.50	352.50
2120	Type DWV, 1-1/4" diameter	10.10	11.90	22
2160	1-1/2" diameter	12.40	13.25	25.65
2180	2" diameter	16.25	16.25	32.50
2200	3" diameter	29	22	51
2220	4" diameter	46.50	32	78.50
2240	5" diameter	114	35.50	149.50
2260	6" diameter	162	47.50	209.50
2280	8" diameter	340	52.50	392.50
2800	Plastic,PVC, DWV, schedule 40, 1-1/4" diameter	6.05	17	23.05
2820	1-1/2" diameter	5.85	19.85	25.70
2830	2" diameter	6.30	22	28.30
2840	3" diameter	9	24.50	33.50
2850	4" diameter	11.25	27	38.25
2890	6" diameter	18.60	33	51.60
3010	Pressure pipe 200 PSI, 1/2" diameter	5.30	13.25	18.55
3030	3/4" diameter	5.70	14	19.70
3040	1" diameter	6.30	15.55	21.85
3050	1-1/4" diameter	7.20	17	24.20
3060	1-1/2" diameter	7.65	19.85	27.50
3070	2" diameter	8.65	22	30.65
3080	2-1/2" diameter	13.50	23	36.50
3090	3" diameter	14.95	24.50	39.45
3100	4" diameter	23	27	50
3110	6" diameter	39	33	72
3120	8" diameter	57.50	41.50	99
4000	Steel, schedule 40, black, threaded, 1/2" diameter	3.19	11.35	14.54
4020	3/4" diameter	3.68	11.70	15.38
4030	1" diameter	4.73	13.50	18.23
4040	1-1/4" diameter	5.70	14.45	20.15
4050	1-1/2" diameter	6.45	16.10	22.55
4060	2" diameter	8.15	20	28.15
4070	2-1/2" diameter	12.70	25.50	38.20
4080	3" diameter	16.05	30	46.05
4090	4" diameter	24.50	35.50	60
4100	Grooved, 5" diameter	44.50	35	79.50
4110	6" diameter	57	47.50	104.50
4120	8" diameter	92.50	54	146.50
4130	10" diameter	125	64.50	189.50
4140	12" diameter	139	74	213
4200	Galvanized, threaded, 1/2" diameter	3.49	11.35	14.84
4220	3/4" diameter	3.96	11.70	15.66
4230	1" diameter	5.30	13.50	18.80
4240	1-1/4" diameter	6.40	14.45	20.85
4250	1-1/2" diameter	7.25	16.10	23.35
4260	2" diameter	9.30	20	29.30
4270	2-1/2" diameter	14.70	25.50	40.20
4280	3" diameter	18.55	30	48.55
4290	4" diameter	27	35.50	62.50

D2090 Other Plumbing Systems

D2090 810	Piping - Installed - Unit Costs	COST PER L.F.		
		MAT.	INST.	TOTAL
4300	Grooved, 5" diameter	30	35	65
4310	6" diameter	37.50	47.50	85
4320	8" diameter	56.50	54	110.50
4330	10" diameter	118	64.50	182.50
4340	12" diameter	144	74	218
5010	Flanged, black, 1" diameter	10.25	19.30	29.55
5020	1-1/4" diameter	11.40	21	32.40
5030	1-1/2" diameter	11.80	23	34.80
5040	2" diameter	13.90	30	43.90
5050	2-1/2" diameter	19.05	37.50	56.55
5060	3" diameter	22	42	64
5070	4" diameter	29.50	52	81.50
5080	5" diameter	57	64.50	121.50
5090	6" diameter	68	82.50	150.50
5100	8" diameter	111	108	219
5110	10" diameter	157	129	286
5120	12" diameter	189	148	337
5720	Grooved joints, black, 3/4" diameter	5.50	10.05	15.55
5730	1" diameter	5.30	11.35	16.65
5740	1-1/4" diameter	6.55	12.30	18.85
5750	1-1/2" diameter	7.25	14	21.25
5760	2" diameter	8.40	17.85	26.25
5770	2-1/2" diameter	12.80	22.50	35.30
5900	3" diameter	15.40	25.50	40.90
5910	4" diameter	21.50	28.50	50
5920	5" diameter	44.50	35	79.50
5930	6" diameter	57	47.50	104.50
5940	8" diameter	92.50	54	146.50
5950	10" diameter	125	64.50	189.50
5960	12" diameter	139	74	213

D2090 820	Standard Pipe Fittings - Installed - Unit Costs	COST EACH		
		MAT.	INST.	TOTAL
0100	Cast iron, soil, B&S service weight, 1/8 Bend, 2" diameter	16.10	80.50	96.60
0110	3" diameter	25	92	117
0120	4" diameter	37	99	136
0130	5" diameter	52	111	163
0140	6" diameter	62.50	118	180.50
0150	8" diameter	187	247	434
0160	10" diameter	269	272	541
0170	12" diameter	510	300	810
0220	1/4 Bend, 2" diameter	22.50	80.50	103
0230	3" diameter	30	92	122
0240	4" diameter	47.50	99	146.50
0250	5" diameter	66	111	177
0260	6" diameter	82	118	200
0270	8" diameter	248	247	495
0280	10" diameter	360	272	632
0290	12" diameter	490	300	790
0360	Tee, 2" diameter	45.50	129	174.50
0370	3" diameter	68	143	211
0380	4" diameter	87	161	248
0390	5" diameter	185	167	352
0400	6" diameter	182	182	364
0410	8" diameter	330	390	720
0810	No hub, 1/8 Bend, 1-1/2" diameter	9.80		9.80
0820	2" diameter	10.95		10.95

D2090 Other Plumbing Systems

D2090 820	Standard Pipe Fittings - Installed - Unit Costs	COST EACH		
		MAT.	INST.	TOTAL
0830	3" diameter	14.65		14.65
0840	4" diameter	19.15		19.15
0850	5" diameter	40		40
0860	6" diameter	42.50		42.50
0870	8" diameter	123		123
0880	10" diameter	233		233
0960	1/4 Bend, 1-1/2" diameter	11.65		11.65
0970	2" diameter	12.70		12.70
0980	3" diameter	17.70		17.70
0990	4" diameter	26		26
1000	5" diameter	63		63
1010	6" diameter	63.50		63.50
1020	8" diameter	178		178
1080	Sanitary tee, 1-1/2" diameter	16.30		16.30
1090	2" diameter	17.50		17.50
1110	3" diameter	21.50		21.50
1120	4" diameter	41		41
1130	5" diameter	95.50		95.50
1140	6" diameter	97.50		97.50
1150	8" diameter	395		395
1300	Coupling clamp & gasket, 1-1/2" diameter	9.05	27	36.05
1310	2" diameter	10.20	29	39.20
1320	3" diameter	12.20	34	46.20
1330	4" diameter	16.80	39	55.80
1340	5" diameter	29.50	45.50	75
1350	6" diameter	30.50	50	80.50
1360	8" diameter	117	82.50	199.50
1370	10" diameter	203	105	308
1510	Steel pipe fitting, CI, threaded, 125 Lb. black, 45° Elbow, 1/2" diameter	10.80	47.50	58.30
1520	3/4" diameter	10.90	51	61.90
1530	1" diameter	12.75	55	67.75
1540	1-1/4" diameter	17.15	58.50	75.65
1550	1-1/2" diameter	28.50	64.50	93
1560	2" diameter	33	71.50	104.50
1570	2-1/2" diameter	85.50	92	177.50
1580	3" diameter	135	129	264
1590	4" diameter	281	214	495
1600	5" diameter	345	129	474
1610	6" diameter	435	160	595
1620	8" diameter	780	191	971
1700	90° Elbow, 1/2" diameter	7.05	47.50	54.55
1710	3/4" diameter	7.35	51	58.35
1720	1" diameter	8.70	55	63.70
1730	1-1/4" diameter	12.40	58.50	70.90
1740	1-1/2" diameter	17.10	64.50	81.60
1750	2" diameter	26.50	71.50	98
1760	2-1/2" diameter	64	92	156
1770	3" diameter	105	129	234
1780	4" diameter	195	214	409
1790	5" diameter	345	129	474
1800	6" diameter	435	160	595
1810	8" diameter	780	191	971
1900	Tee, 1/2" diameter	11	79.50	90.50
1910	3/4" diameter	12.85	79.50	92.35
1920	1" diameter	11.45	89.50	100.95
1930	1-1/4" diameter	21	92	113
1940	1-1/2" diameter	27	99	126
1950	2" diameter	37.50	117	154.50

D2090 Other Plumbing Systems

D2090 820	Standard Pipe Fittings - Installed - Unit Costs	COST EACH		
		MAT.	INST.	TOTAL
1960	2-1/2" diameter	98	143	241
1970	3" diameter	150	214	364
1980	4" diameter	305	320	625
1990	5" diameter	545	195	740
2000	6" diameter	675	238	913
2010	8" diameter	1,250	286	1,536
2300	Copper, wrought, solder joints, 45° Elbow, 1/2" diameter	2.27	35.50	37.77
2310	3/4" diameter	3.87	37.50	41.37
2320	1" diameter	9.70	44.50	54.20
2330	1-1/4" diameter	13	47.50	60.50
2340	1-1/2" diameter	16	55	71
2350	2" diameter	26.50	65	91.50
2360	2-1/2" diameter	52	99	151
2370	3" diameter	72.50	99	171.50
2380	4" diameter	149	143	292
2390	5" diameter	540	214	754
2400	6" diameter	855	222	1,077
2500	90° Elbow, 1/2" diameter	1.24	35.50	36.74
2510	3/4" diameter	2.78	37.50	40.28
2520	1" diameter	6.85	44.50	51.35
2530	1-1/4" diameter	10.10	47.50	57.60
2540	1-1/2" diameter	15.80	55	70.80
2550	2" diameter	29	65	94
2560	2-1/2" diameter	52.50	99	151.50
2610	3" diameter	74.50	117	191.50
2620	4" diameter	163	143	306
2630	5" diameter	640	214	854
2640	6" diameter	855	222	1,077
2700	Tee, 1/2" diameter	2.11	55	57.11
2710	3/4" diameter	5.10	59.50	64.60
2720	1" diameter	15.30	71.50	86.80
2730	1-1/4" diameter	21	79.50	100.50
2740	1-1/2" diameter	32	89.50	121.50
2750	2" diameter	51	102	153
2760	2-1/2" diameter	93.50	161	254.50
2770	3" diameter	134	184	318
2780	4" diameter	315	257	572
2790	5" diameter	965	320	1,285
2800	6" diameter	1,325	335	1,660
2880	Coupling, 1/2" diameter	.94	32.50	33.44
2890	3/4" diameter	1.89	34	35.89
2900	1" diameter	3.73	39.50	43.23
2910	1-1/4" diameter	6.40	42	48.40
2920	1-1/2" diameter	8.65	47.50	56.15
2930	2" diameter	14.45	55	69.45
2940	2-1/2" diameter	28	86	114
2950	3" diameter	39.50	99	138.50
2960	4" diameter	80.50	184	264.50
2970	5" diameter	186	214	400
2980	6" diameter	305	250	555
3200	Malleable iron, 150 Lb. threaded, black, 45° elbow, 1/2" diameter	6.25	47.50	53.75
3210	3/4" diameter	7.75	51	58.75
3220	1" diameter	9.80	55	64.80
3230	1-1/4" diameter	17.30	58.50	75.80
3240	1-1/2" diameter	21.50	64.50	86
3250	2" diameter	32.50	71.50	104
3260	2-1/2" diameter	93.50	92	185.50
3270	3" diameter	122	129	251

D2090 Other Plumbing Systems

D2090 820	Standard Pipe Fittings - Installed - Unit Costs	COST EACH		
		MAT.	INST.	TOTAL
3280	4" diameter	238	214	452
3290	5" diameter	345	129	474
3300	6" diameter	435	160	595
3400	90° Elbow, 1/2" diameter	3.80	47.50	51.30
3500	3/4" diameter	4.59	51	55.59
3510	1" diameter	7.95	55	62.95
3520	1-1/4" diameter	13.15	58.50	71.65
3530	1-1/2" diameter	17.30	64.50	81.80
3540	2" diameter	30.50	71.50	102
3550	2-1/2" diameter	66.50	92	158.50
3560	3" diameter	97	129	226
3570	4" diameter	209	214	423
3580	5" diameter	345	129	474
3590	6" diameter	435	160	595
3700	Tee, 1/2" diameter	5.10	79.50	84.60
3800	3/4" diameter	7.30	79.50	86.80
3810	1" diameter	12.50	89.50	102
3820	1-1/4" diameter	20	92	112
3830	1-1/2" diameter	25	99	124
3910	2" diameter	43	117	160
3920	2-1/2" diameter	92.50	143	235.50
3930	3" diameter	136	214	350
3940	4" diameter	330	320	650
3950	5" diameter	545	195	740
3960	6" diameter	315	120	435
4000	Coupling, 1/2" diameter	5.25	37.50	42.75
4010	3/4" diameter	6.15	39.50	45.65
4020	1" diameter	9.20	47.50	56.70
4030	1-1/4" diameter	11.95	49.50	61.45
4040	1-1/2" diameter	16.15	53.50	69.65
4050	2" diameter	24	61.50	85.50
4060	2-1/2" diameter	66	71.50	137.50
4070	3" diameter	89	92	181
4080	4" diameter	180	129	309
4090	5" diameter	87.50	32	119.50
4100	6" diameter	116	40	156
5100	Plastic, PVC, high impact/pressure sch 40, 45° Elbow, 1/2" diameter	.91	21.50	22.41
5110	3/4" diameter	1.42	25	26.42
5120	1" diameter	1.71	28.50	30.21
5130	1-1/4" diameter	2.35	32	34.35
5140	1-1/2" diameter	2.99	35.50	38.49
5150	2" diameter	3.92	35.50	39.42
5160	3" diameter	15.80	56	71.80
5170	4" diameter	28.50	70.50	99
5180	6" diameter	70	116	186
5190	8" diameter	168	194	362
5260	90° Elbow, 1/2" diameter	.55	21.50	22.05
5270	3/4" diameter	.63	25	25.63
5280	1" diameter	1.11	28.50	29.61
5290	1-1/4" diameter	1.97	32	33.97
5300	1-1/2" diameter	2.13	35.50	37.63
5400	2" diameter	3.33	35.50	38.83
5410	3" diameter	12.15	56	68.15
5420	4" diameter	21.50	70.50	92
5430	6" diameter	69	116	185
5440	8" diameter	178	194	372
5500	Tee, 1/2" diameter	.69	32	32.69
5510	3/4" diameter	.80	37.50	38.30

D2090 Other Plumbing Systems

D2090 820	Standard Pipe Fittings - Installed - Unit Costs	COST EACH		
		MAT.	INST.	TOTAL
5520	1" diameter	1.50	43	44.50
5530	1-1/4" diameter	2.32	48.50	50.82
5540	1-1/2" diameter	2.83	53.50	56.33
5550	2" diameter	4.11	53	57.11
5560	3" diameter	17.85	84.50	102.35
5570	4" diameter	32.50	106	138.50
5580	6" diameter	109	174	283
5590	8" diameter	252	294	546
5680	Coupling, 1/2" diameter	.36	21.50	21.86
5690	3/4" diameter	.51	25	25.51
5700	1" diameter	.88	28.50	29.38
5710	1-1/4" diameter	1.22	32	33.22
5720	1-1/2" diameter	1.22	32	33.22
5730	2" diameter	1.98	35.50	37.48
5740	3" diameter	6.90	56	62.90
5750	4" diameter	10	70.50	80.50
6010	6" diameter	31.50	116	147.50
6020	8" diameter	59	194	253
7810	DWV, socket joints, sch. 40, 1/8 Bend, 1-1/4" diameter	4.75	35.50	40.25
7820	1-1/2" diameter	1.96	39.50	41.46
7830	2" diameter	2.83	39	41.83
7840	3" diameter	8.30	62	70.30
7850	4" diameter	14.55	78	92.55
7860	6" diameter	57	127	184
7960	1/4 bend, 1-1/4" diameter	7.05	35.50	42.55
8000	1-1/2" diameter	2	39.50	41.50
8010	2" diameter	3.16	39	42.16
8020	3" diameter	9.25	62	71.25
8030	4" diameter	18.25	78	96.25
8210	6" diameter	62	127	189
8300	Sanitary tee, 1-1/4" diameter	7.30	53	60.30
8310	1-1/2" diameter	3.49	59	62.49
8320	2" diameter	5.15	64.50	69.65
8330	3" diameter	13.50	92.50	106
8340	4" diameter	25	117	142
8400	Coupling, 1-1/4" diameter	28.50	35.50	64
8410	1-1/2" diameter	30.50	39.50	70
8420	2" diameter	36	39	75
8430	3" diameter	75.50	62	137.50
8440	4" diameter	154	78	232

D3010 Energy Supply

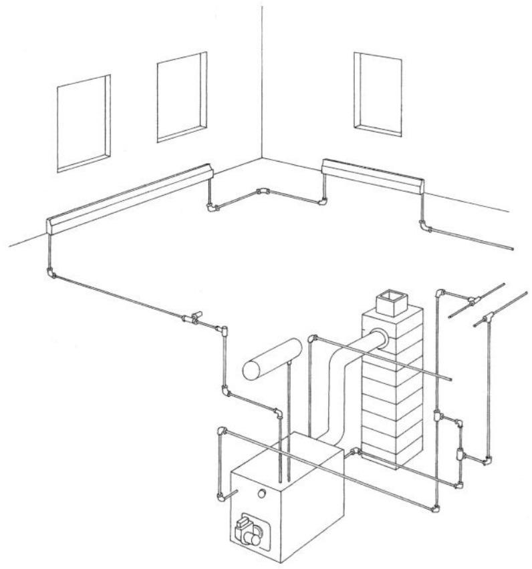

Basis for Heat Loss Estimate, Apartment Type Structures:

1. Masonry walls and flat roof are insulated. U factor is assumed at .08.
2. Window glass area taken as BOCA minimum, 1/10th of floor area. Double insulating glass with 1/4″ air space, U = .65.
3. Infiltration = 0.3 C.F. per hour per S.F. of net wall.
4. Concrete floor loss is 2 BTUH per S.F.
5. Temperature difference taken as 70°F.
6. Ventilating or makeup air has not been included and must be added if desired. Air shafts are not used.

System Components	QUANTITY	UNIT	COST EACH MAT.	COST EACH INST.	COST EACH TOTAL
SYSTEM D3010 510 1760					
HEATING SYSTEM, FIN TUBE RADIATION, FORCED HOT WATER					
1,000 S.F. AREA, 10,000 C.F. VOLUME					
Boiler, oil fired, CI, burner, ctrls/insul/breech/pipe/ftng/valves, 109 MBH	1.000	Ea.	4,462.50	4,068.75	8,531.25
Circulating pump, CI flange connection, 1/12 HP	1.000	Ea.	415	214	629
Expansion tank, painted steel, ASME 18 Gal capacity	1.000	Ea.	815	94	909
Storage tank, steel, above ground, 275 Gal capacity w/supports	1.000	Ea.	540	264	804
Copper tubing type L, solder joint, hanger 10′ OC, 3/4″ diam	100.000	L.F.	575	940	1,515
Radiation, 3/4″ copper tube w/alum fin baseboard pkg, 7″ high	30.000	L.F.	243	675	918
Pipe covering, calcium silicate w/cover, 1′ wall, 3/4′ diam	100.000	L.F.	448	705	1,153
TOTAL			7,498.50	6,960.75	14,459.25
COST PER S.F.			7.50	6.96	14.46

D3010 510	Apartment Building Heating - Fin Tube Radiation	COST PER S.F. MAT.	INST.	TOTAL
1740	Heating systems, fin tube radiation, forced hot water RD3020 -010			
1760	1,000 S.F. area, 10,000 C.F. volume	7.50	6.95	14.45
1800	10,000 S.F. area, 100,000 C.F. volume	3.14	4.18	7.32
1840	20,000 S.F. area, 200,000 C.F. volume	3.43	4.69	8.12
1880	30,000 S.F. area, 300,000 C.F. volume	3.32	4.55	7.87
1890				

D3010 Energy Supply

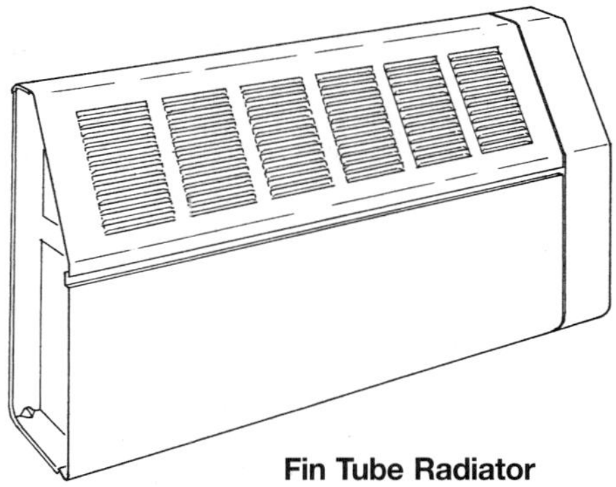

Fin Tube Radiator

Basis for Heat Loss Estimate, Factory or Commercial Type Structures:

1. Walls and flat roof are of lightly insulated concrete block or metal. U factor is assumed at .17.
2. Windows and doors are figured at 1-1/2 S.F. per linear foot of building perimeter. The U factor for flat single glass is 1.13.
3. Infiltration is approximately 5 C.F. per hour per S.F. of wall.
4. Concrete floor loss is 2 BTUH per S.F. of floor.
5. Temperature difference is assumed at 70°F.
6. Ventilation or makeup air has not been included and must be added if desired.

System Components	QUANTITY	UNIT	COST EACH		
			MAT.	INST.	TOTAL
SYSTEM D3010 520 1960					
HEATING SYSTEM, FIN TUBE RADIATION, FORCED HOT WATER					
1,000 S.F. BLDG., ONE FLOOR					
Boiler, oil fired, CI, burner/ctrls/insul/breech/pipe/ftngs/valves, 109 MBH	1.000	Ea.	4,462.50	4,068.75	8,531.25
Expansion tank, painted steel, ASME 18 Gal capacity	1.000	Ea.	2,775	110	2,885
Storage tank, steel, above ground, 550 Gal capacity w/supports	1.000	Ea.	4,325	490	4,815
Circulating pump, CI flanged, 1/8 HP	1.000	Ea.	690	214	904
Pipe, steel, black, sch. 40, threaded, cplg & hngr 10'OC, 1-1/2" diam.	260.000	L.F.	1,677	4,186	5,863
Pipe covering, calcium silicate w/cover , 1" wall, 1-1/2" diam	260.000	L.F.	1,180.40	1,885	3,065.40
Radiation, steel 1-1/4" tube & 4-1/4" fin w/cover & damper, wall hung	42.000	L.F.	1,827	1,533	3,360
Rough in, steel fin tube radiation w/supply & balance valves	4.000	Set	1,300	3,220	4,520
TOTAL			18,236.90	15,706.75	33,943.65
COST PER S.F.			18.24	15.71	33.95

D3010 520	Commercial Building Heating - Fin Tube Radiation	COST PER S.F.		
		MAT.	INST.	TOTAL
1940	Heating systems, fin tube radiation, forced hot water			
1960	1,000 S.F. bldg, one floor	18.20	15.70	33.90
2000	10,000 S.F., 100,000 C.F., total two floors (RD3020-010)	4.70	5.95	10.65
2040	100,000 S.F., 1,000,000 C.F., total three floors	1.90	2.67	4.57
2080	1,000,000 S.F., 10,000,000 C.F., total five floors	1.01	1.40	2.41
2090				

D3010 Energy Supply

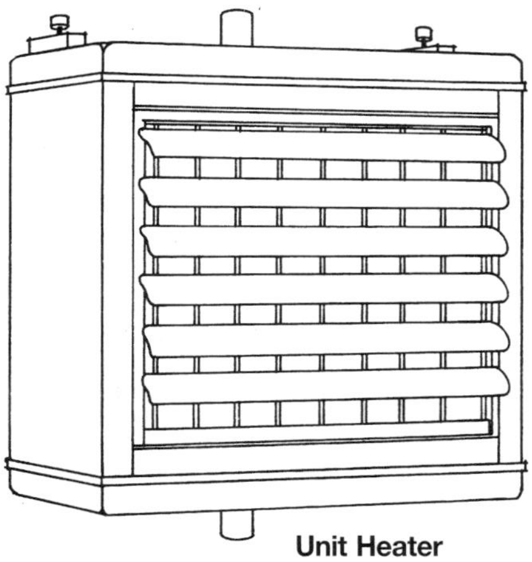

Unit Heater

Basis for Heat Loss Estimate, Factory or Commercial Type Structures:

1. Walls and flat roof are of lightly insulated concrete block or metal. U factor is assumed at .17.
2. Windows and doors are figured at 1-1/2 S.F. per linear foot of building perimeter. The U factor for flat single glass is 1.13.
3. Infiltration is approximately 5 C.F. per hour per S.F. of wall.
4. Concrete floor loss is 2 BTUH per S.F. of floor.
5. Temperature difference is assumed at 70°F.
6. Ventilation or makeup air has not been included and must be added if desired.

System Components	QUANTITY	UNIT	COST EACH MAT.	COST EACH INST.	COST EACH TOTAL
SYSTEM D3010 530 1880					
HEATING SYSTEM, TERMINAL UNIT HEATERS, FORCED HOT WATER					
1,000 S.F. BLDG., ONE FLOOR					
Boiler oil fired, CI, burner/ctrls/insul/breech/pipe/ftngs/valves, 109 MBH	1.000	Ea.	4,462.50	4,068.75	8,531.25
Expansion tank, painted steel, ASME 18 Gal capacity	1.000	Ea.	2,775	110	2,885
Storage tank, steel, above ground, 550 Gal capacity w/supports	1.000	Ea.	4,325	490	4,815
Circulating pump, CI, flanged, 1/8 HP	1.000	Ea.	690	214	904
Pipe, steel, black, schedule 40, threaded, cplg & hngr 10′ OC 1-1/2″ diam	260.000	L.F.	1,677	4,186	5,863
Pipe covering, calcium silicate w/cover, 1″ wall, 1-1/2″ diam	260.000	L.F.	1,180.40	1,885	3,065.40
Unit heater, 1 speed propeller, horizontal, 200° EWT, 26.9 MBH	2.000	Ea.	1,280	330	1,610
Unit heater piping hookup with controls	2.000	Set	1,590	3,100	4,690
TOTAL			17,979.90	14,383.75	32,363.65
COST PER S.F.			17.98	14.38	32.36

D3010 530	Commercial Bldg. Heating - Terminal Unit Heaters	COST PER S.F. MAT.	COST PER S.F. INST.	COST PER S.F. TOTAL
1860	Heating systems, terminal unit heaters, forced hot water			
1880	1,000 S.F. bldg., one floor	18	14.40	32.40
1920	10,000 S.F. bldg., 100,000 C.F. total two floors (RD3020-010)	4.25	4.91	9.16
1960	100,000 S.F. bldg., 1,000,000 C.F. total three floors	1.95	2.43	4.38
2000	1,000,000 S.F. bldg., 10,000,000 C.F. total five floors	1.32	1.57	2.89
2010				

D3010 Energy Supply

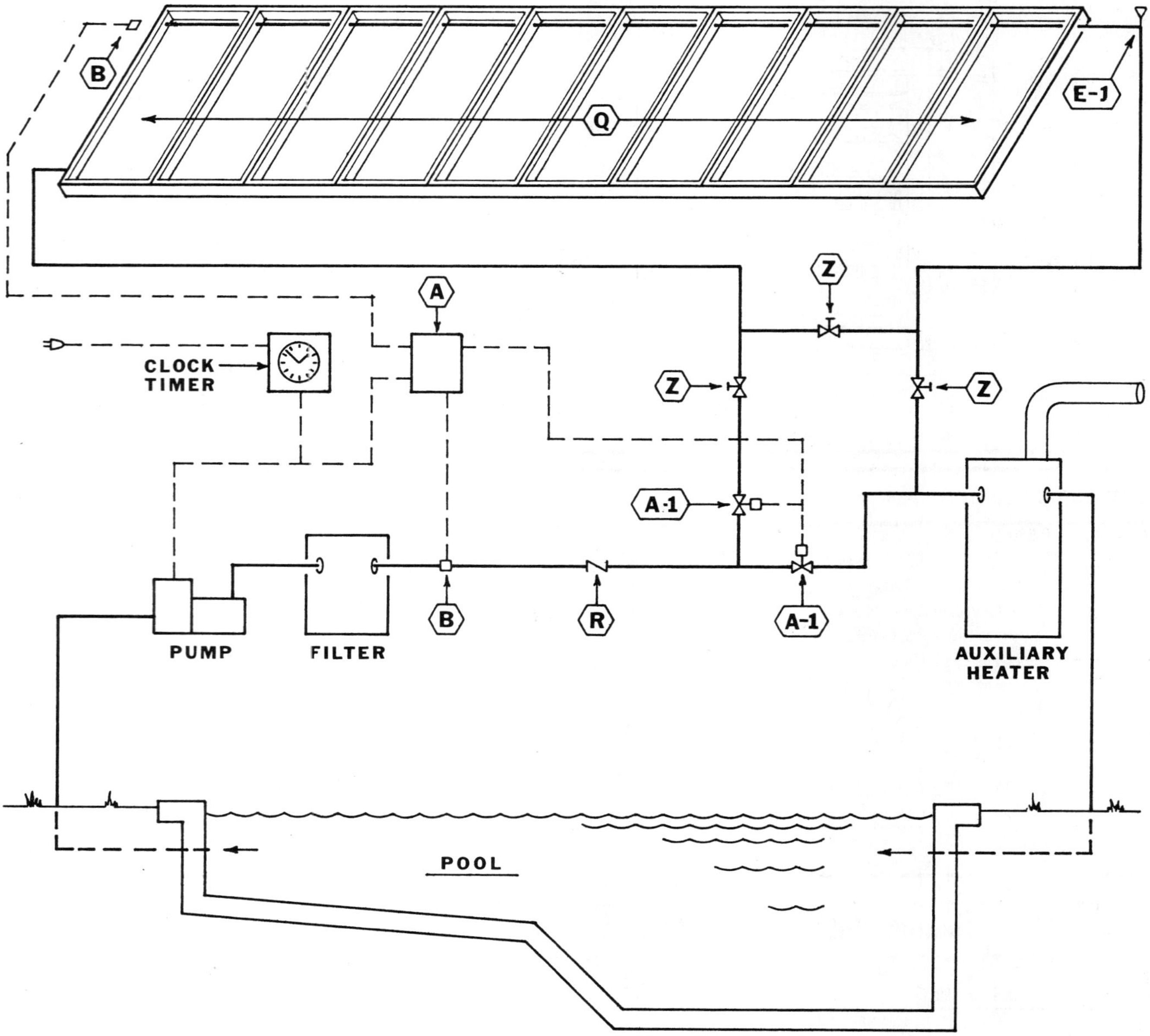

This draindown pool system uses a differential thermostat similar to those used in solar domestic hot water and space heating applications. To heat the pool, the pool water passes through the conventional pump-filter loop and then flows through the collectors. When collection is not possible, or when the pool temperature is reached, all water drains from the solar loop back to the pool through the existing piping. The modes are controlled by solenoid valves or other automatic valves in conjunction with a vacuum breaker relief valve, which facilitates draindown.

D3010 Energy Supply

System Components

System Components		QUANTITY	UNIT	COST EACH MAT.	COST EACH INST.	COST EACH TOTAL
	SYSTEM D3010 660 2640					
	SOLAR SWIMMING POOL HEATER, ROOF MOUNTED COLLECTORS					
	TEN 4′ X 10′ FULLY WETTED UNGLAZED PLASTIC ABSORBERS					
A	Differential thermostat/controller, 110V, adj pool pump system	1.000	Ea.	355	355	710
A-1	Solenoid valve, PVC, normally 1 open 1 closed (included)	2.000	Ea.			
B	Sensor, thermistor type (included)	2.000	Ea.			
E-1	Valve, vacuum relief	1.000	Ea.	32.50	22.50	55
Q	Collector panel, solar energy, plastic, liquid full wetted, 4′ x 10′	10.000	Ea.	3,550	2,570	6,120
R	Valve, ball check, PVC, socket, 1-1/2″ diam	1.000	Ea.	124	35.50	159.50
Z	Valve, ball, PVC, socket, 1-1/2″ diam	3.000	Ea.	324	106.50	430.50
	Pipe, PVC, sch 40, 1-1/2″ diam	80.000	L.F.	616	1,588	2,204
	Pipe fittings, PVC sch 40, socket joint, 1-1/2″ diam	10.000	Ea.	21.30	355	376.30
	Sensor wire, #22-2 conductor, multistranded	.500	C.L.F.	7.83	33	40.83
	Roof clamps for solar energy collector panels	10.000	Set	31.10	184	215.10
	Roof strap, teflon for solar energy collector panels	26.000	L.F.	715	90.74	805.74
	TOTAL			5,776.73	5,340.24	11,116.97

D3010 660	Solar Swimming Pool Heater Systems	COST EACH MAT.	COST EACH INST.	COST EACH TOTAL
2530	Solar swimming pool heater systems, roof mounted collectors			
2540	10 ea. 3′x7′ black chrome absorber, 1/8″ temp. glass	13,200	4,125	17,325
2560	10 ea. 4′x8′ black chrome absorber, 3/16″ temp. glass	15,000	4,900	19,900
2580	10 ea. 3′8″x6′ flat black absorber, 3/16″ temp. glass	11,900	4,200	16,100
2600	10 ea. 4′x9′ flat black absorber, plastic glazing	13,700	5,100	18,800
2620	10 ea. 2′x9′ rubber absorber, plastic glazing	15,100	5,525	20,625
2640	10 ea. 4′x10′ fully wetted unglazed plastic absorber	5,775	5,350	11,125
2660	Ground mounted collectors			
2680	10 ea. 3′x7′ black chrome absorber, 1/8″ temp. glass	13,200	4,600	17,800
2700	10 ea. 4′x8′ black chrome absorber, 3/16″ temp. glass	15,000	5,400	20,400
2720	10 ea. 3′8″x6′ flat blk absorber, 3/16″ temp. glass	11,900	4,700	16,600
2740	10 ea. 4′x9′ flat blk absorber, plastic glazing	13,700	5,600	19,300
2760	10 ea. 2′x9′ rubber absorber, plastic glazing	15,000	5,850	20,850
2780	10 ea. 4′x10′ fully wetted unglazed plastic absorber	5,775	5,825	11,600

RD3010 -600

D3010 Energy Supply

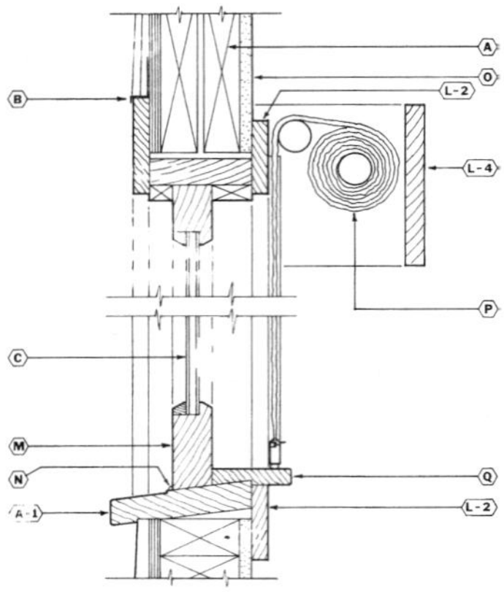

Solar Direct Gain Glazing — Door Panels: In a Direct Gain System the collection, absorption, and storage of solar energy occur directly within the building's living space. In this system, a section of the building's south wall is removed and replaced with glazing. The glazing in the System Component example consists of three standard insulating glass door panels preassembled in wood frames. A direct gain system also includes a thermal storage mass and night insulation.

During the heating season, solar radiation enters the living space directly through the glazing and is converted to heat, which either warms the air or is stored in the mass. The storage mass is typically masonry or water. Masonry often serves as a basic structural component of the building (e.g., a wall or floor). In this system the cost of adding mass to the building is not included. To reduce heat loss through the south facing glass, thereby increasing overall thermal performance, night insulation is included in this system.

System Components		QUANTITY	UNIT	COST EACH		
				MAT.	INST.	TOTAL
	SYSTEM D3010 682 2600 SOLAR PASSIVE HEATING, DIRECT GAIN, 3'X6'8" DOUBLE GLAZED DOOR PANEL 3 PANELS WIDE					
A	Framing headers over openings, fir 2" x 12"	.025	M.B.F.	28.05	51	79.05
A-1	Treated wood sleeper, 1" x 2"	.002	M.B.F.	3.60	6.10	9.70
B	Aluminum flashing, mill finish, .013 thick	5.000	S.F.	7.28	31.36	38.64
	Studs, 8' high wall, 2" x 4"	.020	M.B.F.	13.10	26	39.10
L-2	Interior casing, stock pine, 11/16" x 2-1/2"	25.000	L.F.	38.75	59.50	98.25
L-4	Valance 1" x 8"	12.000	L.F.	15.84	35.64	51.48
M	Vinyl clad sliding door, insulated glass, 6' wide	1.000	Ea.	2,336	380.16	2,716.16
N	Caulking/sealants, silicone rubber, cartridges	.250	Gal.	14.63		14.63
O	Drywall, gypsum plasterboard, nailed to studs, 5/8" standard	39.000	S.F.	14.43	23.01	37.44
O	For taping and finishing, add	39.000	S.F.	1.95	23.01	24.96
P	Mylar shade, dbl. layer heat reflective	49.640	S.F.	332.59	43.19	375.78
Q	Stool cap, pine, 11/16" x 3-1/2"	12.000	L.F.	27.36	35.64	63
	Paint wood trim	40.000	S.F.	5.60	24.40	30
	Demolition, framing	84.000	S.F.		19.74	19.74
	Demolition, shingles	84.000	S.F.		111.72	111.72
	Demolition, drywall	84.000	S.F.		39.48	39.48
	TOTAL			2,839.18	909.95	3,749.13

D3010 682	Solar Direct Gain Glazing, Door Panels	COST EACH		
		MAT.	INST.	TOTAL
2550	Solar passive heating, direct gain, 3'x6'8", double glazed door			
2560	One panel wide	1,425	490	1,915
2580	Two panels wide	2,175	710	2,885
2600	Three panels wide	2,850	910	3,760

D3010 Energy Supply

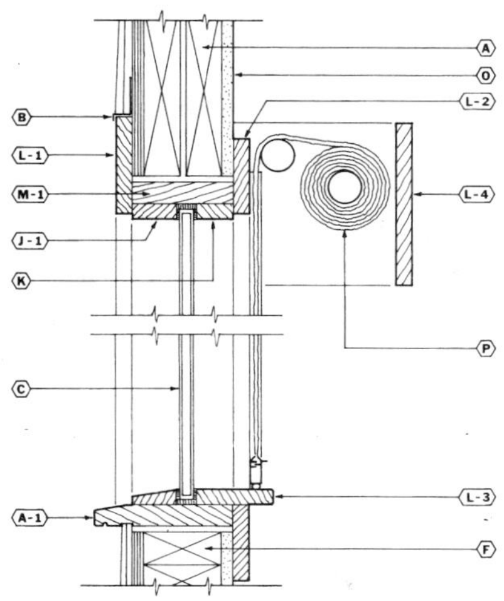

Solar Direct Gain Glazing — Window Units: In a Direct Gain System, collection, absorption, and storage of solar energy occur directly within the building's living space. In this system, a section of the building's south wall is removed and replaced with glazing. The glazing consists of standard insulating replacement glass units in wood site-built frames. A direct gain system also includes a thermal storage mass and night insulation.

During the heating season, solar radiation enters the living space directly through the glazing and is converted to heat, which either warms the air or is stored in the mass. The thermal storage mass is typically masonry or water. Masonry often serves as a basic structural component of the building (e.g., a wall or floor). In this system the cost of adding mass to the building is not included. To reduce heat loss through the south facing glass, thereby increasing overall thermal performance, night insulation is included in this system.

System Components		QUANTITY	UNIT	COST EACH		
				MAT.	INST.	TOTAL
	SYSTEM D3010 684 2600					
	SOLAR PASSIVE HEATING, DIRECT GAIN, 2'-6"X5' DBL. GLAZED WINDOW PANEL					
	3 PANELS WIDE					
A	Framing headers over openings, fir 2" x 8"	.022	M.B.F.	15.40	58.30	73.70
A-1	Wood exterior sill, oak, no horns, 8/4 x 8" deep	8.000	L.F.	58.40	95.20	153.60
B	Aluminum flashing, mill finish, .013 thick	4.150	S.F.	3.79	16.31	20.10
C	Glazing, insul glass 2 lite 3/16" float 5/8" thick unit, 15-30 S.F.	38.000	S.F.	571.90	480.70	1,052.60
F	2"x4" sill blocking	.013	M.B.F.	8.52	45.50	54.02
	Mullion	.027	M.B.F.	17.69	35.10	52.79
J-1	Exterior window stop 1" x 2"	45.000	L.F.	12.60	81	93.60
	Exterior sill stock water repellent 1" x 3"	8.000	L.F.	18.24	23.76	42
K	Interior window stop 1" x 3"	45.000	L.F.	64.35	111.60	175.95
L-1	Exterior trim and molding, corner board, pine, D & better, 1" x 6"	28.000	L.F.	29.52	71.28	100.80
L-2	Interior casing, 1" x 4"	28.000	L.F.	52.92	75.60	128.52
L-3	Apron	8.000	L.F.	16.32	21.60	37.92
L-4	Valance 1" x 8"	10.000	L.F.	13.20	29.70	42.90
M-1	Window frame	52.000	L.F.	296.40	128.96	425.36
O	Drywall, gypsum plasterboard, nailed/screw to studs, 5/8" standard	39.000	S.F.	14.43	23.01	37.44
O	For taping and finishing, add	39.000	S.F.	1.95	23.01	24.96
	Demolition, wall framing	60.000	S.F.		15.98	15.98
	Demolition, shingles	60.000	S.F.		79.80	79.80
P	Mylar triple layer heat reflective shade (see desc. sheet)	37.500	S.F.	320.63	32.63	353.26
	Demolition, drywall	60.000	S.F.		28.20	28.20
	Paint interior & exterior trim	200.000	L.F.	28	122	150
	TOTAL			1,544.26	1,599.24	3,143.50

D3010 684	Solar Direct Gain Glazing-Window Panels		COST EACH		
			MAT.	INST.	TOTAL
2550	Solar passive heating, direct gain, 2'-6"x5', double glazed window				
2560		One panel wide	555	585	1,140
2580		Two panels wide	1,125	1,550	2,675
2600		Three panels wide	1,550	1,600	3,150

D3010 Energy Supply

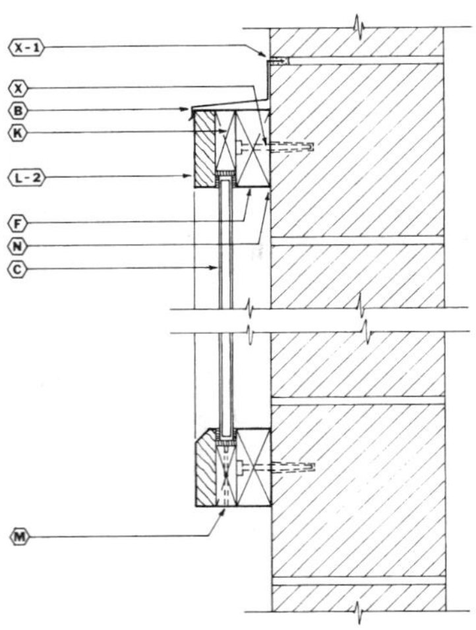

Solar Indirect Gain — Thermal Wall: The thermal storage wall is an Indirect Gain System because collection, absorption, and storage of solar energy occur outside the living space. This system consists of an unvented existing masonry wall located between south facing glazing and the space to be heated. The outside surface of the masonry wall is painted black to enable it to absorb maximum thermal energy. Standard double glazed units in wooden site-built frames are added to the south side of the building.

Thermal energy conducts through the masonry wall and radiates into the living space. A masonry thermal storage wall can also contain vents that allow warm air to flow into the living space through convection.

Indirect gain solar systems can be retrofitted to masonry walls constructed from a wide variety of building materials, including concrete, concrete block (solid and filled), brick, stone and adobe.

System Components		QUANTITY	UNIT	COST EACH MAT.	COST EACH INST.	COST EACH TOTAL
	SYSTEM D3010 686 2600					
	SOLAR PASSIVE HEATING, INDIRECT GAIN, THERMAL STORAGE WALL					
	3′ X 6′-8″ DOUBLE GLAZED PANEL, THREE PANELS WIDE					
B	Aluminum flashing, mill finish, .013 thick	5.500	S.F.	5.01	21.56	26.57
C	Glazing, insul glass 2 lite 3/16″ float 5/8″ thick unit, 15-30 S.F.	54.000	S.F.	812.70	683.10	1,495.80
F	Framing 2″ x 4″ lumber	.030	M.B.F.	19.65	71.25	90.90
K	Molding, miscellaneous, 1″ x 3″	44.000	L.F.	62.92	109.12	172.04
L-2	Wood batten, 1″ x 4″, custom pine or cedar	44.000	L.F.	43.12	130.68	173.80
M	Drill holes, 4 per panel 1/2″ diameter	30.000	Inch		39.60	39.60
N	Caulking/sealants, silicone rubber, cartridges	.250	Gal.	14.63		14.63
X	Lag screws 3/8″ x 3″ long	20.000	Ea.	7	79.20	86.20
X-1	Reglet	8.000	L.F.	.32	35.52	35.84
	Trim, primer & 1 coat	44.000	L.F.	9.68	33.88	43.56
	Layout and drill holes for lag screws	20.000	Ea.	1.20	189	190.20
	Expansion shields for lag screws	20.000	Ea.	9.20	74.40	83.60
	TOTAL			985.43	1,467.31	2,452.74

D3010 686	Passive Solar Indirect Gain Wall	COST EACH MAT.	COST EACH INST.	COST EACH TOTAL
2550	Solar passive heating, indirect gain			
2560	3′x6′8″, double glazed thermal storage wall, one panel wide	350	555	905
2580	Two panels wide	665	990	1,655
2600	Three panels wide	985	1,450	2,435

D3010 Energy Supply

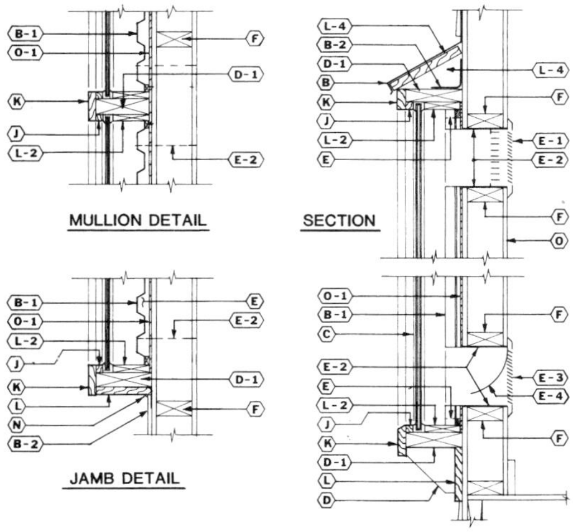

Solar Indirect Gain — Thermosyphon Panel: The Thermosyphon Air Panel (TAP) System is an indirect gain system used primarily for daytime heating because it does not include thermal mass. This site-built TAP System makes use of double glazing and a ribbed aluminum absorber plate. The panel is framed in wood and attached to the south side of a building. Solar radiation passes through the glass and is absorbed by the absorber plate. As the absorber plate heats, the air between it and the glass also heats. Vents cut through the building's south wall and the absorber plate near the top and bottom of the TAP allow hot air to circulate into the building by means of natural convection.

System Components		QUANTITY	UNIT	COST EACH MAT.	INST.	TOTAL
	SYSTEM D3010 688 2600					
	SOLAR PASSIVE HEATING, INDIRECT GAIN, THERMOSYPHON PANEL					
	3'X6'8" DOUBLE GLAZED PANEL, THREE PANELS WIDE					
B	Aluminum flashing and sleeves, mill finish, .013 thick	8.000	S.F.	20.93	90.16	111.09
B-1	Aluminum ribbed, 4" pitch, on steel frame .032" mill finish	45.000	S.F.	126.90	135.45	262.35
B-2	Angle brackets	9.000	Ea.	7.11	30.60	37.71
C	Glazing, insul glass 2 lite 3/16" float 5/8" thick unit, 15-30 S.F.	45.000	S.F.	677.25	569.25	1,246.50
D	Support brackets, fir 2"x6"	.003	M.B.F.	1.97	10.50	12.47
D-1	Framing fir 2" x 6"	.040	M.B.F.	34	119	153
E	Corrugated end closure	15.000	L.F.	15.45	43.65	59.10
E-1	Register, multilouvre operable 6" x 30"	3.000	Ea.	70.50	117	187.50
E-3	Grille 6" x 30"	3.000	Ea.	47.25	91.50	138.75
E-4	Waterproofed kraft with sisal or fiberglass fibers	.038	C.S.F.	.54	.61	1.15
F	House wall blocking 2" x 4"	.012	M.B.F.	26.20	57	83.20
J	Absorber supports & glass stop	54.000	L.F.	25.92	118.80	144.72
K	Exterior & bottom trim & interior glazing stop	40.000	L.F.	50.76	223.72	274.48
L	Exterior trim and molding, cornice board, pine, #2, 1" x 6"	24.000	L.F.	19.44	57.12	76.56
L-2	Trim cap	8.000	L.F.	10.56	23.76	34.32
N	Caulking/sealants, silicone rubber, cartridges	.250	Gal.	14.63		14.63
O	Drywall, nailed & taped, 5/8" thick	60.000	S.F.	22.20	35.40	57.60
O	For taping and finishing, add	60.000	S.F.	3	35.40	38.40
O-1	Insulation foil faced exterior wall	45.000	S.F.	21.60	19.80	41.40
	Demolition, framing	60.000	S.F.		19.74	19.74
	Demolition, drywall	60.000	S.F.		28.20	28.20
	Paint absorber plate black	45.000	S.F.	5.85	54.90	60.75
	Finish paint	125.000	S.F.	17.50	76.25	93.75
	Paint walls & ceiling with roller, primer & 1 coat	60.000	S.F.	8.40	30	38.40
	TOTAL			1,227.96	1,987.81	3,215.77

D3010 688	Passive Solar Indirect Gain Panel	COST EACH MAT.	INST.	TOTAL
2550	Solar passive heating, indirect gain			
2560	3'x6'8", double glazed thermosyphon panel, one panel wide	420	700	1,120
2580	Two panels wide	825	1,350	2,175
2600	Three panels wide	1,225	2,000	3,225

D3010 Energy Supply

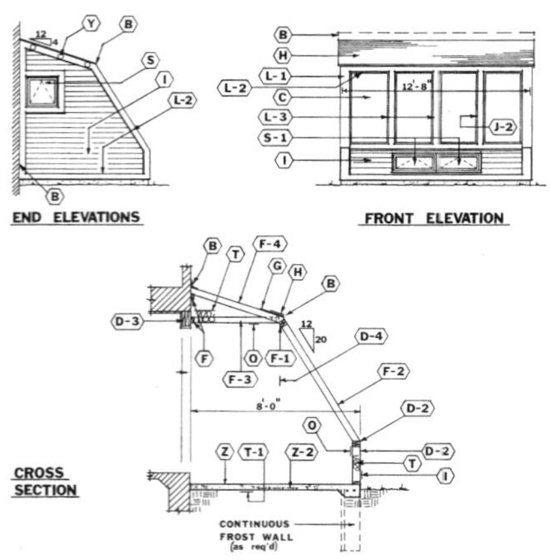

Solar Attached Sunspace: The attached sunspace, a room adjacent to the south side of a building, is designed solely for heat collection. During the heating season, solar radiation enters the sunspace through south facing glass. Radiation is absorbed by elements in the space and converted to heat. This sunspace is thermally isolated from the building. A set of sliding glass doors connecting the sunspace to the building space is closed to prevent night heat losses from the living space. This attached sunspace uses standard size, double glazed units and has a concrete slab floor. Since the required depth of footings (if required) is dependent on local frost conditions, costs of footings at depths of 2′, 3′ and 4′ are provided with the base cost of system as shown.

System Components		QUANTITY	UNIT	COST EACH MAT.	COST EACH INST.	COST EACH TOTAL
	SYSTEM D3010 690 2560					
	SOLAR PASSIVE HEATING, DIRECT GAIN, ATTACHED SUNSPACE					
	WOOD FRAMED, SLAB ON GRADE					
B	Aluminum, flashing, mill finish	21.000	S.F.	19.11	82.32	101.43
C	Glazing, double, tempered glass	72.000	S.F.	1,083.60	910.80	1,994.40
D-2-4	Knee & end wall framing	.188	M.B.F.	125.96	225.60	351.56
F	Ledgers, headers and rafters	.028	M.B.F.	118.56	257.93	376.49
	Building paper, asphalt felt sheathing paper 15 lb	240.000	S.F.	14.04	38.52	52.56
G	Sheathing plywood on roof CDX 1/2″	240.000	S.F.	180	204	384
H	Asphalt shingles inorganic class A 210-235 lb/sq	1.000	Sq.	89	103	192
I	Wood siding, match existing	164.000	S.F.	767.52	331.28	1,098.80
	Window stop, interior or external	96.000	L.F.	53.76	345.60	399.36
L-1	Exterior trim and molding, fascia, pine, D & better, 1″ x 6″	44.000	L.F.	54.12	104.72	158.84
L-2, L-3	Exterior trim 1″ x 4″	206.000	L.F.	201.88	611.82	813.70
O	Drywall, taping & finishing joints	200.000	S.F.	10	118	128
O	Drywall, 1/2″ on walls, standard, no finish	200.000	S.F.	72	118	190
S, S-1	Windows, wood, awning type, double glazed, with screens	4.000	Ea.	2,440	298	2,738
T	Insulation batts, fiberglass, faced	210.000	S.F.	142.80	92.40	235.20
T-1	Polyethylene vapor barrier, standard, 4 mil	2.500	C.S.F.	5.62	32.10	37.72
Y	Button vents	6.000	Ea.	16.38	59.40	75.78
Z	Slab on grade, not including finish, 4″ thick	1.234	C.Y.	175.23	130.29	305.52
	Floor finishing, steel trowel	100.000	S.F.		93	93
Z-2	Mesh, welded wire 6 x 6, #10/10	1.000	C.S.F.	14.85	37.50	52.35
	Paint interior, exterior and trim	330.000	S.F.	49.50	267.30	316.80
	TOTAL			5,633.93	4,461.58	10,095.51

D3010 690	Passive Solar Sunspace	COST EACH MAT.	COST EACH INST.	COST EACH TOTAL
2550	Solar passive heating, direct gain, attached sunspace,100 S.F.			
2560	Wood framed, slab on grade	5,625	4,450	10,075
2580	2′ deep frost wall	6,050	5,475	11,525
2600	3′ deep frost wall	6,200	5,850	12,050
2620	4′ deep frost wall	6,350	6,250	12,600

D3020 Heat Generating Systems

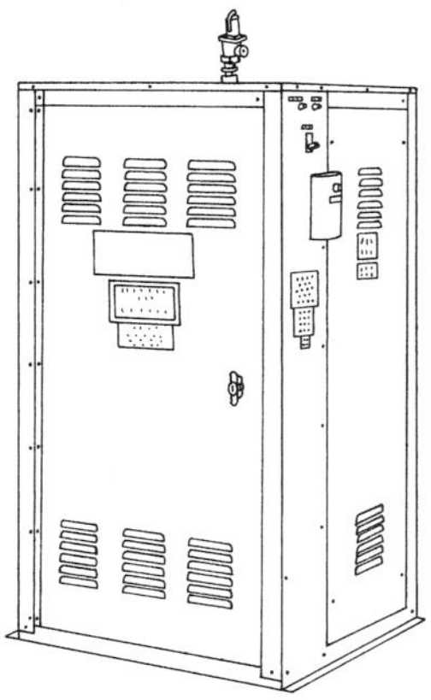

Boiler

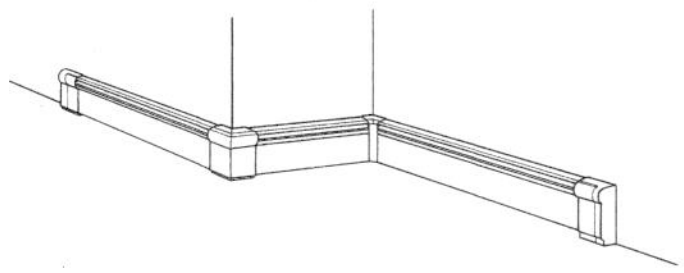

Baseboard Radiation

Small Electric Boiler System Considerations:

1. Terminal units are fin tube baseboard radiation rated at 720 BTU/hr with 200° water temperature or 820 BTU/hr steam.
2. Primary use being for residential or smaller supplementary areas, the floor levels are based on 7-1/2′ ceiling heights.
3. All distribution piping is copper for boilers through 205 MBH. All piping for larger systems is steel pipe.

System Components	QUANTITY	UNIT	COST EACH MAT.	COST EACH INST.	COST EACH TOTAL
SYSTEM D3020 102 1120					
SMALL HEATING SYSTEM, HYDRONIC, ELECTRIC BOILER					
1,480 S.F., 61 MBH, STEAM, 1 FLOOR					
Boiler, electric steam, std cntrls, trim, ftngs and valves, 18 KW, 61.4 MBH	1.000	Ea.	5,225	1,815	7,040
Copper tubing type L, solder joint, hanger 10′OC, 1-1/4″ diam	160.000	L.F.	1,768	1,968	3,736
Radiation, 3/4″ copper tube w/alum fin baseboard pkg 7″ high	60.000	L.F.	486	1,350	1,836
Rough in baseboard panel or fin tube with valves & traps	10.000	Set	2,650	6,900	9,550
Pipe covering, calcium silicate w/cover, 1″ wall 1-1/4″ diam	160.000	L.F.	716.80	1,160	1,876.80
Low water cut-off, quick hookup, in gage glass tappings	1.000	Ea.	289	46	335
TOTAL			11,134.80	13,239	24,373.80
COST PER S.F.			7.52	8.95	16.47

D3020 102	Small Heating Systems, Hydronic, Electric Boilers	COST PER S.F. MAT.	COST PER S.F. INST.	COST PER S.F. TOTAL
1100	Small heating systems, hydronic, electric boilers			
1120	Steam, 1 floor, 1480 S.F., 61 M.B.H.	7.50	8.92	16.42
1160	3,000 S.F., 123 M.B.H. RD3020 -010	5.75	7.85	13.60
1200	5,000 S.F., 205 M.B.H.	4.98	7.25	12.23
1240	2 floors, 12,400 S.F., 512 M.B.H.	3.88	7.20	11.08
1280	3 floors, 24,800 S.F., 1023 M.B.H.	4.32	7.10	11.42
1320	34,750 S.F., 1,433 M.B.H.	3.97	6.90	10.87
1360	Hot water, 1 floor, 1,000 S.F., 41 M.B.H.	12.65	4.95	17.60
1400	2,500 S.F., 103 M.B.H.	8.85	8.95	17.80
1440	2 floors, 4,850 S.F., 205 M.B.H.	8.45	10.75	19.20
1480	3 floors, 9,700 S.F., 410 M.B.H.	8.70	11.10	19.80

D3020 Heat Generating Systems

Boiler

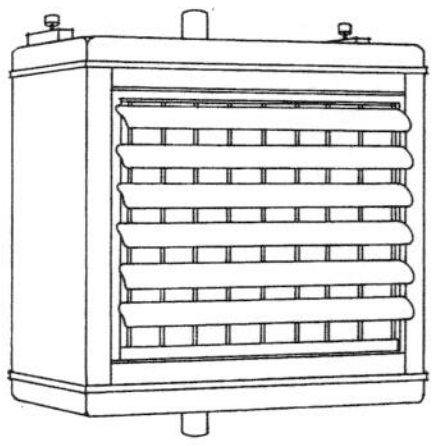
Unit Heater

Large Electric Boiler System Considerations:

1. Terminal units are all unit heaters of the same size. Quantities are varied to accommodate total requirements.
2. All air is circulated through the heaters a minimum of three times per hour.
3. As the capacities are adequate for commercial use, floor levels are based on 10′ ceiling heights.
4. All distribution piping is black steel pipe.

System Components	QUANTITY	UNIT	COST EACH MAT.	COST EACH INST.	COST EACH TOTAL
SYSTEM D3020 104 1240					
LARGE HEATING SYSTEM, HYDRONIC, ELECTRIC BOILER					
9,280 S.F., 135 KW, 461 MBH, 1 FLOOR					
Boiler, electric hot water, std ctrls, trim, ftngs, valves, 135 KW, 461 MBH	1.000	Ea.	12,225	3,975	16,200
Expansion tank, painted steel, 60 Gal capacity ASME	1.000	Ea.	4,350	220	4,570
Circulating pump, CI, close cpld, 50 GPM, 2 HP, 2″ pipe conn	1.000	Ea.	3,050	430	3,480
Unit heater, 1 speed propeller, horizontal, 200° EWT, 72.7 MBH	7.000	Ea.	6,545	1,680	8,225
Unit heater piping hookup with controls	7.000	Set	5,565	10,850	16,415
Pipe, steel, black, schedule 40, welded, 2-1/2″ diam	380.000	L.F.	4,408	10,966.80	15,374.80
Pipe covering, calcium silicate w/cover, 1″ wall, 2-1/2″ diam	380.000	L.F.	2,109	2,850	4,959
TOTAL			38,252	30,971.80	69,223.80
COST PER S.F.			4.12	3.34	7.46

D3020 104	Large Heating Systems, Hydronic, Electric Boilers	COST PER S.F. MAT.	COST PER S.F. INST.	COST PER S.F. TOTAL
1230	Large heating systems, hydronic, electric boilers			
1240	9,280 S.F., 135 K.W., 461 M.B.H., 1 floor	4.13	3.34	7.47
1280	14,900 S.F., 240 K.W., 820 M.B.H., 2 floors	5.35	5.40	10.75
1320	18,600 S.F., 296 K.W., 1,010 M.B.H., 3 floors — RD3020 -010	5.25	5.85	11.10
1360	26,100 S.F., 420 K.W., 1,432 M.B.H., 4 floors	5.15	5.75	10.90
1400	39,100 S.F., 666 K.W., 2,273 M.B.H., 4 floors	4.43	4.81	9.24
1440	57,700 S.F., 900 K.W., 3,071 M.B.H., 5 floors	4.24	4.74	8.98
1480	111,700 S.F., 1,800 K.W., 6,148 M.B.H., 6 floors	3.77	4.07	7.84
1520	149,000 S.F., 2,400 K.W., 8,191 M.B.H., 8 floors	3.71	4.05	7.76
1560	223,300 S.F., 3,600 K.W., 12,283 M.B.H., 14 floors	3.91	4.60	8.51

D3020 Heat Generating Systems

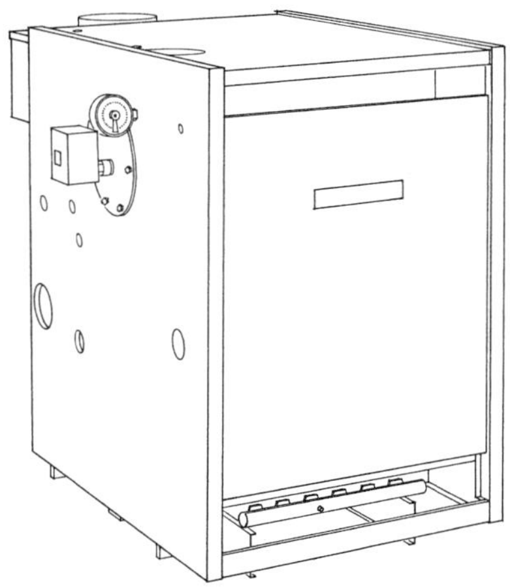

Boiler Selection: The maximum allowable working pressures are limited by ASME "Code for Heating Boilers" to 15 PSI for steam and 160 PSI for hot water heating boilers, with a maximum temperature limitation of 250°F. Hot water boilers are generally rated for a working pressure of 30 PSI. High pressure boilers are governed by the ASME "Code for Power Boilers" which is used almost universally for boilers operating over 15 PSIG. High pressure boilers used for a combination of heating/process loads are usually designed for 150 PSIG.

Boiler ratings are usually indicated as either Gross or Net Output. The Gross Load is equal to the Net Load plus a piping and pickup allowance. When this allowance cannot be determined, divide the gross output rating by 1.25 for a value equal to or greater than the next heat loss requirement of the building.

Table below lists installed cost per boiler and includes insulating jacket, standard controls, burner and safety controls. Costs do not include piping or boiler base pad. Outputs are Gross.

D3020 106	Boilers, Hot Water & Steam	COST EACH		
		MAT.	INST.	TOTAL
0600	Boiler, electric, steel, hot water, 12 K.W., 41 M.B.H.	5,450	1,525	6,975
0620	30 K.W., 103 M.B.H.	5,950	1,650	7,600
0640	60 K.W., 205 M.B.H. RD3020 -010	6,600	1,800	8,400
0660	120 K.W., 410 M.B.H.	7,025	2,200	9,225
0680	210 K.W., 716 M.B.H.	8,375	3,300	11,675
0700	510 K.W., 1,739 M.B.H.	25,400	6,150	31,550
0720	720 K.W., 2,452 M.B.H.	31,400	6,950	38,350
0740	1,200 K.W., 4,095 M.B.H.	39,900	7,975	47,875
0760	2,100 K.W., 7,167 M.B.H.	70,000	10,000	80,000
0780	3,600 K.W., 12,283 M.B.H.	103,500	16,900	120,400
0820	Steam, 6 K.W., 20.5 M.B.H.	4,500	1,650	6,150
0840	24 K.W., 81.8 M.B.H.	5,500	1,800	7,300
0860	60 K.W., 205 M.B.H.	7,600	1,975	9,575
0880	150 K.W., 512 M.B.H.	11,000	3,050	14,050
0900	510 K.W., 1,740 M.B.H.	35,900	7,525	43,425
0920	1,080 K.W., 3,685 M.B.H.	44,700	10,800	55,500
0940	2,340 K.W., 7,984 M.B.H.	95,000	16,900	111,900
0980	Gas, cast iron, hot water, 80 M.B.H.	2,175	1,900	4,075
1000	100 M.B.H.	2,825	2,075	4,900
1020	163 M.B.H.	3,475	2,800	6,275
1040	280 M.B.H.	4,750	3,100	7,850
1060	544 M.B.H.	10,800	5,500	16,300
1080	1,088 M.B.H.	14,900	6,975	21,875
1100	2,000 M.B.H.	23,500	10,900	34,400
1120	2,856 M.B.H.	33,900	14,000	47,900
1140	4,720 M.B.H.	81,500	19,200	100,700
1160	6,970 M.B.H.	144,500	31,400	175,900
1180	For steam systems under 2,856 M.B.H., add 8%			
1520	Oil, cast iron, hot water, 109 M.B.H.	2,550	2,325	4,875
1540	173 M.B.H.	3,225	2,800	6,025
1560	236 M.B.H.	4,100	3,275	7,375
1580	1,084 M.B.H.	12,000	7,425	19,425
1600	1,600 M.B.H.	15,400	10,700	26,100
1620	2,480 M.B.H.	22,600	13,600	36,200
1640	3,550 M.B.H.	30,100	16,300	46,400
1660	Steam systems same price as hot water			

D3020 Heat Generating Systems

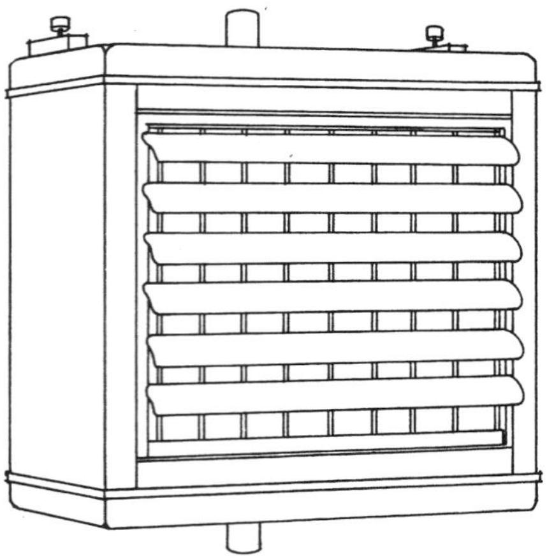

Unit Heater

Fossil Fuel Boiler System Considerations:

1. Terminal units are horizontal unit heaters. Quantities are varied to accommodate total heat loss per building.
2. Unit heater selection was determined by their capacity to circulate the building volume a minimum of three times per hour in addition to the BTU output.
3. Systems shown are forced hot water. Steam boilers cost slightly more than hot water boilers. However, this is compensated for by the smaller size or fewer terminal units required with steam.
4. Floor levels are based on 10′ story heights.
5. MBH requirements are gross boiler output.

System Components	QUANTITY	UNIT	COST EACH MAT.	COST EACH INST.	COST EACH TOTAL
SYSTEM D3020 108 1280					
HEATING SYSTEM, HYDRONIC, FOSSIL FUEL, TERMINAL UNIT HEATERS					
CAST IRON BOILER, GAS, 80 MBH, 1,070 S.F. BUILDING					
Boiler, gas, hot water, CI, burner, controls, insulation, breeching, 80 MBH	1.000	Ea.	2,283.75	1,995	4,278.75
Pipe, steel, black, schedule 40, threaded, cplg & hngr 10′OC, 2″ diam	200.000	L.F.	1,630	4,000	5,630
Unit heater, 1 speed propeller, horizontal, 200° EWT, 72.7 MBH	2.000	Ea.	1,870	480	2,350
Unit heater piping hookup with controls	2.000	Set	1,590	3,100	4,690
Expansion tank, painted steel, ASME, 18 Gal capacity	1.000	Ea.	2,775	110	2,885
Circulating pump, CI, flange connection, 1/12 HP	1.000	Ea.	415	214	629
Pipe covering, calcium silicate w/cover, 1″ wall, 2″ diam	200.000	L.F.	1,040	1,500	2,540
TOTAL			11,603.75	11,399	23,002.75
COST PER S.F.			10.84	10.65	21.49

D3020 108	Heating Systems, Unit Heaters	COST PER S.F. MAT.	COST PER S.F. INST.	COST PER S.F. TOTAL
1260	Heating systems, hydronic, fossil fuel, terminal unit heaters,			
1280	Cast iron boiler, gas, 80 M.B.H., 1,070 S.F. bldg.	10.84	10.65	21.49
1320	163 M.B.H., 2,140 S.F. bldg. RD3020 -010	7.30	7.25	14.55
1360	544 M.B.H., 7,250 S.F. bldg.	5.45	5.15	10.60
1400	1,088 M.B.H., 14,500 S.F. bldg.	4.47	4.85	9.32
1440	3,264 M.B.H., 43,500 S.F. bldg.	3.67	3.66	7.33
1480	5,032 M.B.H., 67,100 S.F. bldg.	4.21	3.75	7.96
1520	Oil, 109 M.B.H., 1,420 S.F. bldg.	12.10	9.50	21.60
1560	235 M.B.H., 3,150 S.F. bldg.	7.50	6.80	14.30
1600	940 M.B.H., 12,500 S.F. bldg.	5.35	4.43	9.78
1640	1,600 M.B.H., 21,300 S.F. bldg.	4.96	4.24	9.20
1680	2,480 M.B.H., 33,100 S.F. bldg.	4.95	3.82	8.77
1720	3,350 M.B.H., 44,500 S.F. bldg.	4.39	3.91	8.30
1760	Coal, 148 M.B.H., 1,975 S.F. bldg.	94.50	6.20	100.70
1800	300 M.B.H., 4,000 S.F. bldg.	51.50	4.86	56.36
1840	2,360 M.B.H., 31,500 S.F. bldg.	11.35	3.89	15.24

D3020 Heat Generating Systems

Fin Tube Radiator

Fossil Fuel Boiler System Considerations:

1. Terminal units are commercial steel fin tube radiation. Quantities are varied to accommodate total heat loss per building.
2. Systems shown are forced hot water. Steam boilers cost slightly more than hot water boilers. However, this is compensated for by the smaller size or fewer terminal units required with steam.
3. Floor levels are based on 10′ story heights.
4. MBH requirements are gross boiler output.

System Components

System Components	QUANTITY	UNIT	COST EACH MAT.	COST EACH INST.	COST EACH TOTAL
SYSTEM D3020 110 3240					
HEATING SYSTEM, HYDRONIC, FOSSIL FUEL, FIN TUBE RADIATION					
CAST IRON BOILER, GAS, 80 MBH, 1,070 S.F. BLDG.					
Boiler, gas, hot water, CI, burner, controls, insulation, breeching, 80 MBH	1.000	Ea.	2,283.75	1,995	4,278.75
Pipe, steel, black, schedule 40, threaded, 2″ diam	200.000	L.F.	1,630	4,000	5,630
Radiation,steel 1-1/4″ tube & 4-1/4″ fin w/cover & damper, wall hung	80.000	L.F.	3,480	2,920	6,400
Rough-in, wall hung steel fin radiation with supply & balance valves	8.000	Set	2,600	6,440	9,040
Expansion tank, painted steel, ASME, 18 Gal capacity	1.000	Ea.	2,775	110	2,885
Circulating pump, CI flanged, 1/12 HP	1.000	Ea.	415	214	629
Pipe covering, calcium silicate w/cover, 1″ wall, 2″ diam	200.000	L.F.	1,040	1,500	2,540
TOTAL			14,223.75	17,179	31,402.75
COST PER S.F.			13.29	16.06	29.35

D3020 110 Heating System, Fin Tube Radiation

D3020 110	Heating System, Fin Tube Radiation	COST PER S.F. MAT.	COST PER S.F. INST.	COST PER S.F. TOTAL
3230	Heating systems, hydronic, fossil fuel, fin tube radiation			
3240	Cast iron boiler, gas, 80 MBH, 1,070 S.F. bldg.	13.27	16.07	29.34
3280	169 M.B.H., 2,140 S.F. bldg. RD3020 -010	8.40	10.20	18.60
3320	544 M.B.H., 7,250 S.F. bldg.	7.20	8.75	15.95
3360	1,088 M.B.H., 14,500 S.F. bldg.	6.40	8.55	14.95
3400	3,264 M.B.H., 43,500 S.F. bldg.	5.75	7.55	13.30
3440	5,032 M.B.H., 67,100 S.F. bldg.	6.30	7.65	13.95
3480	Oil, 109 M.B.H., 1,420 S.F. bldg.	16.45	17.55	34
3520	235 M.B.H., 3,150 S.F. bldg.	9.25	10.30	19.55
3560	940 M.B.H., 12,500 S.F. bldg.	7.30	8.15	15.45
3600	1,600 M.B.H., 21,300 S.F. bldg.	7.05	8.10	15.15
3640	2,480 M.B.H., 33,100 S.F. bldg.	7.05	7.70	14.75
3680	3,350 M.B.H., 44,500 S.F. bldg.	6.45	7.80	14.25
3720	Coal, 148 M.B.H., 1,975 S.F. bldg.	96	9.70	105.70
3760	300 M.B.H., 4,000 S.F. bldg.	53.50	8.35	61.85
3800	2,360 M.B.H., 31,500 S.F. bldg.	13.35	7.70	21.05
4080	Steel boiler, oil, 97 M.B.H., 1,300 S.F. bldg.	12.25	14.40	26.65
4120	315 M.B.H., 4,550 S.F. bldg.	6.30	7.30	13.60
4160	525 M.B.H., 7,000 S.F. bldg.	9.10	8.45	17.55
4200	1,050 M.B.H., 14,000 S.F. bldg.	8.20	8.45	16.65
4240	2,310 M.B.H., 30,800 S.F. bldg.	7.10	7.75	14.85
4280	3,150 M.B.H., 42,000 S.F. bldg.	6.60	7.85	14.45

D3020 Heat Generating Systems

D3020 310	Gas Vent, Galvanized Steel, Double Wall	COST PER V.L.F.		
		MAT.	INST.	TOTAL
0030	Gas vent, galvanized steel, double wall, round pipe diameter, 3″	7.25	17.50	24.75
0040	4″	9	18.55	27.55
0050	5″	10.20	19.70	29.90
0060	6″	12.15	21	33.15
0070	7″	23	22.50	45.50
0080	8″	25	24	49
0090	10″	47.50	26.50	74
0100	12″	56.50	28.50	85
0110	14″	94	30	124
0120	16	135	31.50	166.50
0130	18″	166	33	199
0140	20″	196	54.50	250.50
0150	22″	250	57.50	307.50
0160	24″	310	61.50	371.50
0400				

D3020 310	Gas Vent Fittings, Galvanized Steel, Double Wall	COST EACH		
		MAT.	INST.	TOTAL
1020	Gas vent, fittings, 45° Elbow, adjustable thru 8″ diameter			
1030	Pipe diameter, 3″	12.85	35	47.85
1040	4″	15.50	37	52.50
1050	5″	18.05	39.50	57.55
1060	6″	22.50	42	64.50
1070	7″	35.50	45	80.50
1080	8″	47.50	48.50	96
1090	10″	101	52.50	153.50
1100	12″	108	57.50	165.50
1110	14″	169	60	229
1120	16″	220	63	283
1130	18″	289	66.50	355.50
1140	20″	320	109	429
1150	22″	515	115	630
1170	24″	655	123	778
1410	90° Elbow, adjustable, pipe diameter, 3″	22	35	57
1420	4″	25.50	37	62.50
1430	5″	30.50	39.50	70
1440	6″	37	42	79
1450	7″	63.50	45	108.50
1460	8″	64.50	48.50	113
1510	Tees, standard, pipe diameter, 3″	33.50	46.50	80
1520	4″	36	48.50	84.50
1530	5″	38	50.50	88.50
1540	6″	43	52.50	95.50
1550	7″	62	55	117
1560	8″	67.50	57.50	125
1570	10″	180	60	240
1580	12″	185	63	248
1590	14″	330	70	400
1600	16″	490	79	569
1610	18″	590	90	680
1620	20″	810	115	925
1630	22″	1,050	151	1,201
1640	24″	1,200	163	1,363
1660	30″	3,000	196	3,196
1700	38″	3,525	245	3,770
1750	48″	5,350	390	5,740

D3020 Heat Generating Systems

D3020 310	Gas Vent Fittings, Galvanized Steel, Double Wall	COST EACH MAT.	COST EACH INST.	COST EACH TOTAL
1810	Tee, access caps, pipe diameter, 3″	2.23	28	30.23
1820	4″	2.41	30	32.41
1830	5″	3.09	31.50	34.59
1840	6″	4.48	34	38.48
1850	7″	7.50	36	43.50
1860	8″	8.20	37	45.20
1870	10″	66.50	39.50	106
1880	12″	66	42	108
1890	14″	67	45	112
1900	16″	70.50	50.50	121
1910	18″	83.50	52.50	136
1920	20″	94	72.50	166.50
1930	22″	350	89	439
1940	24″	390	93.50	483.50
2110	Tops, bird proof, pipe diameter, 3″	17.05	27.50	44.55
2120	4″	17.95	28.50	46.45
2130	5″	19.90	30	49.90
2140	6″	22	31.50	53.50
2150	7″	49	33	82
2160	8″	60.50	35	95.50
2170	10″	99.50	37	136.50
2180	12″	121	39.50	160.50
2190	14″	194	42	236
2200	16″	249	45	294
2210	18″	345	48.50	393.50
2220	20″	520	70	590
2230	22″	795	89	884
2240	24″	990	98	1,088
2410	Roof flashing, tall cone/adjustable, pipe diameter, 3″	7.50	35	42.50
2420	4″	8.70	37	45.70
2430	5″	26.50	39.50	66
2440	6″	21	42	63
2450	7″	27.50	45	72.50
2460	8″	29.50	48.50	78
2470	10″	39.50	52.50	92
2480	12″	55.50	57.50	113
2490	14″	126	63	189
2500	16″	146	70	216
2510	18″	208	79	287
2520	20″	275	109	384
2530	22″	340	140	480
2540	24″	400	163	563

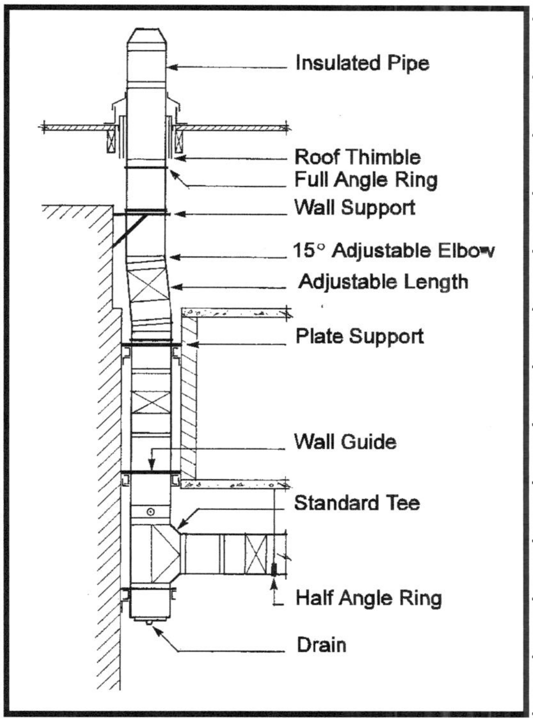

D3020 315	Chimney, Stainless Steel, Insulated, Wood & Oil	COST PER V.L.F. MAT.	COST PER V.L.F. INST.	COST PER V.L.F. TOTAL
0530	Chimney, stainless steel, insulated, diameter, 6″	71.50	21	92.50
0550	8″	82	24	106
0560	10″	91.50	26.50	118
0570	12″	105	28.50	133.50
0580	14″	118	30	148

D3030 Cooling Generating Systems

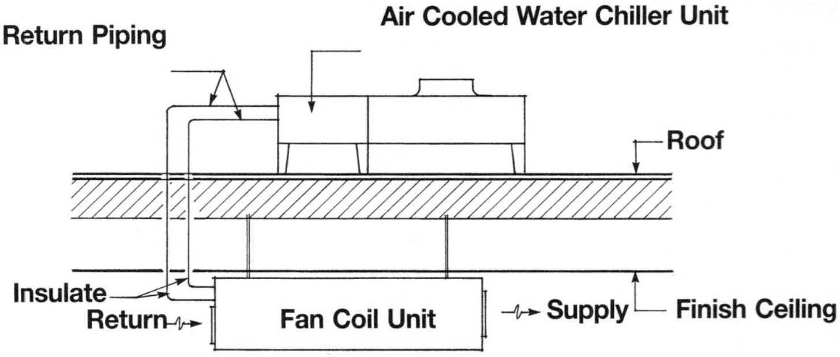

Design Assumptions: The chilled water, air cooled systems priced, utilize reciprocating hermetic compressors and propeller-type condenser fans. Piping with pumps and expansion tanks is included based on a two pipe system. No ducting is included and the fan-coil units are cooling only. Water treatment and balancing are not included. Chilled water piping is insulated. Area distribution is through the use of multiple fan coil units. Fewer but larger fan coil units with duct distribution would be approximately the same S.F. cost.

System Components	QUANTITY	UNIT	COST EACH MAT.	COST EACH INST.	COST EACH TOTAL
SYSTEM D3030 110 1200					
PACKAGED CHILLER, AIR COOLED, WITH FAN COIL UNIT					
APARTMENT CORRIDORS, 3,000 S.F., 5.50 TON					
Fan coil air conditioning unit, cabinet mounted & filters chilled water	1.000	Ea.	3,253.58	604.89	3,858.47
Water chiller, air conditioning unit, air cooled	1.000	Ea.	6,545	2,255	8,800
Chilled water unit coil connections	1.000	Ea.	1,275	1,650	2,925
Chilled water distribution piping	440.000	L.F.	8,426	22,220	30,646
TOTAL			19,499.58	26,729.89	46,229.47
COST PER S.F.			6.50	8.91	15.41

D3030 110	Chilled Water, Air Cooled Condenser Systems	COST PER S.F. MAT.	COST PER S.F. INST.	COST PER S.F. TOTAL
1180	Packaged chiller, air cooled, with fan coil unit			
1200	Apartment corridors, 3,000 S.F., 5.50 ton	6.50	8.90	15.40
1360	40,000 S.F., 73.33 ton (RD3030 -010)	5.10	3.89	8.99
1440	Banks and libraries, 3,000 S.F., 12.50 ton	9.95	10.35	20.30
1560	20,000 S.F., 83.33 ton	8.50	5.25	13.75
1680	Bars and taverns, 3,000 S.F., 33.25 ton	17.55	12.10	29.65
1760	10,000 S.F., 110.83 ton	12.70	2.76	15.46
1920	Bowling alleys, 3,000 S.F., 17.00 ton	12.55	11.60	24.15
2040	20,000 S.F., 113.33 ton	9.45	5.30	14.75
2160	Department stores, 3,000 S.F., 8.75 ton	9.15	9.90	19.05
2320	40,000 S.F., 116.66 ton	6.20	4	10.20
2400	Drug stores, 3,000 S.F., 20.00 ton	14.25	11.95	26.20
2520	20,000 S.F., 133.33 ton	11.10	5.55	16.65
2640	Factories, 2,000 S.F., 10.00 ton	8.75	9.95	18.70
2800	40,000 S.F., 133.33 ton	6.70	4.10	10.80
2880	Food supermarkets, 3,000 S.F., 8.50 ton	9	9.85	18.85
3040	40,000 S.F., 113.33 ton	6.05	4.01	10.06
3120	Medical centers, 3,000 S.F., 7.00 ton	8.10	9.60	17.70
3280	40,000 S.F., 93.33 ton	5.60	3.94	9.54
3360	Offices, 3,000 S.F., 9.50 ton	8.25	9.80	18.05
3520	40,000 S.F., 126.66 ton	6.60	4.14	10.74
3600	Restaurants, 3,000 S.F., 15.00 ton	11.15	10.75	21.90
3720	20,000 S.F., 100.00 ton	9.65	5.55	15.20
3840	Schools and colleges, 3,000 S.F., 11.50 ton	9.45	10.20	19.65
3960	20,000 S.F., 76.66 ton	8.35	5.30	13.65

D3030 Cooling Generating Systems

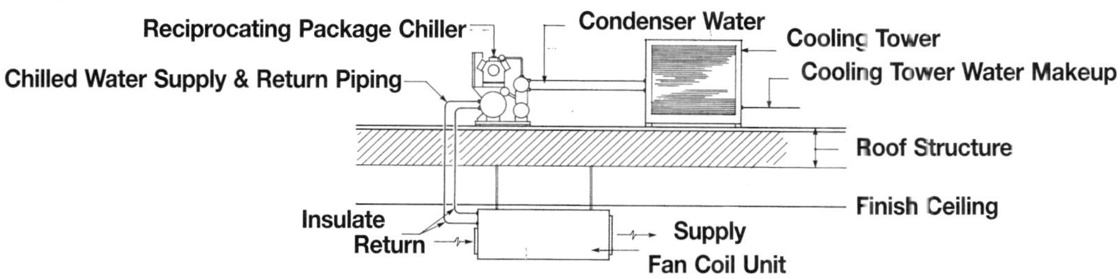

General: Water cooled chillers are available in the same sizes as air cooled units. They are also available in larger capacities.

Design Assumptions: The chilled water systems with water cooled condenser include reciprocating hermetic compressors, water cooling tower, pumps, piping and expansion tanks and are based on a two pipe system. Chilled water piping is insulated. No ducts are included and fan-coil units are cooling only. Area distribution is through use of multiple fan coil units. Fewer but larger fan coil units with duct distribution would be approximately the same S.F. cost. Water treatment and balancing are not included.

System Components	QUANTITY	UNIT	COST EACH		
			MAT.	INST.	TOTAL
SYSTEM D3030 115 1320					
PACKAGED CHILLER, WATER COOLED, WITH FAN COIL UNIT					
APARTMENT CORRIDORS, 4,000 S.F., 7.33 TON					
Fan coil air conditioner unit, cabinet mounted & filters, chilled water	2.000	Ea.	4,336.33	806.19	5,142.52
Water chiller, water cooled, 1 compressor, hermetic scroll,	1.000	Ea.	5,793.70	3,938.80	9,732.50
Cooling tower, draw thru single flow, belt drive	1.000	Ea.	1,737.21	168.59	1,905.80
Cooling tower pumps & piping	1.000	System	864.94	395.82	1,260.76
Chilled water unit coil connections	2.000	Ea.	2,550	3,300	5,850
Chilled water distribution piping	520.000	L.F.	9,958	26,260	36,218
TOTAL			25,240.18	34,869.40	60,109.58
COST PER S.F.			6.31	8.72	15.03
*Cooling requirements would lead to choosing a water cooled unit.					

D3030 115	Chilled Water, Cooling Tower Systems	COST PER S.F.		
		MAT.	INST.	TOTAL
1300	Packaged chiller, water cooled, with fan coil unit			
1320	Apartment corridors, 4,000 S.F., 7.33 ton	6.30	8.73	15.03
1600	Banks and libraries, 4,000 S.F., 16.66 ton (RD3030 -010)	10.55	9.45	20
1800	60,000 S.F., 250.00 ton	9.20	7.60	16.80
1880	Bars and taverns, 4,000 S.F., 44.33 ton	17	11.75	28.75
2000	20,000 S.F., 221.66 ton	20.50	9.70	30.20
2160	Bowling alleys, 4,000 S.F., 22.66 ton	12.35	10.35	22.70
2320	40,000 S.F., 226.66 ton	11.80	7.15	18.95
2440	Department stores, 4,000 S.F., 11.66 ton	7.20	9.40	16.60
2640	60,000 S.F., 175.00 ton	8.10	6.95	15.05
2720	Drug stores, 4,000 S.F., 26.66 ton	13.30	10.65	23.95
2880	40,000 S.F., 266.67 ton	12.10	8.10	20.20
3000	Factories, 4,000 S.F., 13.33 ton	9.10	9	18.10
3200	60,000 S.F., 200.00 ton	8.35	7.30	15.65
3280	Food supermarkets, 4,000 S.F., 11.33 ton	7.10	9.30	16.40
3480	60,000 S.F., 170.00 ton	8.05	6.95	15
3560	Medical centers, 4.000 S.F., 9.33 ton	6.15	8.55	14.70
3760	60,000 S.F., 140.00 ton	6.95	7.05	14
3840	Offices, 4,000 S.F., 12.66 ton	8.80	8.95	17.75
4040	60,000 S.F., 190.00 ton	8.10	7.20	15.30
4120	Restaurants, 4,000 S.F., 20.00 ton	11.05	9.60	20.65
4320	60,000 S.F., 300.00 ton	10.15	7.85	18
4400	Schools and colleges, 4,000 S.F., 15.33 ton	9.95	9.30	19.25
4600	60,000 S.F., 230.00 ton	8.55	7.30	15.85

D3030 Cooling Generating Systems

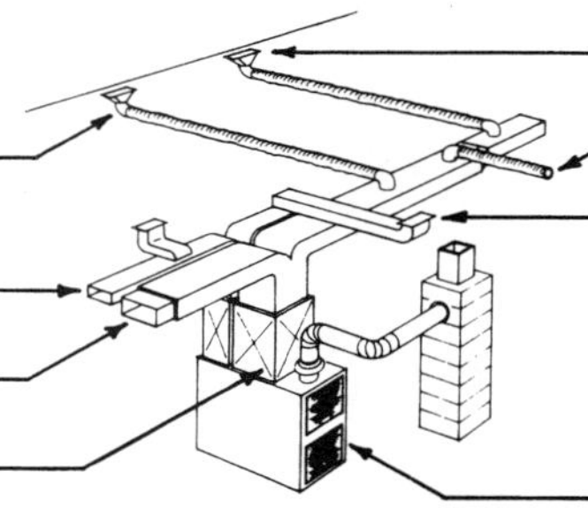

System Components	QUANTITY	UNIT	COST EACH MAT.	COST EACH INST.	COST EACH TOTAL
SYSTEM D3030 214 1200					
HEATING/COOLING, GAS FIRED FORCED AIR,					
ONE ZONE, 1200 S.F. BLDG, SEER 14					
Thermostat manual	1.000	Ea.	58.50	91.50	150
Intermittent pilot	1.000	Ea.	168		168
Furnace, 3 Ton cooling, 115 MBH	1.000	Ea.	2,500	420	2,920
Cooling tubing 25 feet	1.000	Ea.	305		305
Ductwork	158.000	Lb.	99.54	1,319.30	1,418.84
Ductwork connection	12.000	Ea.	195	504	699
Supply ductwork	176.000	SF Surf	153.12	985.60	1,138.72
Supply grill	2.000	Ea.	58	64	122
Duct insulation	1.000	L.F.	465.12	698.40	1,163.52
Return register	1.000	Ea.	123.60	264	387.60
TOTAL			4,125.88	4,346.80	8,472.68

D3030 214	Heating/Cooling System	COST EACH MAT.	COST EACH INST.	COST EACH TOTAL
1200	Heating/Cooling system, gas fired, 3 ton, SEER 14, 1200 S.F. Bldg	4,125	4,350	8,475
1300	5 ton, 2000 S.F. Bldg	4,975	6,900	11,875
1400	Heating/Cooling system, heat pump 3 ton, SEER 14, 1200 S.F. Bldg	5,275	5,575	10,850
1500	5 ton, SEER 14, 2000 S.F. Bldg	6,950	6,475	13,425

D3050 Terminal & Package Units

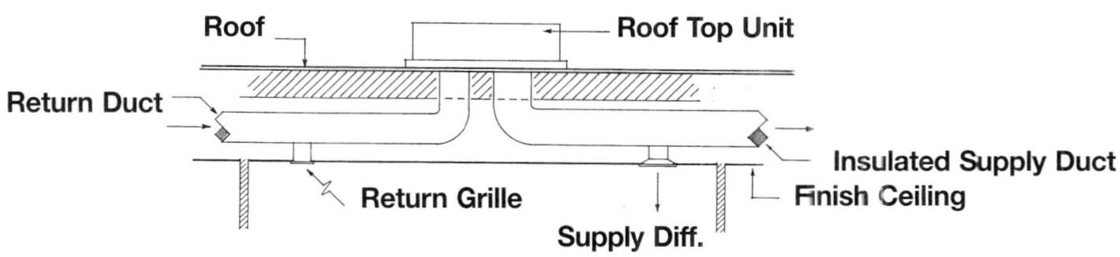

System Description: Rooftop single zone units are electric cooling and gas heat. Duct systems are low velocity, galvanized steel supply and return. Price variations between sizes are due to several factors. Jumps in the cost of the rooftop unit occur when the manufacturer shifts from the largest capacity unit on a small frame to the smallest capacity on the next larger frame, or changes from one compressor to two. As the unit capacity increases for larger areas the duct distribution grows in proportion. For most applications there is a tradeoff point where it is less expensive and more efficient to utilize smaller units with short simple distribution systems. Larger units also require larger initial supply and return ducts which can create a space problem. Supplemental heat may be desired in colder locations. The table below is based on one unit supplying the area listed. The 10,000 S.F. unit for bars and taverns is not listed because a nominal 110 ton unit would be required and this is above the normal single zone rooftop capacity.

System Components	QUANTITY	UNIT	COST EACH		
			MAT.	INST.	TOTAL
SYSTEM D3050 150 1280					
ROOFTOP, SINGLE ZONE, AIR CONDITIONER					
APARTMENT CORRIDORS, 500 S.F., .92 TON					
Rooftop air conditioner, 1 zone, electric cool, standard controls, curb	1.000	Ea.	1,414.50	862.50	2,277
Ductwork package for rooftop single zone units	1.000	System	172.96	1,219	1,391.96
TOTAL			1,587.46	2,081.50	3,668.96
COST PER S.F.			3.17	4.16	7.33

D3050 150	Rooftop Single Zone Unit Systems	COST PER S.F.		
		MAT.	INST.	TOTAL
1260	Rooftop, single zone, air conditioner			
1280	Apartment corridors, 500 S.F., .92 ton	3.15	4.15	7.30
1480	10,000 S.F., 18.33 ton (RD3030 -010)	3.32	2.81	6.13
1560	Banks or libraries, 500 S.F., 2.08 ton	7.20	9.40	16.60
1760	10,000 S.F., 41.67 ton	5.80	6.35	12.15
1840	Bars and taverns, 500 S.F. 5.54 ton	11.80	12.55	24.35
2000	5,000 S.F., 55.42 ton	13.25	9.55	22.80
2080	Bowling alleys, 500 S.F., 2.83 ton	6.10	10.60	16.70
2280	10,000 S.F., 56.67 ton	7.30	8.65	15.95
2360	Department stores, 500 S.F., 1.46 ton	5.05	6.60	11.65
2560	10,000 S.F., 29.17 ton	4.36	4.45	8.81
2640	Drug stores, 500 S.F., 3.33 ton	7.20	12.45	19.65
2840	10,000 S.F., 66.67 ton	8.60	10.15	18.75
2920	Factories, 500 S.F., 1.67 ton	5.75	7.55	13.30
3120	10,000 S.F., 33.33 ton	4.99	5.10	10.09
3200	Food supermarkets, 500 S.F., 1.42 ton	4.88	6.40	11.28
3400	10,000 S.F., 28.33 ton	4.23	4.31	8.54
3480	Medical centers, 500 S.F., 1.17 ton	4.03	5.30	9.33
3680	10,000 S.F., 23.33 ton	4.23	3.58	7.81
3760	Offices, 500 S.F., 1.58 ton	5.45	7.15	12.60
3960	10,000 S.F., 31.67 ton	4.74	4.83	9.57
4000	Restaurants, 500 S.F., 2.50 ton	8.65	11.30	19.95
4200	10,000 S.F., 50.00 ton	6.45	7.60	14.05
4240	Schools and colleges, 500 S.F., 1.92 ton	6.60	8.70	15.30
4440	10,000 S.F., 38.33 ton	5.35	5.85	11.20

D3050 Terminal & Package Units

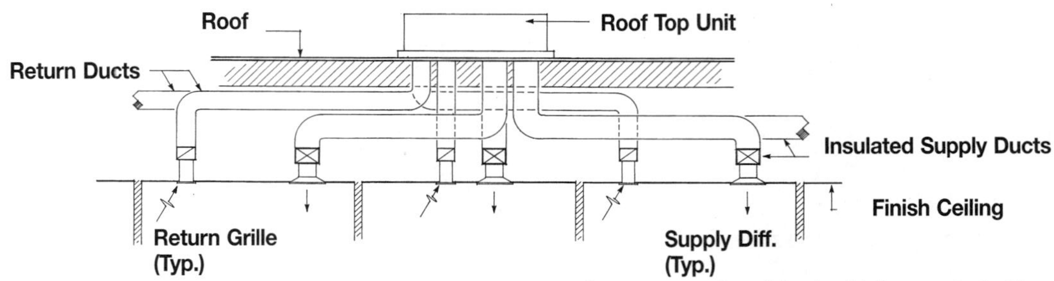

System Description: Rooftop units are multizone with up to 12 zones, and include electric cooling, gas heat, thermostats, filters, supply and return fans complete. Duct systems are low velocity, galvanized steel supply and return with insulated supplies.

Multizone units cost more per ton of cooling than single zone. However, they offer flexibility where load conditions are varied due to heat generating areas or exposure to radiational heating. For example, perimeter offices on the "sunny side" may require cooling at the same time "shady side" or central offices may require heating. It is possible to accomplish similar results using duct heaters in branches of the single zone unit. However, heater location could be a problem and total system operating energy efficiency could be lower.

System Components	QUANTITY	UNIT	COST EACH MAT.	COST EACH INST.	COST EACH TOTAL
SYSTEM D3050 155 1280					
ROOFTOP, MULTIZONE, AIR CONDITIONER					
APARTMENT CORRIDORS, 3,000 S.F., 5.50 TON					
Rooftop multizone unit, standard controls, curb	1.000	Ea.	31,900	2,013	33,913
Ductwork package for rooftop multizone units	1.000	System	1,523.50	13,887.50	15,411
TOTAL			33,423.50	15,900.50	49,324
COST PER S.F.			11.14	5.30	16.44
Note A: Small single zone unit recommended.					

D3050 155	Rooftop Multizone Unit Systems		COST PER S.F. MAT.	COST PER S.F. INST.	COST PER S.F. TOTAL
1240	Rooftop, multizone, air conditioner				
1260	Apartment corridors, 1,500 S.F., 2.75 ton. See Note A.				
1280	3,000 S.F., 5.50 ton	RD3030 -010	11.13	5.30	16.43
1440	25,000 S.F., 45.80 ton		7.30	5.10	12.40
1520	Banks or libraries, 1,500 S.F., 6.25 ton		25.50	12.05	37.55
1640	15,000 S.F., 62.50 ton		12.15	11.55	23.70
1720	25,000 S.F., 104.00 ton		10.80	11.50	22.30
1800	Bars and taverns, 1,500 S.F., 16.62 ton		55	17.35	72.35
1840	3,000 S.F., 33.24 ton		46	16.75	62.75
1880	10,000 S.F., 110.83 ton		27.50	16.65	44.15
2080	Bowling alleys, 1,500 S.F., 8.50 ton		34.50	16.40	50.90
2160	10,000 S.F., 56.70 ton		22.50	15.85	38.35
2240	20,000 S.F., 113.00 ton		14.70	15.65	30.35
2640	Drug stores, 1,500 S.F., 10.00 ton		40.50	19.25	59.75
2680	3,000 S.F., 20.00 ton		28	18.55	46.55
2760	15,000 S.F., 100.00 ton		17.35	18.45	35.80
3760	Offices, 1,500 S.F., 4.75 ton, See Note A.				
3880	15,000 S.F., 47.50 ton		12.65	8.85	21.50
3960	25,000 S.F., 79.16 ton		9.25	8.80	18.05
4000	Restaurants, 1,500 S.F., 7.50 ton		30.50	14.45	44.95
4080	10,000 S.F., 50.00 ton		19.95	13.95	33.90
4160	20,000 S.F., 100.00 ton		13	13.85	26.85
4240	Schools and colleges, 1,500 S.F., 5.75 ton		23.50	11.10	34.60
4360	15,000 S.F., 57.50 ton		11.20	10.65	21.85
4440	25,000 S.F., 95.83 ton		10.50	10.65	21.15

D3050 Terminal & Package Units

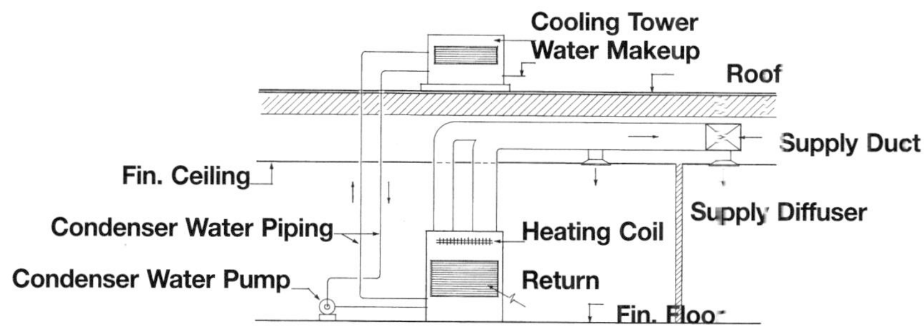

System Description: Self-contained, single package water cooled units include cooling tower, pump, piping allowance. Systems for 1000 S.F. and up include duct and diffusers to provide for even distribution of air. Smaller units distribute air through a supply air plenum, which is integral with the unit.

Returns are not ducted and supplies are not insulated.

Hot water or steam heating coils are included but piping to boiler and the boiler itself is not included.

Where local codes or conditions permit single pass cooling for the smaller units, deduct 10%.

System Components	QUANTITY	UNIT	COST EACH		
			MAT.	INST.	TOTAL
SYSTEM D3050 160 1300					
SELF-CONTAINED, WATER COOLED UNIT					
APARTMENT CORRIDORS, 500 S.F., .92 TON					
Self contained, water cooled, single package air conditioner unit	1.000	Ea.	1,318.80	688.80	2,007.60
Ductwork package for water or air cooled packaged units	1.000	System	61.18	920	981.18
Cooling tower, draw thru single flow, belt drive	1.000	Ea.	218.04	21.16	239.20
Cooling tower pumps & piping	1.000	System	108.56	49.68	158.24
TOTAL			1,706.58	1,679.64	3,386.22
COST PER S.F.			3.41	3.36	6.77

D3050 160	Self-contained, Water Cooled Unit Systems	COST PER S.F.		
		MAT.	INST.	TOTAL
1280	Self-contained, water cooled unit	3.42	3.36	6.78
1300	Apartment corridors, 500 S.F., .92 ton	3.40	3.35	6.75
1440	10,000 S.F., 18.33 ton (RD3030 -010)	3.94	2.32	6.26
1520	Banks or libraries, 500 S.F., 2.08 ton	7.45	3.44	10.89
1680	10,000 S.F., 41.66 ton	7.80	5.30	13.10
1760	Bars and taverns, 500 S.F., 5.54 ton	16.30	5.80	22.10
1920	10,000 S.F., 110.00 ton	19.30	8.20	27.50
2000	Bowling alleys, 500 S.F., 2.83 ton	10.15	4.70	14.85
2160	10,000 S.F., 56.66 ton	10.35	7.10	17.45
2200	Department stores, 500 S.F., 1.46 ton	5.25	2.42	7.67
2360	10,000 S.F., 29.17 ton	5.70	3.57	9.27
2440	Drug stores, 500 S.F., 3.33 ton	11.95	5.50	17.45
2600	10,000 S.F., 66.66 ton	12.15	8.35	20.50
2680	Factories, 500 S.F., 1.66 ton	5.95	2.76	8.71
2840	10,000 S.F., 33.33 ton	6.50	4.08	10.58
2920	Food supermarkets, 500 S.F., 1.42 ton	5.05	2.35	7.40
3080	10,000 S.F., 28.33 ton	5.50	3.46	8.96
3160	Medical centers, 500 S.F., 1.17 ton	4.18	1.93	6.11
3320	10,000 S.F., 23.33 ton	5	2.96	7.96
3400	Offices, 500 S.F., 1.58 ton	5.65	2.62	8.27
3560	10,000 S.F., 31.67 ton	6.15	3.87	10.02
3640	Restaurants, 500 S.F., 2.50 ton	8.95	4.15	13.10
3800	10,000 S.F., 50.00 ton	6.70	5.85	12.55
3880	Schools and colleges, 500 S.F., 1.92 ton	6.85	3.18	10.03
4040	10,000 S.F., 38.33 ton	7.20	4.87	12.07

D3050 Terminal & Package Units

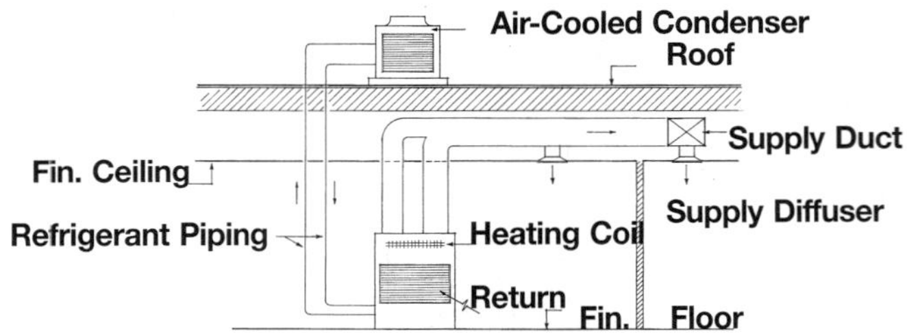

System Description: Self-contained air cooled units with remote air cooled condenser and interconnecting tubing. Systems for 1000 S.F. and up include duct and diffusers. Smaller units distribute air directly.

Returns are not ducted and supplies are not insulated.

Potential savings may be realized by using a single zone rooftop system or through-the-wall unit, especially in the smaller capacities, if the application permits.

Hot water or steam heating coils are included but piping to boiler and the boiler itself is not included.

Condenserless models are available for 15% less where remote refrigerant source is available.

System Components	QUANTITY	UNIT	COST EACH MAT.	COST EACH INST.	COST EACH TOTAL
SYSTEM D3050 165 1320					
SELF-CONTAINED, AIR COOLED UNIT					
APARTMENT CORRIDORS, 500 S.F., .92 TON					
Air cooled, package unit	1.000	Ea.	1,385.50	450.50	1,836
Ductwork package for water or air cooled packaged units	1.000	System	61.18	920	981.18
Refrigerant piping	1.000	System	372.60	607.20	979.80
Air cooled condenser, direct drive, propeller fan	1.000	Ea.	829.25	170.50	999.75
TOTAL			2,648.53	2,148.20	4,796.73
COST PER S.F.			5.30	4.30	9.60

D3050 165	Self-contained, Air Cooled Unit Systems	COST PER S.F. MAT.	COST PER S.F. INST.	COST PER S.F. TOTAL
1300	Self-contained, air cooled unit			
1320	Apartment corridors, 500 S.F., .92 ton	5.30	4.30	9.60
1480	10,000 S.F., 18.33 ton	3.34	3.47	6.81
1560	Banks or libraries, 500 S.F., 2.08 ton	11.50	5.50	17
1720	10,000 S.F., 41.66 ton	8.50	7.70	16.20
1800	Bars and taverns, 500 S.F., 5.54 ton	26.50	12	38.50
1960	10,000 S.F., 110.00 ton	23	14.90	37.90
2040	Bowling alleys, 500 S.F., 2.83 ton	15.80	7.55	23.35
2200	10,000 S.F., 56.66 ton	11.80	10.45	22.25
2240	Department stores, 500 S.F., 1.46 ton	8.15	3.87	12.02
2400	10,000 S.F., 29.17 ton	6.05	5.40	11.45
2480	Drug stores, 500 S.F., 3.33 ton	18.60	8.85	27.45
2640	10,000 S.F., 66.66 ton	14.05	12.35	26.40
2720	Factories, 500 S.F., 1.66 ton	9.45	4.46	13.91
2880	10,000 S.F., 33.33 ton	6.90	6.15	13.05
3200	Medical centers, 500 S.F., 1.17 ton	6.55	3.11	9.66
3360	10,000 S.F., 23.33 ton	4.27	4.42	8.69
3440	Offices, 500 S.F., 1.58 ton	8.85	4.21	13.06
3600	10,000 S.F., 31.66 ton	6.60	5.85	12.45
3680	Restaurants, 500 S.F., 2.50 ton	13.90	6.60	20.50
3840	10,000 S.F., 50.00 ton	10.75	9.20	19.95
3920	Schools and colleges, 500 S.F., 1.92 ton	10.70	5.10	15.80
4080	10,000 S.F., 38.33 ton	7.80	7.10	14.90

RD3030 -010

D3050 Terminal & Package Units

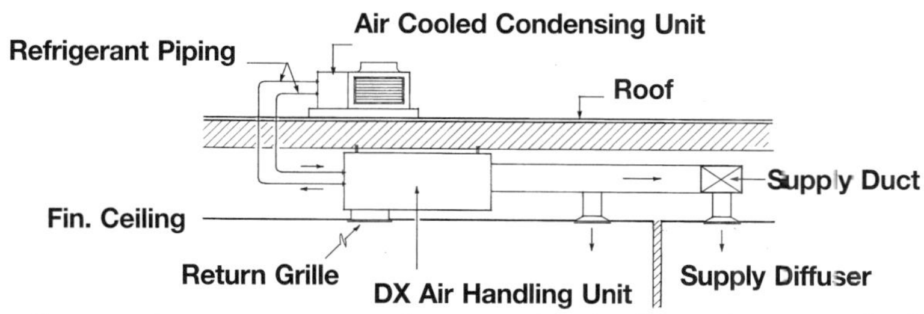

General: Split systems offer several important advantages which should be evaluated when a selection is to be made. They provide a greater degree of flexibility in component selection which permits an accurate match-up of the proper equipment size and type with the particular needs of the building. This allows for maximum use of modern energy saving concepts in heating and cooling. Outdoor installation of the air cooled condensing unit allows space savings in the building and also isolates the equipment operating sounds from building occupants.

Design Assumptions: The systems below are comprised of a direct expansion air handling unit and air cooled condensing unit with interconnecting copper tubing. Ducts and diffusers are also included for distribution of air. Systems are priced for cooling only. Heat can be added as desired either by putting hot water/steam coils into the air unit or into the duct supplying the particular area of need. Gas fired duct furnaces are also available. Refrigerant liquid line is insulated.

System Components	QUANTITY	UNIT	COST EACH MAT.	COST EACH INST.	COST EACH TOTAL
SYSTEM D3050 170 1280					
SPLIT SYSTEM, AIR COOLED CONDENSING UNIT					
APARTMENT CORRIDORS, 1,000 S.F., 1.80 TON					
Fan coil AC unit, cabinet mntd & filters direct expansion air cool	1.000	Ea.	411.75	161.04	572.79
Ductwork package, for split system, remote condensing unit	1.000	System	57.19	873.83	931.02
Refrigeration piping	1.000	System	345.87	1,006.50	1,352.37
Condensing unit, air cooled, incls compressor & standard controls	1.000	Ea.	1,708	646.60	2,354.60
TOTAL			2,522.81	2,687.97	5,210.78
COST PER S.F.			2.52	2.69	5.21

*Cooling requirements would lead to choosing a water cooled unit.

D3050 170	Split Systems With Air Cooled Condensing Units		COST PER S.F. MAT.	COST PER S.F. INST.	COST PER S.F. TOTAL
1260	Split system, air cooled condensing unit				
1280	Apartment corridors, 1,000 S.F., 1.83 ton		2.53	2.70	5.23
1440	20,000 S.F., 36.66 ton	RD3030 -010	2.15	3.76	5.91
1520	Banks and libraries, 1,000 S.F., 4.17 ton		4.07	6.20	10.27
1680	20,000 S.F., 83.32 ton		5.95	8.90	14.85
1760	Bars and taverns, 1,000 S.F., 11.08 ton		8.45	12.15	20.60
1880	10,000 S.F., 110.84 ton		13.10	14.20	27.30
2000	Bowling alleys, 1,000 S.F., 5.66 ton		5.60	11.50	17.10
2160	20,000 S.F., 113.32 ton		8.75	12.60	21.35
2320	Department stores, 1,000 S.F., 2.92 ton		2.88	4.25	7.13
2480	20,000 S.F., 58.33 ton		3.44	6	9.44
2560	Drug stores, 1,000 S.F., 6.66 ton		6.60	13.55	20.15
2720	20,000 S.F., 133.32 ton*				
2800	Factories, 1,000 S.F., 3.33 ton		3.28	4.85	8.13
2960	20,000 S.F., 66.66 ton		4.05	7.15	11.20
3040	Food supermarkets, 1,000 S.F., 2.83 ton		2.80	4.13	6.93
3200	20,000 S.F., 56.66 ton		3.34	5.80	9.14
3280	Medical centers, 1,000 S.F., 2.33 ton		2.51	3.32	5.83
3440	20,000 S.F., 46.66 ton		2.76	4.79	7.55
3520	Offices, 1,000 S.F., 3.17 ton		3.12	4.61	7.73
3680	20,000 S.F., 63.32 ton		3.84	6.80	10.64
3760	Restaurants, 1,000 S.F., 5.00 ton		4.97	10.15	15.12
3920	20,000 S.F., 100.00 ton		7.75	11.15	18.90
4000	Schools and colleges, 1,000 S.F., 3.83 ton		3.74	5.70	9.44
4160	20,000 S.F., 76.66 ton		4.65	8.20	12.85

D3050 Terminal & Package Units

Computer rooms impose special requirements on air conditioning systems. A prime requirement is reliability, due to the potential monetary loss that could be incurred by a system failure. A second basic requirement is the tolerance of control with which temperature and humidity are regulated, and dust eliminated. As the air conditioning system reliability is so vital, the additional cost of reserve capacity and redundant components is often justified.

System Descriptions: Computer areas may be environmentally controlled by one of three methods as follows:

1. **Self-contained Units**
   These are units built to higher standards of performance and reliability. They usually contain alarms and controls to indicate component operation failure, filter change, etc. It should be remembered that these units in the room will occupy space that is relatively expensive to build and that all alterations and service of the equipment will also have to be accomplished within the computer area.
2. **Decentralized Air Handling Units** In operation these are similar to the self-contained units except that their cooling capability comes from remotely located refrigeration equipment as refrigerant or chilled water. As no compressors or refrigerating equipment are required in the air units, they are smaller and require less service than self-contained units. An added plus for this type of system occurs if some of the computer components themselves also require chilled water for cooling.
3. **Central System Supply** Cooling is obtained from a central source which, since it is not located within the computer room, may have excess capacity and permit greater flexibility without interfering with the computer components. System performance criteria must still be met.

Note: The costs shown below do not include an allowance for ductwork or piping.

D3050 185	Computer Room Cooling Units	COST EACH		
		MAT.	INST.	TOTAL
0560	Computer room unit, air cooled, includes remote condenser			
0580	3 ton	21,500	2,650	24,150
0600	5 ton (RD3030 -010)	22,900	2,925	25,825
0620	8 ton	42,800	4,875	47,675
0640	10 ton	44,800	5,275	50,075
0660	15 ton	49,300	6,000	55,300
0680	20 ton	59,500	8,550	68,050
0700	23 ton	73,500	9,775	83,275
0800	Chilled water, for connection to existing chiller system			
0820	5 ton	16,400	1,775	18,175
0840	8 ton	16,700	2,650	19,350
0860	10 ton	16,900	2,700	19,600
0880	15 ton	17,600	2,750	20,350
0900	20 ton	18,700	2,875	21,575
0920	23 ton	19,900	3,250	23,150
1000	Glycol system, complete except for interconnecting tubing			
1020	3 ton	26,900	3,300	30,200
1040	5 ton	29,500	3,475	32,975
1060	8 ton	47,300	5,725	53,025
1080	10 ton	50,500	6,275	56,775
1100	15 ton	62,500	7,900	70,400
1120	20 ton	68,500	8,550	77,050
1140	23 ton	72,000	8,925	80,925
1240	Water cooled, not including condenser water supply or cooling tower			
1260	3 ton	21,800	2,125	23,925
1280	5 ton	23,800	2,450	26,250
1300	8 ton	39,000	4,000	43,000
1320	15 ton	47,200	4,875	52,075
1340	20 ton	50,500	5,400	55,900
1360	23 ton	53,500	5,850	59,350

D4010 Sprinklers

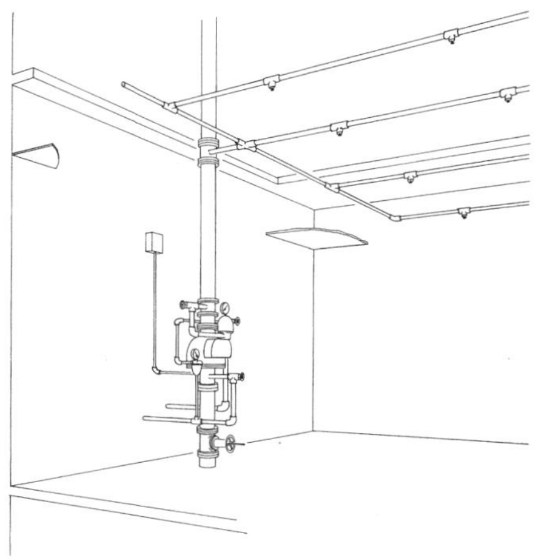

Dry Pipe System: A system employing automatic sprinklers attached to a piping system containing air under pressure, the release of which from the opening of sprinklers permits the water pressure to open a valve known as a "dry pipe valve". The water then flows into the piping system and out the opened sprinklers.

All areas are assumed to be open.

System Components	QUANTITY	UNIT	COST EACH MAT.	INST.	TOTAL
SYSTEM D4010 310 0580					
DRY PIPE SPRINKLER, STEEL, BLACK, SCH. 40 PIPE					
LIGHT HAZARD, ONE FLOOR, 2000 S.F.					
Valve, gate, iron body 125 lb., OS&Y, flanged, 4″ pipe size	1.000	Ea.	667.50	322.50	990
Valve, swing check, bronze, 125 lb, regrinding disc, 2-1/2″ pipe size	1.000	Ea.	693.75	64.50	758.25
Valve, angle, bronze, 150 lb., rising stem, threaded, 2″ pipe size	1.000	Ea.	656.25	48.75	705
*Alarm valve, 2-1/2″ pipe size	1.000	Ea.	1,425	315	1,740
Alarm, water motor, complete with gong	1.000	Ea.	360	131.25	491.25
Fire alarm horn, electric	1.000	Ea.	50.25	73.88	124.13
Valve swing check w/balldrip CI with brass trim, 4″ pipe size	1.000	Ea.	303.75	315	618.75
Pipe, steel, black, schedule 40, 4″ diam.	10.000	L.F.	144.38	275.48	419.86
Dry pipe valve, trim & gauges, 4″ pipe size	1.000	Ea.	2,475	937.50	3,412.50
Pipe, steel, black, schedule 40, threaded, cplg & hngr 10′OC 2-1/2″ diam.	20.000	L.F.	190.50	382.50	573
Pipe, steel, black, schedule 40, threaded, cplg & hngr 10′OC 2″ diam.	12.500	L.F.	76.41	187.50	263.91
Pipe, steel, black, schedule 40, threaded, cplg & hngr 10′OC 1-1/4″ diam.	37.500	L.F.	160.31	406.41	566.72
Pipe, steel, black, schedule 40, threaded, cplg & hngr 10′OC 1″ diam.	112.000	L.F.	397.32	1,134	1,531.32
Pipe Tee, malleable iron black, 150 lb. threaded, 4″ pipe size	2.000	Ea.	495	480	975
Pipe Tee, malleable iron black, 150 lb. threaded, 2-1/2″ pipe size	2.000	Ea.	138.75	214.50	353.25
Pipe Tee, malleable iron black, 150 lb. threaded, 2″ pipe size	1.000	Ea.	32.25	87.75	120
Pipe Tee, malleable iron black, 150 lb. threaded, 1-1/4″ pipe size	5.000	Ea.	75	345	420
Pipe Tee, malleable iron black, 150 lb. threaded, 1″ pipe size	4.000	Ea.	37.50	268.50	306
Pipe 90° elbow malleable iron black, 150 lb. threaded, 1″ pipe size	6.000	Ea.	35.78	247.50	283.28
Sprinkler head dry 1/2″ orifice 1″ NPT, 3″ to 4-3/4″ length	12.000	Ea.	1,740	600	2,340
Air compressor, 200 Gal sprinkler system capacity, 1/3 HP	1.000	Ea.	667.50	401.25	1,068.75
*Standpipe connection, wall, flush, brs. w/plug & chain 2-1/2″x2-1/2″	1.000	Ea.	138	189	327
Valve gate bronze, 300 psi, NRS, class 150, threaded, 1″ pipe size	1.000	Ea.	88.50	28.13	116.63
TOTAL			11,048.70	7,455.90	18,504.60
COST PER S.F.			5.52	3.73	9.25

*Not included in systems under 2000 S.F.

D4010 310	Dry Pipe Sprinkler Systems		COST PER S.F. MAT.	INST.	TOTAL
0520	Dry pipe sprinkler systems, steel, black, sch. 40 pipe				
0530	Light hazard, one floor, 500 S.F.		10.05	6.30	16.35
0560	1000 S.F.	RD4010 -100	5.70	3.72	9.42
0580	2000 S.F.		5.50	3.74	9.24
0600	5000 S.F.	RD4020 -300	2.87	2.54	5.41
0620	10,000 S.F.		2.01	2.09	4.10

D4010 Sprinklers

D4010 310	Dry Pipe Sprinkler Systems	COST PER S.F.		
		MAT.	INST.	TOTAL
0640	50,000 S.F.	1.47	1.88	3.35
0660	Each additional floor, 500 S.F.	2.04	3.09	5.13
0680	1000 S.F.	1.85	2.57	4.42
0700	2000 S.F.	1.81	2.33	4.14
0720	5000 S.F.	1.48	2.03	3.51
0740	10,000 S.F.	1.33	1.86	3.19
0760	50,000 S.F.	1.22	1.68	2.90
1000	Ordinary hazard, one floor, 500 S.F.	10.25	6.40	16.65
1020	1000 S.F.	5.90	3.73	9.63
1040	2000 S.F.	5.70	3.89	9.59
1060	5000 S.F.	3.33	2.72	6.05
1080	10,000 S.F.	2.65	2.75	5.40
1100	50,000 S.F.	2.37	2.60	4.97
1140	Each additional floor, 500 S.F.	2.26	3.17	5.43
1160	1000 S.F.	2.14	2.85	4.99
1180	2000 S.F.	2.18	2.59	4.77
1200	5000 S.F.	2.03	2.25	4.28
1220	10,000 S.F.	1.84	2.17	4.01
1240	50,000 S.F.	1.76	1.89	3.65
1500	Extra hazard, one floor, 500 S.F.	14.40	7.95	22.35
1520	1000 S.F.	8.80	5.75	14.55
1540	2000 S.F.	6.30	5	11.30
1560	5000 S.F.	3.83	3.79	7.62
1580	10,000 S.F.	4.14	3.59	7.73
1600	50,000 S.F.	4.47	3.49	7.96
1660	Each additional floor, 500 S.F.	3.32	3.92	7.24
1680	1000 S.F.	3.28	3.72	7
1700	2000 S.F.	3	3.74	6.74
1720	5000 S.F.	2.55	3.26	5.81
1740	10,000 S.F.	3.14	2.96	6.10
1760	50,000 S.F.	3.16	2.86	6.02
2020	Grooved steel, black, sch. 40 pipe, light hazard, one floor, 2000 S.F.	5.35	3.20	8.55
2060	10,000 S.F.	2.05	1.79	3.84
2100	Each additional floor, 2000 S.F.	1.84	1.90	3.74
2150	10,000 S.F.	1.37	1.56	2.93
2200	Ordinary hazard, one floor, 2000 S.F.	5.65	3.39	9.04
2250	10,000 S.F.	2.55	2.32	4.87
2300	Each additional floor, 2000 S.F.	2.12	2.09	4.21
2350	10,000 S.F.	1.88	2.09	3.97
2400	Extra hazard, one floor, 2000 S.F.	6.35	4.29	10.64
2450	10,000 S.F.	3.64	2.98	6.62
2500	Each additional floor, 2000 S.F.	3.06	3.08	6.14
2550	10,000 S.F.	2.78	2.60	5.38
3050	Grooved steel, black, sch. 10 pipe, light hazard, one floor, 2000 S.F.	5.30	3.18	8.48
3100	10,000 S.F.	2.03	1.77	3.80
3150	Each additional floor, 2000 S.F.	1.79	1.88	3.67
3200	10,000 S.F.	1.35	1.54	2.89
3250	Ordinary hazard, one floor, 2000 S.F.	5.60	3.36	8.96
3300	10,000 S.F.	2.49	2.27	4.76
3350	Each additional floor, 2000 S.F.	2.08	2.06	4.14
3400	10,000 S.F.	1.82	2.04	3.86
3450	Extra hazard, one floor, 2000 S.F.	6.35	4.26	10.61
3500	10,000 S.F.	3.46	2.94	6.40
3550	Each additional floor, 2000 S.F.	3.03	3.05	6.08
3600	10,000 S.F.	2.72	2.57	5.29
4050	Copper tubing, type M, light hazard, one floor, 2000 S.F.	5.65	3.16	8.81
4100	10,000 S.F.	2.51	1.80	4.31
4150	Each additional floor, 2000 S.F.	2.18	1.90	4.08

D4010 Sprinklers

D4010 310	Dry Pipe Sprinkler Systems	COST PER S.F.		
		MAT.	INST.	TOTAL
4200	10,000 S.F.	1.84	1.58	3.42
4250	Ordinary hazard, one floor, 2000 S.F.	6.10	3.53	9.63
4300	10,000 S.F.	3.08	2.14	5.22
4350	Each additional floor, 2000 S.F.	2.84	2.19	5.03
4400	10,000 S.F.	2.37	1.87	4.24
4450	Extra hazard, one floor, 2000 S.F.	7	4.31	11.31
4500	10,000 S.F.	5.15	3.25	8.40
4550	Each additional floor, 2000 S.F.	3.69	3.10	6.79
4600	10,000 S.F.	3.83	2.84	6.67
5050	Copper tubing, type M, T-drill system, light hazard, one floor			
5060	2000 S.F.	5.70	2.95	8.65
5100	10,000 S.F.	2.40	1.50	3.90
5150	Each additional floor, 2000 S.F.	2.20	1.69	3.89
5200	10,000 S.F.	1.73	1.28	3.01
5250	Ordinary hazard, one floor, 2000 S.F.	5.90	3.02	8.92
5300	10,000 S.F.	2.99	1.91	4.90
5350	Each additional floor, 2000 S.F.	2.40	1.72	4.12
5400	10,000 S.F.	2.22	1.59	3.81
5450	Extra hazard, one floor, 2000 S.F.	6.50	3.58	10.08
5500	10,000 S.F.	4.46	2.39	6.85
5550	Each additional floor, 2000 S.F.	3.22	2.37	5.59
5600	10,000 S.F.	3.12	1.98	5.10

D4010 Sprinklers

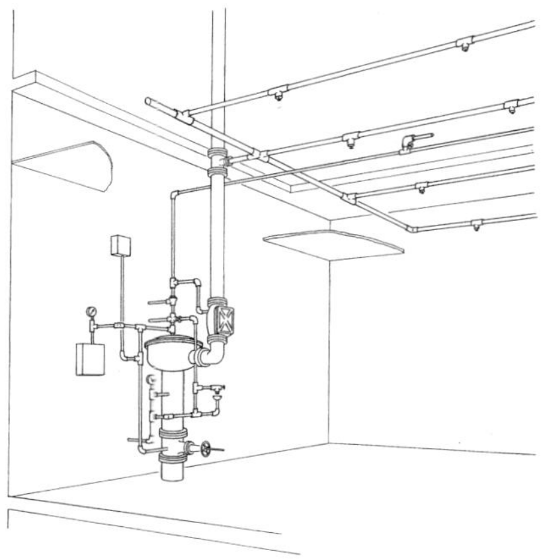

Pre-Action System: A system employing automatic sprinklers attached to a piping system containing air that may or may not be under pressure, with a supplemental heat responsive system of generally more sensitive characteristics than the automatic sprinklers themselves, installed in the same areas as the sprinklers. Actuation of the heat responsive system, as from a fire, opens a valve which permits water to flow into the sprinkler piping system and to be discharged from those sprinklers which were opened by heat from the fire.

All areas are assumed to be open.

System Components	QUANTITY	UNIT	COST EACH MAT.	COST EACH INST.	COST EACH TOTAL
SYSTEM D4010 350 0580					
PREACTION SPRINKLER SYSTEM, STEEL BLACK SCH. 40 PIPE					
LIGHT HAZARD, 1 FLOOR, 2000 S.F.					
Valve, gate, iron body 125 lb., OS&Y, flanged, 4″ pipe size	1.000	Ea.	667.50	322.50	990
*Valve, swing check w/ball drip CI with brass trim 4″ pipe size	1.000	Ea.	303.75	315	618.75
Valve, swing check, bronze, 125 lb, regrinding disc, 2-1/2″ pipe size	1.000	Ea.	693.75	64.50	758.25
Valve, angle, bronze, 150 lb., rising stem, threaded, 2″ pipe size	1.000	Ea.	656.25	48.75	705
*Alarm valve, 2-1/2″ pipe size	1.000	Ea.	1,425	315	1,740
Alarm, water motor, complete with gong	1.000	Ea.	360	131.25	491.25
Fire alarm horn, electric	1.000	Ea.	50.25	73.88	124.13
Thermostatic release for release line	2.000	Ea.	1,170	52.50	1,222.50
Pipe, steel, black, schedule 40, 4″ diam.	10.000	L.F.	144.38	275.48	419.86
Dry pipe valve, trim & gauges, 4″ pipe size	1.000	Ea.	2,475	937.50	3,412.50
Pipe, steel, black, schedule 40, threaded, cplg. & hngr. 10′OC 2-1/2″ diam.	20.000	L.F.	190.50	382.50	573
Pipe steel black, schedule 40, threaded, cplg. & hngr. 10′OC 2″ diam.	12.500	L.F.	76.41	187.50	263.91
Pipe, steel, black, schedule 40, threaded, cplg. & hngr. 10′OC 1-1/4″ diam.	37.500	L.F.	160.31	406.41	566.72
Pipe, steel, black, schedule 40, threaded, cplg. & hngr. 10′OC 1″ diam.	112.000	L.F.	397.32	1,134	1,531.32
Pipe, Tee, malleable iron, black, 150 lb. threaded, 4″ diam.	2.000	Ea.	495	480	975
Pipe, Tee, malleable iron, black, 150 lb. threaded, 2-1/2″ pipe size	2.000	Ea.	138.75	214.50	353.25
Pipe, Tee, malleable iron, black, 150 lb. threaded, 2″ pipe size	1.000	Ea.	32.25	87.75	120
Pipe, Tee, malleable iron, black, 150 lb. threaded, 1-1/4″ pipe size	5.000	Ea.	75	345	420
Pipe, Tee, malleable iron, black, 150 lb. threaded, 1″ pipe size	4.000	Ea.	37.50	268.50	306
Pipe, 90° elbow, malleable iron, blk., 150 lb. threaded, 1″ pipe size	6.000	Ea.	35.78	247.50	283.28
Sprinkler head, std. spray, brass 135°-286°F 1/2″ NPT, 3/8″ orifice	12.000	Ea.	213.60	522	735.60
Air compressor auto complete 200 Gal sprinkler sys. cap., 1/3 HP	1.000	Ea.	667.50	401.25	1,068.75
*Standpipe conn.,wall, flush, brass w/plug & chain 2-1/2″ x 2-1/2″	1.000	Ea.	138	189	327
Valve, gate, bronze, 300 psi, NRS, class 150, threaded, 1″ pipe size	1.000	Ea.	88.50	28.13	116.63
TOTAL			10,692.30	7,430.40	18,122.70
COST PER S.F.			5.35	3.72	9.07
*Not included in systems under 2000 S.F.					

D4010 350	Preaction Sprinkler Systems		COST PER S.F. MAT.	COST PER S.F. INST.	COST PER S.F. TOTAL
0520	Preaction sprinkler systems, steel, black sch. 40 pipe	RD4010 -100			
0530	Light hazard, one floor, 500 S.F.		9.90	5.05	14.95
0560	1000 S.F.	RD4020 -300	5.75	3.80	9.55
0580	2000 S.F.		5.35	3.73	9.08

D4010 Sprinklers

D4010 350	Preaction Sprinkler Systems	COST PER S.F.		
		MAT.	INST.	TOTAL
0600	5000 S.F.	2.73	2.53	5.26
0620	10,000 S.F.	1.88	2.08	3.96
0640	50,000 S.F.	1.34	1.87	3.21
0660	Each additional floor, 500 S.F.	2.40	2.76	5.16
0680	1000 S.F.	1.87	2.57	4.44
0700	2000 S.F.	1.82	2.34	4.16
0720	5000 S.F.	1.34	2.02	3.36
0740	10,000 S.F.	1.20	1.85	3.05
0760	50,000 S.F.	1.17	1.72	2.89
1000	Ordinary hazard, one floor, 500 S.F.	10.20	5.45	15.65
1020	1000 S.F.	5.70	3.72	9.42
1040	2000 S.F.	5.80	3.89	9.69
1060	5000 S.F.	3.04	2.70	5.74
1080	10,000 S.F.	2.30	2.73	5.03
1100	50,000 S.F.	2.01	2.57	4.58
1140	Each additional floor, 500 S.F.	2.67	3.18	5.85
1160	1000 S.F.	1.86	2.57	4.43
1180	2000 S.F.	1.71	2.56	4.27
1200	5000 S.F.	1.85	2.42	4.27
1220	10,000 S.F.	1.63	2.50	4.13
1240	50,000 S.F.	1.53	2.23	3.76
1500	Extra hazard, one floor, 500 S.F.	13.95	7	20.95
1520	1000 S.F.	8	5.25	13.25
1540	2000 S.F.	5.75	4.99	10.74
1560	5000 S.F.	3.48	4.09	7.57
1580	10,000 S.F.	3.56	3.97	7.53
1600	50,000 S.F.	3.81	3.88	7.69
1660	Each additional floor, 500 S.F.	3.35	3.91	7.26
1680	1000 S.F.	2.72	3.69	6.41
1700	2000 S.F.	2.44	3.71	6.15
1720	5000 S.F.	2	3.28	5.28
1740	10,000 S.F.	2.42	2.99	5.41
1760	50,000 S.F.	2.33	2.82	5.15
2020	Grooved steel, black, sch. 40 pipe, light hazard, one floor, 2000 S.F.	5.35	3.21	8.56
2060	10,000 S.F.	1.92	1.78	3.70
2100	Each additional floor of 2000 S.F.	1.85	1.91	3.76
2150	10,000 S.F.	1.24	1.55	2.79
2200	Ordinary hazard, one floor, 2000 S.F.	5.45	3.38	8.83
2250	10,000 S.F.	2.20	2.30	4.50
2300	Each additional floor, 2000 S.F.	1.94	2.08	4.02
2350	10,000 S.F.	1.53	2.07	3.60
2400	Extra hazard, one floor, 2000 S.F.	5.80	4.26	10.06
2450	10,000 S.F.	2.93	2.95	5.88
2500	Each additional floor, 2000 S.F.	2.50	3.05	5.55
2550	10,000 S.F.	2.03	2.56	4.59
3050	Grooved steel, black, sch. 10 pipe light hazard, one floor, 2000 S.F.	5.30	3.19	8.49
3100	10,000 S.F.	1.90	1.76	3.66
3150	Each additional floor, 2000 S.F.	1.80	1.89	3.69
3200	10,000 S.F.	1.22	1.53	2.75
3250	Ordinary hazard, one floor, 2000 S.F.	5.30	3.14	8.44
3300	10,000 S.F.	1.85	2.24	4.09
3350	Each additional floor, 2000 S.F.	1.90	2.05	3.95
3400	10,000 S.F.	1.47	2.02	3.49
3450	Extra hazard, one floor, 2000 S.F.	5.75	4.23	9.98
3500	10,000 S.F.	2.71	2.90	5.61
3550	Each additional floor, 2000 S.F.	2.47	3.02	5.49
3600	10,000 S.F.	1.97	2.53	4.50
4050	Copper tubing, type M, light hazard, one floor, 2000 S.F.	5.70	3.17	8.87

D4010 Sprinklers

D4010 350	Preaction Sprinkler Systems	COST PER S.F.		
		MAT.	INST.	TOTAL
4100	10,000 S.F.	2.38	1.79	4.17
4150	Each additional floor, 2000 S.F.	2.20	1.91	4.11
4200	10,000 S.F.	1.42	1.56	2.98
4250	Ordinary hazard, one floor, 2000 S.F.	5.90	3.52	9.42
4300	10,000 S.F.	2.73	2.12	4.85
4350	Each additional floor, 2000 S.F.	2.19	1.94	4.13
4400	10,000 S.F.	1.84	1.70	3.54
4450	Extra hazard, one floor, 2000 S.F.	6.45	4.28	10.73
4500	10,000 S.F.	4.38	3.20	7.58
4550	Each additional floor, 2000 S.F.	3.13	3.07	6.20
4600	10,000 S.F.	3.08	2.80	5.88
5050	Copper tubing, type M, T-drill system, light hazard, one floor			
5060	2000 S.F.	5.70	2.96	8.66
5100	10,000 S.F.	2.27	1.49	3.76
5150	Each additional floor, 2000 S.F.	2.21	1.70	3.91
5200	10,000 S.F.	1.60	1.27	2.87
5250	Ordinary hazard, one floor, 2000 S.F.	5.75	3.01	8.76
5300	10,000 S.F.	2.64	1.89	4.53
5350	Each additional floor, 2000 S.F.	2.22	1.72	3.94
5400	10,000 S.F.	1.97	1.66	3.63
5450	Extra hazard, one floor, 2000 S.F.	5.95	3.55	9.50
5500	10,000 S.F.	3.67	2.34	6.01
5550	Each additional floor, 2000 S.F.	2.66	2.34	5
5600	10,000 S.F.	2.37	1.94	4.31

D4010 Sprinklers

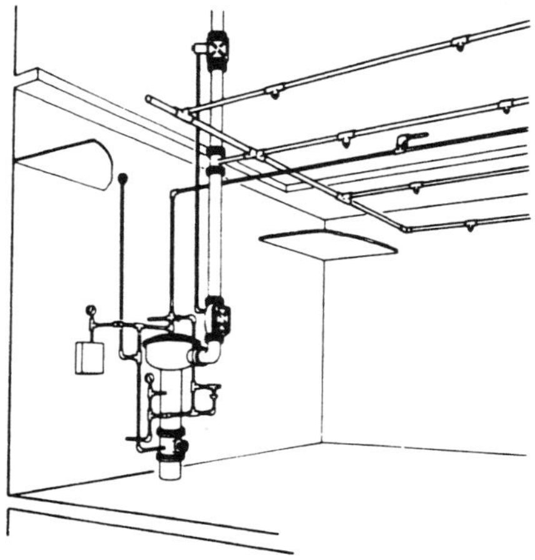

Deluge System: A system employing open sprinklers attached to a piping system connected to a water supply through a valve which is opened by the operation of a heat responsive system installed in the same areas as the sprinklers. When this valve opens, water flows into the piping system and discharges from all sprinklers attached thereto.

All areas are assumed to be open.

System Components	QUANTITY	UNIT	COST EACH MAT.	COST EACH INST.	COST EACH TOTAL
SYSTEM D4010 370 0580					
DELUGE SPRINKLER SYSTEM, STEEL BLACK SCH. 40 PIPE					
LIGHT HAZARD, 1 FLOOR, 2000 S.F.					
Valve, gate, iron body 125 lb., OS&Y, flanged, 4″ pipe size	1.000	Ea.	667.50	322.50	990
Valve, swing check w/ball drip, CI w/brass ftngs., 4″ pipe size	1.000	Ea.	303.75	315	618.75
Valve, swing check, bronze, 125 lb, regrinding disc, 2-1/2″ pipe size	1.000	Ea.	693.75	64.50	758.25
Valve, angle, bronze, 150 lb., rising stem, threaded, 2″ pipe size	1.000	Ea.	656.25	48.75	705
*Alarm valve, 2-1/2″ pipe size	1.000	Ea.	1,425	315	1,740
Alarm, water motor, complete with gong	1.000	Ea.	360	131.25	491.25
Fire alarm horn, electric	1.000	Ea.	50.25	73.88	124.13
Thermostatic release for release line	2.000	Ea.	1,170	52.50	1,222.50
Pipe, steel, black, schedule 40, 4″ diam.	10.000	L.F.	144.38	275.48	419.86
Deluge valve trim, pressure relief, emergency release, gauge, 4″ pipe size	1.000	Ea.	4,068.75	937.50	5,006.25
Deluge system, monitoring panel w/deluge valve & trim	1.000	Ea.	4,912.50	29.25	4,941.75
Pipe, steel, black, schedule 40, threaded, cplg & hngr 10′ OC 2-1/2″ diam.	20.000	L.F.	190.50	382.50	573
Pipe, steel, black, schedule 40, threaded, cplg & hngr 10′ OC 2″ diam.	12.500	L.F.	76.41	187.50	263.91
Pipe, steel, black, schedule 40, threaded, cplg & hngr 10′ OC 1-1/4″ diam.	37.500	L.F.	160.31	406.41	566.72
Pipe, steel, black, schedule 40, threaded, cplg & hngr 10′ OC 1″ diam.	112.000	L.F.	397.32	1,134	1,531.32
Pipe, Tee, malleable iron, black, 150 lb. threaded, 4″ pipe size	2.000	Ea.	495	480	975
Pipe, Tee, malleable iron, black, 150 lb. threaded, 2-1/2″ pipe size	2.000	Ea.	138.75	214.50	353.25
Pipe, Tee, malleable iron, black, 150 lb. threaded, 2″ pipe size	1.000	Ea.	32.25	87.75	120
Pipe, Tee, malleable iron, black, 150 lb. threaded, 1-1/4″ pipe size	5.000	Ea.	75	345	420
Pipe, Tee, malleable iron, black, 150 lb. threaded, 1″ pipe size	4.000	Ea.	37.50	268.50	306
Pipe, 90° elbow, malleable iron, black, 150 lb. threaded 1″ pipe size	6.000	Ea.	35.78	247.50	283.28
Sprinkler head, std spray, brass 135°-286°F 1/2″ NPT, 3/8″ orifice	9.720	Ea.	213.60	522	735.60
Air compressor, auto, complete, 200 Gal sprinkler sys. cap., 1/3 HP	1.000	Ea.	667.50	401.25	1,068.75
*Standpipe connection, wall, flush w/plug & chain 2-1/2″ x 2-1/2″	1.000	Ea.	138	189	327
Valve, gate, bronze, 300 psi, NRS, class 150, threaded, 1″ pipe size	1.000	Ea.	88.50	28.13	116.63
TOTAL			17,198.55	7,459.65	24,658.20
COST PER S.F.			8.60	3.73	12.33

*Not included in systems under 2000 S.F.

D4010 370	Deluge Sprinkler Systems		COST PER S.F. MAT.	COST PER S.F. INST.	COST PER S.F. TOTAL
0520	Deluge sprinkler systems, steel, black, sch. 40 pipe	RD4010 -100			
0530	Light hazard, one floor, 500 S.F.		21.50	5.10	26.60

D4010 Sprinklers

D4010 370	Deluge Sprinkler Systems	COST PER S.F. MAT.	COST PER S.F. INST.	COST PER S.F. TOTAL
0560	1000 S.F. RD4020 -300	11.55	3.66	15.21
0580	2000 S.F.	8.60	3.74	12.34
0600	5000 S.F.	4.03	2.54	6.57
0620	10,000 S.F.	2.53	2.08	4.61
0640	50,000 S.F.	1.47	1.87	3.34
0660	Each additional floor, 500 S.F.	2.40	2.76	5.16
0680	1000 S.F.	1.87	2.57	4.44
0700	2000 S.F.	1.82	2.34	4.16
0720	5000 S.F.	1.34	2.02	3.36
0740	10,000 S.F.	1.20	1.85	3.05
0760	50,000 S.F.	1.17	1.72	2.89
1000	Ordinary hazard, one floor, 500 S.F.	22.50	5.85	28.35
1020	1000 S.F.	11.55	3.75	15.30
1040	2000 S.F.	9.05	3.90	12.95
1060	5000 S.F.	4.34	2.71	7.05
1080	10,000 S.F.	2.95	2.73	5.68
1100	50,000 S.F.	2.20	2.61	4.81
1140	Each additional floor, 500 S.F.	2.67	3.18	5.85
1160	1000 S.F.	1.86	2.57	4.43
1180	2000 S.F.	1.71	2.56	4.27
1200	5000 S.F.	1.74	2.23	3.97
1220	10,000 S.F.	1.62	2.20	3.82
1240	50,000 S.F.	1.46	2.05	3.51
1500	Extra hazard, one floor, 500 S.F.	25.50	7.05	32.55
1520	1000 S.F.	14.30	5.45	19.75
1540	2000 S.F.	9	5	14
1560	5000 S.F.	4.55	3.77	8.32
1580	10,000 S.F.	4.09	3.63	7.72
1600	50,000 S.F.	4.22	3.56	7.78
1660	Each additional floor, 500 S.F.	3.35	3.91	7.26
1680	1000 S.F.	2.72	3.69	6.41
1700	2000 S.F.	2.44	3.71	6.15
1720	5000 S.F.	2	3.28	5.28
1740	10,000 S.F.	2.46	3.10	5.56
1760	50,000 S.F.	2.46	3	5.46
2000	Grooved steel, black, sch. 40 pipe, light hazard, one floor			
2020	2000 S.F.	8.60	3.22	11.82
2060	10,000 S.F.	2.59	1.80	4.39
2100	Each additional floor, 2,000 S.F.	1.85	1.91	3.76
2150	10,000 S.F.	1.24	1.55	2.79
2200	Ordinary hazard, one floor, 2000 S.F.	5.45	3.38	8.83
2250	10,000 S.F.	2.85	2.30	5.15
2300	Each additional floor, 2000 S.F.	1.94	2.08	4.02
2350	10,000 S.F.	1.53	2.07	3.60
2400	Extra hazard, one floor, 2000 S.F.	9.05	4.27	13.32
2450	10,000 S.F.	3.61	2.96	6.57
2500	Each additional floor, 2000 S.F.	2.50	3.05	5.55
2550	10,000 S.F.	2.03	2.56	4.59
3000	Grooved steel, black, sch. 10 pipe, light hazard, one floor			
3050	2000 S.F.	7.90	3.06	10.96
3100	10,000 S.F.	2.55	1.76	4.31
3150	Each additional floor, 2000 S.F.	1.80	1.89	3.69
3200	10,000 S.F.	1.22	1.53	2.75
3250	Ordinary hazard, one floor, 2000 S.F.	8.65	3.36	12.01
3300	10,000 S.F.	2.50	2.24	4.74
3350	Each additional floor, 2000 S.F.	1.90	2.05	3.95
3400	10,000 S.F.	1.47	2.02	3.49
3450	Extra hazard, one floor, 2000 S.F.	9	4.24	13.24

D4010 Sprinklers

D4010 370	Deluge Sprinkler Systems	COST PER S.F.		
		MAT.	INST.	TOTAL
3500	10,000 S.F.	3.36	2.90	6.26
3550	Each additional floor, 2000 S.F.	2.47	3.02	5.49
3600	10,000 S.F.	1.97	2.53	4.50
4000	Copper tubing, type M, light hazard, one floor			
4050	2000 S.F.	8.90	3.18	12.08
4100	10,000 S.F.	3.03	1.79	4.82
4150	Each additional floor, 2000 S.F.	2.19	1.91	4.10
4200	10,000 S.F.	1.42	1.56	2.98
4250	Ordinary hazard, one floor, 2000 S.F.	9.15	3.53	12.68
4300	10,000 S.F.	3.38	2.12	5.50
4350	Each additional floor, 2000 S.F.	2.19	1.94	4.13
4400	10,000 S.F.	1.84	1.70	3.54
4450	Extra hazard, one floor, 2000 S.F.	9.70	4.29	13.99
4500	10,000 S.F.	5.10	3.22	8.32
4550	Each additional floor, 2000 S.F.	3.13	3.07	6.20
4600	10,000 S.F.	3.08	2.80	5.88
5000	Copper tubing, type M, T-drill system, light hazard, one floor			
5050	2000 S.F.	8.95	2.97	11.92
5100	10,000 S.F.	2.92	1.49	4.41
5150	Each additional floor, 2000 S.F.	2.25	1.72	3.97
5200	10,000 S.F.	1.60	1.27	2.87
5250	Ordinary hazard, one floor, 2000 S.F.	9	3.02	12.02
5300	10,000 S.F.	3.29	1.89	5.18
5350	Each additional floor, 2000 S.F.	2.22	1.71	3.93
5400	10,000 S.F.	1.97	1.66	3.63
5450	Extra hazard, one floor, 2000 S.F.	9.20	3.56	12.76
5500	10,000 S.F.	4.32	2.34	6.66
5550	Each additional floor, 2000 S.F.	2.66	2.34	5
5600	10,000 S.F.	2.37	1.94	4.31

D4010 Sprinklers

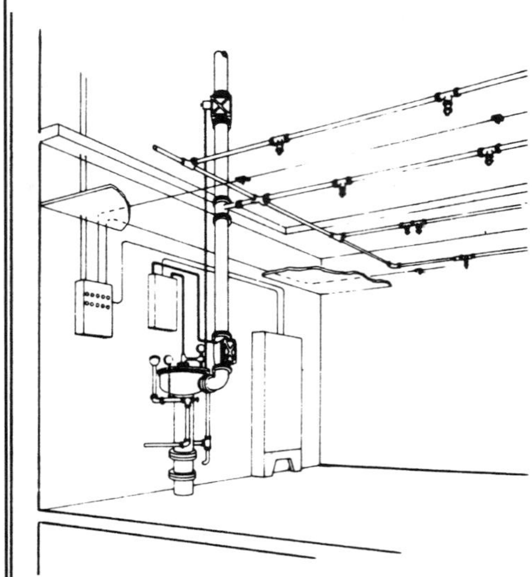

Firecycle is a fixed fire protection sprinkler system utilizing water as its extinguishing agent. It is a time delayed, recycling, preaction type which automatically shuts the water off when heat is reduced below the detector operating temperature and turns the water back on when that temperature is exceeded.

The system senses a fire condition through a closed circuit electrical detector system which controls water flow to the fire automatically. Batteries supply up to 90 hour emergency power supply for system operation. The piping system is dry (until water is required) and is monitored with pressurized air. Should any leak in the system piping occur, an alarm will sound, but water will not enter the system until heat is sensed by a Firecycle detector.

All areas are assumed to be open.

System Components	QUANTITY	UNIT	COST EACH MAT.	COST EACH INST.	COST EACH TOTAL
SYSTEM D4010 390 0580					
FIRECYCLE SPRINKLER SYSTEM, STEEL BLACK SCH. 40 PIPE					
LIGHT HAZARD, ONE FLOOR, 2000 S.F.					
Valve, gate, iron body 125 lb., OS&Y, flanged, 4" pipe size	1.000	Ea.	667.50	322.50	990
Valve, angle, bronze, 150 lb., rising stem, threaded, 2" pipe size	1.000	Ea.	656.25	48.75	705
Valve, swing check, bronze, 125 lb, regrinding disc, 2-1/2" pipe size	1.000	Ea.	693.75	64.50	758.25
*Alarm valve, 2-1/2" pipe size	1.000	Ea.	1,425	315	1,740
Alarm, water motor, complete with gong	1.000	Ea.	360	131.25	491.25
Pipe, steel, black, schedule 40, 4" diam.	10.000	L.F.	144.38	275.48	419.86
Fire alarm, horn, electric	1.000	Ea.	50.25	73.88	124.13
Pipe, steel, black, schedule 40, threaded, cplg & hngr 10' OC 2-1/2" diam.	20.000	L.F.	190.50	382.50	573
Pipe, steel, black, schedule 40, threaded, cplg & hngr 10' OC 2" diam.	12.500	L.F.	76.41	187.50	263.91
Pipe, steel, black, schedule 40, threaded, cplg & hngr 10' OC 1-1/4" diam.	37.500	L.F.	160.31	406.41	566.72
Pipe, steel, black, schedule 40, threaded, cplg & hngr 10' OC 1" diam.	112.000	L.F.	397.32	1,134	1,531.32
Pipe, Tee, malleable iron, black, 150 lb. threaded, 4" pipe size	2.000	Ea.	495	480	975
Pipe, Tee, malleable iron, black, 150 lb. threaded, 2-1/2" pipe size	2.000	Ea.	138.75	214.50	353.25
Pipe, Tee, malleable iron, black, 150 lb. threaded, 2" pipe size	1.000	Ea.	32.25	87.75	120
Pipe, Tee, malleable iron, black, 150 lb. threaded, 1-1/4" pipe size	5.000	Ea.	75	345	420
Pipe, Tee, malleable iron, black, 150 lb. threaded, 1" pipe size	4.000	Ea.	37.50	268.50	306
Pipe, 90° elbow, malleable iron, black, 150 lb. threaded, 1" pipe size	6.000	Ea.	35.78	247.50	283.28
Sprinkler head std spray, brass 135°-286°F 1/2" NPT, 3/8" orifice	12.000	Ea.	213.60	522	735.60
Firecycle controls, incls panel, battery, solenoid valves, press switches	1.000	Ea.	18,150	1,987.50	20,137.50
Detector, firecycle system	2.000	Ea.	1,162.50	65.25	1,227.75
Firecycle pkg, swing check & flow control valves w/trim 4" pipe size	1.000	Ea.	4,893.75	937.50	5,831.25
Air compressor, auto, complete, 200 Gal sprinkler sys. cap., 1/3 HP	1.000	Ea.	667.50	401.25	1,068.75
*Standpipe connection, wall, flush, brass w/plug & chain 2-1/2"x2-1/2"	1.000	Ea.	138	189	327
Valve, gate, bronze 300 psi, NRS, class 150, threaded, 1" diam.	1.000	Ea.	88.50	28.13	116.63
TOTAL			30,949.80	9,115.65	40,065.45
COST PER S.F.			15.47	4.56	20.03
*Not included in systems under 2000 S.F.					

D4010 Sprinklers

D4010 390	Firecycle Sprinkler Systems	COST PER S.F.		
		MAT.	INST.	TOTAL
0520	Firecycle sprinkler systems, steel black sch. 40 pipe (RD4010-100)			
0530	Light hazard, one floor, 500 S.F.	51	9.90	60.90
0560	1000 S.F. (RD4020-300)	26.50	6.20	32.70
0580	2000 S.F.	15.45	4.56	20.01
0600	5000 S.F.	6.80	2.88	9.68
0620	10,000 S.F.	3.97	2.26	6.23
0640	50,000 S.F.	1.79	1.90	3.69
0660	Each additional floor of 500 S.F.	2.39	2.78	5.17
0680	1000 S.F.	1.86	2.57	4.43
0700	2000 S.F.	1.53	2.33	3.86
0720	5000 S.F.	1.33	2.03	3.36
0740	10,000 S.F.	1.26	1.86	3.12
0760	50,000 S.F.	1.20	1.71	2.91
1000	Ordinary hazard, one floor, 500 S.F.	51.50	10.30	61.80
1020	1000 S.F.	26	6.15	32.15
1040	2000 S.F.	15.65	4.71	20.36
1060	5000 S.F.	7.10	3.05	10.15
1080	10,000 S.F.	4.39	2.91	7.30
1100	50,000 S.F.	2.73	2.91	5.64
1140	Each additional floor, 500 S.F.	2.66	3.20	5.86
1160	1000 S.F.	1.85	2.57	4.42
1180	2000 S.F.	1.92	2.36	4.28
1200	5000 S.F.	1.73	2.24	3.97
1220	10,000 S.F.	1.55	2.16	3.71
1240	50,000 S.F.	1.47	1.94	3.41
1500	Extra hazard, one floor, 500 S.F.	55	11.85	66.85
1520	1000 S.F.	28	7.65	35.65
1540	2000 S.F.	15.90	5.80	21.70
1560	5000 S.F.	7.30	4.11	11.41
1580	10,000 S.F.	5.60	4.12	9.72
1600	50,000 S.F.	4.68	4.57	9.25
1660	Each additional floor, 500 S.F.	3.34	3.93	7.27
1680	1000 S.F.	2.71	3.69	6.40
1700	2000 S.F.	2.44	3.71	6.15
1720	5000 S.F.	1.99	3.29	5.28
1740	10,000 S.F.	2.48	3	5.48
1760	50,000 S.F.	2.47	2.89	5.36
2020	Grooved steel, black, sch. 40 pipe, light hazard, one floor			
2030	2000 S.F.	15.50	4.04	19.54
2060	10,000 S.F.	4.34	2.72	7.06
2100	Each additional floor, 2000 S.F.	1.85	1.91	3.76
2150	10,000 S.F.	1.30	1.56	2.86
2200	Ordinary hazard, one floor, 2000 S.F.	15.60	4.21	19.81
2250	10,000 S.F.	4.60	2.61	7.21
2300	Each additional floor, 2000 S.F.	1.94	2.08	4.02
2350	10,000 S.F.	1.59	2.08	3.67
2400	Extra hazard, one floor, 2000 S.F.	15.95	5.10	21.05
2450	10,000 S.F.	4.95	3.10	8.05
2500	Each additional floor, 2000 S.F.	2.50	3.05	5.55
2550	10,000 S.F.	2.09	2.57	4.66
3050	Grooved steel, black, sch. 10 pipe light hazard, one floor,			
3060	2000 S.F.	15.45	4.02	19.47
3100	10,000 S.F.	3.99	1.94	5.93
3150	Each additional floor, 2000 S.F.	1.80	1.89	3.69
3200	10,000 S.F.	1.28	1.54	2.82
3250	Ordinary hazard, one floor, 2000 S.F.	15.55	4.18	19.73

D4010 Sprinklers

D4010 390	Firecycle Sprinkler Systems	COST PER S.F.		
		MAT.	INST.	TOTAL
3300	10,000 S.F.	4.23	2.43	6.66
3350	Each additional floor, 2000 S.F.	1.90	2.05	3.95
3400	10,000 S.F.	1.53	2.03	3.56
3450	Extra hazard, one floor, 2000 S.F.	15.90	5.05	20.95
3500	10,000 S.F.	4.77	3.06	7.83
3550	Each additional floor, 2000 S.F.	2.47	3.02	5.49
3600	10,000 S.F.	2.03	2.54	4.57
4060	Copper tubing, type M, light hazard, one floor, 2000 S.F.	15.80	4	19.80
4100	10,000 S.F.	4.47	1.97	6.44
4150	Each additional floor, 2000 S.F.	2.19	1.91	4.10
4200	10,000 S.F.	1.77	1.58	3.35
4250	Ordinary hazard, one floor, 2000 S.F.	16.05	4.35	20.40
4300	10,000 S.F.	4.82	2.30	7.12
4350	Each additional floor, 2000 S.F.	2.19	1.94	4.13
4400	10,000 S.F.	1.87	1.69	3.56
4450	Extra hazard, one floor, 2000 S.F.	16.55	5.10	21.65
4500	10,000 S.F.	6.50	3.41	9.91
4550	Each additional floor, 2000 S.F.	3.13	3.07	6.20
4600	10,000 S.F.	3.14	2.81	5.95
5060	Copper tubing, type M, T-drill system, light hazard, one floor 2000 S.F.	15.80	3.79	19.59
5100	10,000 S.F.	4.36	1.67	6.03
5150	Each additional floor, 2000 S.F.	2.42	1.79	4.21
5200	10,000 S.F.	1.66	1.28	2.94
5250	Ordinary hazard, one floor, 2000 S.F.	15.85	3.84	19.69
5300	10,000 S.F.	4.73	2.07	6.80
5350	Each additional floor, 2000 S.F.	2.22	1.71	3.93
5400	10,000 S.F.	2.03	1.67	3.70
5450	Extra hazard, one floor, 2000 S.F.	16.10	4.38	20.48
5500	10,000 S.F.	5.75	2.50	8.25
5550	Each additional floor, 2000 S.F.	2.66	2.34	5
5600	10,000 S.F.	2.43	1.95	4.38

D4010 Sprinklers

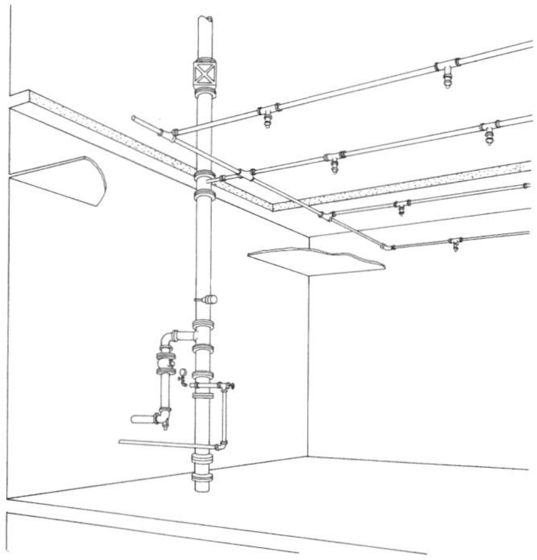

Wet Pipe System. A system employing automatic sprinklers attached to a piping system containing water and connected to a water supply so that water discharges immediately from sprinklers opened by heat from a fire.

All areas are assumed to be open.

System Components	QUANTITY	UNIT	COST EACH MAT.	COST EACH INST.	COST EACH TOTAL
SYSTEM D4010 410 0580					
WET PIPE SPRINKLER, STEEL, BLACK, SCH. 40 PIPE					
LIGHT HAZARD, ONE FLOOR, 2000 S.F.					
Valve, gate, iron body, 125 lb., OS&Y, flanged, 4″ diam.	1.000	Ea.	667.50	322.50	990
Valve, swing check, bronze, 125 lb, regrinding disc, 2-1/2″ pipe size	1.000	Ea.	693.75	64.50	758.25
Valve, angle, bronze, 150 lb., rising stem, threaded, 2″ diam.	1.000	Ea.	656.25	48.75	705
*Alarm valve, 2-1/2″ pipe size	1.000	Ea.	1,425	315	1,740
Alarm, water motor, complete with gong	1.000	Ea.	360	131.25	491.25
Valve, swing check, w/balldrip CI with brass trim 4″ pipe size	1.000	Ea.	303.75	315	618.75
Pipe, steel, black, schedule 40, 4″ diam.	10.000	L.F.	144.38	275.48	419.86
*Flow control valve, trim & gauges, 4″ pipe size	1.000	Set	5,006.25	712.50	5,718.75
Fire alarm horn, electric	1.000	Ea.	50.25	73.88	124.13
Pipe, steel, black, schedule 40, threaded, cplg & hngr 10′ OC, 2-1/2″ diam.	20.000	L.F.	190.50	382.50	573
Pipe, steel, black, schedule 40, threaded, cplg & hngr 10′ OC, 2″ diam.	12.500	L.F.	76.41	187.50	263.91
Pipe, steel, black, schedule 40, threaded, cplg & hngr 10′ OC, 1-1/4″ diam.	37.500	L.F.	160.31	406.41	566.72
Pipe, steel, black, schedule 40, threaded cplg & hngr 10′ OC, 1″ diam.	112.000	L.F.	397.32	1,134	1,531.32
Pipe Tee, malleable iron black, 150 lb. threaded, 4″ pipe size	2.000	Ea.	495	480	975
Pipe Tee, malleable iron black, 150 lb. threaded, 2-1/2″ pipe size	2.000	Ea.	138.75	214.50	353.25
Pipe Tee, malleable iron black, 150 lb. threaded, 2″ pipe size	1.000	Ea.	32.25	87.75	120
Pipe Tee, malleable iron black, 150 lb. threaded, 1-1/4″ pipe size	5.000	Ea.	75	345	420
Pipe Tee, malleable iron black, 150 lb. threaded, 1″ pipe size	4.000	Ea.	37.50	268.50	306
Pipe 90° elbow, malleable iron black, 150 lb. threaded, 1″ pipe size	6.000	Ea.	35.78	247.50	283.28
Sprinkler head, standard spray, brass 135°-286°F 1/2″ NPT, 3/8″ orifice	12.000	Ea.	213.60	522	735.60
Valve, gate, bronze, NRS, class 150, threaded, 1″ pipe size	1.000	Ea.	88.50	28.13	116.63
*Standpipe connection, wall, single, flush w/plug & chain 2-1/2″x2-1/2″	1.000	Ea.	138	189	327
TOTAL			11,386.05	6,751.65	18,137.70
COST PER S.F.			5.69	3.38	9.07
*Not included in systems under 2000 S.F.					

D4010 410	Wet Pipe Sprinkler Systems		COST PER S.F. MAT.	COST PER S.F. INST.	COST PER S.F. TOTAL
0520	Wet pipe sprinkler systems, steel, black, sch. 40 pipe				
0530	Light hazard, one floor, 500 S.F.		2.96	3.26	6.22
0560	1000 S.F.	RD4010-100	6.35	3.37	9.72
0580	2000 S.F.		5.70	3.39	9.09
0600	5000 S.F.	RD4020-300	2.64	2.38	5.02
0620	10,000 S.F.		1.65	2	3.65

D4010 Sprinklers

D4010 410	Wet Pipe Sprinkler Systems	COST PER S.F.		
		MAT.	INST.	TOTAL
0640	50,000 S.F.	1.01	1.83	2.84
0660	Each additional floor, 500 S.F.	1.26	2.77	4.03
0680	1000 S.F.	1.31	2.60	3.91
0700	2000 S.F.	1.24	2.31	3.55
0720	5000 S.F.	.87	2	2.87
0740	10,000 S.F.	.79	1.83	2.62
0760	50,000 S.F.	.66	1.45	2.11
1000	Ordinary hazard, one floor, 500 S.F.	3.26	3.49	6.75
1020	1000 S.F.	6.30	3.29	9.59
1040	2000 S.F.	5.85	3.54	9.39
1060	5000 S.F.	2.95	2.55	5.50
1080	10,000 S.F.	2.07	2.65	4.72
1100	50,000 S.F.	1.64	2.51	4.15
1140	Each additional floor, 500 S.F.	1.50	3.13	4.63
1160	1000 S.F.	1.27	2.54	3.81
1180	2000 S.F.	1.42	2.55	3.97
1200	5000 S.F.	1.40	2.44	3.84
1220	10,000 S.F.	1.22	2.48	3.70
1240	50,000 S.F.	1.16	2.21	3.37
1500	Extra hazard, one floor, 500 S.F.	12.25	5.35	17.60
1520	1000 S.F.	7.55	4.65	12.20
1540	2000 S.F.	6.15	4.78	10.93
1560	5000 S.F.	3.55	4.17	7.72
1580	10,000 S.F.	3.26	3.93	7.19
1600	50,000 S.F.	3.62	3.82	7.44
1660	Each additional floor, 500 S.F.	2.18	3.86	6.04
1680	1000 S.F.	2.13	3.66	5.79
1700	2000 S.F.	1.86	3.68	5.54
1720	5000 S.F.	1.53	3.26	4.79
1740	10,000 S.F.	2.01	2.97	4.98
1760	50,000 S.F.	1.99	2.85	4.84
2020	Grooved steel, black sch. 40 pipe, light hazard, one floor, 2000 S.F.	5.70	2.87	8.57
2060	10,000 S.F.	2.19	1.77	3.96
2100	Each additional floor, 2000 S.F.	1.27	1.88	3.15
2150	10,000 S.F.	.83	1.53	2.36
2200	Ordinary hazard, one floor, 2000 S.F.	5.80	3.04	8.84
2250	10,000 S.F.	1.97	2.22	4.19
2300	Each additional floor, 2000 S.F.	1.36	2.05	3.41
2350	10,000 S.F.	1.12	2.05	3.17
2400	Extra hazard, one floor, 2000 S.F.	6.15	3.92	10.07
2450	10,000 S.F.	2.66	2.86	5.52
2500	Each additional floor, 2000 S.F.	1.92	3.02	4.94
2550	10,000 S.F.	1.62	2.54	4.16
3050	Grooved steel, black sch. 10 pipe, light hazard, one floor, 2000 S.F.	5.65	2.85	8.50
3100	10,000 S.F.	1.67	1.68	3.35
3150	Each additional floor, 2000 S.F.	1.22	1.86	3.08
3200	10,000 S.F.	.81	1.51	2.32
3250	Ordinary hazard, one floor, 2000 S.F.	5.75	3.01	8.76
3300	10,000 S.F.	1.91	2.17	4.08
3350	Each additional floor, 2000 S.F.	1.32	2.02	3.34
3400	10,000 S.F.	1.06	2	3.06
3450	Extra hazard, one floor, 2000 S.F.	6.10	3.89	9.99
3500	10,000 S.F.	2.48	2.82	5.30
3550	Each additional floor, 2000 S.F.	1.89	2.99	4.88
3600	10,000 S.F.	1.56	2.51	4.07
4050	Copper tubing, type M, light hazard, one floor, 2000 S.F.	6	2.83	8.83
4100	10,000 S.F.	2.15	1.71	3.86
4150	Each additional floor, 2000 S.F.	1.61	1.88	3.49

D4010 Sprinklers

D4010 410	Wet Pipe Sprinkler Systems	COST PER S.F.		
		MAT.	INST.	TOTAL
4200	10,000 S.F.	1.30	1.55	2.85
4250	Ordinary hazard, one floor, 2000 S.F.	6.25	3.18	9.43
4300	10,000 S.F.	2.50	2.04	4.54
4350	Each additional floor, 2000 S.F.	1.85	2.08	3.93
4400	10,000 S.F.	1.61	1.83	3.44
4450	Extra hazard, one floor, 2000 S.F.	6.80	3.94	10.74
4500	10,000 S.F.	4.15	3.12	7.27
4550	Each additional floor, 2000 S.F.	2.55	3.04	5.59
4600	10,000 S.F.	2.67	2.78	5.45
5050	Copper tubing, type M, T-drill system, light hazard, one floor			
5060	2000 S.F.	6.05	2.62	8.67
5100	10,000 S.F.	2.04	1.41	3.45
5150	Each additional floor, 2000 S.F.	1.63	1.67	3.30
5200	10,000 S.F.	1.19	1.25	2.44
5250	Ordinary hazard, one floor, 2000 S.F.	6.10	2.67	8.77
5300	10,000 S.F.	2.41	1.81	4.22
5350	Each additional floor, 2000 S.F.	1.64	1.68	3.32
5400	10,000 S.F.	1.56	1.64	3.20
5450	Extra hazard, one floor, 2000 S.F.	6.30	3.21	9.51
5500	10,000 S.F.	3.44	2.26	5.70
5550	Each additional floor, 2000 S.F.	2.18	2.36	4.54
5600	10,000 S.F.	1.96	1.92	3.88

D4020 Standpipes

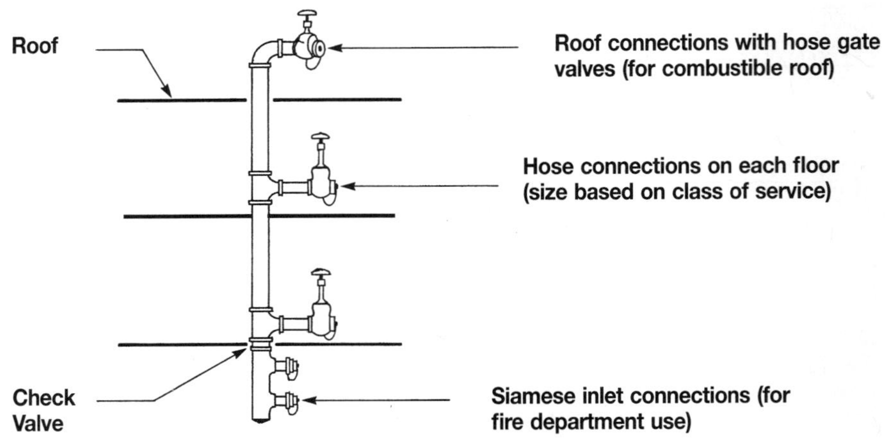

System Components	QUANTITY	UNIT	COST PER FLOOR MAT.	INST.	TOTAL
SYSTEM D4020 310 0560					
WET STANDPIPE RISER, CLASS I, STEEL, BLACK, SCH. 40 PIPE, 10′ HEIGHT					
4″ DIAMETER PIPE, ONE FLOOR					
Pipe, steel, black, schedule 40, threaded, 4″ diam.	20.000	L.F.	490	710	1,200
Pipe, Tee, malleable iron, black, 150 lb. threaded, 4″ pipe size	2.000	Ea.	660	640	1,300
Pipe, 90° elbow, malleable iron, black, 150 lb threaded 4″ pipe size	1.000	Ea.	209	214	423
Pipe, nipple, steel, black, schedule 40, 2-1/2″ pipe size x 3″ long	2.000	Ea.	24.40	161	185.40
Fire valve, gate, 300 lb., brass w/handwheel, 2-1/2″ pipe size	1.000	Ea.	219	100	319
Fire valve, pressure reducing rgh brs, 2-1/2″ pipe size	1.000	Ea.	820	200	1,020
Valve, swing check, w/ball drip, CI w/brs. ftngs., 4″ pipe size	1.000	Ea.	405	420	825
Standpipe conn wall dble. flush brs. w/plugs & chains 2-1/2″x2-1/2″x4″	1.000	Ea.	600	252	852
Valve, swing check, bronze, 125 lb, regrinding disc, 2-1/2″ pipe size	1.000	Ea.	925	86	1,011
Roof manifold, fire, w/valves & caps, horiz/vert brs 2-1/2″x2-1/2″x4″	1.000	Ea.	205	262	467
Fire, hydrolator, vent & drain, 2-1/2″ pipe size	1.000	Ea.	102	58	160
Valve, gate, iron body 125 lb., OS&Y, threaded, 4″ pipe size	1.000	Ea.	890	430	1,320
TOTAL			5,549.40	3,533	9,082.40

D4020 310	Wet Standpipe Risers, Class I	COST PER FLOOR MAT.	INST.	TOTAL
0550	Wet standpipe risers, Class I, steel, black sch. 40, 10′ height			
0560	4″ diameter pipe, one floor	5,550	3,525	9,075
0580	Additional floors	1,325	1,100	2,425
0600	6″ diameter pipe, one floor	9,375	6,200	15,575
0620	Additional floors RD4020 -300	2,375	1,725	4,100
0640	8″ diameter pipe, one floor	14,200	7,475	21,675
0660	Additional floors	3,450	2,100	5,550
0680				

D4020 310	Wet Standpipe Risers, Class II	COST PER FLOOR MAT.	INST.	TOTAL
1030	Wet standpipe risers, Class II, steel, black sch. 40, 10′ height			
1040	2″ diameter pipe, one floor	2,475	1,275	3,750
1060	Additional floors	810	490	1,300
1080	2-1/2″ diameter pipe, one floor	3,425	1,875	5,300
1100	Additional floors	890	570	1,460
1120				

D4020 Standpipes

D4020 310	Wet Standpipe Risers, Class III	COST PER FLOOR		
		MAT.	INST.	TOTAL
1530	Wet standpipe risers, Class III, steel, black sch. 40, 10′ height			
1540	4″ diameter pipe, one floor	5,675	3,525	9,200
1560	Additional floors	1,175	915	2,090
1580	6″ diameter pipe, one floor	9,500	6,200	15,700
1600	Additional floors	2,450	1,725	4,175
1620	8″ diameter pipe, one floor	14,400	7,475	21,875
1640	Additional floors	3,525	2,100	5,625

D4020 Standpipes

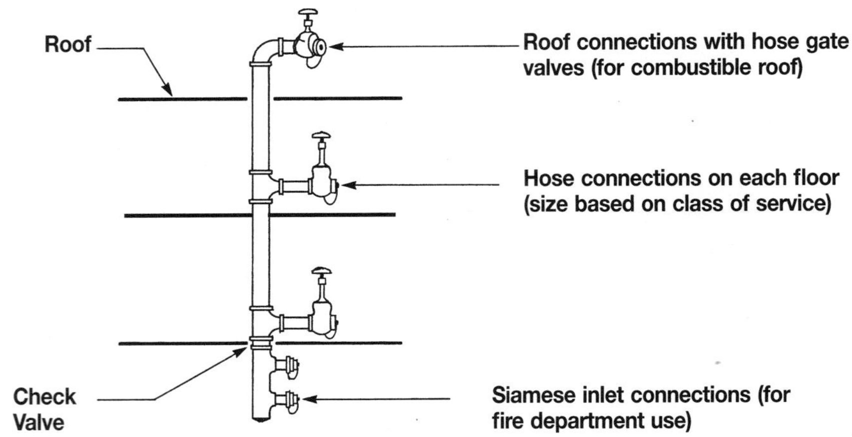

System Components	QUANTITY	UNIT	COST PER FLOOR MAT.	COST PER FLOOR INST.	COST PER FLOOR TOTAL
SYSTEM D4020 330 0540					
DRY STANDPIPE RISER, CLASS I, PIPE, STEEL, BLACK, SCH 40, 10′ HEIGHT					
4″ DIAMETER PIPE, ONE FLOOR					
Pipe, steel, black, schedule 40, threaded, 4″ diam.	20.000	L.F.	490	710	1,200
Pipe, Tee, malleable iron, black, 150 lb. threaded, 4″ pipe size	2.000	Ea.	660	640	1,300
Pipe, 90° elbow, malleable iron, black, 150 lb threaded 4″ pipe size	1.000	Ea.	209	214	423
Pipe, nipple, steel, black, schedule 40, 2-1/2″ pipe size x 3″ long	2.000	Ea.	24.40	161	185.40
Fire valve gate NRS 300 lb., brass w/handwheel, 2-1/2″ pipe size	1.000	Ea.	219	100	319
Fire valve, pressure reducing rgh brs, 2-1/2″ pipe size	1.000	Ea.	410	100	510
Standpipe conn wall dble. flush brs. w/plugs & chains 2-1/2″x2-1/2″x4″	1.000	Ea.	600	252	852
Valve swing check w/ball drip CI w/brs. ftngs., 4″pipe size	1.000	Ea.	405	420	825
Roof manifold, fire, w/valves & caps, horiz/vert brs 2-1/2″x2-1/2″x4″	1.000	Ea.	205	262	467
TOTAL			3,222.40	2,859	6,081.40

D4020 330	Dry Standpipe Risers, Class I	COST PER FLOOR MAT.	COST PER FLOOR INST.	COST PER FLOOR TOTAL
0530	Dry standpipe riser, Class I, steel, black sch. 40, 10′ height			
0540	4″ diameter pipe, one floor	3,225	2,850	6,075
0560	Additional floors	1,225	1,025	2,250
0580	6″ diameter pipe, one floor	6,825	4,925	11,750
0600	Additional floors	2,275	1,675	3,950
0620	8″ diameter pipe, one floor	10,600	5,975	16,575
0640	Additional floors	3,350	2,050	5,400
0660				

D4020 330	Dry Standpipe Risers, Class II	COST PER FLOOR MAT.	COST PER FLOOR INST.	COST PER FLOOR TOTAL
1030	Dry standpipe risers, Class II, steel, black sch. 40, 10′ height			
1040	2″ diameter pipe, one floor	2,075	1,325	3,400
1060	Additional floor	705	435	1,140
1080	2-1/2″ diameter pipe, one floor	2,750	1,550	4,300
1100	Additional floors	790	515	1,305
1120				

D4020 Standpipes

D4020 330	Dry Standpipe Risers, Class III	COST PER FLOOR		
		MAT.	INST.	TOTAL
1530	Dry standpipe risers, Class III, steel, black sch. 40, 10' height			
1540	4" diameter pipe, one floor	3,275	2,800	6,075
1560	Additional floors	1,075	935	2,010
1580	6" diameter pipe, one floor	6,900	4,925	11,825
1600	Additional floors	2,350	1,675	4,025
1620	8" diameter pipe, one floor	10,600	5,975	16,575
1640	Additional floor	3,425	2,050	5,475

D4020 Standpipes

D4020 410	Fire Hose Equipment	MAT.	INST.	TOTAL
		COST EACH		
0100	Adapters, reducing, 1 piece, FxM, hexagon, cast brass, 2-1/2" x 1-1/2"	64.50		64.50
0200	Pin lug, 1-1/2" x 1"	55.50		55.50
0250	3" x 2-1/2"	144		144
0300	For polished chrome, add 75% mat.			
0400	Cabinets, D.S. glass in door, recessed, steel box, not equipped			
0500	Single extinguisher, steel door & frame	134	157	291
0550	Stainless steel door & frame	237	157	394
0600	Valve, 2-1/2" angle, steel door & frame	156	105	261
0650	Aluminum door & frame	189	105	294
0700	Stainless steel door & frame	255	105	360
0750	Hose rack assy, 2-1/2" x 1-1/2" valve & 100' hose, steel door & frame	292	210	502
0800	Aluminum door & frame	430	210	640
0850	Stainless steel door & frame	575	210	785
0900	Hose rack assy & extinguisher,2-1/2"x1-1/2" valve & hose,steel door & frame	305	252	557
0950	Aluminum	550	252	802
1000	Stainless steel	605	252	857
1550	Compressor, air, dry pipe system, automatic, 200 gal., 3/4 H.P.	890	535	1,425
1600	520 gal., 1 H.P.	925	535	1,460
1650	Alarm, electric pressure switch (circuit closer)	102	27	129
2500	Couplings, hose, rocker lug, cast brass, 1-1/2"	61.50		61.50
2550	2-1/2"	81		81
3000	Escutcheon plate, for angle valves, polished brass, 1-1/2"	16.70		16.70
3050	2-1/2"	27.50		27.50
3500	Fire pump, electric, w/controller, fittings, relief valve			
3550	4" pump, 30 H.P., 500 G.P.M.	17,000	3,925	20,925
3600	5" pump, 40 H.P., 1000 G.P.M.	19,000	4,425	23,425
3650	5" pump, 100 H.P., 1000 G.P.M.	25,800	4,925	30,725
3700	For jockey pump system, add	2,975	630	3,605
5000	Hose, per linear foot, synthetic jacket, lined,			
5100	300 lb. test, 1-1/2" diameter	3.74	.48	4.22
5150	2-1/2" diameter	6.45	.57	7.02
5200	500 lb. test, 1-1/2" diameter	3.86	.48	4.34
5250	2-1/2" diameter	6.85	.57	7.42
5500	Nozzle, plain stream, polished brass, 1-1/2" x 10"	55.50		55.50
5550	2-1/2" x 15" x 13/16" or 1-1/2"	116		116
5600	Heavy duty combination adjustable fog and straight stream w/handle 1-1/2"	475		475
5650	2-1/2" direct connection	590		590
6000	Rack, for 1-1/2" diameter hose 100 ft. long, steel	71	63	134
6050	Brass	120	63	183
6500	Reel, steel, for 50 ft. long 1-1/2" diameter hose	151	90	241
6550	For 75 ft. long 2-1/2" diameter hose	245	90	335
7050	Siamese, w/plugs & chains, polished brass, sidewalk, 4" x 2-1/2" x 2-1/2"	730	505	1,235
7100	6" x 2-1/2" x 2-1/2"	945	630	1,575
7200	Wall type, flush, 4" x 2-1/2" x 2-1/2"	600	252	852
7250	6" x 2-1/2" x 2-1/2"	835	273	1,108
7300	Projecting, 4" x 2-1/2" x 2-1/2"	550	252	802
7350	6" x 2-1/2" x 2-1/2"	925	273	1,198
7400	For chrome plate, add 15% mat.			
8000	Valves, angle, wheel handle, 300 Lb., rough brass, 1-1/2"	115	58	173
8050	2-1/2"	202	100	302
8100	Combination pressure restricting, 1-1/2"	98.50	58	156.50
8150	2-1/2"	202	100	302
8200	Pressure restricting, adjustable, satin brass, 1-1/2"	350	58	408
8250	2-1/2"	410	100	510
8300	Hydrolator, vent and drain, rough brass, 1-1/2"	102	58	160
8350	2-1/2"	102	58	160
8400	Cabinet assy, incls. adapter, rack, hose, and nozzle	925	380	1,305

D4090 Other Fire Protection Systems

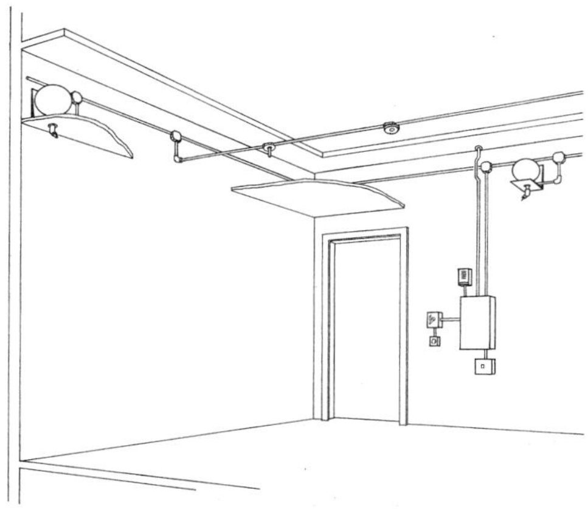

General: Automatic fire protection (suppression) systems other than water sprinklers may be desired for special environments, high risk areas, isolated locations or unusual hazards. Some typical applications would include:

Paint dip tanks
Securities vaults
Electronic data processing
Tape and data storage
Transformer rooms
Spray booths
Petroleum storage
High rack storage

Piping and wiring costs are dependent on the individual application and must be added to the component costs shown below.

All areas are assumed to be open.

D4090 910	Fire Suppression Unit Components	COST EACH		
		MAT.	INST.	TOTAL
0020	Detectors with brackets			
0040	Fixed temperature heat detector	56	82.50	138.50
0060	Rate of temperature rise detector	56	82.50	138.50
0080	Ion detector (smoke) detector	121	106	227
0200	Extinguisher agent			
0240	200 lb FM200, container	8,550	320	8,870
0280	75 lb carbon dioxide cylinder	1,400	214	1,614
0320	Dispersion nozzle			
0340	FM200 1-1/2" dispersion nozzle	73.50	51	124.50
0380	Carbon dioxide 3" x 5" dispersion nozzle	73.50	39.50	113
0420	Control station			
0440	Single zone control station with batteries	1,900	660	2,560
0470	Multizone (4) control station with batteries	3,600	1,325	4,925
0490				
0500	Electric mechanical release	184	345	529
0520				
0550	Manual pull station	66.50	119	185.50
0570				
0640	Battery standby power 10" x 10" x 17"	455	165	620
0700				
0740	Bell signalling device	106	82.50	188.50

D4090 920	FM200 Systems	COST PER C.F.		
		MAT.	INST.	TOTAL
0820	Average FM200 system, minimum			1.94
0840	Maximum			3.85

D5010 Electrical Service/Distribution

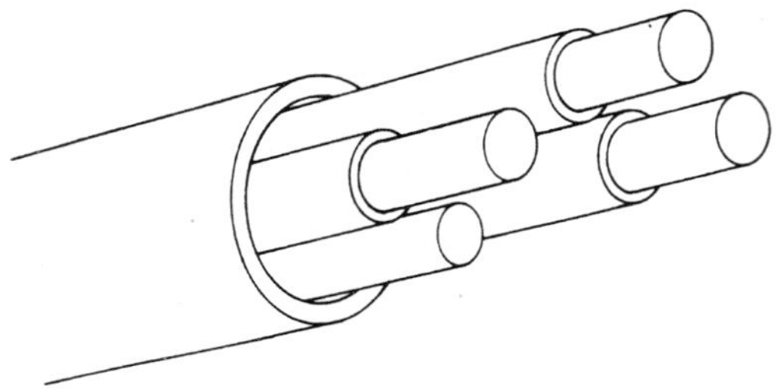

System Components	QUANTITY	UNIT	COST PER L.F. MAT.	INST.	TOTAL
SYSTEM D5010 110 0200					
HIGH VOLTAGE CABLE, NEUTRAL AND CONDUIT INCLUDED, COPPER #2, 5 kV					
Shielded cable, no splice/termn, copper, XLP shielding, 5 kV, #2	.030	C.L.F.	9.15	9.90	19.05
Wire, 600 volt, type THW, copper, stranded, #8	.010	C.L.F.	.36	.83	1.19
Rigid galv steel conduit to 15′ H, 2″ diam, w/term, ftng & support	1.000	L.F.	8.60	14.65	23.25
TOTAL			18.11	25.38	43.49

D5010 110	High Voltage Shielded Conductors	COST PER L.F. MAT.	INST.	TOTAL
0200	High voltage cable, neutral & conduit included, copper #2, 5 kV	18.10	25.50	43.60
0240	Copper #1, 5 kV	25.50	30	55.50
0280	15 kV	26.50	30	56.50
0320	Copper 1/0, 5 kV	30	38	68
0360	15 kV	30.50	38	68.50
0400	25 kV	36	38.50	74.50
0440	35 kV	37	39	76
0480	Copper 2/0, 5 kV	32	38.50	70.50
0520	15 kV	33	38.50	71.50
0560	25 kV	38	39.50	77.50
0600	35 kV	45.50	43.50	89
0640	Copper 4/0, 5 kV	43.50	44	87.50
0680	15 kV	43	44	87
0720	25 kV	48.50	44.50	93
0760	35 kV	52	45.50	97.50
0800	Copper 250 kcmil, 5 kV	46.50	44.50	91
0840	15 kV	45.50	44.50	90
0880	25 kV	54.50	45.50	100
0920	35 kV	59.50	49.50	109
0960	Copper 350 kcmil, 5 kV	56.50	49.50	106
1000	15 kV	54	49.50	103.50
1040	25 kV	63	51	114
1080	35 kV	66	52.50	118.50
1120	Copper 500 kcmil, 5 kV	86	62.50	148.50
1160	15 kV	85	62.50	147.50
1200	25 kV	94	64	158
1240	35 kV	98.50	66	164.50

D5010 Electrical Service/Distribution

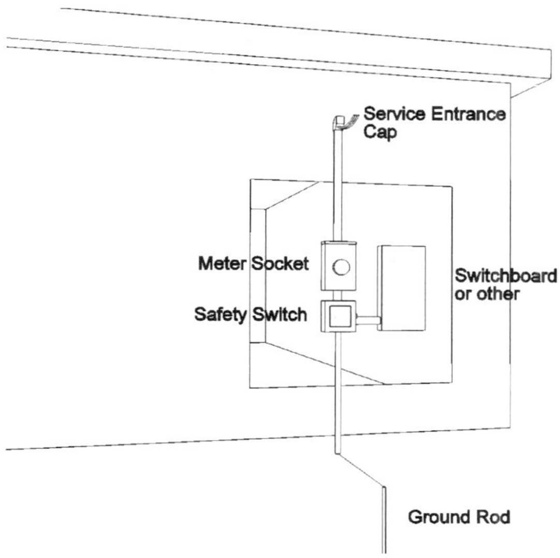

System Components	QUANTITY	UNIT	COST EACH MAT.	COST EACH INST.	COST EACH TOTAL
SYSTEM D5010 120 0220					
SERVICE INSTALLATION, INCLUDES BREAKERS, METERING, 20′ CONDUIT & WIRE					
3 PHASE, 4 WIRE, 60 A					
Wire, 600 volt, copper type XHHW, stranded #6	.600	C.L.F.	36.60	60.60	97.20
Rigid galvanized steel conduit, 3/4″ including fittings	20.000	L.F.	62	165	227
Service entrance cap, 3/4″ diameter	1.000	Ea.	6.55	50.50	57.05
Conduit LB fitting with cover, 3/4″ diameter	1.000	Ea.	12.10	50.50	62.60
Meter socket, three phase, 100 A	1.000	Ea.	181	236	417
Safety switches, heavy duty, 240 volt, 3 pole NEMA 1 fusible, 60 amp	1.000	Ea.	186	287	473
Grounding, wire ground bare armored, #8-1 conductor	.200	C.L.F.	15.30	66	81.30
Grounding, clamp, bronze, 3/4″ diameter	1.000	Ea.	6.50	20.50	27
Grounding, rod, copper clad, 8′ long, 3/4″ diameter	1.000	Ea.	37.50	124	161.50
Wireway w/fittings, 2-1/2″ x 2-1/2″	1.000	L.F.	11.90	14.65	26.55
TOTAL			555.45	1,074.75	1,630.20

D5010 120	Overhead Electric Service, 3 Phase - 4 Wire		COST EACH MAT.	COST EACH INST.	COST EACH TOTAL
0200	Service installation, includes breakers, metering, 20′ conduit & wire				
0220	3 phase, 4 wire, 120/208 volts, 60 A		555	1,075	1,630
0240	100 A		725	1,225	1,950
0245	100 A w/circuit breaker		1,525	1,500	3,025
0280	200 A		1,350	1,675	3,025
0285	200 A w/circuit breaker		3,025	2,100	5,125
0320	400 A	RD5010 -110	2,700	3,350	6,050
0325	400 A w/circuit breaker		5,875	4,175	10,050
0360	600 A		4,600	4,925	9,525
0365	600 A, w/switchboard		9,175	6,250	15,425
0400	800 A		6,300	5,925	12,225
0405	800 A, w/switchboard		10,900	7,425	18,325
0440	1000 A		7,850	7,300	15,150
0445	1000 A, w/switchboard		13,300	8,950	22,250
0480	1200 A		10,400	8,325	18,725
0485	1200 A, w/groundfault switchboard		34,100	10,300	44,400
0520	1600 A		12,800	10,900	23,700
0525	1600 A, w/groundfault switchboard		38,600	13,000	51,600

D5010 Electrical Service/Distribution

D5010 120	Overhead Electric Service, 3 Phase - 4 Wire	COST EACH		
		MAT.	INST.	TOTAL
0560	2000 A	17,100	13,200	30,300
0565	2000 A, w/groundfault switchboard	46,200	16,300	62,500
0610	1 phase, 3 wire, 120/240 volts, 100 A (no safety switch)	182	595	777
0615	100 A w/load center	490	1,150	1,640
0620	200 A	420	815	1,235
0625	200 A w/load center	1,250	1,725	2,975

D5010 Electrical Service/Distribution

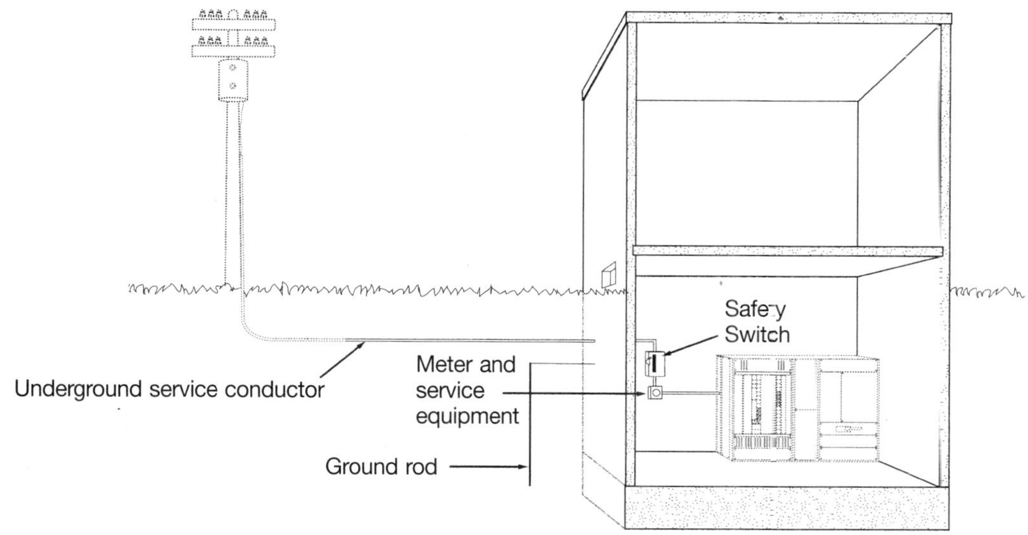

System Components	QUANTITY	UNIT	COST EACH MAT.	COST EACH INST.	COST EACH TOTAL
SYSTEM D5010 130 1000					
2000 AMP UNDERGROUND ELECTRIC SERVICE WITH GROUNDFAULT SWITCHBOARD INCLUDING EXCAVATION, BACKFILL, AND COMPACTION					
Excavate Trench	44.440	B.C.Y.		299.08	299.08
4 inch conduit bank	108.000	L.F.	1,004.40	2,646	3,650.40
4 inch fitting	8.000	Ea.	360	440	800
4 inch bells	4.000	Ea.	7.56	220	227.56
Concrete material	16.580	C.Y.	1,956.44		1,956.44
Concrete placement	16.580	C.Y.		407.54	407.54
Backfill trench	33.710	L.C.Y.		99.44	99.44
Compact fill material in trench	25.930	E.C.Y.		130.17	130.17
Dispose of excess fill material on-site	24.070	L.C.Y.		283.79	283.79
500 kcmil power cable	18.000	C.L.F.	18,450	7,380	25,830
Wire, 600 volt, type THW, copper, stranded, 1/0	6.000	C.L.F.	1,440	1,200	2,640
Saw cutting, concrete walls, plain, per inch of depth	64.000	L.F.	3.20	522.24	525.44
Meter centers and sockets, single pos, 4 terminal, 400 amp	5.000	Ea.	3,975	1,950	5,925
Safety switch, 400 amp	5.000	Ea.	6,500	3,675	10,175
Ground rod clamp	6.000	Ea.	39	123	162
Wireway	1.000	L.F.	11.90	14.65	26.55
600 volt stranded copper wire, 1/0	1.200	C.L.F.	288	240	528
Flexible metallic conduit	120.000	L.F.	76.80	494.40	571.20
Ground rod	6.000	Ea.	336	990	1,326
TOTAL			34,448.30	21,115.31	55,563.61

D5010 130	Underground Electric Service	COST EACH MAT.	COST EACH INST.	COST EACH TOTAL
0950	Underground electric service including excavation,backfill, and compaction			
1000	3 phase, 4 wire, 277/480 volts, 2000 A	34,400	21,100	55,500
1050	2000 A w/groundfault switch	62,000	23,500	85,500
1100	1600 A	27,300	17,000	44,300
1150	1600 A w/groundfault switchboard	53,000	19,100	72,100
1200	1200 A	20,900	13,500	34,400
1250	1200 A w/groundfault switchboard	44,600	15,500	60,100
1400	800 A	14,700	11,200	25,900

D5010 Electrical Service/Distribution

D5010 130	Underground Electric Service	COST EACH		
		MAT.	INST.	TOTAL
1450	800 A w/ switchboard	19,300	12,700	32,000
1500	600 A	10,300	10,300	20,600
1550	600 A w/ switchboard	14,900	11,600	26,500
1600	1 phase, 3 wire, 120/240 volts, 200 A	3,100	3,550	6,650
1650	200 A w/load center	3,925	4,450	8,375
1700	100 A	2,150	2,900	5,050
1750	100 A w/load center	3,000	3,825	6,825

D5010 Electrical Service/Distribution

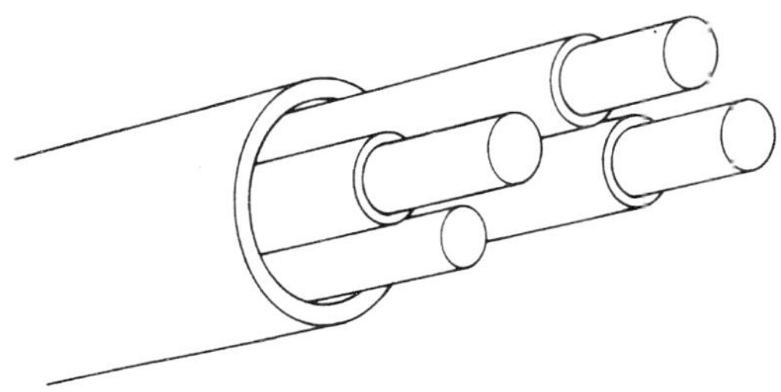

System Components	QUANTITY	UNIT	COST PER L.F. MAT.	INST.	TOTAL
SYSTEM D5010 230 0200					
FEEDERS, INCLUDING STEEL CONDUIT & WIRE, 60 A					
Rigid galvanized steel conduit, 3/4", including fittings	1.000	L.F.	3.10	8.25	11.35
Wire 600 volt, type XHHW copper stranded #6	.030	C.L.F.	1.83	3.03	4.86
Wire 600 volt, type XHHW copper stranded #8	.010	C.L.F.	.37	.83	1.20
TOTAL			5.30	12.11	17.41

D5010 230	Feeder Installation	COST PER L.F. MAT.	INST.	TOTAL
0200	Feeder installation 600 V, including RGS conduit and XHHW wire, 60 A	5.30	12.10	17.40
0240	100 A	9.50	15.80	25.30
0280	200 A	21	24	45
0320	400 A (RD5010 -140)	42.50	47.50	90
0360	600 A	78.50	78	156.50
0400	800 A	102	92	194
0440	1000 A	128	121	249
0480	1200 A	157	156	313
0520	1600 A	204	184	388
0560	2000 A	256	242	498
1200	Branch installation 600 V, including EMT conduit and THW wire, 15 A	1.13	5.90	7.03
1240	20 A	1.32	6.30	7.62
1280	30 A	1.65	6.50	8.15
1320	50 A	3.27	9.05	12.32
1360	65 A	4.02	9.60	13.62
1400	85 A	6.15	11.15	17.30
1440	100 A	6.80	11.40	18.20
1480	130 A	10.20	13.30	23.50
1520	150 A	12.50	15.25	27.75
1560	200 A	17	17.40	34.40

D5010 Electrical Service/Distribution

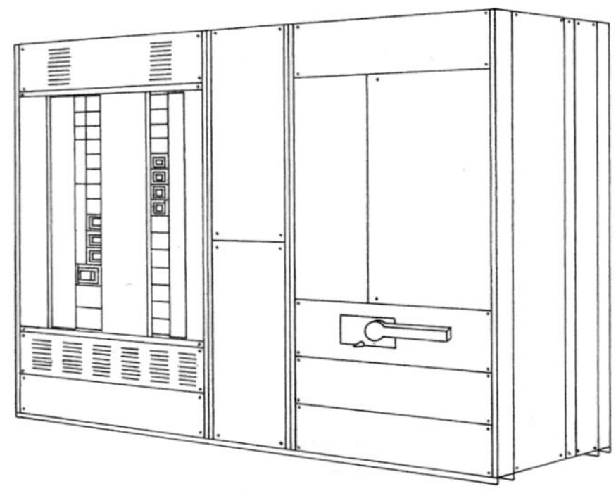

System Components	QUANTITY	UNIT	COST EACH MAT.	COST EACH INST.	COST EACH TOTAL
SYSTEM D5010 240 0240					
SWITCHGEAR INSTALLATION, INCL SWBD, PANELS & CIRC BREAKERS, 600 A					
Switchboards, 120/208 V, 600 amp	1.000	Ea.	4,575	1,325	5,900
Aluminum bus bars, 120/208 V, 600 amp	1.000	Ea.	1,750	1,325	3,075
Feeder section, circuit breakers, KA frame, 70 to 225 amp	1.000	Ea.	1,700	206	1,906
Feeder section, circuit breakers, LA frame, 125 to 400 amp	1.000	Ea.	5,175	287	5,462
TOTAL			13,200	3,143	16,343

D5010 240	Switchgear	COST EACH MAT.	COST EACH INST.	COST EACH TOTAL
0190	Switchgear installation, including switchboard, panels, & circuit breaker			
0200	120/208 V, 1 phase, 400 A	9,225	2,800	12,025
0240	600 A	13,200	3,150	16,350
0280	800 A (RD5010 -110)	15,300	3,625	18,925
0300	1000 A	18,800	4,000	22,800
0320	1200 A	19,800	4,525	24,325
0360	1600 A	32,100	5,225	37,325
0400	2000 A	38,800	5,700	44,500
0500	277/480 V, 3 phase, 400 A	12,600	5,200	17,800
0520	600 A	18,400	6,000	24,400
0540	800 A	19,900	6,625	26,525
0560	1000 A	24,700	7,300	32,000
0580	1200 A	27,100	8,175	35,275
0600	1600 A	40,000	9,025	49,025
0620	2000 A	50,500	9,950	60,450

D5010 Electrical Service/Distribution

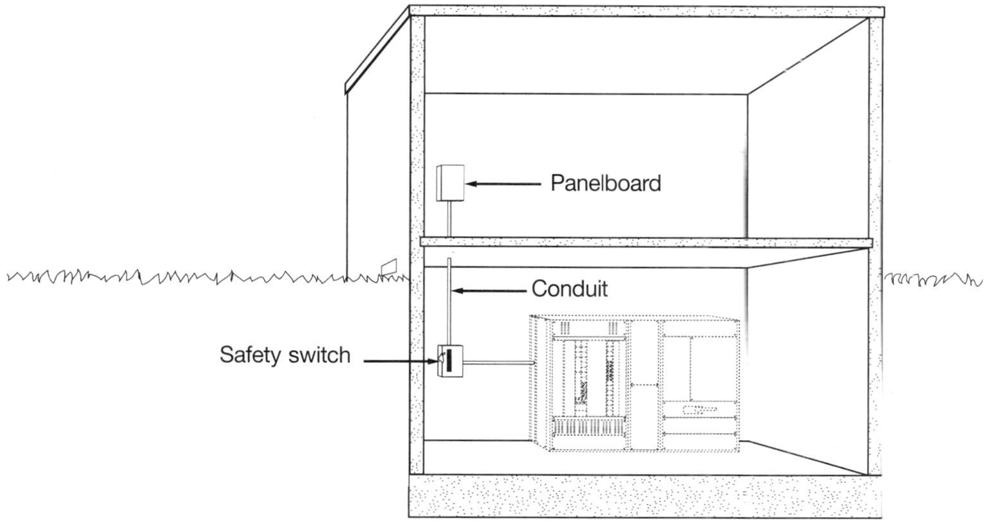

System Components	QUANTITY	UNIT	COST EACH MAT.	COST EACH INST.	COST EACH TOTAL
SYSTEM D5010 250 1020					
PANELBOARD INSTALLATION, INCLUDES PANELBOARD, CONDUCTOR, & CONDUIT					
Conduit, galvanized steel, 1-1/4"dia.	37.000	L.F.	207.20	407	614.20
Panelboards, NQOD, 4 wire, 120/208 volts, 100 amp main, 24 circuits	1.000	Ea.	1,400	1,400	2,800
Wire, 600 volt, type THW, copper, stranded, #3	1.480	C.L.F.	171.68	195.36	367.04
TOTAL			1,778.88	2,002.36	3,781.24

D5010 250	Panelboard	COST EACH MAT.	COST EACH INST.	COST EACH TOTAL
0900	Panelboards, NQOD, 4 wire, 120/208 volts w/conductor & conduit			
1000	100 A, 0 stories, 0′ horizontal	1,400	1,400	2,800
1020	1 stories, 25′ horizontal	1,775	2,000	3,775
1040	5 stories, 50′ horizontal	2,525	3,200	5,725
1060	10 stories, 75′ horizontal	3,400	4,575	7,975
1080	225A, 0 stories, 0′ horizontal	2,625	1,825	4,450
2000	1 stories, 25′ horizontal	3,750	2,900	6,650
2020	5 stories, 50′ horizontal	5,950	5,050	11,000
2040	10 stories, 75′ horizontal	8,500	7,550	16,050
2060	400A, 0 stories, 0′ horizontal	3,875	2,750	6,625
2080	1 stories, 25′ horizontal	5,425	4,500	9,925
3000	5 stories, 50′ horizontal	8,475	8,000	16,475
3020	10 stories, 75′ horizontal	16,900	15,600	32,500
3040	600 A, 0 stories, 0′ horizontal	5,725	3,300	9,025
3060	1 stories, 25′ horizontal	8,650	6,175	14,825
3080	5 stories, 50′ horizontal	14,400	11,900	26,300
4000	10 stories, 75′ horizontal	30,000	23,300	53,300
4010	Panelboards, NEHB, 4 wire, 277/480 volts w/conductor, conduit, & safety switch			
4020	100 A, 0 stories, 0′ horizontal, includes safety switch	3,125	1,925	5,050
4040	1 stories, 25′ horizontal	3,500	2,525	6,025
4060	5 stories, 50′ horizontal	4,250	3,700	7,950
4080	10 stories, 75′ horizontal	5,125	5,100	10,225
5000	225 A, 0 stories, 0′ horizontal	4,700	2,325	7,025
5020	1 stories, 25′ horizontal	5,800	3,425	9,225
5040	5 stories, 50′ horizontal	8,025	5,550	13,575
5060	10 stories, 75′ horizontal	10,600	8,050	18,650

D5010 Electrical Service/Distribution

D5010 250	Panelboard	COST EACH		
		MAT.	INST.	TOTAL
5080	400 A, 0 stories, 0' horizontal	7,450	3,600	11,050
6000	1 stories, 25' horizontal	9,000	5,375	14,375
6020	5 stories, 50' horizontal	12,100	8,850	20,950
6040	10 stories, 75' horizontal	20,400	16,500	36,900
6060	600 A, 0 stories, 0' horizontal	11,500	4,575	16,075
6080	1 stories, 25' horizontal	14,400	7,450	21,850
7000	5 stories, 50' horizontal	20,200	13,100	33,300
7020	10 stories, 75' horizontal	35,800	24,600	60,400

D5020 Lighting and Branch Wiring

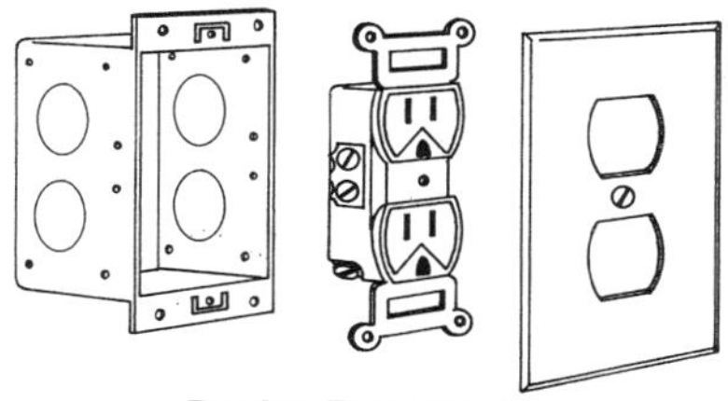

Duplex Receptacle

System Components	QUANTITY	UNIT	COST PER S.F. MAT.	COST PER S.F. INST.	COST PER S.F. TOTAL
SYSTEM D5020 110 0200					
RECEPTACLES INCL. PLATE, BOX, CONDUIT, WIRE & TRANS. WHEN REQUIRED					
2.5 PER 1000 S.F., .3 WATTS PER S.F.					
Steel intermediate conduit, (IMC) 1/2" diam	167.000	L.F.	.40	1.10	1.50
Wire 600V type THWN-THHN, copper solid #12	3.340	C.L.F.	.04	.20	.24
Wiring device, receptacle, duplex, 120V grounded, 15 amp	2.500	Ea.		.04	.04
Wall plate, 1 gang, brown plastic	2.500	Ea.		.02	.02
Steel outlet box 4" square	2.500	Ea.	.01	.08	.09
Steel outlet box 4" plaster rings	2.500	Ea.	.01	.03	.04
TOTAL			.46	1.47	1.93

D5020 110	Receptacle (by Wattage)	COST PER S.F. MAT.	COST PER S.F. INST.	COST PER S.F. TOTAL
0190	Receptacles include plate, box, conduit, wire & transformer when required			
0200	2.5 per 1000 S.F., .3 watts per S.F.	.46	1.47	1.93
0240	With transformer	.54	1.54	2.08
0280	4 per 1000 S.F., .5 watts per S.F. RD5010 -110	.52	1.71	2.23
0320	With transformer	.65	1.83	2.48
0360	5 per 1000 S.F., .6 watts per S.F.	.61	2.03	2.64
0400	With transformer	.78	2.18	2.96
0440	8 per 1000 S.F., .9 watts per S.F.	.64	2.24	2.88
0480	With transformer	.91	2.47	3.38
0520	10 per 1000 S.F., 1.2 watts per S.F.	.65	2.44	3.09
0560	With transformer	.98	2.73	3.71
0600	16.5 per 1000 S.F., 2.0 watts per S.F.	.78	3.04	3.82
0640	With transformer	1.33	3.52	4.85
0680	20 per 1000 S.F., 2.4 watts per S.F.	.81	3.32	4.13
0720	With transformer	1.48	3.91	5.39

D5020 Lighting and Branch Wiring

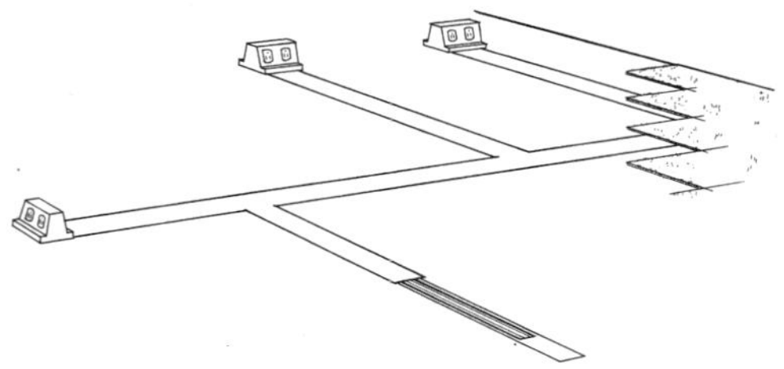

Underfloor Receptacle System

Description: Table D5020 115 includes installed costs of raceways and copper wire from panel to and including receptacle.

National Electrical Code prohibits use of undercarpet system in residential, school or hospital buildings. Can only be used with carpet squares.

Low density = (1) Outlet per 259 S.F. of floor area.

High density = (1) Outlet per 127 S.F. of floor area.

System Components	QUANTITY	UNIT	COST PER S.F. MAT.	INST.	TOTAL
SYSTEM D5020 115 0200					
RECEPTACLE SYSTEMS, UNDERFLOOR DUCT, 5′ ON CENTER, LOW DENSITY					
Underfloor duct 3-1/8″ x 7/8″ w/insert 24″ on center	.190	L.F.	4.18	1.79	5.97
Vertical elbow for underfloor duct, 3-1/8″, included					
Underfloor duct conduit adapter, 2″ x 1-1/4″, included					
Underfloor duct junction box, single duct, 3-1/8″	.003	Ea.	1.47	.50	1.97
Underfloor junction box carpet pan	.003	Ea.	1.22	.02	1.24
Underfloor duct outlet, high tension receptacle	.004	Ea.	.46	.33	.79
Wire 600V type THWN-THHN copper solid #12	.010	C.L.F.	.12	.60	.72
TOTAL			7.45	3.24	10.69

D5020 115	Receptacles, Floor	COST PER S.F. MAT.	INST.	TOTAL
0200	Receptacle systems, underfloor duct, 5′ on center, low density	7.45	3.24	10.69
0240	High density	8	4.17	12.17
0280	7′ on center, low density (RD5010 -110)	5.90	2.78	8.68
0320	High density	6.45	3.71	10.16
0400	Poke thru fittings, low density	1.25	1.59	2.84
0440	High density	2.50	3.17	5.67
0520	Telepoles, using Romex, low density	1.19	.97	2.16
0560	High density	2.37	1.95	4.32
0600	Using EMT, low density	1.23	1.29	2.52
0640	High density	2.51	2.57	5.08
0720	Conduit system with floor boxes, low density	1.18	1.11	2.29
0760	High density	2.38	2.21	4.59
0840	Undercarpet power system, 3 conductor with 5 conductor feeder, low density	1.57	.39	1.96
0880	High density	3.11	.81	3.92

D5020 Lighting and Branch Wiring

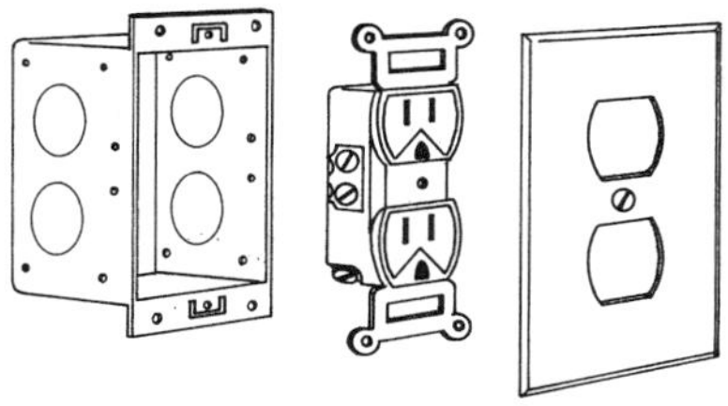

Duplex Receptacle

Wall Switch

System Components	QUANTITY	UNIT	COST PER EACH MAT.	COST PER EACH INST.	COST PER EACH TOTAL
SYSTEM D5020 125 0520					
RECEPTACLES AND WALL SWITCHES, RECEPTICLE DUPLEX 120 V GROUNDED, 15 A					
Electric metallic tubing conduit, (EMT), 3/4" diam	22.000	L.F.	24.20	111.10	135.30
Wire, 600 volt, type THWN-THHN, copper, solid #12	.630	C.L.F.	7.40	37.80	45.20
Steel outlet box 4" square	1.000	Ea.	3.17	33	36.17
Steel outlet box, 4" square, plaster rings	1.000	Ea.	2.24	10.30	12.54
Receptacle, duplex, 120 volt grounded, 15 amp	1.000	Ea.	1.39	16.50	17.89
Wall plate, 1 gang, brown plastic	1.000	Ea.	.42	8.25	8.67
TOTAL			38.82	216.95	255.77

D5020 125	Receptacles & Switches by Each	COST PER EACH MAT.	COST PER EACH INST.	COST PER EACH TOTAL
0460	Receptacles & Switches, with box, plate, 3/4" EMT conduit & wire			
0520	Receptacle duplex 120 V grounded, 15 A	39	217	256
0560	20 A	46	225	271
0600	Receptacle duplex ground fault interrupting, 15 A	52.50	225	277.50
0640	20 A	81.50	225	306.50
0680	Toggle switch single, 15 A	48	217	265
0720	20 A	41	225	266
0760	3 way switch, 15 A	39.50	229	268.50
0800	20 A	41.50	237	278.50
0840	4 way switch, 15 A	45.50	244	289.50
0880	20 A	84.50	260	344.50

D5020 Lighting and Branch Wiring

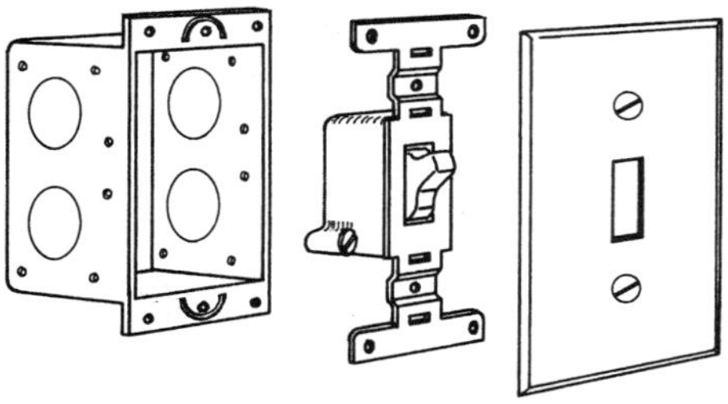

Description: Table D5020 130 includes the cost for switch, plate, box, conduit in slab or EMT exposed and copper wire. Add 20% for exposed conduit.

No power required for switches.

Federal energy guidelines recommend the maximum lighting area controlled per switch shall not exceed 1000 S.F. and that areas over 500 S.F. shall be so controlled that total illumination can be reduced by at least 50%.

System Components	QUANTITY	UNIT	COST PER S.F. MAT.	COST PER S.F. INST.	COST PER S.F. TOTAL
SYSTEM D5020 130 0360					
WALL SWITCHES, 5.0 PER 1000 S.F.					
Steel, intermediate conduit (IMC), 1/2" diameter	88.000	L.F.	.21	.58	.79
Wire, 600V type THWN-THHN, copper solid #12	1.710	C.L.F.	.02	.10	.12
Toggle switch, single pole, 15 amp	5.000	Ea.	.05	.08	.13
Wall plate, 1 gang, brown plastic	5.000	Ea.		.04	.04
Steel outlet box 4" plaster rings	5.000	Ea.	.02	.17	.19
Plaster rings	5.000	Ea.	.01	.05	.06
TOTAL			.31	1.02	1.33

D5020 130	Wall Switch by Sq. Ft.	COST PER S.F. MAT.	COST PER S.F. INST.	COST PER S.F. TOTAL
0200	Wall switches, 1.0 per 1000 S.F.	.06	.25	.31
0240	1.2 per 1000 S.F.	.08	.26	.34
0280	2.0 per 1000 S.F.	.12	.38	.50
0320	2.5 per 1000 S.F.	.15	.48	.63
0360	5.0 per 1000 S.F.	.31	1.02	1.33
0400	10.0 per 1000 S.F.	.62	2.07	2.69

D5020 Lighting and Branch Wiring

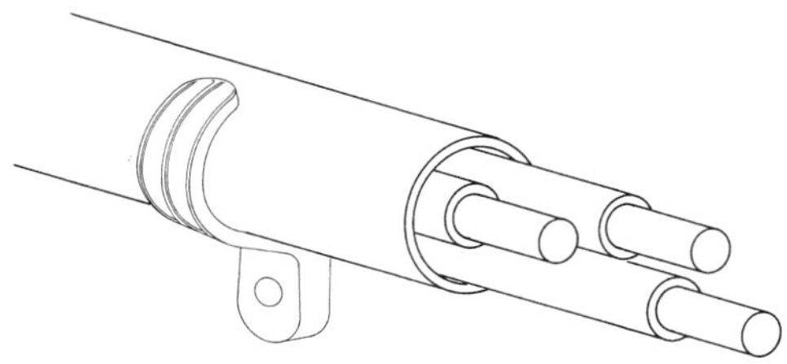

System D5020 135 includes all wiring and connections.

System Components	QUANTITY	UNIT	COST PER S.F. MAT.	COST PER S.F. INST.	COST PER S.F. TOTAL
SYSTEM D5020 135 0200					
MISCELLANEOUS POWER, TO .5 WATTS					
Steel intermediate conduit, (IMC) 1/2" diam	15.000	L.F.	.04	.10	.14
Wire 600V type THWN-THHN, copper solid #12	.325	C.L.F.		.02	.02
TOTAL			.04	.12	.16

D5020 135	Miscellaneous Power	COST PER S.F. MAT.	COST PER S.F. INST.	COST PER S.F. TOTAL
0200	Miscellaneous power, to .5 watts	.04	.12	.16
0240	.8 watts	.06	.17	.23
0280	1 watt	.08	.22	.30
0320	1.2 watts	.09	.26	.35
0360	1.5 watts	.11	.30	.41
0400	1.8 watts	.12	.35	.47
0440	2 watts	.14	.42	.56
0480	2.5 watts	.18	.52	.70
0520	3 watts	.21	.61	.82

D5020 Lighting and Branch Wiring

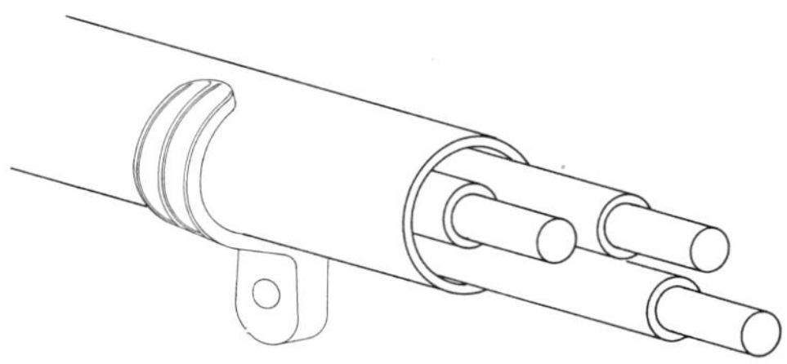

System D5020 140 includes all wiring and connections for central air conditioning units.

System Components	QUANTITY	UNIT	COST PER S.F. MAT.	COST PER S.F. INST.	COST PER S.F. TOTAL
SYSTEM D5020 140 0200					
CENTRAL AIR CONDITIONING POWER, 1 WATT					
Steel intermediate conduit, 1/2" diam.	.030	L.F.	.07	.20	.27
Wire 600V type THWN-THHN, copper solid #12	.001	C.L.F.	.01	.06	.07
TOTAL			.08	.26	.34

D5020 140	Central A. C. Power (by Wattage)	COST PER S.F. MAT.	COST PER S.F. INST.	COST PER S.F. TOTAL
0200	Central air conditioning power, 1 watt	.08	.26	.34
0220	2 watts	.10	.30	.40
0240	3 watts	.20	.44	.64
0280	4 watts	.20	.45	.65
0320	6 watts	.35	.59	.94
0360	8 watts	.45	.65	1.10
0400	10 watts	.63	.75	1.38

D5020 Lighting and Branch Wiring

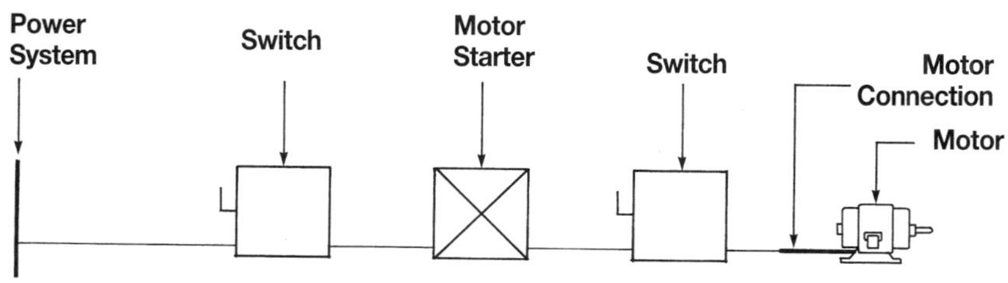

System D5020 145 installed cost of motor wiring using 50′ of rigid conduit and copper wire. **Cost and setting of motor not included.**

System Components	QUANTITY	UNIT	COST EACH MAT.	COST EACH INST.	COST EACH TOTAL
SYSTEM D5020 145 0200					
MOTOR INST., SINGLE PHASE, 115V, TO AND INCLUDING 1/3 HP MOTOR SIZE					
Wire 600V type THWN-THHN, copper solid #12	1.250	C.L.F.	14.69	75	89.69
Steel intermediate conduit, (IMC) 1/2″ diam.	50.000	L.F.	121	330	451
Magnetic FVNR, 115V, 1/3 HP, size 00 starter	1.000	Ea.	197	165	362
Safety switch, fused, heavy duty, 240V 2P 30 amp	1.000	Ea.	85.50	188	273.50
Safety switch, non fused, heavy duty, 600V, 3 phase, 30 A	1.000	Ea.	106	206	312
Flexible metallic conduit, Greenfield 1/2″ diam.	1.500	L.F.	.68	4.95	5.63
Connectors for flexible metallic conduit Greenfield 1/2″ diam.	1.000	Ea.	2.13	8.25	10.38
Coupling for Greenfield to conduit 1/2″ diam. flexible metalic conduit	1.000	Ea.	1.38	13.20	14.58
Fuse cartridge nonrenewable, 250V 30 amp	1.000	Ea.	2.64	13.20	15.84
TOTAL			531.02	1,003.60	1,534.62

D5020 145	Motor Installation	COST EACH MAT.	COST EACH INST.	COST EACH TOTAL
0200	Motor installation, single phase, 115V, 1/3 HP motor size	530	1,000	1,530
0240	1 HP motor size	555	1,000	1,555
0280	2 HP motor size (RD5010 -170)	605	1,075	1,680
0320	3 HP motor size	685	1,075	1,760
0360	230V, 1 HP motor size	535	1,025	1,560
0400	2 HP motor size	570	1,025	1,595
0440	3 HP motor size	640	1,100	1,740
0520	Three phase, 200V, 1-1/2 HP motor size	610	1,125	1,735
0560	3 HP motor size	740	1,225	1,965
0600	5 HP motor size	725	1,350	2,075
0640	7-1/2 HP motor size	755	1,375	2,130
0680	10 HP motor size	1,125	1,725	2,850
0720	15 HP motor size	1,550	1,925	3,475
0760	20 HP motor size	1,950	2,225	4,175
0800	25 HP motor size	2,000	2,225	4,225
0840	30 HP motor size	3,150	2,625	5,775
0880	40 HP motor size	3,900	3,100	7,000
0920	50 HP motor size	6,900	3,625	10,525
0960	60 HP motor size	7,200	3,825	11,025
1000	75 HP motor size	8,875	4,375	13,250
1040	100 HP motor size	26,600	5,150	31,750
1080	125 HP motor size	26,900	5,650	32,550
1120	150 HP motor size	31,300	6,650	37,950
1160	200 HP motor size	32,300	7,850	40,150
1240	230V, 1-1/2 HP motor size	585	1,100	1,685
1280	3 HP motor size	715	1,200	1,915
1320	5 HP motor size	700	1,325	2,025
1360	7-1/2 HP motor size	700	1,325	2,025
1400	10 HP motor size	1,075	1,650	2,725
1440	15 HP motor size	1,200	1,800	3,000
1480	20 HP motor size	1,850	2,150	4,000
1520	25 HP motor size	1,950	2,225	4,175

D5020 Lighting and Branch Wiring

D5020 145	Motor Installation	COST EACH		
		MAT.	INST.	TOTAL
1560	30 HP motor size	2,000	2,225	4,225
1600	40 HP motor size	3,800	3,025	6,825
1640	50 HP motor size	4,075	3,200	7,275
1680	60 HP motor size	6,925	3,625	10,550
1720	75 HP motor size	8,225	4,125	12,350
1760	100 HP motor size	9,475	4,600	14,075
1800	125 HP motor size	27,100	5,325	32,425
1840	150 HP motor size	28,900	6,025	34,925
1880	200 HP motor size	30,300	6,725	37,025
1960	460V, 2 HP motor size	725	1,125	1,850
2000	5 HP motor size	855	1,225	2,080
2040	10 HP motor size	805	1,325	2,130
2080	15 HP motor size	1,075	1,525	2,600
2120	20 HP motor size	1,125	1,650	2,775
2160	25 HP motor size	1,200	1,725	2,925
2200	30 HP motor size	1,550	1,875	3,425
2240	40 HP motor size	1,975	2,000	3,975
2280	50 HP motor size	2,200	2,225	4,425
2320	60 HP motor size	3,350	2,600	5,950
2360	75 HP motor size	3,950	2,875	6,825
2400	100 HP motor size	4,375	3,200	7,575
2440	125 HP motor size	7,225	3,650	10,875
2480	150 HP motor size	8,950	4,075	13,025
2520	200 HP motor size	10,300	4,600	14,900
2600	575V, 2 HP motor size	725	1,125	1,850
2640	5 HP motor size	855	1,225	2,080
2680	10 HP motor size	805	1,325	2,130
2720	20 HP motor size	1,075	1,525	2,600
2760	25 HP motor size	1,125	1,650	2,775
2800	30 HP motor size	1,550	1,875	3,425
2840	50 HP motor size	1,675	1,925	3,600
2880	60 HP motor size	3,300	2,575	5,875
2920	75 HP motor size	3,350	2,600	5,950
2960	100 HP motor size	3,950	2,875	6,825
3000	125 HP motor size	7,050	3,575	10,625
3040	150 HP motor size	7,225	3,650	10,875
3080	200 HP motor size	9,050	4,150	13,200

D5020 Lighting and Branch Wiring

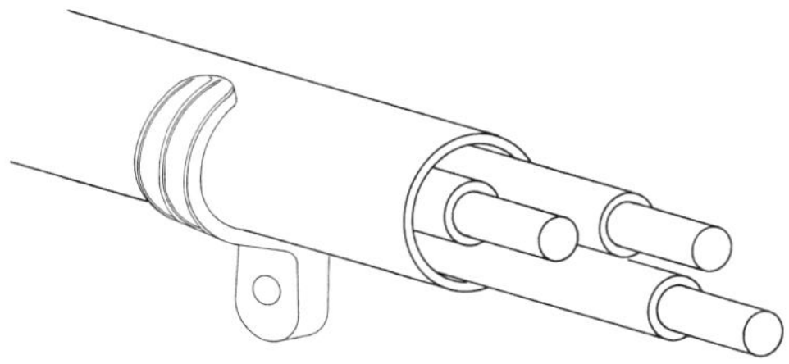

System Components	QUANTITY	UNIT	COST PER L.F. MAT.	COST PER L.F. INST.	COST PER L.F. TOTAL
SYSTEM D5020 155 0200					
MOTOR FEEDER SYSTEMS, SINGLE PHASE, UP TO 115V, 1HP OR 230V, 2HP					
Steel intermediate conduit, (IMC) 1/2" diam	1.000	L.F.	2.42	6.60	9.02
Wire 600V type THWN-THHN, copper solid #12	.020	C.L.F.	.24	1.20	1.44
TOTAL			2.66	7.80	10.46

D5020 155	Motor Feeder	COST PER L.F. MAT.	COST PER L.F. INST.	COST PER L.F. TOTAL
0200	Motor feeder systems, single phase, feed up to 115V 1HP or 230V 2 HP	2.66	7.80	10.46
0240	115V 2HP, 230V 3HP	2.79	7.90	10.69
0280	115V 3HP (RD5010-170)	3.09	8.25	11.34
0360	Three phase, feed to 200V 3HP, 230V 5HP, 460V 10HP, 575V 10HP	2.77	8.40	11.17
0440	200V 5HP, 230V 7.5HP, 460V 15HP, 575V 20HP	2.97	8.60	11.57
0520	200V 10HP, 230V 10HP, 460V 30HP, 575V 30HP	3.43	9.10	12.53
0600	200V 15HP, 230V 15HP, 460V 40HP, 575V 50HP	4.76	10.40	15.16
0680	200V 20HP, 230V 25HP, 460V 50HP, 575V 60HP	6.75	13.10	19.85
0760	200V 25HP, 230V 30HP, 460V 60HP, 575V 75HP	7.65	13.35	21
0840	200V 30HP	8.50	13.80	22.30
0920	230V 40HP, 460V 75HP, 575V 100HP	10.80	15.10	25.90
1000	200V 40HP	12.45	16.15	28.60
1080	230V 50HP, 460V 100HP, 575V 125HP	14.95	17.80	32.75
1160	200V 50HP, 230V 60HP, 460V 125HP, 575V 150HP	17.15	18.90	36.05
1240	200V 60HP, 460V 150HP	21	22	43
1320	230V 75HP, 575V 200HP	24	23	47
1400	200V 75HP	26.50	23.50	50
1480	230V 100HP, 460V 200HP	36.50	27.50	64
1560	200V 100HP	50.50	34.50	85
1640	230V 125HP	50.50	34.50	85
1720	200V 125HP, 230V 150HP	56.50	42.50	99
1800	200V 150HP	68	46.50	114.50
1880	200V 200HP	85	49	134
1960	230V 200HP	81	51	132

D5020 Lighting and Branch Wiring

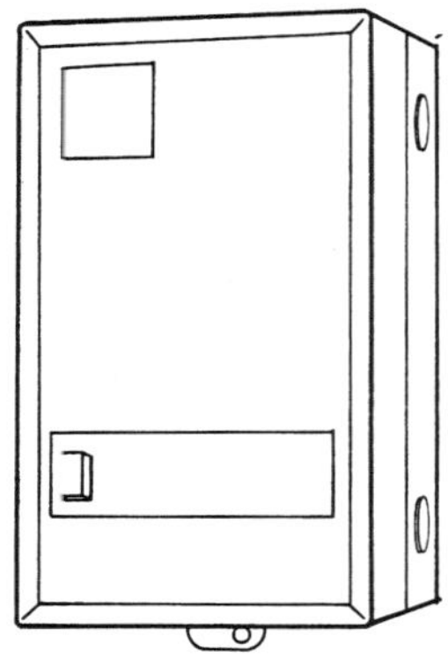

Starters are full voltage, type NEMA 1 for general purpose indoor application with motor overload protection and include mounting and wire connections.

System Components	QUANTITY	UNIT	COST EACH MAT.	COST EACH INST.	COST EACH TOTAL
SYSTEM D5020 160 0200					
MAGNETIC STARTER, SIZE 00 TO 1/3 HP, 1 PHASE 115V OR 1 HP 230V					
Magnetic starter, size 00, to 1/3 HP, 1 phase, 115V or 1 HP 230V	1.000	Ea.	197	165	362
TOTAL			197	165	362

D5020 160	Magnetic Starter	COST EACH MAT.	COST EACH INST.	COST EACH TOTAL
0200	Magnetic starter, size 00, to 1/3 HP, 1 phase, 115V or 1 HP 230V	197	165	362
0280	Size 00, to 1-1/2 HP, 3 phase, 200-230V or 2 HP 460-575V	216	188	404
0360	Size 0, to 1 HP, 1 phase, 115V or 2 HP 230V	219	165	384
0440	Size 0, to 3 HP, 3 phase, 200-230V or 5 HP 460-575V	345	287	632
0520	Size 1, to 2 HP, 1 phase, 115V or 3 HP 230V	252	220	472
0600	Size 1, to 7-1/2 HP, 3 phase, 200-230V or 10 HP 460-575V	294	410	704
0680	Size 2, to 10 HP, 3 phase, 200V, 15 HP-230V or 25 HP 460-575V	550	600	1,150
0760	Size 3, to 25 HP, 3 phase, 200V, 30 HP-230V or 50 HP 460-575V	900	735	1,635
0840	Size 4, to 40 HP, 3 phase, 200V, 50 HP-230V or 100 HP 460-575V	2,000	1,100	3,100
0920	Size 5, to 75 HP, 3 phase, 200V, 100 HP-230V or 200 HP 460-575V	4,675	1,475	6,150
1000	Size 6, to 150 HP, 3 phase, 200V, 200 HP-230V or 400 HP 460-575V	20,900	1,650	22,550

D5020 Lighting and Branch Wiring

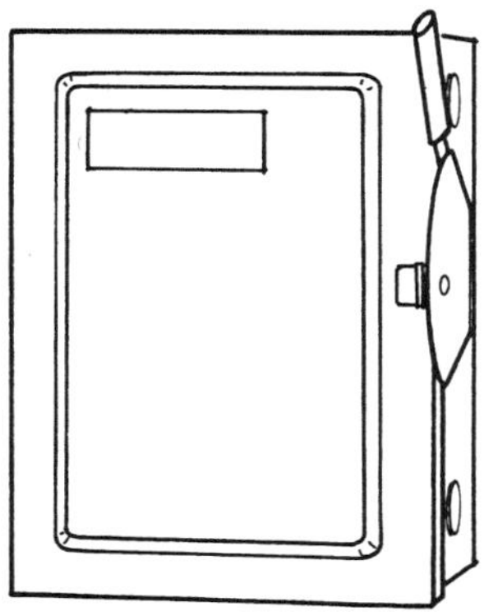

Safety switches are type NEMA 1 for general purpose indoor application, and include time delay fuses, insulation and wire terminations.

System Components	QUANTITY	UNIT	COST EACH MAT.	COST EACH INST.	COST EACH TOTAL
SYSTEM D5020 165 0200					
SAFETY SWITCH, 30A FUSED, 1 PHASE, 115V OR 230V					
Safety switch fused, hvy duty, 240V 2p 30 amp	1.000	Ea.	85.50	188	273.50
Fuse, dual element time delay 250V, 30 amp	2.000	Ea.	21.90	26.40	48.30
TOTAL			107.40	214.40	321.80

D5020 165	Safety Switches	COST EACH MAT.	COST EACH INST.	COST EACH TOTAL
0200	Safety switch, 30 A fused, 1 phase, 2 HP 115 V or 3 HP, 230 V	107	214	321
0280	3 phase, 5 HP, 200 V or 7 1/2 HP, 230 V	143	246	389
0360	15 HP, 460 V or 20 HP, 575 V	259	256	515
0440	60 A fused, 3 phase, 15 HP 200 V or 15 HP 230 V	224	325	549
0520	30 HP 460 V or 40 HP 575 V	330	335	665
0600	100 A fused, 3 phase, 20 HP 200 V or 25 HP 230 V	385	395	780
0680	50 HP 460 V or 60 HP 575 V	635	400	1,035
0760	200 A fused, 3 phase, 50 HP 200 V or 60 HP 230 V	725	560	1,285
0840	125 HP 460 V or 150 HP 575 V	1,025	570	1,595
0920	400 A fused, 3 phase, 100 HP 200 V or 125 HP 230 V	1,675	800	2,475
1000	250 HP 460 V or 350 HP 575 V	2,475	820	3,295
1020	600 A fused, 3 phase, 150 HP 200 V or 200 HP 230 V	3,675	1,175	4,850
1040	400 HP 460 V	3,775	1,200	4,975

D5020 Lighting and Branch Wiring

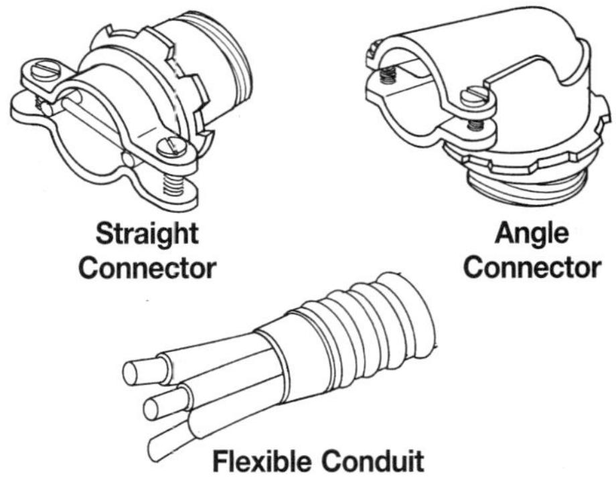

Table below includes costs for the flexible conduit. Not included are wire terminations and testing motor for correct rotation.

System Components	QUANTITY	UNIT	COST EACH MAT.	COST EACH INST.	COST EACH TOTAL
SYSTEM D5020 170 0200					
MOTOR CONNECTIONS, SINGLE PHASE, 115V/230V UP TO 1 HP					
Motor connection, flexible conduit & fittings, 1 HP motor 115V	1.000	Ea.	6.10	82.50	88.60
TOTAL			6.10	82.50	88.60

D5020 170	Motor Connections	COST EACH MAT.	COST EACH INST.	COST EACH TOTAL
0200	Motor connections, single phase, 115/230V, up to 1 HP	6.10	82.50	88.60
0240	Up to 3 HP	12.50	101	113.50
0280	Three phase, 200/230/460/575V, up to 3 HP (RD5010 -170)	7.55	97.50	105.05
0320	Up to 5 HP	6.25	121	127.25
0360	Up to 7-1/2 HF	9.55	143	152.55
0400	Up to 10 HP	21	157	178
0440	Up to 15 HP	21	200	221
0480	Up to 25 HP	27.50	244	271.50
0520	Up to 50 HP	57.50	300	357.50
0560	Up to 100 HP	125	440	565

D5020 Lighting and Branch Wiring

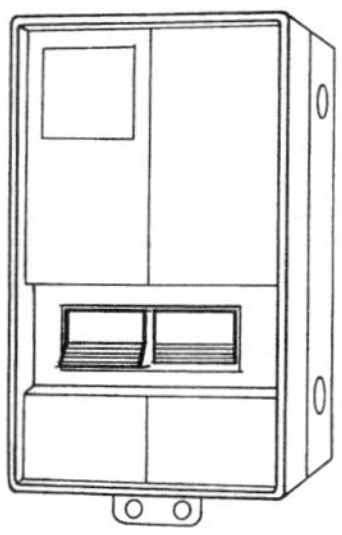

Manual Starter

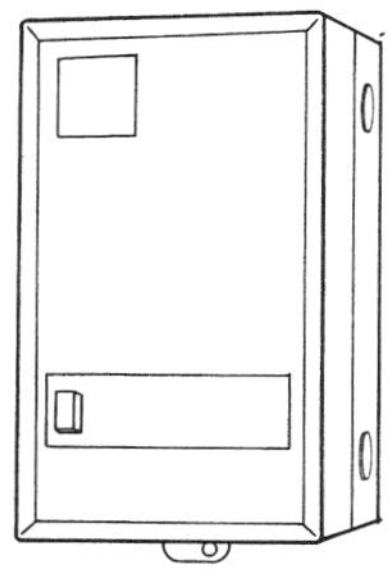

Magnetic Starter

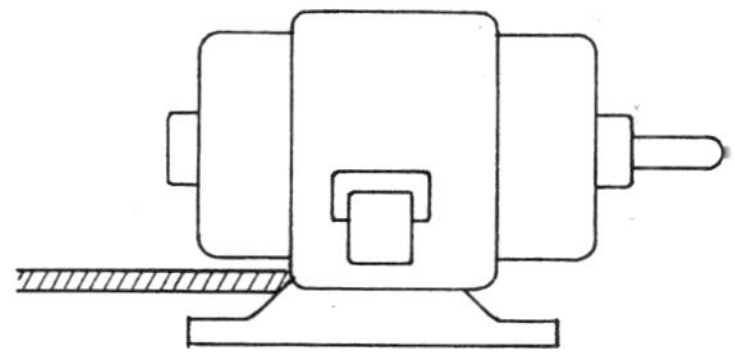

Induction Motor

For 230/460 Volt A.C., 3 phase, 60 cycle ball bearing squirrel cage induction motors, NEMA Class B standard line. Installation included.

No conduit, wire, or terminations included.

System Components	QUANTITY	UNIT	COST EACH MAT.	COST EACH INST.	COST EACH TOTAL
SYSTEM D5020 175 0220					
MOTOR, DRIPPROOF CLASS B INSULATION, 1.15 SERVICE FACTOR, WITH STARTER					
1 H.P., 1200 RPM WITH MANUAL STARTER					
Motor, dripproof, class B insul, 1.15 serv fact, 1200 RPM, 1 HP	1.000	Ea.	640	147	787
Motor starter, manual, 3 phase, 1 HP motor	1.000	Ea.	228	188	416
TOTAL			868	335	1,203

D5020 175	Motor & Starter	COST EACH MAT.	COST EACH INST.	COST EACH TOTAL
0190	Motor, dripproof, premium efficient, 1.15 service factor			
0200	1 HP, 1200 RPM, motor only	640	147	787
0220	With manual starter	870	335	1,205
0240	With magnetic starter (RD5010 -170)	855	335	1,190
0260	1800 RPM, motor only	305	147	452
0280	With manual starter	535	335	870
0300	With magnetic starter	520	335	855
0320	2 HP, 1200 RPM, motor only	710	147	857
0340	With manual starter	940	335	1,275
0360	With magnetic starter	1,050	435	1,485
0380	1800 RPM, motor only	370	147	517
0400	With manual starter	600	335	935
0420	With magnetic starter	715	435	1,150
0440	3600 RPM, motor only	630	147	777
0460	With manual starter	860	335	1,195
0480	With magnetic starter	975	435	1,410
0500	3 HP, 1200 RPM, motor only	975	147	1,122
0520	With manual starter	1,200	335	1,535
0540	With magnetic starter	1,325	435	1,760
0560	1800 RPM, motor only	755	147	902
0580	With manual starter	985	335	1,320
0600	With magnetic starter	1,100	435	1,535
0620	3600 RPM, motor only	640	147	787
0640	With manual starter	870	335	1,205
0660	With magnetic starter	985	435	1,420
0680	5 HP, 1200 RPM, motor only	1,275	147	1,422
0700	With manual starter	1,550	475	2,025
0720	With magnetic starter	1,575	555	2,130
0740	1800 RPM, motor only	670	147	817
0760	With manual starter	940	475	1,415
0780	With magnetic starter	965	555	1,520
0800	3600 RPM, motor only	570	147	717

D5020 Lighting and Branch Wiring

D5020 175	Motor & Starter	COST EACH MAT.	COST EACH INST.	COST EACH TOTAL
0820	With manual starter	840	475	1,315
0840	With magnetic starter	865	555	1,420
0860	7.5 HP, 1800 RPM, motor only	1,025	157	1,182
0880	With manual starter	1,300	485	1,785
0900	With magnetic starter	1,575	755	2,330
0920	10 HP, 1800 RPM, motor only	1,250	165	1,415
0940	With manual starter	1,525	495	2,020
0960	With magnetic starter	1,800	765	2,565
0980	15 HP, 1800 RPM, motor only	1,750	206	1,956
1000	With magnetic starter	2,300	805	3,105
1040	20 HP, 1800 RPM, motor only	2,125	254	2,379
1060	With magnetic starter	3,025	990	4,015
1100	25 HP, 1800 RPM, motor only	2,775	264	3,039
1120	With magnetic starter	3,675	1,000	4,675
1160	30 HP, 1800 RPM, motor only	3,050	275	3,325
1180	With magnetic starter	3,950	1,000	4,950
1220	40 HP, 1800 RPM, motor only	3,975	330	4,305
1240	With magnetic starter	5,975	1,425	7,400
1280	50 HP, 1800 RPM, motor only	4,050	410	4,460
1300	With magnetic starter	6,050	1,500	7,550
1340	60 HP, 1800 RPM, motor only	4,475	470	4,945
1360	With magnetic starter	9,150	1,950	11,100
1400	75 HP, 1800 RPM, motor only	5,250	550	5,800
1420	With magnetic starter	9,925	2,025	11,950
1460	100 HP, 1800 RPM, motor only	6,750	735	7,485
1480	With magnetic starter	11,400	2,200	13,600
1520	125 HP, 1800 RPM, motor only	7,075	940	8,015
1540	With magnetic starter	28,000	2,600	30,600
1580	150 HP, 1800 RPM, motor only	10,700	1,100	11,800
1600	With magnetic starter	31,600	2,750	34,350
1640	200 HP, 1800 RPM, motor only	13,100	1,325	14,425
1660	With magnetic starter	34,000	2,975	36,975
1680	Totally encl, premium efficient, 1.0 ser. fac., 1HP, 1200 RPM, motor only	545	147	692
1700	With manual starter	775	335	1,110
1720	With magnetic starter	760	335	1,095
1740	1800 RPM, motor only	610	147	757
1760	With manual starter	840	335	1,175
1780	With magnetic starter	825	335	1,160
1800	2 HP, 1200 RPM, motor only	630	147	777
1820	With manual starter	860	335	1,195
1840	With magnetic starter	975	435	1,410
1860	1800 RPM, motor only	775	147	922
1880	With manual starter	1,000	335	1,335
1900	With magnetic starter	1,125	435	1,560
1920	3600 RPM, motor only	465	147	612
1940	With manual starter	695	335	1,030
1960	With magnetic starter	810	435	1,245
1980	3 HP, 1200 RPM, motor only	970	147	1,117
2000	With manual starter	1,200	335	1,535
2020	With magnetic starter	1,325	435	1,760
2040	1800 RPM, motor only	675	147	822
2060	With manual starter	905	335	1,240
2080	With magnetic starter	1,025	435	1,460
2100	3600 RPM, motor only	615	147	762
2120	With manual starter	845	335	1,180
2140	With magnetic starter	960	435	1,395
2160	5 HP, 1200 RPM, motor only	1,150	147	1,297
2180	With manual starter	1,425	475	1,900

D5020 Lighting and Branch Wiring

D5020 175	Motor & Starter	COST EACH		
		MAT.	INST.	TOTAL
2200	With magnetic starter	1,450	555	2,005
2220	1800 RPM, motor only	805	147	952
2240	With manual starter	1,075	475	1,550
2260	With magnetic starter	1,100	555	1,655
2280	3600 RPM, motor only	730	147	877
2300	With manual starter	1,000	475	1,475
2320	With magnetic starter	1,025	555	1,580
2340	7.5 HP, 1800 RPM, motor only	1,050	157	1,207
2360	With manual starter	1,325	485	1,810
2380	With magnetic starter	1,600	755	2,355
2400	10 HP, 1800 RPM, motor only	1,500	165	1,665
2420	With manual starter	1,775	495	2,270
2440	With magnetic starter	2,050	765	2,815
2460	15 HP, 1800 RPM, motor only	1,800	206	2,006
2480	With magnetic starter	2,350	805	3,155
2500	20 HP, 1800 RPM, motor only	2,425	254	2,679
2520	With magnetic starter	3,325	990	4,315
2540	25 HP, 1800 RPM, motor only	3,475	264	3,739
2560	With magnetic starter	4,375	1,000	5,375
2580	30 HP, 1800 RPM, motor only	3,425	275	3,700
2600	With magnetic starter	4,325	1,000	5,325
2620	40 HP, 1800 RPM, motor only	4,200	330	4,530
2640	With magnetic starter	6,200	1,425	7,625
2660	50 HP, 1800 RPM, motor only	4,525	410	4,935
2680	With magnetic starter	6,525	1,500	8,025
2700	60 HP, 1800 RPM, motor only	6,150	470	6,620
2720	With magnetic starter	10,800	1,950	12,750
2740	75 HP, 1800 RPM, motor only	8,025	550	8,575
2760	With magnetic starter	12,700	2,025	14,725
2780	100 HP, 1800 RPM, motor only	9,500	735	10,235
2800	With magnetic starter	14,200	2,200	16,400
2820	125 HP, 1800 RPM, motor only	12,300	940	13,240
2840	With magnetic starter	33,200	2,600	35,800
2860	150 HP, 1800 RPM, motor only	14,400	1,100	15,500
2880	With magnetic starter	35,300	2,750	38,050
2900	200 HP, 1800 RPM, motor only	18,200	1,325	19,525
2920	With magnetic starter	39,100	2,975	42,075

D5020 Lighting and Branch Wiring

A. Strip Fixture

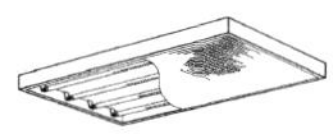

B. Surface Mounted

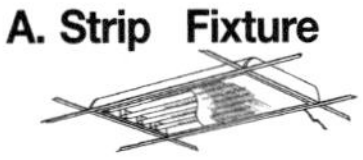

C. Recessed

D. Pendent Mounted

Design Assumptions:

1. A 100 footcandle average maintained level of illumination.
2. Ceiling heights range from 9′ to 11′.
3. Average reflectance values are assumed for ceilings, walls and floors.
4. Cool white (CW) fluorescent lamps with 3150 lumens for 40 watt lamps and 6300 lumens for 8′ slimline lamps.
5. Four 40 watt lamps per 4′ fixture and two 8′ lamps per 8′ fixture.
6. Average fixture efficiency values and spacing to mounting height ratios.
7. Installation labor is average U.S. rate as of January 1.

System Components	QUANTITY	UNIT	COST PER S.F. MAT.	COST PER S.F. INST.	COST PER S.F. TOTAL
SYSTEM D5020 208 0520					
FLUORESCENT FIXTURES MOUNTED 9′-11″ ABOVE FLOOR, 100 FC					
TYPE A, 8 FIXTURES PER 400 S.F.					
Conduit, steel intermediate, 1/2″ diam.	.185	L.F.	.45	1.22	1.67
Wire, 600V, type THWN-THHN, copper, solid, #12	.004	C.L.F.	.04	.22	.26
Fluorescent strip fixture 8 long, surface mounted, two 75W SL	.020	Ea.	1.34	2.12	3.46
Steel outlet box 4″ concrete	.020	Ea.	.23	.66	.89
Steel outlet box plate with stud, 4″ concrete	.020	Ea.	.15	.17	.32
Fixture hangers, flexible, 1/2″ diameter, 4″ long	.040	Ea.	.72	2.20	2.92
Fixture whip, THHN wire, three #12, 3/8″ Greenfield	.040	Ea.	.41	.82	1.23
TOTAL			3.34	7.41	10.75

D5020 208	Fluorescent Fixtures (by Type)	COST PER S.F. MAT.	COST PER S.F. INST.	COST PER S.F. TOTAL
0520	Fluorescent fixtures, type A, 8 fixtures pe· 400 S.F.	3.34	7.40	10.74
0560	11 fixtures per 600 S.F.	3.06	6.75	9.81
0600	17 fixtures per 1000 S.F. RD5020 -200	2.85	6.30	9.15
0640	23 fixtures per 1600 S.F.	2.38	5.30	7.68
0680	28 fixtures per 2000 S.F.	2.35	5.20	7.55
0720	41 fixtures per 3000 S.F.	2.31	5.10	7.41
0800	53 fixtures per 4000 S.F.	2.22	4.92	7.14
0840	64 fixtures per 5000 S.F.	2.18	4.80	6.98
0880	Type B, 11 fixtures per 400 S.F.	5.15	10.55	15.70
0920	15 fixtures per 600 S.F.	4.68	9.50	14.18
0960	24 fixtures per 1000 S.F.	4.52	9.15	13.67
1000	35 fixtures per 1600 S.F.	4.10	8.35	12.45
1040	42 fixtures per 2000 S.F.	3.92	7.95	11.87
1080	61 fixtures per 3000 S F.	3.76	7.70	11.46
1160	80 fixtures per 4000 S F.	2.71	4.89	7.60
1200	98 fixtures per 5000 S F.	3.69	7.50	11.19
1240	Type C, 11 fixtures per 400 S.F.	4.54	11	15.54
1280	14 fixtures per 600 S.F.	3.80	9.20	13
1320	23 fixtures per 1000 S F.	3.76	9.10	12.86
1360	34 fixtures per 1600 S F.	3.49	8.40	11.89
1400	43 fixtures per 2000 S.F.	3.58	8.60	12.18
1440	63 fixtures per 3000 S.F.	3.45	8.30	11.75
1520	81 fixtures per 4000 S.F.	3.31	8.05	11.36
1560	101 fixtures per 5000 S.F.	3.28	7.95	11.23
1600	Type D, 8 fixtures per 400 S.F.	4.16	8.30	12.46
1640	12 fixtures per 600 S.F.	4.16	8.30	12.46
1680	19 fixtures per 1000 S.F.	3.96	7.90	11.86
1720	27 fixtures per 1600 S.F.	3.55	7.05	10.60
1760	34 fixtures per 2000 S.F.	3.55	7.05	10.60
1800	48 fixtures per 3000 S.F.	3.34	6.65	9.99
1880	64 fixtures per 4000 S.F.	3.34	6.65	9.99
1920	79 fixtures per 5000 S.F.	3.32	6.60	9.92

D5020 Lighting and Branch Wiring

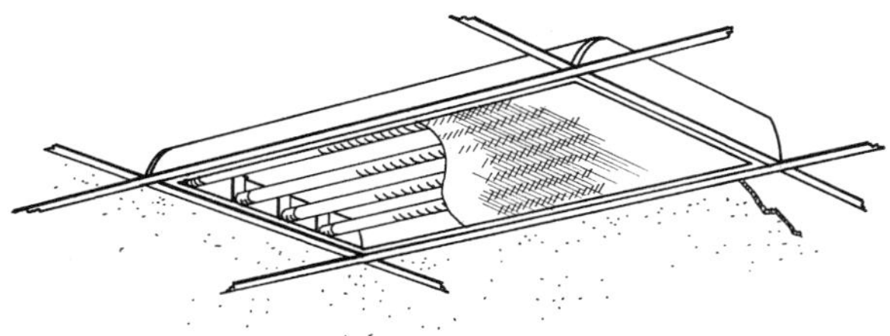

Type C. Recessed, mounted on grid ceiling suspension system, 2′ x 4′, four 40 watt lamps, acrylic prismatic diffusers.

5.3 watts per S.F. for 100 footcandles.

3 watts per S.F. for 57 footcandles.

System Components	QUANTITY	UNIT	COST PER S.F. MAT.	INST.	TOTAL
SYSTEM D5020 210 0200					
FLUORESCENT FIXTURES RECESS MOUNTED IN CEILING					
1 WATT PER S.F., 20 FC, 5 FIXTURES PER 1000 S.F.					
Steel intermediate conduit, (IMC) 1/2″ diam.	.128	L.F.	.31	.84	1.15
Wire, 600 volt, type THW, copper, solid, #12	.003	C.L.F.	.04	.18	.22
Fluorescent fixture, recessed, 2′x 4′, four 40W, w/lens, for grid ceiling	.005	Ea.	.33	.70	1.03
Steel outlet box 4″ square	.005	Ea.	.06	.17	.23
Fixture whip, Greenfield w/#12 THHN wire	.005	Ea.	.04	.04	.08
TOTAL			.78	1.93	2.71

D5020 210	Fluorescent Fixtures (by Wattage)	COST PER S.F. MAT.	INST.	TOTAL
0190	Fluorescent fixtures recess mounted in ceiling			
0195	T12, standard 40 watt lamps			
0200	1 watt per S.F., 20 FC, 5 fixtures @40 watts per 1000 S.F. (RD5020 -200)	.78	1.93	2.71
0240	2 watt per S.F., 40 FC, 10 fixtures @40 watt per 1000 S.F.	1.55	3.80	5.35
0280	3 watt per S.F., 60 FC, 15 fixtures @40 watt per 1000 S.F	2.30	5.75	8.05
0320	4 watt per S.F., 80 FC, 20 fixtures @40 watt per 1000 S.F.	3.07	7.60	10.67
0400	5 watt per S.F., 100 FC, 25 fixtures @40 watt per 1000 S.F.	3.84	9.55	13.39
0450	T8, energy saver 32 watt lamps			
0500	0.8 watt per S.F., 20 FC, 5 fixtures @32 watt per 1000 S.F.	.85	1.93	2.78
0520	1.6 watt per S.F., 40 FC, 10 fixtures @32 watt per 1000 S.F.	1.68	3.80	5.48
0540	2.4 watt per S.F., 60 FC, 15 fixtures @ 32 watt per 1000 S.F	2.49	5.75	8.24
0560	3.2 watt per S.F., 80 FC, 20 fixtures @32 watt per 1000 S.F.	3.33	7.60	10.93
0580	4 watt per S.F., 100 FC, 25 fixtures @32 watt per 1000 S.F.	4.17	9.55	13.72

D5020 Lighting and Branch Wiring

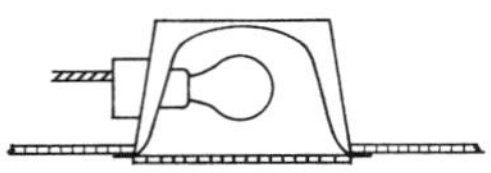

Type A

Type B

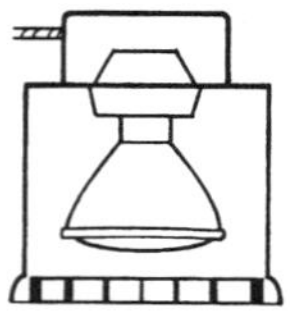

Type C

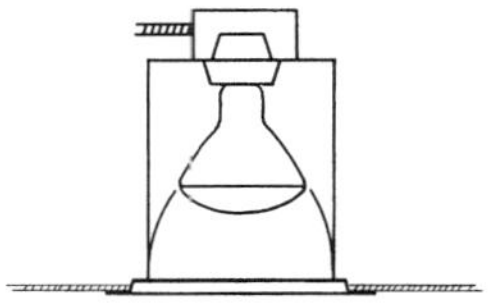

Type D

Type A. Recessed wide distribution reflector with flat glass lens 150 W.

Maximum spacing = 1.2 x mounting height.

13 watts per S.F. for 100 footcandles.

Type B. Recessed reflector down light with baffles 150 W.

Maximum spacing = 0.8 x mounting height.

18 watts per S.F. for 100 footcandles.

Type C. Recessed PAR-38 flood lamp with concentric louver 150 W.

Maximum spacing = 0.5 x mounting height.

19 watts per S.F. for 100 footcandles.

Type D. Recessed R-40 flood lamp with reflector skirt.

Maximum spacing = 0.7 x mounting height.

15 watts per S.F. for 100 footcandles.

System Components	QUANTITY	UNIT	COST PER S.F.		
			MAT.	INST.	TOTAL
SYSTEM D5020 214 0400					
INCANDESCENT FIXTURE RECESS MOUNTED, 100 FC					
TYPE A, 34 FIXTURES PER 400 S.F.					
Steel intermediate conduit (IMC) 1/2" diam	1.060	L.F.	2.57	7	9.57
Wire, 600V, type THWN-THHN, copper, solid, #12	.033	C.L.F.	.39	1.98	2.37
Steel outlet box 4" square	.085	Ea.	.98	2.81	3.79
Fixture whip, Greenfield w/#12 THHN wire	.085	Ea.	.64	.70	1.34
Incandescent fixture, recessed, w/lens, prewired, square trim, 200W	.085	Ea.	9.10	8.37	17.47
TOTAL			13.68	20.86	34.54

D5020 214	Incandescent Fixture (by Type)	COST PER S.F.		
		MAT.	INST.	TOTAL
0380	Incandescent fixture recess mounted, 100 FC			
0400	Type A, 34 fixtures per 400 S.F.	13.70	21	34.70
0440	49 fixtures per 600 S.F. RD5020 -200	13.30	20.50	33.80
0480	63 fixtures per 800 S.F.	13	20.50	33.50
0520	90 fixtures per 1200 S.F.	12.55	19.85	32.40
0560	116 fixtures per 1600 S.F.	12.30	19.70	32
0600	143 fixtures per 2000 S.F.	12.20	19.60	31.80
0640	Type B, 47 fixtures per 400 S.F.	12.75	26.50	39.25
0680	66 fixtures per 600 S.F.	12.25	25.50	37.75
0720	88 fixtures per 800 S.F.	12.25	25.50	37.75
0760	127 fixtures per 1200 S.F.	12.05	25.50	37.55
0800	160 fixtures per 1600 S.F.	11.85	25.50	37.35
0840	206 fixtures per 2000 S.F.	11.80	25	36.80
0880	Type C, 51 fixtures per 400 S.F.	17.35	27.50	44.85
0920	74 fixtures per 600 S.F.	16.90	27	43.90
0960	97 fixtures per 800 S.F.	16.70	27	43.70
1000	142 fixtures per 1200 S.F.	16.45	26.50	42.95
1040	186 fixtures per 1600 S.F.	16.20	26.50	42.70
1080	230 fixtures per 2000 S.F.	16.15	26.50	42.65
1120	Type D, 39 fixtures per 400 S.F.	16.20	21.50	37.70
1160	57 fixtures per 600 S.F.	15.80	21.50	37.30
1200	75 fixtures per 800 S.F.	15.70	21.50	37.20
1240	109 fixtures per 1200 S.F.	15.35	21	36.35
1280	143 fixtures per 1600 S.F.	15.10	21	36.10
1320	176 fixtures per 2000 S.F.	15	21	36

D5020 Lighting and Branch Wiring

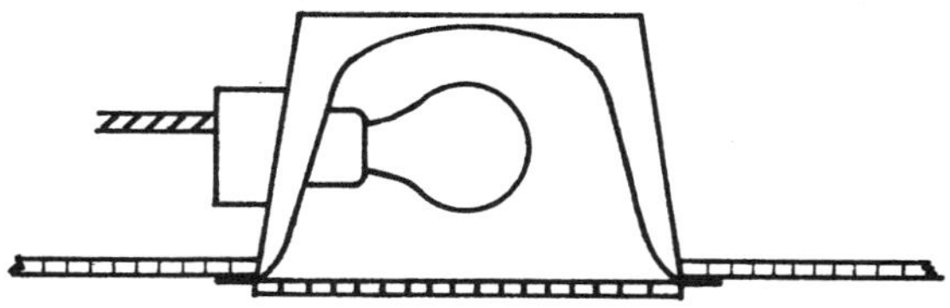

Type A. Recessed, wide distribution reflector with flat glass lens.

150 watt inside frost—2500 lumens per lamp.

PS–25 extended service lamp.

Maximum spacing = 1.2 x mounting height.

13 watts per S.F. for 100 footcandles.

System Components	QUANTITY	UNIT	COST PER S.F. MAT.	COST PER S.F. INST.	COST PER S.F. TOTAL
SYSTEM D5020 216 0200					
INCANDESCENT FIXTURE RECESS MOUNTED, TYPE A					
1 WATT PER S.F., 8 FC, 6 FIXT PER 1000 S.F.					
Steel intermediate conduit, (IMC) 1/2" diam	.091	L.F.	.22	.60	.82
Wire, 600V, type THWN-THHN, copper, solid, #12	.002	C.L.F.	.02	.12	.14
Incandescent fixture, recessed, w/lens, prewired, square trim, 200W	.006	Ea.	.64	.59	1.23
Steel outlet box 4" square	.006	Ea.	.07	.20	.27
Fixture whip, Greenfield w/#12 THHN wire	.006	Ea.	.05	.05	.10
TOTAL			1	1.56	2.56

D5020 216	Incandescent Fixture (by Wattage)	COST PER S.F. MAT.	COST PER S.F. INST.	COST PER S.F. TOTAL
0190	Incandescent fixture recess mounted, type A			
0200	1 watt per S.F., 8 FC, 6 fixtures per 1000 S.F.	1	1.56	2.56
0240	2 watt per S.F., 16 FC, 12 fixtures per 1000 S.F. RD5020-200	2	3.12	5.12
0280	3 watt per S.F., 24 FC, 18 fixtures, per 1000 S.F.	3	4.61	7.61
0320	4 watt per S.F., 32 FC, 24 fixtures per 1000 S.F.	3.99	6.15	10.14
0400	5 watt per S.F., 40 FC, 30 fixtures per 1000 S.F.	5	7.75	12.75

D5020 Lighting and Branch Wiring

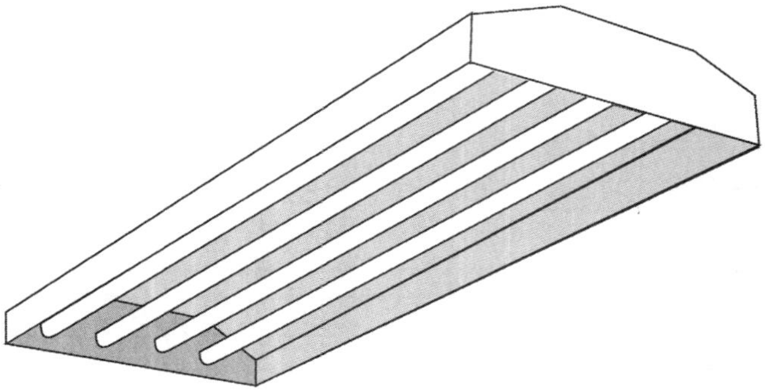

High Bay Fluorescent Fixtures
Four T5 HO/54 watt lamps

System Components	QUANTITY	UNIT	COST PER S.F. MAT.	COST PER S.F. INST.	COST PER S.F. TOTAL
SYSTEM D5020 218 0200					
FLUORESCENT HIGH BAY FIXTURE, 4 LAMP, 8′-10′ ABOVE WORK PLANE					
0.5 WATT/S.F., 29 FC, 2 FIXTURES/1000 S.F.					
Steel intermediate conduit, (IMC) 1/2″ diam	.100	L.F.	.24	.66	.90
Wire, 600V, type THWN-THHN, copper, solid, #10	.002	C.L.F.	.04	.13	.17
Steel outlet box 4″ concrete	.002	Ea.	.02	.07	.09
Steel outlet box plate with stud, 4″ concrete	.002	Ea.	.02	.02	.04
Flourescent , hi bay, 4 lamp, T-5 fixture	.002	Ea.	.50	.30	.80
TOTAL			.82	1.18	2

D5020 218	Fluorescent Fixture, High Bay, 8′-10′ (by Wattage)	COST PER S.F. MAT.	COST PER S.F. INST.	COST PER S.F. TOTAL
0190	Fluorescent high bay-4 lamp fixture, 8′-10′ above work plane			
0200	.5 watt/SF, 29 FC, 2 fixtures per 1000 S.F.	.82	1.18	2
0400	1 watt/SF, 59 FC, 4 fixtures per 1000 S.F.	1.60	2.24	3.84
0600	1.5 watt/SF, 103 FC, 7 fixtures per 1000 S.F.	2.65	3.51	6.16
0800	2 watt/SF, 133 FC, 9 fxtures per 1000 S.F.	3.46	4.62	8.08
1000	2.5 watt/SF, 162 FC, 11 fixtures per 1000 S.F.	4.27	5.75	10.02

D5020 Lighting and Branch Wiring

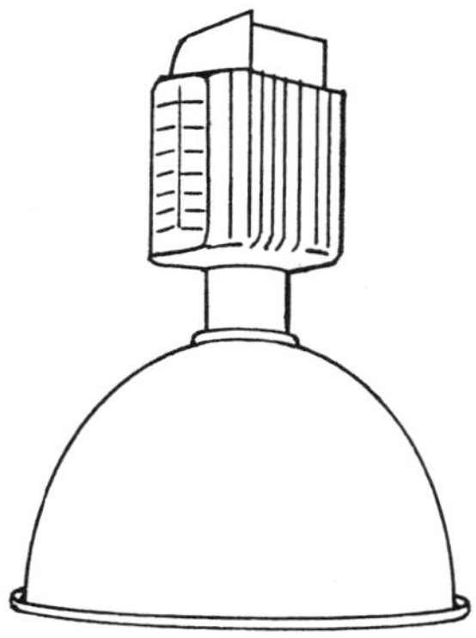

HIGH BAY FIXTURES

B. Metal halide 400 watt
C. High pressure sodium 400 watt
E. Metal halide 1000 watt
F. High pressure sodium 1000 watt
G. Metal halide 1000 watt
125,000 lumen lamp

System Components	QUANTITY	UNIT	COST PER S.F. MAT.	COST PER S.F. INST.	COST PER S.F. TOTAL
SYSTEM D5020 220 0880					
HIGH INTENSITY DISCHARGE FIXTURE, 8'-10' ABOVE WORK PLANE, 100 FC					
TYPE B, 8 FIXTURES PER 900 S.F.					
Steel intermediate conduit, (IMC) 1/2" diam	.460	L.F.	1.11	3.04	4.15
Wire, 600V, type THWN-THHN, copper, solid, #10	.009	C.L.F.	.17	.59	.76
Steel outlet box 4" concrete	.009	Ea.	.10	.30	.40
Steel outlet box plate with stud 4" concrete	.009	Ea.	.07	.07	.14
Metal halide, hi bay, aluminum reflector, 400 W lamp	.009	Ea.	4.05	2.58	6.63
TOTAL			5.50	6.58	12.08

D5020 220	H.I.D. Fixture, High Bay, 8'-10' (by Type)	COST PER S.F. MAT.	COST PER S.F. INST.	COST PER S.F. TOTAL
0500	High intensity discharge fixture, 8'-10' above work plane, 100 FC			
0880	Type B, 8 fixtures per 900 S.F.	5.50	6.60	12.10
0920	15 fixtures per 1800 S.F. (RD5020-200)	5.05	6.35	11.40
0960	24 fixtures per 3000 S.F.	5.05	6.35	11.40
1000	31 fixtures per 4000 S.F. (RD5020-240)	5.05	6.30	11.35
1040	38 fixtures per 5000 S.F.	5.05	6.30	11.35
1080	60 fixtures per 8000 S.F.	5.05	6.30	11.35
1120	72 fixtures per 10000 S.F.	4.54	5.85	10.39
1160	115 fixtures per 16000 S.F.	4.54	5.85	10.39
1200	230 fixtures per 32000 S.F.	4.54	5.85	10.39
1240	Type C, 4 fixtures per 900 S.F.	2.68	4.01	6.69
1280	8 fixtures per 1800 S.F.	2.68	4.04	6.72
1320	13 fixtures per 3000 S.F.	2.68	4.01	6.69
1360	17 fixtures per 4000 S.F.	2.68	4.01	6.69
1400	21 fixtures per 5000 S.F.	2.58	3.69	6.27
1440	33 fixtures per 8000 S.F.	2.56	3.62	6.18
1480	40 fixtures per 10000 S.F.	2.48	3.42	5.90
1520	63 fixtures per 16000 S.F.	2.48	3.42	5.90
1560	126 fixtures per 32000 S.F.	2.48	3.42	5.90

D5020 Lighting and Branch Wiring

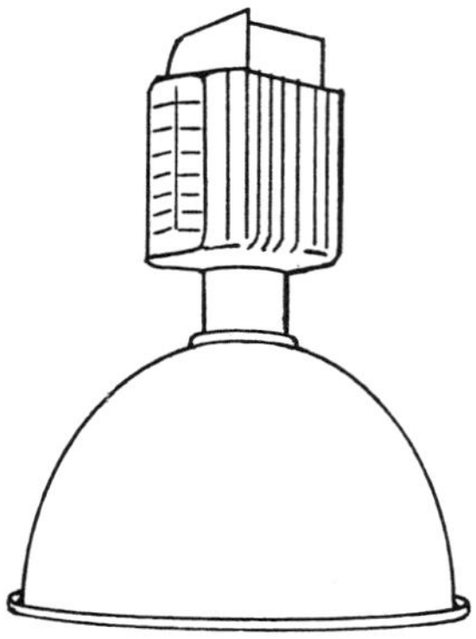

HIGH BAY FIXTURES
B. Metal halide 400 watt
C. High pressure sodium 400 watt
E. Metal halide 1000 watt
F. High pressure sodium 1000 watt
G. Metal halide 1000 watt
125,000 lumen lamp

System Components	QUANTITY	UNIT	COST PER S.F. MAT.	COST PER S.F. INST.	COST PER S.F. TOTAL
SYSTEM D5020 222 0240					
HIGH INTENSITY DISCHARGE FIXTURE, 8′-10′ ABOVE WORK PLANE					
1 WATT/S.F., TYPE B, 29 FC, 2 FIXTURES/1000 S.F.					
Steel intermediate conduit, (IMC) 1/2″ diam	.100	L.F.	.24	.66	.90
Wire, 600V, type THWN-THHN, copper, solid, #10	.002	C.L.F.	.04	.13	.17
Steel outlet box 4″ concrete	.002	Ea.	.02	.07	.09
Steel outlet box plate with stud, 4″ concrete	.002	Ea.	.02	.02	.04
Metal halide, hi bay, aluminum reflector, 400 W lamp	.002	Ea.	.90	.57	1.47
TOTAL			1.22	1.45	2.67

D5020 222	H.I.D. Fixture, High Bay, 8′-10′ (by Wattage)	COST PER S.F. MAT.	COST PER S.F. INST.	COST PER S.F. TOTAL
0190	High intensity discharge fixture, 8′-10′ above work plane			
0240	1 watt/S.F., type B, 29 FC, 2 fixtures/1000 S.F.	1.22	1.45	2.67
0280	Type C, 54 FC, 2 fixtures/1000 S.F. RD5020 -200	1.03	1.11	2.14
0400	2 watt/S.F., type B, 59 FC, 4 fixtures/1000 S.F.	2.40	2.80	5.20
0440	Type C, 108 FC, 4 fixtures/1000 S.F. RD5020 -240	2.03	2.13	4.16
0560	3 watt/S.F., type B, 103 FC, 7 fixtures/1000 S.F.	4.04	4.48	8.52
0600	Type C, 189 FC, 6 fixtures/1000 S.F.	3.20	2.76	5.96
0720	4 watt/S.F., type B, 133 FC, 9 fixtures/1000 S.F.	5.25	5.85	11.10
0760	Type C, 243 FC, 9 fixtures/1000 S.F.	4.54	4.75	9.29
0880	5 watt/S.F., type B, 162 FC, 11 fixtures/1000 S.F.	6.45	7.25	13.70
0920	Type C, 297 FC, 11 fixtures/1000 S.F.	5.60	5.90	11.50

D5020 Lighting and Branch Wiring

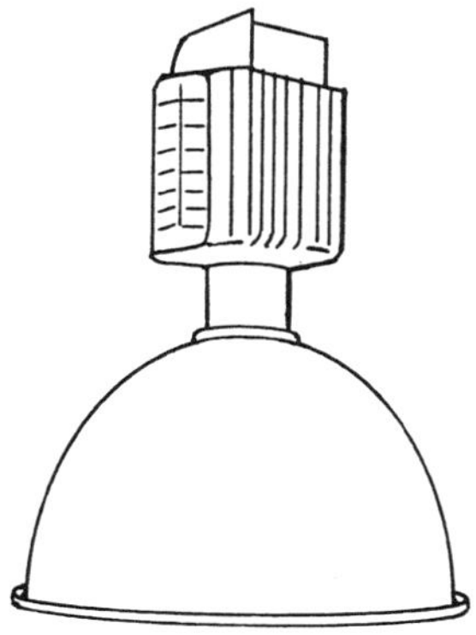

HIGH BAY FIXTURES

B. Metal halide 400 watt
C. High pressure sodium 400 watt
E. Metal halide 1000 watt
F. High pressure sodium 1000 watt
G. Metal halide 1000 watt
125,000 lumen lamp

System Components	QUANTITY	UNIT	COST PER S.F. MAT.	COST PER S.F. INST.	COST PER S.F. TOTAL
SYSTEM D5020 224 1240					
HIGH INTENSITY DISCHARGE FIXTURE, 16′ ABOVE WORK PLANE, 100 FC					
TYPE C, 5 FIXTURES PER 900 S.F.					
Steel intermediate conduit, (IMC) 1/2″ diam	.260	L.F.	.63	1.72	2.35
Wire, 600V, type THWN-THHN, copper, solid, #10	.007	C.L.F.	.13	.46	.59
Steel outlet box 4″ concrete	.006	Ea.	.07	.20	.27
Steel outlet box plate with stud, 4″ concrete	.006	Ea.	.05	.05	.10
High pressure sodium, hi bay, aluminum reflector, 400 W lamp	.006	Ea.	2.49	1.72	4.21
TOTAL			3.37	4.15	7.52

D5020 224	H.I.D. Fixture, High Bay, 16′ (by Type)		COST PER S.F. MAT.	COST PER S.F. INST.	COST PER S.F. TOTAL
0510	High intensity discharge fixture, 16′ above work plane, 100 FC				
1240	Type C, 5 fixtures per 900 S.F.		3.37	4.15	7.52
1280	9 fixtures per 1800 S.F.	RD5020 -200	3.13	4.36	7.49
1320	15 fixtures per 3000 S.F.		3.13	4.36	7.49
1360	18 fixtures per 4000 S.F.	RD5020 -240	2.96	3.90	6.86
1400	22 fixtures per 5000 S.F.		2.96	3.90	6.86
1440	36 fixtures per 8000 S.F.		2.96	3.90	6.86
1480	42 fixtures per 10,000 S.F.		2.69	4.02	6.71
1520	65 fixtures per 16,000 S.F.		2.69	4.02	6.71
1600	Type G, 4 fixtures per 900 S.F.		4.46	6.55	11.01
1640	6 fixtures per 1800 S.F.		3.74	6.15	9.89
1720	9 fixtures per 4000 S.F.		3.17	5.95	9.12
1760	11 fixtures per 5000 S.F.		3.17	5.95	9.12
1840	21 fixtures per 10,000 S.F.		2.95	5.35	8.30
1880	33 fixtures per 16,000 S.F.		2.95	5.35	8.30

D5020 Lighting and Branch Wiring

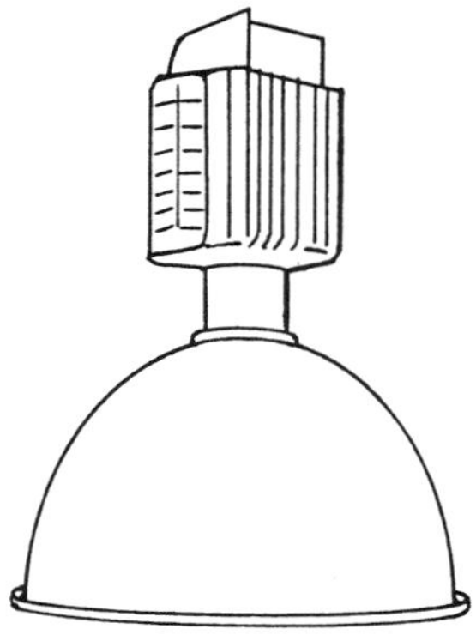

HIGH BAY FIXTURES
B. Metal halide 400 watt
C. High pressure sodium 400 watt
E. Metal halide 1000 watt
F. High pressure sodium 1000 watt
G. Metal halide 1000 watt
125,000 lumen lamp

System Components	QUANTITY	UNIT	COST PER S.F. MAT.	COST PER S.F. INST.	COST PER S.F. TOTAL
SYSTEM D5020 226 0240					
HIGH INTENSITY DISCHARGE FIXTURE, 16′ ABOVE WORK PLANE					
1 WATT/S.F., TYPE E, 42 FC, 1 FIXTURE/1000 S.F.					
Steel intermediate conduit, (IMC) 1/2″ diam	.160	L.F.	.39	1.06	1.45
Wire, 600V, type THWN-THHN, copper, solid, #10	.003	C.L.F.	.06	.20	.26
Steel outlet box 4″ concrete	.001	Ea.	.01	.03	.04
Steel outlet box plate with stud, 4″ concrete	.001	Ea.	.01	.01	.02
Metal halide, hi bay, aluminum reflector, 1000 W lamp	.001	Ea.	.65	.33	.98
TOTAL			1.12	1.63	2.75

D5020 226	H.I.D. Fixture, High Bay, 16′ (by Wattage)		COST PER S.F. MAT.	COST PER S.F. INST.	COST PER S.F. TOTAL
0190	High intensity discharge fixture, 16′ above work plane				
0240	1 watt/S.F., type E, 42 FC, 1 fixture/1000 S.F.		1.12	1.63	2.75
0280	Type G, 52 FC, 1 fixture/1000 S.F.	RD5020 -200	1.12	1.63	2.75
0320	Type C, 54 FC, 2 fixture/1000 S.F.		1.28	1.82	3.10
0440	2 watt/S.F., type E, 84 FC, 2 fixture/1000 S.F.	RD5020 -240	2.24	3.32	5.56
0480	Type G, 105 FC, 2 fixture/1000 S.F.		2.24	3.32	5.56
0520	Type C, 108 FC, 4 fixture/1000 S.F.		2.56	3.64	6.20
0640	3 watt/S.F., type E, 126 FC, 3 fixture/1000 S.F.		3.34	4.94	8.28
0680	Type G, 157 FC, 3 fixture/1000 S.F.		3.34	4.94	8.28
0720	Type C, 162 FC, 6 fixture/1000 S.F.		3.84	5.45	9.29
0840	4 watt/S.F., type E, 168 FC, 4 fixture/1000 S.F.		4.49	6.60	11.09
0880	Type G, 210 FC, 4 fixture/1000 S.F.		4.49	6.60	11.09
0920	Type C, 243 FC, 9 fixture/1000 S.F.		5.55	7.60	13.15
1040	5 watt/S.F., type E, 210 FC, 5 fixture/1000 S.F.		5.60	8.25	13.85
1080	Type G, 262 FC, 5 fixture/1000 S.F.		5.60	8.25	13.85
1120	Type C, 297 FC, 11 fixture/1000 S.F.		6.85	9.40	16.25

D5020 Lighting and Branch Wiring

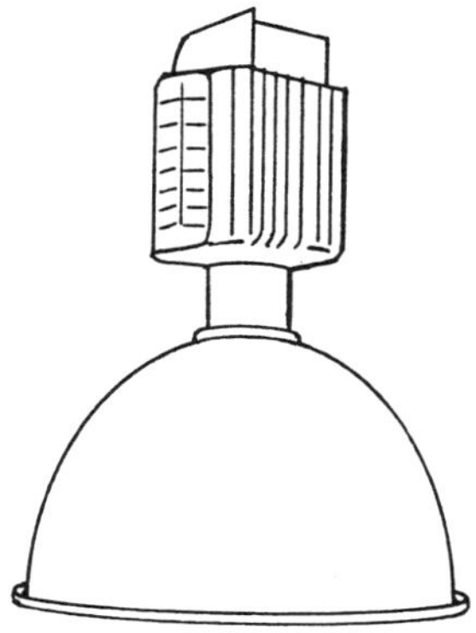

HIGH BAY FIXTURES
B. Metal halide 400 watt
C. High pressure sodium 400 watt
E. Metal halide 1000 watt
F. High pressure sodium 1000 watt
G. Metal halide 1000 watt
125,000 lumen lamp

System Components	QUANTITY	UNIT	COST PER S.F. MAT.	COST PER S.F. INST.	COST PER S.F. TOTAL
SYSTEM D5020 228 1240					
HIGH INTENSITY DISCHARGE FIXTURE, 20′ ABOVE WORK PLANE, 100 FC					
TYPE C, 6 FIXTURES PER 900 S.F.					
Steel intermediate conduit, (IMC) 1/2″ diam	.350	L.F.	.85	2.31	3.16
Wire, 600V, type THWN-THHN, copper, solid, #10.	.011	C.L.F.	.20	.73	.93
Steel outlet box 4″ concrete	.007	Ea.	.08	.23	.31
Steel outlet box plate with stud, 4″ concrete	.007	Ea.	.05	.06	.11
High pressure sodium, hi bay, aluminum reflector, 400 W lamp	.007	Ea.	2.91	2.01	4.92
TOTAL			4.09	5.34	9.43

D5020 228	H.I.D. Fixture, High Bay, 20′ (by Type)	COST PER S.F. MAT.	COST PER S.F. INST.	COST PER S.F. TOTAL
0510	High intensity discharge fixture 20′ above work plane, 100 FC			
1240	Type C, 6 fixtures per 900 S.F.	4.09	5.35	9.44
1280	10 fixtures per 1800 S.F. RD5020 -200	3.66	5	8.66
1320	16 fixtures per 3000 S.F.	3.27	4.82	8.09
1360	20 fixtures per 4000 S.F. RD5020 -240	3.27	4.82	8.09
1400	24 fixtures per 5000 S.F.	3.27	4.82	8.09
1440	38 fixtures per 8000 S.F.	3.20	4.62	7.82
1520	68 fixtures per 16000 S.F.	3.05	5.10	8.15
1560	132 fixtures per 32000 S.F.	3.05	5.10	8.15
1600	Type G, 4 fixtures per 900 S.F.	4.39	6.45	10.84
1640	6 fixtures per 1800 S.F.	3.73	6.15	9.88
1680	7 fixtures per 3000 S.F.	3.66	5.95	9.61
1720	10 fixtures per 4000 S.F.	3.64	5.85	9.49
1760	11 fixtures per 5000 S.F.	3.27	6.30	9.57
1800	18 fixtures per 8000 S.F.	3.27	6.30	9.57
1840	22 fixtures per 10000 S.F.	3.27	6.30	9.57
1880	34 fixtures per 16000 S.F.	3.19	6.10	9.29
1920	66 fixtures per 32000 S.F.	3	5.55	8.55

D5020 Lighting and Branch Wiring

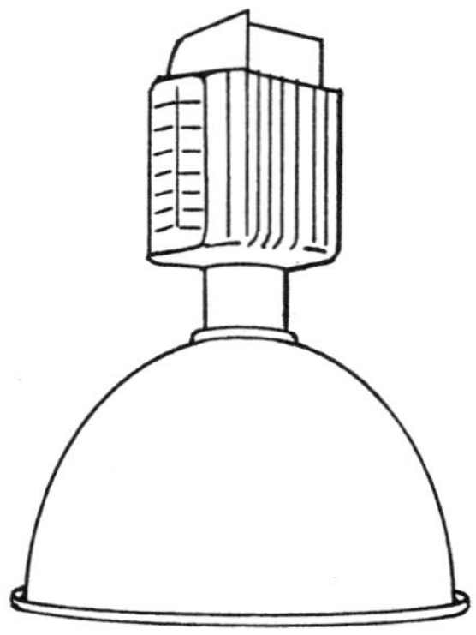

HIGH BAY FIXTURES
B. Metal halide 400 watt
C. High pressure sodium 400 watt
E. Metal halide 1000 watt
F. High pressure sodium 1000 watt
G. Metal halide 1000 watt
125,000 lumen lamp

System Components	QUANTITY	UNIT	COST PER S.F. MAT.	COST PER S.F. INST.	COST PER S.F. TOTAL
SYSTEM D5020 230 0240					
HIGH INTENSITY DISCHARGE FIXTURE, 20′ ABOVE WORK PLANE					
1 WATT/S.F., TYPE E, 40 FC, 1 FIXTURE 1000 S.F.					
Steel intermediate conduit, (IMC) 1/2″ diam	.160	L.F.	.39	1.06	1.45
Wire, 600V, type THWN-THHN, copper, solid, #10	.005	C.L.F.	.09	.33	.42
Steel outlet box 4″ concrete	.001	Ea.	.01	.03	.04
Steel outlet box plate with stud, 4″ concrete	.001	Ea.	.01	.01	.02
Metal halide, hi bay, aluminum reflector, 1000 W lamp	.001	Ea.	.65	.33	.98
TOTAL			1.15	1.76	2.91

D5020 230	H.I.D. Fixture, High Bay, 20′ (by Wattage)		COST PER S.F. MAT.	COST PER S.F. INST.	COST PER S.F. TOTAL
0190	High intensity discharge fixture, 20′ above work plane				
0240	1 watt/S.F., type E, 40 FC, 1 fixture/1000 S.F.		1.15	1.76	2.91
0280	Type G, 50 FC, 1 fixture/1000 S.F.	RD5020 -200	1.15	1.76	2.91
0320	Type C, 52 FC, 2 fixtures/1000 S.F.		1.34	2.03	3.37
0440	2 watt/S.F., type E, 81 FC, 2 fixtures/1000 S.F.	RD5020 -240	2.29	3.52	5.81
0480	Type G, 101 FC, 2 fixtures/1000 S.F.		2.29	3.52	5.81
0520	Type C, 104 FC, 4 fixtures/1000 S.F.		2.67	3.99	6.66
0640	3 watt/S.F., type E, 121 FC, 3 fixtures/1000 S.F.		3.44	5.25	8.69
0680	Type G, 151 FC, 3 fixtures/1000 S.F.		3.44	5.25	8.69
0720	Type C, 155 FC, 6 fixtures/1000 S.F.		4.02	6	10.02
0840	4 watt/S.F., type E, 161 FC, 4 fixtures/1000 S.F.		4.60	7	11.60
0880	Type G, 202 FC, 4 fixtures/1000 S.F.		4.60	7	11.60
0920	Type C, 233 FC, 9 fixtures/1000 S.F.		5.75	8.30	14.05
1040	5 watt/S.F., type E, 202 FC, 5 fixtures/1000 S.F.		5.75	8.80	14.55
1080	Type G, 252 FC, 5 fixtures/1000 S.F.		5.75	8.80	14.55
1120	Type C, 285 FC, 11 fixtures/1000 S.F.		7.10	10.35	17.45

D5020 Lighting and Branch Wiring

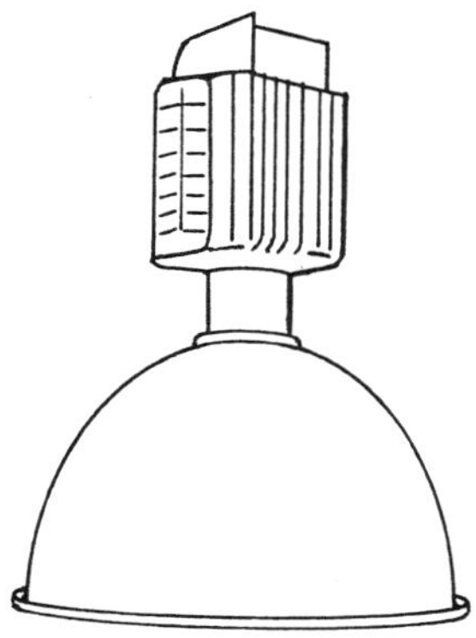

HIGH BAY FIXTURES
B. Metal halide 400 watt
C. High pressure sodium 400 watt
E. Metal halide 1000 watt
F. High pressure sodium 1000 watt
G. Metal halide 1000 watt
125,000 lumen lamp

System Components	QUANTITY	UNIT	COST PER S.F. MAT.	COST PER S.F. INST.	COST PER S.F. TOTAL
SYSTEM D5020 232 1240					
HIGH INTENSITY DISCHARGE FIXTURE, 30′ ABOVE WORK PLANE, 100 FC					
TYPE F, 4 FIXTURES PER 900 S.F.					
Steel intermediate conduit, (IMC) 1/2″ diam	.580	L.F.	1.40	3.83	5.23
Wire, 600V, type THWN-THHN, copper, solid, #10	.018	C.L.F.	.33	1.19	1.52
Steel outlet box 4″ concrete	.004	Ea.	.05	.13	.18
Steel outlet box plate with stud, 4″ concrete	.004	Ea.	.03	.03	.06
High pressure sodium, hi bay, aluminum, 1000 W lamp	.004	Ea.	2.40	1.32	3.72
TOTAL			4.21	6.50	10.71

D5020 232	H.I.D. Fixture, High Bay, 30′ (by Type)	COST PER S.F. MAT.	COST PER S.F. INST.	COST PER S.F. TOTAL
0510	High intensity discharge fixture, 30′ above work plane, 100 FC			
1240	Type F, 4 fixtures per 900 S.F.	4.21	6.50	10.71
1280	6 fixtures per 1800 S.F. RD5020 -200	3.61	6.20	9.81
1320	8 fixtures per 3000 S.F.	3.61	6.20	9.81
1360	9 fixtures per 4000 S.F. RD5020 -240	3.02	5.90	8.92
1400	10 fixtures per 5000 S.F.	3.02	5.90	8.92
1440	17 fixtures per 8000 S.F.	3.02	5.90	8.92
1480	18 fixtures per 10,000 S.F.	2.85	5.45	8.30
1520	27 fixtures per 16,000 S.F.	2.85	5.45	8.30
1560	52 fixtures per 32000 S.F.	2.83	5.35	8.18
1600	Type G, 4 fixtures per 900 S.F.	4.56	6.90	11.46
1640	6 fixtures per 1800 S.F.	3.79	6.25	10.04
1680	9 fixtures per 3000 S.F.	3.71	6.05	9.76
1720	11 fixtures per 4000 S.F.	3.64	5.85	9.49
1760	13 fixtures per 5000 S.F.	3.64	5.85	9.49
1800	21 fixtures per 8000 S.F.	3.64	5.85	9.49
1840	23 fixtures per 10,000 S.F.	3.18	6.25	9.43
1880	36 fixtures per 16,000 S.F.	3.18	6.25	9.43
1920	70 fixtures per 32,000 S.F.	3.18	6.25	9.43

D5020 Lighting and Branch Wiring

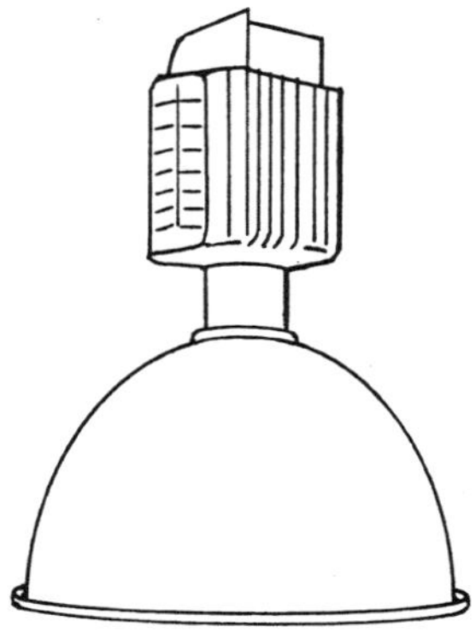

HIGH BAY FIXTURES

B. Metal halide 400 watt
C. High pressure sodium 400 watt
E. Metal halide 1000 watt
F. High pressure sodium 1000 watt
G. Metal halide 1000 watt
125,000 lumen lamp

System Components	QUANTITY	UNIT	COST PER S.F. MAT.	COST PER S.F. INST.	COST PER S.F. TOTAL
SYSTEM D5020 234 0240					
HIGH INTENSITY DISCHARGE FIXTURE, 30′ ABOVE WORK PLANE					
1 WATT/S.F., TYPE E, 37 FC, 1 FIXTURE/1000 S.F.					
Steel intermediate conduit, (IMC) 1/2″ diam	.196	L.F.	.47	1.29	1.76
Wire, 600V type THWN-THHN, copper, solid, #10	.006	C.L.F.	.11	.40	.51
Steel outlet box 4″ concrete	.001	Ea.	.01	.03	.04
Steel outlet box plate with stud, 4″ concrete	.001	Ea.	.01	.01	.02
Metal halide, hi bay, aluminum reflector, 1000 W lamp	.001	Ea.	.65	.33	.98
TOTAL			1.25	2.06	3.31

D5020 234	**H.I.D. Fixture, High Bay, 30′ (by Wattage)**		COST PER S.F. MAT.	COST PER S.F. INST.	COST PER S.F. TOTAL
0190	High intensity discharge fixture, 30′ above work plane				
0240	1 watt/S.F., type E, 37 FC, 1 fixture/1000 S.F.		1.25	2.06	3.31
0280	Type G, 45 FC., 1 fixture/1000 S.F.	RD5020 -200	1.25	2.06	3.31
0320	Type F, 50 FC, 1 fixture/1000 S.F.		1.05	1.61	2.66
0440	2 watt/S.F., type E, 74 FC, 2 fixtures/1000 S.F.	RD5020 -240	2.51	4.13	6.64
0480	Type G, 92 FC, 2 fixtures/1000 S.F.		2.51	4.13	6.64
0520	Type F, 100 FC, 2 fixtures/1000 S.F.		2.13	3.29	5.42
0640	3 watt/S.F., type E, 110 FC, 3 fixtures/1000 S.F.		3.77	6.25	10.02
0680	Type G, 138 FC, 3 fixtures/1000 S.F.		3.77	6.25	10.02
0720	Type F, 150 FC, 3 fixtures/1000 S.F.		3.16	4.90	8.06
0840	4 watt/S.F., type E, 143 FC, 4 fixtures/1000 S.F.		5.05	8.30	13.35
0880	Type G, 185 FC, 4 fixtures/1000 S.F.		5.05	8.30	13.35
0920	Type F, 200 FC, 4 fixtures/1000 S.F.		4.24	6.60	10.84
1040	5 watt/S.F., type E, 185 FC, 5 fixtures/1000 S.F.		6.30	10.40	16.70
1080	Type G, 230 FC, 5 fixtures/1000 S.F.		6.30	10.40	16.70
1120	Type F, 250 FC, 5 fixtures/1000 S.F.		5.30	8.20	13.50

D5020 Lighting and Branch Wiring

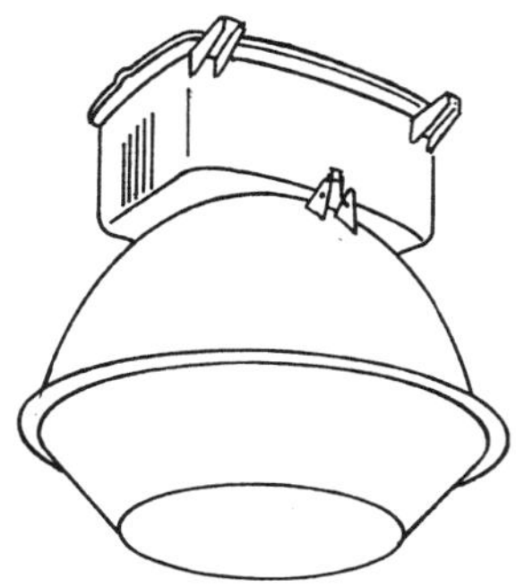

LOW BAY FIXTURES

J. Metal halide 250 watt

K. High pressure sodium 150 watt

System Components	QUANTITY	UNIT	COST PER S.F. MAT.	COST PER S.F. INST.	COST PER S.F. TOTAL
SYSTEM D5020 236 0920					
HIGH INTENSITY DISCHARGE FIXTURE, 8'-10' ABOVE WORK PLANE, 50 FC					
TYPE J, 13 FIXTURES PER 1800 S.F.					
Steel intermediate conduit, (IMC) 1/2" diam	.550	L.F.	1.33	3.63	4.96
Wire, 600V, type THWN-THHN, copper, solid, #10	.012	C.L.F.	.22	.79	1.01
Steel outlet box 4" concrete	.007	Ea.	.08	.23	.31
Steel outlet box plate with stud, 4" concrete	.007	Ea.	.05	.06	.11
Metal halide, lo bay, aluminum reflector, 250 W DX lamp	.007	Ea.	2.80	1.44	4.24
TOTAL			4.48	6.15	10.63

D5020 236	H.I.D. Fixture, Low Bay, 8'-10' (by Type)	COST PER S.F. MAT.	COST PER S.F. INST.	COST PER S.F. TOTAL
0510	High intensity discharge fixture, 8'-10' above work plane, 50 FC			
0880	Type J, 7 fixtures per 900 S.F.	4.83	6.20	11.03
0920	13 fixtures per 1800 S.F. RD5020 -200	4.48	6.15	10.63
0960	21 fixtures per 3000 S.F.	4.48	6.15	10.63
1000	28 fixtures per 4000 S.F.	4.48	6.15	10.63
1040	35 fixtures per 5000 S.F.	4.48	6.15	10.63
1120	62 fixtures per 10,000 S.F.	4.07	5.90	9.97
1160	99 fixtures per 16,000 S.F.	4.07	5.90	9.97
1200	199 fixtures per 32,000 S.F.	4.07	5.90	9.97
1240	Type K, 9 fixtures per 900 S.F.	4.80	5.30	10.10
1280	16 fixtures per 1800 S.F.	4.43	5.15	9.58
1320	26 fixtures per 3000 S.F.	4.41	5.05	9.46
1360	31 fixtures per 4000 S.F.	4.08	4.95	9.03
1400	39 fixtures per 5000 S.F.	4.08	4.95	9.03
1440	62 fixtures per 8000 S.F.	4.08	4.95	9.03
1480	78 fixtures per 10,000 S.F.	4.08	4.95	9.03
1520	124 fixtures per 16,000 S.F.	4.03	4.82	8.85
1560	248 fixtures per 32,000 S.F.	3.93	5.35	9.28

D5020 Lighting and Branch Wiring

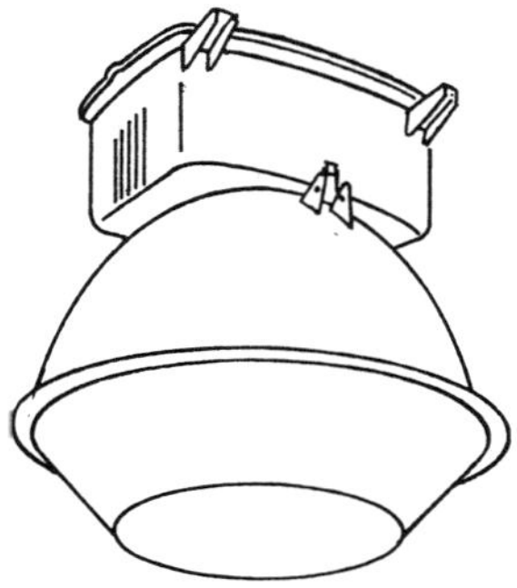

LOW BAY FIXTURES

J. Metal halide 250 watt

K. High pressure sodium 150 watt

System Components	QUANTITY	UNIT	COST PER S.F. MAT.	COST PER S.F. INST.	COST PER S.F. TOTAL
SYSTEM D5020 238 0240 HIGH INTENSITY DISCHARGE FIXTURE, 8′-10′ ABOVE WORK PLANE 1 WATT/S.F., TYPE J, 30 FC, 4 FIXTURES/1000 S.F.					
Steel intermediate conduit, (IMC) 1/2″ diam	.280	L.F.	.68	1.85	2.53
Wire, 600V, type THWN-THHN, copper, solid, #10	.008	C.L.F.	.15	.53	.68
Steel outlet box 4″ concrete	.004	Ea.	.05	.13	.18
Steel outlet box plate with stud, 4″ concrete	.004	Ea.	.03	.03	.06
Metal halide, lo bay, aluminum reflector, 250 W DX lamp	.004	Ea.	1.60	.82	2.42
TOTAL			2.51	3.36	5.87

D5020 238	H.I.D. Fixture, Low Bay, 8′-10′ (by Wattage)	COST PER S.F. MAT.	COST PER S.F. INST.	COST PER S.F. TOTAL
0190	High intensity discharge fixture, 8′-10′ above work plane			
0240	1 watt/S.F., type J, 30 FC, 4 fixtures/1000 S.F.	2.51	3.36	5.87
0280	Type K, 29 FC, 5 fixtures/1000 S.F. (RD5020 -200)	2.52	3.03	5.55
0400	2 watt/S.F., type J, 52 FC, 7 fixtures/1000 S.F.	4.46	6.15	10.61
0440	Type K, 63 FC, 11 fixtures/1000 S.F.	5.45	6.35	11.80
0560	3 watt/S.F., type J, 81 FC, 11 fixtures/1000 S.F.	6.90	9.30	16.20
0600	Type K, 92 FC, 16 fixtures/1000 S.F.	7.95	9.35	17.30
0720	4 watt/S.F., type J, 103 FC, 14 fixtures/1000 S.F.	8.90	12.25	21.15
0760	Type K, 127 FC, 22 fixtures/1000 S.F.	10.85	12.70	23.55
0880	5 watt/S.F., type J, 133 FC, 18 fixtures/1000 S.F.	11.35	15.45	26.80
0920	Type K, 155 FC, 27 fixtures/1000 S.F.	13.35	15.70	29.05

D5020 Lighting and Branch Wiring

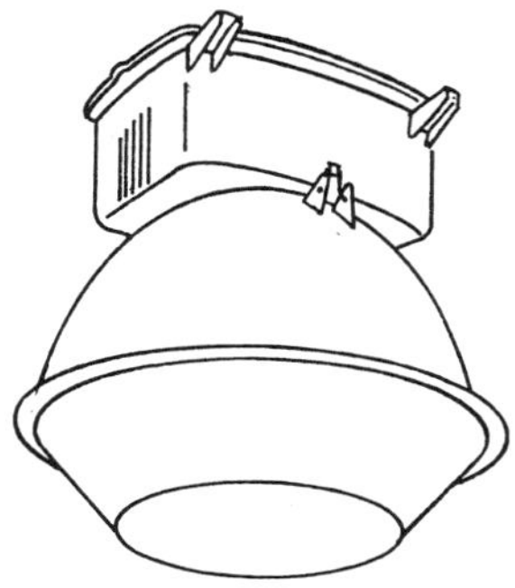

LOW BAY FIXTURES

J. Metal halide 250 watt

K. High pressure sodium 150 watt

System Components	QUANTITY	UNIT	COST PER S.F. MAT.	COST PER S.F. INST.	COST PER S.F. TOTAL
SYSTEM D5020 240 0880					
HIGH INTENSITY DISCHARGE FIXTURE, 16′ ABOVE WORK PLANE, 50 FC					
TYPE J, 9 FIXTURES PER 900 S.F.					
Steel intermediate conduit, (IMC) 1/2″ diam	.630	L.F.	1.52	4.16	5.68
Wire, 600V type, THWN-THHN, copper, solid, #10	.012	C.L.F.	.22	.79	1.01
Steel outlet box 4″ concrete	.010	Ea.	.12	.33	.45
Steel outlet box plate with stud, 4″ concrete	.010	Ea.	.08	.08	.16
Metal halide, lo bay, aluminum reflector, 250 W DX lamp	.010	Ea.	4	2.06	6.06
TOTAL			5.94	7.42	13.36

D5020 240	H.I.D. Fixture, Low Bay, 16′ (by Type)	COST PER S.F. MAT.	COST PER S.F. INST.	COST PER S.F. TOTAL
0510	High intensity discharge fixture, 16′ above work plane, 50 FC			
0880	Type J, 9 fixtures per 900 S.F.	5.95	7.40	13.35
0920	14 fixtures per 1800 S.F.	5.10	7	12.10
0960	24 fixtures per 3000 S.F.	5.15	7.10	12.25
1000	32 fixtures per 4000 S.F.	5.15	7.10	12.25
1040	35 fixtures per 5000 S.F.	4.77	6.95	11.72
1080	56 fixtures per 8000 S.F.	4.77	6.95	11.72
1120	70 fixtures per 10,000 S.F.	4.76	6.95	11.71
1160	111 fixtures per 16,000 S.F.	4.76	6.95	11.71
1200	222 fixtures per 32,000 S.F.	4.76	6.95	11.71
1240	Type K, 11 fixtures per 900 S.F.	5.95	6.95	12.90
1280	20 fixtures per 1800 S.F.	5.50	6.40	11.90
1320	29 fixtures per 3000 S.F.	5.15	6.30	11.45
1360	39 fixtures per 4000 S.F.	5.10	6.25	11.35
1400	44 fixtures per 5000 S.F.	4.77	6.10	10.87
1440	62 fixtures per 8000 S.F.	4.77	6.10	10.87
1480	87 fixtures per 10,000 S.F.	4.77	6.10	10.87
1520	138 fixtures per 16,000 S.F.	4.77	6.10	10.87

D5020 Lighting and Branch Wiring

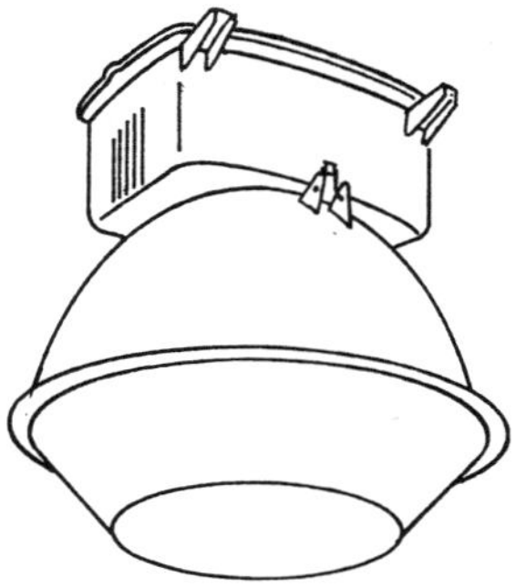

LOW BAY FIXTURES

J. Metal halide 250 watt

K. High pressure sodium 150 watt

System Components	QUANTITY	UNIT	COST PER S.F. MAT.	COST PER S.F. INST.	COST PER S.F. TOTAL
SYSTEM D5020 242 0240					
HIGH INTENSITY DISCHARGE FIXTURE, 16′ ABOVE WORK PLANE					
1 WATT/S.F., TYPE J, 28 FC, 4 FIXTURES/1000 S.F.					
Steel intermediate conduit, (IMC) 1/2″ diam	.328	L.F.	.79	2.16	2.95
Wire, 600V, type THWN-THHN, copper, solid, #10	.010	C.L.F.	.18	.66	.84
Steel outlet box 4″ concrete	.004	Ea.	.05	.13	.18
Steel outlet box plate with stud, 4″ concrete	.004	Ea.	.03	.03	.06
Metal halide, lo bay, aluminum reflector, 250 W DX lamp	.004	Ea.	1.60	.82	2.42
TOTAL			2.65	3.80	6.45

D5020 242	H.I.D. Fixture, Low Bay, 16′ (by Wattage)	COST PER S.F. MAT.	COST PER S.F. INST.	COST PER S.F. TOTAL
0190	High intensity discharge fixture, mounted 16′ above work plane			
0240	1 watt/S.F., type J, 28 FC, 4 fixt./1000 S.F.	2.65	3.80	6.45
0280	Type K, 27 FC, 5 fixt./1000 S.F. RD5020 -200	2.95	4.28	7.23
0400	2 watt/S.F., type J, 48 FC, 7 fixt/1000 S.F.	4.89	7.40	12.29
0440	Type K, 58 FC, 11 fixt/1000 S.F.	6.25	8.75	15
0560	3 watt/S.F., type J, 75 FC, 11 fixt/1000 S.F.	7.55	11.20	18.75
0600	Type K, 85 FC, 16 fixt/1000 S.F.	9.20	13	22.20
0720	4 watt/S.F., type J, 95 FC, 14 fixt/1000 S.F.	9.80	14.75	24.55
0760	Type K, 117 FC, 22 fixt/1000 S.F.	12.50	17.45	29.95
0880	5 watt/S.F., type J, 122 FC, 18 fixt/1000 S.F.	12.45	18.55	31
0920	Type K, 143 FC, 27 fixt/1000 S.F.	15.45	21.50	36.95

D5020 Lighting and Branch Wiring

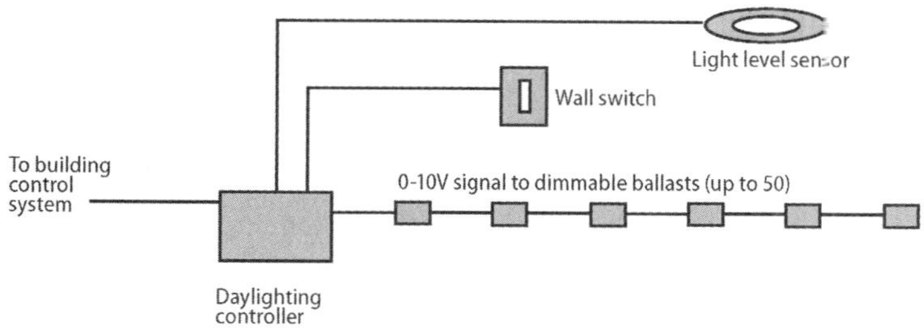

Daylight Dimming System

System Components	QUANTITY	UNIT	COST PER S.F. MAT.	INST.	TOTAL
SYSTEM D5020 290 0800					
DAYLIGHT DIMMING CONTROL SYSTEM					
5 FIXTURES PER 1000 M.S.F.					
Tray cable, type TC, copper #16-4 conductor	.300	C.L.F.	.01	.03	.04
Wire, 600 volt, type THWN-THHN, copper, solid, #12	.320	C.L.F.		.02	.02
Conduit (EMT) , to 15′ H, incl 2 termn,2 elb&11 bm clp per 100′, 3/4″	10.000	L.F.	.01	.05	.06
Cabinet, hinged, steel, NEMA 1, 12″W x 12″H x 4″D	.200	Ea.	.01	.02	.03
Lighting control module	.200	Ea.	.07	.07	.14
Dimmable ballast three-lamp	5.000	Ea.	.68	.48	1.16
Daylight level sensor, wall mounted, on/off or dimming	.200	Ea.	.03	.02	.05
Automatic wall switches	.200	Ea.	.02	.01	.03
Remote power pack	.100	Ea.		.01	.01
TOTAL			.83	.71	1.54

D5020 290	Daylight Dimming System	COST PER S.F. MAT.	INST.	TOTAL
0500	Daylight Dimming System (no fixtures or fixture power)			
0800	5 fixtures per 1000 S.F.	.83	.71	1.54
1000	10 fixtures per 1000 S.F.	1.65	1.30	2.95
2000	12 fixtures per 1000 S.F.	1.98	1.57	3.55
3000	15 fixtures per 1000 S.F.	2.47	1.94	4.41
4000	20 fixtures per 1000 S.F.	3.29	2.56	5.85
5000	25 fixtures per 1000 S.F.	4.11	3.18	7.29
6000	50 fixtures per 1000 S.F.	8.20	6.30	14.50

D5020 Lighting and Branch Wiring

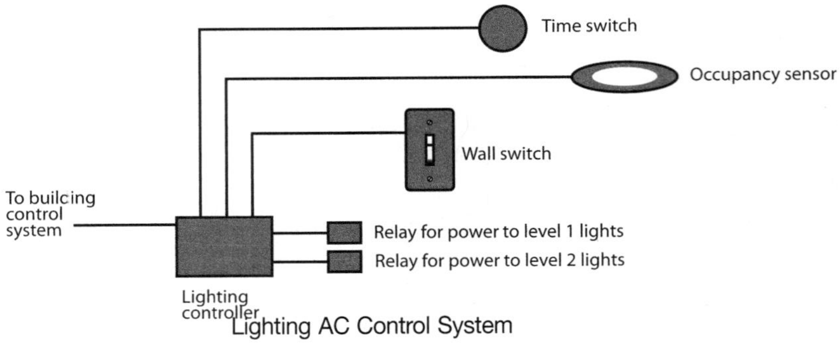

Lighting AC Control System

System Components	QUANTITY	UNIT	COST PER S.F. MAT.	COST PER S.F. INST.	COST PER S.F. TOTAL
SYSTEM D5020 295 0800					
LIGHTING ON/OFF CONTROL SYSTEM					
5 FIXTURES PER 1000 S.F.					
Tray cable, type TC, copper #16-4 conductor	.600	C.L.F.	.03	.05	.08
Wire, 600 volt, type THWN-THHN, copper, solid, #12	.320	C.L.F.		.02	.02
Conduit (EMT) , to 15′ H, incl 2 termn,2 elb&11 bm clp per 100′, 3/4″	10.000	L.F.	.01	.05	.06
Cabinet, hinged, steel, NEMA 1, 12″W x 12″H x 4″D	.400	Ea.	.01	.04	.05
Relays, 120 V or 277 V standard	.800	Ea.	.04	.04	.08
24 hour dial with reserve power	.400	Ea.	.27	.07	.34
Lighting control module	.200	Ea.	.07	.07	.14
Occupancy sensors, passive infrared ceiling mounted	.400	Ea.	.03	.04	.07
Automatic wall switches	.400	Ea.	.03	.01	.04
Remote power pack	.200	Ea.	.01	.01	.02
TOTAL			.50	.40	.90

D5020 295	Lighting On/Off Control System	COST PER S.F. MAT.	COST PER S.F. INST.	COST PER S.F. TOTAL
0500	Includes occupancy and time switching (no fixtures or fixture power)			
0800	5 fixtures per 1000 SF	.50	.40	.90
1000	10 fixtures per 1000 S.F.	.57	.46	1.03
2000	12 fixtures per 1000 S.F.	.68	.57	1.25
3000	15 fixtures per 1000 S.F.	.86	.70	1.56
4000	20 fixtures per 1000 S.F.	1.13	.90	2.03
5000	25 fixtures per 1000 S.F.	1.42	1.09	2.51
6000	50 fixtures per 1000 S.F.	2.82	2.13	4.95

D5030 Communications and Security

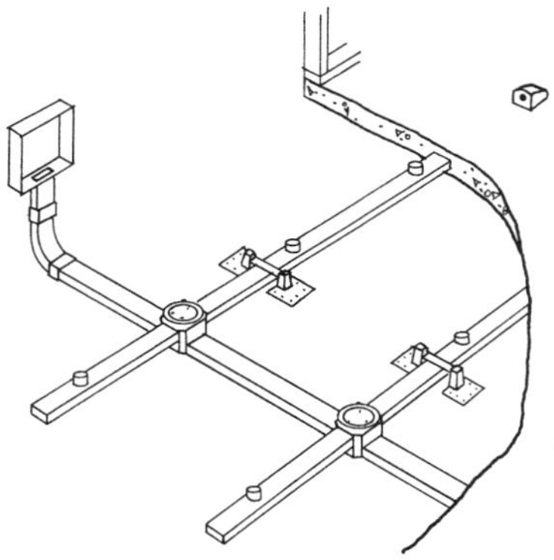

Description: System below includes telephone fitting installed. Does not include cable.

When poke thru fittings and telepoles are used for power, they can also be used for telephones at a negligible additional cost.

System Components	QUANTITY	UNIT	COST PER S.F. MAT.	COST PER S.F. INST.	COST PER S.F. TOTAL
SYSTEM D5030 310 0200					
TELEPHONE SYSTEMS, UNDERFLOOR DUCT, 5′ ON CENTER, LOW DENSITY					
Underfloor duct 7-1/4″ w/insert 2′ O.C. 1-3/8″ x 7-1/4″ super duct	.190	L.F.	7.22	2.51	9.73
Vertical elbow for underfloor superduct, 7-1/4″, included					
Underfloor duct conduit adapter, 2″ x 1-1/4″, included					
Underfloor duct junction box, single duct, 7-1/4″ x 3 1/8″	.003	Ea.	1.80	.50	2.30
Underfloor junction box carpet pan	.003	Ea.	1.22	.02	1.24
Underfloor duct outlet, low tension	.004	Ea.	.46	.33	.79
TOTAL			10.70	3.36	14.06

D5030 310	Telephone Systems	COST PER S.F. MAT.	COST PER S.F. INST.	COST PER S.F. TOTAL
0200	Telephone systems, underfloor duct, 5′ on center, low density	10.70	3.36	14.06
0240	5′ on center, high density	11.15	3.69	14.84
0280	7′ on center, low density	8.55	2.79	11.34
0320	7′ on center, high density	9	3.12	12.12
0400	Poke thru fittings, low density	1.18	1.19	2.37
0440	High density	2.36	2.39	4.75
0520	Telepoles, low density	1.15	.71	1.86
0560	High density	2.30	1.42	3.72
0640	Conduit system with floor boxes, low density	1.34	1.18	2.52
0680	High density	2.67	2.32	4.99
1020	Telephone wiring for offices & laboratories, 8 jacks/M.S.F.	.37	1.81	2.18

D5030 Communications and Security

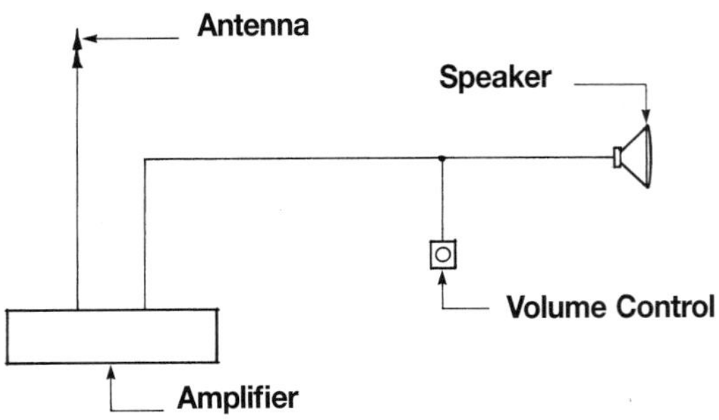

Sound System Includes AM–FM antenna, outlets, rigid conduit, and copper wire.
Fire Detection System Includes pull stations, signals, smoke and heat detectors, rigid conduit, and copper wire.
Intercom System Includes master and remote stations, rigid conduit, and copper wire.
Master Clock System Includes clocks, bells, rigid conduit, and copper wire.
Master TV Antenna Includes antenna, VHF–UHF reception and distribution, rigid conduit, and copper wire.

System Components	QUANTITY	UNIT	COST EACH MAT.	COST EACH INST.	COST EACH TOTAL
SYSTEM D5030 910 0220					
SOUND SYSTEM, INCLUDES OUTLETS, BOXES, CONDUIT & WIRE					
Steel intermediate conduit, (IMC) 1/2″ diam	1200.000	L.F.	2,904	7,920	10,824
Wire sound shielded w/drain, #22-2 conductor	15.500	C.L.F.	237.93	1,278.75	1,516.68
Sound system speakers ceiling or wall	12.000	Ea.	1,608	990	2,598
Sound system volume control	12.000	Ea.	1,146	990	2,136
Sound system amplifier, 250 Watts	1.000	Ea.	1,550	660	2,210
Sound system antenna, AM FM	1.000	Ea.	166	165	331
Sound system monitor panel	1.000	Ea.	445	165	610
Sound system cabinet	1.000	Ea.	965	660	1,625
Steel outlet box 4″ square	12.000	Ea.	38.04	396	434.04
Steel outlet box 4″ plaster rings	12.000	Ea.	26.88	123.60	150.48
TOTAL			9,086.85	13,348.35	22,435.20

D5030 910	Communication & Alarm Systems	COST EACH MAT.	COST EACH INST.	COST EACH TOTAL
0200	Communication & alarm systems, includes outlets, boxes, conduit & wire			
0210	Sound system, 6 outlets	6,275	8,350	14,625
0220	12 outlets (RD5010 -110)	9,075	13,300	22,375
0240	30 outlets	16,100	25,200	41,300
0280	100 outlets	49,300	84,500	133,800
0320	Fire detection systems, non-addressable, 12 detectors	3,550	6,900	10,450
0360	25 detectors	6,200	11,700	17,900
0400	50 detectors	12,200	23,200	35,400
0440	100 detectors	22,600	42,100	64,700
0450	Addressable type, 12 detectors	5,225	6,975	12,200
0452	25 detectors	9,475	11,800	21,275
0454	50 detectors	18,300	23,200	41,500
0456	100 detectors	35,300	42,500	77,800
0458	Fire alarm control panel, 8 zone, excluding wire and conduit	755	1,325	2,080
0459	12 zone	2,625	1,975	4,600
0460	Fire alarm command center, addressable without voice, excl. wire & conduit	4,925	1,150	6,075
0462	Addressable with voice	10,400	1,825	12,225
0480	Intercom systems, 6 stations	4,225	5,700	9,925
0520	12 stations	8,225	11,400	19,625
0560	25 stations	14,000	21,900	35,900
0600	50 stations	27,000	41,100	68,100
0640	100 stations	53,000	80,000	133,000
0680	Master clock systems, 6 rooms	5,575	9,300	14,875
0720	12 rooms	8,975	15,900	24,875
0760	20 rooms	12,500	22,400	34,900
0800	30 rooms	20,800	41,400	62,200

D5030 Communications and Security

D5030 910	Communication & Alarm Systems	COST EACH		
		MAT.	INST.	TOTAL
0840	50 rooms	34,000	70,000	104,000
0880	100 rooms	66,000	138,500	204,500
0920	Master TV antenna systems, 6 outlets	2,775	5,900	8,675
0960	12 outlets	5,150	11,000	16,150
1000	30 outlets	12,700	25,400	38,100
1040	100 outlets	42,300	83,000	125,300

D5030 920	Data Communication	COST PER M.S.F.		
		MAT.	INST.	TOTAL
0100	Data communication system, incl data/voice outlets, boxes, conduit, & cable			
0102	Data and voice system, 2 data/voice outlets per 1000 S.F.	140	490	630
0104	4 data/voice outlets per 1000 S.F.	264	960	1,224
0106	6 data/voice outlets per 1000 S.F.	375	1,425	1,800
0110	8 data/voice outlets per 1000 S.F.	480	1,875	2,355

D5090 Other Electrical Systems

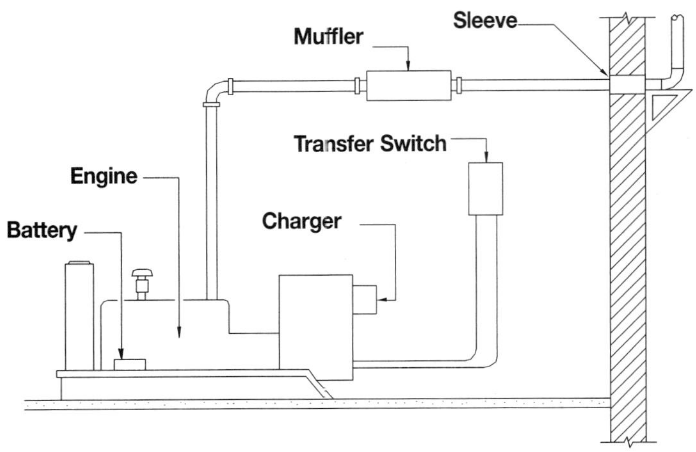

Description: System below tabulates the installed cost for generators by kW. Included in costs are battery, charger, muffler, and transfer switch.

No conduit, wire, or terminations included.

System Components	QUANTITY	UNIT	COST PER kW MAT.	COST PER kW INST.	COST PER kW TOTAL
SYSTEM D5090 210 0200 GENERATOR SET, INCL. BATTERY, CHARGER, MUFFLER & TRANSFER SWITCH GAS/GASOLINE OPER., 3 PHASE, 4 WIRE, 277/480V, 7.5 kW					
Generator set, gas or gasoline operated, 3 ph 4 W, 277/480 V, 7.5 kW	.133	Ea.	1,256.67	288.13	1,544.80
TOTAL			1,256.67	288.13	1,544.80

D5090 210	Generators (by kW)	COST PER kW MAT.	COST PER kW INST.	COST PER kW TOTAL
0190	Generator sets, include battery, charger, muffler & transfer switch			
0200	Gas/gasoline operated, 3 phase, 4 wire, 277/480 volt, 7.5 kW	1,250	288	1,538
0240	11.5 kW	1,150	219	1,369
0280	20 kW (RD5010 -110)	785	142	927
0320	35 kW	535	93	628
0360	80 kW	385	56.50	441.50
0400	100 kW	335	54.50	389.50
0440	125 kW	550	51.50	601.50
0480	185 kW	490	39	529
0560	Diesel engine with fuel tank, 30 kW	410	108	518
0600	50 kW	435	85.50	520.50
0720	125 kW	255	49	304
0760	150 kW	273	46	319
0800	175 kW	261	41	302
0840	200 kW	247	37.50	284.50
0880	250 kW	210	31	241
0920	300 kW	187	27.50	214.50
0960	350 kW	181	25.50	206.50
1000	400 kW	196	23.50	219.50
1040	500 kW	198	19.95	217.95
1200	750 kW	213	12.55	225.55
1400	1000 kW	198	13.20	211.20

D5090 Other Electrical Systems

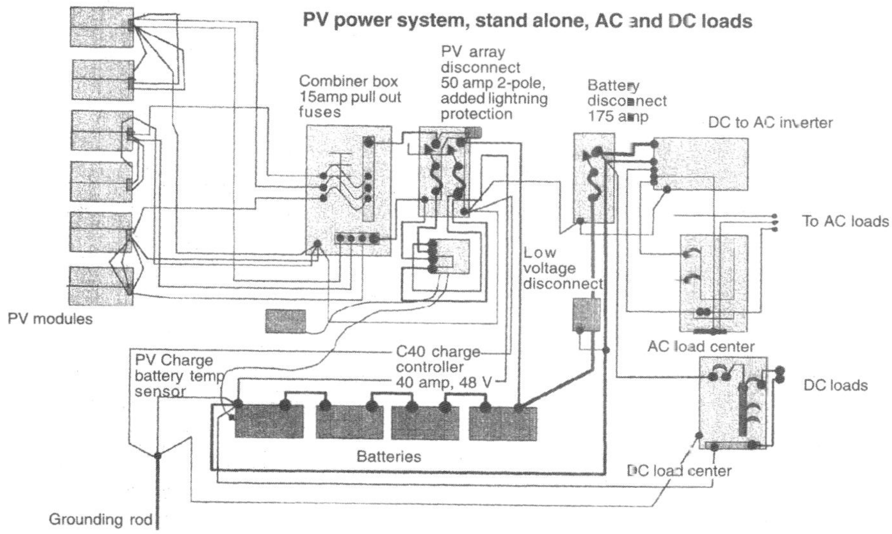

System Components	QUANTITY	UNIT	COST EACH MAT.	INST.	TOTAL
SYSTEM D5090 420 0100					
PHOTOVOLTAIC POWER SYSTEM, STAND ALONE					
Alternative energy sources, photovoltaic module, 75 watt, 17 volts	12.000	Ea.	5,640	990	6,630
PV rack system, roof, penetrating surface mount, on wood framing, 1 panel	12.000	Ea.	726	1,296	2,022
DC to AC inverter for, 24 V, 2,500 watt	1.000	Ea.	2,250	165	2,415
PV components, combiner box, 10 lug, NEMA 3R enclosure	1.000	Ea.	297	165	462
Alternative energy sources, PV components, fuse, 15 A for combiner box	10.000	Ea.	235	165	400
Battery charger controller w/temperature sensor	1.000	Ea.	540	165	705
Digital readout panel, displays hours, volts, amps, etc.	1.000	Ea.	202	165	367
Deep cycle solar battery, 6 V, 180 Ah (C/20)	4.000	Ea.	1,340	330	1,670
Battery interconnection, 15″ AWG #2/0, sealed w/copper ring lugs	3.000	Ea.	60	123	183
Battery interconn, 24″ AWG #2/0, sealed w/copper ring lugs	2.000	Ea.	52	82	134
Battery interconn, 60″ AWG #2/0, sealed w/copper ring lugs	2.000	Ea.	124	82	206
Batt temp computer probe, RJ11 jack, 15′ cord	1.000	Ea.	24.50	41	65.50
System disconnect, DC 175 amp circuit breaker	1.000	Ea.	194	82.50	276.50
Conduit box for inverter	1.000	Ea.	52.50	82.50	135
Low voltage disconnect	1.000	Ea.	60	82.50	142.50
Vented battery enclosure, wood	1.000	Ea.	198	297	495
Grounding, rod, copper clad, 8′ long, 5/8″ diameter	1.000	Ea.	21.50	120	141.50
Grounding, clamp, bronze, 5/8″ dia	1.000	Ea.	4.93	20.50	25.43
Bare copper wire, stranded, #8	1.000	C.L.F.	27.50	60	87.50
Wire, 600 volt, type THW, copper, stranded, #12	3.600	C.L.F.	49.68	216	265.68
Wire, 600 volt, type THW, copper, stranded, #10	1.050	C.L.F.	23.10	69.30	92.40
Cond to 15′ H,incl 2 termn,2 elb&11 bm CLP per 100′,galv stl,1/2″ dia	120.000	L.F.	330	882	1,212
Conduit,to 15′ H,incl 2 termn,2 elb&11 bm clp per 100′,(EMT), 1″ dia	30.000	L.F.	54.90	172.50	227.40
Lightning surge suppressor	1.000	Ea.	56	20.50	76.50
General duty 240 volt, 2 pole, nonfusible, NEMA 3R, 60 amp	1.000	Ea.	217	287	504
Load centers, 3 wire, 120/240V, 100 amp main lugs, indoor, 8 circuits	2.000	Ea.	404	940	1,344
Circuit breaker, Plug-in, 120/240 volt, to 60 amp, 1 pole	5.000	Ea.	53.50	275	328.50
Fuses, dual element, time delay, 250 volt, 50 amp	2.000	Ea.	25.10	26.40	51.50
TOTAL			13,262.21	7,402.70	20,664.91

D5090 Other Electrical Systems

D5090 420	Photovoltaic Power System, Stand Alone	COST EACH		
		MAT.	INST.	TOTAL
0050	24 V capacity, 900 W (~102 SF)			
0100	Roof mounted, on wood framing	13,300	7,400	20,700
0150	with standoff	13,400	7,400	20,800
0200	Ground placement, ballast, fixed	19,500	7,925	27,425
0250	Adjustable	19,800	7,925	27,725
0300	Top of pole, passive tracking	19,600	7,925	27,525
0350	1.8 kW (~204 SF)			
0400	Roof mounted, on wood framing	20,600	11,800	32,400
0450	with standoff	20,900	11,800	32,700
0500	Ground placement, ballast, fixed	32,200	11,100	43,300
0550	Adjustable	32,900	11,100	44,000
0600	Top of pole, passive tracking	33,800	11,100	44,900
0650	3.2 kW (~408 SF)			
0700	Roof mounted, on wood framing	36,600	20,200	56,800
0750	with standoff	37,100	20,200	57,300
0800	Ground placement, ballast, fixed	59,000	17,200	76,200
0850	Adjustable	60,500	17,200	77,700
0900	Top of pole, passive tracking	62,000	17,100	79,100

D5090 Other Electrical Systems

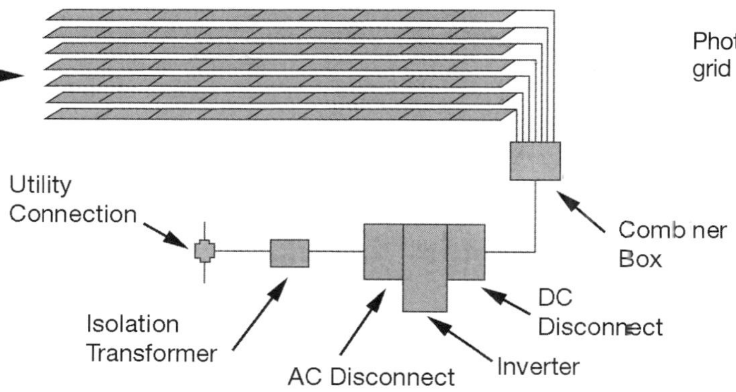

System Components	QUANTITY	UNIT	COST EACH MAT.	COST EACH INST.	COST EACH TOTAL
SYSTEM D5090 430 0100					
PHOTOVOLTAIC POWER SYSTEM, GRID CONNECTED 10 kW					
Photovoltaic module, 150 watt, 33 volts	60.000	Ea.	27,000	4,950	31,950
PV rack system, roof, non-penetrating ballast, 1 panel	60.000	Ea.	63,000	2,340	65,340
DC to AC inverter for, 24 V, 4,000 watt	3.000	Ea.	8,325	990	9,315
Combiner box 10 lug, NEMA 3R	1.000	Ea.	297	165	462
15 amp fuses	10.000	Ea.	235	165	400
Safety switch, 60 amp	1.000	Ea.	310	300	610
Safety switch, 100 amp	1.000	Ea.	580	365	945
Utility connection, 3 pole breaker	1.000	Ea.	260	106	366
Fuse, 60 A	3.000	Ea.	38.40	39.60	78
Fuse, 100 A	3.000	Ea.	93	49.50	142.50
10 kVA isolation tranformer	1.000	Ea.	1,875	825	2,700
EMT Conduit w/fittings & support	1.000	L.F.	329.40	1,035	1,364.40
RGS Conduit w/fittings & support	267.000	L.F.	1,142.76	2,710.05	3,852.81
Enclosure 24" x 24" x 10", NEMA 4	1.000	Ea.	5,475	1,325	6,800
Wire, 600 volt, type THWN-THHN, copper, stranded, #6	8.000	C.L.F.	460	808	1,268
Wire, 600 volt, copper type XLPE-USE(RHW), stranded, #12	18.000	C.L.F.	387	1,080	1,467
Grounding, bare copper wire stranded, 4/0	2.000	C.L.F.	740	462	1,202
Grounding, exothermic weld, 4/0 wire to building steel	4.000	Ea.	45.40	376	421.40
Grounding, brazed connections, #6 wire	2.000	Ea.	35.90	110	145.90
Insulated ground wire, copper, #6	.400	C.L.F.	23	40.40	63.40
TOTAL			110,651.86	18,241.55	128,893.41

D5090 430	Photovoltaic Power System, Grid Connected	COST EACH MAT.	COST EACH INST.	COST EACH TOTAL
0050	10 kW			
0100	Roof mounted, non-penetrating ballast	110,500	18,200	128,700
0200	Penetrating surface mount, on steel framing	51,500	29,800	81,300
0300	on wood framing	51,500	22,400	73,900
0400	with standoff	52,000	22,400	74,400
0500	Ground placement, ballast, fixed	79,000	18,100	97,100
0600	adjustable	81,000	18,100	99,100
0700	Top of pole, passive tracking	81,000	18,600	99,600
1050	20 kW			
1100	Roof mounted, non-penetrating ballast	215,000	30,600	245,600
1200	Penetrating surface mount, on steel framing	96,500	53,500	150,000
1300	on wood framing	96,000	38,900	134,900
1400	with standoff	97,500	38,900	136,400
1500	Ground placement, ballast, fixed	151,000	29,400	180,400
1600	adjustable	154,500	29,400	183,900
1700	Top of pole, passive tracking	155,500	30,300	185,800

D5090 Other Electrical Systems

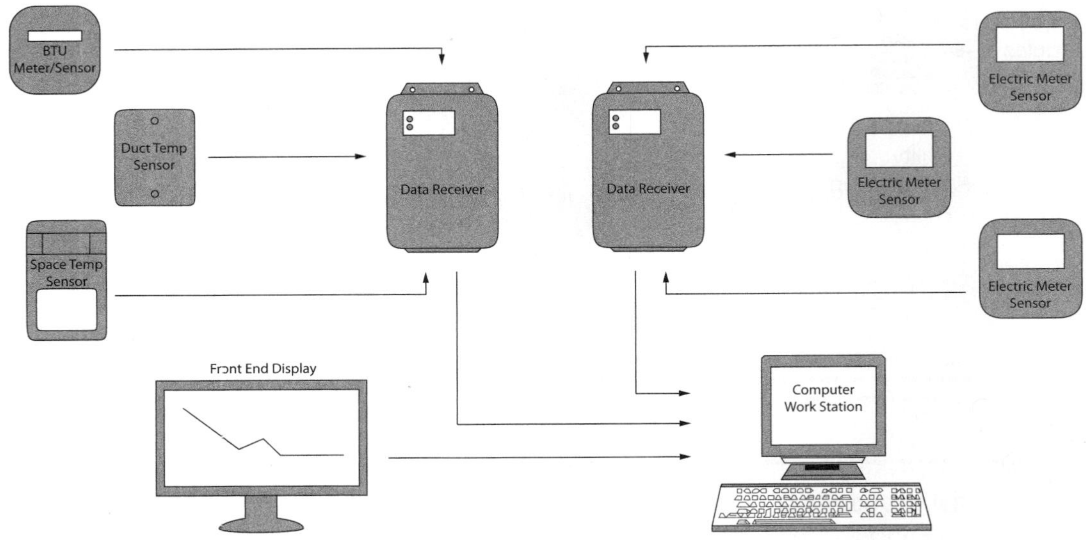

System Components	QUANTITY	UNIT	COST EACH MAT.	COST EACH INST.	COST EACH TOTAL
SYSTEM D5090 480 1100					
ELECTRICAL, SINGLE PHASE, 1 METER					
#18 twisted shielded pair in 1/2" EMT conduit	.050	C.L.F.	8.10	11	19.10
Wire, 600 volt, type THW, copper, solid, #12	.150	C.L.F.	1.76	9	10.76
Conduit (EMT) , to 15' H, incl 2 termn,2 elb&11 bm clp per 100', 3/4"	5.000	L.F.	5.50	25.25	30.75
Outlet boxes, pressed steel, handy box	1.000	Ea.	2.70	24.50	27.20
Outlet boxes, pressed steel, handy box, covers, device	1.000	Ea.	.96	10.30	11.26
Wiring devices, receptacle, duplex, 120 volt, ground, 20 amp	1.000	Ea.	8.70	24.50	33.20
Single phase, 277 volt, 200 amp	1.000	Ea.	450	75	525
Data recorder, 8 meters	1.000	Ea.	1,625	60	1,685
Software package, per meter, premium	1.000	Ea.	750		750
TOTAL			2,852.72	239.55	3,092.27

D5090 480	Energy Monitoring Systems	COST EACH MAT.	COST EACH INST.	COST EACH TOTAL
1000	Electrical			
1100	Single phase, 1 meter	2,850	240	3,090
1110	4 meters	7,250	1,825	9,075
1120	8 meters	14,600	3,725	18,325
1200	Three phase, 1 meter	3,275	305	3,580
1210	5 meters	10,800	2,850	13,650
1220	10 meters	22,200	5,825	28,025
1230	25 meters	50,500	11,600	62,100
2000	Mechanical			
2100	BTU, 1 meter	3,625	890	4,515
2110	w/1 duct sensor	3,825	1,100	4,925
2120	& 1 space sensor	4,350	1,250	5,600
2130	& 5 space sensors	6,425	1,825	8,250
2140	& 10 space sensors	9,625	2,575	12,200
2200	BTU, 3 meters	7,600	2,550	10,150
2210	w/3 duct sensors	8,225	3,175	11,400
2220	& 3 space sensors	9,800	3,625	13,425
2230	& 15 space sensors	16,000	5,375	21,375
2240	& 30 space sensors	24,400	7,600	32,000
9000	Front end display	550	129	679
9100	Computer workstation	1,450	1,725	3,175

D5090 Other Electrical Systems

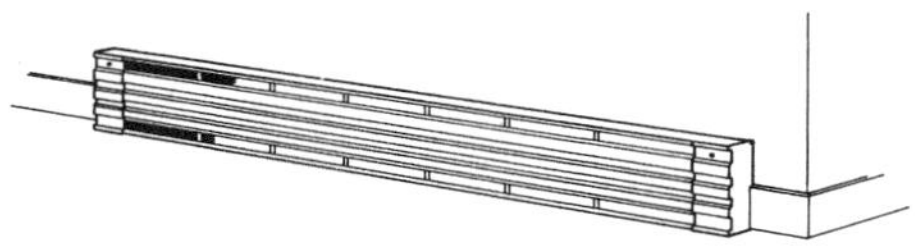

Low Density and Medium Density Baseboard Radiation
The costs shown in Table below are based on the following system considerations:

1. The heat loss per square foot is based on approximately 34 BTU/hr. per S.F. of floor or 10 watts per S.F. of floor.
2. Baseboard radiation is based on the low watt density type rated 187 watts per L.F. and the medium density type rated 250 watts per L.F.
3. Thermostat is not included.
4. Wiring costs include branch circuit wiring.

System Components	QUANTITY	UNIT	COST PER S.F. MAT.	COST PER S.F. INST.	COST PER S.F. TOTAL
SYSTEM D5090 510 1000					
ELECTRIC BASEBOARD RADIATION, LOW DENSITY, 900 S.F., 31 MBH, 9 kW					
Electric baseboard radiator, 5′ long 935 watt	.011	Ea.	.56	1.28	1.84
Steel intermediate conduit, (IMC) 1/2″ diam	.170	L.F.	.41	1.12	1.53
Wire 600 volt, type THW, copper, solid, #12	.005	C.L.F.	.06	.30	.36
TOTAL			1.03	2.70	3.73

D5090 510	Electric Baseboard Radiation (Low Density)	COST PER S.F. MAT.	COST PER S.F. INST.	COST PER S.F. TOTAL
1000	Electric baseboard radiation, low density, 900 S.F., 31 MBH, 9 kW	1.03	2.70	3.73
1200	1500 S.F., 51 MBH, 15 kW	1.01	2.64	3.65
1400	2100 S.F., 72 MBH, 21 kW	.91	2.36	3.27
1600	3000 S.F., 102 MBH, 30 kW	.83	2.14	2.97
2000	Medium density, 900 S.F., 31 MBH, 9 kW	.93	2.46	3.39
2200	1500 S.F., 51 MBH, 15 kW	.91	2.40	3.31
2400	2100 S.F., 72 MBH, 21 kW	.86	2.24	3.10
2600	3000 S.F., 102 MBH, 30 kW	.78	2.03	2.81

D5090 Other Electrical Systems

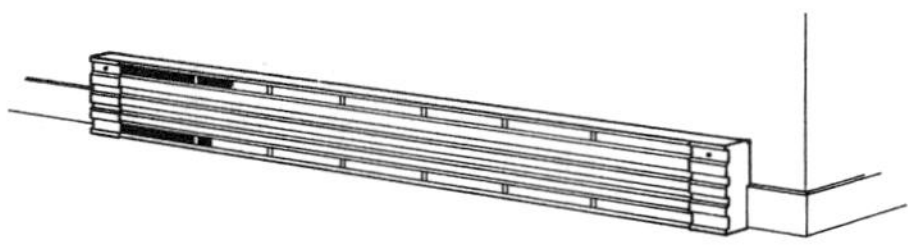

Commercial Duty Baseboard Radiation
The costs shown in Table below are based on the following system considerations:

1. The heat loss per square foot is based on approximately 41 BTU/hr. per S.F. of floor or 12 watts per S.F.
2. The baseboard radiation is of the commercial duty type rated 250 watts per L.F. served by 277 volt, single phase power.
3. Thermostat is not included.
4. Wiring costs include branch circuit wiring.

System Components	QUANTITY	UNIT	COST PER S.F. MAT.	COST PER S.F. INST.	COST PER S.F. TOTAL
SYSTEM D5090 520 1000					
ELECTRIC BASEBOARD RADIATION, MEDIUM DENSITY, 1230 S.F., 51 MBH, 15 kW					
Electric baseboard radiator, 5′ long	.013	Ea.	.66	1.51	2.17
Steel intermediate conduit, (IMC) 1/2″ diam	.154	L.F.	.37	1.02	1.39
Wire 600 volt, type THW, copper, solid, #12	.004	C.L.F.	.05	.24	.29
TOTAL			1.08	2.77	3.85

D5090 520	Electric Baseboard Radiation (Medium Density)	COST PER S.F. MAT.	COST PER S.F. INST.	COST PER S.F. TOTAL
1000	Electric baseboard radiation, medium density, 1230 SF, 51 MBH, 15 kW	1.08	2.77	3.85
1200	2500 S.F. floor area, 106 MBH, 31 kW	1.01	2.59	3.60
1400	3700 S.F. floor area, 157 MBH, 46 kW	.98	2.51	3.49
1600	4800 S.F. floor area, 201 MBH, 59 kW	.91	2.33	3.24
1800	11,300 S.F. floor area, 464 MBH, 136 kW	.85	2.14	2.99
2000	30,000 S.F. floor area, 1229 MBH, 360 kW	.83	2.09	2.92

E Equipment & Furnishings

E1010 Commercial Equipment

E1010 110	Security/Vault, EACH	COST EACH		
		MAT.	INST.	TOTAL
0100	Bank equipment, drive up window, drawer & mike, no glazing, economy	7,725	1,450	9,175
0110	Deluxe	10,100	2,875	12,975
0120	Night depository, economy	8,300	1,450	9,750
0130	Deluxe	11,800	2,875	14,675
0140	Pneumatic tube systems, 2 station, standard	28,800	5,100	33,900
0150	Teller, automated, 24 hour, single unit	49,700	5,100	54,800
0160	Teller window, bullet proof glazing, 44″ x 60″	4,650	945	5,595
0170	Pass through, painted steel, 72″ x 40″	4,800	1,800	6,600
0300	Safe, office type, 1 hr. rating, 34″ x 20″ x 20″	2,500		2,500
0310	4 hr. rating, 62″ x 33″ x 20″	9,750		9,750
0320	Data storage, 4 hr. rating, 63″ x 44″ x 16″	14,800		14,800
0330	Jewelers, 63″ x 44″ x 16″	15,800		15,800
0340	Money, "B" label, 9″ x 14″ x 14″	595		595
0350	Tool and torch resistive, 24″ x 24″ x 20″	8,650	233	8,883
0500	Security gates-scissors type, painted steel, single, 6′ high, 5-1/2′ wide	252	360	612
0510	Double gate, 7-1/2′ high, 14′ wide	695	720	1,415

E1010 510	Mercantile Equipment, EACH	COST EACH		
		MAT.	INST.	TOTAL
0015	Barber equipment, chair, hydraulic, economy	690	25	715
0020	Deluxe	4,375	37	4,412
0030	Console, including mirrors, deluxe	565	130	695
0040	Sink, hair washing basin	575	89.50	664.50
0100	Checkout counter, single belt	3,500	93	3,593
0110	Double belt, power take-away	5,025	103	5,128
0200	Display cases, freestanding, glass and aluminum, 3′-6″ x 3′ x 1′-0″ deep	1,425	149	1,574
0210	5′-10″ x 4′ x 1′-6″ deep	4,300	198	4,498
0220	Wall mounted, glass and aluminum, 3′ x 4′ x 1′-4″ deep	2,375	238	2,613
0230	16′ x 4′ x 1′-4″ deep	4,550	795	5,345
0240	Table exhibit, flat, 3′ x 4′ x 2′-0″ wide	1,350	238	1,588
0250	Sloping, 3′ x 4′ x 3′-0″ wide	1,400	395	1,795
0300	Refrigerated food cases, dairy, multi-deck, rear sliding doors, 6 ft. long	14,400	440	14,840
0310	Delicatessen case, service type, 12 ft. long	9,250	340	9,590
0320	Frozen food, chest type, 12 ft. long	9,050	400	9,450
0330	Glass door, reach-in, 5 door	15,500	440	15,940
0340	Meat case, 12 ft. long, single deck	8,025	400	8,425
0350	Multi-deck	11,400	425	11,825
0360	Produce case, 12 ft. long, single deck	7,550	400	7,950
0370	Multi-deck	9,600	425	10,025

E1010 610	Laundry/Dry Cleaning, EACH	COST EACH		
		MAT.	INST.	TOTAL
0100	Laundry equipment, dryers, gas fired, residential, 16 lb. capacity	755	238	993
0110	Commercial, 30 lb. capacity, single	3,700	238	3,938
0120	Dry cleaners, electric, 20 lb. capacity	37,500	6,875	44,375
0130	30 lb. capacity	57,500	9,175	66,675
0140	Ironers, commercial, 120″ with canopy, 8 roll	182,500	19,600	202,100
0150	Institutional, 110″, single roll	35,900	3,300	39,200
0160	Washers, residential, 4 cycle	975	238	1,213
0170	Commercial, coin operated, deluxe	3,775	238	4,013

E1020 Institutional Equipment

E1020 110	Ecclesiastical Equipment, EACH	COST EACH		
		MAT.	INST.	TOTAL
0090	Church equipment, altar, wood, custom, plain	2,875	425	3,300
0100	Granite, custom, deluxe	40,500	5,600	46,100
0110	Baptistry, fiberglass, economy	6,175	1,550	7,725
0120	Bells & carillons, keyboard operation	20,400	8,650	29,050
0130	Confessional, wood, single, economy	3,575	990	4,565
0140	Double, deluxe	20,900	2,975	23,875
0150	Steeples, translucent fiberglas, 30″ square, 15′ high	9,400	1,850	11,250
0160	Porcelain enamel, custom, 60′ high	23,600	12,400	36,000

E1020 130	Ecclesiastical Equipment, L.F.	COST PER L.F.		
		MAT.	INST.	TOTAL
0100	Arch. equip., church equip. pews, bench type, hardwood, economy	111	29.50	140.50
0110	Deluxe	194	39.50	233.50

E1020 210	Library Equipment, EACH	COST EACH		
		MAT.	INST.	TOTAL
0110	Library equipment, carrels, metal, economy	289	119	408
0120	Hardwood, deluxe	2,250	149	2,399

E1020 230	Library Equipment, L.F.	COST PER L.F.		
		MAT.	INST.	TOTAL
0100	Library equipment, book shelf, metal, single face, 90″ high x 10″ shelf	134	49.50	183.50
0110	Double face, 90″ high x 10″ shelf	490	107	597
0120	Charging desk, built-in, with counter, plastic laminate	335	85	420

E1020 310	Theater and Stage Equipment, EACH	COST EACH		
		MAT.	INST.	TOTAL
0200	Movie equipment, changeover, economy	535		535
0210	Film transport, incl. platters and autowind, economy	5,600		5,600
0220	Lamphouses, incl. rectifiers, xenon, 1000W	7,400	330	7,730
0230	4000W	10,800	440	11,240
0240	Projector mechanisms, 35 mm, economy	12,700		12,700
0250	Deluxe	17,400		17,400
0260	Sound systems, incl. amplifier, single, economy	3,800	735	4,535
0270	Dual, Dolby/super sound	18,900	1,650	20,550
0280	Projection screens, wall hung, manual operation, 50 S.F., economy	355	119	474
0290	Electric operation, 100 S.F., deluxe	2,650	595	3,245
0400	Stage equipment, control boards, incl. dimmers and breakers, economy	13,900	660	14,560
0410	Deluxe	143,500	3,300	146,800
0420	Spotlight, stationary, quartz, 6″ lens	198	165	363
0430	Follow, incl. transformer, 2100W	3,550	165	3,715

E1020 320	Theater and Stage Equipment, S.F.	COST PER S.F.		
		MAT.	INST.	TOTAL
0090	Movie equipment, projection screens, rigid in wall, acrylic, 1/4″ thick	48	5.85	53.85
0100	1/2″ thick	55.50	8.75	64.25
0110	Stage equipment, curtains, velour, medium weight	9.05	1.98	11.03
0120	Silica based yarn, fireproof	16.90	24	40.90
0130	Stages, portable with steps, folding legs, 8″ high	37		37
0140	Telescoping platforms, aluminum, deluxe	55	31	86

E1020 330	Theater and Stage Equipment, L.F.	COST PER L.F.		
		MAT.	INST.	TOTAL
0100	Stage equipment, curtain track, heavy duty	69	66	135
0110	Lights, border, quartz, colored	192	33	225

E1020 Institutional Equipment

E1020 610	Detention Equipment, EACH	COST EACH		
		MAT.	INST.	TOTAL
0110	Detention equipment, cell front rolling door, 7/8" bars, 5' x 7' high	5,900	1,525	7,425
0120	Cells, prefab., including front, 5' x 7' x 7' deep	10,800	2,050	12,850
0130	Doors and frames, 3' x 7', single plate	5,050	770	5,820
0140	Double plate	6,150	770	6,920
0150	Toilet apparatus, incl wash basin	3,775	1,025	4,800
0160	Visitor cubicle, vision panel, no intercom	3,525	1,525	5,050

E1020 710	Laboratory Equipment, EACH	COST EACH		
		MAT.	INST.	TOTAL
0110	Laboratory equipment, glassware washer, distilled water, economy	6,850	765	7,615
0120	Deluxe	15,300	1,375	16,675
0130	Glove box, fiberglass, bacteriological	18,600		18,600
0140	Radio isotope	18,600		18,600
0150	Titration unit, 4-2000 ml. reservoirs	6,100		6,100

E1020 720	Laboratory Equipment, S.F.	COST PER S.F.		
		MAT.	INST.	TOTAL
0100	Arch. equip., lab equip., counter tops, acid proof, economy	47	14.50	61.50
0110	Stainless steel	165	14.50	179.50

E1020 730	Laboratory Equipment, L.F.	COST PER L.F.		
		MAT.	INST.	TOTAL
0110	Laboratory equipment, cabinets, wall, open	203	59.50	262.50
0120	Base, drawer units	580	66	646
0130	Fume hoods, not incl. HVAC, economy	560	220	780
0140	Deluxe incl. fixtures	1,075	495	1,570

E1020 810	Medical Equipment, EACH	COST EACH		
		MAT.	INST.	TOTAL
0100	Dental equipment, central suction system, economy	1,750	595	2,345
0110	Compressor-air, deluxe	9,925	1,225	11,150
0120	Chair, hydraulic, economy	2,450	1,225	3,675
0130	Deluxe	8,600	2,450	11,050
0140	Drill console with accessories, economy	2,375	385	2,760
0150	Deluxe	5,375	385	5,760
0160	X-ray unit, portable	2,625	154	2,779
0170	Panoramic unit	17,600	1,025	18,625
0300	Medical equipment, autopsy table, standard	10,700	715	11,415
0310	Deluxe	17,900	1,200	19,100
0320	Incubators, economy	3,425		3,425
0330	Deluxe	13,100		13,100
0700	Station, scrub-surgical, single, economy	4,825	238	5,063
0710	Dietary, medium, with ice	18,400		18,400
0720	Sterilizers, general purpose, single door, 20" x 20" x 28"	14,200		14,200
0730	Floor loading, double door, 28" x 67" x 52"	227,000		227,000
0740	Surgery tables, standard	13,400	1,025	14,425
0750	Deluxe	28,900	1,450	30,350
0770	Tables, standard, with base cabinets, economy	1,250	395	1,645
0780	Deluxe	6,175	595	6,770
0790	X-ray, mobile, economy	18,500		18,500
0800	Stationary, deluxe	247,000		247,000

E1030 Vehicular Equipment

E1030 110	Vehicular Service Equipment, EACH	MAT.	INST.	TOTAL
		COST EACH		
0110	Automotive equipment, compressors, electric, 1-1/2 H.P., std. controls	515	1,125	1,640
0120	5 H.P., dual controls	3,100	1,675	4,775
0130	Hoists, single post, 4 ton capacity, swivel arms	6,050	4,200	10,250
0140	Dual post, 12 ton capacity, adjustable frame	10,400	880	11,280
0150	Lube equipment, 3 reel type, with pumps	9,675	3,350	13,025
0160	Product dispenser, 6 nozzles, w/vapor recovery, not incl. piping, installed	27,200		27,200
0800	Scales, dial type, built in floor, 5 ton capacity, 8′ x 6′ platform	9,350	3,575	12,925
0810	10 ton capacity, 9′ x 7′ platform	12,100	5,100	17,200
0820	Truck (including weigh bridge), 20 ton capacity, 24′ x 10′	13,900	5,950	19,850
0830	Digital type, truck, 60 ton capacity, 75′ x 10′	46,000	14,900	60,900
0840	Concrete foundations, 8′ x 6′	1,250	4,500	5,750
0850	70′ x 10′	5,400	15,000	20,400

E1030 210	Parking Control Equipment, EACH	MAT.	INST.	TOTAL
		COST EACH		
0110	Parking equipment, automatic gates, 8 ft. arm, one way	4,425	1,200	5,625
0120	Traffic detectors, single treadle	2,225	550	2,775
0130	Booth for attendant, economy	8,000		8,000
0140	Deluxe	29,000		29,000
0150	Ticket printer/dispenser, rate computing	9,450	940	10,390
0160	Key station on pedestal	635	320	955
0200	Concrete filled steel bollard, 8″ diameter, 8′ long	760	129	889

E1030 310	Loading Dock Equipment, EACH	MAT.	INST.	TOTAL
		COST EACH		
0110	Dock bumpers, rubber blocks, 4-1/2″ thick, 10″ high, 14″ long	66	23	89
0120	6″ thick, 20″ high, 11″ long	137	45.50	182.50
0130	Dock boards, H.D., 5′ x 5′, aluminum, 5000 lb. capacity	1,500		1,500
0140	16,000 lb. capacity	1,725		1,725
0150	Dock levelers, hydraulic, 7′ x 8′, 10 ton capacity	6,325	1,500	7,825
0160	Dock lifters, platform, 6′ x 6′, portable, 3000 lb. capacity	10,200		10,200
0170	Dock shelters, truck, scissor arms, economy	2,050	595	2,645
0180	Deluxe	2,750	1,200	3,950

E1090 Other Equipment

E1090 110	Maintenance Equipment, EACH	COST EACH MAT.	INST.	TOTAL
0110	Vacuum cleaning, central, residential, 3 valves	1,200	685	1,885
0120	7 valves	1,925	1,525	3,450

E1090 210	Solid Waste Handling Equipment, EACH	COST EACH MAT.	INST.	TOTAL
0110	Waste handling, compactors, single bag, 250 lbs./hr., hand fed	16,900	700	17,600
0120	Heavy duty industrial, 5 C.Y. capacity	36,500	3,350	39,850
0130	Incinerator, electric, 100 lbs./hr., economy	72,500	3,275	75,775
0140	Gas, 2000 lbs./hr., deluxe	419,500	27,200	446,700
0150	Shredder, no baling, 35 tons/hr.	336,500		336,500
0160	Incl. baling, 50 tons/day	672,000		672,000

E1090 350	Food Service Equipment, EACH	COST EACH MAT.	INST.	TOTAL
0110	Kitchen equipment, bake oven, single deck	6,075	161	6,236
0120	Broiler, without oven	3,925	161	4,086
0130	Commercial dish washer, semiautomatic, 50 racks/hr.	8,100	990	9,090
0140	Automatic, 275 racks/hr.	34,600	4,175	38,775
0150	Cooler, beverage, reach-in, 6 ft. long	3,700	214	3,914
0160	Food warmer, counter, 1.65 kw	770		770
0170	Fryers, with submerger, single	1,375	184	1,559
0180	Double	2,650	257	2,907
0185	Ice maker, 1000 lb. per day, with bin	5,650	1,275	6,925
0190	Kettles, steam jacketed, 20 gallons	9,575	283	9,858
0200	Range, restaurant type, burners, 2 ovens and 24″ griddle	5,225	214	5,439
0210	Range hood, incl. carbon dioxide system, economy	2,075	430	2,505
0220	Deluxe	2,075	430	2,505
1005	Range, restaurant type induction cooker, electric	2,025	283	2,308
2325	Stainless steel shelving, 3′ wide, food service equipment	2,175	77.50	2,252.50

E1090 360	Food Service Equipment, S.F.	COST PER S.F. MAT.	INST.	TOTAL
0110	Refrigerators, prefab, walk-in, 7′-6″ high, 6′ x 6′	181	21.50	202.50
0120	12′ x 20′	114	10.85	124.85

E1090 410	Residential Equipment, EACH	COST EACH MAT.	INST.	TOTAL
0110	Arch. equip., appliances, range, cook top, 4 burner, economy	385	110	495
0120	Built in, single oven 30″ wide, economy	990	110	1,100
0130	Standing, single oven-21″ wide, economy	555	93	648
0135	Free standing 30″ wide, 1 oven, average	1,000	186	1,186
0140	Double oven-30″ wide, deluxe	3,075	93	3,168
0150	Compactor, residential, economy	760	119	880
0160	Deluxe	1,275	198	1,473
0170	Dish washer, built-in, 2 cycles, economy	330	345	675
0180	4 or more cycles, deluxe	1,325	685	2,010
0190	Garbage disposer, sink type, economy	102	137	239
0200	Deluxe	267	137	404
0210	Refrigerator, no frost, 10 to 12 C.F., economy	520	93	613
0220	21 to 29 C.F., deluxe	4,225	310	4,535
0300	Washing machine, automatic	1,825	715	2,540

E1090 550	Darkroom Equipment, EACH	COST EACH MAT.	INST.	TOTAL
0110	Darkroom equipment, developing tanks, 2′ x 4′, 5″ deep	4,575	645	5,220
0120	2′ x 9′, 10″ deep	4,100	860	4,960

E1090 Other Equipment

E1090 550	Darkroom Equipment, EACH	COST EACH		
		MAT.	INST.	TOTAL
0130	Combination, tray & tank sinks, washers & dry tables	11,800	2,850	14,650
0140	Dryers, dehumidified filtered air, 3′ x 2′ x 5′-8″ high	4,900	330	5,230
0150	Processors, manual, 16″ x 20″ print size	10,500	990	11,490
0160	Automatic, color print, deluxe	27,600	3,300	30,900
0170	Washers, round, 11″ x 14″ sheet size	3,575	645	4,220
0180	Square, 50″ x 56″ sheet size	5,400	1,600	7,000

E1090 610	School Equipment, EACH	COST EACH		
		MAT.	INST.	TOTAL
0110	School equipment, basketball backstops, wall mounted, wood, fixed	1,650	1,050	2,700
0120	Suspended type, electrically operated	7,250	1,975	9,225
0130	Bleachers-telescoping, manual operation, 15 tier, economy (per seat)	106	37	143
0140	Power operation, 30 tier, deluxe (per seat)	435	65.50	500.50
0150	Weight lifting gym, universal, economy	385	930	1,315
0160	Deluxe	16,100	1,850	17,950
0170	Scoreboards, basketball, 1 side, economy	2,750	870	3,620
0180	4 sides, deluxe	17,300	11,900	29,200
0800	Vocational shop equipment, benches, metal	630	238	868
0810	Wood	735	238	973
0820	Dust collector, not incl. ductwork, 6′ diam.	5,450	635	6,085
0830	Planer, 13″ x 6″	1,200	297	1,497

E1090 620	School Equipment, S.F.	COST PER S.F.		
		MAT.	INST.	TOTAL
0110	School equipment, gym mats, naugahyde cover, 2″ thick	5.15		5.15
0120	Wrestling, 1″ thick, heavy duty	5.15		5.15

E1090 810	Athletic, Recreational, and Therapeutic Equipment, EACH	COST EACH		
		MAT.	INST.	TOTAL
0050	Bowling alley with gutters	51,000	12,500	63,500
0060	Combo. table and ball rack	1,400		1,400
0070	Bowling alley automatic scorer	11,000		11,000
0110	Sauna, prefabricated, incl. heater and controls, 7′ high, 6′ x 4′	5,850	900	6,750
0120	10′ x 12′	14,100	1,975	16,075
0130	Heaters, wall mounted, to 200 C.F.	805		805
0140	Floor standing, to 1000 C.F., 12500 W	4,275	220	4,495
0610	Shooting range incl. bullet traps, controls, separators, ceilings, economy	40,100	7,675	47,775
0620	Deluxe	59,500	12,900	72,400
0650	Sport court, squash, regulation, in existing building, economy			40,400
0660	Deluxe			45,000
0670	Racketball, regulation, in existing building, economy	43,800	11,200	55,000
0680	Deluxe	47,400	22,500	69,900
0700	Swimming pool equipment, diving stand, stainless steel, 1 meter	10,500	440	10,940
0710	3 meter	17,200	2,975	20,175
0720	Diving boards, 16 ft. long, aluminum	4,475	440	4,915
0730	Fiberglass	3,625	440	4,065
0740	Filter system, sand, incl. pump, 6000 gal./hr.	2,225	795	3,020
0750	Lights, underwater, 12 volt with transformer, 300W	340	660	1,000
0760	Slides, fiberglass with aluminum handrails & ladder, 6′ high, straight	3,975	745	4,720
0780	12′ high, straight with platform	16,000	990	16,990

E1090 820	Athletic, Recreational, and Therapeutic Equipment, S.F.	COST PER S.F.		
		MAT.	INST.	TOTAL
0110	Swimming pools, residential, vinyl liner, metal sides	23.50	8.20	31.70
0120	Concrete sides	28.50	14.85	43.35
0130	Gunite shell, plaster finish, 350 S.F.	52.50	30.50	83
0140	800 S.F.	42.50	17.80	60.30

E1090 Other Equipment

E1090 820	Athletic, Recreational, and Therapeutic Equipment, S.F.	COST PER S.F.		
		MAT.	INST.	TOTAL
0150	Motel, gunite shell, plaster finish	65	39	104
0160	Municipal, gunite shell, tile finish, formed gutters	249	67	316

E1090 910	Other Equipment, EACH	COST EACH		
		MAT.	INST.	TOTAL
0200	Steam bath heater, incl. timer and head, single, to 140 C.F.	2,400	595	2,995
0210	Commercial size, to 2500 C.F.	8,425	895	9,320
0900	Wine Cellar, redwood, refrigerated, 6′-8″ high, 6′W x 6′D	4,675	795	5,470
0910	6′W x 12′D	8,100	1,200	9,300

E2010 Fixed Furnishings

E2010 310	Window Treatment, EACH	COST EACH		
		MAT.	INST.	TOTAL
0110	Furnishings, blinds, exterior, aluminum, louvered, 1′-4″ wide x 3′-0″ long	220	59.50	279.50
0120	1′-4″ wide x 6′-8″ long	390	66	456
0130	Hemlock, solid raised, 1-′4″ wide x 3′-0″ long	79	59.50	138.50
0140	1′-4″ wide x 6′-9″ long	167	66	233
0150	Polystyrene, louvered, 1′-3″ wide x 3′-3″ long	39.50	59.50	99
0160	1′-3″ wide x 6′-8″ long	75	66	141
0200	Interior, wood folding panels, louvered, 7″ x 20″ (per pair)	90.50	35	125.50
0210	18″ x 40″ (per pair)	148	35	183

E2010 320	Window Treatment, S.F.	COST PER S.F.		
		MAT.	INST.	TOTAL
0110	Furnishings, blinds-interior, venetian-aluminum, stock, 2″ slats	6	1.01	7.01
0120	Custom, 1″ slats, deluxe	5.15	1.01	6.16
0130	Vertical, PVC or cloth, T&B track, economy	8.60	1.29	9.89
0140	Deluxe	13.80	1.49	15.29
0150	Draperies, unlined, economy	3.19		3.19
0160	Lightproof, deluxe	10.55		10.55
0510	Shades, mylar, wood roller, single layer, non-reflective	3.52	.87	4.39
0520	Metal roller, triple layer, heat reflective	8.95	.87	9.82
0530	Vinyl, light weight, 4 ga.	.69	.87	1.56
0540	Heavyweight, 6 ga.	2.12	.87	2.99
0550	Vinyl coated cotton, lightproof decorator shades	4.99	.87	5.86
0560	Woven aluminum, 3/8″ thick, light and fireproof	6.65	1.70	8.35

E2010 420	Fixed Floor Grilles and Mats, S.F.	COST PER S.F.		
		MAT.	INST.	TOTAL
0110	Floor mats, recessed, inlaid black rubber, 3/8″ thick, solid	25.50	3	28.50
0120	Colors, 1/2″ thick, perforated	30.50	3	33.50
0130	Link-including nosings, steel-galvanized, 3/8″ thick	27	3	30
0140	Vinyl, in colors	28.50	3	31.50

E2010 510	Fixed Multiple Seating, EACH	COST EACH		
		MAT.	INST.	TOTAL
0110	Seating, painted steel, upholstered, economy	144	34	178
0120	Deluxe	490	42.50	532.50
0400	Seating, lecture hall, pedestal type, economy	221	54	275
0410	Deluxe	540	82	622
0500	Auditorium chair, veneer construction	260	54	314
0510	Fully upholstered, spring seat	266	54	320

E2020 Moveable Furnishings

E2020 210	Furnishings/EACH	COST EACH		
		MAT.	INST.	TOTAL
0200	Hospital furniture, beds, manual, economy	865		865
0210	Deluxe	2,700		2,700
0220	All electric, economy	1,925		1,925
0230	Deluxe	4,175		4,175
0240	Patient wall systems, no utilities, economy, per room	1,575		1,575
0250	Deluxe, per room	2,125		2,125
0300	Hotel furnishings, standard room set, economy, per room	2,775		2,775
0310	Deluxe, per room	9,475		9,475
0500	Office furniture, standard employee set, economy, per person	705		705
0510	Deluxe, per person	2,675		2,675
0550	Posts, portable, pedestrian traffic control, economy	165		165
0560	Deluxe	465		465
0700	Restaurant furniture, booth, molded plastic, stub wall and 2 seats, economy	420	297	717
0710	Deluxe	1,550	395	1,945
0720	Upholstered seats, foursome, single-economy	860	119	979
0730	Foursome, double-deluxe	1,950	198	2,148

E2020 220	Furniture and Accessories, L.F.	COST PER L.F.		
		MAT.	INST.	TOTAL
0210	Dormitory furniture, desk top (built-in),laminated plastc, 24"deep, economy	49.50	24	73.50
0220	30" deep, deluxe	276	29.50	305.50
0230	Dressing unit, built-in, economy	204	99	303
0240	Deluxe	610	149	759
0310	Furnishings, cabinets, hospital, base, laminated plastic	300	119	419
0320	Stainless steel	555	119	674
0330	Countertop, laminated plastic, no backsplash	52	29.50	81.50
0340	Stainless steel	169	29.50	198.50
0350	Nurses station, door type, laminated plastic	350	119	469
0360	Stainless steel	670	119	789
0710	Restaurant furniture, bars, built-in, back bar	228	119	347
0720	Front bar	315	119	434
0910	Wardrobes & coatracks, standing, steel, single pedestal, 30" x 18" x 63"	120		120
0920	Double face rack, 39" x 26" x 70"	136		136
0930	Wall mounted rack, steel frame & shelves, 12" x 15" x 26"	77.50	8.45	85.95
0940	12" x 15" x 50"	42.50	4.43	46.93

F1010 Special Structures

F1010 120	Air-Supported Structures, S.F.	COST PER S.F.		
		MAT.	INST.	TOTAL
0110	Air supported struc., polyester vinyl fabric, 24oz., warehouse, 5000 S.F.	29	.37	29.37
0120	50,000 S.F.	13.55	.30	13.85
0130	Tennis, 7,200 S.F.	26	.31	26.31
0140	24,000 S.F.	18.05	.31	18.36
0150	Woven polyethylene, 6 oz., shelter, 3,000 S.F.	17.60	.62	18.22
0160	24,000 S.F.	13.20	.31	13.51
0170	Teflon coated fiberglass, stadium cover, economy	63	.16	63.16
0180	Deluxe	75	.22	75.22
0190	Air supported storage tank covers, reinf. vinyl fabric, 12 oz., 400 S.F.	25.50	.52	26.02
0200	18,000 S.F.	9.30	.47	9.77

F1010 210	Pre-Engineered Structures, EACH	COST EACH		
		MAT.	INST.	TOTAL
0600	Radio towers, guyed, 40 lb. section, 50′ high, 70 MPH basic wind speed	3,075	1,450	4,525
0610	90 lb. section, 400′ high, wind load 70 MPH basic wind speed	42,400	16,500	58,900
0620	Self supporting, 60′ high, 70 MPH basic wind speed	5,000	2,900	7,900
0630	190′ high, wind load 90 MPH basic wind speed	31,400	11,600	43,000
0700	Shelters, aluminum frame, acrylic glazing, 8′ high, 3′ x 9′	3,375	1,275	4,650
0710	9′ x 12′	8,025	1,975	10,000

F1010 320	Other Special Structures, S.F.	COST PER S.F.		
		MAT.	INST.	TOTAL
0110	Swimming pool enclosure, translucent, freestanding, economy	37	5.95	42.95
0120	Deluxe	102	17	119
0510	Tension structures, steel frame, polyester vinyl fabric, 12,000 S.F.	14.55	2.71	17.26
0520	20,800 S.F.	14.30	2.44	16.74

F1010 330	Special Structures, EACH	COST EACH		
		MAT.	INST.	TOTAL
0110	Kiosks, round, 5′ diam., 7′ high, aluminum wall, illuminated	27,600		27,600
0120	Rectangular, 5′ x 9′, 1″ insulated dbl. wall fiberglass, 7′-6″ high	28,200		28,200
0220	Silos, steel prefab, 30,000 gal., painted, economy	24,300	5,800	30,100
0230	Epoxy-lined, deluxe	50,500	11,600	62,100

F1010 340	Special Structures, S.F.	COST PER S.F.		
		MAT.	INST.	TOTAL
0110	Comfort stations, prefab, mobile on steel frame, economy	196		196
0120	Permanent on concrete slab, deluxe	211	48	259
0210	Domes, bulk storage, wood framing, wood decking, 50′ diam.	75.50	1.87	77.37
0220	116′ diam.	36.50	2.16	38.66
0230	Steel framing, metal decking, 150′ diam.	36.50	11.80	48.30
0240	400′ diam.	29.50	9	38.50
0250	Geodesic, wood framing, wood panels, 30′ diam.	37.50	2.14	39.64
0260	60′ diam.	24.50	1.32	25.82
0270	Aluminum framing, acrylic panels, 40′ diam.			70.50
0280	Aluminum panels, 400′ diam.			22
0310	Garden house, prefab, wood, shell only, 48 S.F.	57	5.95	62.95
0320	200 S.F.	43.50	25	68.50
0410	Greenhouse, shell-stock, residential, lean-to, 8′-6″ long x 3′-10″ wide	47.50	35	82.50
0420	Freestanding, 8′-6″ long x 13′-6″ wide	47.50	11	58.50
0430	Commercial-truss frame, under 2000 S.F., deluxe			14.65
0440	Over 5,000 S.F., economy			13.40
0450	Institutional-rigid frame, under 500 S.F., deluxe			30
0460	Over 2,000 S.F., economy			12.25
0510	Hangar, prefab, galv. roof and walls, bottom rolling doors, economy	13.55	4.69	18.24
0520	Electric bifolding doors, deluxe	13.40	6.10	19.50

F1020 Integrated Construction

F1020 110	Integrated Construction, EACH	COST EACH		
		MAT.	INST.	TOTAL
0110	Integrated ceilings, radiant electric, 2′ x 4′ panel, manila finish	101	26.50	127.50
0120	ABS plastic finish	114	50.50	164.50

F1020 120	Integrated Construction, S.F.	COST PER S.F.		
		MAT.	INST.	TOTAL
0110	Integrated ceilings, Luminaire, suspended, 5′ x 5′ modules, 50% lighted	4.41	14.15	18.56
0120	100% lighted	6.55	25.50	32.05
0130	Dimensionaire, 2′ x 4′ module tile system, no air bar	2.66	2.38	5.04
0140	With air bar, deluxe	3.90	2.38	6.28
0220	Pedestal access floor pkg., inc. stl. pnls, peds.& stringers, w/vinyl cov.	27	3.17	30.17
0230	With high pressure laminate covering	25	3.17	28.17
0240	With carpet covering	26.40	3.17	29.55
0250	Aluminum panels, no stringers, no covering	36	2.38	38.38

F1020 250	Special Purpose Room, EACH	COST EACH		
		MAT.	INST.	TOTAL
0110	Portable booth, acoustical, 27 db 1000 hz., 15 S.F. floor	4,100		4,100
0120	55 S.F. flr.	8,450		8,450

F1020 260	Special Purpose Room, S.F.	COST PER S.F.		
		MAT.	INST.	TOTAL
0110	Anechoic chambers, 7′ high, 100 cps cutoff, 25 S.F.			2,650
0120	200 cps cutoff, 100 S.F.			1,375
0130	Audiometric rooms, under 500 S.F.	59	24.50	83.50
0140	Over 500 S.F.	57.50	19.80	77.30
0300	Darkrooms, shell, not including door, 240 S.F., 8′ high	31	9.90	40.90
0310	64 S.F., 12′ high	76.50	18.60	95.10
0510	Music practice room, modular, perforated steel, under 500 S.F.	38	17	55
0520	Over 500 S.F.	32	14.85	46.85

F1020 330	Special Construction, L.F.	COST PER L.F.		
		MAT.	INST.	TOTAL
0110	Spec. const., air curtains, shipping & receiving, 8′high x 5′wide, economy	785	140	925
0120	20′ high x 8′ wide, heated, deluxe	3,675	350	4,025
0130	Customer entrance, 10′ high x 5′ wide, economy	1,175	140	1,315
0140	12′ high x 4′ wide, heated, deluxe	2,125	175	2,300

F1030 Special Construction Systems

F1030 120	Sound, Vibration, and Seismic Construction, S.F.	COST PER S.F.		
		MAT.	INST.	TOTAL
0020	Special construction, acoustical, enclosure, 4" thick, 8 psf panels	37	25	62
0030	Reverb chamber, 4" thick, parallel walls	52	29.50	81.50
0110	Sound absorbing panels, 2'-6" x 8', painted metal	13.85	8.30	22.15
0120	Vinyl faced	10.80	7.45	18.25
0130	Flexible transparent curtain, clear	8.35	9.75	18.10
0140	With absorbing foam, 75% coverage	11.65	9.75	21.40
0150	Strip entrance, 2/3 overlap	8.35	15.55	23.90
0160	Full overlap	10.55	18.25	28.80
0200	Audio masking system, plenum mounted, over 10,000 S.F.	.81	.30	1.11
0210	Ceiling mounted, under 5,000 S.F.	1.42	.55	1.97

F1030 210	Radiation Protection, EACH	COST EACH		
		MAT.	INST.	TOTAL
0110	Shielding, lead x-ray protection, radiography room, 1/16" lead, economy	11,100	4,475	15,575
0120	Deluxe	13,400	7,450	20,850

F1030 220	Radiation Protection, S.F.	COST PER S.F.		
		MAT.	INST.	TOTAL
0110	Shielding, lead, gypsum board, 5/8" thick, 1/16" lead	12.30	7.10	19.40
0120	1/8" lead	25.50	8.10	33.60
0130	Lath, 1/16" thick	11.55	8.25	19.80
0140	1/8" thick	34.50	9.30	43.80
0150	Radio frequency, galvanized steel, prefab type, economy	5.65	3.17	8.82
0160	Radio frequency, door, copper/wood laminate, 4" x 7'	9.95	8.50	18.45

F1030 910	Other Special Construction Systems, EACH	COST EACH		
		MAT.	INST.	TOTAL
0110	Disappearing stairways, folding, pine, 8'-6" ceiling	194	149	343
0120	9'-6" ceiling	216	149	365
0220	Automatic electric, wood, 8' to 9' ceiling	9,575	1,200	10,775
0230	Aluminum, 14' to 15' ceiling	11,000	1,700	12,700
0300	Fireplace prefabricated, freestanding or wall hung, painted	1,625	455	2,080
0310	Stainless steel	3,350	660	4,010
0320	Woodburning stoves, cast iron, economy, less than 1500 S.F. htg. area	1,425	915	2,340
0330	Greater than 2000 S.F. htg. area	3,075	1,475	4,550

F1040 Special Facilities

F1040 210	Ice Rinks, EACH	COAT EACH		
		MAT.	INST.	TOTAL
0100	Ice skating rink, 85′ x 200′, 55° system, 5 mos., 100 ton			664,000
0110	90° system, 12 mos., 135 ton			751,000
0120	Dash boards, acrylic screens, polyethylene coated plywood	156,000	40,000	196,000
0130	Fiberglass and aluminum construction	179,000	40,000	219,000

F1040 510	Liquid & Gas Storage Tanks, EACH	COST EACH		
		MAT.	INST.	TOTAL
0100	Tanks, steel, ground level, 100,000 gal.			226,500
0110	10,000,000 gal.			5,151,500
0120	Elevated water, 50,000 gal.		196,000	468,000
0130	1,000,000 gal.		913,000	2,073,000
0150	Cypress wood, ground level, 3,000 gal.	9,900	11,800	21,700
0160	Redwood, ground level, 45,000 gal.	77,500	32,100	109,600

F1040 910	Special Construction, EACH	COST EACH		
		MAT.	INST.	TOTAL
0110	Special construction, bowling alley incl. pinsetter, scorer etc., economy	42,500	11,900	54,400
0120	Deluxe	58,500	13,200	71,700
0130	For automatic scorer, economy, add	6,400		6,400
0140	Deluxe, add	11,000		11,000
0300	Control tower, modular, 12′ x 10′, incl. instrumentation, economy			882,000
0310	Deluxe			1,386,000
0400	Garage costs, residential, prefab, wood, single car economy	6,275	1,200	7,475
0410	Two car deluxe	14,200	2,375	16,575
0500	Hangars, prefab, galv. steel, bottom rolling doors, economy (per plane)	18,000	5,200	23,200
0510	Electrical bi-folding doors, deluxe (per plane)	16,000	7,150	23,150

G1030 Site Earthwork

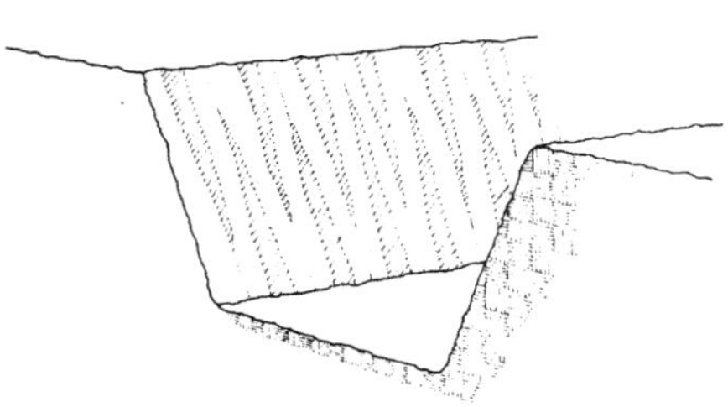

Trenching Systems are shown on a cost per linear foot basis. The systems include: excavation; backfill and removal of spoil; and compaction for various depths and trench bottom widths. The backfill has been reduced to accommodate a pipe of suitable diameter and bedding.

The slope for trench sides varies from none to 1:1.

The Expanded System Listing shows Trenching Systems that range from 2′ to 12′ in width. Depths range from 2′ to 25′.

System Components	QUANTITY	UNIT	COST PER L.F. EQUIP.	COST PER L.F. LABOR	COST PER L.F. TOTAL
SYSTEM G1030 805 1310					
TRENCHING COMMON EARTH, NO SLOPE, 2′ WIDE, 2′ DP, 3/8 C.Y. BUCKET					
Excavation, trench, hyd. backhoe, track mtd., 3/8 C.Y. bucket	.148	B.C.Y.	.40	1.07	1.47
Backfill and load spoil, from stockpile	.153	L.C.Y.	.13	.33	.46
Compaction by vibrating plate, 6″ lifts, 4 passes	.118	E.C.Y.	.03	.39	.42
Remove excess spoil, 8 C.Y. dump truck, 2 mile roundtrip	.040	L.C.Y.	.15	.17	.32
TOTAL			.71	1.96	2.67

G1030 805	Trenching Common Earth	COST PER L.F. EQUIP.	COST PER L.F. LABOR	COST PER L.F. TOTAL
1310	Trenching, common earth, no slope, 2′ wide, 2′ deep, 3/8 C.Y. bucket	.71	1.96	2.67
1320	3′ deep, 3/8 C.Y. bucket	1	2.94	3.94
1330	4′ deep, 3/8 C.Y. bucket	1.29	3.92	5.21
1340	6′ deep, 3/8 C.Y. bucket	1.68	5.10	6.78
1350	8′ deep, 1/2 C.Y. bucket	2.22	6.75	8.97
1360	10′ deep, 1 C.Y. bucket	3.46	8.05	11.51
1400	4′ wide, 2′ deep, 3/8 C.Y. bucket	1.61	3.89	5.50
1410	3′ deep, 3/8 C.Y. bucket	2.21	5.85	8.06
1420	4′ deep, 1/2 C.Y. bucket	2.57	6.55	9.12
1430	6′ deep, 1/2 C.Y. bucket	4.16	10.50	14.66
1440	8′ deep, 1/2 C.Y. bucket	6.65	13.60	20.25
1450	10′ deep, 1 C.Y. bucket	8	16.85	24.85
1460	12′ deep, 1 C.Y. bucket	10.30	21.50	31.80
1470	15′ deep, 1-1/2 C.Y. bucket	9.45	19.20	28.65
1480	18′ deep, 2-1/2 C.Y. bucket	13.05	27	40.05
1520	6′ wide, 6′ deep, 5/8 C.Y. bucket w/trench box	8.55	15.70	24.25
1530	8′ deep, 3/4 C.Y. bucket	11.30	20.50	31.80
1540	10′ deep, 1 C.Y. bucket	11.55	21.50	33.05
1550	12′ deep, 1-1/2 C.Y. bucket	12.40	23	35.40
1560	16′ deep, 2-1/2 C.Y. bucket	16.85	28.50	45.35
1570	20′ deep, 3-1/2 C.Y. bucket	22	34	56
1580	24′ deep, 3-1/2 C.Y. bucket	26	41	67
1640	8′ wide, 12′ deep, 1-1/2 C.Y. bucket w/trench box	17.35	29	46.35
1650	15′ deep, 1-1/2 C.Y. bucket	22.50	38	60.50
1660	18′ deep, 2-1/2 C.Y. bucket	24.50	38	62.50
1680	24′ deep, 3-1/2 C.Y. bucket	35.50	53	88.50
1730	10′ wide, 20′ deep, 3-1/2 C.Y. bucket w/trench box	29	50.50	79.50
1740	24′ deep, 3-1/2 C.Y. bucket	42.50	60.50	103
1780	12′ wide, 20′ deep, 3-1/2 C.Y. bucket w/trench box	45.50	64	109.50
1790	25′ deep, bucket	56	81.50	137.50
1800	1/2 to 1 slope, 2′ wide, 2′ deep, 3/8 C.Y. bucket	1	2.94	3.94
1810	3′ deep, 3/8 C.Y. bucket	1.67	5.15	6.82

G1030 Site Earthwork

G1030 805	Trenching Common Earth	COST PER L.F.		
		EQUIP.	LABOR	TOTAL
1820	4' deep, 3/8 C.Y. bucket	2.50	7.85	10.35
1840	6' deep, 3/8 C.Y. bucket	4.02	12.75	16.77
1860	8' deep, 1/2 C.Y. bucket	6.40	20.50	26.90
1880	10' deep, 1 C.Y. bucket	11.90	28.50	40.40
2300	4' wide, 2' deep, 3/8 C.Y. bucket	1.91	4.87	6.78
2310	3' deep, 3/8 C.Y. bucket	2.89	8.05	10.94
2320	4' deep, 1/2 C.Y. bucket	3.63	9.95	13.58
2340	6' deep, 1/2 C.Y. bucket	6.95	18.65	25.60
2360	8' deep, 1/2 C.Y. bucket	12.80	27.50	40.30
2380	10' deep, 1 C.Y. bucket	17.70	38.50	56.20
2400	12' deep, 1 C.Y. bucket	24.50	52	76.50
2430	15' deep, 1-1/2 C.Y. bucket	26.50	56	82.50
2460	18' deep, 2-1/2 C.Y. bucket	46	87.50	133.50
2840	6' wide, 6' deep, 5/8 C.Y. bucket w/trench box	12.60	23	35.60
2860	8' deep, 3/4 C.Y. bucket	18.25	35	53.25
2880	10' deep, 1 C.Y. bucket	18.65	35.50	54.15
2900	12' deep, 1-1/2 C.Y. bucket	23.50	46.50	70
2940	16' deep, 2-1/2 C.Y. bucket	38.50	68	106.50
2980	20' deep, 3-1/2 C.Y. bucket	55	92	147
3020	24' deep, 3-1/2 C.Y. bucket	78	126	204
3100	8' wide, 12' deep, 1-1/2 C.Y. bucket w/trench box	29	53	82
3120	15' deep, 1-1/2 C.Y. bucket	42.50	76.50	119
3140	18' deep, 2-1/2 C.Y. bucket	53	91	144
3180	24' deep, 3-1/2 C.Y. bucket	87.50	138	225.50
3270	10' wide, 20' deep, 3-1/2 C.Y. bucket w/trench box	56	106	162
3280	24' deep, 3-1/2 C.Y. bucket	96.50	150	246.50
3370	12' wide, 20' deep, 3-1/2 C.Y. bucket w/trench box	81	123	204
3380	25' deep, 3-1/2 C.Y. bucket	113	175	288
3500	1 to 1 slope, 2' wide, 2' deep, 3/8 C.Y. bucket	1.29	3.92	5.21
3520	3' deep, 3/8 C.Y. bucket	3.73	8.95	12.68
3540	4' deep, 3/8 C.Y. bucket	3.69	11.80	15.49
3560	6' deep, 1/2 C.Y. bucket	4.02	12.75	16.77
3580	8' deep, 1/2 C.Y. bucket	7.95	25.50	33.45
3600	10' deep, 1 C.Y. bucket	20.50	48.50	69
3800	4' wide, 2' deep, 3/8 C.Y. bucket	2.21	5.85	8.06
3820	3' deep, 3/8 C.Y. bucket	3.56	10.30	13.86
3840	4' deep, 1/2 C.Y. bucket	4.66	13.35	18.01
3860	6' deep, 1/2 C.Y. bucket	9.80	27	36.80
3880	8' deep, 1/2 C.Y. bucket	19.05	41.50	60.55
3900	10' deep, 1 C.Y. bucket	27.50	60	87.50
3920	12' deep, 1 C.Y. bucket	40.50	87	127.50
3940	15' deep, 1-1/2 C.Y. bucket	44	93	137
3960	18' deep, 2-1/2 C.Y. bucket	63.50	121	184.50
4030	6' wide, 6' deep, 5/8 C.Y. bucket w/trench box	16.50	31.50	48
4040	8' deep, 3/4 C.Y. bucket	23.50	43	66.50
4050	10' deep, 1 C.Y. bucket	27	52	79
4060	12' deep, 1-1/2 C.Y. bucket	36	71	107
4070	16' deep, 2-1/2 C.Y. bucket	60	108	168
4080	20' deep, 3-1/2 C.Y. bucket	89	150	239
4090	24' deep, 3-1/2 C.Y. bucket	130	211	341
4500	8' wide, 12' deep, 1-1/2 C.Y. bucket w/trench box	41	77	118
4550	15' deep, 1-1/2 C.Y. bucket	62	115	177
4600	18' deep, 2-1/2 C.Y. bucket	80.50	141	221.50
4650	24' deep, 3-1/2 C.Y. bucket	139	222	361
4800	10' wide, 20' deep, 3-1/2 C.Y. bucket w/trench box	83.50	161	244.50
4850	24' deep, 3-1/2 C.Y. bucket	148	235	383
4950	12' wide, 20' deep, 3-1/2 C.Y. bucket w/trench box	117	181	298
4980	25' deep, 3-1/2 C.Y. bucket	168	265	433

G1030 Site Earthwork

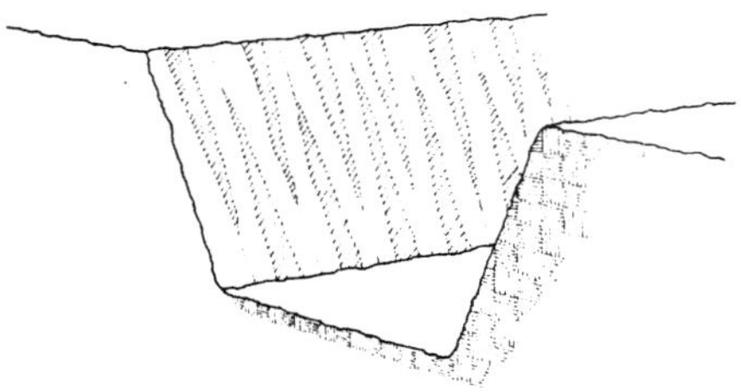

Trenching Systems are shown on a cost per linear foot basis. The systems include: excavation; backfill and removal of spoil; and compaction for various depths and trench bottom widths. The backfill has been reduced to accommodate a pipe of suitable diameter and bedding.

The slope for trench sides varies from none to 1:1.

The Expanded System Listing shows Trenching Systems that range from 2′ to 12′ in width. Depths range from 2′ to 25′.

System Components	QUANTITY	UNIT	COST PER L.F.		
			EQUIP.	LABOR	TOTAL
SYSTEM G1030 806 1310					
TRENCHING LOAM & SANDY CLAY, NO SLOPE, 2′ WIDE, 2′ DP, 3/8 C.Y. BUCKET					
Excavation, trench, hyd. backhoe, track mtd., 3/8 C.Y. bucket	.148	B.C.Y.	.37	.99	1.36
Backfill and load spoil, from stockpile	.165	L.C.Y.	.14	.35	.49
Compaction by vibrating plate 18″ wide, 6″ lifts, 4 passes	.118	E.C.Y.	.03	.39	.42
Remove excess spoil, 8 C.Y. dump truck, 2 mile roundtrip	.042	L.C.Y.	.16	.18	.34
TOTAL			.70	1.91	2.61

G1030 806	Trenching Loam & Sandy Clay	COST PER L.F.		
		EQUIP.	LABOR	TOTAL
1310	Trenching, loam & sandy clay, no slope, 2′ wide, 2′ deep, 3/8 C.Y. bucket	.70	1.91	2.61
1320	3′ deep, 3/8 C.Y. bucket	1.09	3.15	4.24
1330	4′ deep, 3/8 C.Y. bucket	1.27	3.83	5.10
1340	6′ deep, 3/8 C.Y. bucket	1.80	4.56	6.36
1350	8′ deep, 1/2 C.Y. bucket	2.38	6.05	8.43
1360	10′ deep, 1 C.Y. bucket	2.69	6.45	9.14
1400	4′ wide, 2′ deep, 3/8 C.Y. bucket	1.61	3.82	5.43
1410	3′ deep, 3/8 C.Y. bucket	2.19	5.75	7.94
1420	4′ deep, 1/2 C.Y. bucket	2.56	6.45	9.01
1430	6′ deep, 1/2 C.Y. bucket	4.41	9.45	13.86
1440	8′ deep, 1/2 C.Y. bucket	6.45	13.40	19.85
1450	10′ deep, 1 C.Y. bucket	6.45	13.70	20.15
1460	12′ deep, 1 C.Y. bucket	8.15	17.05	25.20
1470	15′ deep, 1-1/2 C.Y. bucket	9.85	19.85	29.70
1480	18′ deep, 2-1/2 C.Y. bucket	11.50	22	33.50
1520	6′ wide, 6′ deep, 5/8 C.Y. bucket w/trench box	8.30	15.40	23.70
1530	8′ deep, 3/4 C.Y. bucket	10.95	20	30.95
1540	10′ deep, 1 C.Y. bucket	10.65	20.50	31.15
1550	12′ deep, 1-1/2 C.Y. bucket	12.10	23	35.10
1560	16′ deep, 2-1/2 C.Y. bucket	16.45	29	45.45
1570	20′ deep, 3-1/2 C.Y. bucket	20.50	34	54.50
1580	24′ deep, 3-1/2 C.Y. bucket	25.50	41.50	67
1640	8′ wide, 12′ deep, 1-1/4 C.Y. bucket w/trench box	17.10	29	46.10
1650	15′ deep, 1-1/2 C.Y. bucket	21	37	58
1660	18′ deep, 2-1/2 C.Y. bucket	25	41.50	66.50
1680	24′ deep, 3-1/2 C.Y. bucket	34.50	54	88.50
1730	10′ wide, 20′ deep, 3-1/2 C.Y. bucket w/trench box	35	54	89
1740	24′ deep, 3-1/2 C.Y. bucket	43.50	67	110.50
1780	12′ wide, 20′ deep, 3-1/2 C.Y. bucket w/trench box	42	64	106
1790	25′ deep, 3-1/2 C.Y. bucket	54.50	83	137.50
1800	1/2:1 slope, 2′ wide, 2′ deep, 3/8 C.Y. bucket	.98	2.88	3.86
1810	3′ deep, 3/8 C.Y. bucket	1.63	5.05	6.68
1820	4′ deep, 3/8 C.Y. bucket	2.44	7.70	10.14
1840	6′ deep, 3/8 C.Y. bucket	4.31	11.40	15.71

G1030 Site Earthwork

G1030 806	Trenching Loam & Sandy Clay	COST PER L.F.		
		EQUIP.	LABOR	TOTAL
1860	8' deep, 1/2 C.Y. bucket	6.85	18.20	25.05
1880	10' deep, 1 C.Y. bucket	9.15	23	32.15
2300	4' wide, 2' deep, 3/8 C.Y. bucket	1.90	4.78	6.68
2310	3' deep, 3/8 C.Y. bucket	2.85	7.90	10.75
2320	4' deep, 1/2 C.Y. bucket	3.59	9.80	13.39
2340	6' deep, 1/2 C.Y. bucket	7.40	16.80	24.20
2360	8' deep, 1/2 C.Y. bucket	12.50	27	39.50
2380	10' deep, 1 C.Y. bucket	14.20	31.50	45.70
2400	12' deep, 1 C.Y. bucket	24	52	76
2430	15' deep, 1-1/2 C.Y. bucket	28	58	86
2460	18' deep, 2-1/2 C.Y. bucket	45.50	88.50	134
2840	6' wide, 6' deep, 5/8 C.Y. bucket w/trench box	12.05	23.50	35.55
2860	8' deep, 3/4 C.Y. bucket	17.60	34	51.60
2880	10' deep, 1 C.Y. bucket	19	38.50	57.50
2900	12' deep, 1-1/2 C.Y. bucket	23.50	47	70.50
2940	16' deep, 2-1/2 C.Y. bucket	37.50	69	106.50
2980	20' deep, 3-1/2 C.Y. bucket	53.50	93	146.50
3020	24' deep, 3-1/2 C.Y. bucket	75.50	128	203.50
3100	8' wide, 12' deep, 1-1/2 C.Y. bucket w/trench box	28.50	53	81.50
3120	15' deep, 1-1/2 C.Y. bucket	39	74.50	113.50
3140	18' deep, 2-1/2 C.Y. bucket	52	92.50	144.50
3180	24' deep, 3-1/2 C.Y. bucket	85	140	225
3270	10' wide, 20' deep, 3-1/2 C.Y. bucket w/trench box	68	113	181
3280	24' deep, 3-1/2 C.Y. bucket	93.50	153	246.50
3320	12' wide, 20' deep, 3-1/2 C.Y. bucket w/trench box	75	123	198
3380	25' deep, 3-1/2 C.Y. bucket w/trench box	103	165	268
3500	1:1 slope, 2' wide, 2' deep, 3/8 C.Y. bucket	1.27	3.84	5.11
3520	3' deep, 3/8 C.Y. bucket	2.29	7.20	9.49
3540	4' deep, 3/8 C.Y. bucket	3.59	11.55	15.14
3560	6' deep, 1/2 C.Y. bucket	4.31	11.40	15.71
3580	8' deep, 1/2 C.Y. bucket	11.35	30.50	41.85
3600	10' deep, 1 C.Y. bucket	15.65	39	54.65
3800	4' wide, 2' deep, 3/8 C.Y. bucket	2.19	5.75	7.94
3820	3' deep, 1/2 C.Y. bucket	3.50	10.10	13.60
3840	4' deep, 1/2 C.Y. bucket	4.60	13.15	17.75
3860	6' deep, 1/2 C.Y. bucket	10.35	24	34.35
3880	8' deep, 1/2 C.Y. bucket	18.50	41	59.50
3900	10' deep, 1 C.Y. bucket	22	49	71
3920	12' deep, 1 C.Y. bucket	31.50	69.50	101
3940	15' deep, 1-1/2 C.Y. bucket	46	96	142
3960	18' deep, 2-1/2 C.Y. bucket	62	122	184
4030	6' wide, 6' deep, 5/8 C.Y. bucket w/trench box	15.80	31	46.80
4040	8' deep, 3/4 C.Y. bucket	24.50	48	72.50
4050	10' deep, 1 C.Y. bucket	27.50	56.50	84
4060	12' deep, 1-1/2 C.Y. bucket	35	71	106
4070	16' deep, 2-1/2 C.Y. bucket	58.50	109	167.50
4080	20' deep, 3-1/2 C.Y. bucket	86.50	152	238.50
4090	24' deep, 3-1/2 C.Y. bucket	126	214	340
4500	8' wide, 12' deep, 1-1/4 C.Y. bucket w/trench box	40	77	117
4550	15' deep, 1-1/2 C.Y. bucket	57	112	169
4600	18' deep, 2-1/2 C.Y. bucket	78.50	143	221.50
4650	24' deep, 3-1/2 C.Y. bucket	135	226	361
4800	10' wide, 20' deep, 3-1/2 C.Y. bucket w/trench box	101	172	273
4850	24' deep, 3-1/2 C.Y. bucket	144	239	383
4950	12' wide, 20' deep, 3-1/2 C.Y. bucket w/trench box	108	182	290
4980	25' deep, 3-1/2 C.Y. bucket	163	269	432

G1030 Site Earthwork

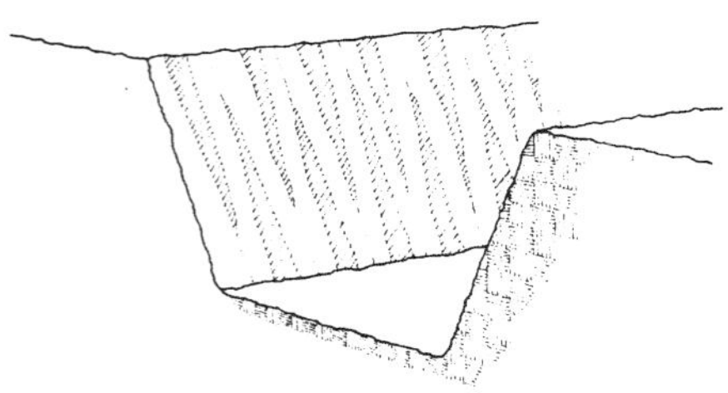

Trenching Systems are shown on a cost per linear foot basis. The systems include: excavation; backfill and removal of spoil; and compaction for various depths and trench bottom widths. The backfill has been reduced to accommodate a pipe of suitable diameter and bedding.

The slope for trench sides varies from none to 1:1.

The Expanded System Listing shows Trenching Systems that range from 2′ to 12′ in width. Depths range from 2′ to 25′.

System Components	QUANTITY	UNIT	COST PER L.F. EQUIP.	LABOR	TOTAL
SYSTEM G1030 807 1310					
TRENCHING SAND & GRAVEL, NO SLOPE, 2′ WIDE, 2′ DEEP, 3/8 C.Y. BUCKET					
Excavation, trench, hyd. backhoe, track mtd., 3/8 C.Y. bucket	.148	B.C.Y.	.36	.97	1.33
Backfill and load spoil, from stockpile	.140	L.C.Y.	.11	.30	.41
Compaction by vibrating plate 18″ wide, 6″ lifts, 4 passes	.118	E.C.Y.	.03	.39	.42
Remove excess spoil, 8 C.Y. dump truck, 2 mile roundtrip	.035	L.C.Y.	.14	.15	.29
TOTAL			.64	1.81	2.45

G1030 807	Trenching Sand & Gravel	COST PER L.F. EQUIP.	LABOR	TOTAL
1310	Trenching, sand & gravel, no slope, 2′ wide, 2′ deep, 3/8 C.Y. bucket	.64	1.81	2.45
1320	3′ deep, 3/8 C.Y. bucket	1.03	3.02	4.05
1330	4′ deep, 3/8 C.Y. bucket	1.19	3.64	4.83
1340	6′ deep, 3/8 C.Y. bucket	1.71	4.33	6.04
1350	8′ deep, 1/2 C.Y. bucket	2.25	5.75	8
1360	10′ deep, 1 C.Y. bucket	2.50	6.05	8.55
1400	4′ wide, 2′ deep, 3/8 C.Y. bucket	1.46	3.59	5.05
1410	3′ deep, 3/8 C.Y. bucket	2.01	5.45	7.46
1420	4′ deep, 1/2 C.Y. bucket	2.34	6.10	8.44
1430	6′ deep, 1/2 C.Y. bucket	4.18	9	13.18
1440	8′ deep, 1/2 C.Y. bucket	6.10	12.70	18.80
1450	10′ deep, 1 C.Y. bucket	6.10	12.95	19.05
1460	12′ deep, 1 C.Y. bucket	7.70	16.15	23.85
1470	15′ deep, 1-1/2 C.Y. bucket	9.25	18.70	27.95
1480	18′ deep, 2-1/2 C.Y. bucket	10.85	20.50	31.35
1520	6′ wide, 6′ deep, 5/8 C.Y. bucket w/trench box	7.80	14.55	22.35
1530	8′ deep, 3/4 C.Y. bucket	10.35	19.15	29.50
1540	10′ deep, 1 C.Y. bucket	10	19.35	29.35
1550	12′ deep, 1-1/2 C.Y. bucket	11.45	21.50	32.95
1560	16′ deep, 2 C.Y. bucket	15.70	27.50	43.20
1570	20′ deep, 3-1/2 C.Y. bucket	19.25	32	51.25
1580	24′ deep, 3-1/2 C.Y. bucket	24	39	63
1640	8′ wide, 12′ deep, 1-1/2 C.Y. bucket w/trench box	16.10	27.50	43.60
1650	15′ deep, 1-1/2 C.Y. bucket	19.55	35	54.55
1660	18′ deep, 2-1/2 C.Y. bucket	23.50	39	62.50
1680	24′ deep, 3-1/2 C.Y. bucket	32.50	50.50	83
1730	10′ wide, 20′ deep, 3-1/2 C.Y. bucket w/trench box	33	50.50	83.50
1740	24′ deep, 3-1/2 C.Y. bucket	41	62.50	103.50
1780	12′ wide, 20′ deep, 3-1/2 C.Y. bucket w/trench box	39.50	60	99.50
1790	25′ deep, 3-1/2 C.Y. bucket	51.50	77.50	129
1800	1/2:1 slope, 2′ wide, 2′ deep, 3/8 C.Y. bucket	.92	2.73	3.65
1810	3′ deep, 3/8 C.Y. bucket	1.54	4.79	6.33
1820	4′ deep, 3/8 C.Y. bucket	2.29	7.30	9.59
1840	6′ deep, 3/8 C.Y. bucket	4.11	10.90	15.01

G1030 Site Earthwork

G1030 807	Trenching Sand & Gravel	COST PER L.F.		
		EQUIP.	LABOR	TOTAL
1860	8' deep, 1/2 C.Y. bucket	6.55	17.40	23.95
1880	10' deep, 1 C.Y. bucket	8.55	21.50	30.05
2300	4' wide, 2' deep, 3/8 C.Y. bucket	1.74	4.51	6.25
2310	3' deep, 3/8 C.Y. bucket	2.63	7.50	10.13
2320	4' deep, 1/2 C.Y. bucket	3.31	9.30	12.61
2340	6' deep, 1/2 C.Y. bucket	7.05	16.05	23.10
2360	8' deep, 1/2 C.Y. bucket	11.85	25.50	37.35
2380	10' deep, 1 C.Y. bucket	13.45	29.50	42.95
2400	12' deep, 1 C.Y. bucket	22.50	49.50	72
2430	15' deep, 1-1/2 C.Y. bucket	26.50	55	81.50
2460	18' deep, 2-1/2 C.Y. bucket	43	83.50	126.50
2840	6' wide, 6' deep, 5/8 C.Y. bucket w/trench box	11.40	22	33.40
2860	8' deep, 3/4 C.Y. bucket	16.70	32.50	49.20
2880	10' deep, 1 C.Y. bucket	17.95	36.50	54.45
2900	12' deep, 1-1/2 C.Y. bucket	22.50	44.50	67
2940	16' deep, 2 C.Y. bucket	36	65	101
2980	20' deep, 3-1/2 C.Y. bucket	50.50	87.50	138
3020	24' deep, 3-1/2 C.Y. bucket	72	120	192
3100	8' wide, 12' deep, 1-1/4 C.Y. bucket w/trench box	27	50	77
3120	15' deep, 1-1/2 C.Y. bucket	36.50	70.50	107
3140	18' deep, 2-1/2 C.Y. bucket	49	87	136
3180	24' deep, 3-1/2 C.Y. bucket	80.50	131	211.50
3270	10' wide, 20' deep, 3-1/2 C.Y. bucket w/trench box	64.50	106	170.50
3280	24' deep, 3-1/2 C.Y. bucket	88.50	143	231.50
3370	12' wide, 20' deep, 3-1/2 C.Y. bucket w/trench box	72	118	190
3380	25' deep, 3-1/2 C.Y. bucket	103	165	268
3500	1:1 slope, 2' wide, 2' deep, 3/8 C.Y. bucket	2.06	4.55	6.61
3520	3' deep, 3/8 C.Y. bucket	2.16	6.85	9.01
3540	4' deep, 3/8 C.Y. bucket	3.38	11	14.38
3560	6' deep, 3/8 C.Y. bucket	4.11	10.90	15.01
3580	8' deep, 1/2 C.Y. bucket	10.80	29	39.80
3600	10' deep, 1 C.Y. bucket	14.65	37	51.65
3800	4' wide, 2' deep, 3/8 C.Y. bucket	2.01	5.45	7.46
3820	3' deep, 3/8 C.Y. bucket	3.25	9.55	12.80
3840	4' deep, 1/2 C.Y. bucket	4.26	12.50	16.76
3860	6' deep, 1/2 C.Y. bucket	9.90	23	32.90
3880	8' deep, 1/2 C.Y. bucket	17.55	38.50	56.05
3900	10' deep, 1 C.Y. bucket	21	46.50	67.50
3920	12' deep, 1 C.Y. bucket	30	66	96
3940	15' deep, 1-1/2 C.Y. bucket	43.50	91	134.50
3960	18' deep, 2-1/2 C.Y. bucket	59	115	174
4030	6' wide, 6' deep, 5/8 C.Y. bucket w/trench box	15	29.50	44.50
4040	8' deep, 3/4 C.Y. bucket	23	45.50	68.50
4050	10' deep, 1 C.Y. bucket	26	53.50	79.50
4060	12' deep, 1-1/2 C.Y. bucket	33.50	67	100.50
4070	16' deep, 2 C.Y. bucket	56.50	103	159.50
4080	20' deep, 3-1/2 C.Y. bucket	82	143	225
4090	24' deep, 3-1/2 C.Y. bucket	119	201	320
4500	8' wide, 12' deep, 1-1/2 C.Y. bucket w/trench box	38	73	111
4550	15' deep, 1-1/2 C.Y. bucket	53.50	106	159.50
4600	18' deep, 2-1/2 C.Y. bucket	74.50	135	209.50
4650	24' deep, 3-1/2 C.Y. bucket	128	212	340
4800	10' wide, 20' deep, 3-1/2 C.Y. bucket w/trench box	95.50	161	256.50
4850	24' deep, 3-1/2 C.Y. bucket	136	224	360
4950	12' wide, 20' deep, 3-1/2 C.Y. bucket w/trench box	103	171	274
4980	25' deep, 3-1/2 C.Y. bucket	155	253	408

G1030 Site Earthwork

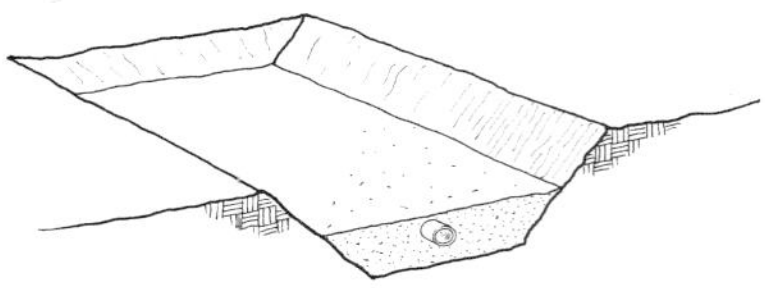

The Pipe Bedding System is shown for various pipe diameters. Compacted bank sand is used for pipe bedding and to fill 12″ over the pipe. No backfill is included. Various side slopes are shown to accommodate different soil conditions. Pipe sizes vary from 6″ to 84″ diameter.

System Components	QUANTITY	UNIT	COST PER L.F. MAT.	COST PER L.F. INST.	COST PER L.F. TOTAL
SYSTEM G1030 815 1440					
PIPE BEDDING, SIDE SLOPE 0 TO 1, 1′ WIDE, PIPE SIZE 6″ DIAMETER					
Borrow, bank sand, 2 mile haul, machine spread	.086	C.Y.	1.88	.71	2.59
Compaction, vibrating plate	.086	C.Y.		.22	.22
TOTAL			1.88	.93	2.81

G1030 815	Pipe Bedding	COST PER L.F. MAT.	COST PER L.F. INST.	COST PER L.F. TOTAL
1440	Pipe bedding, side slope 0 to 1, 1′ wide, pipe size 6″ diameter	1.88	.93	2.81
1460	2′ wide, pipe size 8″ diameter	4.07	1.99	6.06
1480	Pipe size 10″ diameter	4.16	2.04	6.20
1500	Pipe size 12″ diameter	4.25	2.08	6.33
1520	3′ wide, pipe size 14″ diameter	6.90	3.38	10.28
1540	Pipe size 15″ diameter	6.95	3.41	10.36
1560	Pipe size 16″ diameter	7	3.44	10.44
1580	Pipe size 18″ diameter	7.15	3.52	10.67
1600	4′ wide, pipe size 20″ diameter	10.15	4.99	15.14
1620	Pipe size 21″ diameter	10.25	5.05	15.30
1640	Pipe size 24″ diameter	10.45	5.15	15.60
1660	Pipe size 30″ diameter	10.65	5.25	15.90
1680	6′ wide, pipe size 32″ diameter	18.25	8.95	27.20
1700	Pipe size 36″ diameter	18.70	9.15	27.85
1720	7′ wide, pipe size 48″ diameter	29.50	14.55	44.05
1740	8′ wide, pipe size 60″ diameter	36	17.75	53.75
1760	10′ wide, pipe size 72″ diameter	50.50	24.50	75
1780	12′ wide, pipe size 84″ diameter	66.50	32.50	99
2140	Side slope 1/2 to 1, 1′ wide, pipe size 6″ diameter	3.51	1.71	5.22
2160	2′ wide, pipe size 8″ diameter	5.95	2.92	8.87
2180	Pipe size 10″ diameter	6.40	3.14	9.54
2200	Pipe size 12″ diameter	6.80	3.32	10.12
2220	3′ wide, pipe size 14″ diameter	9.80	4.81	14.61
2240	Pipe size 15″ diameter	10	4.91	14.91
2260	Pipe size 16″ diameter	10.25	5.05	15.30
2280	Pipe size 18″ diameter	10.80	5.30	16.10
2300	4′ wide, pipe size 20″ diameter	14.25	7	21.25
2320	Pipe size 21″ diameter	14.55	7.15	21.70
2340	Pipe size 24″ diameter	15.50	7.60	23.10
2360	Pipe size 30″ diameter	17.20	8.45	25.65
2380	6′ wide, pipe size 32″ diameter	25.50	12.40	37.90
2400	Pipe size 36″ diameter	27	13.20	40.20
2420	7′ wide, pipe size 48″ diameter	42	20.50	62.50
2440	8′ wide, pipe size 60″ diameter	53.50	26	79.50
2460	10′ wide, pipe size 72″ diameter	73	36	109
2480	12′ wide, pipe size 84″ diameter	96	47	143
2620	Side slope 1 to 1, 1′ wide, pipe size 6″ diameter	5.15	2.52	7.67
2640	2′ wide, pipe size 8″ diameter	7.90	3.89	11.79

G1030 Site Earthwork

G1030 815	Pipe Bedding	COST PER L.F.		
		MAT.	INST.	TOTAL
2660	Pipe size 10" diameter	8.60	4.22	12.82
2680	Pipe size 12" diameter	9.35	4.60	13.95
2700	3' wide, pipe size 14" diameter	12.70	6.20	18.90
2720	Pipe size 15" diameter	13.10	6.40	19.50
2740	Pipe size 16" diameter	13.55	6.65	20.20
2760	Pipe size 18" diameter	14.45	7.10	21.55
2780	4' wide, pipe size 20" diameter	18.35	9	27.35
2800	Pipe size 21" diameter	18.85	9.25	28.10
2820	Pipe size 24" diameter	20.50	10.05	30.55
2840	Pipe size 30" diameter	23.50	11.60	35.10
2860	6' wide, pipe size 32" diameter	32.50	15.90	48.40
2880	Pipe size 36" diameter	35	17.30	52.30
2900	7' wide, pipe size 48" diameter	54.50	26.50	81
2920	8' wide, pipe size 60" diameter	70.50	34.50	105
2940	10' wide, pipe size 72" diameter	96	47	143
2960	12' wide, pipe size 84" diameter	125	61.50	186.50

G2010 Roadways

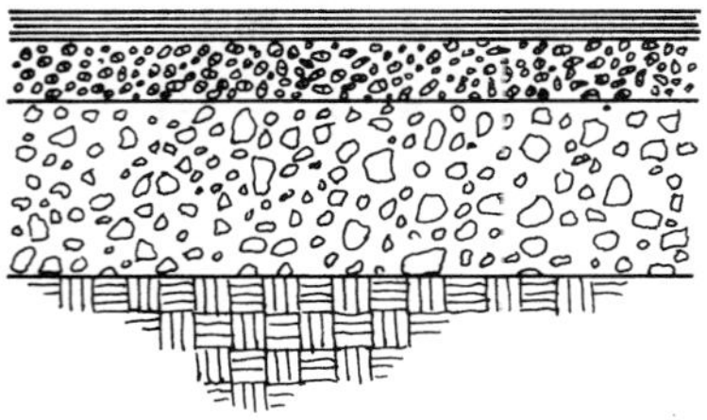

The Bituminous Roadway Systems are listed for pavement thicknesses between 3-1/2" and 7" and crushed stone bases from 3" to 22" in depth. Systems costs are expressed per linear foot for varying widths of two and multi-lane roads. Earth moving is not included. Granite curbs and line painting are added as required system components.

System Components	QUANTITY	UNIT	COST PER L.F. MAT.	INST.	TOTAL
SYSTEM G2010 232 1050					
BITUMINOUS ROADWAY, TWO LANES, 3-1/2" PAVEMENT					
3" THICK CRUSHED STONE BASE, 24' WIDE					
Compact subgrade, 4 passes	.222	E.C.Y.		.48	.48
3/4" crushed stonel, 2 mi haul, dozer spread	.266	C.Y.	7.05	2.20	9.25
Compaction, granular material to 98%	.259	E.C.Y.		.17	.17
Grading, fine grade, 3 passes with grader	3.300	S.Y.		15.41	15.41
Bituminous paving, binder course, 2-1/2" thick	2.670	S.Y.	27.37	4.14	31.51
Bituminous paving, wearing course, 1" thick	2.670	S.Y.	10.87	2.43	13.30
Curbs, granite, split face, straight, 5" x 16"	2.000	L.F.	30.80	27.94	58.74
Thermoplastic, white or yellow, 4" wide, less than 6000 L.F.	4.000	L.F.	1.44	1.08	2.52
TOTAL			77.53	53.85	131.38

G2010 232	Bituminous Roadways Crushed Stone	COST PER L.F. MAT.	INST.	TOTAL
1050	Bitum. roadway, two lanes, 3-1/2" th. pvmt., 3" th. crushed stone,24' wide	77.50	54	131.50
1100	28' wide	85	54.50	139.50
1150	32' wide	92.50	63	155.50
1200	4" thick crushed stone, 24' wide	80	51.50	131.50
1210	28' wide	87.50	55.50	143
1220	32' wide	95.50	59.50	155
1222	36' wide	103	63	166
1224	40' wide	111	66.50	177.50
1230	6" thick crushed stone, 24' wide	84.50	54	138.50
1240	28' wide	93	57.50	150.50
1250	32' wide	102	62	164
1252	36' wide	110	65.50	175.50
1254	40' wide	119	69.50	188.50
1256	8" thick crushed stone, 24' wide	89	55.50	144.50
1258	28' wide	98.50	60	158.50
1260	32' wide	108	64	172
1262	36' wide	117	68.50	185.50
1264	40' wide	127	73	200
1300	9" thick crushed stone, 24' wide	91.50	56.50	148
1350	28' wide	101	61.50	162.50
1400	32' wide	111	66	177
1410	10" thick crushed stone, 24' wide	93.50	57	150.50
1412	28' wide	104	61.50	165.50
1414	32' wide	114	66.50	180.50
1416	36' wide	124	71	195
1418	40' wide	135	76	211

G2010 Roadways

G2010 232	Bituminous Roadways Crushed Stone	COST PER L.F.		
		MAT.	INST.	TOTAL
1420	12″ thick crushed stone, 24′ wide	98.50	58.50	157
1422	28′ wide	109	64	173
1424	32′ wide	120	69	189
1426	36′ wide	131	74	205
1428	40′ wide	142	78.50	220.50
1430	15″ thick crushed stone, 24′ wide	105	61.50	166.50
1432	28′ wide	117	67	184
1434	32′ wide	130	72	202
1436	36′ wide	142	78	220
1438	40′ wide	154	83	237
1440	18″ thick crushed stone, 24′ wide	112	64	176
1442	28′ wide	125	70	195
1444	32′ wide	139	76	215
1446	36′ wide	152	82	234
1448	40′ wide	165	87	252
1550	4″ thick pavement., 4″ thick crushed stone, 24′ wide	85.50	52.50	138
1600	28′ wide	94	56	150
1650	32′ wide	103	60.50	163.50
1652	36′ wide	112	64	176
1654	40′ wide	120	68	188
1700	6″ thick crushed stone , 24′ wide	90	54	144
1710	28′ wide	99.50	58.50	158
1720	32′ wide	109	62.50	171.50
1722	36′ wide	119	66.50	185.50
1724	40′ wide	128	71	199
1726	8″ thick crushed stone , 24′ wide	94.50	56	150.50
1728	28′ wide	105	60	165
1730	32′ wide	115	64.50	179.50
1732	36′ wide	126	69	195
1734	40′ wide	136	74	210
1800	10″ thick crushed stone, 24′ wide	99.50	58	157.50
1850	28′ wide	111	61.50	172.50
1900	32′ wide	122	67.50	189.50
1902	36′ wide	132	72	204
1904	40′ wide	138	75	213
2050	4″ th. pavement, 5″ thick crushed stone, 24′ wide	88.50	56	144.50
2100	28′ wide	98	61	159
2150	32′ wide	108	65	173
2300	12″ thick crushed stone, 24′ wide	105	62	167
2350	28′ wide	117	67	184
2400	32′ wide	130	72.50	202.50
2550	4-1/2″ thick pavement, 5″ thick crushed stone, 24′ wide	94	56.50	150.50
2600	28′ wide	104	61.50	165.50
2650	32′ wide	115	66	181
2800	13″ thick crushed stone, 24′ wide	113	63	176
2850	28′ wide	126	68.50	194.50
2900	32′ wide	140	74.50	214.50
3050	5″ thick pavement, 6″ thick crushed stone, 24′ wide	102	58	160
3100	28′ wide	114	62.50	176.50
3150	32′ wide	125	67.50	192.50
3300	14″ thick crushed stone, 24′ wide	121	64	185
3350	28′ wide	135	70	205
3400	32′ wide	151	76.50	227.50
3550	5-1/2″ thick pavement., 7″ thick crushed stone, 24′ wide	110	59.50	169.50
3600	28′ wide	124	64.50	188.50
3650	32′ wide	137	69.50	206.50
3800	17″ thick crushed stone, 24′ wide	134	67.50	201.50
3850	28′ wide	164	77.50	241.50

G2010 Roadways

G2010 232	Bituminous Roadways Crushed Stone	COST PER L.F.		
		MAT.	INST.	TOTAL
3900	32' wide	195	88.50	283.50
4050	Multi lane, 5" thick pavement, 6" thick crushed stone, 48' wide	172	87.50	259.50
4100	72' wide	243	117	360
4150	96' wide	315	147	462
4300	14" thick crushed stone, 48' wide	210	100	310
4350	72' wide	299	136	435
4400	96' wide	390	172	562
4550	5-1/2" thick pavement, 7" thick crushed stone, 48' wide	190	90.50	280.50
4600	72' wide	269	122	391
4650	96' wide	350	153	503
4800	17" thick crushed stone, 48' wide	237	106	343
4850	72' wide	340	145	485
4900	96' wide	440	184	624
5050	6" thick pavement, 8" thick crushed stone, 48' wide	205	93	298
5100	72' wide	292	125	417
5150	96' wide	380	158	538
5300	20" thick crushed stone, 48' wide	262	112	374
5350	72' wide	375	154	529
5400	96' wide	490	196	686
5550	7" thick pavement, 9" thick crushed stone, 48' wide	233	96.50	329.50
5600	72' wide	335	131	466
5650	96' wide	435	165	600
5800	22" thick crushed stone, 48' wide	294	118	412
5850	72' wide	425	162	587
5900	96' wide	555	206	761

G2010 305	Curbs & Berms	COST PER L.F.		
		MAT.	INST.	TOTAL
1000	Bituminous, curbs, 8" wide, 6" high	1.90	2.22	4.12
1100	8" high	2.54	2.45	4.99
1500	Berm, 12" wide, 3" to 6" high	.04	3.15	3.19
1600	1-1/2" to 4" high	.03	2.24	2.27
2000	Concrete, curb, 6" wide, 18" high, cast-in-place	3.45	6.90	10.35
2100	Precast	9.90	6.45	16.35
2500	Curb and gutter, monolithic, 6" high, 24" wide	17.70	9.10	26.80
2600	30" wide	19.40	10.05	29.45
3000	Granite, curb, 4-1/2" wide, 12" high	7.70	12.80	20.50
3100	5" wide, 16" high	15.40	13.95	29.35
3200	6" wide, 18" high	20	15.35	35.35

G2020 Parking Lots

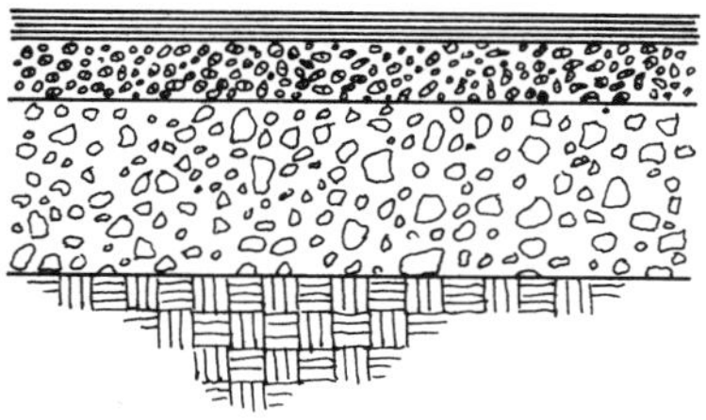

The Parking Lot System includes: compacted bank-run gravel; fine grading with a grader and roller; and bituminous concrete wearing course. All Parking Lot systems are on a cost per car basis. There are three basic types of systems: 90° angle, 60° angle, and 45° angle. The gravel base is compacted to 98%. Final stall design and lay-out of the parking lot with precast bumpers, sealcoating and white paint is also included.

The Expanded System Listing shows the three basic parking lot types with various depths of both gravel base and wearing course. The gravel base depths range from 6″ to 10″. The bituminous paving wearing course varies from a depth of 3″ to 6″.

System Components	QUANTITY	UNIT	COST PER CAR MAT.	COST PER CAR INST.	COST PER CAR TOTAL
SYSTEM G2020 210 1500					
PARKING LOT, 90° ANGLE PARKING, 3″ BITUMINOUS PAVING, 6″ GRAVEL BASE					
Surveying crew for layout, 4 man crew	.020	Day		50.96	50.96
Borrow, bank run gravel, haul 2 mi., spread w/dozer, no compaction	7.223	L.C.Y.	162.52	59.66	222.18
Grading, fine grade 3 passes with motor grader	43.333	S.Y.		102.74	102.74
Compact w/vibrating plate, 8″ lifts, granular mat'l. to 98%	7.223	E.C.Y.		48.61	48.61
Binder course paving 2″ thick	43.333	S.Y.	355.33	59.36	414.69
Wear course paving 1″ thick	43.333	S.Y.	176.37	39.44	215.81
Seal coating, petroleum resistant under 1,000 S.Y.	43.333	S.Y.	63.70	58.50	122.20
Lines on pvmt., parking stall, paint, white, 4″ wide	1.000	Ea.	5.10	3.74	8.84
Precast concrete parking bar, 6″ x 10″ x 6′-0″	1.000	Ea.	44	19.60	63.60
TOTAL			807.02	442.61	1,249.63

G2020 210	Parking Lots Gravel Base	COST PER CAR MAT.	COST PER CAR INST.	COST PER CAR TOTAL
1500	Parking lot, 90° angle parking, 3″ bituminous paving, 6″ gravel base	805	440	1,245
1520	8″ gravel base	860	480	1,340
1540	10″ gravel base (RG2020 -500)	915	515	1,430
1560	4″ bituminous paving, 6″ gravel base	1,025	470	1,495
1580	8″ gravel base	1,075	505	1,580
1600	10″ gravel base	1,150	540	1,690
1620	6″ bituminous paving, 6″ gravel base	1,375	500	1,875
1640	8″ gravel base	1,450	535	1,985
1660	10″ gravel base	1,500	575	2,075
1800	60° angle parking, 3″ bituminous paving, 6″ gravel base	805	440	1,245
1820	8″ gravel base	860	480	1,340
1840	10″ gravel base	915	515	1,430
1860	4″ bituminous paving, 6″ gravel base	1,025	470	1,495
1880	8″ gravel base	1,075	505	1,580
1900	10″ gravel base	1,150	540	1,690
1920	6″ bituminous paving, 6″ gravel base	1,375	500	1,875
1940	8″ gravel base	1,450	535	1,985
1960	10″ gravel base	1,500	575	2,075
2200	45° angle parking, 3″ bituminous paving, 6″ gravel base	825	455	1,280
2220	8″ gravel base	880	490	1,370
2240	10″ gravel base	935	530	1,465
2260	4″ bituminous paving, 6″ gravel base	1,050	480	1,530
2280	8″ gravel base	1,100	520	1,620
2300	10″ gravel base	1,175	555	1,730
2320	6″ bituminous paving, 6″ gravel base	1,425	515	1,940
2340	8″ gravel base	1,475	550	2,025
2360	10″ gravel base	1,525	590	2,115

G2030 Pedestrian Paving

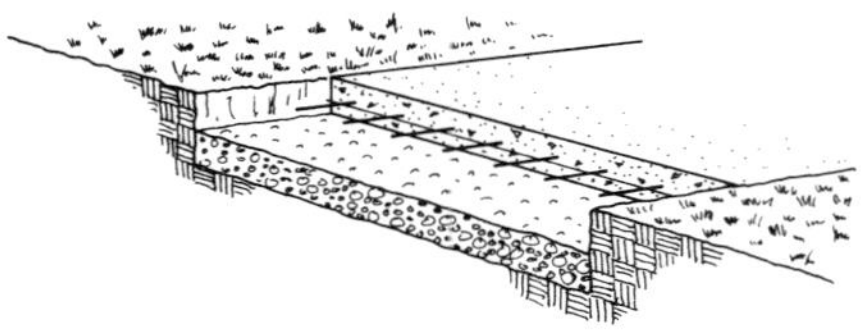

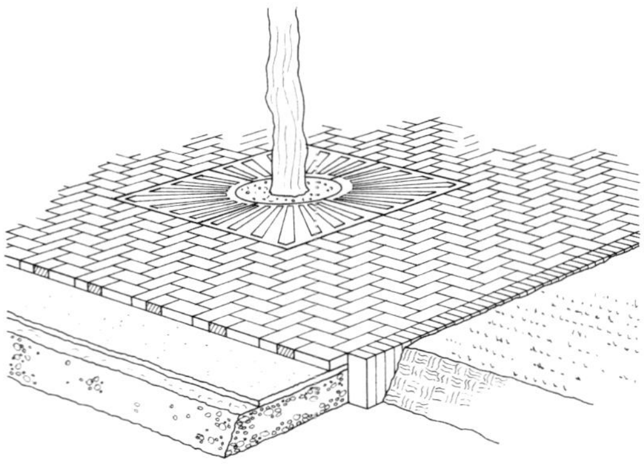

The Bituminous and Concrete Sidewalk Systems include excavation, hand grading and compacted gravel base. Pavements are shown for two conditions of each of the following variables; pavement thickness, gravel base thickness and pavement width. Costs are given on a linear foot basis.

The Plaza Systems listed include several brick and tile paving surfaces on two different bases: gravel and slab on grade. The type of bedding for the pavers depends on the base being used, and alternate bedding may be desirable. Also included in the paving costs are edging and precast grating costs and where concrete bases are involved, expansion joints. Costs are given on a square foot basis.

G2030 110	Bituminous Sidewalks	COST PER L.F.		
		MAT.	INST.	TOTAL
1580	Bituminous sidewalk, 1″ thick paving, 4″ gravel base, 3′ width	2.17	4.65	6.82
1600	4′ width	2.88	5.05	7.93
1640	6″ gravel base, 3′ width	2.60	4.89	7.49
1660	4′ width	3.45	5.35	8.80
2120	2″ thick paving, 4″ gravel base, 3′ width	3.49	5.60	9.09
2140	4′ width	4.65	6.25	10.90
2180	6″ gravel base, 3′ width	3.92	5.85	9.77
2200	4′ width	5.20	6.55	11.75

G2030 120	Concrete Sidewalks	COST PER L.F.		
		MAT.	INST.	TOTAL
1580	Concrete sidewalk, 4″ thick, 4″ gravel base, 3′ wide	6.70	12.20	18.90
1600	4′ wide	8.95	15.05	24
1640	6″ gravel base, 3′ wide	7.15	12.50	19.65
1660	4′ wide	9.50	15.40	24.90
2120	6″ thick concrete, 4″ gravel base, 3′ wide	10	14	24
2140	4′ wide	13.35	17.30	30.65
2180	6″ gravel base, 3′ wide	10.45	14.20	24.65
2200	4′ wide	13.90	17.65	31.55

G2030 150	Brick & Tile Plazas	COST PER S.F.		
		MAT.	INST.	TOTAL
2050	Brick pavers, 4″ x 8″ x 1-3/4″, gravel base, stone dust bedding	5.85	6	11.85
2100	Slab on grade, asphalt bedding	8.15	8.85	17
3550	Concrete paving stone, 4″ x 8″ x 2-1/2″, gravel base, sand bedding	4.51	4.28	8.79
3600	Slab on grade, asphalt bedding	6.35	6.60	12.95
4050	Concrete patio blocks, 8″ x 16″ x 2″, gravel base, sand bedding	4.63	5.50	10.13
4100	Slab on grade, asphalt bedding	6.70	8.20	14.90
6050	Granite pavers, 3-1/2″ x 3-1/2″ x 3-1/2″, gravel base, sand bedding	19.30	13.15	32.45
6100	Slab on grade, mortar bedding	22.50	21	43.50

G2040 Site Development

G2040 105	Fence & Guardrails	COST PER L.F.		
		MAT.	INST.	TOTAL
1000	Fence, chain link, 2" post, 1-5/8" rail, 9 ga. galv. wire, 5' high	21	5.75	26.75
1050	6' high w/barb. wire	21.50	7.20	28.70
1100	6 ga. galv. wire, 6' high	22	6.90	28.90
1150	6' high w/barb. wire	25	7.20	32.20
1200	1-3/8" rail, 11 ga., galv. wire, 10' high	25	15	40
1250	12' high	25	16.80	41.80
1300	9 ga. vinyl covered wire, 10' high	24	15	39
1350	12' high	28	16.80	44.80
1500	1-5/8" post, 1-3/8" rail, 11 ga. galv. wire, 3' high	2.38	3.43	5.81
1550	4' high	8.35	4.30	12.65
1600	9 ga. vinyl covered wire, 3' high	8.80	3.43	12.23
1650	4' high	9.15	4.30	13.45
2000	Guardrail, corrugated steel, galvanized steel posts	26.50	3.36	29.86
2100	Timber, 6" x 8" posts, 4" x 8" rails	13.45	2.97	16.42
2200	Steel box beam, 6" x 6" rails	38	24	62
2250	6" x 8" rails	50	13.30	63.30
3000	Barrier, median, precast concrete, single face	60.50	11.85	72.35
3050	Double face	69.50	13.25	82.75

G2040 810	Flagpoles	COST EACH		
		MAT.	INST.	TOTAL
0110	Flagpoles, on grade, aluminum, tapered, 20' high	1,175	720	1,895
0120	70' high	9,700	1,800	11,500
0130	Fiberglass, tapered, 23' high	625	720	1,345
0140	59' high	5,050	1,600	6,650
0150	Concrete, internal halyard, 20' high	1,500	575	2,075
0160	100' high	19,900	1,450	21,350

G2040 950	Other Site Development, EACH	COST EACH		
		MAT.	INST.	TOTAL
0110	Grandstands, permanent, closed deck, steel, economy (per seat)			24
0120	Deluxe (per seat)			27
0130	Composite design, economy (per seat)			46.50
0140	Deluxe (per seat)			82.50

G2040 Site Development

G2040 990	Site Development Components for Buildings	COST EACH		
		MAT.	INST.	TOTAL
0970	Assume minimal and balanced cut & fill, no rock, no demolition, no haz mat.			
0975	Lines can be adjusted linearly +/- 20% within same use & number of floors.			
0980	60,000 S.F. 2-Story Office Bldg on 3.3 Acres w/20% Green Space			
1000	Site Preparation	225	48,700	48,925
1002	Utilities	90,500	105,500	196,000
1004	Pavement	322,500	201,000	523,500
1006	Stormwater Management	297,000	59,500	356,500
1008	Sidewalks	14,600	19,100	33,700
1010	Exterior lighting	48,800	39,000	87,800
1014	Landscaping	73,500	52,000	125,500
1028	80,000 S.F. 4-Story Office Bldg on 3.5 Acres w/20% Green Space			
1030	Site preparation	230	52,500	52,730
1032	Utilities	84,000	95,500	179,500
1034	Pavement	387,000	240,500	627,500
1036	Stormwater Managmement	313,000	62,500	375,500
1038	Sidewalks	11,900	15,700	27,600
1040	Lighting	48,300	41,200	89,500
1042	Landscaping	75,500	54,000	129,500
1058	16,000 S.F. 1-Story Medical Office Bldg on 1.7 Acres w/25% Green Space			
1060	Site preparation	162	37,400	37,562
1062	Utilities	65,000	77,500	142,500
1064	Pavement	150,000	94,500	244,500
1066	Stormwater Management	145,000	29,100	174,100
1068	Sidewalks	10,600	14,000	24,600
1070	Lighting	30,000	22,600	52,600
1072	Landscaping	49,700	32,000	81,700
1098	36,000 S.F. 3-Story Apartment Bldg on 1.6 Acres w/25% Green Space			
1100	Site preparation	157	37,000	37,157

G2040 Site Development

G2040 990	Site Development Components for Buildings	COST EACH		
		MAT.	INST.	TOTAL
1102	Utilities	60,500	70,500	131,000
1104	Pavement	151,500	95,000	246,500
1106	Stormwater Management	135,000	27,100	162,100
1108	Sidewalks	9,225	12,100	21,325
1110	Lighting	28,000	21,500	49,500
1112	Landscaping	47,600	30,400	78,000
1128	130,000 S.F. 3-Story Hospital Bldg on 5.9 Acres w/17% Green Space			
1130	Site preparation	300	68,500	68,800
1132	Utilities	120,500	138,000	258,500
1134	Pavement	658,500	409,000	1,067,500
1136	Stormwater Management	556,000	111,500	667,500
1138	Sidewalks	17,500	23,000	40,500
1140	Lighting	79,500	69,500	149,000
1142	Landscaping	106,000	82,000	188,000
1158	60,000 S.F. 1-Story Light Manufacturing Bldg on 3.9 Acres w/18% Green Space			
1160	Site preparation	246	52,000	52,246
1162	Utilities	108,500	131,000	239,500
1164	Pavement	311,000	174,000	485,000
1166	Stormwater Management	366,000	73,500	439,500
1168	Sidewalks	20,600	27,100	47,700
1170	Lighting	59,000	44,600	103,600
1172	Landscaping	79,500	58,000	137,500
1198	8,000 S.F. 1-Story Restaurant Bldg on 1.4 Acres w/26% Green Space			
1200	Site preparation	148	36,100	36,248
1202	Utilities	54,500	61,500	116,000
1204	Pavement	143,500	91,500	235,000
1206	Stormwater Management	119,000	23,900	142,900
1208	Sidewalks	7,525	9,875	17,400
1210	Lighting	24,900	19,800	44,700
1212	Landscaping	44,700	27,900	72,600
1228	20,000 S.F. 1-Story Retail Store Bldg on 2.1 Acres w/23% Green Space			
1230	Site preparation	181	43,200	43,381
1232	Utilities	72,500	85,500	158,000
1234	Pavement	195,000	122,500	317,500
1236	Stormwater Management	185,500	37,300	222,800
1238	Sidewalks	11,900	15,700	27,600
1240	Lighting	34,400	26,200	60,600
1242	Landscaping	56,500	37,700	94,200
1258	60,000 S.F. 1-Story Warehouse on 3.1 Acres w/19% Green Space			
1260	Site preparation	219	47,800	48,019
1262	Utilities	98,000	123,000	221,000
1264	Pavement	189,500	98,000	287,500
1266	Stormwater Management	285,000	57,500	342,500
1268	Sidewalks	20,600	27,100	47,700
1270	Lighting	49,700	34,700	84,400
1272	Landscaping	68,500	48,100	116,600

G3010 Water Supply

G3010 110	Water Distribution Piping	MAT.	INST.	TOTAL
		COST PER L.F.		
2000	Piping, excav. & backfill excl., ductile iron class 250, mech. joint			
2130	4″ diameter	33.50	17	50.50
2150	6″ diameter	35	21.50	56.50
2160	8″ diameter	49	25.50	74.50
2170	10″ diameter	64	29.50	93.50
2180	12″ diameter	87	32.50	119.50
2210	16″ diameter	104	47	151
2220	18″ diameter	139	49	188
3000	Tyton joint			
3130	4″ diameter	17.05	8.50	25.55
3150	6″ diameter	20	10.20	30.20
3160	8″ diameter	30.50	17	47.50
3170	10″ diameter	40	18.70	58.70
3180	12″ diameter	42	21.50	63.50
3210	16″ diameter	45	29.50	74.50
3220	18″ diameter	50	34	84
3230	20″ diameter	52	38.50	90.50
3250	24″ diameter	65	44.50	109.50
4000	Copper tubing, type K			
4050	3/4″ diameter	7.30	3.22	10.52
4060	1″ diameter	9.70	4.02	13.72
4080	1-1/2″ diameter	15.50	4.85	20.35
4090	2″ diameter	24	5.60	29.60
4110	3″ diameter	49	9.60	58.60
4130	4″ diameter	74.50	13.55	88.05
4150	6″ diameter	161	25	186
5000	Polyvinyl chloride class 160, S.D.R. 26			
5130	1-1/2″ diameter	.55	1.20	1.75
5150	2″ diameter	.80	1.31	2.11
5160	4″ diameter	3	5.90	8.90
5170	6″ diameter	5.75	7.10	12.85
5180	8″ diameter	8.85	8.50	17.35
6000	Polyethylene 160 psi, S.D.R. 7			
6050	3/4″ diameter	.37	1.71	2.08
6060	1″ diameter	.58	1.85	2.43
6080	1-1/2″ diameter	1.16	2	3.16
6090	2″ diameter	1.93	2.46	4.39

G3020 Sanitary Sewer

G3020 110	Drainage & Sewage Piping	COST PER L.F.		
		MAT.	INST.	TOTAL
2000	Piping, excavation & backfill excluded, PVC, plain			
2130	4″ diameter	1.64	4.19	5.83
2150	6″ diameter	3.61	4.48	8.09
2160	8″ diameter	6.75	4.69	11.44
2900	Box culvert, precast, 8′ long			
3000	6′ x 3′	275	35	310
3020	6′ x 7′	345	39	384
3040	8′ x 3′	345	43	388
3060	8′ x 8′	420	49	469
3080	10′ x 3′	480	44.50	524.50
3100	10′ x 8′	545	61	606
3120	12′ x 3′	710	49	759
3140	12′ x 8′	895	73	968
4000	Concrete, nonreinforced			
4150	6″ diameter	8	12.60	20.60
4160	8″ diameter	8.75	14.95	23.70
4170	10″ diameter	9.75	15.50	25.25
4180	12″ diameter	10.95	16.75	27.70
4200	15″ diameter	15.40	18.60	34
4220	18″ diameter	18.65	23.50	42.15
4250	24″ diameter	21	33.50	54.50
4400	Reinforced, no gasket			
4580	12″ diameter	13.35	22.50	35.85
4600	15″ diameter	17.25	22.50	39.75
4620	18″ diameter	22	25.50	47.50
4650	24″ diameter	24	33.50	57.50
4670	30″ diameter	42	49.50	91.50
4680	36″ diameter	65.50	60.50	126
4690	42″ diameter	89	66	155
4700	48″ diameter	108	75	183
4720	60″ diameter	162	99.50	261.50
4730	72″ diameter	243	120	363
4740	84″ diameter	335	150	485
4800	With gasket			
4980	12″ diameter	14.70	12.15	26.85
5000	15″ diameter	19	12.75	31.75
5020	18″ diameter	24.50	13.40	37.90
5050	24″ diameter	31.50	15	46.50
5070	30″ diameter	50.50	49.50	100
5080	36″ diameter	75.50	60.50	136
5090	42″ diameter	97.50	60	157.50
5100	48″ diameter	122	75	197
5120	60″ diameter	194	96	290
5130	72″ diameter	263	120	383
5140	84″ diameter	400	199	599
5700	Corrugated metal, alum. or galv. bit. coated			
5760	8″ diameter	9.65	10.10	19.75
5770	10″ diameter	10	12.90	22.90
5780	12″ diameter	12.20	15.95	28.15
5800	15″ diameter	16.85	16.75	33.60
5820	18″ diameter	22	17.60	39.60
5850	24″ diameter	27	21	48
5870	30″ diameter	32.50	36.50	69
5880	36″ diameter	38	36.50	74.50
5900	48″ diameter	57	44	101
5920	60″ diameter	87.50	63.50	151
5930	72″ diameter	105	106	211
6000	Plain			

G3020 Sanitary Sewer

G3020 110	Drainage & Sewage Piping	COST PER L.F.		
		MAT.	INST.	TOTAL
6060	8″ diameter	8.65	9.45	18.10
6070	10″ diameter	9.90	11.95	21.85
6080	12″ diameter	11	15.25	26.25
6100	15″ diameter	13.75	15.25	29
6120	18″ diameter	16.60	16.30	32.90
6140	24″ diameter	24	19.15	43.15
6170	30″ diameter	29.50	33.50	63
6180	36″ diameter	34	33.50	67.50
6200	48″ diameter	51	39.50	90.50
6220	60″ diameter	78.50	61.50	140
6230	72″ diameter	178	156	334
6300	Steel or alum. oval arch, coated & paved invert			
6400	15″ equivalent diameter	15.95	16.75	32.70
6420	18″ equivalent diameter	21	22.50	43.50
6450	24″ equivalent diameter	27	26.50	53.50
6470	30″ equivalent diameter	32.50	33.50	66
6480	36″ equivalent diameter	39	44	83
6490	42″ equivalent diameter	47.50	48.50	96
6500	48″ equivalent diameter	61.50	58	119.50
6600	Plain			
6700	15″ equivalent diameter	14.50	14.90	29.40
6720	18″ equivalent diameter	17.65	19.15	36.80
6750	24″ equivalent diameter	24.50	22.50	47
6770	30″ equivalent diameter	29.50	40.50	70
6780	36″ equivalent diameter	35.50	40.50	76
6790	42″ equivalent diameter	42	47.50	89.50
6800	48″ equivalent diameter	59	58	117
8000	Polyvinyl chloride SDR 35			
8130	4″ diameter	1.64	4.19	5.83
8150	6″ diameter	3.61	4.48	8.09
8160	8″ diameter	6.75	4.69	11.44
8170	10″ diameter	12.20	6.15	18.35
8180	12″ diameter	14	6.40	20.40
8200	15″ diameter	14.30	8.50	22.80

G3030 Storm Sewer

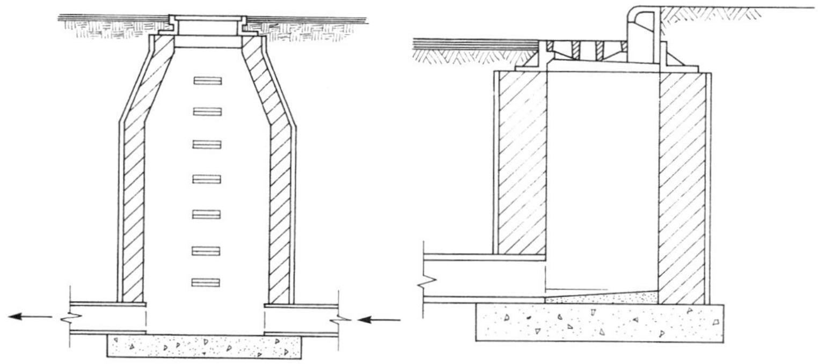

The Manhole and Catch Basin System includes: excavation with a backhoe; a formed concrete footing; frame and cover; cast iron steps and compacted backfill.

The Expanded System Listing shows manholes that have a 4′, 5′ and 6′ inside diameter riser. Depths range from 4′ to 14′. Construction material shown is either concrete, concrete block, precast concrete, or brick.

System Components	QUANTITY	UNIT	COST PER EACH		
			MAT.	INST.	TOTAL
SYSTEM G3030 210 1920					
MANHOLE/CATCH BASIN, BRICK, 4′ I.D. RISER, 4′ DEEP					
Excavation, hydraulic backhoe, 3/8 C.Y. bucket	14.815	B.C.Y.		119.71	119.71
Trim sides and bottom of excavation	64.000	S.F.		62.72	62.72
Forms in place, manhole base, 4 uses	20.000	SFCA	15.40	109	124.40
Reinforcing in place footings, #4 to #7	.019	Ton	19.95	23.75	43.70
Concrete, 3000 psi	.925	C.Y.	109.15		109.15
Place and vibrate concrete, footing, direct chute	.925	C.Y.		49.73	49.73
Catch basin or MH, brick, 4′ ID, 4′ deep	1.000	Ea.	590	1,025	1,615
Catch basin or MH steps; heavy galvanized cast iron	1.000	Ea.	21	14.10	35.10
Catch basin or MH frame and cover	1.000	Ea.	270	247.50	517.50
Fill, granular	12.954	L.C.Y.	310.90		310.90
Backfill, spread with wheeled front end loader	12.954	L.C.Y.		34.46	34.46
Backfill compaction, 12″ lifts, air tamp	12.954	E.C.Y.		118.27	118.27
TOTAL			1,336.40	1,804.24	3,140.64

G3030 210	Manholes & Catch Basins	COST PER EACH		
		MAT.	INST.	TOTAL
1920	Manhole/catch basin, brick, 4′ I.D. riser, 4′ deep	1,325	1,800	3,125
1940	6′ deep	1,900	2,500	4,400
1960	8′ deep	2,525	3,425	5,950
1980	10′ deep	2,975	4,225	7,200
3000	12′ deep	3,725	4,600	8,325
3020	14′ deep	4,575	6,425	11,000
3200	Block, 4′ I.D. riser, 4′ deep	1,175	1,450	2,625
3220	6′ deep	1,650	2,050	3,700
3240	8′ deep	2,200	2,850	5,050
3260	10′ deep	2,600	3,500	6,100
3280	12′ deep	3,325	4,450	7,775
3300	14′ deep	4,150	5,450	9,600
4620	Concrete, cast-in-place, 4′ I.D. riser, 4′ deep	1,325	2,525	3,850
4640	6′ deep	1,900	3,375	5,275
4660	8′ deep	2,675	4,875	7,550
4680	10′ deep	3,225	6,050	9,275
4700	12′ deep	4,050	7,500	11,550
4720	14′ deep	5,000	8,975	13,975
5820	Concrete, precast, 4′ I.D. riser, 4′ deep	1,550	1,350	2,900
5840	6′ deep	2,550	1,800	4,350

G3030 Storm Sewer

G3030 210	Manholes & Catch Basins	COST PER EACH MAT.	COST PER EACH INST.	COST PER EACH TOTAL
5860	8′ deep	3,425	2,500	5,925
5880	10′ deep	4,050	3,100	7,150
5900	12′ deep	4,975	3,825	8,800
5920	14′ deep	6,000	4,850	10,850
6000	5′ I.D. riser, 4′ deep	2,575	1,525	4,100
6020	6′ deep	3,250	2,125	5,375
6040	8′ deep	4,500	2,825	7,325
6060	10′ deep	5,825	3,575	9,400
6080	12′ deep	7,250	4,550	11,800
6100	14′ deep	8,800	5,550	14,350
6200	6′ I.D. riser, 4′ deep	3,425	2,000	5,425
6220	6′ deep	4,500	2,650	7,150
6240	8′ deep	5,600	3,700	9,300
6260	10′ deep	7,225	4,700	11,925
6280	12′ deep	8,975	5,875	14,850
6300	14′ deep	10,900	7,125	18,025

G3060 Fuel Distribution

G3060 110	Gas Service Piping	COST PER L.F.		
		MAT.	INST.	TOTAL
2000	Piping, excavation & backfill excluded, polyethylene			
2070	1-1/4" diam, SDR 10	1.78	4.12	5.90
2090	2" diam, SDR 11	2.48	4.59	7.07
2110	3" diam, SDR 11.5	5.15	5.50	10.65
2130	4" diam,SDR 11	14.85	9.65	24.50
2150	6" diam, SDR 21	37	10.35	47.35
2160	8" diam, SDR 21	51	12.50	63.50
3000	Steel, schedule 40, plain end, tarred & wrapped			
3060	1" diameter	5.50	9.25	14.75
3090	2" diameter	8.65	9.95	18.60
3110	3" diameter	14.30	10.70	25
3130	4" diameter	18.60	16.65	35.25
3140	5" diameter	27	19.35	46.35
3150	6" diameter	33	23.50	56.50
3160	8" diameter	52.50	30.50	83
3170	10" diameter	139	42.50	181.50
3180	12" diameter	154	53.50	207.50
3190	14" diameter	164	57	221
3200	16" diameter	180	60.50	240.50
3210	18" diameter	231	65.50	296.50
3220	20" diameter	360	71	431
3230	24" diameter	410	85	495

G4020 Site Lighting

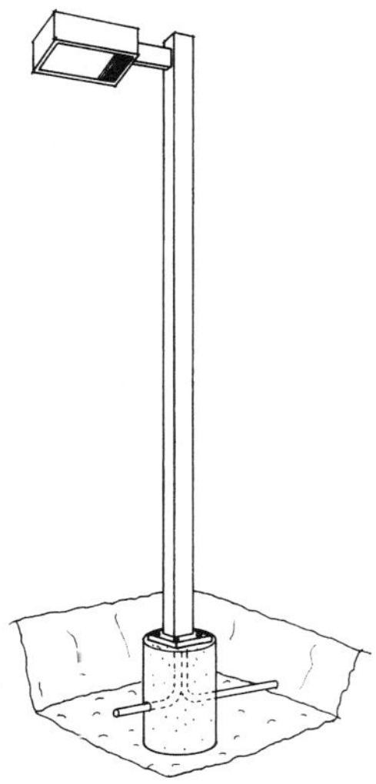

The Site Lighting System includes the complete unit from foundation to electrical fixtures. Each system includes: excavation; concrete base; backfill by hand; compaction with a plate compacter; pole of specified material; all fixtures; and lamps.

The Expanded System Listing shows Site Lighting Systems that use one of two types of lamps: high pressure sodium; and metal halide. Systems are listed for 400-watt and 1000-watt lamps. Pole height varies from 20′ to 40′. There are four types of poles possibly listed: aluminum, fiberglass, steel and wood.

G4020 110	Site Lighting	COST EACH		
		MAT.	INST.	TOTAL
2320	Site lighting, high pressure sodium, 400 watt, aluminum pole, 20′ high	1,800	1,425	3,225
2360	40′ high	3,250	2,275	5,525
2920	Wood pole, 20′ high	995	1,400	2,395
2960	40′ high	1,575	2,175	3,750
3120	1000 watt, aluminum pole, 20′ high	1,875	1,475	3,350
3160	40′ high	3,350	2,300	5,650
3520	Wood pole, 20′ high	1,075	1,425	2,500
3560	40′ high	1,675	2,200	3,875
5820	Metal halide, 400 watt, aluminum pole, 20′ high	1,775	1,425	3,200
5860	40′ high	3,225	2,275	5,500
7620	1000 watt, aluminum pole, 20′ high	1,850	1,475	3,325
7660	40′ high	3,300	2,300	5,600

G4020 Site Lighting

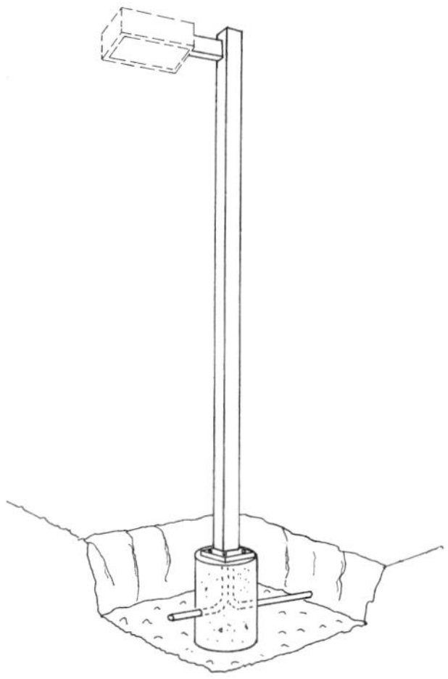

Table G4020 210 Procedure for Calculating Floodlights Required for Various Footcandles
Poles should not be spaced more than 4 times the fixture mounting height for good light distribution.

Estimating Chart
Select Lamp type.

Determine total square feet.

Chart will show quantity of fixtures to provide 1 footcandle initial, at intersection of lines. Multiply fixture quantity by desired footcandle level.

Chart based on use of wide beam luminaires in an area whose dimensions are large compared to mounting height and is approximate only.

To maintain 1 footcandle over a large area use these watts per square foot:

Incandescent	0.15
Metal Halide	0.032
Mercury Vapor	0.05
High Pressure Sodium	0.024

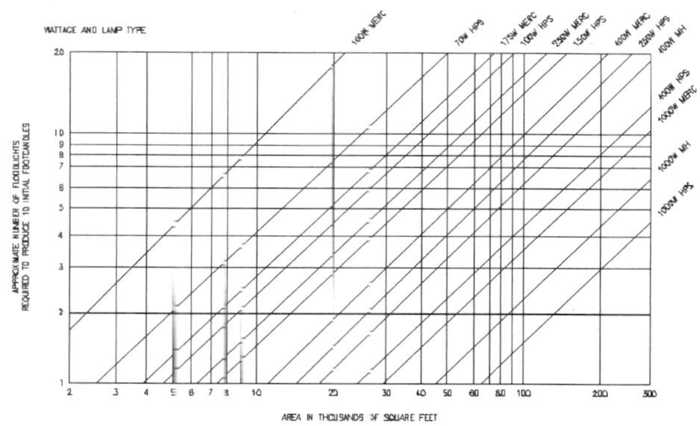

System Components	QUANTITY	UNIT	COST EACH MAT.	INST.	TOTAL
SYSTEM G4020 210 0200					
LIGHT POLES, ALUMINUM, 20′ HIGH, 1 ARM BRACKET					
Aluminum light pole, 20′, no concrete base	1.000	Ea.	1,050	618	1,668
Bracket arm for Aluminum light pole	1.000	Ea.	135	82.50	217.50
Excavation by hand, pits to 6′ deep, heavy soil or clay	2.368	C.Y.		274.69	274.69
Footing, concrete incl forms, reinforcing, spread, under 1 C.Y.	.465	C.Y.	87.42	131.71	219.13
Backfill by hand	1.903	C.Y.		80.88	80.88
Compaction vibrating plate	1.903	C.Y.		10.88	10.88
TOTAL			1,272.42	1,198.66	2,471.08

G4020 210	Light Pole (Installed)	COST EACH MAT.	INST.	TOTAL
0200	Light pole, aluminum, 20′ high, 1 arm bracket	1,275	1,200	2,475
0240	2 arm brackets	1,400	1,200	2,600
0280	3 arm brackets	1,550	1,225	2,775
0320	4 arm brackets	1,675	1,225	2,900
0360	30′ high, 1 arm bracket	2,225	1,500	3,725
0400	2 arm brackets	2,350	1,500	3,850
0440	3 arm brackets	2,500	1,550	4,050
0480	4 arm brackets	2,625	1,550	4,175
0680	40′ high, 1 arm bracket	2,725	2,025	4,750
0720	2 arm brackets	2,850	2,025	4,875
0760	3 arm brackets	2,975	2,050	5,025
0800	4 arm brackets	3,125	2,050	5,175
0840	Steel, 20′ high, 1 arm bracket	1,425	1,250	2,675
0880	2 arm brackets	1,475	1,250	2,725
0920	3 arm brackets	1,450	1,300	2,750
0960	4 arm brackets	1,550	1,300	2,850
1000	30′ high, 1 arm bracket	1,750	1,625	3,375
1040	2 arm brackets	1,800	1,625	3,425
1080	3 arm brackets	1,800	1,650	3,450
1120	4 arm brackets	1,900	1,650	3,550
1320	40′ high, 1 arm bracket	2,300	2,175	4,475
1360	2 arm brackets	2,350	2,175	4,525
1400	3 arm brackets	2,350	2,225	4,575
1440	4 arm brackets	2,450	2,225	4,675

Reference Section

All the reference information is in one section, making it easy to find what you need to know . . . and easy to use the data set on a daily basis. This section is visually identified by a vertical black bar on the page edges.

In this section, you'll see the background that relates to the reference numbers that appeared in the Assembly Cost Sections. You'll find reference tables, explanations, and estimating information that support how we derive the systems data. Also included are alternate pricing methods, technical data, and estimating procedures, along with information on design and economy in construction.

Also in this Reference Section, we've included Historical Cost Indexes for cost comparison over time; City Cost Indexes and Location Factors for adjusting costs to the region you are in; Reference Aids; Estimating Forms; and an explanation of all Abbreviations used in the data set.

Table of Contents

General: A spread footing is used to convert a concentrated load (from one superstructure column or substructure grade beams) into an allowable area load on supporting soil.

Because of punching action from the column load, a spread footing is usually thicker than strip footings which support wall loads. One or two story commercial or residential buildings should have no less than 1′ thick spread footings. Heavier loads require no less than 2′ thick. Spread footings may be square, rectangular or octagonal in plan.

Spread footings tend to minimize excavation and foundation materials, as well as labor and equipment. Another advantage is that footings and soil conditions can be readily examined. They are the most widely used type of footing, especially in mild climates and for buildings of four stories or under. This is because they are usually more economical than other types, if suitable soil and site conditions exist.

They are used when suitable supporting soil is located within several feet of the surface or line of subsurface excavation. Suitable soil types include sands and gravels, gravels with a small amount of clay or silt, hardpan, chalk, and rock. Pedestals may be used to bring the column base load down to the top of the footing. Alternately, undesirable soil between the underside of the footing and the top of the bearing level can be removed and replaced with lean concrete mix or compacted granular material.

Depth of footing should be below topsoil, uncompacted fill, muck, etc. It must be lower than frost penetration (see local code or Table L1030-502 in Section L) but should be above the water table. It must not be at the ground surface because of potential surface erosion. If the ground slopes, approximately three horizontal feet of edge protection must remain. Differential footing elevations may overlap soil stresses or cause excavation problems if clear spacing between footings is less than the difference in depth.

Other footing types are usually used for the following reasons:

A. Bearing capacity of soil is low.
B. Very large footings are required, at a cost disadvantage.
C. Soil under footing (shallow or deep) is very compressible, with probability of causing excessive or differential settlement.
D. Good bearing soil is deep.
E. Potential for scour action exists.
F. Varying subsoil conditions within building perimeter.

Cost of spread footings for a building is determined by:

1. The soil bearing capacity.
2. Typical bay size.
3. Total load (live plus dead) per S.F. for roof and elevated floor levels.
4. The size and shape of the building.
5. Footing configuration. Does the building utilize outer spread footings or are there continuous perimeter footings only or a combination of spread footings plus continuous footings?

COST DETERMINATION

1. Determine Soil Bearing Capacity by a known value or by using Table A1010-121 as a guide.

Table A1010-121 Soil Bearing Capacity in Kips per S.F.

Bearing Material	Typical Allowable Bearing Capacity
Hard sound rock	120 KSF
Medium hard rock	80
Hardpan overlaying rock	24
Compact gravel and boulder-gravel; very compact sandy gravel	20
Soft rock	16
Loose gravel; sandy gravel; compact sand; very compact sand-inorganic silt	12
Hard dry consolidated clay	10
Loose coarse to medium sand; medium compact fine sand	8
Compact sand-clay	6
Loose fine sand; medium compact sand-inorganic silts	4
Firm or stiff clay	3
Loose saturated sand-clay; medium soft clay	2

Table A1010-122 Working Load Determination

2. Determine Bay Size in S.F.

3. Determine Load to the Footing

Type Load	Reference	Working Loads			
		Roof	Floor	Column	Total
1. Total Load	Floor System	PSF	PSF		
2. Whole Bay Load/Level	(Line 1) x Bay Area ÷ 1000	Kips	Kips		
3. Load to Column	[(Roof + (Floor x No. Floors)]	Kips	Kips		Kips
4. Column Weight	Systems B1010 201 to B1010 208			Kips	Kips
5. Fireproofing	System B1010 720			Kips	Kips
6. Total Load to Footing	Roof + Floor + Column				Kips

Working loads represent the highest actual load a structural member is designed to support.

4. Determine Size and Shape of Building

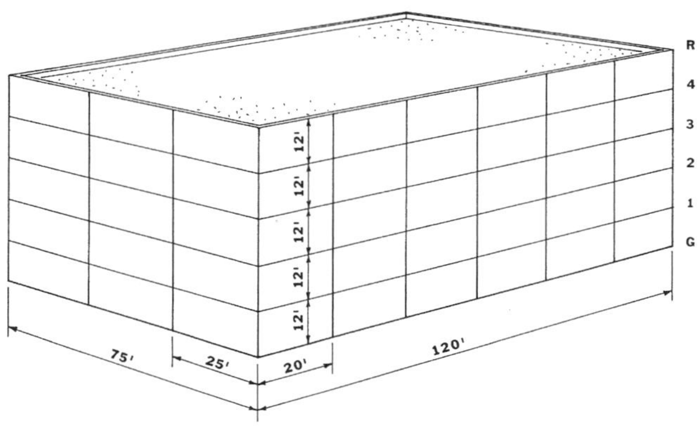

Figure A1010-123

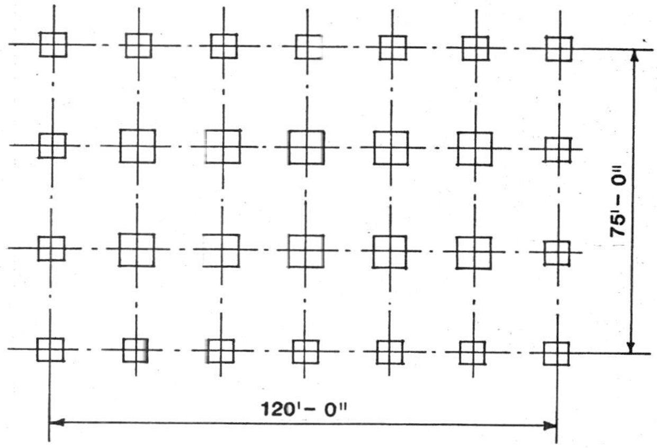

Figure A1010-124

5. Footing Configuration Possibilities

a. Building has spread footings throughout (including exterior and corners).
b. Building has continuous perimeter footings to support exterior columns, corners and exterior walls (as well as interior spread footings).
c. Building has both perimeter continuous footings for walls plus exterior spread footings for column loads. If there are interior spread footings also, use "b."
   1. Figure continuous footings from tables in Section A1010-110.
   2. Add total dollars for spread footings and continuous footings to determine total foundation dollars.

Table A1010-126 Working Load Determination

Tabulate interior footing costs and add continuous footing costs from Systems A1010 210 and A1010 110.

Example: Spread Footing Cost Determination

Office Building	5 Story
Story Height	12′
Dimensions	75′ x 120′
Bay Size	20′ x 25′
Soil Bearing Capacity	6 KSF
Foundation	Spread Footings
Slab on Grade	5″ Reinforced
Code	IBC
Construction	From Example RB1010-112 Using concrete flat plate

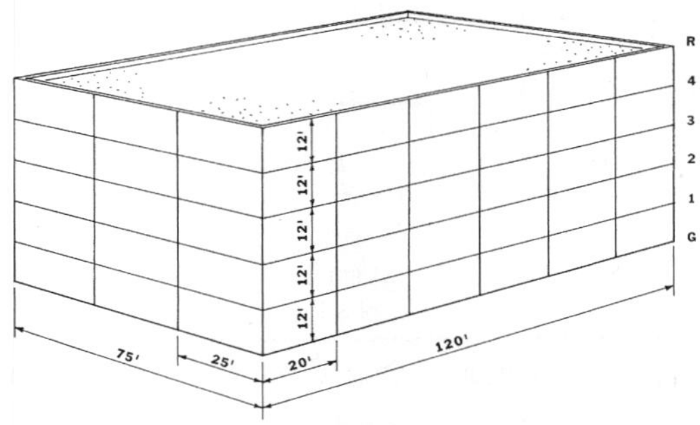

Figure A1010-125

Determine load to footings Table A1010-122 or Example RB1010-112.

Type Load	Reference	Working Loads			
		Roof	Floor	Column	Total
Determine floor system	System in Div. B1010 (Example: Flat Plate)				
1. Total Load	System # B1010 223	146PSF	188PSF		
2. Whole Bay Load/Level	Bay Area x Line 1/1000	73K	94K		
3. Load To Column	Roof + (Floor x No. Floors)	73K	276K		449K
4. Column Weight	Systems B1010 201 to B1010 208			25K	25K
5. Fireproofing	N/A			—	
6. Load to Footing	Roof + Floor + Column				474K

1. Total load working from System B1010 223: 40# superimposed load for roof and 75# superimposed load for floors.
2. Bay Area 20′ x 25′ = 500 S.F. x Line 1
   500 x 146/1000 = 73K roof
   500 x 188/1000 = 94K floor
3. Roof + (floor x no. floors)
   73K + (4 x 94K) = 449K
4. Enter System B1010 203 with a total working load of 474K and 12′ story height. A 16″ square tied column with 6 KSI concrete is most economical. Minimum column size from System B1010 223 is 20″. Use the larger. Column weight = 394 #/V.L.F. x 12′ = approx. 5K.
5. No fireproofing is required for concrete column.
6. Add roof, floors and column loads for a total load to each interior footing.

Footing Cost: Enter System A1010 210 with total working load and allowable soil pressure. Determine the cost per footing using the closest higher load on the table. Add the total costs for interior, exterior, and corner footings.

Table A1010-127 Typical Spread Footing, Square

EXTERIOR FOOTINGS

Exterior vs. Interior Footings

System A1010 210 contains the cost for individual spread interior footings for various loading and soil bearing conditions. To determine the loads for exterior footings:

1. Multiply whole bay working load by 60% for half bay footings and by 45% for quarter bay (or corner) footings. This will give the approximate column load for half and quarter bay footings.
2. Enter System A1010 210 and find cost of footing with loads determined in step 1.
3. Determine the number of each type of footing on the job and multiply the cost per footing by the appropriate quantity. This will give the total dollar cost for exterior and interior spread footings.

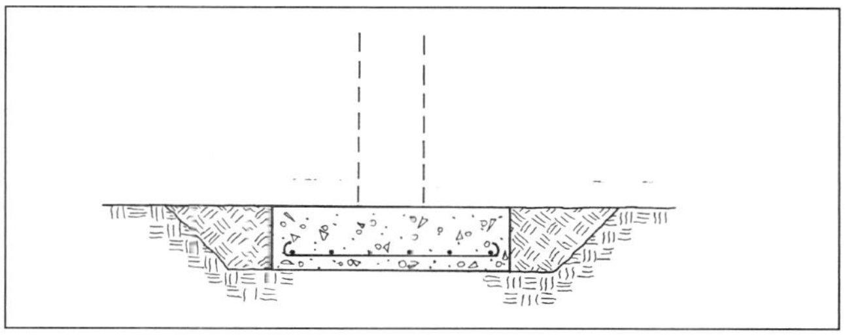

Example: Using soil capacity, 5KSF
Total working load from Table A1010-126 . . . . = 474K
Half Bay Footings . . . . 474K x 60% = 284K
Corner Footings . . . . 474K x 45% = 213K

Table A1010-128 Spread Footing Cost

Footing Type	Working Load (Kips)	Total Quantity	System A1010 210 Unit Cost	Total
Interior	474	10	$2,650	$26,500
Exterior	284	14	1,475	20,650
Corner	213	4	1,475	5,900
Total				$53,050

Table A1010-141 Section Through Typical Strip Footing

General: Strip (continuous) footings are a type of spread footing used to convert the lineal type wall into an allowable area load on supporting soil.

Many of the general comments on isolated spread footings (Section A1010 210) apply here as well.

Strip footings may be used under walls of concrete, brick, block or stone. They are constructed of continuously placed concrete, 2000 psi or greater. Normally they are not narrower than twice the wall thickness, nor are they thinner than the wall thickness. Plain concrete footings should be at least 8″ thick, and reinforced footings require at least 6″ of thickness above the bottom reinforcement. Where there is no basement, the bottom of footing is usually 3′ to 5′ below finished exterior grade, and 12″ below the average frost penetration (see Figure A1010-141), resting on undisturbed soil. In some cases strip footings serve as leveling pads (stone walls, etc).

Method of construction is affected by reinforcement and soil type. If steel mats and/or dowels are present and require accurate placement (frequently for reinforced concrete or reinforced masonry walls), then side forms are required, as they are in wet or sandy soil. If there is no reinforcement or only longitudinal reinforcement, and the soil will "stand" forms might be eliminated and economically should be.

Preferably, strip footings are used on sand and gravel. The presence of a small amount of clay or dense silty sand in gravel is acceptable, as is bearing on rock or chalk. They are sometimes used in uniform, firm or stiff clays with little nearby ground vegetation, and placed at least 3-1/2′ into the clay. If the clay is sloped, there is potential for downhill creep. Footings up to 3-1/2′ wide are sometimes used in soft or silty clays, but settlement must be expected and provision made.

Sloping terrain requires stepping the footing to maintain depth, since a sloped footing will create horizontal thrust which may distress the structure. Steps should be at least 2′ long horizontally and each vertical step no greater than three-fourths the horizontal distance between steps. Vertical risers must be at least 6″ thick and of footing width. Potential for erosion of surrounding soil must be considered.

Strip footings should be used under walls constructed of unreinforced concrete, or of units bonded with mortar. They are also used when bearing capacity of soil is inadequate to support wall thickness alone.

Footings require horizontal reinforcement at right angles to the wall line when side projection of footings beyond the face of the wall is greater than 1/2 to 2/3 footing thickness. Horizontal reinforcement parallel to the wall line should be used if soil bearing capacity is low, or soil compressibility is variable, or the footing spans pipe trenches, etc. Vertical reinforcing dowels depend upon wall design.

Alternate foundation types include bored piles, grade beams, and mat. Short, bored compacted concrete piles are sometimes competitive with strip footings, especially in shrinkable clay soils where they may provide an added factor of safety. When soil bearing capacity is inadequate, grade beams can be used to transfer wall loads across the inadequate ground to column support foundations. Mat footings replace both strip and isolated spread footings. They may eliminate differential settlement problems.

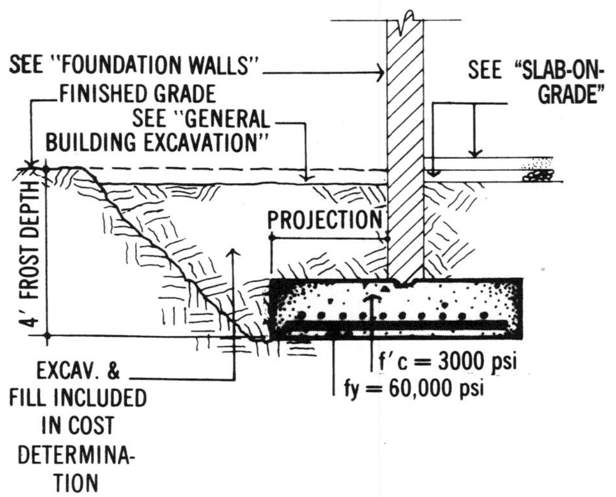

Figure A1010-141
Section Through Typical Strip Footing

Example: Strip Footing Cost Determination for Bearing Wall

Office Building . . . . 5 Story
Story Height . . . . 12 VLF
Dimensions . . . . 50′ x 80′ = 4000 S.F.G.
Exterior Wall . . . . 12″ Brick and Block
Interior Wall . . . . 8″ Block
Floor . . . . Prestressed Concrete Slabs + 2″ Topping
Roof . . . . Prestressed Concrete Slabs, No Topping
Soil Bearing Capacity . . . . 6 KSF
Superimposed Load:
floor . . . . 100 PSF
roof . . . . 40 PSF

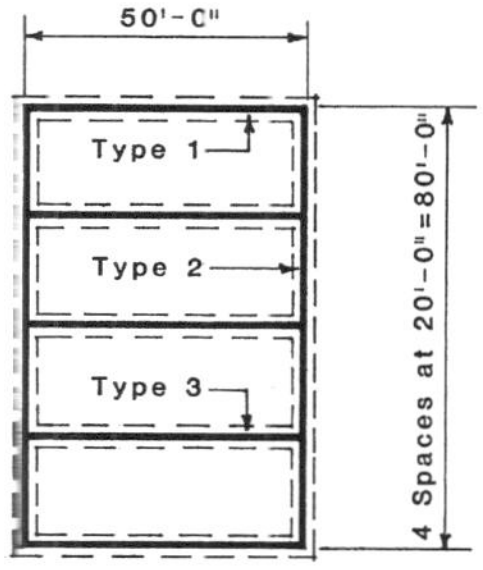

Figure A1010-142 Footing Plan

A. Total load to each footing

Table A1010-143 Strip Footing, Load Determination

Type Load	Reference	Working Loads Roof	Floor	Wall	Total
1. Total Load	System B1010 229	90 PSF	175 PSF	95 PSF	
2. Load Span or Height	Building Design	10 LF	10 LF	12 VLF	
3. Whole Bay Load/Level	(Line 1 x Line 2) ÷ 1000	.9 KIPS	1.75 KIPS	1.1 KIPS	
4. Load to Footing/L.F.	Roof + (Floor + Wall)	.9 KLF	7.0 KLF	5.9 KLF	13.8 KLF

1. Total load from System B1010-229.
2. Roof and floor load spans for exterior strip footings are 1/2 total span of 20′ or 10′ (wall height for our example is 12′).
3. Multiply Line 1 by Line 2 in each column and divide by 1,000 Lb/Kip.
4. Add 1 roof + 4 elevated floors + 5.33 stories of wall (includes .33 story for a frost wall) for the total load to the footing.

B. Enter System A1010-110 with a load of 13.8 KLF and interpolate between line items 2700 and 3100.

$$\text{Interpolating: } \$41.15 + \left[\left[\frac{13.8 - 11.1}{14.8 - 11.1} \right] \times (\$46.50 - \$41.15) \right] = \$45.00$$

C. If the building design incorporates other strip footing loads, repeat steps A and B for each type of footing load.

Table A1010-144 Cost Determination, Strip Footings

Description	Strip Footing Type or Size	Total Length Each Type (Ft.)	Cost From System A1010 110 ($/L.F.)	Cost x Length ($/Each Type)	Total Cost All Types
End Bearing Wall	1	100	$45.00	$ 4,500	
Non Bearing Wall	2	160	41.15	$ 6,584	
Interior Bearing Wall	3	150	92.50	$13,875	$24,959

Note: Type 2 wall: use "wall" loads only in Step A above.
Type 3 wall: same as Step A above with load span increased to 20′.
Interpolated values from System A1010 110 are rounded to $.05.

A10 Foundations | RA1010-330 Pile Caps

General: The function of a reinforced concrete pile cap is to transfer superstructure load from isolated column or pier to each pile in its supporting cluster. To do this, the cap must be thick and rigid, with all piles securely embedded into and bonded to it.

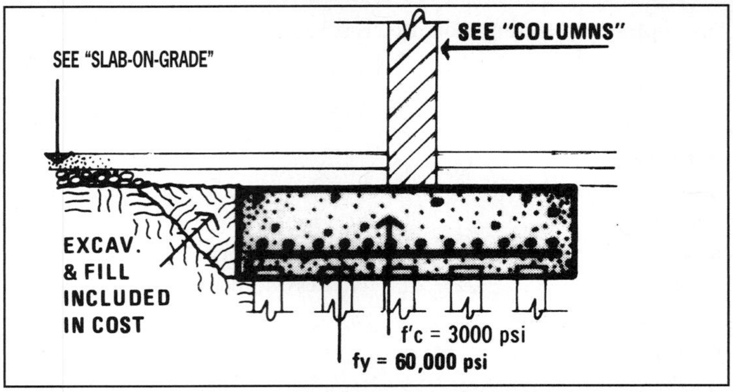

Figure A1010-331 Section Through Pile Cap

Table A1010-332 Concrete Quantities for Pile Caps

Load	Number of Piles @ 3'-0" O.C. Per Footing Cluster									
Working (K)	2 (CY)	4 (CY)	6 (CY)	8 (CY)	10 (CY)	12 (CY)	14 (CY)	16 (CY)	18 (CY)	20 (CY)
50	(.9)	(1.9)	(3.3)	(4.9)	(5.7)	(7.8)	(9.9)	(11.1)	(14.4)	(16.5)
100	(1.0)	(2.2)	(3.3)	(4.9)	(5.7)	(7.8)	(9.9)	(11.1)	(14.4)	(16.5)
200	(1.0)	(2.2)	(4.0)	(4.9)	(5.7)	(7.8)	(9.9)	(11.1)	(14.4)	(16.5)
400	(1.1)	(2.6)	(5.2)	(6.3)	(7.4)	(8.2)	(13.7)	(11.1)	(14.4)	(16.5)
800		(2.9)	(5.8)	(7.5)	(9.2)	(13.6)	(17.6)	(15.9)	(19.7)	(22.1)
1200			(5.8)	(8.3)	(9.7)	(14.2)	(18.3)	(20.4)	(21.2)	(22.7)
1600				(9.8)	(11.4)	(14.5)	(19.5)	(20.4)	(24.6)	(27.2)
2000				(9.8)	(11.4)	(16.6)	(24.1)	(21.7)	(26.0)	(28.8)
3000						(17.5)		(26.5)	(30.3)	(32.9)
4000								(30.2)	(30.7)	(36.5)

Table A1010-333 Concrete Quantities for Pile Caps

Load	Number of Piles @ 4'-6" O.C. Per Footing					
Working (K)	2 (CY)	3 (CY)	4 (CY)	5 (CY)	6 (CY)	7 (CY)
50	(2.3)	(3.6)	(5.6)	(11.0)	(13.7)	(12.9)
100	(2.3)	(3.6)	(5.6)	(11.0)	(13.7)	(12.9)
200	(2.3)	(3.6)	(5.6)	(11.0)	(13.7)	(12.9)
400	(3.0)	(3.6)	(5.6)	(11.0)	(13.7)	(12.9)
800			(6.2)	(11.5)	(14.0)	(12.9)
1200				(13.0)	(13.7)	(13.4)
1600						(14.0)

General: Piles are column-like shafts which receive superstructure loads, overturning forces, or uplift forces. They receive these loads from isolated column or pier foundations (pile caps), foundation walls, grade beams, or foundation mats. The piles then transfer these loads through shallower poor soil strata to deeper soil of adequate support strength and acceptable settlement with load.

Be sure that other foundation types aren't better suited to the job. Consider ground and settlement, as well as loading, when reviewing. Piles usually are associated with difficult foundation problems and substructure condition. Ground conditions determine type of pile (different pile types have been developed to suit ground conditions). Decide each case by technical study, experience, and sound engineering judgment—not rules of thumb. A full investigation of ground conditions, early, is essential to provide maximum information for professional foundation engineering and an acceptable structure.

Piles support loads by end bearing and friction. Both are generally present; however, piles are designated by their principal method of load transfer to soil.

Boring should be taken at expected pile locations. Ground strata (to bedrock or depth of 1-1/2 building width) must be located and identified with appropriate strengths and compressibilities. The sequence of strata determines if end-bearing or friction piles are best suited. See **Table A1020-105** for site investigation costs.

End-bearing piles have shafts which pass through soft strata or thin hard strata and tip bear on bedrock or penetrate some distance into a dense, adequate soil (sand or gravel).

Friction piles have shafts which may be entirely embedded in cohesive soil (moist clay). They develop required support mainly by adhesion or "skin-friction" between soil and shaft area.

Piles pass through soil by either one of two ways:

1. Displacement piles force soil out of the way. This may cause compaction, ground heaving, remolding of sensitive soils, damage to adjacent structures, or hard driving.
2. Non-displacement piles have either a hole bored and the pile cast or placed in the hole, or open-ended pipe (casing) driven and the soil core removed. They tend to eliminate heaving or lateral pressure damage to adjacent structures of piles. Steel "HP" piles are considered of small displacement.

Placement of piles (attitude) is most often vertical; however, they are sometimes battered (placed at a small angle from vertical) to advantageously resist lateral loads. Seldom are piles installed singly but rather in clusters (or groups). Codes require a minimum of three piles per major column load or two per foundation wall or grade beam. Single pile capacity is limited by pile structural strength or support strength of soil. Support capacity of a pile cluster is almost always less than the sum of its individual pile capacities due to overlapping of bearing the friction stresses. See **Table A1020-101** for minimum pile spacing requirements.

Large rigs for heavy, long piles create large soil surface loads and additional expense on weak ground. See **Table A1020-103** for percent of cost increases.

Fewer piles create higher costs per pile. See **Table A1020-102** for effect of mobilization.

Pile load tests are frequently required by code, ground situation, or pile type. See **Table A1020-104** for costs. Test load is twice the design load.

Table A1020-101 Min. Pile Spacing by Pile Type

Type Support		Min. Pile Spacing	
		X's Butt Diameter	Foot
End	Bedrock	2	2'-0"
Bearing	Hard Strata	2.5	2'-6"
Friction		3 to 5	3'-6"

Table A1020-102 Add Cost for Mobilization (Set Up and Removal)

Job Size	Cost	
	$/Job	$/L.F. Pile
Small (12,000 L.F.)	$15,100	$1.26
Large (25,000 L.F.)	25,200	1.01

Table A1020-103 Add Costs for Special Soil Conditions

Special Conditions	Add % of Total
Soft, damp ground	40%
Swampy, wet ground	40%
Barge mounted drilling rig	30%

Table A1020-104 Testing Costs, if Required, Any Pile Type

Test Weight (Tons)	$/Test
50-100T	$15,500
100-200T	22,000
150-300T	28,500
200-400T	31,000

Table A1020-105 Cost/L.F. of Boring Types (4"), and Per Job Cost of Other Items

Type Soil	Type Boring	Sample	$/L.F. Boring	Total $
Earth	Auger	None	$ 33.50	
	Cased	Yes	64.00	
Rock, "BX"	Core	None	80.50	
	Cased	Yes	106.00	
Filed survey, mobilization, demobilization & engineering report $3,280				

A10 Foundations RA1020-200 Caissons

General: Caissons, as covered in this section, are drilled cylindrical foundation shafts which function primarily as short column-like compression members. They transfer superstructure loads through inadequate soils to bedrock or hard stratum. They may be either reinforced or unreinforced and either straight or belled out at the bearing level.

Shaft diameters range in size from 20″ to 84″ with the most usual sizes beginning at 34″. If inspection of bottom is required, the minimum diameter practical is 30″. If handwork is required (in addition to mechanical belling, etc.) the minimum diameter is 32″. The most frequently used shaft diameter is probably 36″ with a 5′ or 6′ bell diameter. The maximum bell diameter practical is three times the shaft diameter.

Plain concrete is commonly used, poured directly against the excavated face of soil. Permanent casings add to cost and economically should be avoided. Wet or loose strata are undesirable. The associated installation sometimes involves a mudding operation with bentonite clay slurry to keep walls of excavation stable (costs not included here).

Reinforcement is sometimes used, especially for heavy loads. It is required if uplift, bending moment, or lateral loads exist. A small amount of reinforcement is desirable at the top portion of each caisson, even if the above conditions theoretically are not present. This will provide for construction eccentricities and other possibilities. Reinforcement, if present, should extend below the soft strata. Horizontal reinforcement is not required for belled bottoms.

There are three basic types of caisson bearing details:

1. Belled, which are generally recommended to provide reduced bearing pressure on soil. These are not for shallow depths or poor soils. Good soils for belling include most clays, hardpan, soft shale, and decomposed rock.

   Soils requiring handwork include hard shale, limestone, and sandstone.

   Soils not recommended include sand, gravel, silt, and igneous rock. Compact sand and gravel above water table may stand. Water in the bearing strata is undesirable.
2. Straight shafted, which have no bell but the entire length is enlarged to permit safe bearing pressures. They are most economical for light loads on high bearing capacity soil.
3. Socketed (or keyed), which are used for extremely heavy loads. They involve sinking the shaft into rock for combined friction and bearing support action. Reinforcement of shaft is usually necessary. Wide flange cores are frequently used here.

Advantages include:

A. Shafts can pass through soils that piles cannot
B. No soil heaving or displacement during installation
C. No vibration during installation
D. Less noise than pile driving
E. Bearing strata can be visually inspected & tested

Uses include:

A. Situations where unsuitable soil exists to moderate depth
B. Tall structures
C. Heavy structures
D. Underpinning (extensive use)

See **Table A1010-121** for Soil Bearing Capacities.

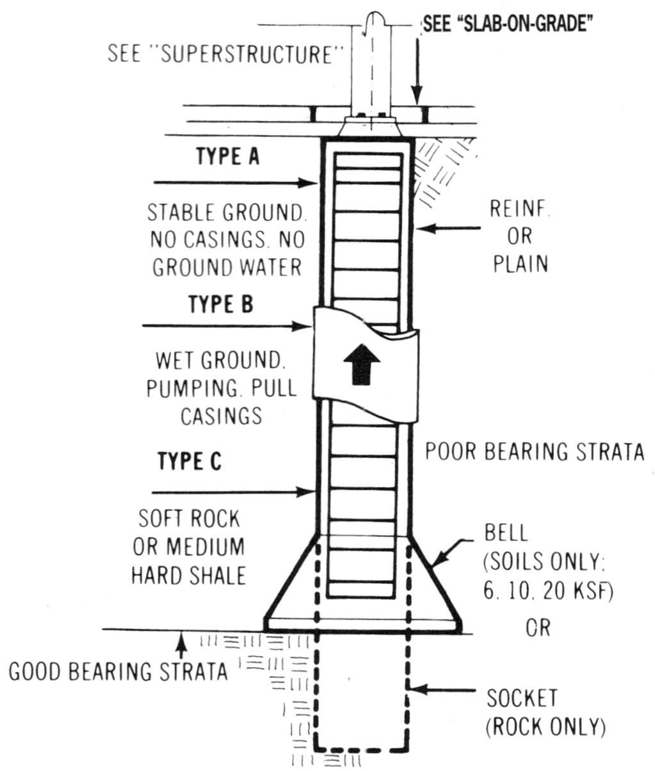

Figure A1020-201 Design Assumptions

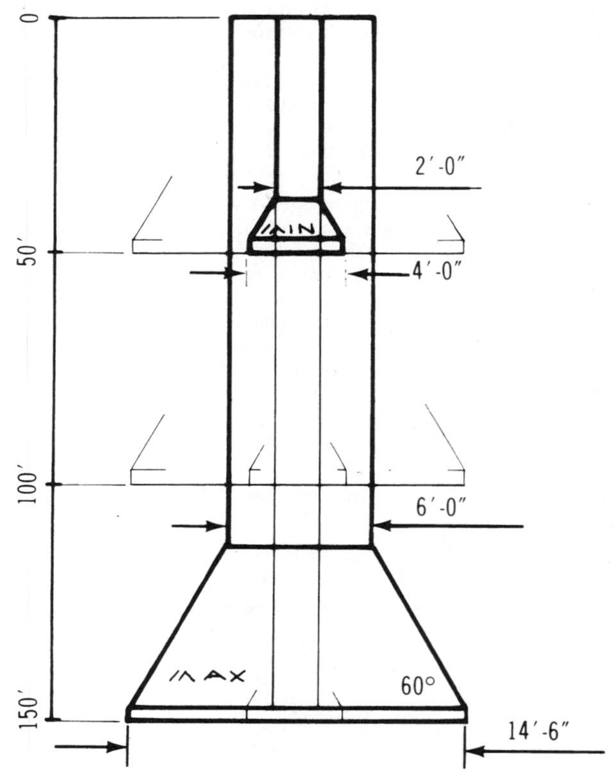

Figure A1020-202 Size Range

Table A1020-231 Grade Beam Detail

General: Grade beams are stiff self-supporting structural members, partly exposed above grade. They support wall loads and carry them across unacceptable soil to column footings, support piles or caissons.

They should be deep enough to be below frost depth. Therefore, they are more frequently used in mild climates.

They must be stiff enough to prevent cracking of the supported wall. Usually, they are designed as simply supported beams, so as to minimize effects of unequal footing settlement.

Heavy column loads and light wall loads tend to make grade beams an economical consideration. Light column loads but heavy wall loads tend to make strip footings more economical.

Conditions which favor other foundation types include:

A. Deep frost penetration requiring deep members
B. Good soil bearing near surface
C. Large live loads to grade beam
D. Ground floor 2′ or 3′ above finished grade requiring wall depth beams (not included here).
E. Varying ground elevations
F. Basement, requiring walls

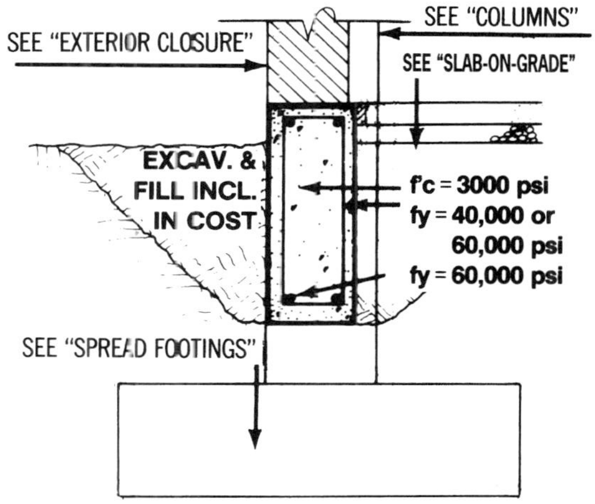

Figure A1020-231 Grade Beam Detail

Design Assumption: See Figure A1020-231
Concrete placed by chute
Forms, four uses
Simply supported
Max. deflection = L/480
Design span = Bay width less 2′

Table A1030-202 Thickness and Loading Assumptions by Type of Use

General: Grade slabs are classified on the basis of use. Thickness is generally controlled by the heaviest concentrated load supported. If load area is greater than 80 sq. in., soil bearing may be important. The base granular fill must be a uniformly compacted material of limited capillarity, such as gravel or crushed rock. Concrete is placed on this surface or the vapor barrier on top of the base.

Grade slabs are either single or two course floors. Single course are widely used. Two course floors have a subsequent wear resistant topping.

Reinforcement is provided to maintain tightly closed cracks.

Control joints limit crack locations and provide for differential horizontal movement only. Isolation joints allow both horizontal and vertical differential movements.

Use of Table: Determine the appropriate type of slab (A, B, C, or D) by considering the type of use or amount of abrasive wear of traffic type.

Determine thickness by maximum allowable wheel load or uniform load, opposite 1st column thickness. Increase the controlling thickness if details require, and select either plain or reinforced slab thickness and type.

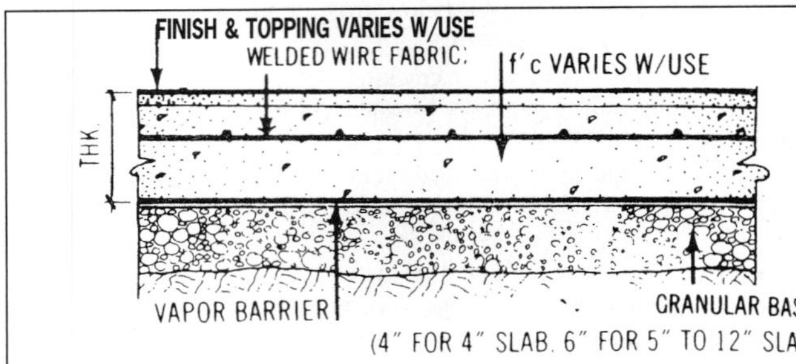

Figure A1030-201 Section, Slab-on-Grade

Table A1030-202 Thickness and Loading Assumptions by Type of Use

		A	B	C	D	◄ Slab I.D.
		Non	Light	Normal	Heavy	◄ Industrial
		Little	Light	Moderate	Severe	◄ Abrasion
		Foot Only	Pneumatic Wheels	Solid Rubber Wheels	Steel Tires	◄ Type of Traffic
SLAB THICKNESS (IN.)	TYPE	Load* (K)	Load* (K)	Load* (K)	Load* (K)	Max. Uniform Load to Slab ▼ (PSF)
4"	Reinf. Plain	4K				100
5"	Reinf. Plain	6K	4K			200
6"	Reinf. Plain		8K	6K	6K	500 to 800
7"	Reinf. Plain			9K	8K	1,500
8"	Reinf. Plain				11K	* Max. Wheel Load in Kips (incl. impact)
10"	Reinf. Plain				14K	
12"	Reinf. Plain					
DESIGN ASSUMPTIONS	Concrete, Chuted	f'c = 3.5 KSI	4 KSI	4.5 KSI	Slab @ 3.5 KSI	ASSUMPTIONS BY SLAB TYPE
	Topping			1" Integral	1" Bonded	
	Finish	Steel Trowel	Steel Trowel	Steel Trowel	Screed & Steel Trowel	
	Compacted Granular Base	4" deep for 4" slab thickness 6" deep for 5" slab thickness & greater				ASSUMPTIONS FOR ALL SLAB TYPES
	Vapor Barrier	6 mil polyethylene				
	Forms & Joints	Allowances included				
	Rein-forcement	WWF As required ≥ 60,000 psi				

Table A2020-212 Minimum Wall Reinforcement Weight (PSF)

General: Foundation wall heights depend upon depth of frost penetration, basement configuration or footing requirements. For certain light load conditions, good soil bearing and adequate wall thickness, strip footings may not be required.

Thickness requirements are based upon code requirements (usually 8″ min. for foundation, exposed basement, fire and party walls), unsupported height or length, configuration of wall above, and structural load considerations.

Shrinkage and temperature forces make longitudinal reinforcing desirable, while high vertical and lateral loads make vertical reinforcing necessary. Earthquake design also necessitates reinforcing.

Figure A2020-211 shows the range in wall sizes for which costs are provided in System A2020 110. Other pertinent information is also given.

Design Assumptions: Reinforced concrete wall designs are based upon ACI 318-71-14, Empirical Design of Walls. Plain concrete wall designs are based upon nonreinforced load bearing masonry criteria (fm = 1700 psi).

Earthquake, lateral or shear load requirements may add to reinforcement costs and are not included.

Minimum Wall Reinforcement: The table below lists the weight of wall for minimum quantities of reinforcing steel for various wall thicknesses.

Design Assumptions:

Reinforcing steel is A 615, grade 60 (fy = 60 k).
Total percent of steel area is 0.20% for horizontal steel, 0.12% for vertical steel.
For steel at both faces, mats are identical.
For single layer, mat is in center of wall.
For grade 40 steel, add 25% to weights and costs.
If other than the minimum amount of steel area is required, factor weights directly with increased percentage.

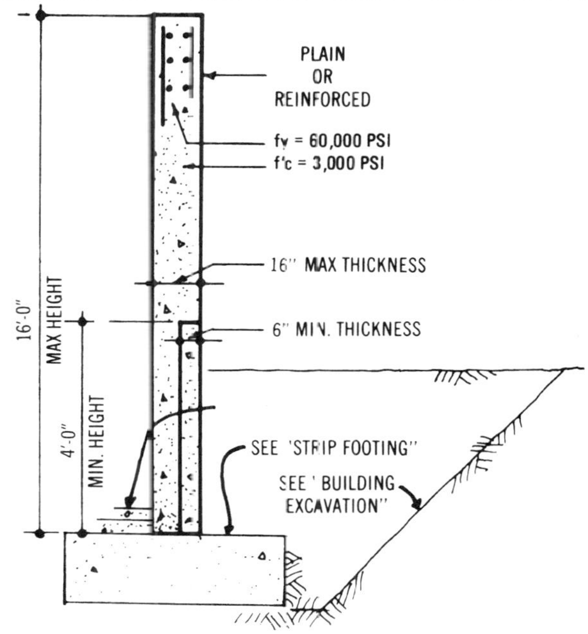

Figure A2020-211 Concrete Foundation Walls Size Range for System A2020 110

Note: Excavation and fill costs are not included in System A2020 110, but are provided in STRIP FOOTING COST, System A1010 110

Table A2020-212 Minimum Wall Reinforcement Weight (PSF)

Location	Wall Thickness	Horizontal Steel				Vertical Steel				Horizontal & Vertical Steel
		Bar Size	Spacing C/C	Sq. In. per L.F.	Total Wt. per S.F.	Bar Size	Spacing C/C	Sq. In. per L.F.	Total Wt. per S.F.	Total Wt. per S.F.
Both Faces	10″	#4	18″	.13	.891#	#3	18″	.07	.501#	1.392#
	12″	#4	16″	.15	1.002	#3	15″	.09	.602	1.604
	14″	#4	14″	.17	1.145	#3	13″	.10	.694	1.839
	16″	#4	12″	.20	1.336	#3	11″	.12	.820	2.156
	18″	#5	17″	.22	1.472	#4	18″	.13	.891	2.363
One Face	6″	#3	9″	.15	.501	#3	18″	.07	.251	.752
	8″	#4	12″	.20	.668	#3	11″	.12	.410	1.078
	10″	#5	15″	.25	.834	#4	16″	.15	.501	1.335

Table A2020-213 Weight of Steel Reinforcing per S.F. in Walls (PSF)

Reinforced Weights: The table below lists the weight per S.F. for reinforcing steel in walls.

Design Assumptions: Reinforcing Steel is A615 grade 60 (for costs).

Weights will be correct for any grade steel reinforcing. For bars in two directions, add weights for each size and spacing.

C/C Spacing in Inches	Bar size								
	#3 Wt. (PSF)	#4 Wt. (PSF)	#5 Wt. (PSF)	#6 Wt. (PSF)	#7 Wt. (PSF)	#8 Wt. (PSF)	#9 Wt. (PSF)	#10 Wt. (PSF)	#11 Wt. (PSF)
2″	2.26	4.01	6.26	9.01	12.27				
3″	1.50	2.67	4.17	6.01	8.18	0.68	13.60	17.21	21.25
4″	1.13	2.01	3.13	4.51	6.13	8.10	10.20	12.91	15.94
5″	.90	1.60	2.50	3.60	4.91	6.41	8.16	10.33	12.75
6″	.752	1.34	2.09	3.00	4.09	5.34	6.80	8.61	10.63
8″	.564	1.00	1.57	2.25	3.07	4.01	5.10	6.46	7.97
10″	.451	.802	1.25	1.80	2.45	3.20	4.08	5.16	6.38
12″	.376	.668	1.04	1.50	2.04	2.67	3.40	4.30	5.31
18″	.251	.445	.695	1.00	1.32	1.78	2.27	2.86	3.54
24″	.188	.334	.522	.751	1.02	1.34	1.70	2.15	2.66
30″	.150	.267	.417	.600	.817	1.07	1.36	1.72	2.13
36″	.125	.223	.348	.501	.681	.890	1.13	1.43	1.77
42″	.107	.191	.298	.429	.584	.763	.97	1.17	1.52
48″	.094	.167	.261	.376	.511	.668	.85	1.08	1.33

Selecting a Floor System:

1. Determine size of building — total S.F. and occupancy (i.e., office building, dormitory, etc.).
2. Determine number of floors and S.F./floor.
3. Select a possible bay size and layout.
4. From Table L1010-101 determine the minimum design live load. Add partition load, ceiling load and miscellaneous loads such as mechanical, light fixtures and flooring.

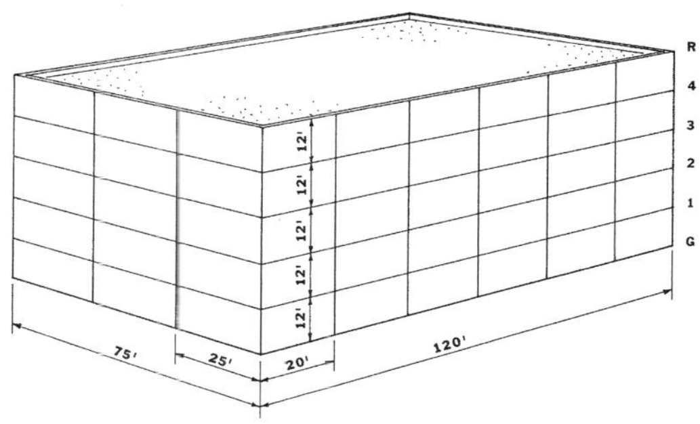

Example:

Live Load Office Building	50	PSF	BOCA Code Table L1010-101
Partitions	20		Assume using Tables L1010-201
Ceiling	5		through L1010-226
	75	PSF	Total Superimposed Load

Suspended Floors					
Building Dimension (Ft.)	**Suspended Floor (S.F.)**	**Number of Suspended Floors**	**Total Suspended Floors (S.F.)**	**Bay Size (Ft.)**	**Superimposed Load PSF**
75 x 120	9,000	4	36,000	20 x 25	75

5. Enter Tables with 75 PSF superimposed load and 20 x 25 bay size.

Table B1010-223	C.I.P. Flat Plate	$15.80 /S.F.
Table B1010-256	Composite Beam, Deck & Slab	19.65 /S.F.
Table B1010-250	Steel Joists & Beams on Col. (col. not included)	16.75 /S.F.

It appears, from the costs listed above, that the most economical system for framing the suspended floors with 20′ x 25′ bays, 75 PSF superimposed load, is the open web joists, beams, slab form and 2-1/2″ concrete slabs. However, the system will require a fireproof ceiling for fireproofing the structure, which will ultimately raise the cost.

The concrete flat plate may be more economical if no hung ceiling is required and the soil capacity is sufficient to justify the additional load to the foundations.

Some of the factors affecting the selection of a floor system are:

1. Location
2. Owner's preference
3. Fire code
4. Availability of materials
5. Subsoil conditions
6. Economy
7. Clear span
8. Acoustical characteristics
9. Contractor capability

Table B1010-101 Comparative Costs ($/S.F.) of Floor Systems/Type (Table Number), Bay Size, & Load

	Cast-In-Place Concrete						Precast Concrete			Structural Steel					Wood	
Bay Size	1 Way BM & Slab B1010-219	2 Way BM & Slab B1010-220	Flat Slab B1010-222	Flat Plate B1010-223	Joist Slab B1010-226	Waffle Slab B1010-227	Beams & Hollow Core Slabs B1010-236	Beams & Hollow Core Slabs Topped B1010-238	Beams & Double Tees Topped B1010-239	Bar Joists on Cols. & Brg. Walls B1010-248	Bar Joists & Beams on cols. B1010-250	Composite Beams & C.I.P. Slab B1010-252	Composite Deck & Slab, W Shapes B1010-254	Composite Beam & DK., Lt. Wt. Slab B1010-256	Wood Beams & Joists B1010-264	Laminated Wood Beams & Joists B1010-265
Superimposed Load = 40 PSF																
15 x 15	16.65	15.70	13.95	13.45	18.00	—	—	—	—	—	—	—	—	—	13.75	15.00
15 x 20	16.65	16.30	14.40	14.25	18.15	—	—	—	—	11.10	13.05	—	18.75	—	16.20	14.95
20 x 20	16.80	17.10	14.85	14.25	18.10	19.65	26.70	29.70	—	11.35	14.00	—	20.05	—	16.60	15.15
20 x 25	17.05	18.00	15.85	15.50	18.30	19.90	24.65	28.05	—	11.85	14.10	21.75	20.25	19.15	—	—
25 x 25	17.10	18.40	16.30	15.80	18.05	20.25	25.45	28.30	—	12.70	15.95	22.35	22.60	18.65	—	—
25 x 30	17.75	19.15	17.20	—	18.90	20.50	23.35	26.25	26.40	14.05	17.60	24.10	22.60	18.85	—	—
30 x 30	19.20	20.55	18.10	—	19.40	21.40	23.75	26.65	26.25	13.90	17.60	24.10	24.75	19.15	—	—
30 x 35	20.05	22.10	19.20	—	20.10	21.70	22.25	26.25		15.45	19.65	25.05	26.75	19.90	—	—
35 x 35	21.20	22.95	19.50	—	20.15	22.55	23.10	26.70	—	15.85	20.10	26.00	27.25	20.80	—	—
35 x 40	21.65	24.20	—	—	20.40	23.00	23.10	26.95	24.30	—	—	28.10	28.55	22.80	—	—
40 x 40	—	—	—	—	21.40	23.65	24.15	26.40	26.30	—	—	—	—	—	—	—
40 x 45	—	—	—	—	22.15	24.25	—	—	—	—	—	—	—	—	—	—
40 x 50	—	—	—	—	—	—	—	—	23.25	—	—	—	—	—	—	—
Superimposed Load = 75 PSF																
15 x 15	16.80	16.00	14.15	13.40	18.05	—	—	—	—	—	—	—	—	—	17.70	18.55
15 x 20	17.25	17.55	14.90	14.80	18.90	—	—	—	—	11.95	14.75	—	20.35	—	21.00	20.45
20 x 20	18.15	18.45	15.45	14.85	19.10	19.95	21.35	30.30	—	12.80	16.00	—	22.35	—	22.10	20.35
20 x 25	18.55	19.95	16.70	15.80	19.10	20.25	24.65	28.10	—	13.35	16.75	24.15	24.65	19.65	—	—
25 x 25	18.45	19.85	16.95	16.00	19.05	20.70	25.45	23.40	—	14.55	18.40	25.45	25.65	20.10	—	—
25 x 30	18.65	20.65	18.05	—	19.50	20.95	24.05	26.45	26.40	14.80	18.75	27.00	26.80	19.90	—	—
30 x 30	20.60	22.30	19.00	—	19.95	21.75	23.75	28.15	26.24	15.90	20.15	26.75	28.55	20.40	—	—
30 x 35	20.90	23.10	20.15	—	20.30	21.70	23.15	26.00	—	18.05	22.55	28.70	30.60	21.30	—	—
35 x 35	23.10	23.85	20.70	—	21.00	22.95	24.10	28.00	—	19.35	24.15	29.60	31.15	23.25	—	—
35 x 40	23.40	25.00	—	—	21.75	23.65	—	—	24.10	—	—	31.50	32.65	24.35	—	—
40 x 40	—	—	—	—	21.90	24.40	—	—	25.30	—	—	—	—	—	—	—
40 x 45	—	—	—	—	22.35	25.00	—	—	—	—	—	—	—	—	—	—
40 x 50	—	—	—	—	—	—	—	—	21.25	—	—	—	—	—	—	—
Superimposed Load = 125 PSF																
15 x 15	17.05	16.65	14.50	13.70	18.35	—	—	—	—	—	—	—	—	—	26.05	27.00
15 x 20	17.90	18.95	15.45	15.60	19.30	—	—	—	—	12.85	16.25	—	23.25	—	31.45	31.15
20 x 20	19.05	19.10	16.40	15.55	19.35	20.25	—	—	—	14.45	15.20	—	25.00	—	41.40	30.10
20 x 25	19.70	20.45	17.70	16.65	20.15	20.60	—	—	—	15.70	19.45	26.95	28.60	23.60	—	—
25 x 25	21.25	21.50	17.85	17.05	21.05	21.15	—	—	—	16.90	21.30	28.85	30.20	21.20	—	—
25 x 30	21.35	22.35	18.70	—	20.80	21.40	—	—	—	17.90	22.50	31.50	30.00	22.90	—	—
30 x 30	21.90	23.35	19.60	—	21.00	22.05	—	—	—	19.80	25.05	31.40	33.25	24.05	—	—
30 x 35	23.30	25.15	20.80	—	21.05	22.55	—	—	—	20.35	25.55	35.25	35.25	24.70	—	—
35 x 35	24.50	26.05	21.15	—	21.10	23.35	—	—	—	22.60	26.70	34.05	37.00	27.10	—	—
35 x 40	24.75	26.20	—	—	21.60	24.45	—	—	—	—	—	36.65	37.95	27.70	—	—
40 x 40	—	—	—	—	22.75	24.80	—	—	—	—	—	—	—	—	—	—
40 x 45	—	—	—	—	23.10	25.35	—	—	—	—	—	—	—	—	—	—
40 x 50	—	—	—	—	—	—	—	—	—	—	—	—	—	—	—	—
Superimposed Load = 200 PSF																
15 x 15	17.75	17.75	15.05	—	18.90	—	—	—	—	—	—	—	—	—	49.05	47.30
15 x 20	19.30	20.35	15.75	—	19.85	—	—	—	—	—	—	—	27.90	—	43.15	41.75
20 x 20	20.70	20.65	16.65	—	20.15	21.05	—	—	—	—	—	—	29.25	—	—	43.90
20 x 25	21.20	22.25	18.20	—	21.10	21.30	—	—	—	—	—	32.80	31.95	26.20	—	—
25 x 25	23.05	24.10	18.45	—	21.65	21.65	—	—	—	—	—	34.05	34.65	27.45	—	—
25 x 30	23.20	24.25	19.55	—	21.90	22.85	—	—	—	—	—	36.20	37.00	27.60	—	—
30 x 30	24.30	24.95	20.70	—	22.10	23.55	—	—	—	—	—	38.15	44.30	28.30	—	—
30 x 35	24.45	26.40	—	—	22.60	24.40	—	—	—	—	—	41.40	42.85	28.80	—	—
35 x 35	26.40	27.35	—	—	22.60	24.75	—	—	—	—	—	42.10	46.50	30.60	—	—
35 x 40	26.65	28.00	—	—	23.05	26.20	—	—	—	—	—	45.35	47.70	32.95	—	—
40 x 40	—	—	—	—	—	—	—	—	—	—	—	—		—	—	—
40 x 45	—	—	—	—	—	—	—	—	—	—	—	—	—	—	—	—
40 x 50	—	—	—	—	—	—	—	—	—	—	—	—	—	—	—	—

Design Assumptions:

Bay Size 20′ x 25′ = 500 S.F.

Use Concrete Flat Plate

Roof Load:	Superimposed Load	40 PSF
	Total Load	146 PSF
Floor Load:	Superimposed Load	75 PSF
	Total Load	188 PSF

Minimum Column size 20″

ROOF	Total Roof Load .146 KSF x 500 S.F. =	73 K
	(top of column)	73 K
	Column Load 1.67′ x 1.67′ x .15 KCF x 12′ =	5 K
		78 K
FOURTH	Total Floor Load .188 K x 500 S.F. =	94 K
	(top of column)	172 K
	Column	5 K
		177 K
THIRD	Total Floor Load	94 K
	(top of column)	271 K
	Column	5 K
		276 K
SECOND	Total Floor Load	94 K
	(top of column)	370 K
	Column	5 K
		375 K
FIRST	Total Floor Load	94 K
	(top of column)	469 K
	Column	5 K
		474 K
GROUND		
	Total Load	474 K

Description:

1. Multiply roof load x bay area.
2. Show total load at top of column.
3. Multiply est. column weight x floor to floor height.
4. Add to roof load.
5. Multiply floor load x bay area.
6. Add to roof and column load.
7. Multiply est. column weight x floor to floor height.
8. Add to total loads above.
9. Repeat steps above for remainder of floors & columns.
10. Enter the minimum reinforced portion of the table with total load on the column and the minimum allowable column size for the selected cast-in-place floor system.

    If the total load on the column does not exceed the allowable load shown, use the cost per L.F. multiplied by the length of columns required to obtain the column cost.
11. If the total load on the column exceeds the allowable working load shown in the minimum reinforced portion of the table enter the first portion of the table with the total load on the column and the minimum allowable column size from the selected cast-in-place floor system.

    Select a cost per L.F. for bottom level columns by total load or minimum allowable column size.

    Select a cost per L.F. for top level columns using the column size required for bottom level columns from the minimum reinforced portion of the table.

$$\frac{\text{Bottom \& Top Col. Costs/L.F.}}{2} = \text{Average Column Cost/L.F.}$$

Column Cost = Average Col. Cost/L.F. x Length of Cols. Req'd.

Description: Below is an example of steel column determination when roof and floor loads are known.

Calculation of Total Loads

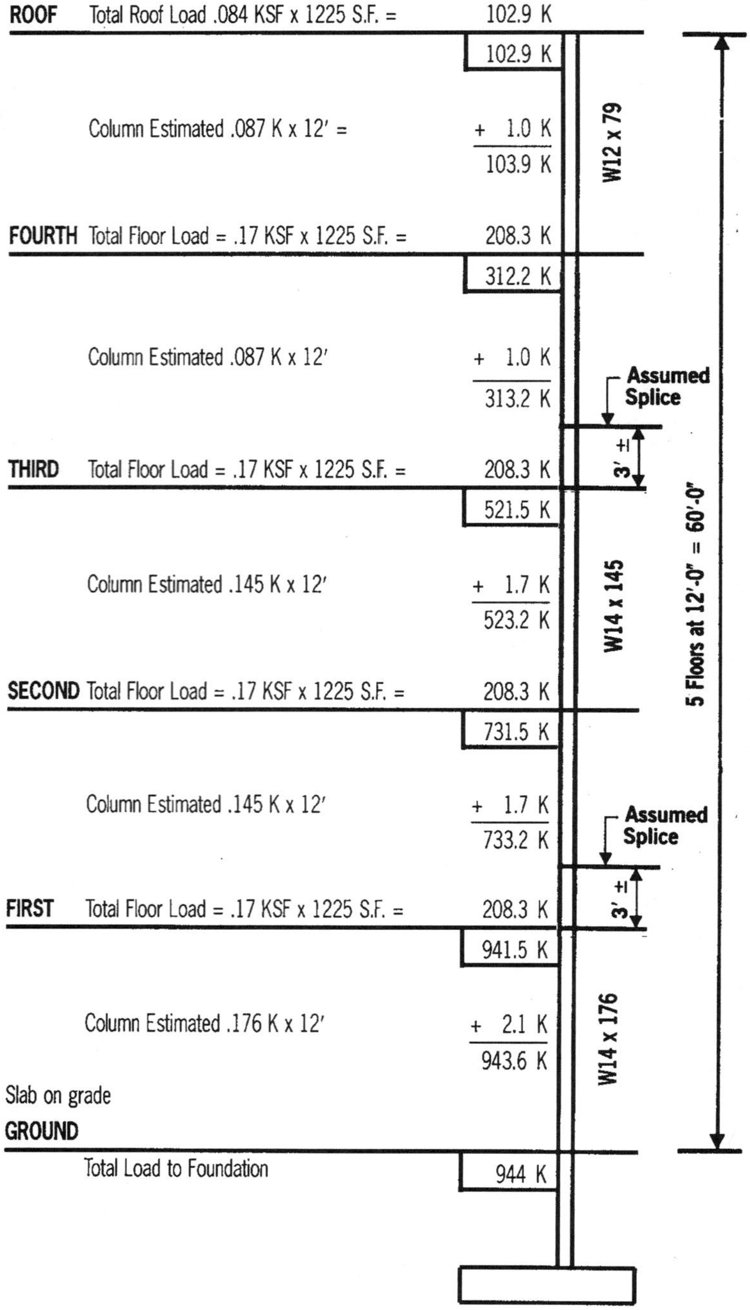

Design Assumptions:

Bay Size	35′ x 35′ = 1,225 S.F.		
Roof Load:	Superimposed Load	40	PSF
	Dead Load	44	PSF
	Total Load	84	PSF
Floor Load:	Superimposed Load	125	PSF
	Dead Load	45	PSF
	Total Load	170	PSF

Description:

1. Multiply roof load x bay area.
2. Show total load at top of column.
3. Multiply estimated column weight* x floor to floor height.
4. Add to roof load.
5. Multiply floor load x bay area.
6. Add to roof and column load.
7. Multiply estimated column weight x floor to floor height.
8. Add to total loads above.
9. Choose column from **System B1010 208** using unsupported height.
10. Interpolate or use higher loading to obtain cost/L.F.
11. Repeat steps above for remainder of floors and columns.
12. Multiply number of columns by the height of the column times the cost per foot to obtain the cost of each type of column.

*** To Estimate Column Weight**

Roof Load		.084 KSF
Floor Load x No Floors above Splice		
170 x 1		.170 KSF
	Total	.254 KSF

Total Load (KSF) x Bay Area (S.F.) = Load to Col.
.254 KSF x 1,225 KSF = 311 K

From **System B1010 208**, choose a column by:

Load,	Height,	Weight
400 K	10′	79 lb.

Table B1010-241 One, Two and Three Member Beams, Maximum Load for Various Spans

Description: The table below lists the maximum uniform load (W) or concentrated load (P) allowable for various size beams at various spans.

Design Assumptions: Fiber strength (f) is 1,000 psi.

Maximum deflection does not exceed 1/360 the span of the beam.

Modulus of elasticity (E) is 1,100,000 psi. Anything less will result in excessive deflection in the longer members so that spans must be reduced or member size increased.

The uniform loads (W) are in pounds per foot; concentrated loads (P), at the midpoint, are in pounds.

The span is in feet and is the unsupported clear span.

The member sizes are from 2″ x 6″ to 4″ x 12″ and include one, two and three pieces of each member size to arrive at spans and loading.

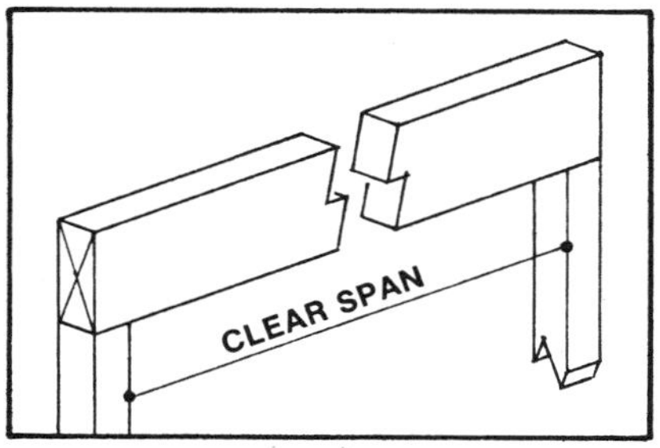

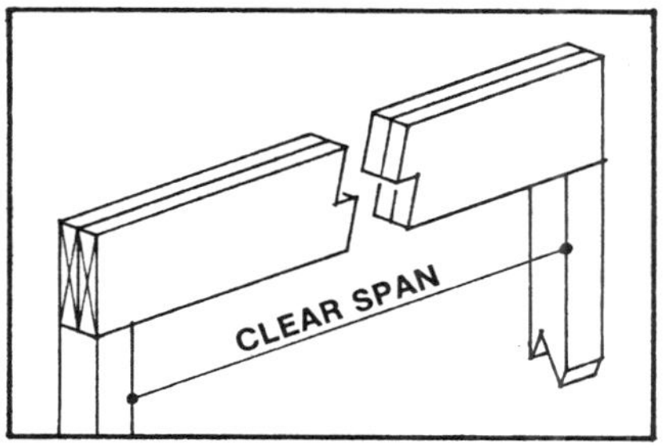

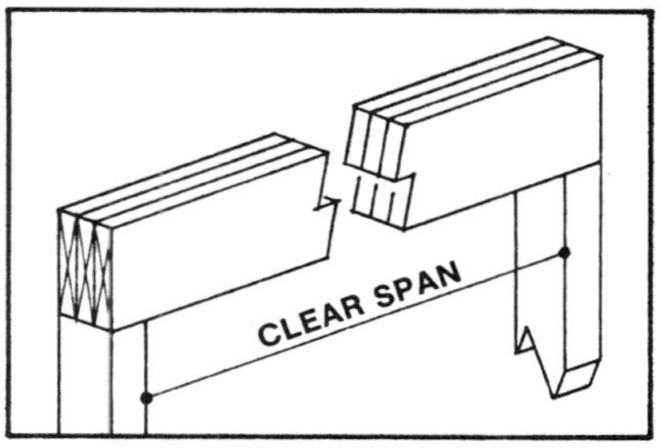

Individual Member		Span and Load													
		6′		8′		10′		12′		14′		16′		18′	
		W	P	W	P	W	P	W	P	W	P	W	P	W	P
Size	No.	#/L.F.	#	#/L.F.	#	#/L.F.	#	#/L.F.	#	#/L.F.	#	#/L.F.	#	#/L.F.	#
2″ x 6″	2	238	840	—	—										
	3	357	1260	—	—										
2″ x 8″	2	314	1460	235	1095										
	3	471	2190	353	1642	232	1164								
2″ x 10″	2	400	2376	300	1782	240	1426	—	—						
	3	600	3565	450	2673	360	2139	279	1679						
2″ x 12″	2	487	2925	365	2636	292	2109	243	1757	208	1479	—	—		
	3	731	4387	548	3955	548	3164	365	2636	313	2219	212	1699		
2″ x 14″	2	574	3445	430	3445	344	2926	287	2438	246	2090	215	1828	—	—
	3	861	5167	645	5167	516	4389	430	3657	369	3135	322	2743	243	2193
3″ x 6″	1	—	—	—	—										
	2	397	1400	200	882										
	3	595	2100	300	1323										
3″ x 8″	1	261	1216	—	—	—	—	—	—						
	2	523	2433	392	1825	258	1293	—	—						
	3	785	3650	589	2737	388	1940	224	1347						
3″ x 10″	1	334	1980	250	1485	200	1188	—	—	—	—				
	2	668	3961	500	2970	400	2376	310	1865	—	—	—	—		
	3	1002	5941	750	4456	600	3565	466	2798	293	2056				
3″ x 12″	1	406	2437	304	2197	243	1757	203	1464	—	—	—	—	—	—
	2	812	4875	608	4394	487	3515	406	2929	348	2466	236	1888	—	—
	3	1218	7312	912	6591	731	5273	609	4394	522	3699	354	2832	248	2237
4″ x 6″	1	278	980	—	—	—	—	—	—						
	2	556	1960	308	1235	—	—								
4″ x 8″	1	366	1703	274	1277	—	—	—	—						
	2	733	3406	549	2555	362	1811	209	1257						
4″ x 10″	1	467	2772	350	2079	280	1663	217	1306	—	—				
	2	935	5545	700	4159	560	3327	435	2612	274	1919				
4″ x 12″	1	568	3412	426	3076	341	2460	284	2050	243	1726	—	—		
	2	1137	6825	853	6152	682	4921	568	4100	487	3452	330	2643	232	2088

Table B1010-711 Maximum Floor Joist Spans

Description: The table below lists the maximum clear spans and the framing lumber quantity required per S.F. of floor.

Design Assumptions: Dead load = joist weight plus floor weight of 5 psf plus ceiling or partition load of 10 psf.

Maximum L.L. deflection is 1/360 of the clear span.

Modulus of elasticity is 1,100,000 psi.

Fiber strength (f_b) is 1,000 psi.

10% allowance has been added to framing quantities for overlaps, waste, double joists at openings, etc. 5% added to subfloor for waste.

Maximum span is in feet and is the unsupported clear span.

Floor Joist		Framing Lumber	Live Load in Pounds per Square Foot							
Size in Inches	Spacing C/C	B.F./S.F.	30 Span	40 Span	50 Span	60 Span	70 Span	80 Span	90 Span	100 Span
2″ x 6″	12″	1.10	9′-0″	8′-6″	8′-0″	7′-8″	7′-4″	7′-0″	6′-10″	6′-7″
	16″	.83	8′-4″	7′-9″	7′-4″	7′-0″	6′-8″	6′-4″	6′-0″	—
	24″	.55	7′-2″	6′-9″	6′-2″	5′-9″	—	—	—	—
2″ x 8″	12″	1.46	12′-0″	11′-2″	10′-7″	10′-0″	9′-8″	9′-4″	9′-0″	8′-8″
	16″	1.10	10′-10″	10′-2″	9′-7″	9′-2″	8′-9″	8′-4″	7′-10″	7′-6″
	24″	.74	9′-6″	8′-10″	8′-2″	7′-6″	7′-0″	6′-9″	6′-5″	6′-0″
2″ x 10″	12″	1.84	15′-4″	14′-4″	13′-6″	12′-10″	12′-4″	11′-10″	11′-6″	11′-2″
	16″	1.38	13′-10″	13′-0″	12′-4″	11′-8″	11′-2″	10′-7″	10′-0″	9′-7″
	24″	.91	12′-0″	11′-4″	10′-5″	9′-9″	9′-0″	8′-7″	8′-3″	7′-10″
2″ x 12″	12″	2.20	18′-7″	17′-4″	16′-5″	15′-8″	15′-0″	14′-6″	14′-0″	13′-6″
	16″	1.65	16′-10″	15′-9″	14′-10″	14′-3″	13′-7″	12′-10″	12′-3″	11′-8″
	24″	1.10	14′-9″	13′-9″	12′-8″	11′-9″	11′-2″	10′-6″	10′-0″	9′-6″
2″ x 14″	12″	2.56	21′-10″	20′-6″	19′-4″	18′-5″	17′-9″	17′-0″	16′-6″	15′-10″
	16″	1.93	19′-10″	18′-7″	17′-7″	16′-9″	16′-0″	15′-2″	14′-5″	13′-9″
	24″	1.29	17′-4″	16′-3″	15′-0″	13′-10″	13′-2″	12′-4″	11′-9″	11′-3″
3″ x 6″	12″	1.65	10′-9″	10′-0″	9′-6″	9′-0″	8′-8″	8′-4″	8′-0″	7′-10″
	16″	1.24	9′-9″	9′-0″	8′-8″	8′-3″	7′-10″	7′-7″	7′-4″	7′-0″
	24″	.83	8′-6″	8′-0″	7′-6″	7′-2″	6′-10″	6′-8″	6′-4″	6′-0″
3″ x 8″	12″	2.20	14′-2″	13′-3″	12′-6″	11′-10″	11′-6″	11′-0″	10′-8″	10′-4″
	16″	1.65	12′-10″	12′-0″	11′-4″	10′-10″	10′-5″	10′-0″	9′-9″	9′-5″
	24″	1.10	11′-3″	10′-6″	9′-10″	9′-6″	9′-0″	8′-9″	8′-3″	7′-10″
3″ x 10″	12″	2.75	18′-0″	16′-10″	16′-0″	15′-4″	14′-8″	14′-0″	13′-8″	13′-3″
	16″	2.07	16′-5″	15′-4″	14′-6″	13′-10″	13′-3″	12′-10″	12′-4″	12′-0″
	24″	1.38	14′-4″	13′-5″	12′-8″	12′-0″	11′-7″	11′-0″	10′-7″	10′-0″
3″ x 12″	12″	3.30	22′-0″	20′-7″	19′-6″	18′-7″	17′-10″	17′-0″	16′-7″	16′-0″
	16″	2.48	20′-0″	18′-9″	17′-8″	16′-10″	16′-2″	15′-7″	15′-0″	14′-7″
	24″	1.65	17′-6″	16′-4″	15′-6″	14′-9″	14′-0″	13′-7″	12′-10″	12′-4″
4″ x 6″	12″	2.20	—	—	—	10′-0″	9′-9″	9′-4″	9′-0″	8′-9″
	16″	1.65	—	—	—	9′-2″	8′-10″	8′-6″	8′-3″	8′-0″
	24″	1.10	—	—	—	8′-0″	7′-9″	7′-4″	7′-2″	7′-0″
4″ x 8″	12″	2.93	—	—	—	13′-4″	12′-10″	12′-4″	12′-0″	11′-7″
	16″	2.20	—	—	—	12′-0″	11′-8″	11′-3″	10′-10″	10′-6″
	24″	1.47	—	—	—	10′-7″	9′-2″	9′-9″	9′-6″	9′-2″
4″ x 10″	12″	3.67	—	—	—	17′-0″	16′-4″	15′-9″	15′-3″	14′-9″
	16″	2.75	—	—	—	15′-6″	14′-7″	14′-4″	13′-10″	13′-5″
	24″	1.83	—	—	—	13′-6″	13′-0″	12′-6″	12′-0″	11′-9″

Table B1010-781 Decking Material Characteristics

Description: The table below lists the maximum spans for commonly used wood decking materials for various loading conditions.

Design Assumptions: Applied load is the total load, live plus dead.

Maximum deflection is 1/180 of the clear span which is not suitable if plaster ceilings will be supported by the roof or floor.

Modulus of elasticity (E) and fiber strength (F) are as shown in the table to the right.

No allowance for waste has been included.

Deck Material	Modulus of Elasticity	Fiber Strength
Cedar	1,100,000 psi	1,000 psi
Douglas Fir	1,760,000	1,200
Hemlock	1,600,000	1,200
White Spruce	1,320,000	1,200

Assume Dead Load of wood plank is 33.3 lbs. per C.F.

Table B1010-782 Maximum Spans for Wood Decking

Nominal Thickness	Type Wood	B.F. per S.F.	Total Uniform Load per S.F. 40	50	70	80	100	150	200	250
2″	Cedar	2.40	6.5′	6′	5.7′	5.3′	5	4.5′	4′	3.5′
	Douglas fir	2.40	8′	7′	6.7′	6.3′	6	5′	4.5′	4′
	Hemlock	2.40	7.5′	7′	6.5′	6′	5.5′	5′	4.5′	4′
	White spruce	2.40	7′	6.5′	6′	5.5′	5′	4.5′	4′	3.5′
3″	Cedar	3.65	11′	10′	9′	8.5′	8′	7′	6.5′	6′
	Douglas fir	3.65	13′	12′	11′	10.5′	10′	8′	7.5′	7′
	Hemlock	3.65	12′	11′	10.5′	10′	9′	8′	7.5′	7′
	White spruce	3.65	11′	10′	9.5′	9′	8′	7.5′	7′	6′
4″	Cedar	4.65	15′	14′	13′	12′	11′	10′	9′	8′
	Douglas fir	4.65	18′	17′	15′	14′	13′	11′	10.5′	10′
	Hemlock	4.65	17′	16′	14′	13.5′	13′	11′	10′	9′
	White spruce	4.65	16′	15′	13′	12.5′	12′	10′	9′	8′
6″	Cedar	6.65	23′	21′	19′	18′	17′	15′	14′	13′
	Douglas fir	6.65	24′+	24′	23′	22′	21′	18′	16′	15′
	Hemlock	6.65	24′+	24′	22′	21′	20′	17′	16′	15′
	White spruce	6.65	24′	23′	21′	20′	18′	16′	14′	13′

Table B1020-511 Maximum Roof Joist or Rafter Spans

Description: The table below lists the maximum clear spans and the framing lumber quantity required per S.F. of roof. Spans and loads are based on flat roofs.

Design Assumptions: Dead load = Joist weight plus weight of roof sheathing at 2.5 psf plus either of two alternate roofing weights.

Maximum deflection is 1/360 of the clear span.

Modulus of elasticity is 1,100,000 psi.

Fiber strength (f) is 1,000 psi.

Allowance: 10% has been added to framing quantities for overlaps, waste, double joists at openings, etc.

Maximum span is measured horizontally in feet and is the clear span.

Note: To convert to inclined measurements or to add for overhangs, multiply quantities in Table B1020-511 by factors in Table B1020-512.

Joist			Live Load in Pounds per Square Foot							
			Group I Covering (2.5 psf)				Group II Covering (8.0 psf)			
Size in Inches	Center to Center Spacing	Framing B.F. per S.F.	15 Span	20 Span	30 Span	40 Span	20 Span	30 Span	40 Span	50 Span
2" x 4"	12"	0.73	7'-11"	7'-3"	6'-5"	5'-10"	6'-9"	6'-1"	5'-7"	5'-3"
	16"	0.55	7'-2"	6'-7"	5'-10"	5'-4"	6'-1"	5'-6"	5'-1"	4'-10"
	24"	0.37	6'-3"	5'-9"	5'-0"	4'-8"	5'-4"	4'-10"	4'-6"	4'-2"
2" x 6"	12"	1.10	12'-5"	11'-5"	10'-0"	9'-3"	10'-7"	9'-7"	8'-10"	8'-4"
	16"	0.83	11'-3"	10'-4"	9'-2"	8'-5"	9'-8"	8'-9"	8'-0"	7'-6"
	24"	0.55	9'-10"	9'-0"	8'-0"	7'-4"	8'-5"	7'-7"	7'-0"	6'-6"
2" x 8"	12"	1.46	16'-5"	15'-0"	13'-4"	12'-2"	14'-0"	12'-8"	11'-8"	11'-0"
	16"	1.10	14'-11"	13'-8"	12'-0"	11'-0"	12'-9"	11'-6"	10'-7"	11'-0"
	24"	0.74	13'-0"	12'-0"	10'-7"	9'-8"	11'-1"	10'-0"	9'-3"	10'-0"
2" x 10"	12"	1.84	20'-11"	19'-3"	17'-0"	19'-7"	17'-11"	16'-2"	14'-11"	14'-0"
	16"	1.38	19'-0"	17'-6"	15'-6"	14'-2"	16'-3"	14'-8"	13'-7"	12'-9"
	24"	0.91	16'-7"	15'-3"	13'-6"	12'-4"	14'-2"	12'-10"	11'-10"	11'-0"
2" x 12"	12"	2.20	25'-6"	23'-5"	20'-9"	18'-11"	21'-9"	19'-8"	18'-2"	17'-1"
	16"	1.65	23'-1"	21'-3"	18'-10"	17'-2"	19'-9"	17'-10"	16'-6"	15'-6"
	24"	1.10	20'-2"	18'-7"	16'-5"	15'-0"	17'-3"	15'-7"	14'-5"	13'-5"
2" x 14"	12"	2.56	30'-0"	27'-7"	24'-5"	22'-4"	25'-8"	23'-2"	21'- 5"	20'-1"
	16"	1.93	27'-3"	25'-0"	22'-2"	20'-3"	23'-3"	21'-0"	19'-6"	18'-3"
	24"	1.29	23'-10"	21'-11"	19'-4"	17'-8"	20'-4"	18'-4"	17'-0"	15'-10"
3" x 6"	12"	1.65	14'-9"	13'-7"	12'-0"	11'-0"	12'-7"	11'-4"	10'-6"	9'-11"
	16"	1.24	13'-4"	12'-4"	10'-11"	9'-11"	11'-5"	10'-4"	9'-7"	9'-0"
	24"	.83	11'-9"	10'-9"	9'-6"	8'-8"	10'-0"	9'-0"	8'-4"	7'-10"
3" x 8"	12"	2.20	19'-5"	17'-11"	15'-10"	14'-6"	16'-7"	15'-0"	13'-11"	13'-0"
	16"	1.65	17'-8"	16'-9"	14'-4"	13'-2"	15'-1"	13'-7"	12'-7"	11'-10"
	24"	1.10	15'-5"	14'-2"	12'-6"	11'-6"	13'-2"	11'-11"	11'-0"	10'-4"
3" x 10"	12"	2.75	24'-10"	22'-10"	22'-2"	18'-6"	21'-9"	19'-2"	17'-9"	16'-8"
	16"	2.07	22'-6"	20'-9"	18'-4"	16'-9"	19'-9"	17'-4"	16'-1"	15'-1"
	24"	1.38	19'-8"	18'-1"	16'-0"	14'-8"	16'-10"	15'-2"	14'-0"	13'-2"
3" x 12"	12"	3.30	30'-2"	27'-9"	24'-6"	22'-5"	25'-10"	23'-4"	21'-6"	20'-3"
	16"	2.48	27'-5"	25'-3"	22'-3"	20'-4"	23'-5"	21'-2"	19'-7"	18'-4"
	24"	1.65	24'-0"	22'-0"	19'-6"	17'-10"	20'-6"	18'-6"	17'-1"	16'-0"

Table B1020-512 Factors for Converting Inclined to Horizontal

Rafters: The quantities shown in Table B1020-511 are for flat roofs. For inclined roofs using rafters, the quantities should be multiplied by the factors in Table B1020-512 to allow for the increased area of the inclined roofs.

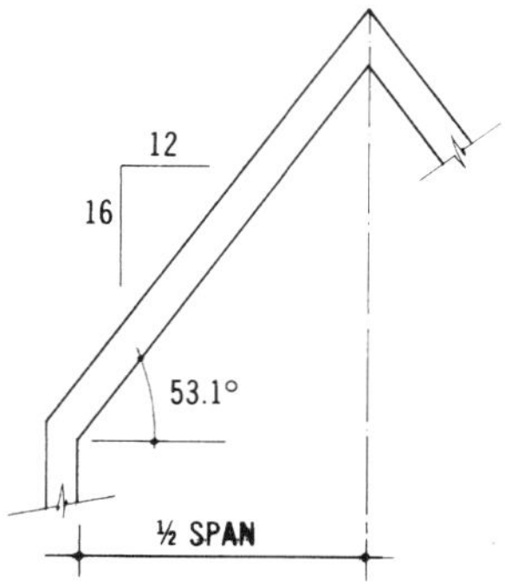

INCLINED TO HORIZONTAL SECTIONS

Roof or Deck Overhangs: The quantities shown in the table do not include cantilever overhangs. The S.F. quantities should be multiplied by the factors in Table B1020-513 to allow for the increased area of the overhang. All dimensions are horizontal. For inclined overhangs also multiply by factors in Table B1020-512 above.

Roof Slope	Approx. Angle	Factor	Roof Slope	Approx. Angle	Factor
Flat	0°	1.000	12 in 12	45.0	1.414
1 in 12	4.8°	1.003	13 in 12	47.3	1.474
2 in 12	9.5°	1.014	14 in 12	49.4	1.537
3 in 12	14.0°	1.031	15 in 12	51.3	1.601
4 in 12	18.4°	1.054	16 in 12	53.1	1.667
5 in 12	22.6°	1.083	17 in 12	54.8	1.734
6 in 12	26.6°	1.118	18 in 12	56.3	1.803
7 in 12	30.3°	1.158	19 in 12	57.7	1.873
8 in 12	33.7°	1.202	20 in 12	59.0	1.943
9 in 12	36.9°	1.250	21 in 12	60.3	2.015
10 in 12	39.8°	1.302	22 in 12	61.4	2.088
11 in 12	42.5°	1.357	23 in 12	62.4	2.162

Table B1020-513 below can be used two ways:

1. Use 1/2 span with overhang on one side only.
2. Use whole span with total overhang from both sides.

Table B1020-513 Allowance Factors for Including Roof or Deck Overhangs

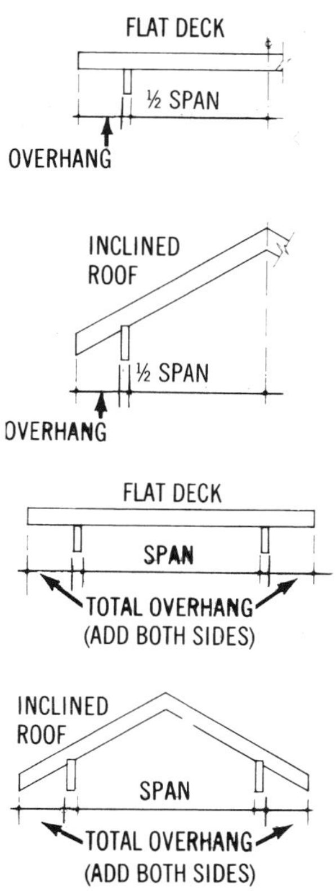

OVERHANG SECTIONS

Horizontal Span	Roof overhang measured horizontally							
	0′-6″	1′-0″	1′-6″	2′-0″	2′-6″	3′-0″	3′-6″	4′-0″
6′	1.083	1.167	1.250	1.333	1.417	1.500	1.583	1.667
7′	1.071	1.143	1.214	1.286	1.357	1.429	1.500	1.571
8′	1.063	1.125	1.188	1.250	1.313	1.375	1.438	1.500
9′	1.056	1.111	1.167	1.222	1.278	1.333	1.389	1.444
10′	1.050	1.100	1.150	1.200	1.250	1.300	1.350	1.400
11′	1.045	1.091	1.136	1.182	1.227	1.273	1.318	1.364
12′	1.042	1.083	1.125	1.167	1.208	1.250	1.292	1.333
13′	1.038	1.077	1.115	1.154	1.192	1.231	1.269	1.308
14′	1.036	1.071	1.107	1.143	1.179	1.214	1.250	1.286
15′	1.033	1.067	1.100	1.133	1.167	1.200	1.233	1.267
16′	1.031	1.063	1.094	1.125	1.156	1.188	1.219	1.250
17′	1.029	1.059	1.088	1.118	1.147	1.176	1.206	1.235
18′	1.028	1.056	1.083	1.111	1.139	1.167	1.194	1.222
19′	1.026	1.053	1.079	1.105	1.132	1.158	1.184	1.211
20′	1.025	1.050	1.075	1.100	1.125	1.150	1.175	1.200
21′	1.024	1.048	1.071	1.095	1.119	1.143	1.167	1.190
22′	1.023	1.045	1.068	1.091	1.114	1.136	1.159	1.182
23′	1.022	1.043	1.065	1.087	1.109	1.130	1.152	1.174
24′	1.021	1.042	1.063	1.083	1.104	1.125	1.146	1.167
25′	1.020	1.040	1.060	1.080	1.100	1.120	1.140	1.160
26′	1.019	1.038	1.058	1.077	1.096	1.115	1.135	1.154
27′	1.019	1.037	1.056	1.074	1.093	1.111	1.130	1.148
28′	1.018	1.036	1.054	1.071	1.089	1.107	1.125	1.143
29′	1.017	1.034	1.052	1.069	1.086	1.103	1.121	1.138
30′	1.017	1.033	1.050	1.067	1.083	1.100	1.117	1.133
32′	1.016	1.031	1.047	1.063	1.078	1.094	1.109	1.125

B20 Exterior Enclosure — RB2010-010 Exterior Walls

The "Exterior Closure" Division has many types of wall systems which are most commonly used in current United States and Canadian building construction.

Systems referenced have an illustration as well as a description and cost breakdown. Total costs per square foot of each system are provided. Each system is itemized with quantities, extended and summed for material, labor, or total.

Exterior walls usually consist of several component systems. The weighted square foot cost of the enclosure is made up of individual component systems costs, prorated for their percent of wall area. For instance, a given wall system may take up 70% of the enclosure wall's composite area. The fenestration may make up the balance of 30%. The components should be factored by the proportion of area each contributes to the total.

Example: An enclosure wall is 100′ long and 15′ high.
The total wall area is therefore 1,500 S.F.
70% of the wall area is a brick & block system.
@ \$32.90/S.F. (B2010-132-1240)

30% is a window system:

Framing	\$30.00	(B2020-210-1700)
Glazing	34.00	(B2020-220-1400)
	\$64.00	

Solution: Brick & block = 1,500 S.F. x .70 x \$32.90 = \$34,545
Windows = 1,500 S.F. x .30 x \$64.00 = \$28,800
Total enclosure wall cost = \$63,345
or \$42.23 per S.F. of wall

The square foot unit costs include the installing contractor's markup for overhead and profit. However, fees for professional services must be added along with other items. See H1010, General Conditions, for these costs. Note that interior finishes (drywall, etc.) are included elsewhere.

There may be times when the wall systems costs will need to be expressed in units of square feet of floor area.

Example: A building 25′ x 25′ requires a wall 100′ long and 15′ high.

Cost of the wall is \$42.23 per S.F. of wall.
The floor area is 625 S.F.

Solution: Wall cost per S.F. of floor area

$$\frac{100 \text{ L.F.} \times 15' \times \$42.23}{625 \text{ S.F.}} = \$101.35 \text{ per S.F. of floor}$$

B20 Exterior Enclosure — RB2010-020 Masonry Block Lintels

Concrete Block Lintels

Design Assumptions:
Bearing length each end = 8 inches
f′c = 2,000 p.s.i.
Wall load = 300 lbs. per linear foot (p.l.f.)
Floor & roof load (includes wall load) = 100 p.l.f.
Weight of 8″ deep lintel (included in wall load) = 50 p.l.f.
Weight of 16″ deep lintel (included in wall load) = 100 p.l.f.

Load to be supported by the lintel is shown below. The weight of the masonry above the opening can be safely assumed as the weight of a triangular section whose height is one-half the clear span of the opening. Corbelling action of the masonry above the top of the opening may be counted upon to support the weight of the masonry outside the triangle. The dead load of a wall must be added to the uniform dead & live loads of floors & roof that frame into the wall above the opening and below the apex of the 45° triangle. Any load above the apex may be neglected.

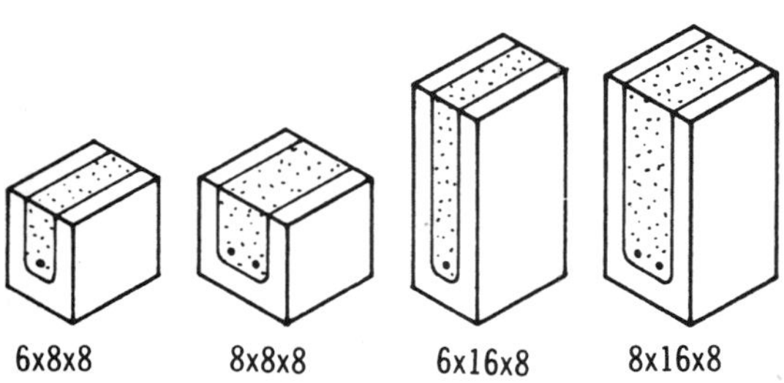

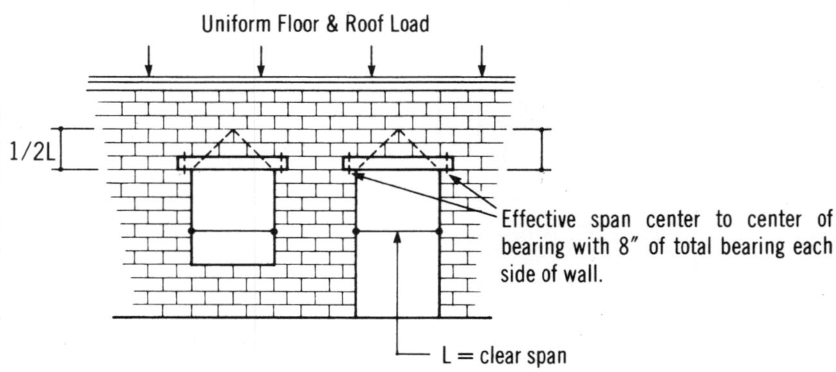

Table B2010-031 Example Using 14″ Masonry Cavity Wall with 1″ Plaster for the Exterior Closure

Total Heat Transfer is found using the equation
$Q = AU (T_2 - T_1)$ where:
Q = Heat flow, BTU per hour
A = Area, square feet
U = Overall heat transfer coefficient
$(T_2 - T_1)$ = Difference in temperature of the air on each side of the construction component in degrees Fahrenheit

Coefficients of Transmission ("U") are expressed in BTU per (hour) (square foot) (Fahrenheit degree difference in temperature between the air on two sides) and are based on 15 mph outside wind velocity.

The lower the U-value the higher the insulating value.
U = 1/R where "R" is the summation of the resistances of air films, materials and air spaces that make up the assembly.

Figure B2010-031 Example Using 14″ Masonry Cavity Wall with 1″ Plaster for the Exterior Closure

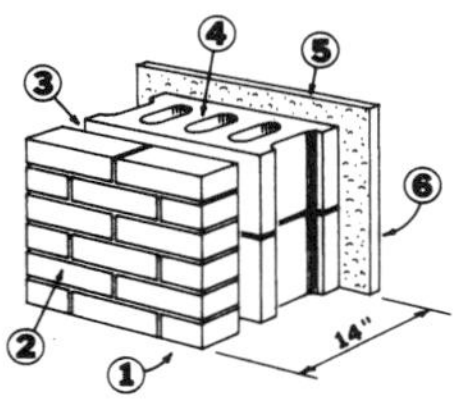

Construction	Resistance R.
1. Outside surface (15 mph wind)	0.17
2. Face Brick (4″)	0.44
3. Air space (2″, 50° mean temp, 10° diff)	1.02
4. Concrete block (8″, lightweight)	2.12
5. Plaster (1″, sand aggregate)	0.18
6. Inside Surface (still air)	0.68
Total resistance	4.61
U = 1/R = 1/4.61 =	0.22

Weight of system = 77 psf
I.S. = Initial System

Assume 20,000 S.F. of wall in Boston, MA (5600 degree days) Table L1030-501
Initial system effective U = Uw x M = 0.22 x 0.905 = 0.20
(M = weight correction factor from Table L1030-203)
Modified system effective U = Uw x M = 0.067 x 0.915 = 0.06
Initial systems: Q = 20,000 x 0.20 x (72° — 10°) = 248,000 BTU/hr.
Modified system: Q = 20,000 x 0.06 x (72° — 10°) = 74,400 BTU/hr.

Reduction in BTU/hr. heat loss; $\frac{\text{I.S. - M.S.}}{\text{I.S.}} \times 100 = \frac{248{,}000 - 74{,}400}{248{,}000} \times 100 = 70\%$

Replace item 3 with 2″ smooth rigid polystyrene insulation and item 5 with 3/4″ furring and 1/2″ drywall.

Total resistance			4.61
Deduct	3. Airspace	1.02	
	5. Plaster	0.18	
		1.20	
Difference 4.61 — 1.20 =			3.41
Add	rigid polystyrene insulation		10.00
	3/4″ air space		1.01
	1/2″ gypsum board		0.45
	Total resistance		14.87
	U = 1/R = 1/14.87 =		0.067

Weight of system = 70 psf
M.S. = Modified System

Table B2010-032 Thermal Coefficients of Exterior Closures

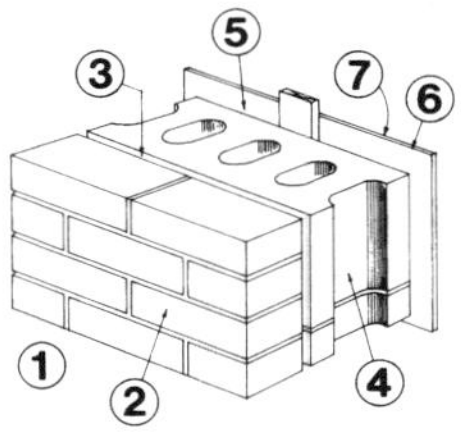

EXAMPLE:

Construction	Resistance R
1. Outside surface (15 MPH wind)	0.17
2. Common brick, 4″	0.80
3. Nonreflective air space, 0.75″	1.01
4. Concrete block, S&G aggregate, 8″	1.46
5. Nonreflective air space, 0.75″	1.01
6. Gypsum board, 0.5″	0.45
7. Inside surface (still air)	0.68
Total resistance	5.58

U = 1/R = 1/5.58 = 0.18

System weight = 97 psf

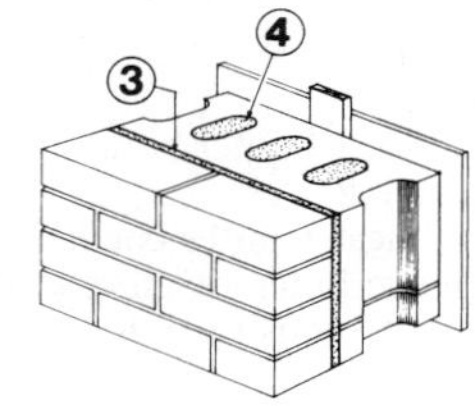

Substitution

Replace item 3 with perlite loose fill insulation and fill block cavities with the same.

Total resistance (Example)		R = 5.58
Deduct 3. Air space	1.01	
4. Concrete block	1.46	
	2.47	– 2.47
Difference		3.11
Add 0.75″ perlite cavity fill		2.03
Perlite filled 8″ block		2.94
Net total resistance		8.08
U = 1/R = 1/8.08 =		0.12

System weight = 97 psf

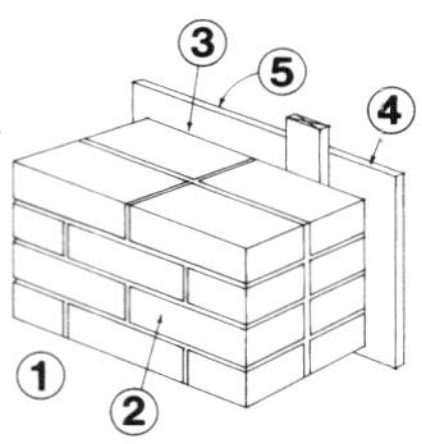

EXAMPLE:

Construction	Resistance R
1. Outside surface (15 MPH wind)	0.17
2. Common brick, 8″	1.60
3. Nonreflective air space, 0.75″	1.01
4. Gypsum board, 0.625″	0.56
5. Inside surface (still air)	0.68
Total resistance	4.02

U = 1/R = 1/4.02 = 0.25

System weight = 82 psf

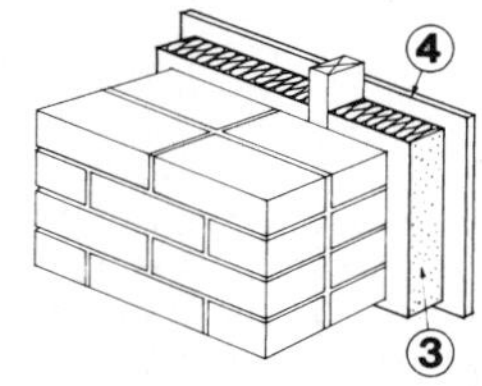

Substitution

Replace item 3 with 4″ blanket insulation and item 4 with 0.75″ Gypsum plaster (sand agg.).

Total resistance (Example)		R = 4.02
Deduct 3. Air space	1.01	
4. Gypsum board	0.56	
	1.57	– 1.57
Difference		2.45
ADD Blanket insulation		13.00
Gypsum plaster		0.14
Net total resistance		15.59
U = 1/R = 1/15.59 =		0.06

System weight = 87 psf

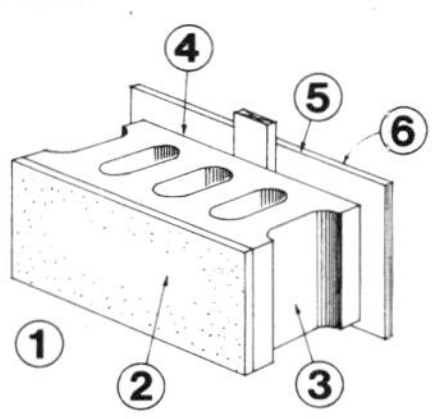

EXAMPLE:

Construction	Resistance R
1. Outside surface (15 MPH wind)	0.17
2. Cement stucco, 0.75″	0.15
3. Concrete block, 8″ light weight	2.12
4. Reflective air space, 0.75″	2.77
5. Gypsum board, 0.5″	0.45
6. Inside surface (still air)	0.68
Total resistance	6.34

U = 1/R = 1/6.34 = 0.16

System weight = 47 psf

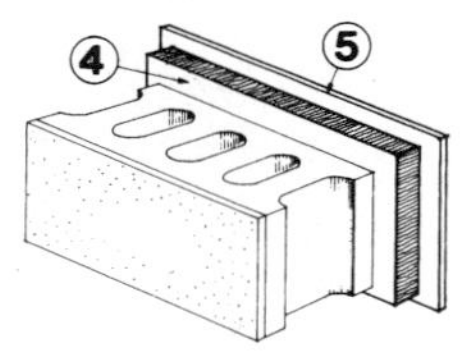

Substitution

Replace item 4 with 2″ insulation board and item 5 with 0.75″ Perlite plaster.

Total resistance (Example)		R = 6.34
Deduct 4. Air space	2.77	
5. Gypsum board	0.45	
	3.22	– 3.22
Difference		3.12
Add Insulation board 2″ (polyurethane)		12.50
Perlite plaster		1.34
Net total resistance		16.96
U = 1/R = 1/16.96 =		0.06

System weight = 48 psf

Table B2010-112 Partially and Fully Grouted Reinforced Concrete Masonry Wall Capacities Per L.F. (Kips & In-Kips)

Earthquake Zones 1, 2 & 3					Allowable Vertical Wall Loads									Allowable Wall Moments (Without Vertical Wall Loads)	
Thk.	Length Or Height		Grouted Core & Rebar		Eccentric Loads				Without Wind or Eccentric Loads		With Wind			Not Wind or Earthquake	
					7.0 in-K/Ft.		3.5 in-K/Ft.				Inspection				
					Inspection		Inspection		Inspection		No		Yes	Inspection	
T (Nom.) (in)	h′ (Ft.)	h′/T (in/in)	(spacing) (in O.C.)	Rebar Size (@ d.)	No (K/Ft.)	Yes (K/Ft.)	No (K/Ft.)	Yes (K/Ft.)	No (K/Ft.)	Yes (K/Ft.)	15 psf (K/Ft.)	30 psf (K/Ft.)	15 & 30 (K/Ft.)	No (in.-K/Ft.)	Yes (in.-K/Ft.)
8″ Conc. Block	8′	12	48″	#8	5.10	12.55	6.25	13.70	7.75	14.90	7.45	7.45	14.90	7.55	12.20
			32″	#5	5.45	13.35	6.65	14.60	7.90	15.80	7.90	7.90	15.80	6.45	9.60
			16″		6.50	15.80	7.90	17.15	9.25	18.50	9.25	9.25	18.50	7.95	12.85
			8″	↓	10.10	23.40	11.70	25.00	13.30	26.55	13.30	13.30	26.55	10.20	17.15
	12′	18	48″	#8	4.70	11.60	5.80	12.65	6.85	13.75	6.85	6.85	13.75	7.55	12.20
			32″	#5	5.05	12.35	6.15	13.45	7.30	14.60	7.30	7.30	14.60	6.45	9.60
			16″		6.00	14.55	7.30	15.85	8.55	17.10	8.55	8.55	17.10	7.95	12.85
			8″	↓	9.30	21.60	10.80	23.05	12.25	24.55	12.25	12.25	24.55	10.20	17.15
	16′	24	48″	#8	3.95	9.70	4.85	10.60	5.75	11.55	5.75	—	11.55	7.55	12.20
			32″	#5	4.20	12.25	5.15	11.30	6.10	12.25	6.10	—	12.25	6.45	9.60
			16″		5.05	12.20	6.10	13.30	7.15	14.35	7.15	—	14.35	7.95	12.85
			8″	↓	7.80	18.10	9.05	19.35	10.30	20.60	10.30	9.65	20.60	10.20	17.15
10″ Conc. Block	8′	9.6	48″	#8	7.25	16.45	8.20	17.40	9.15	—	9.15	9.15	18.35	13.10	21.05
			32″	↓	7.80	17.60	8.80	18.60	9.80	19.55	9.80	9.80	19.55	14.55	24.15
			16″	#5	9.45	21.15	10.55	22.25	11.65	23.35	11.65	11.65	23.35	13.90	22.35
			8″	↓	14.50	31.65	15.80	32.95	17.10	34.25	17.10	17.10	34.25	18.45	30.15
	12	14.4	48″	#8	7.05	16.00	8.00	16.90	8.90	17.85	8.90	8.90	17.85	13.10	21.05
			32″	↓	7.60	17.10	8.55	18.05	9.50	19.05	9.50	9.50	19.05	14.55	24.15
			16″	#5	9.20	20.55	10.25	21.60	11.35	22.70	11.35	11.35	22.70	13.90	22.35
			8″	↓	14.10	30.75	15.35	32.05	16.65	33.00	16.65	16.65	33.30	18.35	30.15
	16′	19.2	48″	#8	6.70	15.10	7.55	16.00	8.40	16.85	8.40	8.40	16.85	13.10	21.05
			32″	↓	7.15	16.15	8.05	17.05	9.00	17.95	9.00	9.00	17.95	14.55	24.15
			16″	#5	8.70	19.40	9.70	20.40	10.70	21.45	10.70	10.70	21.45	13.90	22.35
			8″	↓	13.35	29.05	14.50	30.25	15.70	31.45	15.70	15.70	31.45	18.35	30.15
	20′	24	48″	#8	6.05	13.65	6.80	14.45	7.60	15.20	7.60	—	15.20	13.10	21.05
			32″	↓	6.45	14.60	7.30	15.40	8.10	16.25	8.10	0.70	16.25	14.55	24.15
			16″	#5	7.85	17.50	8.75	18.45	9.65	19.35	9.65	0.25	19.35	13.90	22.35
			8″	↓	12.05	26.25	13.10	27.30	14.20	28.40	14.20	13.40	28.40	18.35	30.15
12″ Conc. Block	8′	8	48″	#8	9.20	20.00	10.00	20.75	10.75	21.55	10.75	10.75	21.55	15.30	24.40
			32″	↓	9.90	21.50	10.75	22.30	11.55	23.15	11.55	11.55	23.15	17.10	28.10
			16″	#5	12.10	26.05	13.00	26.95	13.90	27.85	13.90	13.90	27.85	16.20	25.90
			8″	↓	18.60	39.35	19.65	40.40	20.75	41.50	20.75	20.75	41.50	21.45	35.10
	12′	12	48″	#8	9.00	19.55	9.75	20.35	10.55	21.10	10.55	10.55	21.10	15.30	24.40
			32″	↓	9.70	21.05	10.50	21.85	11.30	22.65	11.30	11.30	22.65	17.10	28.10
			16″	#5	11.85	25.50	12.75	26.40	13.65	27.30	13.65	13.65	27.30	16.20	25.90
			8″	↓	18.20	38.50	19.25	39.55	20.30	40.60	20.30	20.30	40.60	21.45	35.10
	16′	16	48″	#8	8.65	18.75	9.35	19.50	10.10	20.20	10.10	10.10	20.20	15.30	24.40
			32″	↓	9.30	20.15	10.05	20.90	10.85	21.70	10.85	10.85	21.70	17.10	28.10
			16″	#5	11.35	24.40	12.20	25.25	13.05	26.15	13.05	13.05	26.15	16.20	25.90
			8″	↓	17.45	36.90	18.45	37.90	19.45	38.90	19.45	19.45	38.90	21.45	35.10
	24′	24	48″	#8	7.05	15.35	7.65	15.95	8.30	16.55	8.30	—	16.55	15.30	24.40
			32″	↓	7.60	16.55	8.25	17.15	8.90	17.80	8.90	—	17.80	17.10	28.10
			16″	#5	9.30	20.00	10.00	20.70	10.70	21.45	10.40*	—	21.45	16.20	25.90
			8″	↓	14.30	30.25	15.10	31.10	15.95	31.90	15.55*	15.15	31.90	21.45	35.10

*Zone 3 only

Table B2010-202 Mix Proportions by Volume, Compressive Strength of Mortar

Where Used	Allowable Proportions by Volume					Compressive Strength @ 28 days
	Mortar Type	Portland Cement	Masonry Cement	Hydrated Lime	Masonry Sand	
Plain Masonry	M	1	1	—	6	2500 psi
		1	—	1/4	3	
	S	1/2	1	—	4	1800 psi
		1	—	1/4 to 1/2	4	
	N	—	1	—	3	750 psi
		1	—	1/2 to 1-1/4	6	
	O	—	1	—	3	350 psi
		1	—	1-1/4 to 2-1/2	9	
	K	1	—	2-1/2 to 4	12	75 psi
Reinforced Masonry	PM	1	1	—	6	2500 psi
	PL	1	—	1/4 to 1/2	4	2500 psi

Note: The total aggregate should be between 2.25 to 3 times the sum of the cement and lime used.

Table B2010-203 Volume and Cost Per S.F. of Grout Fill for Concrete Block Walls

Center to Center Spacing Grouted Cores	6″ C.M.U. Per S.F.				8″ C.M.U. Per S.F.				12″ C.M.U. Per S.F.			
	Volume in C.F.		Cost		Volume in C.F.		Cost		Volume in C.F.		Cost	
	40% Solid	75% Solid	40% Solid	75% Solid	40% Solid	75% Solid	40% Solid	75% Solid	40% Solid	75% Solid	40% Solid	75% Solid
All cores grouted solid	0.27	0.11	$3.09	$1.26	0.36	0.15	$4.12	$1.72	0.55	0.23	$6.30	$2.63
cores grouted 16″ O.C.	0.14	0.06	1.60	0.69	0.18	0.08	2.06	0.92	0.28	0.12	3.21	1.37
cores grouted 24″ O.C.	0.09	0.04	1.03	0.46	0.12	0.05	1.37	0.57	0.18	0.08	2.06	0.92
cores grouted 32″ O.C.	0.07	0.03	0.80	0.34	0.09	0.04	1.03	0.46	0.14	0.06	1.60	0.69
cores grouted 40″ O.C.	0.05	0.02	0.57	0.23	0.07	0.03	0.80	0.34	0.11	0.05	1.26	0.57
cores grouted 48″ O.C.	0.04	0.02	0.46	0.23	0.06	0.03	0.69	0.34	0.09	0.04	1.03	0.46

Installed costs, including O&P, based on High-Lift Grouting method

Low-Lift Grouting is used when the wall is built to a maximum height of 5′. The grout is pumped or poured into the cores of the concrete block. The operation is repeated after each five additional feet of wall height has been completed.

High-Lift Grouting is used when the wall has been built to the full-story height. Some of the advantages are: the vertical reinforcing steel can be placed after the wall is completed and the grout can be supplied by a ready-mix concrete supplier so that it may be pumped in a continuous operation.

Pre-engineered Steel Buildings

These buildings are manufactured by many companies and normally erected by franchised dealers throughout the U.S. The four basic types are: Rigid Frames, Truss type, Post and Beam and the Sloped Beam type. The most popular roof slope is a low pitch of 1″ in 12″. The minimum economical area of these buildings is about 3000 S.F. of floor area. Bay sizes are usually 20′ to 24′ but can go as high as 30′ with heavier girts and purlins. Eave heights are usually 12′ to 24′ with 18′ to 20′ most typical.

Material prices generally do not include floors, foundations, interior finishes or utilities. Costs assume at least three bays of 24′ each, a 1″ in 12″ roof slope, and they are based on 30 psf roof load and 20 psf wind load (wind load is a function of wind speed, building height, and terrain characteristics; this should be determined by a registered structural engineer) and no unusual requirements. Costs include the structural frame, 26 ga. non-insulated colored corrugated or ribbed roofing or siding panels, fasteners, closures, trim and flashing but no allowance for insulation, doors, windows, skylights, gutters or downspouts. Very large projects would generally cost less for materials than the prices shown.

Conditions at the site, weather, shape and size of the building, and labor availability will affect the erection cost of the building.

Table B3010-011 Thermal Coefficients for Roof Systems

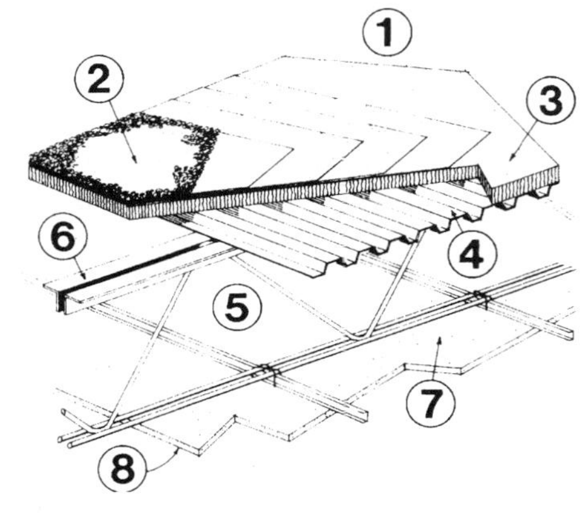

EXAMPLE:

Construction	**Resistance R**
1. Outside surface (15 MPH wind)	0.17
2. Built up roofing .375″ thick	0.33
3. Rigid roof insulation, 2″	5.26
4. Metal decking	0.00
5. Air space (non reflective)	1.14
6. Structural members	0.00
7. Gypsum board ceiling 5/8″ thick	0.56
8. Inside surface (still air)	0.61
Total resistance	8.07
U = 1/R = 1/8.07 =	0.12

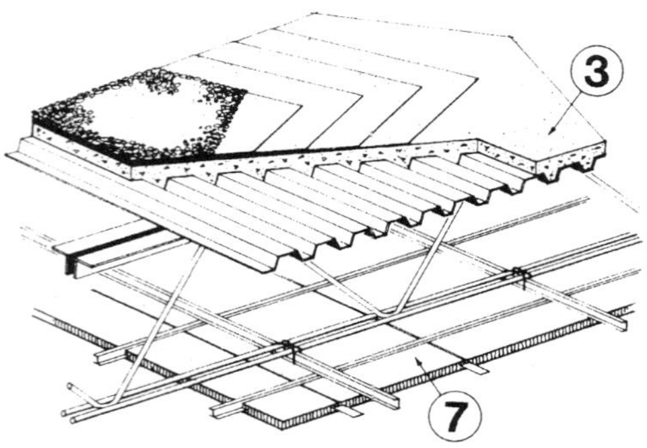

Substitution

Replace item 3 with Perlite lightweight concrete and item 7 with acoustical tile.

Total resistance (Example)		**R =** 8.07
Deduct 3. Rigid insulation	5.26	
7. Gypsum board ceiling	0.05	
	5.31	-5.31
Difference		2.76
Add Perlite concrete, 4″		4.32
Acoustical tile, 3/4″		1.56
Total resistance		8.64
U = 1/R = 1/8.64 =		0.12

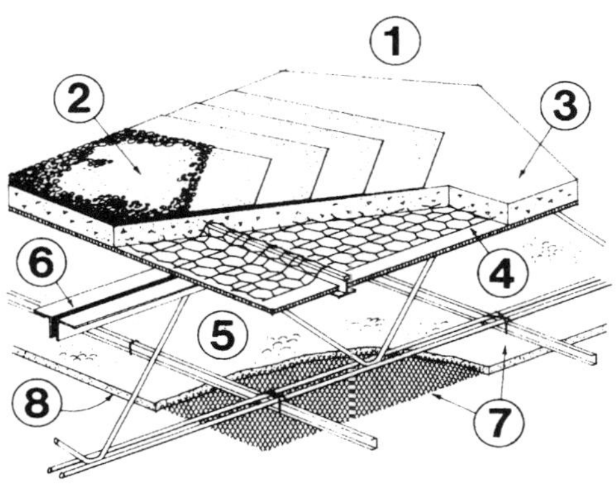

EXAMPLE

Construction	**Resistance R**
1. Outside surface (15 MPH wind)	0.17
2. Built up roofing .375″ thick	0.33
3. Rigid roof insulation, 2″ thick	1.20
4. Form bd., 1/2″ thick	0.45
5. Air space (non reflective)	1.14
6. Structural members	0.00
7. Gypsum board ceiling 3/4″ thick	0.14
8. Inside surface (still air)	0.61
Total resistance	4.04
U = 1/R = 1/4.04 =	0.25

(for modified system see next page)

Table B3010-011 Thermal Coefficients for Roof Systems

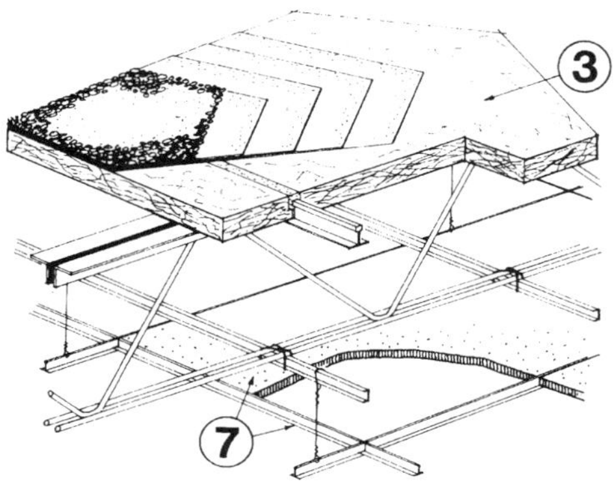

Substitution
Replace item 3 with Tectum board and item 7 with acoustical tile.

Total resistance (Example)		**R** =	4.04
Deduct 3. Gypsum concrete	1.20		
4. Form bd.	0.45		
7. Plaster	0.14		
	1.79		<-1.79>
Difference			2.25
Add Tectum bd. 3″ thick			5.25
Acoustical tile, 3/4″			1.56
		Total resistance	9.06
		U = 1/R = 1/9.06 =	0.11

(for limited system see previous page)

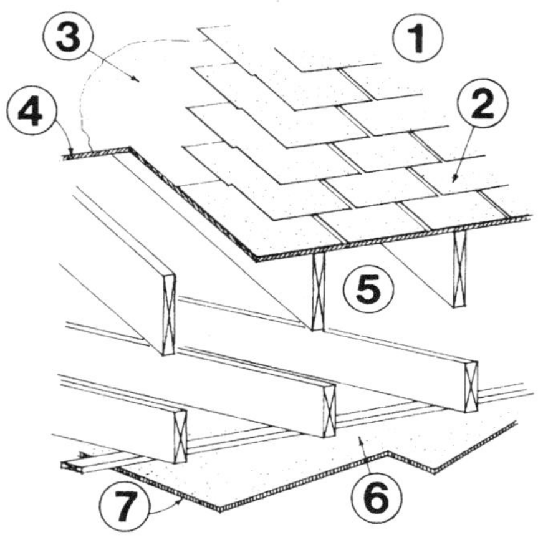

EXAMPLE:

Construction	**Resistance R**
1. Outside surface (15 MPH wind)	0.17
2. Asphalt shingle roof	0.44
3. Felt building paper	0.06
4. Plywood sheathing, 5/8″ thick	0.78
5. Reflective air space, 3-1/2″	2.17
6. Gypsum board, 1/2″ foil backed	0.45
7. Inside surface (still air)	0.61
Total resistance	4.68
U = 1/R = 1/4.68 =	0.21

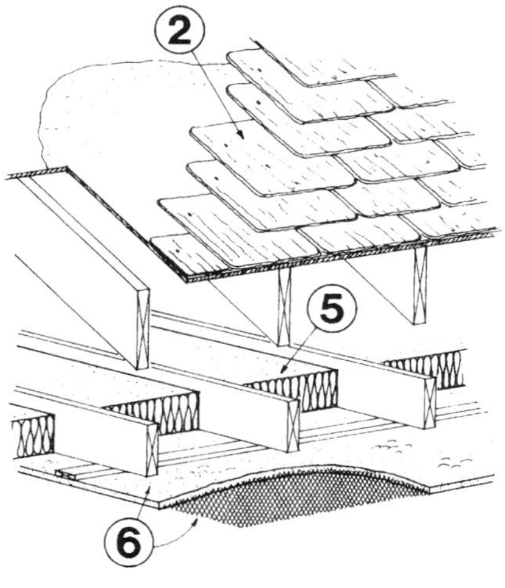

Substitution
Replace item 2 with slate, item 5 with insulation and item 6 with gypsum plaster

Total resistance (Example)		**R** =	4.68
Deduct 2. Asphalt shingles	0.44		
5. Air space	2.17		
6. Gypsum board	0.45		
	3.06		<-3.06>
Difference			1.62
Add Slate, 1/4″ thick			0.03
Insulation, 6″ blanket			19.00
Gypsum sand plaster 3/4″			0.14
		Total resistance	20.79
		U = 1/R = 1/20.79 =	0.05

Table C2010-101 Typical Range of Risers for Various Story Heights

General Design: See Table L1010-101 for code requirements.
Maximum height between landings is 12′; usual stair angle is 20° to 50° with 30° to 35° best. Usual relation of riser to treads is:

Riser + tread = 17.5
2x (Riser) + tread = 25
Riser x tread = 70 or 75

Maximum riser height is 7″ for commercial, 8-1/4″ for residential.
Usual riser height is 6-1/2″ to 7-1/4″.
Minimum tread width is 11″ for commercial, 9″ for residential.

Story Height	Minimum Risers	Maximum Riser Height	Tread Width	Maximum Risers	Minimum Riser Height	Tread Width	Average Risers	Average Riser Height	Tread Width
7′-6″	12	7.50″	10.00″	14	6.43″	11.07″	13	6.92″	10.58″
8′-0″	13	7.38	10.12	15	6.40	11.10	14	6.86	10.64
8′-6″	14	7.29	10.21	16	6.38	11.12	15	6.80	10.70
9′-0″	15	7.20	10.30	17	6.35	11.15	16	6.75	10.75
9′-6″	16	7.13	10.37	18	6.33	11.17	17	6.71	10.79
10′-0″	16	7.50	10.00	19	6.32	11.18	18	6.67	10.83
10′-6″	17	7.41	10.09	20	6.30	11.20	18	7.00	10.50
11′-0″	18	7.33	10.17	21	6.29	11.21	19	6.95	10.55
11′-6″	19	7.26	10.24	22	6.27	11.23	20	6.90	10.60
12′-0″	20	7.20	10.30	23	6.26	11.24	21	6.86	10.64
12′-6″	20	7.50	10.00	24	6.25	11.25	22	6.82	10.68
13′-0″	21	7.43	10.07	25	6.24	11.26	22	7.09	10.41
13′-6″	22	7.36	10.14	25	6.48	11.02	23	7.04	10.46
14′-0″	23	7.30	10.20	26	6.46	11.04	24	7.00	10.50

Table D1010-011 Elevator Hoistway Sizes

Elevator Type	Floors	Building Type	Capacity Lbs.	Capacity Passengers	Entry	Hoistway Width	Hoistway Depth	S.F. Area per Floor
Hydraulic	5	Apt./Small office	1500	10	S	6′-7″	4′-6″	29.6
			2000	13	S	7′-8″	4′-10″	37.4
	7	Average office/Hotel	2500	16	S	8′-4″	5′-5″	45.1
			3000	20	S	8′-4″	5′-11″	49.3
		Large office/Store	3500	23	S	8′-4″	6′-11″	57.6
		Freight light duty	2500		D	7′-2″	7′-10″	56.1
		Heavy duty	5000		D	1C′-2″	10′-10″	110.1
			7500		D	1C′-2″	12′-10″	131
			5000		D	1C′-2″	10′-10″	110.1
			7500		D	1C′-2″	12′-10″	131
			10000		D	1C′-4″	14′-10″	153.3
		Hospital	3500		D	6′-10″	9′-2″	62.7
			4000		D	7′-4″	9′-6″	69.6
Electric								
Traction, high speed	High Rise	Apt./Small office	2000	13	S	7′-8″	5′-10″	44.8
			2500	16	S	8′-4″	6′-5″	54.5
			3000	20	S	8′-4″	6′-11″	57.6
			3500	23	S	8′-4″	7′-7″	63.1
		Store	3500	23	S	9′-5″	6′-10″	64.4
		Large office	4000	26	S	9′-4″	7′-6″	70
		Hospital	3500		D	7′-6″	9′-2″	69.4
			4000		D	7′-10″	9′-6″	58.8
Geared, low speed		Apartment	1200	8	S	6′-4″	5′-3″	33.2
			2000	13	S	7′-8″	5′-8″	43.5
			2500	16	S	8′-4″	6′-3″	52
		Office	3000	20	S	8′-4″	6′-9″	56
		Store	3500	23	S	9′-5″	6′-10″	64.4
						Add 4″ width for multiple units		

Figure D1010-012 Elevator Passenger Capacity During Peak Periods

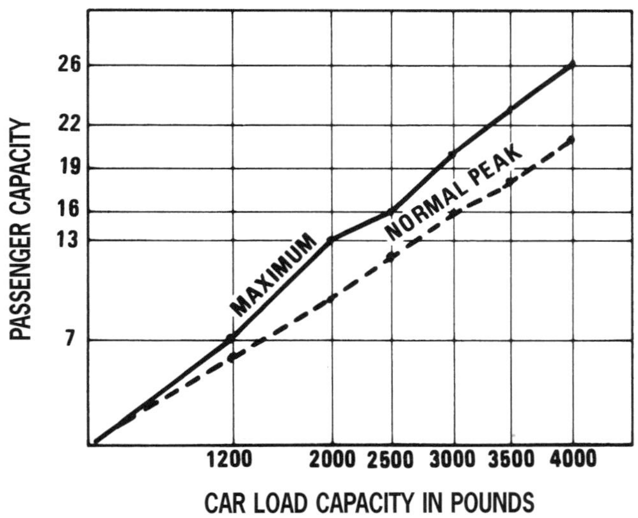

Figure D1010-013 Speed & Travel Chart

All speeds are available with any capacity car

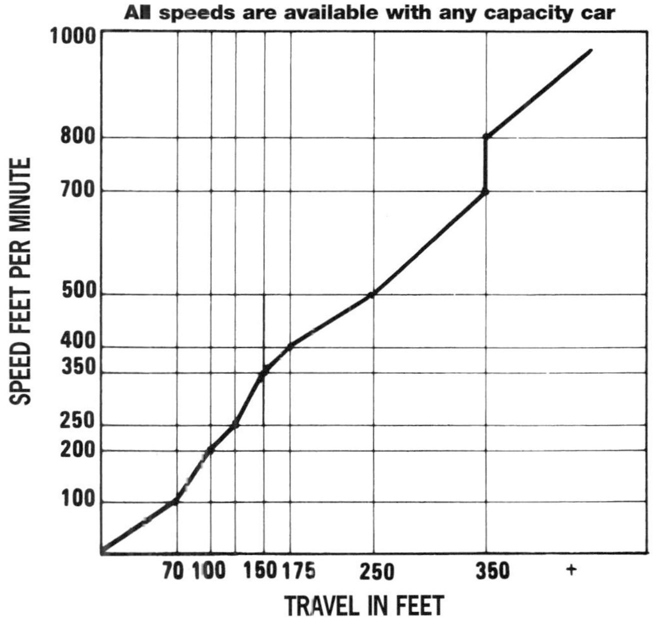

Table D1010-014 Elevator Speed vs. Height Requirements

Building Type and Elevator Capacities	Travel Speeds in Feet per Minute								
	100 fpm	200 fpm	250 fpm	350 fpm	400 fpm	500 fpm	700 fpm	800 fpm	1000 fpm
Apartments 1200 lb. to 2500 lb.	to 70	to 100	to 125	to 150	to 175	to 250	to 350		
Department Stores 1200 lb. to 2500 lb.		100		125	175	250	350		
Hospitals 3500 lb. to 4000 lb.	70	100	125	150	175	250	350		
Office Buildings 2000 lb. to 4000 lb.		100	125	150		175	250	to 350	over 350

Note: Vertical transportation capacity may be determined by code occupancy requirements of the number of square feet per person divided into the total square feet of building type. If we are contemplating an office building, we find that the Occupancy Code Requirement is 100 S.F. per person. For a 20,000 S.F. building, we would have a legal capacity of two hundred people. Elevator handling capacity is subject to the five minute evacuation recommendation, but it may vary from 11 to 18%. Speed required is a function of the travel height, number of stops and capacity of the elevator.

Table D2010-031 Plumbing Approximations for Quick Estimating

Water Control
Water Meter; Backflow Preventer,
Shock Absorbers; Vacuum
Breakers; Mixer .... 10 to 15% of Fixtures
Pipe And Fittings .... 30 to 60% of Fixtures

Note: Lower percentage for compact buildings or larger buildings with plumbing in one area.
Larger percentage for large buildings with plumbing spread out.
In extreme cases pipe may be more than 100% of fixtures.
Percentages **do not** include special purpose or process piping.

Plumbing Labor
1 & 2 Story Residential .... Rough-in Labor = 80% of Materials
Apartment Buildings .... Rough-in Labor = 90 to 100% of Materials
Labor for handling and placing fixtures is approximately 25 to 30% of fixtures
Quality/Complexity Multiplier (for all installations)
Economy installation, add .... 0 to 5%
Good quality, medium complexity, add .... 5 to 15%
Above average quality and complexity, add .... 15 to 25%

Table D2010-032 Pipe Material Considerations

1. Malleable fittings should be used for gas service.
2. Malleable fittings are used where there are stresses/strains due to expansion and vibration.
3. Cast fittings may be broken as an aid to disassembling of heating lines frozen by long use, temperature and minerals.
4. Cast iron pipe is extensively used for underground and submerged service.
5. Type M (light wall) copper tubing is available in hard temper only and is used for nonpressure and less severe applications than K and L.
6. Type L (medium wall) copper tubing is available hard or soft for interior service.
7. Type K (heavy wall) copper tubing, available in hard or soft temper for use where conditions are severe. For underground and interior service.
8. Hard drawn tubing requires fewer hangers or supports but should not be bent. Silver brazed fittings are recommended, but soft solder is normally used.
9. Type DMV (very light wall) copper tubing is designed for drainage, waste and vent plus other non-critical pressure services.

Table D2010-033 Domestic/Imported Pipe and Fittings Cost

The prices shown in this publication for steel/cast iron pipe and steel, cast iron, and malleable iron fittings are based on domestic production sold at the normal trade discounts. The above listed items of foreign manufacture may be available at prices 1/3 to 1/2 of those shown. Some imported items after minor machining or finishing operations are being sold as domestic to further complicate the system.

Caution: Most pipe prices in this data set also include a coupling and pipe hangers which for the larger sizes can add significantly to the per foot cost and should be taken into account when comparing "book cost" with quoted supplier's cost

Table D2010-401 Minimum Plumbing Fixture Requirements

Classification	Occupancy	Description	Water Closet		Lavatories		Bathtubs/Showers	Drinking Fountains	Other
			Male	Female	Male	Female			
Assembly	A-1	Theaters and other buildings for the performing arts and motion pictures	1:125	1:65	1:200			1:500	1 Service Sink
	A-2	Nightclubs, bars, taverns, dance halls	1:40		1:75			1:500	1 Service Sink
		Restaurants, banquet halls, food courts	1:75		1:200			1:500	1 Service Sink
	A-3	Auditorium w/o permanent seating, art galleries, exhibition halls, museums, lecture halls, libraries, arcades & gymnasiums	1:125	1:65	1:200			1:500	1 Service Sink
		Passenger terminals and transportation facilities	1:500		1:750			1:1000	1 Service Sink
		Places of worship and other religious services	1:150	1:75	1:200			1:1000	1 Service Sink
	A-4	Indoor sporting events and activities, coliseums, arenas, skating rinks, pools, and tennis courts	1:75 for the first 1500, then 1:120 for the remainder	1:40 for the first 1520, then 1:60 for the remainder	1:200	1:150		1:1000	1 Service Sink
	A-5	Outdoor sporting events and activities, stadiums, amusement parks, bleachers, grandstands	1:75 for the first 1500, then 1:120 for the remainder	1:40 for the first 1520, then 1:60 for the remainder	1:200	1:150		1:1000	1 Service Sink
Business	B	Buildings for the transaction of business, professional services, other services involving merchandise, office buildings, banks, light industrial	1:25 for the first 50, then 1:50 for the remainder		1:40 for the first 80, then 1:80 for the remainder			1:100	1 Service Sink
Educational	E	Educational facilities	1:50		1:50			1:100	1 Service Sink
Factory and industrial	F-1 and F-2	Structures in which occupants are engaged in work fabricating, assembly or processing of products or materials	1:100		1:100		See *International Plumbing Code*	1:400	1 Service Sink
Institutional	I-1	Residential care	1:10		1:10		1:8	1:100	1 Service Sink
	I-2	Hospitals, ambulatory nursing home care recipient	1 per room		1 per room		1:15	1:100	1 Service Sink
		Employees, other than residential care	1:25		1:35			1:100	
		Visitors, other than residential care	1:75		1:100			1:500	
	I-3	Prisons	1 per cell		1 per cell		1:15	1:100	1 Service Sink
		Reformatories, detention and correction centers	1:15		1:15		1:15	1:100	1 Service Sink
		Employees	1:25		1:35			1:100	
	I-4	Adult and child day care	1:15		1:15		1	1:100	1 Service Sink
Mercantile	M	Retail stores, service stations, shops, salesrooms, markets and shopping centers	1:500		1:750			1:1000	1 Service Sink
Residential	R-1	Hotels, motels, boarding houses (transient)	1 per sleeping unit		1 per sleeping unit		1 per sleeping unit		1 Service Sink
	R-2	Dormitories, fraternities, sororities and boarding houses (not transient)	1:10		1:10		1:8	1:100	1 Service Sink
		Apartment house	1 per dwelling unit		1 per dwelling unit		1 per dwelling unit		1 Kitchen sink per dwelling; 1 clothes washer connection per 20 dwellings
	R-3	1 and 2 family dwellings	1 per dwelling unit		1:10		1 per dwelling unit		1 Kitchen sink per dwelling; 1 clothes washer connection per dwelling
	R-3	Congregate living facilities w/<16 people	1:10		1:10		1:8	1:100	1 Service Sink
	R-4	Congregate living facilities w/<16 people	1:10		1:10		1:8	1:100	1 Service Sink
Storage	S-1 and S-2	Structures for the storage of goods, warehouses, storehouses and freight depots, low and moderate hazard	1:100		1:100		See *International Plumbing Code*	1;1000	1 Service Sink

Table 2902.1

Excerpted from the 2012 *International Building Code*, Copyright 2011. Washington, D.C.: International Code Council. Reproduced with permission. www.ICCSAFE.org

Table D2020-101 Hot Water Consumption Rates

Fixture Type	Water Supply Fixture Unit Value		
	Hot Water	Cold Water	Combined
Bathtub	1.0	1.0	1.4
Clothes Washer	1.0	1.0	1.4
Dishwasher	1.4	0.0	1.4
Kitchen Sink	1.0	1.0	1.4
Laundry Tub	1.0	1.0	1.4
Lavatory	0.5	0.5	0.7
Shower Stall	1.0	1.0	1.4
Water Closet (tank type)	0.0	2.2	2.2
Full Bath Group w/bathtub or shower stall	1.5	2.7	3.6
Half Bath Group w/W.C. and Lavatory	0.5	2.5	2.6
Kitchen Group w/Dishwasher and Sink	1.9	1.0	2.5
Laundry Group w/Clothes Washer and Laundry Tub	1.8	1.8	2.5
Hose bibb (sillcock)	0.0	2.5	2.5

Notes:
Typically, WSFU = 1GPM

Supply loads in the building water-distribution system shall be determined by total load on the pipe being sized, in terms of water supply fixture units (WSFU) and gallons per minute (GPM) flow rates. For fixtures not listed, choose a WSFU value of a fixture with similar flow characteristics. Water Fixture Supply Units determined the required water supply to fixtures and their service systems. Fixture units are equal to one (1) cubic foot of water drained in a 1-1/4″ pipe per minute. It is not a flow rate unit but a design factor.

Excerpted from the 2012 *International Plumbing Code*, Copyright 2011.
Washington, D.C.: International Code Council. Reproduced with permission.

Table D2020-102 Fixture Demands in Gallons per Fixture per Hour

The table below is based on a 140°F final temperature except for dishwashers in public places where 180°F water is mandatory.

Supply Systems for Flush Tanks			Supply Systems for Flushometer Valves		
Load	Demand		Load	Demand	
WSFU	GPM	CU. FT.	WSFU	GPM	CU. FT.
1.0	3.0	0.041040	-	-	-
2.0	5.0	0.068400	-	-	-
3.0	6.5	0.868920	-	-	-
4.0	8.0	1.069440	-	-	-
5.0	9.4	1.256592	5.0	15	2.0052
6.0	10.7	1.430376	6.0	17.4	2.326032
7.0	11.8	1.577424	7.0	19.8	2.646364
8.0	12.8	1.711104	8.0	22.2	2.967696
9.0	13.7	1.831416	9.0	24.6	3.288528
10.0	14.6	1.951728	10.0	27	3.60936
11.0	15.4	2.058672	11.0	27.8	3.716304
12.0	16.0	2.138880	12.0	28.6	3.823248
13.0	16.5	2.205720	13.0	29.4	3.930192
14.0	17.0	2.272560	14.0	30.2	4.037136
15.0	17.5	2.339400	15.0	31	4.14408
16.0	18.0	2.906240	16.0	31.8	4.241024
17.0	18.4	2.459712	17.0	32.6	4.357968
18.0	18.8	2.513184	18.0	33.4	4.464912
19.0	19.2	2.566656	19.0	34.2	4.571856
20.0	19.6	2.620218	20.0	35.0	4.678800
25.0	21.5	2.874120	25.0	38.0	5.079840
30.0	23.3	3.114744	30.0	42.0	5.611356
35.0	24.9	3.328632	35.0	44.0	5.881920
40.0	26.3	3.515784	40.0	46.0	6.149280
45.0	27.7	3.702936	45.0	48.0	6.416640
50.0	29.1	3.890088	50.0	50.0	6.684000

Notes:
When designing a plumbing system that utilizes fixtures other than, or in addition to, water closets, use the data provided in the Supply Systems for Flush Tanks section of the above table.

To obtain the probable maximum demand, multiply the total demands for the fixtures (gal./fixture/hour) by the demand factor. The heater should have a heating capacity in gallons per hour equal to this maximum. The storage tank should have a capacity in gallons equal to the probable maximum demand multiplied by the storage capacity factor.

Table D3010-601 Collector Tilt for Domestic Hot Water

Optimum collector tilt is usually equal to the site latitude. Variations of plus or minus 10 degrees are acceptable, and orientation of 20 degrees on either side of true south is acceptable. However, local climate and collector type may influence the choice between east or west deviations.

Flat plate collectors consist of a number of components as follows: Insulation to reduce heat loss through the bottom and sides of the collector. The enclosure which contains all the components in this assembly is usually weatherproof and prevents dust, wind and water from coming in contact with the absorber plate. The cover plate usually consists of one or more layers of a variety of glass or plastic and reduces the reradiation. It creates an air space which traps the heat by reducing radiation losses between the cover plate and the absorber plate.

The absorber plate must have a good thermal bond with the fluid passages. The absorber plate is usually metallic and treated with a surface coating which improves absorptivity. Black or dark paints or selective coatings are used for this purpose, and the design of this passage and plate combination helps determine a solar system's effectiveness.

Heat transfer fluid passage tubes are attached above and below or integral with an absorber plate for the purpose of transferring thermal energy from the absorber plate to a heat transfer medium. The heat exchanger is a device for transferring thermal energy from one fluid to another. The rule of thumb of space heating sizing is one S.F. of collector per 2.5 S.F. of floor space.

For domestic hot water the rule of thumb is 3/4 S.F. of collector for one gallon of water used per day, with an average use of twenty-five gallons per day per person, plus ten gallons per dishwasher or washing machine.

Table D3020-011

Heating Systems

The basic function of a heating system is to bring an enclosed volume up to a desired temperature and maintain that temperature within a reasonable range. To accomplish this, the selected system must have sufficient capacity to offset transmission losses resulting from the temperature difference on the interior and exterior of the enclosing walls in addition to losses due to cold air infiltration through cracks, crevices and around doors and windows. The amount of heat to be furnished is dependent upon the building size, construction, temperature difference, air leakage, use, shape, orientation and exposure. Air circulation is also an important consideration. Circulation will prevent stratification which could result in heat losses through uneven temperatures at various levels. For example, the most efficient use of unit heaters can usually be achieved by circulating the space volume through the total number of units once every 20 minutes or 3 times an hour. This general rule must, of course, be adapted for special cases such as large buildings with low ratios of heat transmitting surface to cubical volume. The type of occupancy of a building will have considerable bearing on the number of heat transmitting units and the location selected. It is axiomatic, however, that the basis of any successful heating system is to provide the maximum amount of heat at the points of maximum heat loss such as exposed walls, windows, and doors. Large roof areas, wind direction, and wide doorways create problems of excessive heat loss and require special consideration and treatment.

Heat Transmission

Heat transfer is an important parameter to consider during selection of the exterior wall style, material and window area. A high rate of transfer will permit greater heat loss during the wintertime with the resultant increase in heating energy costs and a greater rate of heat gain in the summer with proportionally greater cooling cost. Several terms are used to describe various aspects of heat transfer. However, for general estimating purposes this book lists U valves for systems of construction materials. U is the "overall heat transfer coefficient." It is defined as the heat flow per hour through one square foot when the temperature difference in the air on either side of the structure wall, roof, ceiling or floor is one degree Fahrenheit. The structural segment may be a single homogeneous material or a composite.

Total heat transfer is found using the following equation:

$Q = AU(T_2 - T_1)$ where

- Q = Heat flow, BTU per hour
- A = Area, square feet
- U = Overall heat transfer coefficient
- $(T_2 - T_1)$ = Difference in temperature of air on each side of the construction component. (Also abbreviated TD)

Note that heat can flow through all surfaces of any building and this flow is in addition to heat gain or loss due to ventilation, infiltration and generation (appliances, machinery, people).

Table D3020-012 Heating Approximations for Quick Estimating

Oil Piping & Boiler Room Piping:	
Small System	20 to 30% of Boiler
Complex System with Pumps, Headers, Etc.	80 to 110% of Boiler
Breeching With Insulation:	
Small	10 to 15% of Boiler
Large	15 to 25% of Boiler
Coils:	15 to 30% of Containing Unit
Balancing (Independent)	1/2% of H.V.A.C. Estimating

Quality/Complexity Adjustment: For all heating installations add these adjustments to the estimate to more closely allow for the equipment and conditions of the particular job under consideration.

Economy installation, add	0 to 5% of System
Good quality, medium complexity, add	5 to 15% of System
Above average quality and complexity, add	15 to 25% of System

Table D3030-012 Air Conditioning Requirements

BTU's per hour per S.F. of floor area and S.F. per ton of air conditioning.

Type of Building	BTU/Hr per S.F.	S.F. per Ton	Type of Building	BTU/Hr per S.F.	S.F. per Ton	Type of Building	BTU/Hr per S.F.	S.F. per Ton
Apartments, Individual	26	450	Dormitory, Rooms	40	300	Libraries	50	240
Corridors	22	550	Corridors	30	400	Low Rise Office, Exterior	38	320
Auditoriums & Theaters	40	300/18*	Dress Shops	43	280	Interior	33	360
Banks	50	240	Drug Stores	80	150	Medical Centers	28	425
Barber Shops	48	250	Factories	40	300	Motels	28	425
Bars & Taverns	133	90	High Rise Office—Ext. Rms.	46	263	Office (small suite)	43	280
Beauty Parlors	66	180	Interior Rooms	37	325	Post Office Individual Office	42	285
Bowling Alleys	68	175	Hospitals, Core	43	280	Central Area	46	260
Churches	36	330/20*	Perimeter	46	260	Residences	20	600
Cocktail Lounges	68	175	Hotel, Guest Rooms	44	275	Restaurants	60	200
Computer Rooms	141	85	Corridors	30	400	Schools & Colleges	46	260
Dental Offices	52	230	Public Spaces	55	220	Shoe Stores	55	220
Dept. Stores, Basement	34	350	Industrial Plants, Offices	38	320	Shop'g. Ctrs., Supermarkets	34	350
Main Floor	40	300	General Offices	34	350	Retail Stores	48	250
Upper Floor	30	400	Plant Areas	40	300	Specialty	60	200

*Persons per ton
12,000 BTU = 1 ton of air conditioning

Table D3030-013 Psychrometric Table

Dew Point or Saturation Temperature (F)

100	32	35	40	45	50	55	60	65	70	75	80	85	90	95	100
90	30	33	37	42	47	52	57	62	67	72	77	82	87	92	97
80	27	30	34	39	44	49	54	58	64	68	73	78	83	88	93
70	24	27	31	36	40	45	50	55	60	64	69	74	79	84	88
60	20	24	28	32	36	41	46	51	55	60	65	69	74	79	83
50	16	20	24	28	33	36	41	46	50	55	60	64	69	73	78
40	12	15	18	23	27	31	35	40	45	49	53	58	62	67	71
30	8	10	14	18	21	25	29	33	37	42	46	50	54	59	62
20	6	7	8	9	13	16	20	24	28	31	35	40	43	48	52
10	4	4	5	5	6	8	9	10	13	17	20	24	27	30	34
	32	**35**	**40**	**45**	**50**	**55**	**60**	**65**	**70**	**75**	**80**	**85**	**90**	**95**	**100**

Dry bulb temperature (F)

This table shows the relationship between RELATIVE HUMIDITY, DRY BULB TEMPERATURE AND DEW POINT. As an example, assume that the thermometer in a room reads 75°F, and we know that the relative humidity is 50%. The chart shows the dew point temperature to be 55°. That is, any surface colder than 55°F will "sweat" or collect condensing moisture. This surface could be the outside of an uninsulated chilled water pipe in the summertime or the inside surface of a wall or deck in the wintertime. After determining the extreme ambient parameters, the table at the left is useful in determining which surfaces need insulation or vapor barrier protection.

Table D3030-017 Ductwork Packages (per Ton of Cooling)

System	Sheet Metal	Insulation	Diffusers	Return Register
Roof Top Unit Single Zone	120 Lbs.	52 S.F.	1	1
Roof Top Unit Multizone	240 Lbs.	104 S.F.	2	1
Self-contained Air or Water Cooled	108 Lbs.	—	2	—
Split System Air Cooled	102 Lbs.	—	2	—

Systems reflect most common usage. Refer to system graphics for duct layout.

Table D3030-018 Quality/Complexity Adjustment for Air Conditioning Systems

Economy installation, add .... 0 to 5%
Good quality, medium complexity, add .... 5 to 15%
Above average quality and complexity, add .... 15 to 25%
Add the above adjustments to the estimate to more closely allow for the equipment and conditions of the particular job under consideration.

Table D3030-019 Sheet Metal Calculator (Weight in Lb./Ft. of Length)

Gauge	26	24	22	20	18	16
Wt.-Lb./S.F.	.906	1.156	1.406	1.656	2.156	2.656
SMACNA Max. Dimension – Long Side		30″	54″	84″	85″ Up	
Sum-2 sides						
2	.3	.40	.50	.60	.80	.90
3	.5	.65	.80	.90	1.1	1.4
4	.7	.85	1.0	1.2	1.5	1.8
5	.8	1.1	1.3	1.5	1.9	2.3
6	1.0	1.3	1.5	1.7	2.3	2.7
7	1.2	1.5	1.8	2.0	2.7	3.2
8	1.3	1.7	2.0	2.3	3.0	3.6
9	1.5	1.9	2.3	2.6	3.4	4.1
10	1.7	2.2	2.5	2.9	3.8	4.5
11	1.8	2.4	2.8	3.2	4.2	5.0
12	2.0	2.6	3.0	3.5	4.6	5.4
13	2.2	2.8	3.3	3.8	4.9	5.9
14	2.3	3.0	3.5	4.1	5.3	6.3
15	2.5	3.2	3.8	4.4	5.7	6.8
16	2.7	3.4	4.0	4.6	6.1	7.2
17	2.8	3.7	4.3	4.9	6.5	7.7
18	3.0	3.9	4.5	5.2	6.8	8.1
19	3.2	4.1	4.8	5.5	7.2	8.6
20	3.3	4.3	5.0	5.8	7.6	9.0
21	3.5	4.5	5.3	6.1	8.0	9.5
22	3.7	4.7	5.5	6.4	8.4	9.9
23	3.8	5.0	5.8	6.7	8.7	10.4
24	4.0	5.2	6.0	7.0	9.1	10.8
25	4.2	5.4	6.3	7.3	9.5	11.3
26	4.3	5.6	6.5	7.5	9.9	11.7
27	4.5	5.8	6.8	7.8	10.3	12.2
28	4.7	6.0	7.0	8.1	10.6	12.6
29	4.8	6.2	7.3	8.4	11.0	13.1
30	5.0	6.5	7.5	8.7	11.4	13.5
31	5.2	6.7	7.8	9.0	11.8	14.0
32	5.3	6.9	8.0	9.3	12.2	14.4
33	5.5	7.1	8.3	9.6	12.5	14.9
34	5.7	7.3	8.5	9.9	12.9	15.3
35	5.8	7.5	8.8	10.2	13.3	15.8
36	6.0	7.8	9.0	10.4	13.7	16.2
37	6.2	8.0	9.3	10.7	14.1	16.7
38	6.3	8.2	9.5	11.0	14.4	17.1
39	6.5	8.4	9.8	11.3	14.8	17.6
40	6.7	8.6	10.0	11.6	15.2	18.0
41	6.8	8.8	10.3	11.9	15.6	18.5
42	7.0	9.0	10.5	12.2	16.0	18.9
43	7.2	9.2	10.8	12.5	16.3	19.4
44	7.3	9.5	11.0	12.8	16.7	19.8
45	7.5	9.7	11.3	13.1	17.1	20.3
46	7.7	9.9	11.5	13.3	17.5	20.7
47	7.8	10.1	11.8	13.6	17.9	21.2
48	8.0	10.3	12.0	13.9	18.2	21.6
49	8.2	10.5	12.3	14.2	18.6	22.1
50	8.3	10.7	12.5	14.5	19.0	22.5
51	8.5	11.0	12.8	14.8	19.4	23.0
52	8.7	11.2	13.0	15.1	19.8	23.4
53	8.8	11.4	13.3	15.4	20.1	23.9
54	9.0	11.6	13.5	15.7	20.5	24.3
55	9.2	11.8	13.8	16.0	20.9	24.8
56	9.3	12.0	14.0	16.2	21.3	25.2
57	9.5	12.3	14.3	16.5	21.7	25.7
58	9.7	12.5	14.5	16.8	22.0	26.1
59	9.8	12.7	14.8	17.1	22.4	26.6
60	10.0	12.9	15.0	17.4	22.8	27.0
61	10.2	13.1	15.3	17.7	23.2	27.5
62	10.3	13.3	15.5	18.0	23.6	27.9
63	10.5	13.5	15.8	18.3	24.0	28.4
64	10.7	13.7	16.0	18.6	24.3	28.8
65	10.8	13.9	16.3	18.9	24.7	29.3
66	11.0	14.1	16.5	19.1	25.1	29.7
67	11.2	14.3	16.8	19.4	25.5	30.2
68	11.3	14.6	17.0	19.7	25.8	30.6
69	11.5	14.8	17.3	20.0	26.2	31.1
70	11.7	15.0	17.5	20.3	26.6	31.5
71	11.8	15.2	17.8	20.6	27.0	32.0
72	12.0	15.4	18.0	20.9	27.4	32.4
73	12.2	15.6	18.3	21.2	27.7	32.9
74	12.3	15.8	18.5	21.5	28.1	33.3
75	12.5	16.1	18.8	21.8	28.5	33.8
76	12.7	16.3	19.0	22.0	28.9	34.2
77	12.8	16.5	19.3	22.3	29.3	34.7
78	13.0	16.7	19.5	22.6	29.6	35.1
79	13.2	16.9	19.8	22.9	30.0	35.6
80	13.3	17.1	20.0	23.2	30.4	36.0
81	13.5	17.3	20.3	23.5	30.8	36.5
82	13.7	17.5	20.5	23.8	31.2	36.9
83	13.8	17.8	20.8	24.1	31.5	37.4
84	14.0	18.0	21.0	24.4	31.9	37.8
85	14.2	18.2	21.3	24.7	32.3	38.3
86	14.3	18.4	21.5	24.9	32.7	38.7
87	14.5	18.6	21.8	25.2	33.1	39.2
88	14.7	18.8	22.0	25.5	33.4	39.6
89	14.8	19.0	22.3	25.8	33.8	40.1
90	15.0	19.3	22.5	26.1	34.2	40.5
91	15.2	19.5	22.8	26.4	34.6	41.0
92	15.3	19.7	23.0	26.7	35.0	41.4
93	15.5	19.9	23.3	27.0	35.3	41.9
94	15.7	20.1	23.5	27.3	35.7	42.3
95	15.8	20.3	23.8	27.6	36.1	42.8
96	16.0	20.5	24.0	27.8	36.5	43.2
97	16.2	20.3	24.3	28.1	36.9	43.7
98	16.3	21.0	24.5	28.4	37.2	44.1
99	16.5	21.2	24.8	28.7	37.6	44.6
100	16.7	21.4	25.0	29.0	38.0	45.0
101	16.8	21.6	25.3	29.3	38.4	45.5
102	17.0	21.8	25.5	29.6	38.8	45.9
103	17.2	22.0	25.8	29.9	39.1	46.4
104	17.3	22.3	26.0	30.2	39.5	46.8
105	17.5	22.5	26.3	30.5	39.9	47.3
106	17.7	22.7	26.5	30.7	40.3	47.7
107	17.8	22.9	26.8	31.0	40.7	48.2
108	18.0	23.1	27.0	31.3	41.0	48.6
109	18.2	23.3	27.3	31.6	41.4	49.1
110	18.3	23.5	27.5	31.9	41.8	49.5

Example: If the duct is 34″ x 20″ x 15′ long, 34″ is greater than 30″ maximum, for 24 ga. so it must be 22 ga. 34″ + 20″ = 54″ going across from 54″ find 13.5 lb. per foot. 13.5 x 15′ = 202.5 lbs. For S.F. of surface area 202.5 ÷ 1.406 = 144 S.F.

Note: Figures include an allowance for scrap.
***Do not** use unless engineer specified. 26GA is very light and sometimes used for toilet exhaust 16GA is heavy plate and mostly specified for plenums and hoods.

Table D4010-102 System Classification

System Classification

Rules for installation of sprinkler systems vary depending on the classification of occupancy falling into one of three categories as follows:

Light Hazard Occupancy

The protection area allotted per sprinkler should not exceed 225 S.F., with the maximum distance between lines and sprinklers on lines being 15′. The sprinklers do not need to be staggered. Branch lines should not exceed eight sprinklers on either side of a cross main. Each large area requiring more than 100 sprinklers and without a sub-dividing partition should be supplied by feed mains or risers sized for ordinary hazard occupancy.
Maximum system area = 52,000 S.F.

Included in this group are:

Churches
Clubs
Educational
Hospitals
Institutional
Libraries
(except large stack rooms)
Museums
Nursing Homes
Offices
Residential
Restaurants
Theaters and Auditoriums
(except stages and prosceniums)
Unused Attics

Ordinary Hazard Occupancy

The protection area allotted per sprinkler shall not exceed 130 S.F. of noncombustible ceiling and 130 S.F. of combustible ceiling. The maximum allowable distance between sprinkler lines and sprinklers on line is 15′. Sprinklers shall be staggered if the distance between heads exceeds 12′. Branch lines should not exceed eight sprinklers on either side of a cross main.
Maximum system area = 52,000 S.F.

Included in this group are:

Group 1
Automotive Parking and Showrooms
Bakeries
Beverage manufacturing
Canneries
Dairy Products Manufacturing/Processing
Electronic Plans
Glass and Glass Products Manufacturing
Laundries
Restaurant Service Areas

Group 2
Cereal Mills
Chemical Plants—Ordinary
Confectionery Products
Distilleries
Dry Cleaners
Feed Mills
Horse Stables
Leather Goods Manufacturing
Libraries—Large Stack Room Areas
Machine Shops
Metal Working
Mercantile
Paper and Pulp Mills
Paper Process Plants
Piers and Wharves
Post Offices
Printing and Publishing
Repair Garages
Stages
Textile Manufacturing
Tire Manufacturing
Tobacco Products Manufacturing
Wood Machining
Wood Product Assembly

Extra Hazard Occupancy

The protection area allotted per sprinkler shall not exceed 100 S.F. of noncombustible ceiling and 100 S.F. of combustible ceiling. The maximum allowable distance between lines and sprinklers on lines is 12′. Sprinklers on alternate lines shall be staggered if the distance between sprinklers on lines exceeds 8′. Branch lines should not exceed six sprinklers on either side of a cross main.
Maximum system area:
Design by pipe schedule = 25,000 S.F.
Design by hydraulic calculation = 40,000 S.F.

Included in this group are:

Group 1
Aircraft hangars
Combustible Hydraulic Fluid Use Area
Die Casting
Metal Extruding
Plywood/Particle Board Manufacturing
Printing (inks with flash points < 100 degrees F)
Rubber Reclaiming, Compounding, Drying, Milling, Vulcanizing
Saw Mills
Textile Picking, Opening, Blending, Garnetting, Carding, Combing of Cotton, Synthetics, Wood Shoddy, or Burlap
Upholstering with Plastic Foams

Group 2
Asphalt Saturating
Flammable Liquids Spraying
Flow Coating
Manufactured/Modular Home Building Assemblies (where finished enclosure is present and has combustible interiors)
Open Oil Quenching
Plastics Processing
Solvent Cleaning
Varnish and Paint Dipping

Table D4020-301 Standpipe Systems

The basis for standpipe system design is National Fire Protection Association NFPA 14. However, the authority with jurisdiction should be consulted for special conditions, local requirements and approval.

Standpipe systems, properly designed and maintained, are an effective and valuable time saving aid for extinguishing fires, especially in the upper stories of tall buildings, the interior of large commercial or industrial malls, or other areas where construction features or access make the laying of temporary hose lines time consuming and/or hazardous. Standpipes are frequently installed with automatic sprinkler systems for maximum protection.

There are three general classes of service for standpipe systems:
Class I is for use by fire departments and personnel with special training for heavy streams (2-1/2″ hose connections).
Class II is for use by building occupants until the arrival of the fire department (1-1/2″ hose connector with hose).
Class III is for use by either fire departments and trained personnel or by the building occupants (both 2-1/2″ and 1-1/2″ hose connections or one 2-1/2″ hose valve with an easily removable 2-1/2″ by 1-1/2″ adapter).

Standpipe systems are also classified by the way water is supplied to the system. The four basic types are:
Type 1: Wet standpipe system having supply valve open and water pressure maintained at all times.
Type 2: Standpipe system arranged through the use of approved devices to admit water to the system automatically by opening a hose valve.
Type 3: Standpipe system arranged to admit water to the system through manual operation of approved remote control devices located at each hose station.
Type 4: Dry standpipe having no permanent water supply.

Table D4020-302 NFPA 14 Basic Standpipe Design

Class	Design-Use	Pipe Size Minimums	Water Supply Minimums
Class I	2 1/2″ hose connection on each floor All areas within 150′ of an exit in every exit stairway Fire Department Trained Personnel	Height to 100′, 4″ dia. Heights above 100′, 6″ dia. (275′ max. except with pressure regulators 400′ max.)	For each standpipe riser 500 GPM flow For common supply pipe allow 500 GPM for first standpipe plus 250 GPM for each additional standpipe (2500 GPM max. total) 30 min. duration 65 PSI at 500 GPM
Class II	1 1/2″ hose connection with hose on each floor All areas within 130′ of hose connection measured along path of hose travel Occupant personnel	Height to 50′, 2″ dia. Height above 50′, 2 1/2″ dia.	For each standpipe riser 100 GPM flow For multiple riser common supply pipe 100 GPM 300 min. duration, 65 PSI at 100 GPM
Class III	Both of above. Class I valved connections will meet Class III with additional 2 1/2″ by 1 1/2″ adapter and 1 1/2″ hose.	Same as Class I	Same as Class I

*Note: Where 2 or more standpipes are installed in the same building or section of the building they shall be interconnected at the bottom.

Combined Systems

Combined systems are systems where the risers supply both automatic sprinklers and 2-1/2″ hose connection outlets for fire department use. In such a system the sprinkler spacing pattern shall be in accordance with NFPA 13 while the risers and supply piping will be sized in accordance with NFPA 14. When the building is completely sprinklered the risers may be sized by hydraulic calculation. The minimum size riser for buildings not completely sprinklered is 6″.
The minimum water supply of a completely sprinklered, light hazard, high-rise occupancy building will be 500 GPM while the supply required for other types of completely sprinklered high-rise buildings is 1000 GPM.

General System Requirements

1. Approved valves will be provided at the riser for controlling branch lines to hose outlets.
2. A hose valve will be provided at each outlet for hose attachment.
3. Where pressure at any standpipe outlet exceeds 100 PSI a pressure reducer must be installed to limit the pressure to 100 PSI. Note that the pressure head due to gravity in 100′ of riser is 43.4 PSI. This must be overcome by city pressure, fire pumps, or gravity tanks to provide adequate pressure at the top of the riser.
4. Each hose valve on a wet system having a linen hose shall have an automatic drip connection to prevent valve leakage from entering the hose.
5. Each riser will have a valve to isolate it from the rest of the system.
6. One or more fire department connections as an auxiliary supply shall be provided for each Class I or Class III standpipe system. In buildings having two or more zones, a connection will be provided for each zone.
7. There will be no shutoff valve in the fire department connection, but a check valve will be located in the line before it joins the system.
8. All hose connections street side will be identified on a cast plate or fitting as to purpose.

Table D4020-303 Quality/Complexity Adjustment for Sprinkler/Standpipe Systems

Economy installation, add	0 to 5%
Good quality, medium complexity, add	5 to 15%
Above average quality and complexity, add	15 to 25%

Figure D5010-111 Typical Overhead Service Entrance

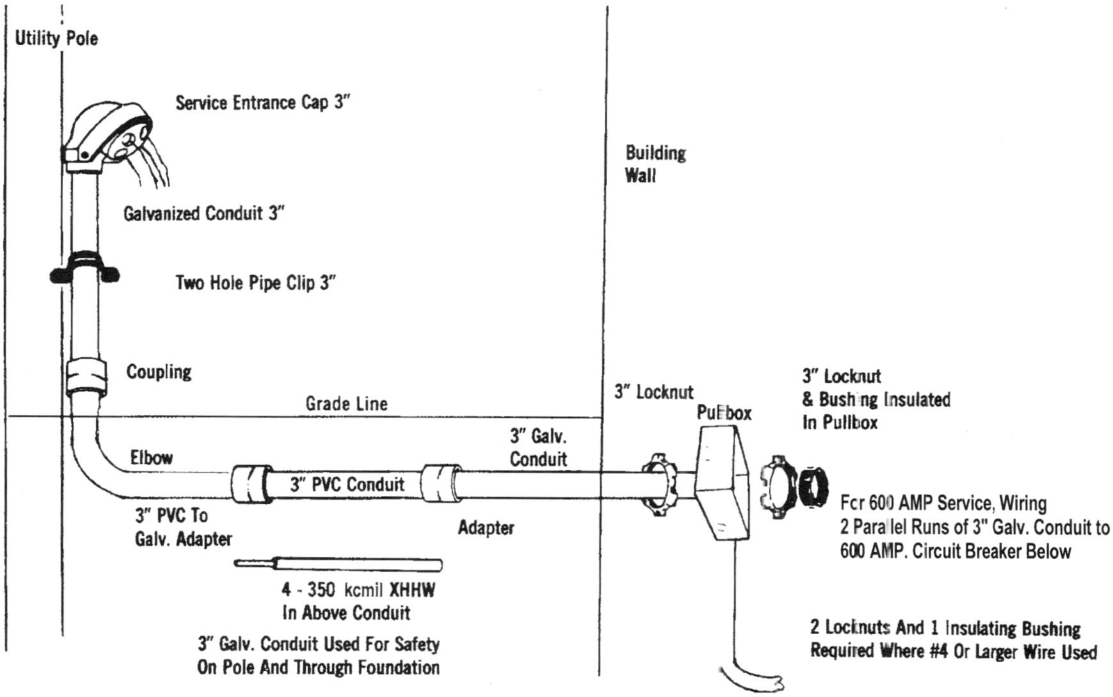

Figure D5010-112 Typical Commercial Electric System

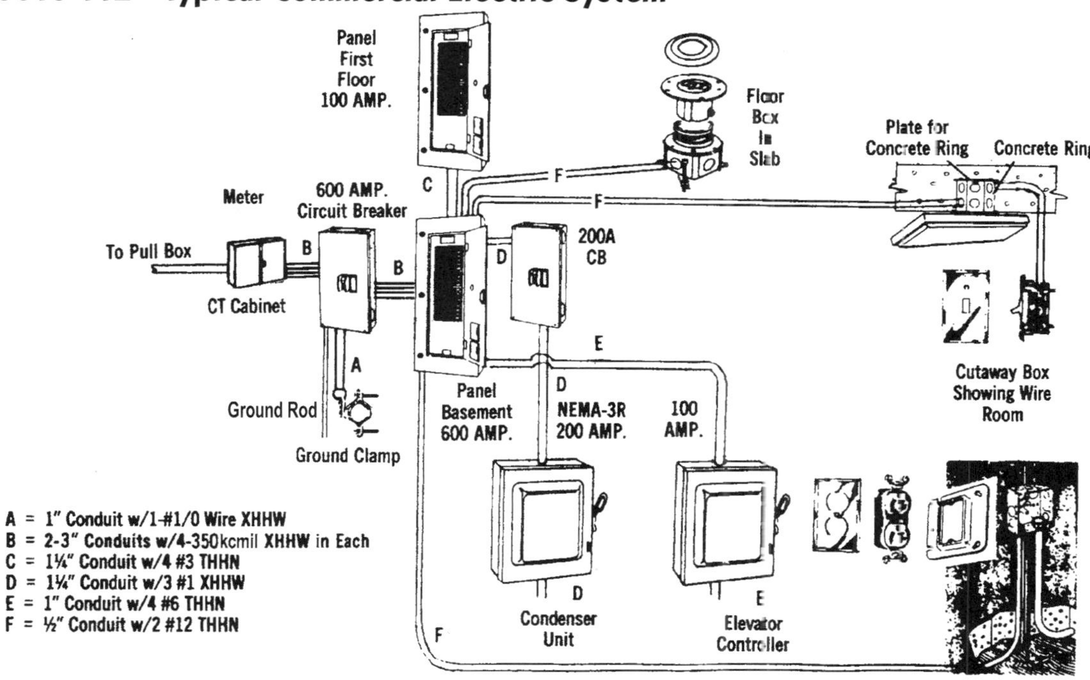

Figure D5010-113 Preliminary Procedure

1. Determine building size and use.
2. Develop total load in watts.
   a. Lighting
   b. Receptacle
   c. Air Conditioning
   d. Elevator
   e. Other power requirements
3. Determine best voltage available from utility company.
4. Determine cost from tables for loads (a) through (e) above.
5. Determine size of service from formulas (D5010-116).
6. Determine costs for service, panels, and feeders from tables.

Figure D5010-114 Office Building 90′ x 210′, 3 story, w/garage

Garage Area = 18,900 S.F.
Office Area = 56,700 S.F.
Elevator = 2 @ 125 FPM

Tables	Power Required		Watts	
D5010-1151	Garage Lighting .5 Watt/S.F.		9,450	
	Office Lighting 3 Watts/S.F.		170,100	
D5010-1151	Office Receptacles 2 Watts/S.F.		113,400	
D5010-1151, R262213-27	Low Rise Office A.C. 4.3 Watts/S.F.		243,810	
D5010-1152, 1153	Elevators - 2 @ 20 HP = 2 @ 17,404 Watts/Ea.		34,808	
D5010-1151	Misc. Motors + Power 1.2 Watts/S.F.		68,040	
		Total	639,608	Watts

Voltage Available

277/480V, 3 Phase, 4 Wire

Formula

D5010-116

$$\text{Amperes} = \frac{\text{Watts}}{\text{Volts x Power Factor x 1.73}} = \frac{639{,}608}{480\text{V x .8 x 1.73}} = 963 \text{ Amps}$$

Use 1200 Amp Service

System	Description	Unit Cost	Unit	Total
D5020 210 0200	Garage Lighting (Interpolated)	$2.72	S.F.	$ 51,408
D5020 210 0540	Office Lighting (Using 32 W lamps)	8.24	S.F.	467,208
D5020 115 0880	Receptacle-Undercarpet	3.92	S.F.	222,264
D5020 140 0280	Air Conditioning	0.65	S.F.	36,855
D5020 135 0320	Misc. Pwr.	0.35	S.F.	19,845
D5020 145 2120	Elevators - 2 @ 20HP	2,775.00	Ea.	5,550
D5010 130 1250	Underground Electric Service-1200 Amp	60,100.00	Ea.	60,100
D5010 230 0480	Feeder - Assume 200 feet	313.00	L.F.	62,600
D5010 240 0580	Panels - 277/480V, 1200 Amp	35,275.00	Ea.	35,275
D5030 910 0400	Fire Detection	35,400.00	Ea.	35,400
		Total		$996,505
		or		$996,500

Table D5010-1151 Nominal Watts Per S.F. for Electric Systems for Various Building Types

Type Construction	1. Lighting	2. Devices	3. HVAC	4. Misc.	5. Elevator	Total Watts
Apartment, luxury high rise	2	2.2	3	1		
Apartment, low rise	2	2	3	1		
Auditorium	2.5	1	3.3	.8		
Bank, branch office	3	2.1	5.7	1.4		
Bank, main office	2.5	1.5	5.7	1.4		
Church	1.8	.8	3.3	.8		
College, science building	3	3	5.3	1.3		
College, library	2.5	.8	5.7	1.4		
College, physical education center	2	1	4.5	1.1		
Department store	2.5	.9	4	1		
Dormitory, college	1.5	1.2	4	1		
Drive-in doughnut shop	3	4	6.8	1.7		
Garage, commercial	.5	.5	0	.5		
Hospital, general	2	4.5	5	1.3		
Hospital, pediatric	3	3.8	5	1.3		
Hotel, airport	2	1	5	1.3		
Housing for the elderly	2	1.2	4	1		
Manufacturing, food processing	3	1	4.5	1.1		
Manufacturing, apparel	2	1	4.5	1.1		
Manufacturing, tools	4	1	4.5	1.1		
Medical clinic	2.5	1.5	3.2	1		
Nursing home	2	1.6	4	1		
Office building, high rise	3	2	4.7	1.2		
Office building, low rise	3	2	4.3	1.2		
Radio-TV studio	3.8	2.2	7.6	1.9		
Restaurant	2.5	2	6.8	1.7		
Retail store	2.5	.9	5.5	1.4		
School, elementary	3	1.9	5.3	1.3		
School, junior high	3	1.5	5.3	1.3		
School, senior high	2.3	1.7	5.3	1.3		
Supermarket	3	1	4	1		
Telephone exchange	1	.6	4.5	1.1		
Theater	2.5	1	3.3	.8		
Town Hall	2	1.9	5.3	1.3		
U.S. Post Office	3	2	5	1.3		
Warehouse, grocery	1	.6	0	.5		

Rule of Thumb: 1 KVA = 1 HP (Single Phase)

Three Phase:

Watts = 1.73 x Volts x Current x Power Factor x Efficiency

$$\text{Horsepower} = \frac{\text{Volts x Current x 1.73 x Power Factor}}{\text{746 Watts}}$$

Table D5010-1152 Horsepower Requirements for Elevators with 3 Phase Motors

Type	Maximum Travel Height in Ft.	Travel Speeds in FPM	Capacity of Cars in Lbs.		
			1200	1500	1800
Hydraulic	70	70	10	15	15
		85	15	15	15
		100	15	15	20
		110	20	20	20
		125	20	20	20
		150	25	25	25
		175	25	30	30
		200	30	30	40
Geared Traction	300	200			
		350			
			2000	2500	3000
Hydraulic	70	70	15	20	20
		85	20	20	25
		100	20	25	30
		110	20	25	30
		125	25	30	40
		150	30	40	50
		175	40	50	50
		200	40	50	60
Geared Traction	300	200	10	10	15
		350	15	15	23
			3500	4000	4500
Hydraulic	70	70	20	25	30
		85	25	30	30
		100	30	40	40
		110	40	40	50
		125	40	50	50
		150	50	50	60
		175	60		
		200	60		
Geared Traction	300	200	15		23
		350	23		35

The power factor of electric motors varies from 80% to 90% in larger size motors. The efficiency likewise varies from 80% on a small motor to 90% on a large motor.

Table D5010-1153 Watts per Motor

90% Power Factor & Efficiency @ 200 or 460V			
HP	Watts	HP	Watts
10	9024	30	25784
15	13537	40	33519
20	17404	50	41899
25	21916	60	49634

Table D5010-116 Electrical Formulas

Ohm's Law

Ohm's Law is a method of explaining the relation between voltage, current, and resistance in an electrical circuit. It is practically the basis of all electrical calculations. The term "electromotive force" is often used to designate pressure in volts. This formula can be expressed in various forms.

To find the current in amperes:

$$\text{Current} = \frac{\text{Voltage}}{\text{Resistance}} \quad \text{or} \quad \text{Amperes} = \frac{\text{Volts}}{\text{Ohms}} \quad \text{or} \quad I = \frac{E}{R}$$

The flow of current in amperes through any circuit is equal to the voltage or electromotive force divided by the resistance of that circuit.

To find the pressure or voltage:

$$\text{Voltage} = \text{Current} \times \text{Resistance} \quad \text{or} \quad \text{Volts} = \text{Amperes} \times \text{Ohms} \quad \text{or} \quad E = I \times R$$

The voltage required to force a current through a circuit is equal to the resistance of the circuit multiplied by the current.

To find the resistance:

$$\text{Resistance} = \frac{\text{Voltage}}{\text{Current}} \quad \text{or} \quad \text{Ohms} = \frac{\text{Volts}}{\text{Amperes}} \quad \text{or} \quad R = \frac{E}{I}$$

The resistance of a circuit is equal to the voltage divided by the current flowing through that circuit.

Power Formulas

One horsepower = 746 watts One kilowatt = 1000 watts

The power factor of electric motors varies from 80% to 90% in the larger size motors.

Single-Phase Alternating Current Circuits

Power in Watts = Volts x Amperes x Power Factor

To find current in amperes:

$$\text{Current} = \frac{\text{Watts}}{\text{Volts} \times \text{Power Factor}} \quad \text{or}$$

$$\text{Amperes} = \frac{\text{Watts}}{\text{Volts} \times \text{Power Factor}} \quad \text{or} \quad I = \frac{W}{E \times PF}$$

To find current of a motor, single phase:

$$\text{Current} = \frac{\text{Horsepower} \times 746}{\text{Volts} \times \text{Power Factor} \times \text{Efficiency}} \quad \text{or}$$

$$I = \frac{HP \times 746}{E \times PF \times Eff.}$$

To find horsepower of a motor, single phase:

$$\text{Horsepower} = \frac{\text{Volts} \times \text{Current} \times \text{Power Factor} \times \text{Efficiency}}{746 \text{ Watts}}$$

$$HP = \frac{E \times I \times PF \times Eff.}{746}$$

To find power in watts of a motor, single phase:

Watts = Volts x Current x Power Factor x Efficiency or
Watts = E x I x PF x Eff.

To find single phase kVA:

$$1 \text{ Phase kVA} = \frac{\text{Volts} \times \text{Amps}}{1000}$$

Three-Phase Alternating Current Circuits

Power in Watts = Volts x Amperes x Power Factor x 1.73

To find current in amperes in each wire:

$$\text{Current} = \frac{\text{Watts}}{\text{Voltage} \times \text{Power Factor} \times 1.73} \quad \text{or}$$

$$\text{Amperes} = \frac{\text{Watts}}{\text{Volts} \times \text{Power Factor} \times 1.73} \quad \text{or} \quad I = \frac{W}{E \times PF \times 1.73}$$

To find current of a motor, 3 phase:

$$\text{Current} = \frac{\text{Horsepower} \times 746}{\text{Volts} \times \text{Power Factor} \times \text{Efficiency} \times 1.73} \quad \text{or}$$

$$I = \frac{HP \times 746}{E \times PF \times Eff. \times 1.73}$$

To find horsepower of a motor, 3 phase:

$$\text{Horsepower} = \frac{\text{Volts} \times \text{Current} \times 1.73 \times \text{Power Factor}}{746 \text{ Watts}}$$

$$HP = \frac{E \times I \times 1.73 \times PF}{746}$$

To find power in watts of a motor, 3 phase:

Watts = Volts x Current x 1.73 x Power Factor x Efficiency or
Watts = E x I x 1.73 x PF x Eff.

To find 3 phase kVA:

$$3 \text{ phase kVA} = \frac{\text{Volts} \times \text{Amps} \times 1.73}{1000} \quad \text{or}$$

$$\text{kVA} = \frac{V \times A \times 1.73}{1000}$$

Power Factor (PF) is the percentage ratio of the measured watts (effective power) to the volt-amperes (apparent watts)

$$\text{Power Factor} = \frac{\text{Watts}}{\text{Volts} \times \text{Amperes}} \times 100\%$$

Table RD5010-1181 ASHRAE Climate Zone Map

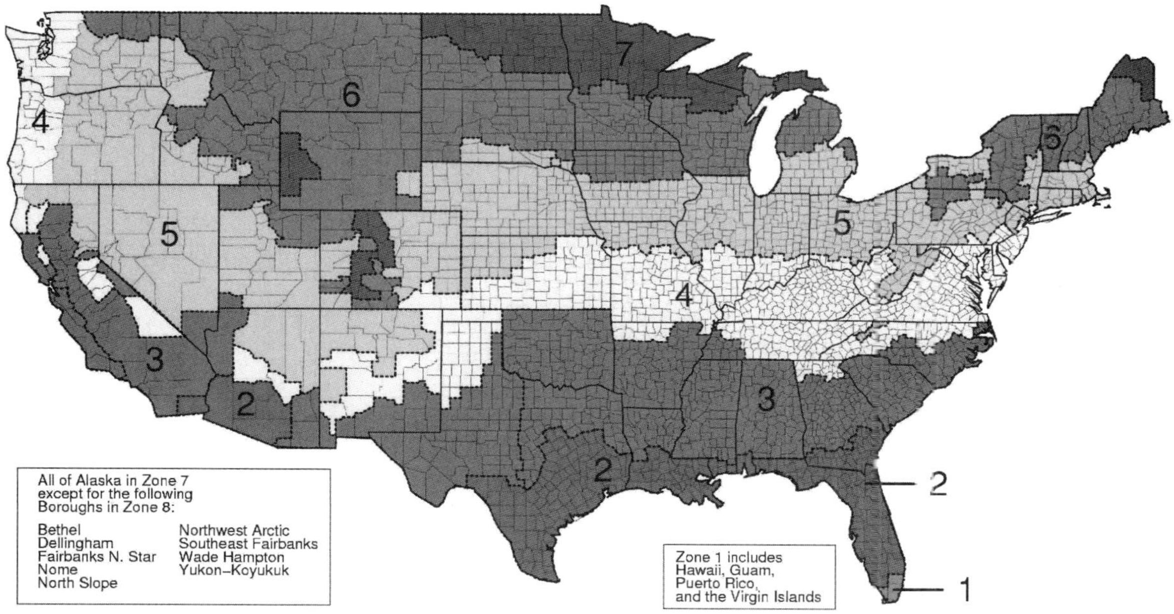

Table RD5010-1182 Watts per Square Foot for Lighting and Misc. Power by Building Type & Space

Building / Space	Percent	Lighting	Misc
Apartment, Low: Area = 20,000 SF; Floors = 2; Heating = DX Coils; Cooling = DX Coils			
Room	85%	0.50	0.10
Storage	8%	1.19	0.00
Laundry	7%	1.28	0.15
Weighted Average		**0.61**	**0.10**
Apartment, Mid: Area = 40,000 SF; Floors = 5; Heating = DX Coils; Cooling = DX Coils			
Room	71%	0.50	0.30
Corridor	16%	0.57	0.00
Storage	7%	1.19	0.00
Laundry	6%	1.28	0.15
Average		**0.61**	**0.22**
Apartment, High: Area = 72,000 SF; Floors = 8; Heating = DX Coils; Cooling = DX Coils			
Room	71%	0.50	0.30
Corridor	16%	0.57	0.00
Storage	7%	1.19	0.00
Laundry	6%	1.28	0.15
Weighted Average		**0.61**	**0.22**
Assisted Living: Area = 10,000 SF; Floors = 1; Heating = Furnace; Cooling = DX Coils			
Room	70%	1.46	0.25
Dining	15%	1.85	0.10
Food Prep	10%	1.19	1.00
Care	5%	1.34	1.00
Weighted Average		**1.49**	**0.34**

Building / Space	Percent	Lighting	Misc
Auditorum: Area = 70,000 SF; Floors = 1; Heating = Hot Water Coils; Cooling = Chilled Water Coils			
Theater	70%	0.90	
Lobby	15%	1.50	
Office	5%	1.30	
Corridor	5%	0.60	
MEP	5%	0.70	
Weighted Average		**0.99**	
Bank: Area = 25,000 SF; Floors = 2; Heating = Furnace; Cooling = DX Coils			
Bank	60%	2.33	0.45
Corridor	10%	0.57	0.00
Office	10%	1.49	0.75
Lobby	5%	1.52	0.25
Restrooms	5%	0.77	0.10
Conference	4%	0.92	0.10
MEP	4%	0.81	0.10
Production	2%	1.64	0.70
Weighted Average		**1.82**	**0.38**
Community Center: Area = 35,000 SF; Floors = 1; Heating = Furnace; Cooling = DX Coils			
Convention	50%	1.27	
Dining	15%	1.85	
Food Prep	15%	1.19	
Storage	10%	1.19	
Corridor	5%	0.57	
Restrooms	5%	0.77	
Weighted Average		**1.28**	
Convention Center: Area = 150,000 SF; Floors = 2; Heating = Hot Water Coils; Cooling = Chilled Water Coils			
Convention	65%	1.27	0.25
Conference	20%	0.92	0.10
Dining	5%	1.85	0.10
Corridor	4%	0.57	0.00
Food Prep	3%	1.19	1.00
Restrooms	3%	0.77	0.10
Weighted Average		**1.18**	**0.22**

Building / Space	Percent	Lighting	Misc
Day Care: Area = 2,500 SF; Floors = 1; Heating = Furnace; Cooling = DX Coils			
Classroom	45%	1.45	
Corridor	15%	0.57	
Library	15%	1.67	
Food Prep	10%	1.19	
Restrooms	7%	0.77	
Office	5%	1.49	
Storage	3%	1.19	
Weighted Average		**1.27**	
Fitness Center: Area = 20,000 SF; Floors = 1; Heating = Furnace; Cooling = DX Coils			
Exercise	80%	1.13	0.10
Corridor	5%	0.57	0.00
MEP	5%	0.81	0.10
Office	5%	1.24	0.75
Restrooms	7%	0.77	0.10
Weighted Average		**1.07**	**0.13**
Hospital: Area = 250,000 SF; Floors = 4; Heating = Hot Water Coils; Cooling = Chilled Water Coils			
Medical	60%	1.34	1.00
Laboratory	15%	3.19	1.00
Corridor	10%	0.57	0.00
Laundry	5%	1.28	0.15
MEP	5%	0.81	0.10
Restrooms	5%	0.77	0.10
Weighted Average		**1.48**	**0.77**
Hotel: Area = 180,000 SF; Floors = 10; Heating = Furnace; Cooling = DX Coils			
Room	75%	1.46	
Meeting Center	15%	1.27	
Corridor	8%	0.57	
MEP	2%	0.81	
Weighted Average		**1.35**	

Steps:

1. Determine climate zone using RD5010-1181
2. Use RD5010-1182 to determine lighting and miscellaneous power for appropriate building type
3. Use corresponding building type and climate zone in RD5010-1183 to determine other systems

*Values for RD5010-1182 & -1183 were derived by performing an energy analysis using the eQUEST® software from James J. Hirsch & Associates. eQUEST is a trademark of James J. Hirsch

**Additional systems may be necessary

Table RD5010-1182 Watts per Square Foot for Lighting and Misc. Power by Building Type & Space (cont.)

Building / Space	Percent	Lighting	Misc
Manufacturing, General: Area = 100,000 SF; Floors = 10; Heating = Furnace; Cooling = DX Coils			
Work	60%	2.66	1.00
Storage	25%	1.19	0.00
Office	10%	1.24	0.75
Restrooms	3%	0.77	0.10
MEP	2%	0.81	0.10
Weighted Average		**2.06**	**0.68**
Manufacturing, High Tech: Area = 200,000 SF; Floors = 1; Heating = Furnace; Cooling = DX Coils			
Work	45%	3.19	1.00
Office	27%	1.24	0.75
Corridor	18%	0.57	0.00
Dining	3%	1.85	0.10
Conference	2%	0.92	0.10
Server	2%	1.24	5.00
MEP	2%	0.81	0.10
Food Prep	1%	1.19	1.00
Weighted Average		**2.00**	**0.77**
Medical Clinic: Area = 20,000 SF; Floors = 1; Heating = Furnace; Cooling = DX Coils			
Clinic	50%	1.40	
General Office	15%	1.30	
Private Office	10%	1.30	
Laboratory	10%	1.30	
Storage	5%	0.60	
Restrooms	5%	0.60	
MEP	5%	0.70	
Weighted Average		**1.25**	
Motel: Area = 10,000 SF; Floors = 1; Heating = DX Coils; Cooling = DX Coils			
Room	95%	1.46	0.25
Office	3%	1.24	0.75
MEP	2%	0.81	0.10
Weighted Average		**1.44**	**0.26**
Museum: Area = 20,000 SF; Floors = 1; Heating = Hot Water Coils; Cooling = Chilled Water Coils			
Exhibit Area	65%	1.03	0.25
Theater	15%	1.02	0.25
Lobby	5%	1.77	0.25
MEP	5%	0.81	0.10
Office	5%	1.49	0.75
Restrooms	5%	0.77	0.10
Weighted Average		**1.06**	**0.26**
Office, Low: Area = 25,000 SF; Floors = 2; Heating = Furnace; Cooling = DX Coils			
Office	70%	1.30	0.75
Corridor	10%	0.60	0.00
Lobby	5%	1.10	0.25
Restrooms	5%	0.60	0.10
Conference	4%	1.60	0.10
MEP	4%	0.70	0.10
Production	2%	1.50	0.70
Weighted Average		**1.18**	**0.56**
Office, Mid: Area = 125,000 SF; Floors = 4; Heating = Hot Water Coils; Cooling = Chilled Water Coils			
Office Open	40%	1.24	0.75
Office Private	30%	1.49	0.75
Corridor	10%	0.57	0.00
Lobby	5%	1.52	0.25
Restrooms	5%	0.77	0.10
Conference	4%	0.92	0.10
MEP	4%	0.81	0.10
Production	2%	1.64	0.70
Weighted Average		**1.22**	**0.56**
Office, High: Area = 250,000 SF; Floors = 8; Heating = Hot Water Coils; Cooling = Chilled Water Coils			
Office Open	45%	1.24	0.75
Office Private	25%	1.49	0.75
Corridor	10%	0.57	0.00
Lobby	5%	1.52	0.25
Conference	4%	0.92	0.10
Production	2%	1.64	0.70
Restrooms	5%	0.77	0.10
MEP	4%	0.81	0.10
Weighted Average		**1.20**	**0.56**

Building / Space	Percent	Lighting	Misc
Restaurant, Bar/Lounge: Area = 4,000 SF; Floors = 1; Heating = Furnace; Cooling = DX Coils			
Lounge	35%	1.85	0.10
Dining	30%	1.85	0.10
Food Prep	25%	1.19	1.00
MEP	5%	0.81	0.10
Restrooms	5%	0.77	0.10
Weighted Average		**1.58**	**0.33**
Restaurant, Full: Area = 5,000 SF; Floors = 1; Heating = Furnace; Cooling = DX Coils			
Dining	60%	1.85	0.10
Food Prep	30%	1.19	1.00
Restrooms	5%	0.77	0.10
MEP	5%	0.81	0.10
Weighted Average		**1.55**	**0.37**
Restaurant, Quick: Area = 2,500 SF; Floors = 1; Heating = Furnace; Cooling = DX Coils			
Dining	50%	1.85	0.10
Food Prep	35%	1.19	1.00
Restrooms	10%	0.77	0.10
MEP	5%	0.81	0.10
Weighted Average		**1.46**	**0.42**
Store, Convenience: Area = 3,000 SF; Floors = 1; Heating = Furnace; Cooling = DX Coils			
Retail	80%	2.38	0.25
Office	10%	1.49	0.75
Restrooms	10%	0.77	0.10
Weighted Average		**2.13**	**0.29**
Store, Department: Area = 120,000 SF; Floors = 2; Heating = Hot Water Coils; Cooling = Chilled Water Coils			
Retail	55%	2.38	0.25
Exhibit Area	15%	1.03	0.25
Storage	16%	1.19	0.00
Office	5%	1.24	0.75
Smoking Lounge	2%	1.85	0.10
Restrooms	2%	0.77	0.10
MEP	5%	0.81	0.10
Weighted Average		**1.81**	**0.22**
Store, Large: Area = 70,000 SF; Floors = 1; Heating = Furnace; Cooling = DX Coils			
Retail	65%	2.38	0.25
Storage	20%	1.19	0.00
Office	5%	1.24	0.75
Dining	3%	1.85	0.10
Food Prep	1%	1.19	1.00
Lounge	2%	1.85	0.10
Restrooms	2%	0.77	0.10
MEP	2%	0.81	0.10
Weighted Average		**1.98**	**0.22**
Store, Small: Area = 5,000 SF; Floors = 1; Heating = Furnace; Cooling = DX Coils			
Retail	58%	2.38	0.25
Exhibit Area	15%	1.03	0.25
Storage	15%	1.19	0.00
Office	5%	1.24	0.75
Restrooms	5%	0.77	0.10
MEP	2%	0.81	0.10
Weighted Average		**1.83**	**0.23**
Store, Strip Mall: Area = 15,000 SF; Floors = 1; Heating = Furnace; Cooling = DX Coils			
Retail	38%	2.38	0.25
Atrium	20%	1.55	0.13
Exhibit Area	15%	1.03	0.25
Storage	15%	1.19	0.00
Office	5%	1.24	0.75
Restrooms	5%	0.77	0.10
MEP	2%	0.81	0.10
Weighted Average		**1.66**	**0.20**

Building / Space	Percent	Lighting	Misc
Store, Warehouse: Area = 150,000 SF; Floors = 1; Heating = Furnace; Cooling = DX Coils			
Retail	75%	2.38	0.25
Storage	20%	1.19	0.00
Office	4%	1.24	0.75
Restrooms	1%	0.77	0.10
Weighted Average		**2.08**	**0.22**
School, Elementary: Area = 35,000 SF; Floors = 1; Heating = Furnace; Cooling = DX Coils			
Classroom	50%	1.45	0.50
Corridor	15%	0.57	0.00
Gym	10%	1.13	0.10
Dining	5%	1.85	0.10
Food Prep	5%	1.19	1.00
Library	5%	1.78	0.25
Auditorium	5%	1.62	0.25
Restrooms	5%	0.77	0.10
Weighted Average		**1.28**	**0.35**
School, Middle: Area = 75,000 SF; Floors = 1; Heating = Furnace; Cooling = DX Coils			
Classroom	50%	1.45	0.50
Corridor	14%	0.57	0.00
Gym	9%	1.13	0.10
Auditorium	7%	1.62	0.25
Dining	6%	1.85	0.10
Food Prep	5%	1.19	1.00
Library	5%	1.78	0.25
Restrooms	4%	0.77	0.10
Weighted Average		**1.31**	**0.35**
School, High: Area = 150,000 SF; Floors = 2; Heating = Furnace; Cooling = DX Coils			
Classroom	45%	1.45	0.50
Corridor	10%	0.57	0.00
Gym	10%	1.13	0.10
Dining	10%	1.85	0.10
Food Prep	5%	1.19	1.00
Library	5%	1.78	0.25
Auditorium	10%	1.62	0.25
Restrooms	5%	0.77	0.10
Weighted Average		**1.36**	**0.34**
School, College: Area = 250,000 SF; Floors = 4; Heating = Furnace; Cooling = DX Coils			
Classroom	50%	1.45	0.50
Office	15%	1.49	0.75
Corridor	12%	0.57	0.00
Gym	10%	1.13	0.10
Library	5%	1.78	0.25
Dining	3%	1.85	0.10
Restrooms	3%	0.77	0.10
Food Prep	2%	1.19	1.00
Weighted Average		**1.32**	**0.41**
Service Station: Area = 2,000 SF; Floors = 1; Heating = Furnace; Cooling = DX Coils			
Workshop	70%	1.31	0.50
Office	10%	1.49	0.75
Restrooms	10%	0.77	0.10
Retail	10%	2.38	0.25
Weighted Average		**1.38**	**0.46**
Storage, High Bay: Area = 200,000 SF; Floors = 1; Heating = Furnace; Cooling = DX Coils			
Storage	85%	1.19	0.00
Office	10%	1.24	0.75
Restrooms	3%	0.77	0.10
MEP	2%	0.81	0.10
Weighted Average		**1.17**	**0.08**
Storage, Low Bay: Area = 70,000 SF; Floors = 1; Heating = Furnace; Cooling = DX Coils			
Storage	85%	1.19	0.00
Office	10%	1.24	0.75
Restrooms	3%	0.77	0.10
MEP	2%	0.81	0.10
Weighted Average		**1.17**	**0.08**

Table RD5010-1183 Watts per S.F. for Various Systems by Building Type & Climate Zone

Building / Zone	Heating	Cooling	Heat rejection	Auxiliary (pumps)	Vent fan	Supp heat pump
Apartment, Low						
Zone 1	0.42	3.65	0.00	0.00	0.12	1.47
Zone 2	0.63	3.53	0.00	0.00	0.12	4.01
Zone 3	0.64	3.45	0.00	0.00	0.13	4.63
Zone 4	0.65	3.09	0.00	0.00	0.13	7.41
Zone 5	0.67	2.88	0.00	0.00	0.12	10.30
Zone 6	0.69	2.66	0.00	0.00	0.13	10.68
Zone 7	0.85	2.11	0.00	0.00	0.12	11.76
Average	**0.65**	**3.05**	**0.00**	**0.00**	**0.13**	**7.18**
Apartment, Mid						
Zone 1	0.00	1.90	0.00	0.00	0.11	0.00
Zone 2	0.43	1.82	0.00	0.00	0.11	2.12
Zone 3	0.48	1.69	0.00	0.00	0.11	2.32
Zone 4	0.67	1.67	0.00	0.00	0.12	4.25
Zone 5	0.70	1.54	0.00	0.00	0.12	4.79
Zone 6	0.72	1.45	0.00	0.00	0.12	5.07
Zone 7	0.93	1.39	0.00	0.00	0.12	5.11
Average	**0.56**	**1.63**	**0.00**	**0.00**	**0.12**	**3.38**
Apartment, High						
Zone 1	0.00	2.11	0.00	0.00	0.14	0.00
Zone 2	0.00	2.03	0.00	0.00	0.13	0.00
Zone 3	0.48	1.88	0.00	0.00	0.14	1.75
Zone 4	0.57	1.77	0.00	0.00	0.14	4.22
Zone 5	0.66	1.64	0.00	0.00	0.14	5.40
Zone 6	0.71	1.63	0.00	0.00	0.15	5.64
Zone 7	0.91	1.18	0.00	0.00	0.14	5.75
Average	**0.48**	**1.75**	**0.00**	**0.00**	**0.14**	**3.25**
Assisted Living						
Zone 1	0.00	2.65	0.00	0.04	0.31	0.00
Zone 2	0.00	2.63	0.00	0.04	0.31	0.00
Zone 3	0.00	2.57	0.00	0.04	0.31	0.00
Zone 4	0.00	2.50	0.00	0.04	0.31	0.00
Zone 5	0.00	2.42	0.00	0.04	0.30	0.00
Zone 6	0.00	1.99	0.00	0.04	0.32	0.00
Zone 7	0.00	1.90	0.00	0.04	0.30	0.00
Average	**0.00**	**2.38**	**0.00**	**0.04**	**0.31**	**0.00**
Auditorium						
Zone 1	0.00	2.32	0.34	0.37	1.03	0.00
Zone 2	0.00	2.32	0.34	0.38	1.03	0.00
Zone 3	0.00	2.29	0.33	0.38	1.04	0.00
Zone 4	0.00	2.26	0.29	0.35	1.05	0.00
Zone 5	0.00	2.20	0.34	0.39	1.04	0.00
Zone 6	0.00	1.71	0.14	0.34	1.05	0.00
Zone 7	0.00	1.38	0.23	0.36	1.04	0.00
Average	**0.00**	**2.07**	**0.29**	**0.37**	**1.04**	**0.00**
Bank						
Zone 1	0.00	3.32	0.00	0.00	0.60	0.00
Zone 2	0.00	3.28	0.00	0.00	0.61	0.00
Zone 3	0.00	3.28	0.00	0.00	0.61	0.00
Zone 4	0.00	3.21	0.00	0.00	0.60	0.00
Zone 5	0.00	3.06	0.00	0.00	0.59	0.00
Zone 6	0.00	2.82	0.00	0.00	0.61	0.00
Zone 7	0.00	2.66	0.00	0.00	0.60	0.00
Average	**0.00**	**3.09**	**0.00**	**0.00**	**0.60**	**0.00**

Building / Zone	Heating	Cooling	Heat rejection	Auxiliary (pumps)	Vent fan	Supp heat pump
Community Center						
Zone 1	0.00	2.70	0.00	0.00	0.44	0.00
Zone 2	0.00	2.66	0.00	0.00	0.44	0.00
Zone 3	0.00	2.66	0.00	0.00	0.45	0.00
Zone 4	0.00	2.44	0.00	0.00	0.45	0.00
Zone 5	0.00	2.42	0.00	0.00	0.44	0.00
Zone 6	0.00	2.16	0.00	0.00	0.44	0.00
Zone 7	0.00	1.91	0.00	0.00	0.44	0.00
Average	**0.00**	**2.42**	**0.00**	**0.00**	**0.45**	**0.00**
Convention Center						
Zone 1	0.00	1.46	0.25	0.27	0.31	0.00
Zone 2	0.00	1.44	0.25	0.27	0.29	0.00
Zone 3	0.00	1.43	0.24	0.26	0.30	0.00
Zone 4	0.00	1.42	0.23	0.26	0.28	0.00
Zone 5	0.00	1.30	0.29	0.31	0.22	0.00
Zone 6	0.00	1.17	0.12	0.29	0.16	0.00
Zone 7	0.00	0.90	0.20	0.31	0.16	0.00
Average	**0.00**	**1.30**	**0.22**	**0.28**	**0.25**	**0.00**
Day Care						
Zone 1	0.00	3.47	0.00	0.02	0.29	0.00
Zone 2	0.00	3.42	0.00	0.02	0.28	0.00
Zone 3	0.00	3.40	0.00	0.02	0.29	0.00
Zone 4	0.00	3.29	0.00	0.02	0.29	0.00
Zone 5	0.00	3.08	0.00	0.02	0.29	0.00
Zone 6	0.00	2.29	0.00	0.02	0.29	0.00
Zone 7	0.00	2.12	0.00	0.02	0.24	0.00
Average	**0.05**	**3.01**	**0.05**	**0.07**	**0.29**	**0.05**
Fitness Center						
Zone 1	0.00	4.97	0.00	0.00	0.59	0.00
Zone 2	0.00	4.94	0.00	0.00	0.59	0.00
Zone 3	0.00	4.80	0.00	0.00	0.60	0.00
Zone 4	0.00	4.76	0.00	0.00	0.60	0.00
Zone 5	0.00	4.75	0.00	0.00	0.59	0.00
Zone 6	0.00	4.13	0.00	0.00	0.60	0.00
Zone 7	0.00	3.27	0.00	0.00	0.60	0.00
Average	**0.00**	**4.52**	**0.00**	**0.00**	**0.59**	**0.00**
Hospital						
Zone 1	0.00	1.54	0.23	0.26	0.80	0.00
Zone 2	0.00	1.47	0.22	0.25	0.80	0.00
Zone 3	0.00	1.44	0.22	0.25	0.81	0.00
Zone 4	0.00	1.42	0.22	0.25	0.81	0.00
Zone 5	0.00	1.37	0.23	0.27	0.81	0.00
Zone 6	0.00	1.26	0.15	0.25	0.82	0.00
Zone 7	0.00	1.07	0.19	0.26	0.82	0.00
Average	**0.00**	**1.36**	**0.21**	**0.26**	**0.81**	**0.00**
Hotel						
Zone 1	0.00	1.77	0.00	0.01	0.26	0.00
Zone 2	0.00	1.64	0.00	0.01	0.26	0.00
Zone 3	0.00	1.62	0.00	0.01	0.27	0.00
Zone 4	0.00	1.62	0.00	0.01	0.27	0.00
Zone 5	0.00	1.51	0.00	0.01	0.27	0.00
Zone 6	0.00	1.25	0.00	0.01	0.28	0.00
Zone 7	0.00	1.23	0.00	0.01	0.28	0.00
Average	**0.00**	**1.52**	**0.00**	**0.01**	**0.27**	**0.00**

Table RD5010-1183 Watts per S.F. for Various Systems by Building Type & Climate Zone (cont.)

Building / Zone	Heating	Cooling	Heat rejection	Auxiliary (pumps)	Vent fan	Supp heat pump
Manufacturing, General						
Zone 1	0.00	1.46	0.00	0.00	0.13	0.00
Zone 2	0.00	1.45	0.00	0.00	0.13	0.00
Zone 3	0.00	1.43	0.00	0.00	0.13	0.00
Zone 4	0.00	1.39	0.00	0.00	0.13	0.00
Zone 5	0.00	1.39	0.00	0.00	0.13	0.00
Zone 6	0.00	1.21	0.00	0.00	0.13	0.00
Zone 7	0.00	1.00	0.00	0.00	0.13	0.00
Average	**0.00**	**1.33**	**0.00**	**0.00**	**0.13**	**0.00**
Manufacturing, High Tech						
Zone 1	0.00	2.15	0.00	0.00	0.17	0.00
Zone 2	0.00	2.12	0.00	0.00	0.17	0.00
Zone 3	0.00	2.04	0.00	0.00	0.17	0.00
Zone 4	0.00	2.01	0.00	0.00	0.17	0.00
Zone 5	0.00	1.88	0.00	0.00	0.17	0.00
Zone 6	0.00	1.63	0.00	0.00	0.17	0.00
Zone 7	0.00	1.50	0.00	0.00	0.17	0.00
Average	**0.00**	**1.90**	**0.00**	**0.00**	**0.17**	**0.00**
Medical Clinic						
Zone 1	0.00	3.15	0.00	0.00	0.54	0.00
Zone 2	0.00	3.09	0.00	0.00	0.53	0.00
Zone 3	0.00	3.07	0.00	0.00	0.54	0.00
Zone 4	0.00	3.02	0.00	0.00	0.54	0.00
Zone 5	0.00	2.99	0.00	0.00	0.54	0.00
Zone 6	0.00	2.58	0.00	0.00	0.55	0.00
Zone 7	0.00	2.45	0.00	0.00	0.52	0.00
Average	**0.00**	**2.91**	**0.00**	**0.00**	**0.54**	**0.00**
Motel						
Zone 1	0.97	2.68	0.00	0.00	0.19	2.15
Zone 2	1.19	2.52	0.00	0.00	0.18	4.43
Zone 3	1.19	2.52	0.00	0.00	0.19	4.81
Zone 4	1.19	2.42	0.00	0.00	0.21	7.39
Zone 5	1.28	2.26	0.00	0.00	0.21	7.82
Zone 6	1.29	2.07	0.00	0.00	0.23	8.89
Zone 7	1.39	1.97	0.00	0.00	0.23	9.51
Average	**1.21**	**2.35**	**0.00**	**0.00**	**0.21**	**6.43**
Museum						
Zone 1	0.00	3.77	0.00	0.11	0.67	0.00
Zone 2	0.00	3.71	0.00	0.12	0.77	0.00
Zone 3	0.00	3.59	0.00	0.12	0.72	0.00
Zone 4	0.00	3.47	0.00	0.14	0.74	0.00
Zone 5	0.00	3.42	0.00	0.16	0.51	0.00
Zone 6	0.00	2.39	0.00	0.15	0.23	0.00
Zone 7	0.00	1.69	0.00	0.17	0.24	0.00
Average	**0.00**	**3.15**	**0.00**	**0.14**	**0.55**	**0.00**
Office, Low						
Zone 1	0.00	2.90	0.00	0.02	0.29	0.00
Zone 2	0.00	2.90	0.00	0.02	0.30	0.00
Zone 3	0.00	2.81	0.00	0.02	0.30	0.00
Zone 4	0.00	2.70	0.00	0.02	0.29	0.00
Zone 5	0.00	2.66	0.00	0.02	0.29	0.00
Zone 6	0.00	2.04	0.00	0.02	0.30	0.00
Zone 7	0.00	1.82	0.00	0.02	0.29	0.00
Average	**0.00**	**2.55**	**0.00**	**0.02**	**0.29**	**0.00**

Building / Zone	Heating	Cooling	Heat rejection	Auxiliary (pumps)	Vent fan	Supp heat pump
Office, Mid						
Zone 1	0.00	1.61	0.31	0.33	0.45	0.00
Zone 2	0.00	1.60	0.29	0.32	0.46	0.00
Zone 3	0.00	1.59	0.28	0.34	0.42	0.00
Zone 4	0.00	1.56	0.27	0.33	0.43	0.00
Zone 5	0.00	1.55	0.34	0.36	0.37	0.00
Zone 6	0.00	1.29	0.16	0.34	0.31	0.00
Zone 7	0.00	1.10	0.22	0.35	0.32	0.00
Average	**0.00**	**1.47**	**0.27**	**0.34**	**0.40**	**0.00**
Office, High						
Zone 1	0.00	1.70	0.30	0.33	0.44	0.00
Zone 2	0.00	1.59	0.29	0.32	0.41	0.00
Zone 3	0.00	1.59	0.32	0.34	0.42	0.00
Zone 4	0.00	1.58	0.28	0.35	0.42	0.00
Zone 5	0.00	1.58	0.32	0.36	0.41	0.00
Zone 6	0.00	1.11	0.16	0.29	0.33	0.00
Zone 7	0.00	0.86	0.23	0.31	0.37	0.00
Average	**0.00**	**1.43**	**0.27**	**0.33**	**0.40**	**0.00**
Restaurant, Bar/Lounge						
Zone 1	0.00	3.07	0.00	0.06	0.24	0.00
Zone 2	0.00	2.87	0.00	0.06	0.25	0.00
Zone 3	0.00	2.85	0.00	0.06	0.24	0.00
Zone 4	0.00	2.83	0.00	0.06	0.24	0.00
Zone 5	0.00	2.80	0.00	0.06	0.24	0.00
Zone 6	0.00	1.88	0.00	0.06	0.24	0.00
Zone 7	0.00	1.87	0.00	0.06	0.23	0.00
Average	**0.00**	**2.59**	**0.00**	**0.06**	**0.24**	**0.00**
Restaurant, Full						
Zone 1	0.00	3.37	0.00	0.05	0.38	0.00
Zone 2	0.00	3.30	0.00	0.05	0.39	0.00
Zone 3	0.00	3.25	0.00	0.05	0.39	0.00
Zone 4	0.00	3.20	0.00	0.05	0.39	0.00
Zone 5	0.00	3.03	0.00	0.05	0.38	0.00
Zone 6	0.00	2.35	0.00	0.05	0.39	0.00
Zone 7	0.00	2.33	0.00	0.05	0.38	0.00
Average	**0.00**	**2.98**	**0.00**	**0.05**	**0.38**	**0.00**
Restaurant, Quick						
Zone 1	0.00	3.78	0.00	0.08	0.48	0.00
Zone 2	0.00	3.72	0.00	0.08	0.50	0.00
Zone 3	0.00	3.64	0.00	0.08	0.50	0.00
Zone 4	0.00	3.60	0.00	0.08	0.50	0.00
Zone 5	0.00	3.42	0.00	0.08	0.49	0.00
Zone 6	0.00	2.74	0.00	0.08	0.51	0.00
Zone 7	0.00	2.69	0.00	0.08	0.49	0.00
Average	**0.00**	**3.37**	**0.00**	**0.08**	**0.50**	**0.00**
Store, Convenience						
Zone 1	0.00	4.08	0.00	0.02	0.29	0.00
Zone 2	0.00	3.66	0.00	0.02	0.30	0.00
Zone 3	0.00	3.60	0.00	0.02	0.30	0.00
Zone 4	0.00	3.46	0.00	0.02	0.29	0.00
Zone 5	0.00	3.28	0.00	0.02	0.29	0.00
Zone 6	0.00	2.66	0.00	0.02	0.27	0.00
Zone 7	0.00	2.50	0.00	0.02	0.24	0.00
Average	**0.00**	**3.32**	**0.00**	**0.02**	**0.28**	**0.00**

Table RD5010-1183 Watts per S.F. for Various Systems by Building Type & Climate Zone (cont.)

Building / Zone	Heating	Cooling	Heat rejection	Auxiliary (pumps)	Vent fan	Supp heat pump
Store, Department						
Zone 1	0.00	1.65	0.20	0.23	0.63	0.00
Zone 2	0.00	1.61	0.20	0.23	0.63	0.00
Zone 3	0.00	1.59	0.19	0.22	0.64	0.00
Zone 4	0.00	1.52	0.19	0.23	0.64	0.00
Zone 5	0.00	1.51	0.21	0.24	0.63	0.00
Zone 6	0.00	1.32	0.13	0.21	0.63	0.00
Zone 7	0.00	0.98	0.17	0.23	0.63	0.00
Average	**0.00**	**1.45**	**0.18**	**0.23**	**0.63**	**0.00**
Store, Large						
Zone 1	0.00	2.68	0.00	0.00	0.16	0.00
Zone 2	0.00	2.61	0.00	0.00	0.17	0.00
Zone 3	0.00	2.61	0.00	0.00	0.17	0.00
Zone 4	0.00	2.56	0.00	0.00	0.16	0.00
Zone 5	0.00	2.53	0.00	0.00	0.16	0.00
Zone 6	0.00	1.84	0.00	0.00	0.16	0.00
Zone 7	0.00	1.50	0.00	0.00	0.16	0.00
Average	**0.00**	**2.33**	**0.00**	**0.00**	**0.16**	**0.00**
Store, Small						
Zone 1	0.00	3.60	0.00	0.01	0.25	0.00
Zone 2	0.00	3.51	0.00	0.01	0.26	0.00
Zone 3	0.00	3.38	0.00	0.01	0.25	0.00
Zone 4	0.00	3.32	0.00	0.01	0.25	0.00
Zone 5	0.00	3.23	0.00	0.01	0.24	0.00
Zone 6	0.00	2.49	0.00	0.01	0.23	0.00
Zone 7	0.00	2.17	0.00	0.01	0.21	0.00
Average	**0.00**	**3.10**	**0.00**	**0.01**	**0.24**	**0.00**
Store, Strip Mall						
Zone 1	0.00	3.36	0.00	0.02	0.25	0.00
Zone 2	0.00	3.30	0.00	0.02	0.25	0.00
Zone 3	0.00	3.14	0.00	0.02	0.25	0.00
Zone 4	0.00	3.14	0.00	0.02	0.25	0.00
Zone 5	0.00	2.24	0.00	0.02	0.24	0.00
Zone 6	0.00	1.96	0.00	0.02	0.25	0.00
Zone 7	0.00	1.64	0.00	0.02	0.24	0.00
Average	**0.00**	**2.52**	**0.00**	**0.02**	**0.25**	**0.00**
Store, Warehouse						
Zone 1	0.00	2.99	0.00	0.00	0.21	0.00
Zone 2	0.00	2.97	0.00	0.00	0.21	0.00
Zone 3	0.00	2.94	0.00	0.00	0.21	0.00
Zone 4	0.00	2.83	0.00	0.00	0.21	0.00
Zone 5	0.00	2.70	0.00	0.00	0.21	0.00
Zone 6	0.00	2.10	0.00	0.00	0.21	0.00
Zone 7	0.00	1.78	0.00	0.00	0.21	0.00
Average	**0.00**	**2.61**	**0.00**	**0.00**	**0.21**	**0.00**
School, Elementary						
Zone 1	0.00	2.73	0.00	0.01	0.21	0.00
Zone 2	0.00	2.69	0.00	0.01	0.21	0.00
Zone 3	0.00	2.63	0.00	0.01	0.21	0.00
Zone 4	0.00	2.61	0.00	0.01	0.20	0.00
Zone 5	0.00	2.51	0.00	0.01	0.20	0.00
Zone 6	0.00	1.71	0.00	0.01	0.20	0.00
Zone 7	0.00	1.64	0.00	0.01	0.20	0.00
Average	**0.00**	**2.36**	**0.00**	**0.01**	**0.20**	**0.00**

Building / Zone	Heating	Cooling	Heat rejection	Auxiliary (pumps)	Vent fan	Supp heat pump
School, Middle						
Zone 1	0.00	2.48	0.00	0.00	0.17	0.00
Zone 2	0.00	2.45	0.00	0.00	0.17	0.00
Zone 3	0.00	2.41	0.00	0.00	0.18	0.00
Zone 4	0.00	2.35	0.00	0.00	0.17	0.00
Zone 5	0.00	2.26	0.00	0.00	0.17	0.00
Zone 6	0.00	1.61	0.00	0.00	0.17	0.00
Zone 7	0.00	1.49	0.00	0.00	0.17	0.00
Average	**0.00**	**2.15**	**0.00**	**0.00**	**0.17**	**0.00**
School, High						
Zone 1	0.00	3.24	0.00	0.01	0.22	0.00
Zone 2	0.00	3.08	0.00	0.01	0.22	0.00
Zone 3	0.00	3.05	0.00	0.01	0.23	0.00
Zone 4	0.00	3.00	0.00	0.01	0.23	0.00
Zone 5	0.00	2.94	0.00	0.01	0.23	0.00
Zone 6	0.00	1.98	0.00	0.01	0.23	0.00
Zone 7	0.00	1.81	0.00	0.01	0.23	0.00
Average	**0.00**	**2.73**	**0.00**	**0.01**	**0.23**	**0.00**
School, College						
Zone 1	0.00	2.92	0.00	0.00	0.20	0.00
Zone 2	0.00	2.91	0.00	0.00	0.20	0.00
Zone 3	0.00	2.78	0.00	0.00	0.20	0.00
Zone 4	0.00	2.72	0.00	0.00	0.20	0.00
Zone 5	0.00	2.43	0.00	0.00	0.20	0.00
Zone 6	0.00	1.77	0.00	0.00	0.20	0.00
Zone 7	0.00	1.51	0.00	0.00	0.19	0.00
Average	**0.00**	**2.43**	**0.00**	**0.00**	**0.20**	**0.00**
Service Station						
Zone 1	0.00	10.42	0.00	0.03	0.47	0.00
Zone 2	0.00	9.01	0.00	0.03	0.46	0.00
Zone 3	0.00	7.82	0.00	0.03	0.49	0.00
Zone 4	0.00	7.02	0.00	0.03	0.47	0.00
Zone 5	0.00	6.13	0.00	0.03	0.46	0.00
Zone 6	0.00	4.32	0.00	0.03	0.45	0.00
Zone 7	0.00	4.18	0.00	0.03	0.40	0.00
Average	**0.00**	**6.99**	**0.00**	**0.03**	**0.46**	**0.00**
Storage, High Bay						
Zone 1	0.00	1.98	0.00	0.00	0.17	0.00
Zone 2	0.00	1.97	0.00	0.00	0.18	0.00
Zone 3	0.00	1.90	0.00	0.00	0.18	0.00
Zone 4	0.00	1.89	0.00	0.00	0.18	0.00
Zone 5	0.00	1.85	0.00	0.00	0.17	0.00
Zone 6	0.00	1.29	0.00	0.00	0.17	0.00
Zone 7	0.00	1.10	0.00	0.00	0.17	0.00
Average	**0.00**	**1.71**	**0.00**	**0.00**	**0.17**	**0.00**
Storage, Low Bay						
Zone 1	0.00	1.87	0.00	0.00	0.16	0.00
Zone 2	0.00	1.85	0.00	0.00	0.16	0.00
Zone 3	0.00	1.82	0.00	0.00	0.16	0.00
Zone 4	0.00	1.77	0.00	0.00	0.16	0.00
Zone 5	0.00	1.76	0.00	0.00	0.16	0.00
Zone 6	0.00	1.20	0.00	0.00	0.16	0.00
Zone 7	0.00	1.00	0.00	0.00	0.16	0.00
Average	**0.00**	**1.61**	**0.00**	**0.00**	**0.16**	**0.00**

D5010-120 Electric Circuit Voltages

General: The following method provides the user with a simple non-technical means of obtaining comparative costs of wiring circuits. The circuits considered serve the electrical loads of motors, electric heating, lighting and transformers, for example, that require low voltage 60 Hertz alternating current.

The method used here is suitable only for obtaining estimated costs. It is **not** intended to be used as a substitute for electrical engineering design applications.

Conduit and wire circuits can represent from twenty to thirty percent of the total building electrical cost. By following the described steps and using the tables the user can translate the various types of electric circuits into estimated costs.

Wire Size: Wire size is a function of the electric load which is usually listed in one of the following units:

1. Amperes (A)
2. Watts (W)
3. Kilowatts (kW)
4. Volt amperes (VA)
5. Kilovolt amperes (kVA)
6. Horsepower (HP)

These units of electric load must be converted to amperes in order to obtain the size of wire necessary to carry the load. To convert electric load units to amperes one must have an understanding of the voltage classification of the power source and the voltage characteristics of the electrical equipment or load to be energized. The seven A.C. circuits commonly used are illustrated in Figures D5010-121 through D5010-127 which show the tranformer load voltage and the point of use voltage at the point on the circuit where the load is connected. The difference between the source and point of use voltage is attributed to the circuit voltage drop and is considered to be approximately 4%.

Motor Voltages: Motor voltages are listed by their point of use voltage and not the power source voltage.

For example: 460 volts instead of 480 volts
200 instead of 208 volts
115 volts instead of 120 volts

Lighting and Heating Voltages: Lighting and heating equipment voltages are listed by the power source voltage and not the point of wire voltage.

For example: 480, 277, 120 volt lighting
480 volt heating or air conditioning unit
208 volt heating unit

Transformer Voltages: Transformer primary (input) and secondary (output) voltages are listed by the power source voltage.

For example: Single phase 10 kVA
Primary 240/480 volts
Secondary 120/240 volts

In this case, the primary voltage may be 240 volts with a 120 volts secondary or 480 volts with either a 120V or a 240V secondary.

For example: Three phase 10 kVA
Primary 480 volts
Secondary 208Y/120 volts

In this case the transformer is suitable for connection to a circuit with a 3-phase, 3-wire or a 3-phase, 4-wire circuit with a 480 voltage. This application will provide a secondary circuit of 3-phase, 4-wire with 208 volts between phase wires and 120 volts between any phase wire and the neutral (white) wire.

Figure D5010-121

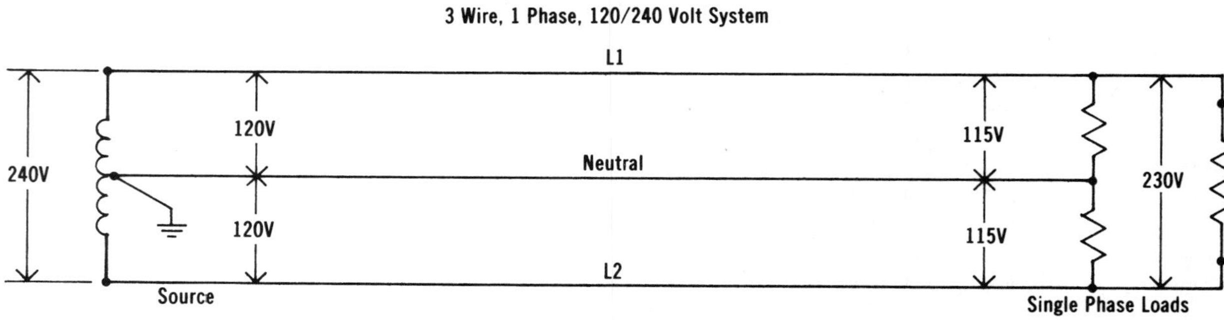

Figure D5010-122

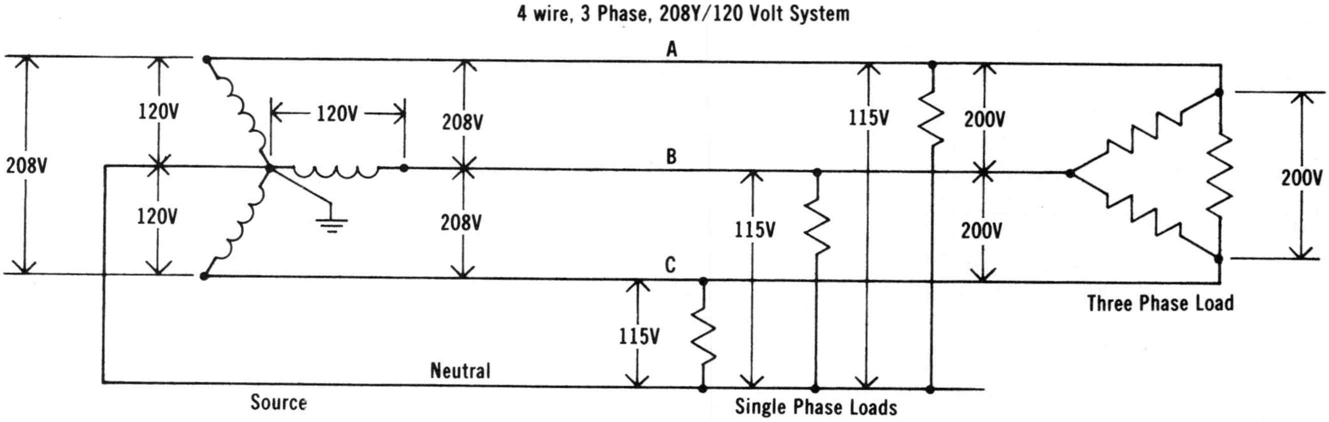

Figure D5010-123 Electric Circuit Voltages (cont.)

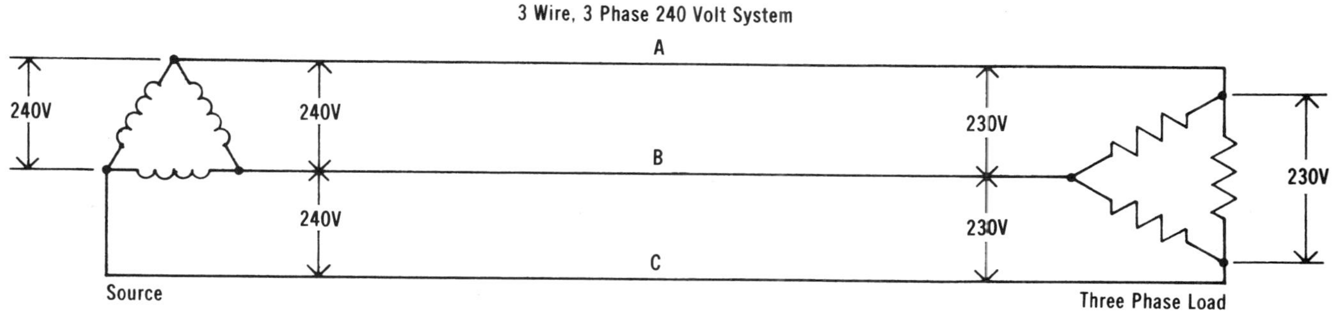

Figure D5010-124

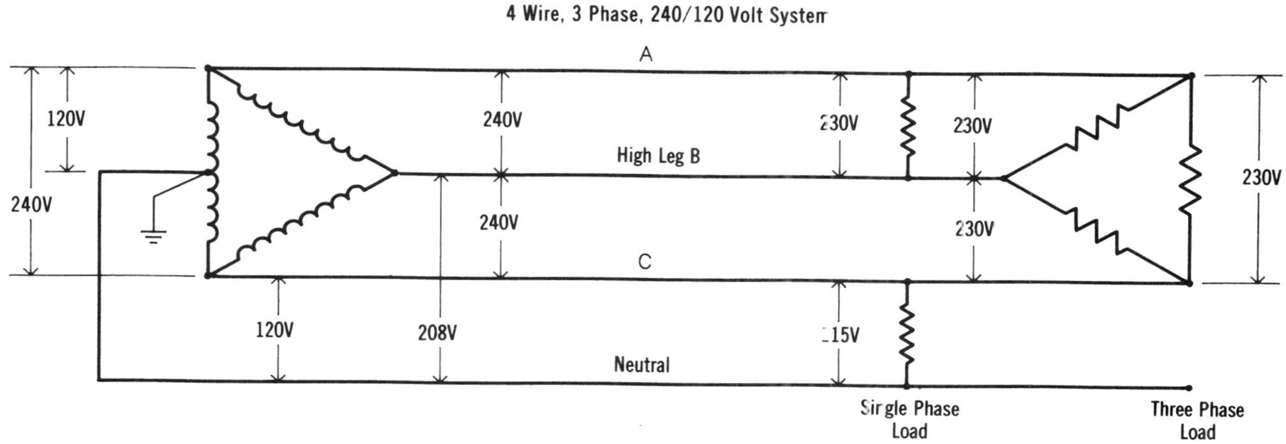

Figure D5010-125

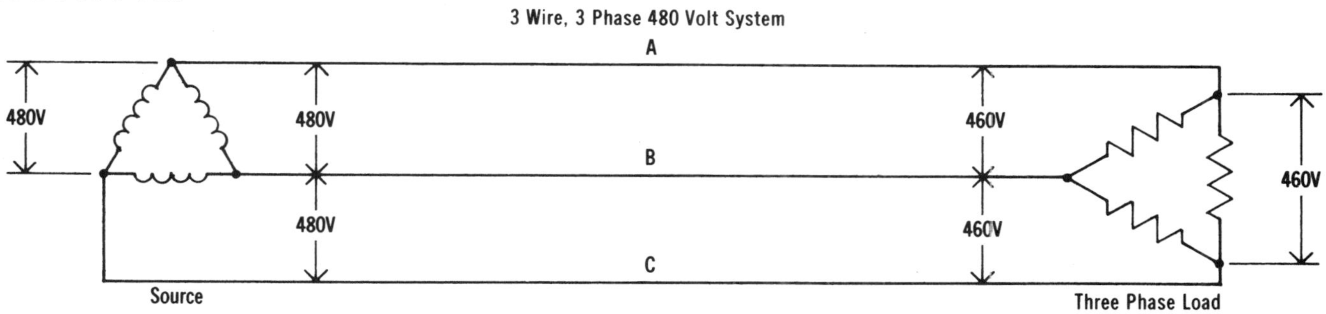

Figure D5010-126

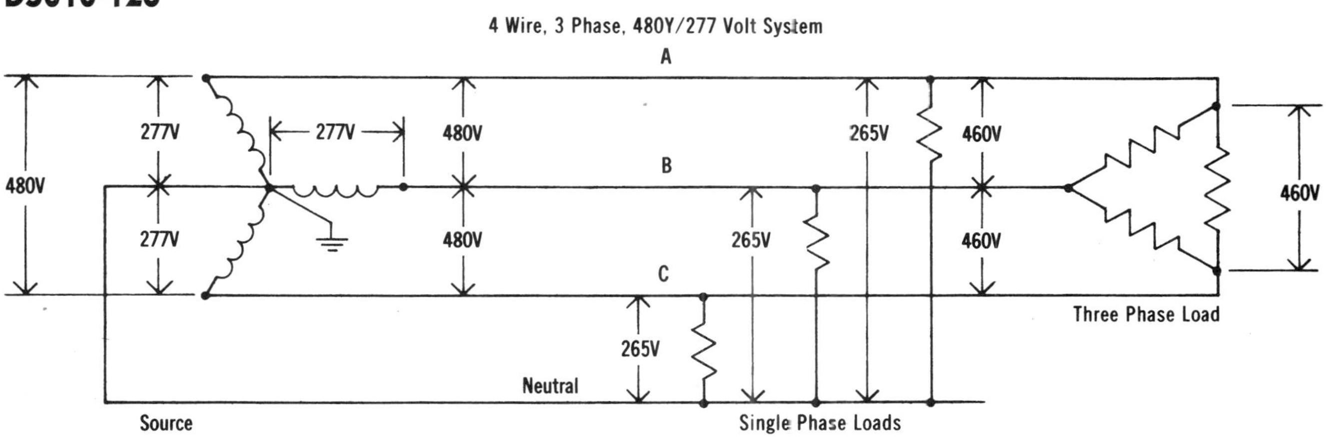

Figure D5010-127 Electric Circuit Voltages (cont.)

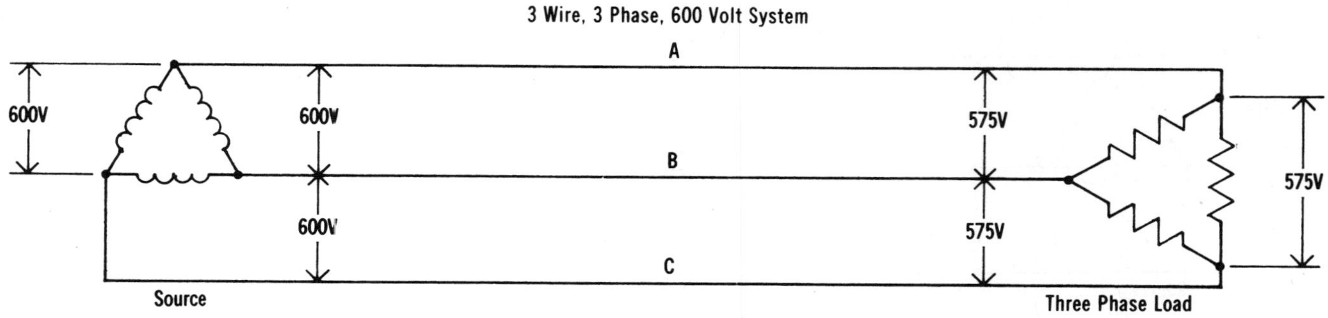

RD5010-141 Minimum Copper and Aluminum Wire Size Allowed for Various Types of Insulation

Minimum Wire Sizes									
	Copper		Aluminum			Copper		Aluminum	
Amperes	THW THWN or XHHW	THHN XHHW *	THW XHHW	THHN XHHW *	Amperes	THW THWN or XHHW	THHN XHHW *	THW XHHW	THHN XHHW *
15A	#14	#14	#12	#12	195	3/0	2/0	250kcmil	4/0
20	#12	#12	#10	#10	200	3/0	3/0	250kcmil	4/0
25	#10	#10	#10	#10	205	4/0	3/0	250kcmil	4/0
30	#10	#10	# 8	# 8	225	4/0	3/0	300kcmil	250kcmil
40	# 8	# 8	# 8	# 8	230	4/0	4/0	300kcmil	250kcmil
45	# 8	# 8	# 6	# 8	250	250kcmil	4/0	350kcmil	300kcmil
50	# 8	# 8	# 6	# 6	255	250kcmil	4/0	400kcmil	300kcmil
55	# 6	# 8	# 4	# 6	260	300kcmil	4/0	400kcmil	350kcmil
60	# 6	# 6	# 4	# 6	270	300kcmil	250kcmil	400kcmil	350kcmil
65	# 6	# 6	# 4	# 4	280	300kcmil	250kcmil	500kcmil	350kcmil
75	# 4	# 6	# 3	# 4	285	300kcmil	250kcmil	500kcmil	400kcmil
85	# 4	# 4	# 2	# 3	290	350kcmil	250kcmil	500kcmil	400kcmil
90	# 3	# 4	# 2	# 2	305	350kcmil	300kcmil	500kcmil	400kcmil
95	# 3	# 4	# 1	# 2	310	350kcmil	300kcmil	500kcmil	500kcmil
100	# 3	# 3	# 1	# 2	320	400kcmil	300kcmil	600kcmil	500kcmil
110	# 2	# 3	1/0	# 1	335	400kcmil	350kcmil	600kcmil	500kcmil
115	# 2	# 2	1/0	# 1	340	500kcmil	350kcmil	600kcmil	500kcmil
120	# 1	# 2	1/0	1/0	350	500kcmil	350kcmil	700kcmil	500kcmil
130	# 1	# 2	2/0	1/0	375	500kcmil	400kcmil	700kcmil	600kcmil
135	1/0	# 1	2/0	1/0	380	500kcmil	400kcmil	750kcmil	600kcmil
150	1/0	# 1	3/0	2/0	385	600kcmil	500kcmil	750kcmil	600kcmil
155	2/0	1/0	3/0	3/0	420	600kcmil	500kcmil		700kcmil
170	2/0	1/0	4/0	3/0	430		500kcmil		750kcmil
175	2/0	2/0	4/0	3/0	435		600kcmil		750kcmil
180	3/0	2/0	4/0	4/0	475		600kcmil		

*Dry Locations Only

Notes:

1. Size #14 to 4/0 is in AWG units (American Wire Gauge).
2. Size 250 to 750 is in kcmil units (Thousand Circular Mils).
3. Use next higher ampere value if exact value is not listed in table.
4. For loads that operate continuously increase ampere value by 25% to obtain proper wire size.
5. Table D5010-141 has been written for estimating purpose only, based on an ambient temperature of 30°C (86° F); for an ambient temperature other than 30°C (86° F), ampacity correction factors will be applied.

Table D5010-142 Conductors in Conduit

The table below lists the maximum number of conductors for various sized conduit using THW, TW or THWN insulations.

Copper Wire Size	Conduit Size																										
	½″			¾″			1″			1-¼″			1-½″			2″			2-½″			3″		3-½″		4″	
	TW	THW	THWN	TW	THW	THWN	TW	THW	THWN	TW	THW	THWN	TW	THW	THWN	TW	THW	THWN	TW	THW	THWN	THW	THWN	THW	THWN	THW	THWN
#14	9	6	13	15	10	24	25	16	39	44	29	69	60	40	94	99	65	154	142	93		143		192			
#12	7	4	10	12	8	18	19	13	29	35	24	51	47	32	70	78	53	114	111	76	164	117		157			
#10	5	4	6	9	6	11	15	11	18	26	19	32	36	26	44	60	43	73	85	61	104	95	160	127		163	
#8	2	1	3	4	3	5	7	5	9	12	10	16	17	13	22	28	22	36	40	32	51	49	79	66	106	85	136
#6		1	1		2	4		4	6		7	11		10	15		16	26		23	37	36	57	48	76	62	98
#4		1	1		1	2		3	4		5	7		7	9		12	16		17	22	27	35	36	47	47	60
#3		1	1		1	1		2	3		4	6		6	8		10	13		15	19	23	29	31	39	40	51
#2		1	1		1	1		2	3		4	5		5	7		9	11		13	16	20	25	27	33	34	43
#1					1	1		1	1		3	3		4	5		6	8		9	12	14	18	19	25	25	32
1/0					1	1		1	1		2	3		3	4		5	7		8	10	12	15	16	21	21	27
2/0					1	1		1	1		1	2		3	3		5	6		7	8	10	13	14	17	18	22
3/0					1	1		1	1		1	1		2	3		4	5		6	7	9	11	12	14	15	18
4/0						1		1	1		1	1		1	2		3	4		5	6	7	9	10	12	13	15
250 MCM								1	1		1	1		1	1		2	3		4	4	6	7	8	10	10	12
300								1	1		1	1		1	1		2	3		3	4	5	6	7	8	9	11
350									1		1	1		1	1		1	2		3	3	4	5	6	7	8	9
400											1	1		1	1		1	1		2	3	4	5	5	6	7	8
500											1	1		1	1		1	1		1	2	3	4	4	5	6	7
600												1		1	1		1	1		1	1	3	3	4	4	5	5
700														1	1		1	1		1	1	2	3	3	4	4	5
750 ▼														1	1		1	1		1	1	2	2	3	3	4	4

Table D5010-146 Metric Equivalent, Wire

U.S. vs. European Wire – Approximate Equivalents			
United States		European	
Size AWG or MCM	Area Cir. Mils. (CM) MM²	Size MM²	Area Cir. Mils.
18	1620/.82	.75	1480
16	2580/1.30	1.0	1974
14	4110/2.08	1.5	2961
12	6530/3.30	2.5	4935
10	10,380/5.25	4	7896
8	16,510/8.36	6	11,844
6	26,240/13.29	10	19,740
4	41,740/21.14	16	31,584
3	52,620/26.65	25	49,350
2	66,360/33.61	–	–
1	83,690/42.39	35	69,090
1/0	105,600/53.49	50	98,700
2/0	133,100/67.42	–	–
3/0	167,800/85.00	70	138,180
4/0	211,600/107.19	95	187,530
250	250,000/126.64	120	236,880
300	300,000/151.97	150	296,100
350	350,000/177.30	–	–
400	400,000/202.63	185	365,190
500	500,000/253.29	240	473,760
600	600,000/303.95	300	592,200
700	700,000/354.60	–	–
750	750,000/379.93	–	–

D5010-147 Concrete for Conduit Encasement

The table below lists C.Y. of concrete for 100 L.F. of trench. Conduits separation center to center should meet 7.5″ (N.E.C.).

Number of Conduits	1	2	3	4	6	8	9	Number of Conduits
Trench Dimension	11.5″ x 11.5″	11.5″ x 19″	11.5″ x 27″	19″ x 19″	19″ x 27″	19″ x 38″	27″ x 27″	Trench Dimension
Conduit Diameter 2.0″	3.29	5.39	7.64	8.83	12.51	17.66	17.72	Conduit Diameter 2.0″
2.5″	3.23	5.29	7.49	8.62	12.19	17.23	17.25	2.5″
3.0″	3.15	5.13	7.24	8.29	11.71	16.59	16.52	3.0″
3.5″	3.08	4.97	7.02	7.99	11.26	15.98	15.84	3.5″
4.0″	2.99	4.80	6.76	7.65	10.74	15.30	15.07	4.0″
5.0″	2.78	4.37	6.11	6.78	9.44	13.57	13.12	5.0″
6.0″	2.52	3.84	5.33	5.74	7.87	11.48	10.77	6.0″

Table D5010-148 Size Required and Weight (Lbs./1000 L.F.) of Aluminum and Copper THW Wire by Ampere Load

Amperes	Copper Size	Aluminum Size	Copper Weight	Aluminum Weight
15	14	12	24	11
20	12	10	33	17
30	10	8	48	39
45	8	6	77	52
65	6	4	112	72
85	4	2	167	101
100	3	1	205	136
115	2	1/0	252	162
130	1	2/0	324	194
150	1/0	3/0	397	233
175	2/0	4/0	491	282
200	3/0	250	608	347
230	4/0	300	753	403
255	250	400	899	512
285	300	500	1068	620
310	350	500	1233	620
335	400	600	1396	772
380	500	750	1732	951

Table D5010-152 Conduit Weight Comparisons (Lbs. per 100 ft.) with Maximum Cable Fill*

Type	1/2″	3/4″	1″	1-1/4″	1-1/2″	2″	2-1/2″	3″	3-1/2″	4″	5″	6″
Rigid Galvanized Steel (RGS)	104	140	235	358	455	721	1022	1451	1749	2148	3083	4343
Intermediate Steel (IMC)	84	113	186	293	379	611	883	1263	1501	1830		
Electrical Metallic Tubing (EMT)	54	116	183	296	368	445	641	930	1215	1540		

*Conduit & Heaviest Conductor Combination

General: Lighting and electric heating loads are expressed in watts and kilowatts.

Cost Determination:

The proper ampere values can be obtained as follows:

1. Convert watts to kilowatts (watts 1000 ÷ kilowatts)
2. Determine voltage rating of equipment.
3. Determine whether equipment is single phase or three phase.
4. Refer to Table D5010-158 to find ampere value from kW, Ton and BTU/hr. values.
5. Determine type of wire insulation – TW, THW, THWN.
6. Determine if wire is copper or aluminum.
7. Refer to Table D5010-141 to obtain copper or aluminum wire size from ampere values.
8. Next refer to Table D5010-142 for the proper conduit size to accommodate the number and size of wires in each particular case.
9. Next refer to unit price data for the per linear foot cost of the conduit.
10. Next refer to unit price data for the per linear foot cost of the wire. Multiply cost of wire per L.F. x number of wires in the circuits to obtain total wire cost per L.F..
11. Add values obtained in Steps 9 and 10 for total cost per linear foot for conduit and wire x length of circuit = Total Cost.

Notes:

1. Phase refers to single phase, 2 wire circuits.
2. Phase refers to three phase, 3 wire circuits.
3. For circuits which operate continuously for 3 hours or more, multiply the ampere values by 1.25 for a given kW requirement.
4. For kW ratings not listed, add ampere values.

   For example: Find the ampere value of 9 kW at 208 volt, single phase.

   4 kW = 19.2A
   5 kW = 24.0A
   9 kW = 43.2A

5. "Length of Circuit" refers to the one way distance of the run, not to the total sum of wire lengths.

Table D5010-158 Ampere Values as Determined by kW Requirements, BTU/HR or Ton, Voltage and Phase Values

			Ampere Values						
			120V	208V		240V		277V	480V
kW	Ton	BTU/HR	1 Phase	1 Phase	3 Phase	1 Phase	3 Phase	1 Phase	3 Phase
0.5	.1422	1,707	4.2A	2.4A	1.4A	2.1A	1.2A	1.8A	0.6A
0.75	.2133	2,560	6.2	3.6	2.1	3.1	1.9	2.7	.9
1.0	.2844	3,413	8.3	4.9	2.8	4.2	2.4	3.6	1.2
1.25	.3555	4,266	10.4	6.0	3.5	5.2	3.0	4.5	1.5
1.5	.4266	5,120	12.5	7.2	4.2	6.3	3.1	5.4	1.8
2.0	.5688	6,826	16.6	9.7	5.6	8.3	4.8	7.2	2.4
2.5	.7110	8,533	20.8	12.0	7.0	10.4	6.1	9.1	3.1
3.0	.8532	10,239	25.0	14.4	8.4	12.5	7.2	10.8	3.6
4.0	1.1376	13,652	33.4	19.2	11.1	16.7	9.6	14.4	4.8
5.0	1.4220	17,065	41.6	24.0	13.9	20.8	12.1	18.1	6.1
7.5	2.1331	25,598	62.4	36.0	20.8	31.2	18.8	27.0	9.0
10.0	2.8441	34,130	83.2	48.0	27.7	41.6	24.0	36.5	12.0
12.5	3.5552	42,663	104.2	60.1	35.0	52.1	30.0	45.1	15.0
15.0	4.2662	51,195	124.8	72.0	41.6	62.4	37.6	54.0	18.0
20.0	5.6883	68,260	166.4	96.0	55.4	83.2	48.0	73.0	24.0
25.0	7.1104	85,325	208.4	120.2	70.0	104.2	60.0	90.2	30.0
30.0	8.5325	102,390		144.0	83.2	124.8	75.2	108.0	36.0
35.0	9.9545	119,455		168.0	97.1	145.6	87.3	126.0	42.1
40.0	11.3766	136,520		192.0	110.8	166.4	96.0	146.0	48.0
45.0	12.7987	153,585			124.8	187.5	112.8	162.0	54.0
50.0	14.2208	170,650			140.0	208.4	120.0	180.4	60.0
60.0	17.0650	204,780			166.4		150.4	216.0	72.0
70.0	19.9091	238,910			194.2		174.6		84.2
80.0	22.7533	273,040			221.6		192.0		96.0
90.0	25.5975	307,170					225.6		108.0
100.0	28.4416	341,300							120.0

General: Control transformers are listed in VA. Step-down and power transformers are listed in kVA.

Cost Determination:

1. Convert VA to kVA. Volt amperes (VA) ÷ 1000 = Kilovolt amperes (kVA).
2. Determine voltage rating of equipment.
3. Determine whether equipment is single phase or three phase.
4. Refer to Table D5010-161 to find ampere value from kVA value.
5. Determine type of wire insulation – TW, THW, THWN.
6. Determine if wire is copper or aluminum.
7. Refer to Table D5010-161 to obtain copper or aluminum wire size from ampere values.
8. Next refer to Table D5010-142 for the proper conduit size to accommodate the number and size of wires in each particular case.
9. Next refer to unit price data for the per linear foot cost of the conduit.
10. Next refer to unit price data for the per linear foot cost of the wire. Multiply cost of wire per L.F. x number of wires in the circuits to obtain total wire cost.
11. Add values obtained in Steps 9 and 10 for total cost per linear foot for conduit and wire x length of circuit = Total Cost.

Example: A transformer rated 10 kVA 480 volts primary, 240 volts secondary, 3 phase has the capacity to furnish the following:

1. Primary amperes = 10 kVA x 1.20 = 12 amperes (from Table D5010-161)
2. Secondary amperes = 10 kVA x 2.40 = 24 amperes (from Table D5010-161)

Note: Transformers can deliver generally 125% of their rated kVA. For instance, a 10 kVA rated transformer can safely deliver 12.5 kVA.

Table D5010-161 Multiplier Values for kVA to Amperes Determined by Voltage and Phase Values

Volts	Multiplier for Circuits	
	2 Wire, 1 Phase	3 Wire, 3 Phase
115	8.70	
120	8.30	
230	4.30	2.51
240	4.16	2.40
200	5.00	2.89
208	4.80	2.77
265	3.77	2.18
277	3.60	2.08
460	2.17	1.26
480	2.08	1.20
575	1.74	1.00
600	1.66	0.96

General: Motors can be powered by any of the seven systems shown in Figure D5010-121 through Figure D5010-127 provided the motor voltage characteristics are compatible with the power system characteristics.

Cost Determination:

Motor Amperes for the various size H.P. and voltage are listed in Table D5010-171. To find the amperes, locate the required H.P. rating and locate the amperes under the appropriate circuit characteristics.

For example:

A. 100 H.P., 3 phase, 460 volt motor = 124 amperes (Table D5010-171)

B. 10 H.P., 3 phase, 200 volt motor = 32.2 amperes (Table D5010-171)

Motor Wire Size: After the amperes are found in Table D5010-171 the amperes must be increased 25% to compensate for power losses. Next refer to Table D5010-141. Find the appropriate insulation column for copper or aluminum wire to determine the proper wire size.

For example:

A. 100 H.P., 3 phase, 460 volt motor has an ampere value of 124 amperes from Table D5010-171

B. 124A x 1.25 = 155 amperes

C. Refer to Table D5010-141 for THW or THWN wire insulations to find the proper wire size. For a 155 ampere load using copper wire a size 2/0 wire is needed.

D. For the 3 phase motor three wires of 2/0 size are required.

Conduit Size: To obtain the proper conduit size for the wires and type of insulation used, refer to Table D5010-142.

For example: For the 100 H.P., 460V, 3 phase motor, it was determined that three 2/0 wires are required. Assuming THWN insulated copper wire, use Table D5010-142 to determine that three 2/0 wires require 1-1/2″ conduit.

Material Cost of the conduit and wire system depends on:

1. Wire size required
2. Copper or aluminum wire
3. Wire insulation type selected
4. Steel or plastic conduit
5. Type of conduit raceway selected.

Labor Cost of the conduit and wire system depends on:

1. Type and size of conduit
2. Type and size of wires installed
3. Location and height of installation in building or depth of trench
4. Support system for conduit.

Table D5010-171 Ampere Values Determined by Horsepower, Voltage and Phase Values

H.P.	Amperes					
	Single Phase		Three Phase			
	115V	230V	200 V	230V	460V	575V
1/6	4.4A	2.2A				
1/4	5.8	2.9				
1/3	7.2	3.6				
1/2	9.8	4.9	2.3A	2.0A	1.0A	0.8A
3/4	13.8	6.9	3.2	2.8	1.4	1.1
1	16	8	4.1	3.6	1.8	1.4
1-1/2	20	10	6.0	5.2	2.6	2.1
2	24	12	7.8	6.8	3.4	2.7
3	34	17	11.0	9.6	4.8	3.9
5			17.5	15.2	7.6	6.1
7-1/2			25.3	22	11	9
10			32.2	28	14	11
15			48.3	42	21	17
20			62.1	54	27	22
25			78.2	68	34	27
30			92.0	80	40	32
40			119.6	104	52	41
50			149.5	130	65	52
60			177	154	77	62
75			221	192	96	77
100			285	248	124	99
125			359	312	156	125
150			414	360	180	144
200			552	480	240	192

Cost Determination (cont.)

Magnetic starters, switches, and motor connection:

To complete the cost picture from H.P. to Costs additional items must be added to the cost of the conduit and wire system to arrive at a total cost.

1. Assembly D5020 160 Magnetic Starters Installed Cost lists the various size starters for single phase and three phase motors.
2. Assembly D5020 165 Heavy Duty Safety Switches Installed Cost lists safety switches required at the beginning of a motor circuit and one required in the vicinity of the motor location.
3. Assembly D5020 170 Motor Connection lists the various costs for single and three phase motors.

Table D5010-172

Worksheet for Motor Circuits					
				Cost	
Item	Type	Size	Quantity	Unit	Total
Wire					
Conduit					
Switch					
Starter					
Switch					
Motor Connection					
Other					
Total Cost					

Worksheet to obtain total motor wiring costs:

It is assumed that the motors or motor driven equipment are furnished and installed under other sections for this estimate and the following work is done under this section:

1. Conduit
2. Wire (add 10% for additional wire beyond conduit ends for connections to switches, boxes, starters, etc.)
3. Starters
4. Safety switches
5. Motor connections

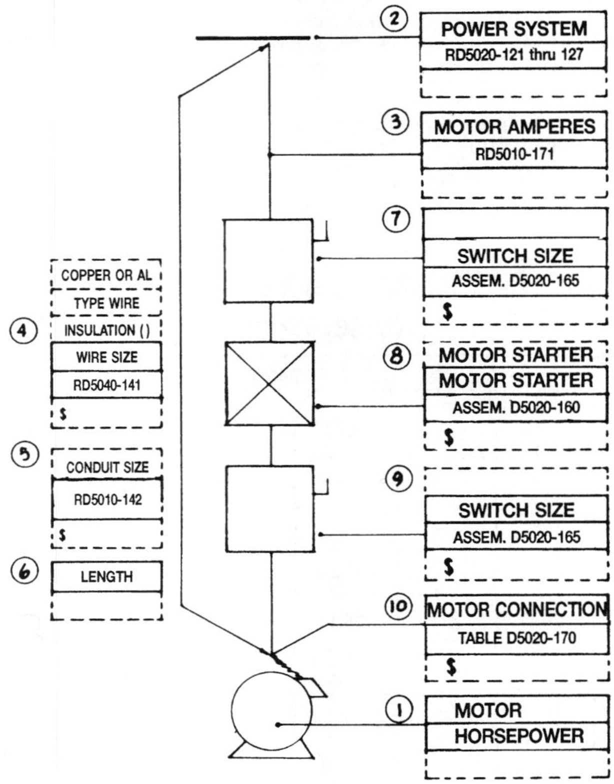

Table D5010-173 Maximum Horsepower for Starter Size by Voltage

Starter	Maximum HP (3ϕ)			
Size	208V	240V	480V	600V
00	1½	1½	2	2
0	3	3	5	5
1	7½	7½	10	10
2	10	15	25	25
3	25	30	50	50
4	40	50	100	100
5		100	200	200
6		200	300	300
7		300	600	600
8		450	900	900
8L		700	1500	1500

General: The cost of the lighting portion of the electrical costs is dependent upon:
1. The footcandle requirement of the proposed building.
2. The type of fixtures required.
3. The ceiling heights of the building.
4. Reflectance value of ceilings, walls and floors.
5. Fixture efficiencies and spacing vs. mounting height ratios.

Footcandle Requirements: See Table D5020-204 for Footcandle and Watts per S.F. determination.

Table D5020-201 IESNA* Recommended Illumination Levels in Footcandles

Commercial Buildings			Industrial Buildings		
Type	Description	Footcandles	Type	Description	Footcandles
Bank	Lobby	50	Assembly Areas	Rough bench & machine work	50
	Customer Areas	70		Medium bench & machine work	100
	Teller Stations	150		Fine bench & machine work	500
	Accounting Areas	150	Inspection Areas	Ordinary	50
Offices	Routine Work	100		Difficult	100
	Accounting	150		Highly Difficult	200
	Drafting	200	Material Handling	Loading	20
	Corridors, Halls, Washrooms	30		Stock Picking	30
Schools	Reading or Writing	70		Packing, Wrapping	50
	Drafting, Labs, Shops	100	Stairways	Service Areas	20
	Libraries	70	Washrooms	Service Areas	20
	Auditoriums, Assembly	15	Storage Areas	Inactive	5
	Auditoriums, Exhibition	30		Active, Rough, Bulky	10
Stores	Circulation Areas	30		Active, Medium	20
	Stock Rooms	30		Active, Fine	50
	Merchandise Areas, Service	100	Garages	Active Traffic Areas	20
	Self-Service Areas	200		Service & Repair	100

*IESNA - Illuminating Engineering Society of North America

Table D5020-202 General Lighting Loads by Occupancies

Type of Occupancy	Unit Load per S.F. (Watts)
Armories and Auditoriums	1
Banks	5
Barber Shops and Beauty Parlors	3
Churches	1
Clubs	2
Court Rooms	2
*Dwelling Units	3
Garages — Commercial (storage)	½
Hospitals	2
*Hotels and Motels, including apartment houses without provisions for cooking by tenants	2
Industrial Commercial (Loft) Buildings	2
Lodge Rooms	1½
Office Buildings	5
Restaurants	2
Schools	3
Stores	3
Warehouses (storage)	¼
*In any of the above occupancies except one-family dwellings and individual dwelling units of multi-family dwellings:	
Assembly Halls and Auditoriums	1
Halls, Corridors, Closets	½
Storage Spaces	¼

Table D5020-203 Lighting Limit (Connected Load) for Listed Occupancies: New Building Proposed Energy Conservation Guideline

Type of Use	Maximum Watts per S.F.
Interior	
Category A: Classrooms, office areas, automotive mechanical areas, museums, conference rooms, drafting rooms, clerical areas, laboratories, merchandising areas, kitchens, examining rooms, book stacks, athletic facilities.	3.00
Category B: Auditoriums, waiting areas, spectator areas, restrooms, dining areas, transportation terminals, working corridors in prisons and hospitals, book storage areas, active inventory storage, hospital bedrooms, hotel and motel bedrooms, enclosed shopping mall concourse areas, stairways.	1.00
Category C: Corridors, lobbies, elevators, inactive storage areas.	0.50
Category D: Indoor parking.	0.25
Exterior	
Category E: Building perimeter: wall-wash, facade, canopy.	5.00 (per linear foot)
Category F: Outdoor parking.	0.10

Table D5020-204 Procedure for Calculating Footcandles and Watts Per Square Foot

1. Initial footcandles = No. of fixtures × lamps per fixture × lumens per lamp × coefficient of utilization ÷ square feet
2. Maintained footcandles = initial footcandles × maintenance factor
3. Watts per square foot = No. of fixtures × lamps × (lamp watts + ballast watts) ÷ square feet

Example: To find footcandles and watts per S.F. for an office 20′ x 20′ with 11 fluorescent fixtures each having 4–40 watt C.W. lamps.

Based on good reflectance and clean conditions:

Lumens per lamp = 40 watt cool white at 3150 lumens per lamp

Coefficient of utilization = .42 (varies from .62 for light colored areas to .27 for dark)

Maintenance factor = .75 (varies from .80 for clean areas with good maintenance to .50 for poor)

Ballast loss = 8 watts per lamp (Varies with manufacturer. See manufacturers' catalog.)

1. Initial footcandles:

$$\frac{11 \times 4 \times 3150 \times .42}{400} = \frac{58{,}212}{400} = 145 \text{ footcandles}$$

2. Maintained footcandles:

145 × .75 = 109 footcandles

3. Watts per S.F.

$$\frac{11 \times 4\ (40 + 8)}{400} = \frac{2{,}112}{400} = 5.3 \text{ watts per S.F.}$$

Table D5020-205 Approximate Watts Per Square Foot for Popular Fixture Types

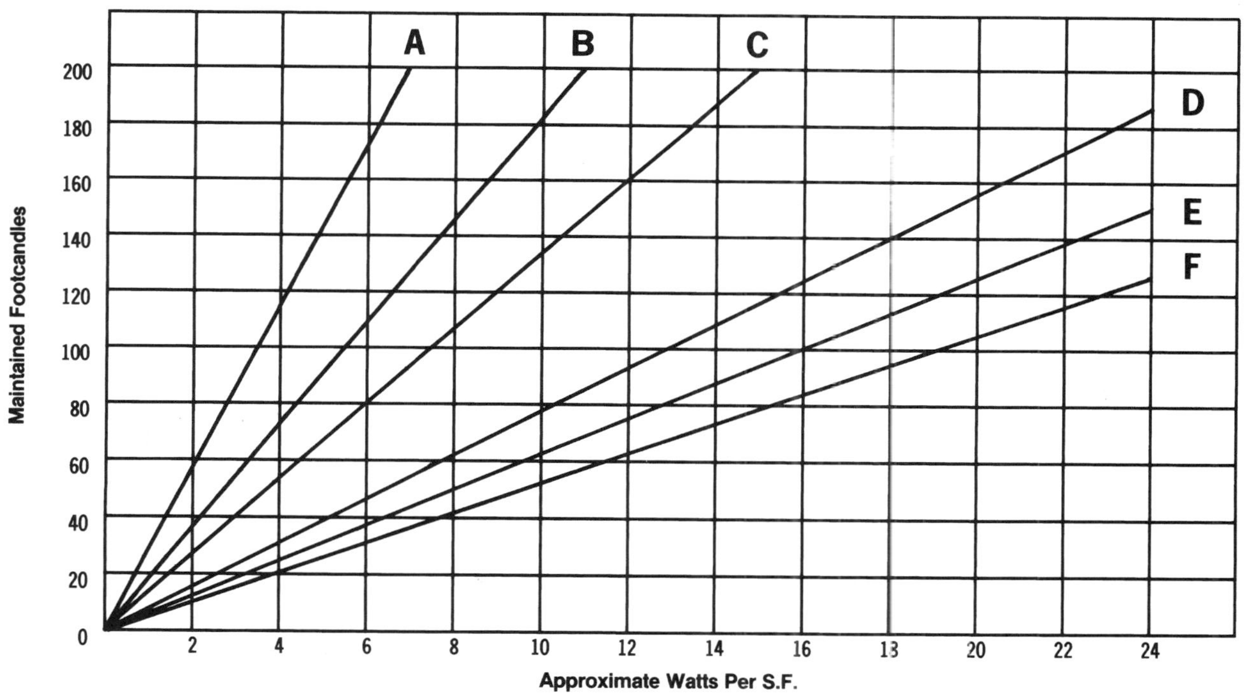

Due to the many variables involved, use for preliminary estimating only:

a. Fluorescent – industrial System D5020 208
b. Fluorescent – lens unit System D5020 208 Fixture types B & C
c. Fluorescent – louvered unit
d. Incandescent – open reflector System D5020 214, Type D
e. Incandescent – lens unit System D5020 214, Type A
f. Incandescent – down light System D5020 214, Type B

Table D5020-242 For Other than Regular Cool White (CW) Lamps

Multiply Material Costs as Follows:					
Regular Lamps	Cool white deluxe (CWX) Warm white deluxe (WWX) Warm white (WW) Natural (N)	x 1.35 x 1.35 x 1.30 x 2.05	Energy Saving Lamps	Cool white (CW/ES) Cool white deluxe (CWX/ES) Warm white (WW/ES) Warm white deluxe (WWX/ES)	x 1.35 x 1.65 x 1.55 x 1.65

G10 Site Preparation — RG1010-010 Information

The Building Site Work Division is divided into five sections:

G10 Site Preparation
G20 Site Improvements
G30 Site Mechanical Utilities
G40 Site Electrical Utilities
G90 Other Site Construction

Clear and Grub, Demolition and Landscaping must be added to these costs as they apply to individual sites.

The Site Preparation section includes typical bulk excavation, trench excavation, and backfill. Sheeting, dewatering, drilling and blasting, traffic control, and any unusual items relating to a particular site must be added to this.

The Site Improvements section, including Parking Lot and Roadway construction, has quantities and prices for typical designs with gravel base and asphaltic concrete pavement. Different climate, soil conditions, material availability, and owner's requirements can easily be adapted to these parameters.

The Site Utilities sections tabulate the installed price for pipe. Excavation and backfill costs for the pipe are derived from the Site Preparation section.

By following the examples and illustrations of typical site work conditions in this division and adapting them to your particular site, an accurate price, including O&P, can be determined.

Table G1010-011 Site Preparation

Description	Unit	Total Costs
Clear and Grub Average	Acre	$11,175
Bulk Excavation with Front End Loader	C.Y.	2.55
Spread and Compact Dumped Material	C.Y.	3.15
Fine Grade and Seed	S.Y.	3.85

G10 Site Preparation — RG1030-400 Excavation, General

Bulk excavation on building jobs today is done entirely by machine. Hand work is used only for squaring out corners, fine grading, cutting out trenches and footings below the general excavating grade, and for minor miscellaneous cuts. Hand trimming of material other than sand and gravel is generally done with picks and shovels.

The following figures on machine excavating apply to basement pits, large footings and other excavation for buildings. On grading excavation this performance can be bettered by 33% to 50%. The figures presume loading directly into trucks. Allowance has been made for idle time due to truck delay and moving about on the job. Figures include no water problems or slow down due to sheeting and bracing operations. They also do not include truck spotters or labor for cleanup.

Table G1030-401 Bulk Excavation Cost, 15′ Deep

1½ C.Y. Hydraulic Excavator Backhoe, Four Tandem Trucks with a Two Mile Round Trip Haul Distance

Item	Daily Cost	Labor & Equipment	Unit Price (C.Y.)
Equipment operator @ $ 79.05 per hr. x 8 hrs. Laborer @ $ 58.15 per hr. x 8 hrs. Four truck drivers @ $ 65.25 per hr. x 8 hrs.	$ 632.40 465.20 2,088.00	$3,185.60 Labor	$4.42 Labor
1½ C.Y. backhoe @ $1,029.00 per day Four 16 ton trucks @ $ 690.80 per day	$1,029.00 2,763.20	$3,792.20 Equipment	$5.27 Equipment
Production = 720 C.Y. per day	$6,977.80	$6,977.80 Labor & Equip.	$9.69 Labor & Equip.

Table G1030-402 Total Cost Per C.Y. Excavation for Various Size Jobs

No. of C.Y. to be Excavated	Cost Per Cubic Yard	Mobilizing & Demobilizing Cost*	Total Cost Per Cubic Yard
500 C.Y.	$9.69	$3.94	$13.63
1,000	↓	1.97	$11.66
1,500		1.31	$11.00
2,000		0.99	$10.68
3,000		0.66	$10.35
5,000		0.39	$10.08
10,000		0.20	$ 9.89
15,000	▼	0.13	$ 9.82

*$985.00 for Mobilizing or Demobilizing 1½ C.Y. Hydraulic Excavator

Table G1030-403 Trench Bottom Widths for Various Outside Diameters of Buried Pipe

Outside Diameter	Trench Bottom Width
24 inches	4.1 feet
30	4.9
36	5.6
42	6.3
48	6.3
60	8.5
72	10.0
84 ▼	11.4 ▼

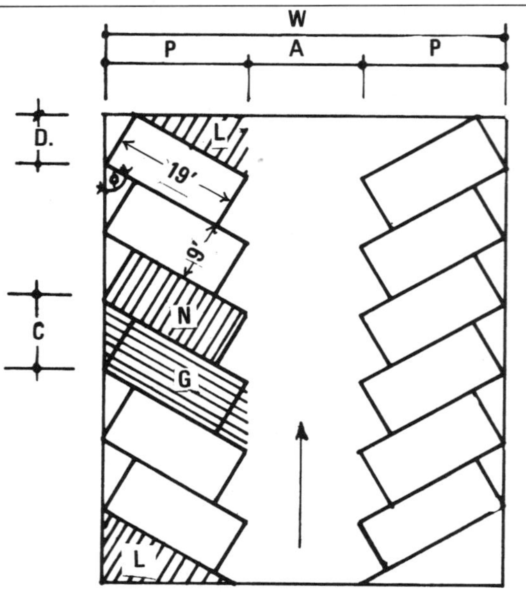

SINGLE UNIT

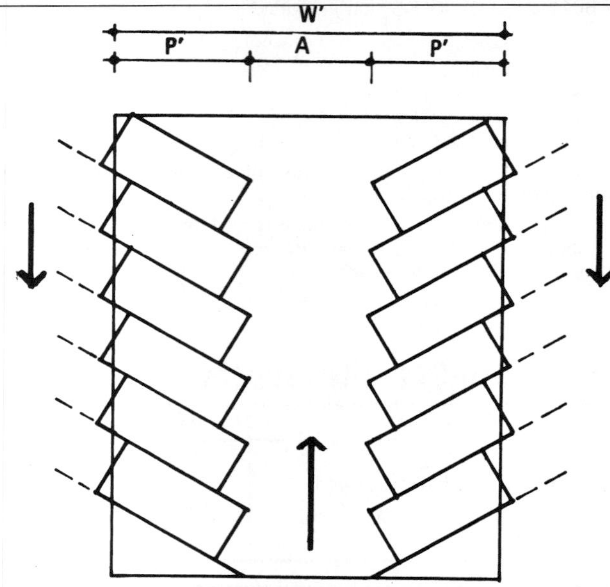

OVERLAPPING UNITS

Table G2020-502 Layout Data Based on 9′ x 19′ Parking Stall Size

φ	P	A	C	W	N	G	D	L	P′	W′
Angle of Stall	Parking Depth	Aisle Width	Curb Length	Width Overall	Net Car Area	Gross Car Area	Distance Last Car	Lost Area	Parking Depth	Width Overall
90°	19′.	24′	9′.	62′.	171 S.F.	171 S.F.	9′.	0	19′.	62′.
60°	21′.8	18′	10.4′	60′.	171 S.F.	217 S.F.	7.8′	205 S.F.	18.8′	55.5′
45°	19.8′	13′	12.8′	52.7′	171 S.F.	252 S.F.	6.4′	286 S.F.	16.6′	46.2′

Note: Square foot per car areas do not include the area of the travel lane.

90° Stall Angle: The main reason for use of this stall angle is to achieve the highest car capacity. This may be sound reasoning for employee lots with all day parking, but in most (in & out) lots there is difficulty in entering the stalls and no traffic lane direction. This may outweigh the advantage of high capacity.

60° Stall Angle: This layout is used most often due to the ease of entering and backing out. Also the traffic aisle may be smaller.

45° Stall Angle: This requires a small change of direction from the traffic aisle to the stall, so the aisle may be reduced in width.

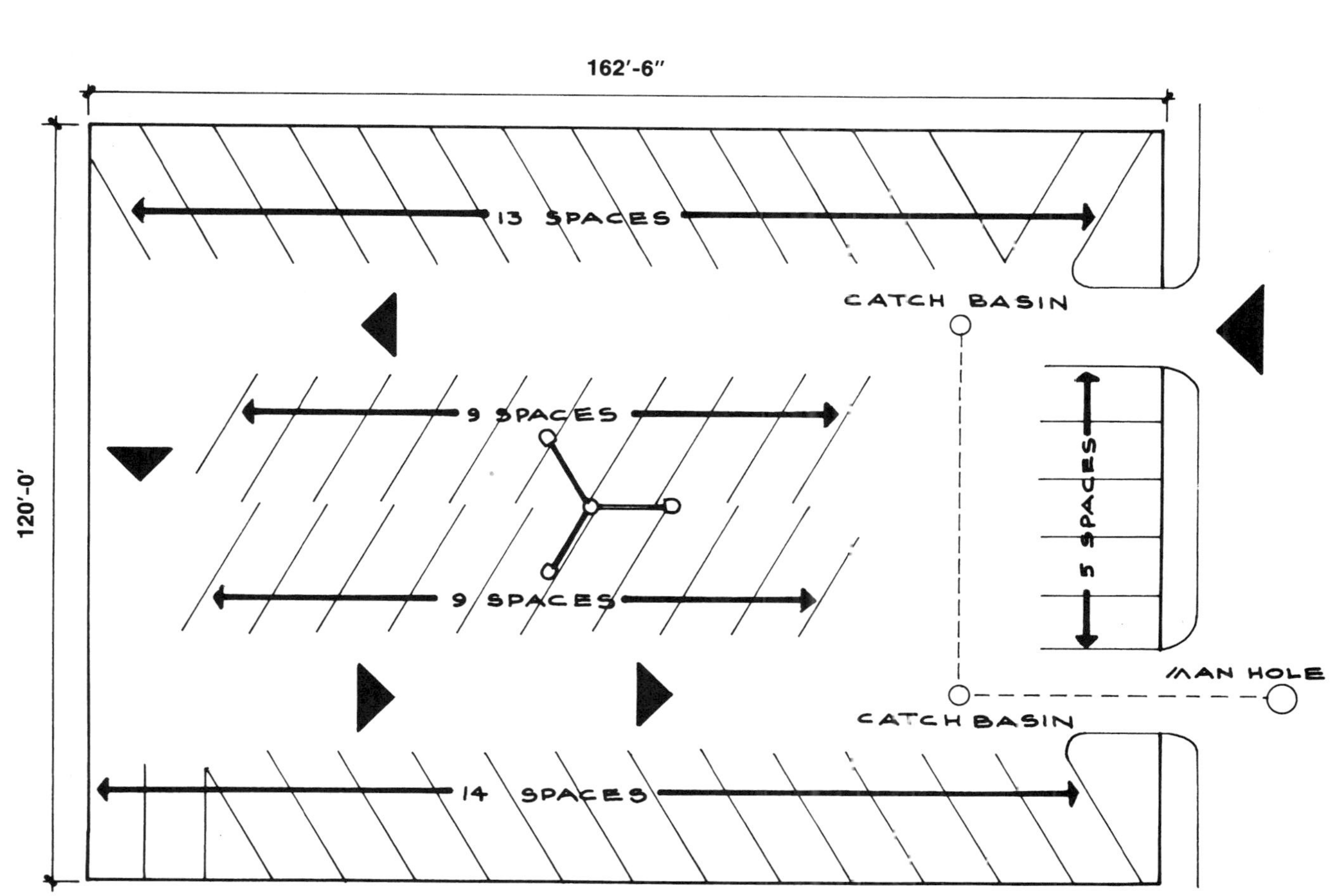

PARKING LOT PLAN (50 CAR)

Preliminary Design Data: The space required for parking and maneuvering is between 300 and 400 S.F. per car, depending upon engineering layout and design.

Ninety degree (90°) parking, with a central driveway and two rows of parked cars, will provide the best economy.

Diagonal parking is easier than 90° for the driver and reduces the necessary driveway width, but requires more total space.

Table H1010-101 General Contractor's Overhead

There are two distinct types of overhead on a construction project: project overhead and main office overhead. Project overhead includes those costs at a construction site not directly associated with the installation of construction materials. Examples of project overhead costs include the following:

1. Superintendent
2. Construction office and storage trailers
3. Temporary sanitary facilities
4. Temporary utilities
5. Security fencing
6. Photographs
7. Clean up
8. Performance and payment bonds

The above project overhead items are also referred to as General Requirements and therefore are estimated in Division 1. Division 1 is the first division listed in the CSI MasterFormat but it is usually the last division estimated. The sum of the costs in Divisions 1 through 49 is referred to as the sum of the direct costs.

All construction projects also include indirect costs. The primary components of indirect costs are the contractor's main office overhead and profit. The amount of the main office overhead expense varies depending on the following:

1. Owner's compensation
2. Project managers and estimator's wages
3. Clerical support wages
4. Office rent and utilities
5. Corporate legal and accounting costs
6. Advertising
7. Automobile expenses
8. Association dues
9. Travel and entertainment expenses

These costs are usually calculated as a percentage of annual sales volume. This percentage can range from 35% for a small contractor doing less than $500,000 to 5% for a large contractor with sales in excess of $100 million.

Table H1010-102 Main Office Expense

A general contractor's main office expense consists of many items not detailed in the front portion of the data set. The percentage of main office expense declines with increased annual volume of the contractor. Typical main office expense ranges from 2% to 20% with the median about 7.2% of total volume. This equals about 7.7% of direct costs. The following are approximate percentages of total overhead for different items usually included in a general contractor's main office overhead. With different accounting procedures, these percentages may vary.

Item	Typical Range	Average
Managers', clerical and estimators' salaries	40 % to 55 %	48%
Profit sharing, pension and bonus plans	2 to 20	12
Insurance	5 to 8	6
Estimating and project management (not inclucing salaries)	5 to 9	7
Legal, accounting and data processing	0.5 to 5	3
Automobile and light truck expense	2 to 8	5
Depreciation of overhead capital expenditures	2 to 6	4
Maintenance of office equipment	0.1 to 1.5	1
Office rental	3 to 5	4
Utilities including phone and light	1 to 3	2
Miscellaneous	5 to 15	8
Total		100%

Table H1010-201 Architectural Fees

Tabulated below are typical percentage fees by project size for good professional architectural service. Fees may vary from those listed depending upon degree of design difficulty and economic conditions in any particular area.

Rates can be interpolated horizontally and vertically. Various portions of the same project requiring different rates should be adjusted proportionately. For alterations, add 50% to the fee for the first $500,000 of project cost and add 25% to the fee for project cost over $500,000.

Architectural fees tabulated below include Structural, Mechanical and Electrical Engineering Fees. They do not include the fees for special consultants such as kitchen planning, security, acoustical, interior design, etc.

Building Types	Total Project Size in Thousands of Dollars						
	100	250	500	1,000	5,000	10,000	50,000
Factories, garages, warehouses, repetitive housing	9.0%	8.0%	7.0%	6.2%	5.3%	4.9%	4.5%
Apartments, banks, schools, libraries, offices, municipal buildings	12.2	12.3	9.2	8.0	7.0	6.6	6.2
Churches, hospitals, homes, laboratories, museums, research	15.0	13.6	12.7	11.9	9.5	8.8	8.0
Memorials, monumental work, decorative furnishings	—	16.0	14.5	13.1	10.0	9.0	8.3

Table H1010-202 Engineering Fees

Typical **Structural Engineering Fees** based on type of construction and total project size. These fees are included in Architectural Fees.

Type of Construction	Total Project Size (in thousands of dollars)			
	$500	$500-$1,000	$1,000-$5,000	Over $5000
Industrial buildings, factories & warehouses	Technical payroll times 2.0 to 2.5	1.60%	1.25%	1.00%
Hotels, apartments, offices, dormitories, hospitals, public buildings, food stores	↓	2.00%	1.70%	1.20%
Museums, banks, churches and cathedrals	↓	2.00%	1.75%	1.25%
Thin shells, prestressed concrete, earthquake resistive	↓	2.00%	1.75%	1.50%
Parking ramps, auditoriums, stadiums, convention halls, hangars & boiler houses	↓	2.50%	2.00%	1.75%
Special buildings, major alterations, underpinning & future expansion	↓	Add to above 0.5%	Add to above 0.5%	Add to above 0.5%

For complex reinforced concrete or unusually complicated structures, add 20% to 50%.

Table H1010-203 Mechanical and Electrical Fees

Typical **Mechanical and Electrical Engineering Fees** based on the size of the subcontract. The fee structure for both is shown below. These fees are included in Architectural Fees.

Type of Construction	Subcontract Size							
	$25,000	$50,000	$100,000	$225,000	$350,000	$500,000	$750,000	$1,000,000
Simple structures	6.4%	5.7%	4.8%	4.5%	4.4%	4.3%	4.2%	4.1%
Intermediate structures	8.0	7.3	6.5	5.6	5.1	5.0	4.9	4.8
Complex structures	10.1	9.0	9.0	8.0	7.5	7.5	7.0	7.0

For renovations, add 15% to 25% to applicable fee.

Table H1010-204 Construction Time Requirements

The table below is average construction time in months for different types of building projects. The table at right is the construction time in months for different size projects. Design time runs 25% to 40% of construction time.

Type Building	Construction Time
Industrial Buildings	12 Months
Commercial Buildings	15 Months
Research & Development	18 Months
Institutional Buildings	20 Months

Project Value	Construction Time
Under $1,400,000	10 Months
Up to $3,800,000	15 Months
Up to $19,000,000	21 Months
over $19,000,000	28 Months

Table H1010-301 Builder's Risk Insurance

Builder's Risk Insurance is insurance on a building during construction. Premiums are paid by the owner or the contractor. Blasting, collapse and underground insurance would raise total insurance costs above those listed. Floater policy for materials delivered to the job runs $.75 to $1.25 per $100 value. Contractor equipment insurance runs $.50 to $1.50 per $100 value. Insurance for miscellaneous tools to $1,500 value runs from $3.00 to $7.50 per $100 value.

Tabulated below are New England Builder's Risk insurance rates in dollars per $100 value for $1,000 deductible. For $25,000 deductible, rates can be reduced 13% to 34%. On contracts over $1,000,000, rates may be lower than those tabulated. Policies are written annually for the total completed value in place. For "all risk" insurance (excluding flood, earthquake and certain other perils) add $.025 to total rates below.

Coverage	Frame Construction (Class 1)		Brick Construction (Class 4)		Fire Resistive (Class 6)	
	Range	Average	Range	Average	Range	Average
Fire Insurance	$.350 to $ .850	$.600	$.158 to $.189	$.174	$.052 to $.080	$.070
Extended Coverage	.115 to .200	.158	.080 to .105	.101	.081 to .105	.100
Vandalism	.012 to .016	.014	.008 to .011	.011	.008 to .011	.010
Total Annual Rate	$.477 to $1.066	$.772	$.246 to $.305	$.286	$.141 to $.196	$.180

Table H1010-302 Performance Bond

This table shows the cost of a performance bond for a construction job scheduled to be completed in 12 months. Add 1% of the premium cost per month for jobs requiring more than 12 months to complete. The rates are "standard" rates offered to contractors that the bonding company considers financially sound and capable of doing the work. Preferred rates are offered by some bonding companies based upon financial strength of the contractor. Actual rates vary from contractor to contractor and from bonding company to bonding company. Contractors should prequalify through a bonding agency before submitting a bid on a contract that requires a bond.

Contract Amount	Building Construction Class B Projects	Highways & Bridges	
		Class A New Construction	Class A-1 Highway Resurfacing
First $ 100,000 bid	$25.00 per M	$15.00 per M	$9.40 per M
Next 400,000 bid	$ 2,500 plus $15.00 per M	$ 1,500 plus $10.00 per M	$ 940 plus $7.20 per M
Next 2,000,000 bid	8,500 plus 10.00 per M	5,500 plus 7.00 per M	3,820 plus 5.00 per M
Next 2,500,000 bid	28,500 plus 7.50 per M	19,500 plus 5.50 per M	15,820 plus 4.50 per M
Next 2,500,000 bid	47,250 plus 7.00 per M	33,250 plus 5.00 per M	28,320 plus 4.50 per M
Over 7,500,000 bid	64,750 plus 6.00 per M	45,750 plus 4.50 per M	39,570 plus 4.00 per M

Table H1010-401 Workers' Compensation Insurance Rates by Trade

The table below tabulates the national averages for workers' compensation insurance rates by trade and type of building. The average "Insurance Rate" is multiplied by the "% of Building Cost" for each trade. This produces the "Workers' Compensation" cost by % of total labor cost, to be added for each trade by building type to determine the weighted average workers' compensation rate for the building types analyzed.

Trade	Insurance Rate (% Labor Cost)				% of Building Cost			Workers' Compensation		
	Range			Average	Office Bldgs.	Schools & Apts.	Mfg.	Office Bldgs.	Schools & Apts.	Mfg.
Excavation, Grading, etc.	3.1 %	to	21.5%	9.3%	4.8%	4.9%	4.5%	0.45%	0.46%	0.42%
Piles & Foundations	6.4	to	32.8	14.6	7.1	5.2	8.7	1.04	0.76	1.27
Concrete	4.3	to	32.5	12.6	5.0	14.8	3.7	0.63	1.86	0.47
Masonry	4.6	to	36.8	13.7	6.9	7.5	1.9	0.95	1.03	0.26
Structural Steel	6.5	to	95.8	27.5	10.7	3.9	17.6	2.94	1.07	4.84
Miscellaneous & Ornamental Metals	4.1	to	28.9	11.7	2.8	4.0	3.6	0.33	0.47	0.42
Carpentry & Millwork	5.0	to	42.8	14.4	3.7	4.0	0.5	0.53	0.58	0.07
Metal or Composition Siding	6.5	to	75.9	18.7	2.3	0.3	4.3	0.43	0.06	0.80
Roofing	6.5	to	107.1	31.4	2.3	2.6	3.1	0.72	0.82	0.97
Doors & Hardware	4.1	to	42.8	11.4	0.9	1.4	0.4	0.10	0.16	0.05
Sash & Glazing	4.8	to	28.0	13.1	3.5	4.0	1.0	0.46	0.52	0.13
Lath & Plaster	2.8	to	41.3	10.5	3.3	6.9	0.8	0.35	0.72	0.08
Tile, Marble & Floors	2.9	to	22.4	8.8	2.6	3.0	0.5	0.23	0.26	0.04
Acoustical Ceilings	3.1	to	33.8	8.6	2.4	0.2	0.3	0.21	0.02	0.03
Painting	4.1	to	34.3	11.7	1.5	1.6	1.6	0.18	0.19	0.19
Interior Partitions	5.0	to	42.8	14.4	3.9	4.3	4.4	0.56	0.62	0.63
Miscellaneous Items	2.5	to	105.9	12.6	5.2	3.7	9.7	0.65	0.47	1.22
Elevators	1.4	to	13.6	5.2	2.1	1.1	2.2	0.11	0.06	0.11
Sprinklers	2.8	to	25.9	7.6	0.5	—	2.0	0.04	—	0.15
Plumbing	1.9	to	19.5	6.9	4.9	7.2	5.2	0.34	0.50	0.36
Heat., Vent., Air Conditioning	3.8	to	18.4	8.8	13.5	11.0	12.9	1.19	0.97	1.14
Electrical	2.3	to	14.4	5.6	10.1	8.4	11.1	0.57	0.47	0.62
Total	1.4 %	to	107.1%	—	100.0%	100.0%	100.0%	13.01%	12.07%	14.27%
Overall Weighted Average 13.12%										

Table H1010-402 Workers' Compensation Insurance Rates by States

The table below lists the weighted average workers' compensation base rate for each state with a factor comparing this with the national average of 13.0%.

State	Weighted Average	Factor	State	Weighted Average	Factor	State	Weighted Average	Factor
Alabama	19.5%	150	Kentucky	14.3%	110	North Dakota	9.2%	71
Alaska	13.7	105	Louisiana	18.6	143	Ohio	8.2	63
Arizona	12.6	97	Maine	12.2	94	Oklahoma	12.5	96
Arkansas	8.3	64	Maryland	14.1	108	Oregon	11.8	91
California	25.8	198	Massachusetts	11.6	89	Pennsylvania	17.4	134
Colorado	7.2	55	Michigan	16.3	125	Rhode Island	12.1	93
Connecticut	23.7	182	Minnesota	22.6	174	South Carolina	18.9	145
Delaware	11.9	92	Mississippi	13.2	102	South Dakota	14.9	115
District of Columbia	11.0	85	Missouri	14.8	114	Tennessee	12.0	92
Florida	10.7	82	Montana	8.5	65	Texas	9.1	70
Georgia	28.1	216	Nebraska	17.7	136	Utah	8.8	68
Hawaii	8.1	62	Nevada	8.7	67	Vermont	13.1	101
Idaho	10.3	79	New Hampshire	20.0	154	Virginia	9.5	73
Illinois	27.1	208	New Jersey	14.4	111	Washington	10.5	81
Indiana	5.5	42	New Mexico	14.4	111	West Virginia	8.0	62
Iowa	14.6	112	New York	17.2	132	Wisconsin	13.1	101
Kansas	8.4	65	North Carolina	16.7	128	Wyoming	6.1	47
Weighted Average for U.S. is 13.7% of payroll = 100%								

The weighted average skilled worker rate for 35 trades is 13.0%. For bidding purposes, apply the full value of Workers' Compensation directly to total labor costs, or if labor is 38%, materials 42% and overhead and profit 20% of total cost, carry 38/80 x 13.0% = 6.2% of cost (before overhead and profit) into overhead. Rates vary not only from state to state but also with the experience rating of the contractor.

Rates are the most current available at the time of publication.

General Conditions — RH1020-300 Overhead & Miscellaneous Data

Unemployment Taxes and Social Security Taxes

State unemployment tax rates vary not only from state to state, but also with the experience rating of the contractor. The federal unemployment tax rate is 6.0% of the first $7,000 of wages. This is reduced by a credit of up to 5.4% for timely payment to the state. The minimum federal unemployment tax is 0.6% after all credits.

Social Security (FICA) for 2016 is estimated at time of publication to be 7.65% of wages up to $118,500.

General Conditions — RH1020-400 Overtime

Overtime

One way to improve the completion date of a project or eliminate negative float from a schedule is to compress activity duration times. This can be achieved by increasing the crew size or working overtime with the proposed crew.

To determine the costs of working overtime to compress activity duration times, consider the following examples. Below is an overtime efficiency and cost chart based on a five, six, or seven day week with an eight through twelve hour day. Payroll percentage increases for time and one half and double time are shown for the various working days.

Days per Week	Hours per Day	Production Efficiency					Payroll Cost Factors	
		1st Week	2nd Week	3rd Week	4th Week	Average 4 Weeks	@ 1-1/2 Times	@ 2 Times
5	8	100%	100%	100%	100%	100%	1.000	1.000
	9	100	100	95	90	96	1.056	1.111
	10	100	95	90	85	93	1.100	1.200
	11	95	90	75	65	81	1.136	1.273
	12	90	85	70	60	76	1.167	1.333
6	8	100	100	95	90	96	1.083	1.167
	9	100	95	90	85	93	1.130	1.259
	10	95	90	85	80	88	1.167	1.333
	11	95	85	70	65	79	1.197	1.394
	12	90	80	65	60	74	1.222	1.444
7	8	100	95	85	75	89	1.143	1.286
	9	95	90	80	70	84	1.183	1.365
	10	90	85	75	65	79	1.214	1.429
	11	85	80	65	60	73	1.240	1.481
	12	85	75	60	55	69	1.262	1.524

General Conditions — RH1030-100 General

Sales Tax by State

State sales tax on materials is tabulated below (5 states have no sales tax). Many states allow local jurisdictions, such as a county or city, to levy additional sales tax.

Some projects may be sales tax exempt, particularly those constructed with public funds.

State	Tax (%)	State	Tax (%)	State	Tax (%)	State	Tax (%)
Alabama	4	Illinois	6.25	Montana	0	Rhode Island	7
Alaska	0	Indiana	7	Nebraska	5.5	South Carolina	6
Arizona	5.6	Iowa	6	Nevada	6.85	South Dakota	4
Arkansas	6	Kansas	6.15	New Hampshire	0	Tennessee	7
California	7.5	Kentucky	6	New Jersey	7	Texas	6.25
Colorado	2.9	Louisiana	4	New Mexico	5.125	Utah	5.95
Connecticut	6.35	Maine	5.5	New York	4	Vermont	6
Delaware	0	Maryland	6	North Carolina	4.75	Virginia	5.3
District of Columbia	5.75	Massachusetts	6.25	North Dakota	5	Washington	6.5
Florida	6	Michigan	6	Ohio	5.75	West Virginia	6
Georgia	4	Minnesota	6.875	Oklahoma	4.5	Wisconsin	5
Hawaii	4	Mississippi	7	Oregon	0	Wyoming	4
Idaho	6	Missouri	4.225	Pennsylvania	6	Average	5.08 %

Table H1040-101 Steel Tubular Scaffolding

On new construction, tubular scaffolding is efficient up to 60′ high or five stories. Above this it is usually better to use hung scaffolding if construction permits. Swing scaffolding operations may interfere with tenants. In this case, the tubular is more practical at all heights.

In repairing or cleaning the front of an existing building, the cost of tubular scaffolding per S.F. of building front goes up as the height increases above the first tier. The first tier cost is relatively high due to leveling and alignment.

The minimum efficient crew for erection is three workers. For heights over 50′, a crew of four is more efficient. Use two or more on top and two at the bottom for handing up or hoisting. Four workers can erect and dismantle about nine frames per hour up to five stories. From five to eight stories, they will average six frames per hour. With 7′ horizontal spacing, this will run about 400 S.F. and 265 S.F. of wall surface, respectively. Time for placing planks must be added to the above. On heights above 50′, five planks can be placed per labor-hour.

The cost per 1,000 S.F. of building front in the table below was developed by pricing the materials required for a typical tubular scaffolding system eleven frames long and two frames high. Planks were figured five wide for standing plus two wide for materials.

Frames are 5′ wide and usually spaced 7′ O.C. horizontally. Sidewalk frames are 6′ wide. Rental rates will be lower for jobs over three months' duration.

For jobs under twenty-five frames, add 50% to rental cost. These figures do not include accessories which are listed separately below. Large quantities for long periods can reduce rental rates by 20%.

Item	Unit	Monthly Rent	Per 1,000 S.F. of Building Front	
			No. of Pieces	Rental per Month
5′ Wide Standard Frame, 6′-4″ High	Ea.	$ 5.90	24	$141.60
Leveling Jack & Plate	↓	2.20	24	52.80
Cross Brace	↓	1.10	44	48.40
Side Arm Bracket, 21″	↓	2.20	12	26.40
Guardrail Post	↓	1.10	12	13.20
Guardrail, 7′ section	↓	1.10	22	24.20
Stairway Section	↓	38.50	2	77.00
Walk-Through Frame Guardrail	↓	2.75	2	5.50
			Total	$389.10
			Per C.S.F., 1 Use/Mo.	$ 38.91

Scaffolding is often used as falsework over 15′ high during construction of cast-in-place concrete beams and slabs. Two-foot wide scaffolding is generally used for heavy beam construction. The span between frames depends upon the load to be carried, with a maximum span of 5′.

Heavy duty shoring frames with a capacity of 10,000#/leg can be spaced up to 10′ O. C., depending upon form support design and loading.

Scaffolding used as horizontal shoring requires less than half the material required with conventional shoring.

On new construction, erection is done by carpenters.

Rolling towers supporting horizontal shores can reduce labor and speed the job. For maintenance work, catwalks with spans up to 70′ can be supported by the rolling towers.

City Adjustments RJ1010-010

General: The following information on current city cost indexes is calculated for over 930 zip code locations in the United States and Canada. Index figures for both material and installation are based upon the 30 major city average of 100 and represent the cost relationship on July 1, 2015.

In addition to index adjustment, the user should consider:

1. productivity
2. management efficiency
3. competitive conditions
4. automation
5. restrictive union practices
6. unique local requirements
7. regional variations due to specific building codes

The weighted-average index is calculated from about 66 materials, 6 equipment types and 21 building trades. The component contribution of these in a model building is tabulated in Table J1010-011 below.

If the systems component distribution of a building is unknown, the weighted-average index can be used to adjust the cost for any city.

Table J1010-011 Labor, Material, and Equipment Cost Distribution for Weighted Average Listed by System Division

Division No.	Building System	Percentage	Division No.	Building System	Percentage
A	Substructure	6.2%	D10	Services: Conveying	4.2%
B10	Shell: Superstructure	18.9	D20-40	Mechanical	22.2
B20	Exterior Closure	11.5	D50	Electrical	11.9
B30	Roofing	2.8	E	Equipment & Furnishings	2.2
C	Interior Construction	16.2	G	Site Work	3.9
				Total weighted average (Div. A-G)	100.0%

How to Use the Component Indexes: Table J1010-012 below shows how the average costs obtained from this data set for each division should be adjusted for your particular building in your city. The example is a building adjusted for New York, NY. Indexes for other cities are tabulated in Division RJ1030.

Table J1010-012 Adjustment of "Book" Costs to a Particular City

Systems Division Number	System Description	New York, NY City Cost Index	Cost Factor	Book Cost	Adjusted Cost	City Cost Index	Cost Factor	Book Cost	Adjusted Cost
A	Substructure	144.3	1.443	$ 103,100	$ 148,800				
B10	Shell: Superstructure	128.9	1.289	284,700	367,000				
B20	Exterior Closure	136.2	1.362	193,000	262,900				
B30	Roofing	130.1	1.301	42,500	55,300				
C	Interior Construction	133.8	1.338	27,900	37,300				
D10	Services: Conveying	108.5	1.085	55,100	59,800				
D20-40	Mechanical	128.9	1.289	341,000	439,500				
D50	Electrical	142.7	1.427	190,800	272,300				
E	Equipment	104.8	1.048	28,800	30,200				
G	Site Work	111.7	1.117	81,100	90,600				
Total Cost, (Div. A-G)				$1,348,000	$1,763,700				

Alternate Method: Use the weighted-average index for your city rather than the component indexes. **For example:**

(Total weighted-average index A-G)/100 x (total cost) = Adjusted city cost

New York, NY 1.311 x $1,348,000 = $1,767,200

City to City: To convert known or estimated costs from one city to another, cost indexes can be used as follows:

$$\text{Unknown City Cost} = \text{Known Cost} \times \frac{\text{Unknown City Index}}{\text{Known City Index}}$$

For example: If the building cost in Boston, MA, is $2,000,000, how much would a duplicated building cost in Los Angeles, CA?

$$\text{L.A. Cost} = \$2{,}000{,}000 \times \frac{\text{(Los Angeles) } 107.7}{\text{(Boston) } 117.5} = \$1{,}833{,}191$$

The table below lists both the RSMeans City Cost Index based on January 1, 1993 = 100 as well as the computed value of an index based on January 1, 2016 costs. Since the January 1, 2016 figure is estimated, space is left to write in the actual index figures as they become available through the quarterly *RSMeans Construction Cost Indexes*. To compute the actual index based on January 1, 2016 = 100, divide the Quarterly City Cost Index for a particular year by the actual January 1, 2016 Quarterly City Cost Index. Space has been left to advance the index figures as the year progresses.

Table J1020-011 Historical Cost Indexes

Year	Historical Cost Index Jan. 1, 1993 = 100		Current Index Based on Jan. 1, 2016 = 100		Year	Historical Cost Index Jan. 1, 1993 = 100	Current Index Based on Jan. 1, 2016 = 100		Year	Historical Cost Index Jan. 1, 1993 = 100	Current Index Based on Jan. 1, 2016 = 100	
	Est.	Actual	Est.	Actual		Actual	Est.	Actual		Actual	Est.	Actual
Oct 2016*					July 2001	125.1	60.4		July 1983	80.2	38.7	
July 2016*					2000	120.9	58.3		1982	76.1	36.8	
April 2016*					1999	117.6	56.8		1980	62.9	30.4	
Jan 2016*	207.2		100.0	100.0	1998	115.1	55.6		1979	57.8	27.9	
July 2015		206.2	99.5		1997	112.8	54.4		1978	53.5	25.8	
2014		204.9	98.9		1996	110.2	53.2		1977	49.5	23.9	
2013		201.2	97.1		1995	107.6	51.9		1976	46.9	22.6	
2012		194.6	93.9		1994	104.4	50.4		1975	44.8	21.6	
2011		191.2	92.3		1993	101.7	49.1		1974	41.4	20.0	
2010		183.5	88.6		1992	99.4	48.0		1973	37.7	18.2	
2009		180.1	86.9		1991	96.8	46.7		1972	34.8	16.8	
2008		180.4	87.1		1990	94.3	45.5		1971	32.1	15.5	
2007		169.4	81.8		1989	92.1	44.5		1970	28.7	13.9	
2006		162.0	78.2		1988	89.9	43.4		1969	26.9	13.0	
2005		151.6	73.2		1987	87.7	42.3		1968	24.9	12.0	
2004		143.7	69.4		1986	84.2	40.7		1967	23.5	11.3	
2003		132.0	63.7		1985	82.6	39.9		1966	22.7	11.0	
2002		128.7	62.1		1984	82.0	39.6		1965	21.7	10.5	

To find the **current cost** from a project built previously in either the same city or a different city, the following formula is used:

$$\text{Present Cost (City X)} = \frac{\text{Current HCI} \times \text{CCI (City X)}}{\text{Previous HCI} \times \text{CCI (City Y)}} \times \text{Former Cost (City Y)}$$

For example: Find the construction cost of a building to be built in San Francisco, CA, as of January 1, 2016 when the identical building cost $500,000 in Boston, MA, on July 1, 1968.

$$\text{Jan. 1, 2016 (San Francisco)} = \frac{\text{(San Francisco) } 207.2 \times 122.5}{\text{(Boston) } 24.9 \times 117.6} \times \$500{,}000 = \$4{,}334{,}000$$

Note: The City Cost Indexes for Canada can be used to convert U.S. national averages to local costs in Canadian dollars.

To Project Future Construction Costs: Using the results of the last five years average percentage increase as a basis, an average increase of 2.4% could be used.

The historical index figures above are compiled from the Means Construction Index Service.

Example:

To estimate and compare the cost of a building in Toronto, ON in 2016 with the known cost of $600,000 (US$) in New York, NY in 2016:

INDEX Toronto = 109.9

INDEX New York = 131.1

$$\frac{\text{INDEX Toronto}}{\text{INDEX New York}} \times \text{Cost New York} = \text{Cost Toronto}$$

$$\frac{109.9}{131.1} \times \$600{,}000 = .838 \times \$600{,}000 = \$502{,}975$$

The construction cost of the building in Toronto is $502,975 (CN$).

*Historical Cost Index updates and other resources are provided on the following website:
http://info.thegordiangroup.com/RSMeans.html

DIV. NO.	BUILDING SYSTEMS	ALABAMA														
		BIRMINGHAM			HUNTSVILLE			MOBILE			MONTGOMERY			TUSCALOOSA		
		MAT.	INST.	TOTAL	MAT.	INST.	TOTAL	MAT.	INST.	TOTAL	MAT.	INST.	TOTAL	MAT.	INST.	TOTAL
A	Substructure	100.4	82.1	89.4	95.5	78.6	85.3	99.4	81.4	88.6	98.6	80.8	87.8	98.1	81.7	88.2
B10	Shell: Superstructure	96.2	88.1	93.0	97.0	85.3	92.3	96.8	88.7	93.5	95.9	87.9	92.7	96.9	88.1	93.4
B20	Exterior Closure	101.1	80.1	91.2	100.5	76.1	89.0	102.9	76.1	90.2	101.8	76.8	90.0	100.0	79.0	90.1
B30	Roofing	102.4	85.0	95.6	100.7	82.6	93.7	97.0	84.3	92.0	95.7	84.6	91.4	100.8	84.7	94.5
C	Interior Construction	100.2	76.4	90.2	100.8	72.3	88.8	96.5	77.6	88.5	96.3	77.0	88.2	100.8	77.5	91.0
D10	Services: Conveying	100.0	91.4	97.5	100.0	90.0	97.1	100.0	91.5	97.5	100.0	90.9	97.3	100.0	90.9	97.3
D20 - 40	Mechanical	100.0	67.3	86.2	100.0	65.6	85.5	99.9	69.2	86.9	100.0	68.8	86.8	100.0	71.3	87.9
D50	Electrical	99.1	63.4	81.1	95.8	67.0	81.3	94.6	59.1	76.6	95.1	64.6	79.6	95.4	63.4	79.2
E	Equipment & Furnishings	100.0	77.9	98.8	100.0	71.3	98.4	100.0	79.4	98.8	100.0	77.9	98.8	100.0	77.8	98.8
G	Site Work	94.6	92.7	93.3	88.0	92.2	90.9	95.3	89.5	91.3	93.4	88.0	89.8	88.6	92.4	91.2
A-G	**WEIGHTED AVERAGE**	99.3	77.3	89.6	98.7	75.3	88.4	98.3	76.6	88.7	97.9	77.0	88.7	98.7	78.1	89.6

DIV. NO.	BUILDING SYSTEMS	ALASKA									ARIZONA					
		ANCHORAGE			FAIRBANKS			JUNEAU			FLAGSTAFF			MESA/TEMPE		
		MAT.	INST.	TOTAL	MAT.	INST.	TOTAL	MAT.	INST.	TOTAL	MAT.	INST.	TOTAL	MAT.	INST.	TOTAL
A	Substructure	126.8	120.4	122.9	128.9	120.7	124.0	132.4	120.4	125.2	93.2	77.3	83.6	98.7	76.8	85.5
B10	Shell: Superstructure	128.0	110.8	121.2	132.7	111.0	124.0	131.9	110.8	123.5	100.1	71.8	88.8	99.2	71.9	88.3
B20	Exterior Closure	161.2	123.4	143.3	153.5	124.0	139.6	158.0	123.4	141.6	126.8	64.8	97.5	114.6	64.1	90.7
B30	Roofing	157.2	117.6	141.8	172.9	117.1	151.2	178.9	117.6	155.1	96.3	68.4	85.5	99.2	64.4	85.7
C	Interior Construction	128.9	119.8	125.1	133.9	120.4	128.2	130.6	119.8	126.1	101.0	62.9	85.0	94.8	64.1	81.9
D10	Services: Conveying	100.0	111.5	103.4	100.0	111.8	103.4	100.0	111.5	103.4	100.0	85.8	95.8	100.0	86.2	96.0
D20 - 40	Mechanical	100.5	105.8	102.7	100.5	109.7	104.4	100.6	105.8	102.8	100.2	79.1	91.3	100.1	79.1	91.2
D50	Electrical	107.1	117.3	112.3	117.5	117.3	117.4	109.5	117.3	113.4	102.0	61.7	81.6	97.1	61.7	79.2
E	Equipment & Furnishings	100.0	117.9	101.0	100.0	118.0	101.0	100.0	117.9	101.0	100.0	69.0	98.3	100.0	66.9	98.1
G	Site Work	128.6	130.6	130.0	123.8	130.6	128.5	142.3	130.6	134.3	90.2	95.6	93.9	95.7	96.7	96.4
A-G	**WEIGHTED AVERAGE**	121.7	115.8	119.1	124.2	116.9	120.9	123.9	115.8	120.3	102.7	71.9	89.2	100.1	72.0	87.7

DIV. NO.	BUILDING SYSTEMS	ARIZONA									ARKANSAS					
		PHOENIX			PRESCOTT			TUCSON			FORT SMITH			JONESBORO		
		MAT.	INST.	TOTAL	MAT.	INST.	TOTAL	MAT.	INST.	TOTAL	MAT.	INST.	TOTAL	MAT.	INST.	TOTAL
A	Substructure	98.6	77.0	85.6	91.4	70.8	79.0	96.3	76.7	84.5	87.7	72.8	78.7	85.7	73.8	78.5
B10	Shell: Superstructure	100.3	72.4	89.1	99.9	66.7	86.6	99.3	71.7	88.3	99.2	69.4	87.3	93.2	74.4	85.7
B20	Exterior Closure	108.1	64.8	87.6	110.9	62.6	88.1	106.8	64.7	86.9	97.0	53.7	76.5	95.0	48.4	73.0
B30	Roofing	98.8	66.4	86.2	95.1	62.8	82.5	100.2	63.6	86.0	100.1	56.9	83.3	104.4	59.1	86.8
C	Interior Construction	97.0	67.0	84.4	99.8	52.3	79.8	92.0	62.8	79.7	97.5	59.8	81.7	97.0	48.9	76.8
D10	Services: Conveying	100.0	86.2	96.0	100.0	85.8	95.8	100.0	86.2	96.0	100.0	83.3	95.1	100.0	72.2	91.9
D20 - 40	Mechanical	100.0	79.2	91.2	100.2	79.0	91.3	100.1	74.4	89.2	100.1	51.5	79.5	100.4	53.5	80.6
D50	Electrical	102.5	64.1	83.1	101.7	61.7	81.5	99.3	61.7	80.3	96.3	59.2	77.5	102.1	63.0	82.3
E	Equipment & Furnishings	100.0	67.0	98.1	100.0	46.9	97.0	100.0	66.9	98.1	100.0	61.3	97.8	100.0	47.1	97.0
G	Site Work	96.1	96.8	96.6	78.5	95.5	90.1	91.5	96.7	95.0	82.2	89.7	87.3	102.3	106.9	105.5
A-G	**WEIGHTED AVERAGE**	100.5	73.0	88.4	100.4	68.4	86.3	98.9	70.8	86.5	97.8	62.3	82.2	97.4	62.6	82.1

DIV. NO.	BUILDING SYSTEMS	ARKANSAS									CALIFORNIA					
		LITTLE ROCK			PINE BLUFF			TEXARKANA			ANAHEIM			BAKERSFIELD		
		MAT.	INST.	TOTAL	MAT.	INST.	TOTAL	MAT.	INST.	TOTAL	MAT.	INST.	TOTAL	MAT.	INST.	TOTAL
A	Substructure	88.1	73.3	79.2	84.3	73.1	77.5	91.0	66.6	76.3	93.6	117.2	107.9	93.4	117.2	107.7
B10	Shell: Superstructure	98.8	71.4	87.9	99.7	70.9	88.2	94.9	63.1	82.2	105.1	109.6	106.9	104.6	109.8	106.6
B20	Exterior Closure	97.4	70.1	84.5	111.0	58.6	86.2	97.0	45.8	72.8	101.2	117.2	108.8	100.0	116.0	107.6
B30	Roofing	94.7	57.9	80.4	96.1	57.2	81.0	96.7	54.0	80.1	102.0	122.2	109.8	96.8	116.5	104.5
C	Interior Construction	96.1	62.6	82.0	99.4	59.5	82.7	103.2	42.9	77.9	103.6	116.8	109.1	96.6	117.7	105.5
D10	Services: Conveying	100.0	83.3	95.1	100.0	83.3	95.1	100.0	68.2	90.7	100.0	110.3	103.0	100.0	107.7	102.2
D20 - 40	Mechanical	99.9	53.9	80.4	100.2	53.8	80.6	100.2	59.5	83.0	100.0	119.4	108.2	100.2	109.9	104.3
D50	Electrical	103.0	71.1	86.8	96.7	69.7	83.1	98.4	61.0	79.5	90.6	105.3	98.0	103.4	102.8	103.1
E	Equipment & Furnishings	100.0	61.9	97.9	100.0	61.8	97.9	100.0	41.6	96.7	100.0	118.9	101.1	100.0	121.4	101.2
G	Site Work	90.6	90.7	90.7	86.5	90.8	89.5	96.6	91.5	93.2	101.3	108.0	105.8	99.4	105.8	103.7
A-G	**WEIGHTED AVERAGE**	98.2	67.4	84.6	99.6	65.2	84.5	98.5	58.7	80.9	100.6	114.0	106.5	100.4	111.4	105.2

DIV. NO.	BUILDING SYSTEMS	CALIFORNIA														
		FRESNO			LOS ANGELES			OAKLAND			OXNARD			REDDING		
		MAT.	INST.	TOTAL	MAT.	INST.	TOTAL	MAT.	INST.	TOTAL	MAT.	INST.	TOTAL	MAT.	INST.	TOTAL
A	Substructure	95.3	121.0	110.8	94.4	117.0	108.0	109.2	128.1	120.6	99.0	116.1	109.3	122.1	124.0	123.2
B10	Shell: Superstructure	105.5	113.0	108.5	103.3	110.3	106.1	100.7	118.1	107.7	102.3	109.3	105.1	114.3	115.4	114.8
B20	Exterior Closure	105.9	128.2	116.4	98.9	121.6	109.6	110.9	135.7	122.6	103.3	115.2	108.9	136.6	124.3	130.8
B30	Roofing	88.5	117.1	99.6	101.2	120.4	108.7	98.9	137.4	113.8	97.2	120.8	106.4	127.4	122.7	125.6
C	Interior Construction	98.4	126.1	110.0	100.0	117.1	107.2	98.1	143.0	117.0	95.8	116.4	104.5	116.3	135.6	124.4
D10	Services: Conveying	100.0	121.5	106.3	100.0	110.7	103.1	100.0	122.5	106.6	100.0	110.6	103.1	100.0	121.5	106.3
D20 - 40	Mechanical	100.2	116.3	107.0	100.0	119.5	108.2	100.1	149.6	121.0	100.1	119.4	108.3	100.4	110.0	104.5
D50	Electrical	94.3	100.7	97.5	98.0	123.5	110.9	103.8	138.2	121.2	98.1	106.5	102.3	99.7	106.5	103.1
E	Equipment & Furnishings	100.0	132.3	101.8	100.0	119.5	101.1	100.0	149.8	102.8	100.0	118.8	101.1	100.0	143.1	102.4
G	Site Work	103.7	105.5	104.9	100.5	107.2	105.1	113.6	108.1	109.9	103.9	103.8	103.8	134.8	104.9	114.4
A-G	**WEIGHTED AVERAGE**	100.5	116.5	107.5	100.2	117.2	107.7	102.1	134.5	116.4	99.9	113.5	105.9	112.3	117.8	114.7

DIV. NO.	BUILDING SYSTEMS	CALIFORNIA														
		RIVERSIDE			SACRAMENTO			SAN DIEGO			SAN FRANCISCO			SAN JOSE		
		MAT.	INST.	TOTAL	MAT.	INST.	TOTAL	MAT.	INST.	TOTAL	MAT.	INST.	TOTAL	MAT.	INST.	TOTAL
A	Substructure	97.6	116.8	109.2	94.5	124.3	112.5	97.6	109.1	104.6	114.3	130.3	124.0	113.0	126.0	120.8
B10	Shell: Superstructure	106.0	109.6	107.5	94.8	111.8	101.6	104.0	105.7	104.6	105.8	122.8	112.6	103.8	120.6	110.5
B20	Exterior Closure	97.8	116.9	106.8	101.1	127.7	113.7	98.9	115.3	106.6	112.0	138.9	124.7	121.8	132.9	127.0
B30	Roofing	102.2	122.1	110.0	107.3	123.8	113.7	100.3	106.1	102.6	100.5	140.5	116.0	101.2	138.3	115.6
C	Interior Construction	103.6	116.8	109.1	104.4	131.6	115.8	100.4	112.2	105.3	101.2	144.4	119.4	97.3	142.8	116.4
D10	Services: Conveying	100.0	110.3	103.0	100.0	122.2	106.5	100.0	109.6	102.8	100.0	123.0	106.7	100.0	121.8	106.4
D20 - 40	Mechanical	100.0	119.4	108.2	99.9	120.1	108.4	100.0	118.6	107.8	100.0	178.0	133.0	100.0	149.8	121.1
D50	Electrical	90.4	105.6	98.1	99.4	109.3	104.4	105.8	98.8	102.3	103.8	163.2	133.8	98.6	160.2	129.8
E	Equipment & Furnishings	100.0	119.0	101.1	100.0	135.4	102.0	100.0	110.9	100.6	100.0	151.0	102.9	100.0	150.8	102.9
G	Site Work	100.1	105.9	104.1	93.9	114.5	107.9	104.5	100.7	101.9	114.8	114.1	114.3	133.3	99.6	110.4
A-G	**WEIGHTED AVERAGE**	100.5	113.9	106.4	99.6	120.1	108.6	101.4	109.8	105.1	104.1	146.0	122.5	103.9	136.9	118.4

DIV. NO.	BUILDING SYSTEMS	CALIFORNIA									COLORADO					
		SANTA BARBARA			STOCKTON			VALLEJO			COLORADO SPRINGS			DENVER		
		MAT.	INST.	TOTAL	MAT.	INST.	TOTAL	MAT.	INST.	TOTAL	MAT.	INST.	TOTAL	MAT.	INST.	TOTAL
A	Substructure	98.8	116.5	109.5	100.0	121.7	113.1	100.8	128.7	117.6	104.0	77.0	87.7	100.8	80.3	88.4
B10	Shell: Superstructure	102.6	109.4	105.3	103.3	113.7	107.4	98.6	115.3	105.4	102.2	78.6	92.8	103.9	79.4	94.1
B20	Exterior Closure	102.3	116.7	109.1	108.4	124.0	115.8	89.1	134.2	110.5	106.2	73.9	90.9	103.4	74.4	89.7
B30	Roofing	94.4	120.5	104.6	105.1	121.0	111.3	100.9	135.7	114.4	97.8	79.4	90.7	96.3	82.0	90.7
C	Interior Construction	96.6	116.4	104.9	104.1	129.9	114.9	105.3	142.3	121.1	101.0	66.5	86.5	103.0	70.8	89.5
D10	Services: Conveying	100.0	110.6	103.1	100.0	121.5	106.3	100.0	121.9	106.4	100.0	90.2	97.1	100.0	90.1	97.1
D20 - 40	Mechanical	100.1	119.4	108.3	100.0	115.0	106.3	100.0	133.5	114.2	100.2	83.8	93.3	100.0	82.5	92.6
D50	Electrical	91.8	109.7	100.8	96.0	110.2	103.2	96.9	124.3	110.8	99.4	81.8	90.5	101.0	83.9	92.3
E	Equipment & Furnishings	100.0	118.9	101.1	100.0	135.4	102.0	100.0	149.1	102.8	100.0	65.2	98.0	100.0	68.8	98.2
G	Site Work	103.8	105.8	105.2	102.7	105.1	104.3	95.5	114.2	108.2	97.1	91.6	93.4	96.2	101.0	99.4
A-G	WEIGHTED AVERAGE	99.2	114.3	105.9	102.1	117.9	109.0	99.1	128.9	112.2	101.3	78.7	91.3	101.6	80.4	92.3

DIV. NO.	BUILDING SYSTEMS	COLORADO												CONNECTICUT		
		FORT COLLINS			GRAND JUNCTION			GREELEY			PUEBLO			BRIDGEPORT		
		MAT.	INST.	TOTAL	MAT.	INST.	TOTAL	MAT.	INST.	TOTAL	MAT.	INST.	TOTAL	MAT.	INST.	TOTAL
A	Substructure	112.0	79.5	92.4	110.7	79.1	91.6	98.3	75.2	84.4	102.1	76.4	86.6	101.8	121.7	113.8
B10	Shell: Superstructure	102.8	77.2	92.5	105.6	76.8	94.1	99.9	74.0	89.5	104.9	79.1	94.6	96.2	126.1	108.1
B20	Exterior Closure	114.9	75.8	96.4	131.2	72.7	103.5	105.4	53.5	80.8	107.8	73.8	91.7	110.5	137.7	123.4
B30	Roofing	97.4	69.2	86.4	104.0	65.3	88.9	96.8	67.7	85.5	103.5	77.8	93.5	98.5	128.9	110.3
C	Interior Construction	99.7	67.3	86.1	105.5	69.9	90.5	99.1	69.6	86.7	99.9	67.2	86.1	96.6	127.2	109.5
D10	Services: Conveying	100.0	89.9	97.0	100.0	90.5	97.2	100.0	89.9	97.0	100.0	91.1	97.4	100.0	111.8	103.5
D20 - 40	Mechanical	100.1	76.1	89.9	99.9	86.6	94.3	100.1	74.9	89.4	99.9	71.1	87.8	99.9	118.8	107.9
D50	Electrical	96.2	83.8	89.9	97.3	53.4	75.1	96.2	83.8	89.9	98.2	72.5	85.2	95.9	111.4	103.7
E	Equipment & Furnishings	100.0	74.9	98.6	100.0	73.3	98.5	100.0	80.1	98.9	100.0	64.5	98.0	100.0	121.7	101.2
G	Site Work	108.3	95.3	99.4	135.4	95.7	108.4	94.7	95.2	95.1	126.3	89.3	101.1	106.5	103.4	104.4
A-G	WEIGHTED AVERAGE	102.4	77.6	91.5	106.5	75.6	92.9	99.7	74.0	88.4	102.4	74.7	90.2	99.5	122.0	109.4

DIV. NO.	BUILDING SYSTEMS	CONNECTICUT														
		BRISTOL			HARTFORD			NEW BRITAIN			NEW HAVEN			NORWALK		
		MAT.	INST.	TOTAL	MAT.	INST.	TOTAL	MAT.	INST.	TOTAL	MAT.	INST.	TOTAL	MAT.	INST.	TOTAL
A	Substructure	97.9	121.6	112.2	98.5	121.6	112.4	98.8	121.6	112.6	99.8	122.0	113.2	100.8	122.2	113.7
B10	Shell: Superstructure	95.3	126.0	107.6	99.1	126.0	109.9	92.5	126.0	105.8	92.9	126.1	106.1	96.0	126.6	108.2
B20	Exterior Closure	106.3	137.7	121.1	108.7	137.7	122.4	107.4	137.7	121.7	120.3	137.7	128.5	106.1	138.9	121.6
B30	Roofing	98.6	127.3	109.8	104.0	127.3	113.1	98.6	124.1	108.5	98.7	126.6	109.5	98.6	129.2	110.5
C	Interior Construction	96.6	126.7	109.3	95.2	126.7	108.4	96.6	126.7	109.3	96.6	127.2	109.5	96.6	126.7	109.3
D10	Services: Conveying	100.0	111.8	103.5	100.0	111.8	103.5	100.0	111.8	103.5	100.0	111.8	103.5	100.0	111.8	103.5
D20 - 40	Mechanical	99.9	118.8	107.9	99.9	118.8	107.9	99.9	118.7	107.9	99.9	118.8	107.9	99.9	118.8	107.9
D50	Electrical	95.9	113.4	104.8	95.4	113.8	104.7	95.9	113.4	104.8	95.8	112.8	104.4	95.9	165.9	131.3
E	Equipment & Furnishings	100.0	121.8	101.2	100.0	121.8	101.2	100.0	121.8	101.2	100.0	121.8	101.2	100.0	124.3	101.4
G	Site Work	105.6	103.4	104.1	102.1	103.4	103.0	105.8	103.4	104.2	105.6	104.2	104.6	106.3	103.4	104.3
A-G	WEIGHTED AVERAGE	98.7	122.2	109.0	99.6	122.2	109.6	98.3	122.1	108.8	99.8	122.3	109.7	99.0	129.7	112.5

DIV. NO.	BUILDING SYSTEMS	CONNECTICUT						D.C.			DELAWARE			FLORIDA		
		STAMFORD			WATERBURY			WASHINGTON			WILMINGTON			DAYTONA BEACH		
		MAT.	INST.	TOTAL	MAT.	INST.	TOTAL	MAT.	INST.	TOTAL	MAT.	INST.	TOTAL	MAT.	INST.	TOTAL
A	Substructure	101.8	122.2	114.1	101.8	121.6	113.7	108.3	92.9	99.0	97.0	104.6	101.6	93.9	76.2	83.3
B10	Shell: Superstructure	96.2	126.6	108.3	96.2	126.0	108.1	102.8	101.3	102.2	98.7	110.3	103.3	97.6	82.8	91.7
B20	Exterior Closure	106.5	138.9	121.8	106.5	137.7	121.2	98.6	88.5	93.8	106.2	104.7	105.5	98.1	67.7	83.8
B30	Roofing	98.6	129.2	110.5	98.6	126.4	109.4	104.3	88.9	98.3	104.9	116.3	109.3	98.0	75.2	89.1
C	Interior Construction	96.6	126.7	109.3	96.6	126.7	109.2	97.2	93.8	95.8	88.8	102.6	94.6	94.5	68.7	83.7
D10	Services: Conveying	100.0	111.8	103.5	100.0	111.8	103.5	100.0	101.6	100.5	100.0	101.3	100.4	100.0	88.8	96.7
D20 - 40	Mechanical	99.9	118.8	107.9	99.9	118.7	107.9	100.0	92.6	96.9	100.2	116.9	107.3	100.0	78.0	90.7
D50	Electrical	95.9	165.9	131.3	95.4	111.4	103.5	100.5	105.0	102.8	100.2	111.1	105.7	94.2	68.1	81.0
E	Equipment & Furnishings	100.0	124.3	101.4	100.0	121.7	101.2	100.0	96.3	99.8	100.0	100.4	100.0	100.0	69.8	98.3
G	Site Work	106.9	103.4	104.5	106.4	103.4	104.4	103.5	97.3	99.2	102.4	112.2	109.0	101.2	89.4	93.2
A-G	WEIGHTED AVERAGE	99.1	129.7	112.6	99.0	121.9	109.1	100.6	95.9	98.5	98.7	109.4	103.4	97.5	75.5	87.8

DIV. NO.	BUILDING SYSTEMS	FLORIDA														
		FORT LAUDERDALE			JACKSONVILLE			MELBOURNE			MIAMI			ORLANDO		
		MAT.	INST.	TOTAL	MAT.	INST.	TOTAL	MAT.	INST.	TOTAL	MAT.	INST.	TOTAL	MAT.	INST.	TOTAL
A	Substructure	95.0	70.8	80.4	94.5	71.2	80.4	106.0	77.0	88.5	94.8	73.3	81.8	101.8	76.3	86.4
B10	Shell: Superstructure	96.7	79.7	89.9	96.6	77.9	89.2	106.4	83.4	97.2	97.1	80.6	90.5	98.6	83.0	92.3
B20	Exterior Closure	102.0	71.2	87.5	98.0	63.5	81.7	99.8	69.3	85.4	102.6	74.9	89.5	100.0	67.8	84.8
B30	Roofing	104.7	69.9	91.2	98.3	68.9	86.9	98.5	73.3	88.7	104.3	74.2	92.6	98.9	75.2	89.7
C	Interior Construction	94.2	68.3	83.3	94.6	60.7	80.3	93.7	71.1	84.2	97.4	69.0	85.4	99.3	68.8	86.5
D10	Services: Conveying	100.0	89.2	96.8	100.0	85.4	95.7	100.0	89.5	96.9	100.0	92.5	97.8	100.0	88.8	96.7
D20 - 40	Mechanical	100.0	62.4	84.1	100.0	65.5	85.4	100.0	80.4	91.7	100.0	60.0	83.1	100.0	61.2	83.6
D50	Electrical	94.8	72.2	83.3	93.9	68.7	81.2	95.2	67.3	81.1	98.3	75.3	86.6	98.6	67.8	83.0
E	Equipment & Furnishings	100.0	67.4	98.2	100.0	59.5	97.7	100.0	69.8	98.3	100.0	68.0	98.2	100.0	69.8	98.3
G	Site Work	94.4	76.7	82.4	101.2	89.2	93.0	108.4	89.5	95.5	95.7	77.3	83.2	101.2	89.4	93.2
A-G	WEIGHTED AVERAGE	97.8	71.2	86.1	97.3	69.6	85.1	100.1	76.6	89.8	98.9	72.3	87.2	99.5	72.0	87.4

DIV. NO.	BUILDING SYSTEMS	FLORIDA														
		PANAMA CITY			PENSACOLA			ST. PETERSBURG			TALLAHASSEE			TAMPA		
		MAT.	INST.	TOTAL	MAT.	INST.	TOTAL	MAT.	INST.	TOTAL	MAT.	INST.	TOTAL	MAT.	INST.	TOTAL
A	Substructure	99.3	67.9	80.3	113.6	79.5	93.0	102.3	75.3	86.0	100.5	69.4	81.7	100.7	77.5	86.7
B10	Shell: Superstructure	98.3	77.7	90.1	102.3	86.3	95.9	100.1	85.4	94.3	96.4	76.8	88.6	99.2	86.3	94.0
B20	Exterior Closure	106.2	52.2	80.7	114.5	67.4	92.2	126.5	66.6	98.2	101.4	57.5	80.6	103.8	76.7	91.0
B30	Roofing	98.6	62.8	84.7	98.5	74.1	89.0	104.5	67.4	90.1	100.1	65.7	86.7	104.8	71.6	91.9
C	Interior Construction	93.9	49.5	75.3	93.3	74.1	85.3	94.1	68.0	83.2	98.4	59.1	81.9	95.4	67.1	83.5
D10	Services: Conveying	100.0	78.8	93.8	100.0	82.5	94.9	100.0	81.6	94.6	100.0	80.9	94.4	100.0	86.9	96.2
D20 - 40	Mechanical	100.0	52.9	80.1	100.0	58.1	82.3	100.0	59.5	82.9	100.1	56.3	81.6	100.0	63.5	84.6
D50	Electrical	93.0	60.6	76.6	96.5	55.4	75.7	95.0	60.4	77.5	97.9	60.9	79.1	94.7	60.4	77.4
E	Equipment & Furnishings	100.0	47.5	97.1	100.0	77.2	98.7	100.0	68.0	98.2	100.0	58.8	97.7	100.0	68.4	98.2
G	Site Work	112.1	88.7	96.2	113.2	89.1	96.8	107.6	89.0	94.9	99.3	88.8	92.1	107.7	89.2	95.1
A-G	WEIGHTED AVERAGE	98.8	62.0	82.6	101.4	71.0	88.0	101.8	70.2	87.9	98.9	65.0	84.0	99.3	72.8	87.6

DIV. NO.	BUILDING SYSTEMS	GEORGIA														
		ALBANY			ATLANTA			AUGUSTA			COLUMBUS			MACON		
		MAT.	INST.	TOTAL	MAT.	INST.	TOTAL	MAT.	INST.	TOTAL	MAT.	INST.	TOTAL	MAT.	INST.	TOTAL
A	Substructure	93.1	63.7	75.4	103.8	81.8	90.6	99.7	79.1	87.3	92.9	66.4	76.9	92.8	69.4	78.7
B10	Shell: Superstructure	96.9	72.8	87.3	96.8	77.5	89.1	94.9	74.3	86.7	96.5	75.0	87.9	93.0	74.2	85.5
B20	Exterior Closure	99.7	54.9	78.5	93.2	68.0	81.3	90.7	57.6	75.0	99.7	64.8	83.2	104.3	55.1	81.0
B30	Roofing	98.8	61.8	84.4	95.1	70.8	85.6	94.8	68.4	84.5	98.7	63.5	85.0	97.1	64.8	84.5
C	Interior Construction	92.2	43.1	71.5	96.6	79.4	89.4	93.5	73.2	85.0	92.1	49.4	74.2	87.9	47.8	71.0
D10	Services: Conveying	100.0	86.8	96.1	100.0	87.1	96.2	100.0	86.6	96.1	100.0	86.8	96.1	100.0	86.9	96.2
D20 - 40	Mechanical	100.0	69.3	87.0	100.0	73.6	88.9	100.1	61.4	83.7	100.0	64.1	84.8	100.0	70.3	87.5
D50	Electrical	93.3	60.4	76.6	100.2	70.9	85.4	101.5	65.4	83.2	93.4	69.6	81.4	92.5	63.0	77.6
E	Equipment & Furnishings	100.0	36.1	96.4	100.0	84.7	99.1	100.0	80.8	98.9	100.0	45.9	97.0	100.0	43.5	96.8
G	Site Work	98.3	78.1	84.5	98.9	96.0	96.9	95.8	94.5	94.9	98.2	77.9	84.4	99.2	93.2	95.2
A-G	**WEIGHTED AVERAGE**	96.9	63.1	82.0	98.1	76.5	88.6	96.8	69.9	85.0	96.8	66.1	83.3	95.8	66.2	82.7

DIV. NO.	BUILDING SYSTEMS	GEORGIA						HAWAII						IDAHO		
		SAVANNAH			VALDOSTA			HONOLULU			STATES & POSS.GUAM			BOISE		
		MAT.	INST.	TOTAL	MAT.	INST.	TOTAL	MAT.	INST.	TOTAL	MAT.	INST.	TOTAL	MAT.	INST.	TOTAL
A	Substructure	96.8	69.3	80.2	92.5	63.5	75.0	157.7	123.5	137.1	187.3	80.3	122.7	92.3	84.2	87.4
B10	Shell: Superstructure	94.9	76.5	87.6	95.3	72.0	86.0	138.5	116.7	129.8	159.7	74.3	125.6	104.4	80.0	94.7
B20	Exterior Closure	99.5	64.7	83.1	108.2	53.8	82.5	151.1	128.0	140.2	173.7	44.9	112.8	121.6	82.7	103.2
B30	Roofing	96.5	64.0	83.9	99.0	60.9	84.2	145.4	122.9	136.6	150.0	68.4	118.3	94.7	79.9	89.0
C	Interior Construction	96.7	59.8	81.2	89.1	43.9	70.1	131.5	136.3	133.5	155.8	57.2	114.4	95.4	72.2	85.6
D10	Services: Conveying	100.0	86.3	96.0	100.0	86.3	96.0	100.0	113.7	104.0	100.0	76.4	93.1	100.0	89.5	96.9
D20 - 40	Mechanical	100.1	64.0	84.8	100.0	69.4	87.1	100.5	107.4	103.4	103.3	37.1	75.3	100.0	72.7	88.5
D50	Electrical	97.2	64.2	80.5	91.7	55.8	73.6	107.7	120.2	114.0	149.9	40.9	94.8	97.6	70.8	84.0
E	Equipment & Furnishings	100.0	57.2	97.6	100.0	37.3	96.5	100.0	137.6	102.1	100.0	55.8	97.5	100.0	72.6	98.5
G	Site Work	98.5	79.2	85.4	107.5	78.1	87.4	167.3	106.2	125.6	195.6	102.0	131.9	89.4	95.3	93.4
A-G	**WEIGHTED AVERAGE**	97.8	67.6	84.5	97.0	62.3	81.7	125.0	119.7	122.7	143.0	57.6	105.4	101.5	77.8	91.1

DIV. NO.	BUILDING SYSTEMS	IDAHO									ILLINOIS					
		LEWISTON			POCATELLO			TWIN FALLS			CHICAGO			DECATUR		
		MAT.	INST.	TOTAL	MAT.	INST.	TOTAL	MAT.	INST.	TOTAL	MAT.	INST.	TOTAL	MAT.	INST.	TOTAL
A	Substructure	101.8	86.9	92.9	94.5	84.1	88.2	97.1	69.1	80.2	103.3	143.6	127.6	96.6	110.0	104.7
B10	Shell: Superstructure	101.0	86.0	95.0	111.4	79.7	98.7	111.9	69.4	94.9	97.3	144.7	116.2	96.9	110.5	102.4
B20	Exterior Closure	129.7	85.0	108.6	118.1	78.7	99.5	127.2	75.1	102.5	98.2	158.9	126.9	92.7	116.4	103.9
B30	Roofing	149.3	86.1	124.7	94.6	72.1	85.9	95.4	72.2	86.4	100.2	143.7	117.1	104.2	111.4	107.0
C	Interior Construction	132.1	81.6	110.9	96.1	72.9	86.3	97.8	49.2	77.4	105.2	155.2	126.2	101.6	115.7	107.5
D10	Services: Conveying	100.0	88.3	96.6	100.0	89.5	96.9	100.0	42.4	83.1	100.0	120.9	106.1	100.0	103.2	100.9
D20 - 40	Mechanical	100.9	85.3	94.3	99.9	72.7	88.4	99.9	68.5	86.7	100.0	134.8	114.7	99.9	98.1	99.2
D50	Electrical	89.0	81.9	85.4	95.5	70.5	82.8	92.1	59.7	75.7	96.5	128.1	112.5	101.0	89.8	95.3
E	Equipment & Furnishings	100.0	78.4	98.8	100.0	72.6	98.5	100.0	47.0	97.0	100.0	153.9	103.0	100.0	115.0	100.8
G	Site Work	96.5	91.6	93.2	89.2	95.3	93.4	96.2	96.1	96.1	100.8	102.9	102.2	94.2	100.4	98.4
A-G	**WEIGHTED AVERAGE**	109.4	85.0	98.6	102.5	77.2	91.3	103.8	66.3	87.3	99.9	140.4	117.8	98.8	105.8	101.9

DIV. NO.	BUILDING SYSTEMS	ILLINOIS														
		EAST ST. LOUIS			JOLIET			PEORIA			ROCKFORD			SPRINGFIELD		
		MAT.	INST.	TOTAL	MAT.	INST.	TOTAL	MAT.	INST.	TOTAL	MAT.	INST.	TOTAL	MAT.	INST.	TOTAL
A	Substructure	93.1	111.3	104.1	103.4	140.1	125.6	97.3	113.0	106.8	98.1	121.8	112.4	94.8	108.9	103.3
B10	Shell: Superstructure	93.8	117.6	103.3	95.8	138.8	113.0	97.8	115.8	105.0	98.0	127.2	109.7	94.4	110.5	100.8
B20	Exterior Closure	86.6	117.1	101.0	99.6	151.4	124.1	110.5	118.3	114.2	98.8	133.3	115.1	98.5	116.7	107.1
B30	Roofing	98.0	108.7	102.2	100.1	142.7	116.7	99.2	112.5	104.4	101.7	127.9	111.9	106.4	110.8	108.2
C	Interior Construction	95.0	112.3	102.3	103.2	158.6	126.5	100.1	117.1	107.3	100.1	125.6	110.8	101.9	113.7	106.9
D10	Services: Conveying	100.0	100.5	100.1	100.0	119.5	105.7	100.0	104.3	101.3	100.0	112.0	103.5	100.0	103.0	100.9
D20 - 40	Mechanical	99.9	98.5	99.3	100.0	133.5	114.2	99.9	104.4	101.8	100.0	115.4	106.5	99.9	103.7	101.5
D50	Electrical	98.1	102.7	100.4	95.8	133.3	114.7	98.2	97.6	97.9	98.5	132.9	115.9	103.6	92.0	97.7
E	Equipment & Furnishings	100.0	111.0	100.6	100.0	158.4	103.3	100.0	116.9	100.9	100.0	125.3	101.4	100.0	115.0	100.8
G	Site Work	100.1	101.0	100.7	100.7	101.9	101.5	99.2	99.4	99.4	98.8	100.7	100.1	97.6	100.3	99.5
A-G	**WEIGHTED AVERAGE**	95.9	108.4	101.4	99.4	139.0	116.8	100.3	109.8	104.5	99.3	123.5	109.9	99.3	106.9	102.6

DIV. NO.	BUILDING SYSTEMS	INDIANA														
		ANDERSON			BLOOMINGTON			EVANSVILLE			FORT WAYNE			GARY		
		MAT.	INST.	TOTAL	MAT.	INST.	TOTAL	MAT.	INST.	TOTAL	MAT.	INST.	TOTAL	MAT.	INST.	TOTAL
A	Substructure	102.7	83.6	91.2	96.3	82.3	87.8	95.0	91.6	93.0	106.5	79.7	90.3	105.7	110.2	108.4
B10	Shell: Superstructure	96.4	85.8	92.2	97.1	78.4	89.6	91.4	83.8	88.4	97.3	81.3	90.9	96.9	111.6	102.8
B20	Exterior Closure	93.4	81.7	87.9	105.0	76.5	91.5	102.5	83.9	93.7	95.0	79.7	87.8	94.0	116.7	104.7
B30	Roofing	109.1	74.7	95.8	97.1	77.3	89.4	101.5	82.2	94.0	108.9	72.0	94.6	107.6	107.5	107.6
C	Interior Construction	96.2	80.8	89.7	97.3	79.2	89.7	93.4	81.5	88.4	96.1	75.4	87.4	95.7	117.2	104.7
D10	Services: Conveying	100.0	93.7	98.2	100.0	89.5	96.9	100.0	98.4	99.5	100.0	92.4	97.8	100.0	105.1	101.5
D20 - 40	Mechanical	99.9	78.5	90.9	99.6	78.8	90.8	99.9	80.6	91.7	99.9	72.6	88.4	99.9	106.1	102.5
D50	Electrical	88.3	89.2	88.8	99.0	88.3	93.6	95.5	87.6	91.5	88.9	78.6	83.7	99.1	106.8	103.0
E	Equipment & Furnishings	100.0	81.3	98.9	100.0	80.8	98.9	100.0	79.9	98.9	100.0	73.8	98.5	100.0	114.9	100.8
G	Site Work	99.6	93.1	95.2	87.7	92.1	90.7	93.1	121.8	112.7	101.2	93.0	95.6	100.4	96.7	97.8
A-G	**WEIGHTED AVERAGE**	97.1	83.6	91.1	98.8	81.2	91.0	96.6	86.6	92.2	97.6	78.6	89.3	98.4	110.0	103.5

DIV. NO.	BUILDING SYSTEMS	INDIANA												IOWA		
		INDIANAPOLIS			MUNCIE			SOUTH BEND			TERRE HAUTE			CEDAR RAPIDS		
		MAT.	INST.	TOTAL	MAT.	INST.	TOTAL	MAT.	INST.	TOTAL	MAT.	INST.	TOTAL	MAT.	INST.	TOTAL
A	Substructure	99.2	87.8	92.4	100.4	84.2	90.6	100.2	82.5	89.5	93.2	91.5	92.2	105.7	84.0	92.6
B10	Shell: Superstructure	96.0	83.2	90.9	99.1	86.1	93.9	95.8	91.8	94.2	91.6	85.0	89.0	91.8	87.0	89.9
B20	Exterior Closure	97.2	81.5	89.8	100.8	81.9	91.8	96.3	80.0	88.6	110.9	80.9	96.7	103.8	78.1	91.6
B30	Roofing	102.4	80.3	93.8	99.7	76.3	90.6	102.4	83.3	95.0	101.6	79.8	93.1	106.0	80.6	96.1
C	Interior Construction	99.6	85.1	93.5	91.7	80.5	87.0	95.3	81.4	89.5	93.6	82.9	89.1	104.6	80.1	94.3
D10	Services: Conveying	100.0	94.7	98.5	100.0	92.5	97.8	100.0	95.3	98.6	100.0	95.7	98.7	100.0	92.9	97.9
D20 - 40	Mechanical	99.8	81.7	92.1	99.6	78.4	90.7	99.9	76.8	90.1	99.9	79.2	91.2	100.1	79.7	91.4
D50	Electrical	102.2	89.2	95.6	91.6	87.8	89.7	98.5	88.1	93.2	93.9	90.2	92.0	99.3	80.0	89.5
E	Equipment & Furnishings	100.0	85.3	99.2	100.0	81.4	99.0	100.0	79.0	98.8	100.0	82.7	99.0	100.0	79.4	98.8
G	Site Work	98.8	96.5	97.2	87.9	92.2	90.8	99.1	92.8	94.8	95.3	122.0	113.5	100.8	95.6	97.3
A-G	**WEIGHTED AVERAGE**	99.0	85.2	93.0	97.3	83.4	91.2	97.8	84.1	91.8	97.4	86.6	92.6	99.9	82.6	92.3

DIV. NO.	BUILDING SYSTEMS	IOWA														
		COUNCIL BLUFFS			DAVENPORT			DES MOINES			DUBUQUE			SIOUX CITY		
		MAT.	INST.	TOTAL	MAT.	INST.	TOTAL	MAT.	INST.	TOTAL	MAT.	INST.	TOTAL	MAT.	INST.	TOTAL
A	Substructure	108.3	79.1	90.7	103.0	94.5	97.9	95.7	85.1	89.3	102.1	86.4	92.6	106.7	70.7	85.0
B10	Shell: Superstructure	95.0	82.9	90.2	91.2	99.5	94.5	94.6	88.4	92.1	89.3	88.0	88.7	91.7	76.8	85.7
B20	Exterior Closure	105.3	76.0	91.4	102.3	88.0	95.5	100.7	82.7	92.2	103.2	75.5	90.1	100.7	59.0	81.0
B30	Roofing	105.4	66.8	90.4	105.4	90.0	99.4	97.6	78.1	90.1	105.6	79.8	95.6	105.4	65.5	89.9
C	Interior Construction	99.8	73.0	88.5	102.8	92.5	98.5	95.1	81.6	89.4	101.7	76.2	91.0	103.6	63.9	86.9
D10	Services: Conveying	100.0	91.5	97.5	100.0	94.9	98.5	100.0	93.7	98.1	100.0	92.3	97.8	100.0	91.0	97.4
D20 - 40	Mechanical	100.1	77.2	90.4	100.1	91.9	96.6	99.7	79.0	91.0	100.1	74.5	89.3	100.1	75.0	89.5
D50	Electrical	104.5	80.9	92.6	97.3	92.4	94.8	104.8	80.7	92.6	103.1	79.1	91.0	99.3	72.4	85.7
E	Equipment & Furnishings	100.0	70.7	98.4	100.0	90.4	99.5	100.0	78.1	98.8	100.0	73.4	98.5	100.0	63.9	98.0
G	Site Work	103.0	91.6	95.3	99.7	99.2	99.4	101.6	99.4	100.1	99.1	92.9	94.8	110.1	95.5	100.1
A-G	WEIGHTED AVERAGE	100.6	79.0	91.1	99.0	93.6	96.6	98.4	83.9	92.0	99.0	80.6	90.9	99.6	72.3	87.6

DIV. NO.	BUILDING SYSTEMS	IOWA			KANSAS											
		WATERLOO			DODGE CITY			KANSAS CITY			SALINA			TOPEKA		
		MAT.	INST.	TOTAL	MAT.	INST.	TOTAL	MAT.	INST.	TOTAL	MAT.	INST.	TOTAL	MAT.	INST.	TOTAL
A	Substructure	107.1	73.7	87.0	115.2	65.6	85.3	93.0	96.5	95.1	103.5	67.0	81.5	96.8	67.2	78.9
B10	Shell: Superstructure	94.3	79.1	88.2	100.0	68.2	87.3	101.1	99.5	100.5	97.4	73.7	87.9	98.2	79.7	90.8
B20	Exterior Closure	101.3	74.0	88.4	120.0	60.2	91.7	106.2	97.9	102.3	121.2	60.4	92.4	102.9	64.5	84.7
B30	Roofing	105.4	71.2	92.1	98.9	60.8	84.1	96.0	98.9	97.1	98.4	62.0	84.2	99.9	70.2	88.3
C	Interior Construction	95.4	58.3	79.8	101.0	52.0	80.4	94.3	93.6	94.0	100.4	53.8	80.8	99.5	54.7	80.7
D10	Services: Conveying	100.0	92.9	97.9	100.0	38.4	82.0	100.0	76.5	93.1	100.0	91.3	97.5	100.0	42.5	83.2
D20 - 40	Mechanical	99.9	79.0	91.1	99.9	69.3	87.0	99.8	95.4	98.0	99.9	69.4	87.0	99.9	71.5	87.9
D50	Electrical	95.9	62.9	79.3	97.3	73.2	85.1	102.9	96.4	99.6	97.1	73.3	85.0	101.7	74.1	87.7
E	Equipment & Furnishings	100.0	46.9	97.0	100.0	56.1	97.5	100.0	96.9	99.8	100.0	51.4	97.3	100.0	44.5	96.9
G	Site Work	107.7	95.8	99.6	115.7	96.0	102.3	92.1	94.1	93.5	106.0	96.5	99.5	93.9	92.5	93.0
A-G	WEIGHTED AVERAGE	98.4	73.7	87.5	103.0	66.1	86.8	99.6	95.9	98.0	101.8	68.9	87.3	99.8	69.8	86.6

DIV. NO.	BUILDING SYSTEMS	KANSAS			KENTUCKY											
		WICHITA			BOWLING GREEN			LEXINGTON			LOUISVILLE			OWENSBORO		
		MAT.	INST.	TOTAL	MAT.	INST.	TOTAL	MAT.	INST.	TOTAL	MAT.	INST.	TOTAL	MAT.	INST.	TOTAL
A	Substructure	99.4	64.7	78.5	86.4	90.0	88.6	91.8	86.1	88.3	90.5	85.3	87.4	89.3	95.4	93.0
B10	Shell: Superstructure	98.6	71.3	87.7	92.9	87.7	90.8	91.7	85.8	89.3	92.0	86.5	89.8	87.3	88.5	87.8
B20	Exterior Closure	106.5	53.4	81.4	104.1	82.2	93.8	96.9	79.9	88.8	95.3	80.4	88.3	109.6	88.1	99.5
B30	Roofing	98.0	54.8	81.2	88.3	81.2	85.5	106.8	81.0	96.8	103.3	79.2	93.9	101.5	85.8	95.4
C	Interior Construction	103.8	53.6	82.7	92.1	84.0	88.7	91.7	70.3	82.7	88.7	84.2	86.8	91.2	85.2	88.7
D10	Services: Conveying	100.0	88.1	96.5	100.0	57.8	87.6	100.0	100.4	100.1	100.0	97.1	99.2	100.0	106.3	101.9
D20 - 40	Mechanical	99.8	64.9	85.0	99.9	83.0	92.8	100.0	81.2	92.0	100.0	81.8	92.3	99.9	78.2	90.7
D50	Electrical	100.7	73.2	86.8	94.4	80.9	87.6	93.0	82.4	87.7	97.3	82.4	90.0	93.9	82.8	88.3
E	Equipment & Furnishings	100.0	49.8	97.2	100.0	84.9	99.2	100.0	66.7	98.1	100.0	84.9	99.2	100.0	82.1	99.0
G	Site Work	101.1	94.8	96.9	79.7	94.5	89.8	91.4	98.2	96.1	85.4	94.8	91.8	93.4	122.4	113.2
A-G	WEIGHTED AVERAGE	101.0	66.1	85.6	95.7	84.1	90.6	95.5	82.2	89.7	95.1	84.3	90.4	95.8	88.0	92.4

DIV. NO.	BUILDING SYSTEMS	LOUISIANA														
		ALEXANDRIA			BATON ROUGE			LAKE CHARLES			MONROE			NEW ORLEANS		
		MAT.	INST.	TOTAL	MAT.	INST.	TOTAL	MAT.	INST.	TOTAL	MAT.	INST.	TOTAL	MAT.	INST.	TOTAL
A	Substructure	94.5	66.7	77.7	99.2	77.1	85.8	98.8	72.3	82.8	94.2	63.9	75.9	95.8	72.8	81.9
B10	Shell: Superstructure	94.2	68.7	84.1	99.1	76.3	90.0	92.3	70.0	83.4	94.1	67.2	83.4	100.7	70.6	88.7
B20	Exterior Closure	109.3	62.3	87.1	97.3	64.5	81.8	95.2	64.5	80.7	106.8	50.1	80.0	96.3	62.7	80.4
B30	Roofing	97.2	67.4	85.6	96.4	72.5	87.1	97.6	72.0	87.6	97.2	62.1	83.5	97.1	70.6	86.8
C	Interior Construction	103.6	50.3	81.2	100.8	65.5	86.0	99.7	65.8	85.4	103.5	45.5	79.2	100.9	67.3	86.8
D10	Services: Conveying	100.0	84.1	95.3	100.0	87.4	96.3	100.0	87.1	96.2	100.0	80.4	94.3	100.0	85.5	95.8
D20 - 40	Mechanical	100.2	59.7	83.1	100.0	64.8	85.2	100.1	68.3	86.7	100.2	54.2	80.7	100.0	66.2	85.7
D50	Electrical	95.4	57.5	76.2	99.4	61.1	80.0	94.4	66.8	80.5	97.1	59.2	77.9	96.5	74.6	85.4
E	Equipment & Furnishings	100.0	41.5	96.7	100.0	66.4	98.1	100.0	66.6	98.1	100.0	41.5	96.7	100.0	69.1	98.3
G	Site Work	102.4	93.0	96.0	102.9	91.4	95.1	101.9	91.3	94.7	102.4	92.3	95.5	98.9	93.0	94.9
A-G	WEIGHTED AVERAGE	99.7	63.3	83.7	99.5	69.8	86.4	97.2	69.9	85.2	99.6	59.3	81.8	99.2	70.7	86.6

DIV. NO.	BUILDING SYSTEMS	LOUISIANA			MAINE											
		SHREVEPORT			AUGUSTA			BANGOR			LEWISTON			PORTLAND		
		MAT.	INST.	TOTAL	MAT.	INST.	TOTAL	MAT.	INST.	TOTAL	MAT.	INST.	TOTAL	MAT.	INST.	TOTAL
A	Substructure	98.3	62.6	76.8	97.4	99.6	98.8	82.6	103.0	94.9	88.3	103.0	97.1	94.6	103.0	99.6
B10	Shell: Superstructure	98.7	64.1	84.9	99.7	92.3	96.8	88.5	94.4	90.9	92.8	94.4	93.4	99.4	94.4	97.4
B20	Exterior Closure	102.8	50.8	78.3	111.2	67.4	90.5	116.7	100.0	108.8	105.8	100.0	103.1	113.1	100.0	106.9
B30	Roofing	95.3	62.4	82.5	99.5	71.1	88.5	97.9	80.2	91.0	97.7	80.2	90.9	103.6	80.2	94.5
C	Interior Construction	104.2	45.6	79.6	101.6	85.6	94.9	96.7	91.6	94.5	99.8	91.6	96.4	101.7	91.6	97.5
D10	Services: Conveying	100.0	83.8	95.3	100.0	97.3	99.2	100.0	111.5	103.4	100.0	111.5	103.4	100.0	111.5	103.4
D20 - 40	Mechanical	99.9	58.4	82.4	99.9	63.6	84.6	100.1	73.5	88.9	100.1	73.5	88.9	99.9	73.5	88.8
D50	Electrical	102.5	66.1	84.1	101.0	79.5	90.1	99.5	76.1	87.7	101.5	80.4	90.8	103.5	80.4	91.8
E	Equipment & Furnishings	100.0	42.3	96.8	100.0	98.7	99.9	100.0	91.4	99.5	100.0	91.6	99.5	100.0	91.6	99.5
G	Site Work	104.3	92.3	96.1	91.2	98.8	96.4	93.9	99.1	97.4	90.6	99.1	96.4	91.7	99.1	96.7
A-G	WEIGHTED AVERAGE	100.9	60.7	83.2	101.2	81.0	92.3	97.9	88.8	93.9	98.5	89.4	94.5	101.6	89.4	96.3

DIV. NO.	BUILDING SYSTEMS	MARYLAND						MASSACHUSETTS								
		BALTIMORE			HAGERSTOWN			BOSTON			BROCKTON			FALL RIVER		
		MAT.	INST.	TOTAL	MAT.	INST.	TOTAL	MAT.	INST.	TOTAL	MAT.	INST.	TOTAL	MAT.	INST.	TOTAL
A	Substructure	107.8	81.6	92.0	90.3	83.8	86.3	100.8	137.6	123.0	97.1	131.8	118.0	95.1	132.8	117.9
B10	Shell: Superstructure	101.9	87.2	96.0	94.9	87.4	91.9	100.0	137.6	115.0	95.9	132.4	110.5	95.5	134.7	111.1
B20	Exterior Closure	102.2	76.5	90.1	99.7	89.1	94.7	104.3	161.2	131.2	104.1	148.5	125.1	104.4	153.8	127.8
B30	Roofing	104.2	80.4	94.9	102.9	76.7	92.7	97.8	146.7	116.8	98.5	141.3	115.2	98.4	138.5	114.0
C	Interior Construction	98.3	74.5	88.3	96.8	79.1	89.4	99.7	153.3	122.2	97.8	144.3	117.3	97.6	144.8	117.4
D10	Services: Conveying	100.0	88.7	96.7	100.0	92.4	97.8	100.0	116.5	104.8	100.0	115.1	104.4	100.0	115.8	104.6
D20 - 40	Mechanical	100.0	81.8	92.3	99.9	85.9	94.0	99.9	131.8	113.4	100.1	110.9	104.7	100.1	111.0	104.7
D50	Electrical	99.7	91.8	95.7	97.9	82.1	89.9	102.1	134.3	118.4	98.4	99.3	98.9	98.3	99.3	98.8
E	Equipment & Furnishings	100.0	74.7	98.6	100.0	74.8	98.6	100.0	144.9	102.5	100.0	131.9	101.8	100.0	131.8	101.8
G	Site Work	103.2	98.1	99.8	94.1	94.6	94.4	97.9	107.0	104.1	94.4	103.5	100.6	93.2	103.6	100.3
A-G	WEIGHTED AVERAGE	100.9	83.4	93.2	97.7	85.2	92.2	100.6	139.1	117.5	98.8	125.1	110.4	98.6	126.3	110.8

DIV. NO.	BUILDING SYSTEMS	MASSACHUSETTS														
		HYANNIS			LAWRENCE			LOWELL			NEW BEDFORD			PITTSFIELD		
		MAT.	INST.	TOTAL	MAT.	INST.	TOTAL	MAT.	INST.	TOTAL	MAT.	INST.	TOTAL	MAT.	INST.	TOTAL
A	Substructure	86.3	132.6	114.2	97.2	131.6	117.9	91.8	131.5	115.8	88.1	132.8	115.1	93.3	112.5	104.9
B10	Shell: Superstructure	90.4	134.3	107.9	95.7	132.0	110.2	94.4	130.6	108.8	94.0	134.7	110.2	94.5	110.5	100.9
B20	Exterior Closure	101.2	153.9	126.1	112.6	155.3	132.8	102.1	154.5	126.9	103.6	153.8	127.3	102.4	129.1	115.0
B30	Roofing	97.9	139.3	114.0	99.7	141.5	116.0	99.5	142.6	116.3	98.3	138.5	113.9	99.6	112.3	104.5
C	Interior Construction	93.9	144.6	115.2	95.3	144.1	115.8	100.0	144.1	118.5	97.6	144.8	117.4	100.1	116.6	107.0
D10	Services: Conveying	100.0	115.1	104.4	100.0	115.1	104.4	100.0	115.1	104.4	100.0	115.8	104.6	100.0	104.8	101.4
D20 - 40	Mechanical	100.1	110.0	104.3	100.1	124.4	110.3	100.1	124.4	110.3	100.1	111.0	104.7	100.1	101.0	100.5
D50	Electrical	96.2	99.3	97.8	97.3	127.9	112.8	97.8	127.9	113.0	99.1	99.3	99.2	97.8	99.0	98.4
E	Equipment & Furnishings	100.0	131.8	101.8	100.0	133.2	101.9	100.0	133.2	101.9	100.0	131.8	101.8	100.0	106.4	100.4
G	Site Work	92.1	103.4	99.8	95.0	103.5	100.8	93.6	103.5	100.3	91.7	103.6	99.8	94.7	101.9	99.6
A-G	WEIGHTED AVERAGE	95.9	125.9	109.1	99.2	132.6	113.9	98.4	132.3	113.3	97.9	126.3	110.4	98.5	109.7	103.4

DIV. NO.	BUILDING SYSTEMS	MASSACHUSETTS						MICHIGAN								
		SPRINGFIELD			WORCESTER			ANN ARBOR			DEARBORN			DETROIT		
		MAT.	INST.	TOTAL	MAT.	INST.	TOTAL	MAT.	INST.	TOTAL	MAT.	INST.	TOTAL	MAT.	INST.	TOTAL
A	Substructure	94.1	114.2	106.2	93.8	128.9	114.9	89.1	107.6	100.2	87.9	107.3	99.6	99.1	107.2	104.0
B10	Shell: Superstructure	97.2	112.3	103.2	97.2	128.7	109.8	94.7	115.6	103.1	94.5	115.6	103.0	97.1	104.9	100.2
B20	Exterior Closure	102.2	127.9	114.4	102.0	153.5	126.4	97.6	109.7	103.3	97.5	113.1	104.9	93.8	111.6	102.2
B30	Roofing	99.5	113.8	105.1	99.5	133.7	112.8	103.0	106.9	104.6	101.4	112.9	105.9	100.0	112.9	105.0
C	Interior Construction	100.0	116.2	106.8	100.0	137.2	115.7	96.4	113.0	103.3	96.3	113.4	103.5	99.3	113.4	105.2
D10	Services: Conveying	100.0	106.2	101.8	100.0	108.5	102.5	100.0	104.7	101.4	100.0	105.3	101.5	100.0	105.3	101.5
D20 - 40	Mechanical	100.1	102.3	101.0	100.1	112.5	105.3	100.1	98.2	99.3	100.1	108.1	103.5	99.9	109.1	103.8
D50	Electrical	97.8	99.0	98.4	97.8	105.5	101.7	96.9	108.0	102.5	96.9	106.2	101.6	97.8	106.1	102.0
E	Equipment & Furnishings	100.0	106.6	100.4	100.0	123.0	101.3	100.0	114.1	100.8	100.0	114.1	100.8	100.0	114.1	100.8
G	Site Work	94.1	102.1	99.6	94.0	103.5	100.5	82.9	96.9	92.4	82.6	97.0	92.4	94.6	99.3	97.8
A-G	WEIGHTED AVERAGE	99.1	110.3	104.0	99.0	124.5	110.2	97.0	107.4	101.6	96.8	109.9	102.6	98.2	108.2	102.6

DIV. NO.	BUILDING SYSTEMS	MICHIGAN														
		FLINT			GRAND RAPIDS			KALAMAZOO			LANSING			MUSKEGON		
		MAT.	INST.	TOTAL	MAT.	INST.	TOTAL	MAT.	INST.	TOTAL	MAT.	INST.	TOTAL	MAT.	INST.	TOTAL
A	Substructure	88.1	95.0	92.3	97.3	88.5	92.0	94.9	88.8	91.2	97.5	94.8	95.9	93.6	86.7	89.5
B10	Shell: Superstructure	95.1	106.0	99.4	96.4	86.4	92.4	98.2	86.8	93.7	96.1	105.7	99.9	96.2	85.5	91.9
B20	Exterior Closure	97.6	97.9	97.8	100.5	86.8	94.0	102.8	87.4	95.5	100.4	96.3	98.5	97.5	87.1	92.6
B30	Roofing	100.7	93.3	97.8	99.3	73.2	89.2	97.7	82.1	91.6	99.7	86.8	94.7	96.6	74.8	88.1
C	Interior Construction	96.1	89.7	93.4	99.8	82.6	92.6	93.9	82.3	89.0	98.1	88.9	94.2	91.6	80.7	87.0
D10	Services: Conveying	100.0	93.3	98.1	100.0	95.9	98.8	100.0	95.7	98.7	100.0	102.8	100.8	100.0	95.1	98.6
D20 - 40	Mechanical	100.1	88.7	95.3	100.1	82.4	92.6	100.2	80.7	91.9	100.0	88.5	95.1	99.9	81.3	92.1
D50	Electrical	96.9	99.0	97.9	97.7	78.5	88.0	93.0	78.2	85.5	99.8	92.6	96.2	93.3	78.5	85.8
E	Equipment & Furnishings	100.0	87.5	99.3	100.0	78.4	98.8	100.0	78.7	98.8	100.0	88.0	99.3	100.0	76.8	98.7
G	Site Work	73.5	96.0	88.8	93.5	85.3	87.9	95.3	85.3	88.5	94.2	96.0	95.5	92.5	85.1	87.5
A-G	WEIGHTED AVERAGE	96.7	95.6	96.2	98.8	83.9	92.3	97.8	83.9	91.7	98.7	94.4	96.8	96.3	83.2	90.5

DIV. NO.	BUILDING SYSTEMS	MICHIGAN			MINNESOTA											
		SAGINAW			DULUTH			MINNEAPOLIS			ROCHESTER			SAINT PAUL		
		MAT.	INST.	TOTAL	MAT.	INST.	TOTAL	MAT.	INST.	TOTAL	MAT.	INST.	TOTAL	MAT.	INST.	TOTAL
A	Substructure	87.3	93.0	90.7	105.6	108.9	107.6	99.4	120.1	111.8	103.1	106.9	105.4	108.4	119.1	114.9
B10	Shell: Superstructure	94.7	104.4	98.6	97.1	118.9	105.8	96.6	127.8	109.1	96.9	119.7	106.0	97.8	127.9	109.8
B20	Exterior Closure	97.7	89.7	93.9	106.8	120.3	113.2	104.0	131.3	116.9	102.0	119.2	110.1	106.5	131.3	118.2
B30	Roofing	101.6	87.0	95.9	103.9	114.8	108.2	102.4	131.0	113.5	107.6	102.3	105.5	102.9	130.6	113.7
C	Interior Construction	95.0	86.3	91.4	104.3	116.0	109.2	104.2	129.8	115.0	101.8	112.8	106.4	99.6	129.9	112.4
D10	Services: Conveying	100.0	92.1	97.7	100.0	94.8	98.5	100.0	107.8	102.3	100.0	103.2	100.9	100.0	107.5	102.2
D20 - 40	Mechanical	100.1	82.3	92.5	99.8	103.5	101.3	99.8	107.2	102.9	99.8	96.4	98.4	99.7	113.6	105.6
D50	Electrical	94.9	89.8	92.3	100.9	102.4	101.6	102.6	113.2	108.0	101.3	88.5	94.9	101.4	113.2	107.4
E	Equipment & Furnishings	100.0	86.3	99.2	100.0	110.9	100.6	100.0	126.8	101.5	100.0	106.2	100.3	100.0	127.9	101.6
G	Site Work	76.1	95.7	89.5	99.4	102.7	101.7	99.6	108.3	105.5	99.4	102.1	101.3	96.1	103.4	101.1
A-G	WEIGHTED AVERAGE	96.3	90.8	93.9	101.3	110.5	105.3	100.7	119.9	109.1	100.3	106.3	103.0	100.7	120.9	109.6

DIV. NO.	BUILDING SYSTEMS	MINNESOTA			MISSISSIPPI											
		ST. CLOUD			BILOXI			GREENVILLE			JACKSON			MERIDIAN		
		MAT.	INST.	TOTAL	MAT.	INST.	TOTAL	MAT.	INST.	TOTAL	MAT.	INST.	TOTAL	MAT.	INST.	TOTAL
A	Substructure	96.5	113.7	106.8	107.3	64.4	81.4	104.7	69.2	83.3	99.4	71.7	82.7	102.8	70.9	83.6
B10	Shell: Superstructure	91.3	122.7	103.8	95.1	72.5	86.1	93.9	76.3	86.9	98.5	77.0	89.9	92.2	76.8	86.0
B20	Exterior Closure	100.6	119.7	109.6	103.7	56.4	81.4	124.4	66.7	97.1	107.6	66.6	88.2	101.4	64.8	84.1
B30	Roofing	105.5	116.9	109.9	97.3	60.1	82.9	97.9	66.5	85.7	96.0	68.1	85.2	97.0	66.7	85.2
C	Interior Construction	90.7	120.9	103.3	94.1	54.5	77.4	94.4	63.4	81.4	95.5	66.2	83.2	91.5	68.0	81.6
D10	Services: Conveying	100.0	99.4	99.8	100.0	71.1	91.6	100.0	75.5	92.8	100.0	75.5	92.8	100.0	74.5	92.5
D20 - 40	Mechanical	99.4	103.5	101.2	100.0	49.7	78.7	100.0	51.0	79.3	100.0	62.8	84.3	100.0	61.4	83.7
D50	Electrical	102.8	113.2	108.1	100.4	56.5	78.2	96.3	58.4	77.1	101.8	58.4	79.9	99.4	57.4	78.2
E	Equipment & Furnishings	100.0	115.3	100.9	100.0	53.6	97.4	100.0	67.6	98.2	100.0	67.6	98.2	100.0	67.4	98.2
G	Site Work	94.8	105.3	101.9	104.8	87.8	93.2	103.3	88.3	93.1	99.3	88.3	91.8	100.9	88.1	92.2
A-G	WEIGHTED AVERAGE	96.8	114.0	104.4	98.8	60.5	81.9	100.3	65.1	84.8	99.8	68.4	86.0	97.1	67.9	84.2

DIV. NO.	BUILDING SYSTEMS	MISSOURI														
		CAPE GIRARDEAU			COLUMBIA			JOPLIN			KANSAS CITY			SPRINGFIELD		
		MAT.	INST.	TOTAL	MAT.	INST.	TOTAL	MAT.	INST.	TOTAL	MAT.	INST.	TOTAL	MAT.	INST.	TOTAL
A	Substructure	95.6	87.5	90.7	92.1	93.3	92.8	101.8	83.7	90.9	95.1	105.2	101.2	98.1	88.1	92.1
B10	Shell: Superstructure	95.3	97.8	96.3	95.8	108.1	100.7	94.6	90.7	93.1	99.2	112.2	104.4	101.1	98.0	99.9
B20	Exterior Closure	106.9	80.9	94.6	121.9	93.8	108.6	97.0	85.0	91.3	100.3	108.0	103.9	107.9	88.1	98.5
B30	Roofing	100.0	86.1	94.6	94.7	87.7	92.0	91.7	77.7	86.2	91.7	105.4	97.0	99.1	79.6	91.5
C	Interior Construction	98.0	81.4	91.0	96.8	85.3	92.0	102.3	74.4	90.6	103.4	104.9	104.1	103.4	79.2	93.2
D10	Services: Conveying	100.0	91.4	97.5	100.0	98.7	99.6	100.0	87.6	96.4	100.0	100.3	100.1	100.0	97.0	99.1
D20 - 40	Mechanical	99.9	95.9	98.2	99.8	98.9	99.4	100.1	72.2	88.3	100.0	101.8	100.7	99.9	74.0	88.9
D50	Electrical	104.7	110.1	107.4	99.0	119.2	109.2	93.2	76.9	85.0	102.5	101.6	102.0	102.9	69.1	85.9
E	Equipment & Furnishings	100.0	83.6	99.1	100.0	81.7	99.0	100.0	71.6	98.4	100.0	102.3	100.1	100.0	75.5	98.6
G	Site Work	94.2	93.3	93.6	99.4	98.3	98.6	106.6	97.5	100.4	99.8	96.0	97.2	99.4	95.9	97.0
A-G	WEIGHTED AVERAGE	99.6	92.8	96.6	100.3	99.7	100.0	98.2	81.0	90.7	100.2	104.8	102.2	101.8	83.3	93.7

DIV. NO.	BUILDING SYSTEMS	MISSOURI						MONTANA								
		ST. JOSEPH			ST. LOUIS			BILLINGS			BUTTE			GREAT FALLS		
		MAT.	INST.	TOTAL	MAT.	INST.	TOTAL	MAT.	INST.	TOTAL	MAT.	INST.	TOTAL	MAT.	INST.	TOTAL
A	Substructure	94.9	98.3	96.9	95.4	103.5	100.3	115.8	79.1	93.7	125.3	77.4	96.4	130.2	77.3	98.3
B10	Shell: Superstructure	96.2	106.8	100.5	99.5	113.9	105.3	108.5	83.2	98.4	105.3	82.2	96.1	109.4	82.3	98.6
B20	Exterior Closure	100.0	99.0	99.5	97.8	111.6	104.3	111.8	75.8	94.8	107.2	76.5	92.7	112.0	76.5	95.3
B30	Roofing	92.2	96.8	94.0	100.1	105.0	102.0	111.2	70.8	95.5	111.0	68.1	94.3	111.6	70.2	95.5
C	Interior Construction	102.8	92.9	98.6	100.6	102.4	101.4	99.8	73.7	88.8	98.9	69.2	86.4	102.5	74.3	90.6
D10	Services: Conveying	100.0	98.2	99.5	100.0	101.0	100.3	100.0	96.9	99.1	100.0	96.9	99.1	100.0	96.9	99.1
D20 - 40	Mechanical	100.1	88.1	95.1	99.9	105.5	102.3	100.0	75.4	89.6	100.1	74.9	89.4	100.1	69.3	87.1
D50	Electrical	100.9	100.8	100.9	107.2	90.6	98.8	97.9	76.6	87.1	104.4	72.6	88.3	97.3	72.6	84.8
E	Equipment & Furnishings	100.0	89.9	99.4	100.0	101.4	100.1	100.0	69.1	98.3	100.0	68.9	98.3	100.0	68.9	98.3
G	Site Work	101.5	91.4	94.6	94.1	96.7	95.9	96.7	96.9	96.8	102.3	96.6	98.4	105.9	96.7	99.7
A-G	WEIGHTED AVERAGE	99.4	96.7	98.2	100.2	104.3	102.0	103.7	78.8	92.7	103.7	77.1	92.0	105.2	76.8	92.7

DIV. NO.	BUILDING SYSTEMS	MONTANA						NEBRASKA								
		HELENA			MISSOULA			GRAND ISLAND			LINCOLN			NORTH PLATTE		
		MAT.	INST.	TOTAL	MAT.	INST.	TOTAL	MAT.	INST.	TOTAL	MAT.	INST.	TOTAL	MAT.	INST.	TOTAL
A	Substructure	108.3	73.1	87.1	101.1	76.1	86.0	110.4	75.6	89.4	93.4	78.2	84.2	111.3	74.2	88.9
B10	Shell: Superstructure	104.0	78.6	93.9	98.0	81.4	91.3	96.1	79.3	89.4	94.2	80.7	88.8	97.0	77.5	89.2
B20	Exterior Closure	109.8	75.9	93.8	113.2	75.8	95.5	111.6	75.4	94.5	105.2	71.5	89.3	101.3	74.2	88.5
B30	Roofing	105.4	68.0	90.9	110.2	77.2	97.4	101.8	76.3	91.9	100.0	76.5	90.9	98.7	77.6	90.5
C	Interior Construction	100.9	61.3	84.3	98.6	70.2	86.7	94.1	73.7	85.5	103.1	75.8	91.7	95.9	72.2	86.0
D10	Services: Conveying	100.0	97.1	99.1	100.0	96.9	99.1	100.0	91.4	97.5	100.0	91.4	97.5	100.0	64.1	89.5
D20 - 40	Mechanical	99.9	69.3	87.0	100.1	68.5	86.7	100.0	77.3	90.4	99.8	77.3	90.3	99.9	75.1	89.4
D50	Electrical	103.5	72.6	87.9	102.3	72.0	87.0	92.3	66.5	79.3	103.9	66.5	85.0	94.8	76.5	85.5
E	Equipment & Furnishings	100.0	60.6	97.8	100.0	68.8	98.2	100.0	69.1	98.3	100.0	71.5	98.4	100.0	69.4	98.3
G	Site Work	95.8	96.2	96.1	84.8	96.3	92.6	106.2	93.8	97.8	90.9	93.8	92.9	105.6	93.0	97.0
A-G	WEIGHTED AVERAGE	102.8	73.6	90.0	101.1	75.7	89.9	99.3	76.6	89.3	99.8	76.9	89.7	98.9	75.9	88.7

DIV. NO.	BUILDING SYSTEMS	NEBRASKA			NEVADA									NEW HAMPSHIRE		
		OMAHA			CARSON CITY			LAS VEGAS			RENO			MANCHESTER		
		MAT.	INST.	TOTAL	MAT.	INST.	TOTAL	MAT.	INST.	TOTAL	MAT.	INST.	TOTAL	MAT.	INST.	TOTAL
A	Substructure	93.8	80.3	85.6	96.4	91.1	93.2	95.1	108.9	103.4	100.2	91.6	95.0	101.1	101.3	101.2
B10	Shell: Superstructure	94.2	77.4	87.5	101.5	94.9	98.8	110.4	109.4	110.0	106.3	96.0	102.2	100.1	96.5	98.6
B20	Exterior Closure	105.8	79.0	93.1	117.1	81.5	100.3	114.6	107.2	111.1	117.1	77.0	98.2	106.4	102.2	104.4
B30	Roofing	95.8	80.6	89.9	104.8	84.2	96.8	115.2	102.8	110.4	101.1	84.2	94.6	105.3	100.6	103.4
C	Interior Construction	102.7	73.6	90.5	96.6	81.7	90.4	96.1	109.8	101.8	95.5	82.0	89.8	98.5	96.6	97.7
D10	Services: Conveying	100.0	90.2	97.1	100.0	87.1	96.2	100.0	101.6	100.5	100.0	87.1	96.2	100.0	108.3	102.4
D20 - 40	Mechanical	99.8	78.6	90.8	100.0	75.2	89.5	100.2	107.7	103.4	100.1	75.2	89.5	99.9	83.6	93.0
D50	Electrical	99.0	81.7	90.2	104.3	97.8	101.0	106.3	120.7	113.6	102.7	97.8	100.2	94.3	81.9	88.0
E	Equipment & Furnishings	100.0	72.1	98.4	100.0	81.5	99.0	100.0	107.5	100.4	100.0	81.5	99.0	100.0	89.5	99.4
G	Site Work	90.1	93.2	92.2	87.3	97.9	94.5	80.6	100.1	93.9	77.1	97.9	91.2	93.0	101.7	98.9
A-G	WEIGHTED AVERAGE	99.1	79.5	90.5	101.8	86.7	95.1	103.6	109.4	106.1	102.2	86.5	95.3	99.9	93.6	97.1

DIV. NO.	BUILDING SYSTEMS	NEW HAMPSHIRE						NEW JERSEY								
		NASHUA			PORTSMOUTH			CAMDEN			ELIZABETH			JERSEY CITY		
		MAT.	INST.	TOTAL	MAT.	INST.	TOTAL	MAT.	INST.	TOTAL	MAT.	INST.	TOTAL	MAT.	INST.	TOTAL
A	Substructure	92.5	101.3	97.8	83.4	101.3	94.2	89.0	121.4	108.6	85.3	125.3	109.6	84.3	124.2	108.4
B10	Shell: Superstructure	96.8	96.5	96.7	91.3	96.6	93.4	97.4	114.9	104.4	92.1	122.0	104.0	94.9	120.1	104.9
B20	Exterior Closure	105.5	102.2	103.9	101.4	98.8	100.1	106.0	134.9	119.6	109.8	134.5	121.5	99.1	134.3	115.7
B30	Roofing	102.2	100.6	101.6	101.8	114.8	106.9	103.1	127.8	112.7	102.8	131.7	114.1	102.1	131.1	113.4
C	Interior Construction	100.9	96.6	99.1	98.2	96.6	97.5	96.7	129.4	110.4	99.1	138.8	115.7	96.3	139.3	114.4
D10	Services: Conveying	100.0	108.3	102.4	100.0	108.3	102.4	100.0	110.4	103.0	100.0	114.8	104.3	100.0	114.8	104.3
D20 - 40	Mechanical	100.1	83.6	93.1	100.1	83.6	93.1	100.0	124.2	110.2	99.3	126.5	111.2	99.9	129.1	112.2
D50	Electrical	95.6	81.9	88.7	93.9	82.9	88.4	93.5	137.0	115.5	90.5	140.1	115.6	94.1	139.0	116.8
E	Equipment & Furnishings	100.0	89.5	99.4	100.0	89.6	99.4	100.0	123.5	101.3	100.0	133.0	101.9	100.0	133.0	101.9
G	Site Work	95.5	101.7	99.7	89.9	101.7	97.9	94.9	103.7	100.9	101.2	104.3	103.3	88.2	104.3	99.2
A-G	WEIGHTED AVERAGE	99.3	93.6	96.8	96.5	93.7	95.3	98.4	124.7	110.0	97.3	128.8	111.5	96.7	128.9	110.9

DIV. NO.	BUILDING SYSTEMS	NEW JERSEY									NEW MEXICO					
		NEWARK			PATERSON			TRENTON			ALBUQUERQUE			FARMINGTON		
		MAT.	INST.	TOTAL	MAT.	INST.	TOTAL	MAT.	INST.	TOTAL	MAT.	INST.	TOTAL	MAT.	INST.	TOTAL
A	Substructure	103.5	125.4	116.7	97.6	125.3	114.3	101.4	120.7	113.1	96.4	75.2	83.6	98.9	75.2	84.6
B10	Shell: Superstructure	100.1	122.2	108.9	94.2	122.1	105.4	99.9	112.2	104.9	105.7	78.1	94.7	104.4	78.1	93.9
B20	Exterior Closure	105.2	134.5	119.0	101.6	134.5	117.1	107.9	129.0	117.9	111.5	62.3	88.3	118.5	62.3	92.0
B30	Roofing	104.4	131.7	115.1	102.7	124.7	111.2	104.0	125.2	112.2	98.5	71.7	88.2	98.9	71.7	88.3
C	Interior Construction	97.9	139.8	115.5	101.1	139.8	117.3	98.1	136.7	114.3	96.2	63.6	82.5	96.9	63.6	82.9
D10	Services: Conveying	100.0	114.8	104.3	100.0	114.8	104.3	100.0	108.9	102.6	100.0	84.1	95.3	100.0	84.1	95.3
D20 - 40	Mechanical	100.0	128.4	112.0	99.9	129.1	112.3	100.0	127.9	111.8	100.2	68.4	86.8	100.1	68.4	86.7
D50	Electrical	97.1	139.1	118.3	94.1	140.1	117.4	96.6	134.4	115.7	89.2	71.3	80.2	87.1	71.3	79.1
E	Equipment & Furnishings	100.0	133.0	101.9	100.0	133.0	101.9	100.0	132.5	101.8	100.0	63.7	98.0	100.0	63.7	98.0
G	Site Work	102.4	104.3	103.7	99.9	104.3	102.9	93.9	103.9	100.7	90.5	102.1	98.4	98.6	102.1	101.0
A-G	WEIGHTED AVERAGE	100.3	129.3	113.1	98.5	129.4	112.1	100.2	124.9	111.1	100.3	72.1	87.9	101.0	72.1	88.2

DIV. NO.	BUILDING SYSTEMS	NEW MEXICO									NEW YORK					
		LAS CRUCES			ROSWELL			SANTA FE			ALBANY			BINGHAMTON		
		MAT.	INST.	TOTAL	MAT.	INST.	TOTAL	MAT.	INST.	TOTAL	MAT.	INST.	TOTAL	MAT.	INST.	TOTAL
A	Substructure	93.5	67.7	77.9	97.5	75.2	84.1	99.1	75.2	84.7	87.2	104.1	97.4	103.1	101.8	102.3
B10	Shell: Superstructure	100.9	72.8	89.7	103.8	78.1	93.6	101.2	78.1	92.0	97.3	107.5	101.4	96.0	111.0	102.0
B20	Exterior Closure	94.3	61.6	78.8	122.9	62.3	94.3	109.4	62.3	87.1	101.9	110.0	105.7	97.5	101.7	99.5
B30	Roofing	86.4	66.2	78.5	99.2	71.7	88.5	100.9	71.7	89.6	110.7	104.0	108.1	109.9	95.5	104.3
C	Interior Construction	96.7	62.8	82.4	95.1	63.6	81.9	99.3	63.6	84.3	98.0	100.0	98.8	93.8	89.9	92.2
D10	Services: Conveying	100.0	81.2	94.5	100.0	84.1	95.3	100.0	84.1	95.3	100.0	99.3	99.8	100.0	98.6	99.6
D20 - 40	Mechanical	100.5	68.1	86.8	99.9	68.4	86.6	100.1	68.4	86.7	100.0	104.9	102.1	100.7	91.7	96.9
D50	Electrical	91.9	66.3	78.9	91.5	71.3	81.3	99.4	71.3	85.2	96.0	103.3	99.7	98.2	101.5	99.9
E	Equipment & Furnishings	100.0	62.7	97.9	100.0	63.7	98.0	100.0	63.7	98.0	100.0	97.9	99.9	100.0	85.2	99.2
G	Site Work	98.5	81.1	86.7	101.4	102.1	101.9	100.3	102.1	101.5	84.0	103.3	97.2	96.0	91.4	92.8
A-G	WEIGHTED AVERAGE	97.5	68.1	84.6	101.4	72.1	88.5	101.1	72.1	88.3	98.3	104.6	101.1	98.2	98.4	98.3

City Cost Indexes — RJ1030-010 Building Systems

DIV. NO.	BUILDING SYSTEMS	NEW YORK														
		BUFFALO			HICKSVILLE			NEW YORK			RIVERHEAD			ROCHESTER		
		MAT.	INST.	TOTAL	MAT.	INST.	TOTAL	MAT.	INST.	TOTAL	MAT.	INST.	TOTAL	MAT.	INST.	TOTAL
A	Substructure	102.7	111.0	107.7	98.7	156.1	133.3	106.3	169.2	144.3	100.4	155.8	133.8	97.2	102.4	100.4
B10	Shell: Superstructure	101.2	104.1	102.4	101.2	154.9	122.6	103.1	167.7	128.9	102.1	154.5	123.0	99.2	105.0	101.5
B20	Exterior Closure	102.9	117.6	109.9	108.4	165.0	135.2	99.7	177.0	136.2	110.4	165.0	136.2	105.2	106.3	105.7
B30	Roofing	101.9	110.3	105.2	109.3	152.6	126.1	107.1	166.2	130.1	110.3	152.6	126.8	104.6	100.5	103.0
C	Interior Construction	98.7	116.0	106.0	93.3	157.4	120.2	98.1	183.1	133.8	93.6	150.8	117.6	104.1	100.0	102.4
D10	Services: Conveying	100.0	104.8	101.4	100.0	123.3	106.8	100.0	129.1	108.5	100.0	123.3	106.8	100.0	100.6	100.2
D20 - 40	Mechanical	100.0	97.7	99.0	99.7	155.2	123.2	100.1	168.3	128.9	100.0	152.0	122.0	100.0	90.1	95.9
D50	Electrical	98.7	101.9	100.3	97.4	146.6	122.2	97.6	186.8	142.7	98.8	146.6	123.0	99.3	97.7	98.5
E	Equipment & Furnishings	100.0	115.6	100.9	100.0	150.3	102.8	100.0	185.0	104.8	100.0	150.3	102.8	100.0	97.4	99.9
G	Site Work	97.7	97.8	97.8	115.1	123.5	120.8	106.0	114.3	111.7	116.8	123.5	121.4	90.6	109.1	103.2
A-G	WEIGHTED AVERAGE	100.3	106.4	103.0	100.3	152.7	123.4	100.7	169.8	131.1	101.1	150.9	123.0	100.8	100.0	100.5

DIV. NO.	BUILDING SYSTEMS	NEW YORK														
		SCHENECTADY			SYRACUSE			UTICA			WATERTOWN			WHITE PLAINS		
		MAT.	INST.	TOTAL	MAT.	INST.	TOTAL	MAT.	INST.	TOTAL	MAT.	INST.	TOTAL	MAT.	INST.	TOTAL
A	Substructure	94.9	104.1	100.5	98.4	98.6	98.5	91.5	97.6	95.2	100.4	100.5	100.5	95.1	143.0	124.0
B10	Shell: Superstructure	97.9	107.5	101.7	97.6	102.3	99.5	94.8	103.0	98.0	95.9	103.5	98.9	91.6	146.5	113.5
B20	Exterior Closure	100.2	110.0	104.8	101.5	104.6	103.0	98.2	101.8	99.9	105.3	106.8	106.0	93.0	154.7	122.2
B30	Roofing	105.3	104.0	104.8	104.2	96.5	101.2	92.3	95.9	93.7	92.6	99.0	95.1	107.8	145.8	122.6
C	Interior Construction	96.3	100.0	97.8	93.9	92.4	93.3	95.7	89.6	93.2	94.4	94.4	94.4	92.0	146.2	114.7
D10	Services: Conveying	100.0	99.3	99.8	100.0	98.7	99.6	100.0	97.4	99.2	100.0	99.3	99.8	100.0	121.6	106.3
D20 - 40	Mechanical	100.2	104.5	102.0	100.3	92.0	96.8	100.3	93.2	97.3	100.3	85.8	94.2	100.3	136.4	115.6
D50	Electrical	95.5	103.3	99.5	98.3	101.4	99.8	96.6	101.4	99.0	98.2	92.4	95.3	89.9	154.0	122.3
E	Equipment & Furnishings	100.0	97.9	99.9	100.0	89.4	99.4	100.0	85.0	99.2	100.0	92.1	99.6	100.0	133.7	101.9
G	Site Work	84.9	103.3	97.4	94.6	102.5	100.0	72.9	101.0	92.0	80.5	102.5	95.5	100.4	110.7	107.4
A-G	WEIGHTED AVERAGE	98.1	104.5	101.0	98.5	98.2	98.3	96.5	97.5	97.0	98.0	96.7	97.4	95.2	143.1	116.3

DIV. NO.	BUILDING SYSTEMS	NEW YORK			NORTH CAROLINA											
		YONKERS			ASHEVILLE			CHARLOTTE			DURHAM			FAYETTEVILLE		
		MAT.	INST.	TOTAL	MAT.	INST.	TOTAL	MAT.	INST.	TOTAL	MAT.	INST.	TOTAL	MAT.	INST.	TOTAL
A	Substructure	104.7	145.7	129.5	107.9	66.6	83.0	110.8	66.9	84.3	94.8	67.7	78.5	112.0	67.5	85.1
B10	Shell: Superstructure	99.9	148.7	119.4	96.7	75.1	88.1	98.2	74.3	88.6	106.1	73.3	93.0	112.3	73.1	96.7
B20	Exterior Closure	99.5	155.0	125.7	100.3	52.4	77.7	101.1	58.5	81.0	99.5	46.3	74.4	100.6	45.0	74.3
B30	Roofing	108.1	147.6	123.5	103.3	53.5	83.9	98.1	55.5	81.5	104.8	54.6	85.3	102.7	52.8	83.3
C	Interior Construction	95.6	153.8	120.0	94.1	59.4	79.5	93.9	58.9	79.2	100.5	58.9	83.0	94.5	58.1	79.2
D10	Services: Conveying	100.0	123.1	106.8	100.0	83.4	95.1	100.0	84.6	95.5	100.0	83.2	95.1	100.0	83.0	95.0
D20 - 40	Mechanical	100.3	136.5	115.6	100.5	55.6	81.5	99.9	57.1	81.8	100.6	55.2	81.4	100.2	54.9	81.1
D50	Electrical	96.1	163.1	129.9	100.5	57.5	78.7	99.5	60.5	79.8	93.9	56.3	74.9	99.5	51.3	75.1
E	Equipment & Furnishings	100.0	143.4	102.4	100.0	62.8	97.9	100.0	63.1	97.9	100.0	62.7	97.9	100.0	62.4	97.9
G	Site Work	105.4	110.7	109.0	99.8	78.3	85.1	101.5	78.5	85.8	98.6	87.6	91.1	99.5	87.5	91.4
A-G	WEIGHTED AVERAGE	99.4	146.3	120.1	99.0	62.4	82.9	99.1	63.8	83.6	100.7	61.7	83.5	102.2	60.6	83.9

DIV. NO.	BUILDING SYSTEMS	NORTH CAROLINA												NORTH DAKOTA		
		GREENSBORO			RALEIGH			WILMINGTON			WINSTON-SALEM			BISMARCK		
		MAT.	INST.	TOTAL	MAT.	INST.	TOTAL	MAT.	INST.	TOTAL	MAT.	INST.	TOTAL	MAT.	INST.	TOTAL
A	Substructure	94.1	67.8	78.2	96.3	67.9	79.2	107.9	66.8	83.1	95.7	67.9	78.9	98.5	69.7	81.1
B10	Shell: Superstructure	100.6	73.2	89.7	95.5	73.3	86.6	96.5	73.7	87.4	98.9	73.3	88.7	97.0	76.2	88.7
B20	Exterior Closure	97.6	46.6	73.5	95.5	47.3	72.7	95.6	49.3	73.7	97.7	46.9	73.7	108.7	66.9	89.0
B30	Roofing	104.6	52.4	84.3	99.0	54.6	81.7	103.3	54.1	84.1	104.6	52.9	84.5	110.8	72.5	95.9
C	Interior Construction	100.6	57.8	82.6	98.7	58.7	81.9	94.5	60.0	80.0	100.6	59.1	83.2	106.0	44.5	80.1
D10	Services: Conveying	100.0	83.3	95.1	100.0	83.5	95.2	100.0	84.6	95.5	100.0	83.5	95.2	100.0	86.7	96.1
D20 - 40	Mechanical	100.5	55.3	81.4	100.1	55.6	81.3	100.5	57.1	82.1	100.5	55.5	81.5	99.9	71.7	88.0
D50	Electrical	93.1	57.5	75.1	95.7	40.1	67.6	100.7	51.3	75.7	93.1	58.0	75.3	104.6	76.2	90.3
E	Equipment & Furnishings	100.0	62.8	97.9	100.0	62.0	97.9	100.0	62.7	97.9	100.0	62.8	97.9	100.0	35.6	96.4
G	Site Work	98.5	87.6	91.1	98.2	87.6	91.0	101.3	78.5	85.7	98.8	87.6	91.2	99.2	97.6	98.1
A-G	WEIGHTED AVERAGE	99.2	61.7	82.7	97.7	59.7	81.0	98.5	61.5	82.2	98.9	62.1	82.7	102.1	70.0	87.9

DIV. NO.	BUILDING SYSTEMS	NORTH DAKOTA									OHIO					
		FARGO			GRAND FORKS			MINOT			AKRON			CANTON		
		MAT.	INST.	TOTAL	MAT.	INST.	TOTAL	MAT.	INST.	TOTAL	MAT.	INST.	TOTAL	MAT.	INST.	TOTAL
A	Substructure	100.0	77.2	86.2	105.4	66.5	81.9	105.4	82.1	91.4	94.2	95.7	95.1	94.8	89.1	91.4
B10	Shell: Superstructure	95.3	81.1	89.6	94.6	74.5	86.6	94.7	84.7	90.7	98.2	87.0	93.7	98.3	79.5	90.8
B20	Exterior Closure	109.5	81.1	96.1	110.1	66.8	89.7	108.1	78.1	93.9	99.1	95.8	97.5	98.7	83.6	91.6
B30	Roofing	107.7	76.8	95.7	110.0	69.8	94.4	109.8	82.1	99.0	111.8	95.9	105.6	113.0	91.9	104.8
C	Interior Construction	102.3	51.1	80.8	103.7	42.7	78.1	103.2	68.0	88.4	100.8	93.4	97.7	97.4	81.1	90.6
D10	Services: Conveying	100.0	88.3	96.6	100.0	48.7	85.0	100.0	90.7	97.3	100.0	99.1	99.7	100.0	97.3	99.2
D20 - 40	Mechanical	100.0	79.0	91.1	100.2	64.7	85.2	100.2	65.7	85.6	100.1	95.6	98.2	100.1	98.7	99.5
D50	Electrical	104.5	69.2	86.7	104.2	67.0	85.4	107.5	77.2	92.2	101.4	93.5	97.4	100.7	94.9	97.7
E	Equipment & Furnishings	100.0	42.7	96.8	100.0	31.9	96.2	100.0	68.2	98.2	100.0	92.2	99.6	100.0	82.1	99.0
G	Site Work	98.8	97.9	98.2	109.2	95.1	99.6	106.7	98.3	101.0	99.3	99.4	99.4	99.5	99.2	99.3
A-G	WEIGHTED AVERAGE	101.1	75.0	89.6	101.9	65.1	85.7	101.9	76.9	90.9	99.9	93.9	97.3	99.3	89.3	94.9

DIV. NO.	BUILDING SYSTEMS	OHIO														
		CINCINNATI			CLEVELAND			COLUMBUS			DAYTON			LORAIN		
		MAT.	INST.	TOTAL	MAT.	INST.	TOTAL	MAT.	INST.	TOTAL	MAT.	INST.	TOTAL	MAT.	INST.	TOTAL
A	Substructure	94.8	86.8	90.0	93.3	100.0	97.3	95.8	87.2	90.6	90.5	85.1	87.3	91.5	94.9	93.5
B10	Shell: Superstructure	97.3	83.0	91.6	99.4	91.1	96.1	97.3	82.1	91.2	95.8	78.7	89.0	98.2	86.4	93.5
B20	Exterior Closure	97.3	84.1	91.1	100.1	102.9	101.4	99.5	90.0	95.0	97.0	83.6	90.6	96.7	101.6	99.0
B30	Roofing	101.5	83.4	94.4	110.5	108.5	109.7	97.6	92.8	95.7	107.2	86.4	99.1	112.9	102.0	108.6
C	Interior Construction	97.6	79.1	89.8	95.2	98.9	96.7	95.9	82.4	90.2	97.7	80.0	90.2	97.4	88.8	93.8
D10	Services: Conveying	100.0	90.9	97.4	100.0	103.0	100.9	100.0	94.5	98.4	100.0	90.9	97.3	100.0	102.7	100.8
D20 - 40	Mechanical	100.0	83.1	92.8	100.0	99.4	99.7	100.0	90.4	95.9	100.9	86.1	94.6	100.1	91.1	96.3
D50	Electrical	96.5	80.2	88.3	101.0	104.4	102.7	98.2	87.1	92.6	95.2	82.3	88.7	100.8	87.3	94.0
E	Equipment & Furnishings	100.0	76.3	98.7	100.0	96.2	99.8	100.0	80.0	98.9	100.0	74.9	98.6	100.0	80.5	98.9
G	Site Work	97.7	99.7	99.1	99.2	99.8	99.6	98.6	95.6	96.5	96.6	99.3	98.4	98.7	100.6	100.0
A-G	WEIGHTED AVERAGE	98.2	83.7	91.8	99.2	99.4	99.3	98.2	87.4	93.5	97.9	83.9	91.7	98.9	92.2	96.0

DIV. NO.	BUILDING SYSTEMS	OHIO									OKLAHOMA					
		SPRINGFIELD			TOLEDO			YOUNGSTOWN			ENID			LAWTON		
		MAT.	INST.	TOTAL	MAT.	INST.	TOTAL	MAT.	INST	TOTAL	MAT.	INST.	TOTAL	MAT.	INST.	TOTAL
A	Substructure	92.0	86.4	88.6	95.7	96.0	95.9	93.7	92.2	92.8	97.3	62.0	76.0	95.5	65.9	77.6
B10	Shell: Superstructure	96.1	79.8	89.6	97.2	90.1	94.4	98.1	83.4	92.3	101.2	60.5	84.9	104.0	63.3	87.7
B20	Exterior Closure	96.6	82.8	90.1	102.9	98.8	100.9	98.6	91.7	95.3	104.7	68.6	87.6	101.0	69.0	85.9
B30	Roofing	107.0	87.1	99.3	99.1	101.1	99.9	113.1	92.8	105.2	109.4	62.7	91.2	109.1	64.3	91.7
C	Interior Construction	96.9	83.9	91.4	94.1	97.0	95.3	97.4	86.7	92.9	98.6	44.7	76.0	100.5	53.0	80.6
D10	Services: Conveying	100.0	90.9	97.3	100.0	100.0	100.0	100.0	97.6	99.3	100.0	86.7	96.1	100.0	86.7	96.1
D20 - 40	Mechanical	100.9	80.8	92.4	100.0	99.7	99.9	100.1	86.1	94.2	100.2	69.2	87.1	100.2	69.2	87.1
D50	Electrical	95.2	87.1	91.1	98.3	103.9	101.1	100.8	85.3	93.0	94.4	70.1	82.2	96.0	70.4	83.0
E	Equipment & Furnishings	100.0	80.7	98.9	100.0	96.1	99.8	100.0	84.9	99.1	100.0	30.4	96.1	100.0	42.9	96.8
G	Site Work	96.9	97.9	97.6	97.9	96.4	96.8	99.2	99.7	99.6	102.4	95.3	97.6	97.9	96.9	97.2
A-G	**WEIGHTED AVERAGE**	97.8	84.1	91.8	98.3	97.6	98.0	99.2	88.1	94.4	100.2	65.2	84.8	100.7	67.5	86.1

DIV. NO.	BUILDING SYSTEMS	OKLAHOMA									OREGON					
		MUSKOGEE			OKLAHOMA CITY			TULSA			EUGENE			MEDFORD		
		MAT.	INST.	TOTAL	MAT.	INST.	TOTAL	MAT.	INST	TOTAL	MAT.	INST.	TOTAL	MAT.	INST.	TOTAL
A	Substructure	90.9	59.2	71.7	97.3	75.1	83.9	96.7	60.3	74.7	101.7	100.7	101.1	103.8	100.6	101.9
B10	Shell: Superstructure	93.8	65.3	82.4	98.6	69.3	86.9	97.3	67.3	85.4	95.5	97.6	96.3	95.2	97.5	96.1
B20	Exterior Closure	102.2	56.4	80.6	102.8	69.5	87.1	97.0	69.8	84.2	103.9	103.0	103.5	108.7	103.0	106.0
B30	Roofing	100.1	60.7	84.8	99.4	69.1	87.6	100.1	64.4	86.2	108.7	95.0	103.4	109.4	92.4	102.8
C	Interior Construction	96.8	38.9	72.5	99.7	64.2	84.8	97.6	44.9	75.5	99.2	98.0	98.7	101.8	97.5	100.0
D10	Services: Conveying	100.0	87.2	96.2	100.0	86.7	96.1	100.0	87.2	96.2	100.0	103.1	100.9	100.0	103.1	100.9
D20 - 40	Mechanical	100.2	65.3	85.5	100.0	69.3	87.0	100.2	65.2	85.4	99.9	99.8	99.9	99.9	99.8	99.9
D50	Electrical	95.5	75.3	85.3	100.9	70.1	85.3	97.4	64.1	80.5	98.4	95.4	96.9	101.8	80.1	90.8
E	Equipment & Furnishings	100.0	31.8	96.2	100.0	59.8	97.7	100.0	36.7	96.4	100.0	99.2	100.0	100.0	98.5	99.9
G	Site Work	92.7	91.2	91.6	95.5	97.4	96.8	98.7	91.4	93.7	97.3	102.7	101.0	105.0	102.7	103.4
A-G	**WEIGHTED AVERAGE**	97.5	62.9	82.3	99.8	71.3	87.3	98.3	64.5	83.4	99.5	99.2	99.3	101.0	96.9	99.2

DIV. NO.	BUILDING SYSTEMS	OREGON						PENNSYLVANIA								
		PORTLAND			SALEM			ALLENTOWN			ALTOONA			ERIE		
		MAT.	INST.	TOTAL	MAT.	INST.	TOTAL	MAT.	INST	TOTAL	MAT.	INST.	TOTAL	MAT.	INST.	TOTAL
A	Substructure	103.7	100.8	102.0	98.8	100.8	100.0	92.4	108.3	102.0	97.9	88.3	92.1	98.2	93.6	95.4
B10	Shell: Superstructure	96.8	97.7	97.2	100.2	97.7	99.2	96.2	119.1	105.3	92.0	101.9	96.0	93.0	104.9	97.7
B20	Exterior Closure	104.1	103.0	103.6	106.2	103.0	104.7	98.5	105.5	101.8	91.7	82.2	87.2	85.3	93.5	89.2
B30	Roofing	108.6	96.8	104.0	105.3	96.8	102.0	104.2	120.0	110.4	103.0	90.1	98.0	103.5	94.6	100.0
C	Interior Construction	98.0	97.7	97.9	98.4	98.0	98.2	92.9	112.0	100.9	88.0	86.8	87.5	89.5	91.2	90.2
D10	Services: Conveying	100.0	103.1	100.9	100.0	103.1	100.9	100.0	104.6	101.3	100.0	97.1	99.1	100.0	102.6	100.8
D20 - 40	Mechanical	99.9	99.9	99.9	100.0	99.9	99.9	100.3	115.8	106.9	99.7	86.8	94.3	99.7	97.6	98.8
D50	Electrical	98.6	109.2	103.9	105.2	95.4	100.2	97.6	100.0	98.8	88.8	111.3	100.2	90.4	99.2	94.9
E	Equipment & Furnishings	100.0	99.8	100.0	100.0	99.2	100.0	100.0	114.5	100.8	100.0	82.8	99.0	100.0	86.1	99.2
G	Site Work	100.0	102.7	101.8	92.6	102.7	99.5	93.2	101.7	99.0	95.0	101.1	99.2	92.0	101.9	98.7
A-G	**WEIGHTED AVERAGE**	99.7	101.1	100.3	100.9	99.2	100.2	97.3	110.6	103.2	94.1	93.5	93.9	94.0	97.5	95.5

DIV. NO.	BUILDING SYSTEMS	PENNSYLVANIA														
		HARRISBURG			PHILADELPHIA			PITTSBURGH			READING			SCRANTON		
		MAT.	INST.	TOTAL	MAT.	INST.	TOTAL	MAT.	INST	TOTAL	MAT.	INST.	TOTAL	MAT.	INST.	TOTAL
A	Substructure	95.8	96.7	96.3	94.4	127.3	114.2	107.9	101.6	104.1	83.2	100.0	93.3	94.9	97.6	96.6
B10	Shell: Superstructure	99.0	110.3	103.5	99.8	132.5	112.8	102.5	113.0	106.7	94.5	110.9	101.1	98.5	109.6	102.9
B20	Exterior Closure	102.4	93.0	97.9	101.1	133.4	116.3	96.9	101.7	99.2	102.9	98.0	100.6	98.7	98.6	98.6
B30	Roofing	98.9	110.6	103.5	102.4	138.1	116.3	102.0	101.5	101.8	108.3	113.8	110.4	104.1	97.4	101.5
C	Interior Construction	96.9	90.5	94.2	99.8	141.7	117.4	98.9	101.4	100.0	90.7	99.3	94.3	93.9	92.4	93.2
D10	Services: Conveying	100.0	99.5	99.8	100.0	117.6	105.2	100.0	103.4	101.0	100.0	101.9	100.6	100.0	103.3	101.0
D20 - 40	Mechanical	99.9	92.9	97.0	100.3	133.2	114.2	100.1	99.7	99.9	100.3	111.1	104.8	100.3	99.8	100.1
D50	Electrical	94.6	89.9	92.2	96.2	156.2	126.5	96.5	111.3	104.0	99.3	97.1	98.2	97.6	97.5	97.5
E	Equipment & Furnishings	100.0	86.7	99.3	100.0	141.2	102.3	100.0	98.1	99.9	100.0	87.1	99.3	100.0	85.8	99.2
G	Site Work	90.2	100.5	97.2	98.9	98.8	98.8	105.3	103.8	104.3	99.4	111.6	107.7	93.8	101.8	99.3
A-G	**WEIGHTED AVERAGE**	98.5	96.5	97.6	99.5	134.7	115.0	100.2	104.6	102.1	97.1	104.5	100.4	98.1	99.8	98.8

DIV. NO.	BUILDING SYSTEMS	PENNSYLVANIA			PUERTO RICO			RHODE ISLAND			SOUTH CAROLINA					
		YORK			SAN JUAN			PROVIDENCE			CHARLESTON			COLUMBIA		
		MAT.	INST.	TOTAL	MAT.	INST.	TOTAL	MAT.	INST	TOTAL	MAT.	INST.	TOTAL	MAT.	INST.	TOTAL
A	Substructure	88.4	97.4	93.8	118.3	36.1	68.7	97.9	120.4	111.5	97.4	71.2	81.6	98.5	74.2	83.8
B10	Shell: Superstructure	93.4	110.6	100.3	115.5	27.8	80.5	99.3	124.2	109.2	97.2	76.6	89.0	95.4	79.1	88.9
B20	Exterior Closure	105.7	95.0	100.6	104.7	17.6	63.5	107.2	131.3	118.6	101.8	61.6	82.8	97.1	61.9	80.5
B30	Roofing	95.9	111.8	102.1	127.9	21.5	86.5	102.4	117.7	108.3	99.6	61.2	84.7	95.8	62.5	82.8
C	Interior Construction	91.8	92.5	92.1	176.6	16.6	109.4	98.2	126.3	110.0	100.4	64.0	85.1	97.6	70.8	86.3
D10	Services: Conveying	100.0	100.4	100.1	100.0	16.8	75.7	100.0	106.8	102.0	100.0	73.1	92.1	100.0	73.1	92.1
D20 - 40	Mechanical	100.1	94.2	97.6	103.8	13.8	65.7	100.0	110.1	104.3	100.5	57.3	82.2	99.9	57.8	82.1
D50	Electrical	89.5	84.6	87.0	120.4	18.1	68.7	96.6	99.6	98.1	100.9	59.2	79.8	100.6	61.8	81.0
E	Equipment & Furnishings	100.0	86.6	99.2	100.0	15.5	95.3	100.0	117.9	101.0	100.0	64.8	98.0	100.0	75.1	98.6
G	Site Work	86.0	100.7	96.0	134.0	89.7	103.8	92.2	102.8	99.4	100.2	88.5	92.2	99.6	88.5	92.1
A-G	**WEIGHTED AVERAGE**	95.9	96.8	96.3	122.0	24.5	79.0	99.8	116.8	107.3	99.8	66.1	84.9	98.2	68.3	85.0

DIV. NO.	BUILDING SYSTEMS	SOUTH CAROLINA									SOUTH DAKOTA					
		FLORENCE			GREENVILLE			SPARTANBURG			ABERDEEN			PIERRE		
		MAT.	INST.	TOTAL	MAT.	INST.	TOTAL	MAT.	INST	TOTAL	MAT.	INST.	TOTAL	MAT.	INST.	TOTAL
A	Substructure	88.5	74.1	79.8	88.9	71.0	78.1	89.1	73.1	79.4	105.8	51.9	73.7	101.9	56.9	74.7
B10	Shell: Superstructure	93.6	78.8	87.7	94.3	73.4	85.9	94.5	78.2	88.0	95.7	53.2	78.7	96.9	63.6	83.6
B20	Exterior Closure	101.9	61.9	83.0	98.4	61.1	80.8	99.5	61.9	81.7	109.9	54.0	83.5	110.4	56.0	84.7
B30	Roofing	99.9	62.5	85.3	99.8	62.5	85.3	99.8	62.5	85.3	100.7	46.8	79.8	102.4	46.6	80.7
C	Interior Construction	96.8	67.5	84.5	97.7	68.4	85.4	98.1	69.3	86.0	100.1	37.2	73.6	102.7	38.5	75.7
D10	Services: Conveying	100.0	73.1	92.1	100.0	73.1	92.1	100.0	73.1	92.1	100.0	39.8	82.4	100.0	79.6	94.0
D20 - 40	Mechanical	100.5	57.4	82.2	100.5	57.1	82.1	100.5	57.3	82.2	100.0	43.4	76.1	99.9	68.2	86.5
D50	Electrical	98.8	61.8	80.1	101.0	58.4	79.4	101.0	58.4	79.4	103.5	76.4	89.8	110.2	51.4	80.5
E	Equipment & Furnishings	100.0	75.1	98.6	100.0	74.8	98.6	100.0	74.8	98.6	100.0	36.0	96.4	100.0	35.7	96.4
G	Site Work	108.3	88.5	94.8	103.8	88.0	93.0	103.6	88.0	92.9	102.1	94.1	96.6	100.6	94.3	96.3
A-G	**WEIGHTED AVERAGE**	98.0	67.7	84.7	98.1	66.0	83.9	98.3	67.3	84.6	101.0	53.7	80.1	102.2	59.3	83.3

DIV. NO.	BUILDING SYSTEMS	SOUTH DAKOTA						TENNESSEE								
		RAPID CITY			SIOUX FALLS			CHATTANOOGA			JACKSON			JOHNSON CITY		
		MAT.	INST.	TOTAL	MAT.	INST.	TOTAL	MAT.	INST.	TOTAL	MAT.	INST.	TOTAL	MAT.	INST.	TOTAL
A	Substructure	103.7	55.7	74.7	95.0	60.5	74.2	99.3	70.0	81.6	99.3	68.2	80.5	86.2	58.1	69.3
B10	Shell: Superstructure	96.6	63.1	83.2	94.5	66.1	83.2	96.3	76.8	88.5	94.9	75.8	87.3	90.6	69.1	82.0
B20	Exterior Closure	108.6	56.3	83.9	103.9	57.9	82.2	105.5	54.0	81.2	112.2	50.3	83.0	132.0	47.1	91.9
B30	Roofing	101.2	46.3	79.8	99.0	49.4	79.7	100.6	62.7	85.9	97.8	60.5	83.3	95.4	54.0	79.3
C	Interior Construction	102.2	42.7	77.2	101.0	51.1	80.0	97.7	59.2	81.6	93.5	45.7	73.4	97.1	40.3	73.2
D10	Services: Conveying	100.0	78.9	93.8	100.0	79.6	94.0	100.0	38.7	82.1	100.0	50.4	85.5	100.0	81.4	94.5
D20 - 40	Mechanical	100.0	67.3	86.2	99.9	42.7	75.7	100.2	62.0	84.1	100.2	68.1	86.6	99.9	58.5	82.4
D50	Electrical	99.9	51.4	75.4	99.3	76.7	87.9	103.1	71.2	87.0	99.5	58.1	78.6	94.1	46.0	69.8
E	Equipment & Furnishings	100.0	32.5	96.2	100.0	45.9	97.0	100.0	59.3	97.7	100.0	44.1	96.9	100.0	35.6	96.4
G	Site Work	100.0	94.2	96.0	92.6	96.0	94.9	102.4	97.8	99.3	97.4	98.2	98.0	109.2	84.7	92.5
A-G	WEIGHTED AVERAGE	100.8	59.6	82.7	99.0	60.5	82.0	99.9	66.6	85.2	99.0	63.5	83.4	99.9	56.4	80.7

DIV. NO.	BUILDING SYSTEMS	TENNESSEE									TEXAS					
		KNOXVILLE			MEMPHIS			NASHVILLE			ABILENE			AMARILLO		
		MAT.	INST.	TOTAL	MAT.	INST.	TOTAL	MAT.	INST.	TOTAL	MAT.	INST.	TOTAL	MAT.	INST.	TOTAL
A	Substructure	93.4	69.1	78.8	94.3	76.5	83.6	92.9	73.5	81.2	97.8	69.3	80.6	94.3	65.8	77.1
B10	Shell: Superstructure	95.8	76.5	88.1	97.1	82.8	91.4	95.8	80.3	89.6	101.4	66.9	87.7	96.7	63.9	83.6
B20	Exterior Closure	93.9	55.6	75.8	102.6	59.2	82.1	97.2	60.7	79.9	97.7	63.9	81.7	101.1	58.3	80.9
B30	Roofing	93.4	65.8	82.7	98.3	68.9	86.9	95.0	66.7	84.0	100.3	66.3	87.1	96.5	63.2	83.6
C	Interior Construction	93.1	60.2	79.3	97.1	64.8	83.5	99.2	64.9	84.8	98.4	64.6	84.2	102.1	56.0	82.8
D10	Services: Conveying	100.0	85.3	95.7	100.0	84.5	95.5	100.0	85.2	95.7	100.0	84.6	95.5	100.0	82.3	94.8
D20 - 40	Mechanical	99.9	67.8	86.3	100.0	73.4	88.8	99.9	83.2	92.9	100.2	49.9	78.9	100.0	54.3	80.7
D50	Electrical	99.7	57.0	78.1	101.5	64.8	82.9	100.1	64.3	82.0	101.7	48.3	74.7	101.4	63.7	82.3
E	Equipment & Furnishings	100.0	60.7	97.8	100.0	67.0	98.1	100.0	66.0	98.1	100.0	65.1	98.0	100.0	55.3	97.5
G	Site Work	90.4	87.4	88.4	92.3	94.3	93.6	98.1	97.1	97.4	102.7	93.1	96.1	98.6	92.0	94.1
A-G	WEIGHTED AVERAGE	96.5	66.8	83.5	98.9	72.5	87.3	98.2	74.1	87.6	100.0	62.3	83.4	99.6	62.3	83.1

DIV. NO.	BUILDING SYSTEMS	TEXAS														
		AUSTIN			BEAUMONT			CORPUS CHRISTI			DALLAS			EL PASO		
		MAT.	INST.	TOTAL	MAT.	INST.	TOTAL	MAT.	INST.	TOTAL	MAT.	INST.	TOTAL	MAT.	INST.	TOTAL
A	Substructure	93.5	66.7	77.3	94.5	70.9	80.3	114.3	65.6	84.9	108.1	70.5	85.4	95.3	68.0	78.8
B10	Shell: Superstructure	98.9	63.9	84.9	97.3	70.0	86.4	99.3	70.2	87.7	102.7	74.4	91.4	96.9	65.0	84.2
B20	Exterior Closure	102.2	57.5	81.0	102.9	65.1	85.0	94.8	57.7	77.3	96.9	59.0	79.0	96.0	58.3	78.2
B30	Roofing	100.1	64.9	86.4	102.3	67.3	88.7	105.1	66.8	90.2	92.0	67.0	82.3	98.8	64.5	85.4
C	Interior Construction	101.0	56.5	82.3	99.3	61.8	83.6	107.2	56.7	86.0	95.7	63.2	82.1	95.3	61.3	81.0
D10	Services: Conveying	100.0	84.3	95.4	100.0	86.7	96.1	100.0	86.3	96.0	100.0	85.4	95.7	100.0	84.0	95.3
D20 - 40	Mechanical	99.9	59.5	82.8	100.2	63.5	84.7	100.2	58.3	82.5	99.9	62.7	84.2	99.9	49.9	78.8
D50	Electrical	100.7	63.4	81.9	96.6	70.1	83.2	97.7	57.2	77.2	94.4	65.5	79.8	103.7	56.9	80.1
E	Equipment & Furnishings	100.0	57.1	97.6	100.0	60.2	97.8	100.0	57.3	97.6	100.0	66.0	98.1	100.0	62.6	97.9
G	Site Work	101.4	90.9	94.2	90.0	97.4	95.1	141.8	86.5	104.1	108.2	92.7	97.7	102.7	91.0	94.7
A-G	WEIGHTED AVERAGE	100.0	63.4	83.9	98.9	68.9	85.7	102.0	63.2	84.9	99.2	67.9	85.4	98.3	61.6	82.2

DIV. NO.	BUILDING SYSTEMS	TEXAS														
		FORT WORTH			HOUSTON			LAREDO			LUBBOCK			ODESSA		
		MAT.	INST.	TOTAL	MAT.	INST.	TOTAL	MAT.	INST.	TOTAL	MAT.	INST.	TOTAL	MAT.	INST.	TOTAL
A	Substructure	99.6	69.6	81.5	98.3	70.1	81.3	91.2	65.7	75.8	102.2	66.2	80.5	98.0	68.7	80.3
B10	Shell: Superstructure	100.2	67.3	87.1	100.4	75.3	90.4	97.3	62.8	83.5	104.5	72.1	91.5	100.9	66.5	87.2
B20	Exterior Closure	98.1	58.1	79.2	99.4	63.6	82.5	101.1	57.3	80.4	97.2	64.8	81.9	97.7	58.7	79.3
B30	Roofing	92.3	66.2	82.2	92.8	67.7	83.0	100.8	64.9	86.9	89.6	64.6	79.9	100.3	66.0	87.0
C	Interior Construction	98.7	63.0	83.7	106.7	61.8	87.9	100.4	56.5	82.0	103.5	56.2	83.6	98.4	62.9	83.5
D10	Services: Conveying	100.0	84.6	95.5	100.0	87.1	96.2	100.0	83.9	95.3	100.0	85.0	95.6	100.0	83.4	95.1
D20 - 40	Mechanical	99.9	59.2	82.7	100.0	65.4	85.4	100.1	63.5	84.6	99.7	50.2	78.8	100.2	49.9	78.9
D50	Electrical	94.7	65.5	79.9	98.1	69.7	83.8	100.0	60.5	80.0	100.5	63.7	81.9	101.8	63.7	82.5
E	Equipment & Furnishings	100.0	65.9	98.1	100.0	62.9	97.9	100.0	57.3	97.6	100.0	55.3	97.5	100.0	65.4	98.1
G	Site Work	101.2	93.1	95.7	101.9	95.1	97.3	104.8	90.9	95.3	124.4	89.5	100.6	102.7	91.4	95.0
A-G	WEIGHTED AVERAGE	98.8	65.7	84.2	100.7	69.8	87.1	99.4	63.5	83.6	101.5	63.6	84.8	99.9	63.2	83.7

DIV. NO.	BUILDING SYSTEMS	TEXAS									UTAH					
		SAN ANTONIO			WACO			WICHITA FALLS			LOGAN			OGDEN		
		MAT.	INST.	TOTAL	MAT.	INST.	TOTAL	MAT.	INST.	TOTAL	MAT.	INST.	TOTAL	MAT.	INST.	TOTAL
A	Substructure	95.5	67.1	78.4	92.5	69.8	78.8	96.3	69.2	80.0	92.0	74.5	81.4	91.3	74.5	81.2
B10	Shell: Superstructure	99.9	63.9	85.6	99.6	66.8	86.5	100.3	67.0	87.0	103.9	73.6	91.8	104.4	73.6	92.1
B20	Exterior Closure	97.0	57.5	78.3	97.7	58.6	79.2	97.9	63.9	81.8	123.2	63.1	94.8	108.1	63.1	86.8
B30	Roofing	98.0	66.2	85.7	96.1	65.9	84.4	96.1	66.3	84.5	97.3	66.9	85.5	96.2	66.9	84.8
C	Interior Construction	103.4	56.8	83.8	82.1	62.3	73.8	82.3	64.6	74.9	92.7	60.6	79.2	91.8	60.6	78.7
D10	Services: Conveying	100.0	84.1	95.4	100.0	84.6	95.5	100.0	82.3	94.8	100.0	89.6	96.9	100.0	89.6	96.9
D20 - 40	Mechanical	100.1	63.4	84.6	100.2	54.2	80.7	100.2	53.7	80.5	99.9	69.0	86.9	99.9	69.0	86.9
D50	Electrical	102.0	60.5	81.0	97.2	59.3	78.0	98.8	60.0	79.2	96.7	75.7	86.1	97.0	75.7	86.2
E	Equipment & Furnishings	100.0	57.4	97.6	100.0	65.6	98.1	100.0	65.6	98.1	100.0	59.0	97.7	100.0	59.0	97.7
G	Site Work	104.0	94.2	97.4	99.4	93.0	95.0	100.3	93.1	95.4	99.6	94.5	96.2	88.2	94.5	92.5
A-G	WEIGHTED AVERAGE	100.3	64.1	84.4	95.9	63.7	81.7	96.5	64.6	82.5	101.3	71.2	88.0	99.3	71.2	86.9

DIV. NO.	BUILDING SYSTEMS	UTAH						VERMONT						VIRGINIA		
		PROVO			SALT LAKE CITY			BURLINGTON			RUTLAND			ALEXANDRIA		
		MAT.	INST.	TOTAL	MAT.	INST.	TOTAL	MAT.	INST.	TOTAL	MAT.	INST.	TOTAL	MAT.	INST.	TOTAL
A	Substructure	93.2	74.1	81.7	96.9	74.4	83.4	101.2	95.4	97.7	90.3	95.4	93.4	100.2	87.2	92.3
B10	Shell: Superstructure	102.8	73.6	91.2	108.5	73.6	94.6	100.0	89.6	95.8	96.3	89.6	93.6	101.1	92.4	97.6
B20	Exterior Closure	124.9	63.1	95.7	131.8	63.1	99.3	113.1	91.8	103.1	102.0	91.8	97.2	102.1	79.4	91.4
B30	Roofing	99.1	66.9	86.6	103.6	66.9	89.3	104.5	75.7	93.3	97.3	75.7	88.9	102.2	84.9	95.5
C	Interior Construction	95.0	60.6	80.6	93.8	60.6	79.9	103.6	87.7	97.0	100.9	87.7	95.4	99.3	86.3	93.9
D10	Services: Conveying	100.0	89.6	96.9	100.0	89.6	96.9	100.0	96.2	98.9	100.0	96.2	98.9	100.0	91.1	97.4
D20 - 40	Mechanical	99.9	69.0	86.9	100.1	69.0	87.0	100.1	71.1	87.9	100.3	71.1	88.0	100.4	87.4	94.9
D50	Electrical	97.4	75.7	86.5	99.4	75.7	87.4	100.4	60.7	80.3	100.0	60.7	80.1	97.6	101.6	99.6
E	Equipment & Furnishings	100.0	59.0	97.7	100.0	59.0	97.7	100.0	82.7	99.0	100.0	82.7	99.0	100.0	89.5	99.4
G	Site Work	96.3	92.9	94.0	87.9	94.5	92.4	96.7	98.0	97.6	93.0	97.8	96.3	115.7	92.1	99.6
A-G	WEIGHTED AVERAGE	101.8	71.1	88.2	103.8	71.2	89.4	102.2	82.5	93.5	99.0	82.5	91.7	100.6	89.4	95.6

DIV. NO.	BUILDING SYSTEMS	VIRGINIA														
		ARLINGTON			NEWPORT NEWS			NORFOLK			PORTSMOUTH			RICHMOND		
		MAT.	INST.	TOTAL	MAT.	INST.	TOTAL	MAT.	INST.	TOTAL	MAT.	INST.	TOTAL	MAT.	INST.	TOTAL
A	Substructure	101.8	86.7	92.7	100.2	71.2	82.7	103.2	71.5	84.1	98.7	67.3	79.7	98.8	66.0	79.0
B10	Shell: Superstructure	99.9	93.5	97.3	99.4	79.1	91.3	101.4	79.2	92.5	97.8	76.2	89.2	100.5	77.0	91.1
B20	Exterior Closure	113.4	79.9	97.6	102.3	54.9	79.9	102.1	54.9	79.8	105.1	55.5	81.7	102.3	55.2	80.0
B30	Roofing	104.2	85.6	97.0	102.1	68.9	89.2	98.9	68.9	87.2	102.1	66.5	88.3	100.7	67.7	87.8
C	Interior Construction	98.3	86.3	93.3	99.7	61.2	83.5	97.2	63.1	82.9	98.2	53.9	79.6	98.4	58.5	81.7
D10	Services: Conveying	100.0	88.5	96.6	100.0	90.5	97.2	100.0	90.5	97.2	100.0	72.9	92.1	100.0	88.8	96.7
D20 - 40	Mechanical	100.4	87.6	95.0	100.5	63.9	85.0	100.2	64.3	85.0	100.5	64.1	85.1	99.9	67.1	86.0
D50	Electrical	95.4	101.6	98.5	94.3	65.3	79.6	97.7	61.9	79.6	92.8	61.9	77.2	97.1	73.1	85.0
E	Equipment & Furnishings	100.0	89.4	99.4	100.0	62.2	97.9	100.0	62.1	97.9	100.0	48.7	97.1	100.0	55.6	97.5
G	Site Work	125.1	89.7	101.0	108.2	90.7	96.3	106.9	91.7	96.5	106.6	90.5	95.6	103.4	90.8	94.8
A-G	**WEIGHTED AVERAGE**	101.5	89.4	96.2	99.8	68.3	85.9	100.1	68.3	86.1	99.3	65.4	84.4	99.8	68.7	86.1

DIV. NO.	BUILDING SYSTEMS	VIRGINIA			WASHINGTON											
		ROANOKE			EVERETT			RICHLAND			SEATTLE			SPOKANE		
		MAT.	INST.	TOTAL	MAT.	INST.	TOTAL	MAT.	INST.	TOTAL	MAT.	INST.	TOTAL	MAT.	INST.	TOTAL
A	Substructure	109.3	68.6	84.8	102.8	102.6	102.7	97.0	84.1	89.2	105.1	102.5	103.5	99.9	83.9	90.2
B10	Shell: Superstructure	102.9	76.9	92.5	102.2	95.0	99.3	99.8	83.1	93.1	101.9	96.1	99.6	102.8	82.7	94.7
B20	Exterior Closure	101.6	58.9	81.5	100.0	100.2	100.1	113.9	83.0	99.3	100.8	101.6	101.2	114.4	82.3	99.2
B30	Roofing	103.6	70.6	90.8	109.0	100.7	105.8	152.3	80.8	124.5	106.6	100.9	104.3	148.6	82.7	122.9
C	Interior Construction	98.8	60.2	82.6	107.9	98.1	103.8	116.1	77.5	99.9	107.6	97.0	103.2	115.1	78.9	99.9
D10	Services: Conveying	100.0	71.1	91.5	100.0	97.7	99.3	100.0	86.5	96.0	100.0	100.3	100.1	100.0	86.5	96.0
D20 - 40	Mechanical	100.4	67.4	86.4	100.2	98.1	99.3	100.7	111.7	105.4	100.1	114.2	106.0	100.6	84.0	93.6
D50	Electrical	97.3	54.0	75.4	101.8	97.7	99.7	84.9	96.8	90.9	101.8	107.1	104.5	83.7	80.1	81.9
E	Equipment & Furnishings	100.0	59.1	97.7	100.0	96.9	99.8	100.0	76.1	98.7	100.0	97.3	99.9	100.0	75.3	98.6
G	Site Work	106.7	88.2	94.1	91.7	110.6	104.6	107.9	90.4	96.0	96.1	108.8	104.7	107.0	90.3	95.7
A-G	**WEIGHTED AVERAGE**	101.0	66.6	85.8	102.2	99.0	100.8	104.4	90.7	98.4	102.3	103.8	103.0	104.8	82.6	95.0

DIV. NO.	BUILDING SYSTEMS	WASHINGTON									WEST VIRGINIA					
		TACOMA			VANCOUVER			YAKIMA			CHARLESTON			HUNTINGTON		
		MAT.	INST.	TOTAL	MAT.	INST.	TOTAL	MAT.	INST.	TOTAL	MAT.	INST.	TOTAL	MAT.	INST.	TOTAL
A	Substructure	104.0	102.7	103.2	113.1	95.8	102.7	108.6	93.7	99.6	98.0	102.9	100.9	104.9	96.2	99.6
B10	Shell: Superstructure	103.4	95.1	100.1	103.2	92.9	99.1	103.0	87.6	96.8	99.5	101.8	100.4	103.7	99.5	102.0
B20	Exterior Closure	100.9	101.5	101.2	105.1	95.5	100.6	98.8	80.7	90.3	97.9	96.0	97.0	100.6	99.4	100.1
B30	Roofing	108.7	100.3	105.4	108.9	93.5	102.9	108.8	83.4	98.9	100.6	87.9	95.7	103.6	88.6	97.8
C	Interior Construction	107.7	98.5	103.8	102.8	91.1	97.9	107.0	86.4	98.4	98.4	94.9	96.9	98.6	95.7	97.4
D10	Services: Conveying	100.0	100.3	100.1	100.0	97.0	99.1	100.0	97.2	99.2	100.0	102.5	100.7	100.0	103.3	101.0
D20 - 40	Mechanical	100.2	98.2	99.4	100.3	92.1	96.8	100.2	111.1	104.9	100.2	99.3	99.8	100.6	100.3	100.5
D50	Electrical	101.6	100.2	100.9	106.4	100.2	103.3	104.3	96.8	100.5	103.2	92.6	97.8	102.1	96.3	99.2
E	Equipment & Furnishings	100.0	97.0	99.8	100.0	92.1	99.6	100.0	96.3	99.8	100.0	90.3	99.5	100.0	90.8	99.5
G	Site Work	94.6	110.6	105.5	103.6	97.6	99.5	97.2	110.0	106.0	99.1	92.8	94.8	104.4	93.0	96.6
A-G	**WEIGHTED AVERAGE**	102.6	99.6	101.3	103.4	94.4	99.4	102.8	94.9	99.3	99.7	97.4	98.7	101.4	97.8	99.8

DIV. NO.	BUILDING SYSTEMS	WEST VIRGINIA						WISCONSIN								
		PARKERSBURG			WHEELING			EAU CLAIRE			GREEN BAY			KENOSHA		
		MAT.	INST.	TOTAL	MAT.	INST.	TOTAL	MAT.	INST.	TOTAL	MAT.	INST.	TOTAL	MAT.	INST.	TOTAL
A	Substructure	100.0	91.3	94.8	100.0	91.9	95.2	99.4	99.6	99.5	102.8	101.9	102.3	106.4	102.3	103.9
B10	Shell: Superstructure	99.3	94.5	97.4	99.5	95.2	97.8	93.1	100.0	95.9	96.2	100.3	97.8	97.6	100.7	98.9
B20	Exterior Closure	99.4	91.7	95.8	111.0	90.5	101.3	97.5	100.7	99.0	114.1	116.0	115.0	99.3	114.5	106.5
B30	Roofing	103.7	88.6	97.8	104.1	87.5	97.7	104.3	99.4	102.4	106.5	87.6	99.1	103.0	105.8	104.1
C	Interior Construction	98.0	91.4	95.2	98.5	92.9	96.1	102.0	99.5	101.0	100.6	103.1	101.6	95.8	100.4	97.7
D10	Services: Conveying	100.0	102.4	100.7	100.0	93.4	98.1	100.0	97.6	99.3	100.0	100.4	100.1	100.0	100.5	100.1
D20 - 40	Mechanical	100.3	90.8	96.3	100.4	98.2	99.5	100.0	86.7	94.4	100.2	88.5	95.3	100.2	96.6	98.7
D50	Electrical	97.7	93.9	95.8	95.1	97.2	96.1	104.9	86.8	95.7	99.4	81.8	90.5	99.9	99.2	99.5
E	Equipment & Furnishings	100.0	86.3	99.2	100.0	91.1	99.5	100.0	97.6	99.9	100.0	97.8	99.9	100.0	94.2	99.7
G	Site Work	109.5	92.8	98.1	110.3	92.4	98.1	98.0	103.9	102.0	100.4	100.2	100.2	102.5	103.5	103.2
A-G	**WEIGHTED AVERAGE**	99.6	92.5	96.5	100.8	94.5	98.0	99.2	95.5	97.6	101.2	97.5	99.5	99.2	101.7	100.3

DIV. NO.	BUILDING SYSTEMS	WISCONSIN												WYOMING		
		LA CROSSE			MADISON			MILWAUKEE			RACINE			CASPER		
		MAT.	INST.	TOTAL	MAT.	INST.	TOTAL	MAT.	INST	TOTAL	MAT.	INST.	TOTAL	MAT.	INST.	TOTAL
A	Substructure	91.0	97.6	95.0	94.1	100.2	97.8	96.8	108.8	104.0	98.1	103.2	101.2	98.3	65.0	78.2
B10	Shell: Superstructure	90.8	97.3	93.4	94.7	97.5	95.8	96.3	102.3	98.7	96.0	100.7	97.9	102.6	64.2	87.2
B20	Exterior Closure	94.1	101.6	97.6	104.5	104.3	104.4	101.6	116.7	108.7	100.5	114.5	107.1	111.8	53.3	84.2
B30	Roofing	103.8	101.2	102.8	100.5	102.3	101.2	102.0	110.8	105.4	102.7	106.6	104.2	109.2	51.8	86.9
C	Interior Construction	100.4	98.3	99.5	100.8	101.5	101.1	99.3	118.3	107.3	97.6	100.4	98.8	104.3	50.6	81.8
D10	Services: Conveying	100.0	97.6	99.3	100.0	100.3	100.1	100.0	101.1	100.3	100.0	100.5	100.1	100.0	90.2	97.1
D20 - 40	Mechanical	100.0	86.6	94.3	99.9	94.5	97.6	99.9	102.4	100.9	100.0	96.6	98.6	99.9	68.1	86.5
D50	Electrical	105.1	86.8	95.9	101.1	96.3	98.7	98.6	101.0	99.8	99.3	98.8	99.1	98.2	64.4	81.1
E	Equipment & Furnishings	100.0	97.5	99.9	100.0	97.9	99.9	100.0	120.2	101.1	100.0	94.1	99.7	100.0	48.3	97.1
G	Site Work	91.8	103.7	99.9	95.4	106.5	103.0	93.9	97.2	96.1	97.2	107.4	104.1	99.3	93.0	95.0
A-G	**WEIGHTED AVERAGE**	97.6	94.8	96.4	99.3	99.1	99.2	98.9	106.9	102.4	98.7	102.0	100.2	102.5	63.8	85.5

DIV. NO.	BUILDING SYSTEMS	WYOMING						CANADA								
		CHEYENNE			ROCK SPRINGS			CALGARY, ALBERTA			EDMONTON, ALBERTA			HALIFAX, NOVA SCOTIA		
		MAT.	INST.	TOTAL	MAT.	INST.	TOTAL	MAT.	INST.	TOTAL	MAT.	INST.	TOTAL	MAT.	INST.	TOTAL
A	Substructure	93.7	69.1	78.8	94.1	68.8	78.9	157.7	96.3	120.7	163.3	96.3	122.9	126.1	80.7	98.7
B10	Shell: Superstructure	103.7	67.4	89.2	101.6	67.1	87.8	143.9	91.2	122.9	144.5	91.2	123.2	131.1	78.7	110.2
B20	Exterior Closure	113.3	53.8	85.2	141.4	51.8	99.0	150.1	85.8	119.7	146.9	85.8	118.0	143.4	82.2	114.5
B30	Roofing	103.2	53.6	83.9	105.1	58.5	87.0	122.4	91.9	110.6	125.6	91.9	112.5	120.6	76.8	103.6
C	Interior Construction	102.2	60.1	84.5	103.9	59.2	85.1	107.4	94.1	101.8	103.8	94.1	99.7	104.2	78.4	93.3
D10	Services: Conveying	100.0	90.2	97.1	100.0	80.5	94.3	131.1	95.1	120.6	131.1	95.1	120.6	131.1	66.8	112.3
D20 - 40	Mechanical	99.9	68.1	86.5	99.9	67.3	86.1	105.5	90.0	98.9	105.4	90.0	98.9	104.9	76.8	93.0
D50	Electrical	100.0	65.9	82.8	97.2	85.4	91.2	103.7	100.5	102.1	106.4	100.6	103.5	105.2	80.6	92.8
E	Equipment & Furnishings	100.0	62.6	97.9	100.0	64.6	98.0	131.1	94.1	129.1	131.1	94.1	129.1	131.1	75.3	128.0
G	Site Work	94.8	93.1	93.6	93.8	92.7	93.1	124.8	104.3	111.2	139.7	104.8	115.9	111.5	97.6	102.0
A-G	**WEIGHTED AVERAGE**	102.3	66.5	86.5	104.9	68.4	88.8	123.8	93.4	110.4	123.9	93.4	110.5	118.2	79.9	101.3

DIV. NO.	BUILDING SYSTEMS	CANADA														
		HAMILTON, ONTARIO			KITCHENER, ONTARIO			LAVAL, QUEBEC			LONDON, ONTARIO			MONTREAL, QUEBEC		
		MAT.	INST.	TOTAL	MAT.	INST.	TOTAL	MAT.	INST.	TOTAL	MAT.	INST.	TOTAL	MAT.	INST.	TOTAL
A	Substructure	131.5	98.6	111.7	115.8	92.3	101.6	124.2	90.6	104.0	133.3	94.1	109.7	138.8	95.8	112.9
B10	Shell: Superstructure	127.8	93.0	113.9	117.4	90.0	106.5	113.4	87.2	102.9	128.3	90.7	113.3	132.5	93.0	116.7
B20	Exterior Closure	135.4	99.3	118.4	112.4	95.6	104.5	132.8	83.5	109.5	135.6	96.7	117.2	131.5	90.9	112.3
B30	Roofing	118.7	92.1	108.4	111.2	89.2	102.7	111.0	88.2	102.1	120.2	90.2	108.6	119.9	94.6	110.1
C	Interior Construction	105.1	93.5	100.2	96.3	87.0	92.4	103.0	87.0	96.3	101.7	88.5	96.2	104.0	93.4	99.5
D10	Services: Conveying	131.1	99.1	121.8	131.1	98.1	121.5	131.1	82.6	116.9	131.1	98.6	121.6	131.1	85.9	117.9
D20 - 40	Mechanical	105.7	88.2	98.3	104.7	86.7	97.1	104.7	90.0	98.5	105.6	85.7	97.2	105.5	93.6	100.4
D50	Electrical	101.9	99.7	100.8	106.3	97.3	101.7	107.0	72.1	89.3	100.2	96.8	98.5	105.7	86.1	95.8
E	Equipment & Furnishings	131.1	91.3	128.9	131.1	83.3	128.5	131.1	84.3	128.5	131.1	84.4	128.5	131.1	90.1	128.8
G	Site Work	114.9	112.5	113.3	101.8	104.5	103.6	103.1	99.0	100.3	110.9	104.7	106.7	114.5	98.6	103.7
A-G	**WEIGHTED AVERAGE**	116.9	95.6	107.5	109.9	91.8	101.9	112.8	86.1	101.1	116.3	92.2	105.7	117.9	92.4	106.7

DIV. NO.	BUILDING SYSTEMS	CANADA														
		OSHAWA, ONTARIO			OTTAWA, ONTARIO			QUEBEC CITY, QUEBEC			REGINA, SASKATCHEWAN			SASKATOON, SASKATCHEWAN		
		MAT.	INST.	TOTAL	MAT.	INST.	TOTAL	MAT.	INST.	TOTAL	MAT.	INST.	TOTAL	MAT.	INST.	TOTAL
A	Substructure	140.3	92.0	111.2	127.7	95.4	108.2	132.8	96.0	110.6	140.5	91.9	111.2	121.9	91.9	103.8
B10	Shell: Superstructure	115.9	90.5	105.7	128.2	91.6	113.6	131.1	93.3	116.0	139.9	88.1	119.2	109.9	88.1	101.2
B20	Exterior Closure	120.8	95.5	108.8	132.4	96.9	115.6	130.9	93.5	113.3	145.3	92.1	120.2	130.2	92.1	112.2
B30	Roofing	112.1	86.3	102.1	126.2	90.3	112.2	120.4	94.8	110.4	126.2	82.0	109.0	109.9	82.4	99.2
C	Interior Construction	99.6	92.2	96.5	108.8	89.8	100.8	106.7	93.4	101.1	109.3	91.7	101.9	102.5	91.7	97.9
D10	Services: Conveying	131.1	98.5	121.6	131.1	96.0	120.9	131.1	86.1	118.0	131.1	69.3	113.1	131.1	69.8	113.2
D20 - 40	Mechanical	104.7	102.6	103.8	105.7	86.2	97.5	105.3	93.6	100.4	105.0	84.0	96.1	104.9	84.0	96.0
D50	Electrical	107.4	92.6	99.9	105.1	98.0	101.5	106.1	86.1	96.0	112.7	91.7	102.1	110.1	91.7	100.8
E	Equipment & Furnishings	131.1	88.0	128.7	131.1	88.1	128.7	131.1	90.1	128.8	131.1	89.0	128.8	131.1	89.0	128.8
G	Site Work	113.6	103.8	106.9	108.8	104.3	105.8	116.4	98.7	104.3	124.1	97.4	105.9	115.4	97.4	103.2
A-G	**WEIGHTED AVERAGE**	112.5	95.3	104.9	117.6	92.9	106.7	117.9	92.8	106.8	123.0	89.0	108.0	112.3	89.0	102.0

DIV. NO.	BUILDING SYSTEMS	CANADA														
		ST. CATHARINES, ONTARIO			ST. JOHN'S, NEWFOUNDLAND			THUNDER BAY, ONTARIO			TORONTO, ONTARIO			VANCOUVER, BRITISH COLUMBIA		
		MAT.	INST.	TOTAL	MAT.	INST.	TOTAL	MAT.	INST.	TOTAL	MAT.	INST.	TOTAL	MAT.	INST.	TOTAL
A	Substructure	112.0	95.9	102.2	150.4	88.7	113.2	118.8	94.7	104.2	126.1	103.4	112.4	126.1	94.9	107.3
B10	Shell: Superstructure	108.9	92.7	102.4	141.2	84.2	118.5	110.5	91.3	102.9	127.6	97.8	115.7	129.6	91.0	114.2
B20	Exterior Closure	112.1	98.5	105.7	147.3	88.2	119.3	115.2	98.0	107.1	134.6	110.6	123.2	137.3	92.6	116.2
B30	Roofing	111.2	93.1	104.2	137.7	89.6	119.0	111.4	90.5	103.3	128.2	99.8	117.2	127.8	85.7	111.5
C	Interior Construction	94.9	94.1	94.6	106.9	78.6	95.0	96.6	91.5	94.5	102.2	99.9	101.3	104.6	89.6	98.3
D10	Services: Conveying	131.1	72.2	113.9	131.1	69.2	113.0	131.1	72.6	114.0	131.1	101.5	122.5	131.1	95.4	120.7
D20 - 40	Mechanical	104.7	86.9	97.2	105.2	81.2	95.0	104.7	87.0	97.2	105.3	96.8	101.7	105.5	82.0	95.6
D50	Electrical	107.7	98.1	102.8	111.9	82.5	97.0	106.3	96.3	101.2	105.1	100.3	102.7	103.8	82.6	93.1
E	Equipment & Furnishings	131.1	91.1	128.9	131.1	80.2	128.3	131.1	88.3	128.7	131.1	97.2	129.2	131.1	87.8	128.7
G	Site Work	102.6	100.9	101.5	127.0	98.9	107.9	107.6	100.8	103.0	129.4	105.4	113.1	116.5	107.2	110.1
A-G	**WEIGHTED AVERAGE**	107.8	93.3	101.5	123.9	83.9	106.3	109.1	92.3	101.7	116.9	100.9	109.9	117.7	89.2	105.1

DIV. NO.	BUILDING SYSTEMS	CANADA														
		WINDSOR, ONTARIO			WINNIPEG, MANITOBA											
		MAT.	INST.	TOTAL	MAT.	INST.	TOTAL	MAT.	INST.	TOTAL	MAT.	INST.	TOTAL	MAT.	INST.	TOTAL
A	Substructure	114.1	94.9	102.5	138.4	75.5	100.5									
B10	Shell: Superstructure	109.6	91.6	102.4	146.8	71.8	116.9									
B20	Exterior Closure	111.9	97.6	105.2	140.9	69.3	107.0									
B30	Roofing	111.3	90.3	103.1	125.9	69.9	104.1									
C	Interior Construction	95.4	90.0	93.1	104.2	66.7	88.5									
D10	Services: Conveying	131.1	72.3	113.9	131.1	66.8	112.3									
D20 - 40	Mechanical	104.7	87.0	97.2	105.4	66.2	88.8									
D50	Electrical	110.4	98.2	104.3	109.9	67.5	88.5									
E	Equipment & Furnishings	131.1	86.7	128.6	131.1	67.1	127.5									
G	Site Work	98.2	100.8	100.0	117.1	100.3	105.6									
A-G	**WEIGHTED AVERAGE**	108.3	92.3	101.3	122.6	70.8	99.8									

Costs shown in RSMeans cost data publications are based on national averages for materials and installation. To adjust these costs to a specific location, simply multiply the base cost by the factor and divide by 100 for that city. The data is arranged alphabetically by state and postal zip code numbers. For a city not listed, use the factor for a nearby city with similar economic characteristics.

STATE/ZIP	CITY	MAT.	INST.	TOTAL
ALABAMA				
350-352	Birmingham	99.3	77.3	89.6
354	Tuscaloosa	98.7	78.1	89.6
355	Jasper	99.2	71.8	87.1
356	Decatur	98.7	76.0	88.7
357-358	Huntsville	98.7	75.3	88.4
359	Gadsden	98.8	67.1	84.8
360-361	Montgomery	97.9	77.0	88.7
362	Anniston	97.3	77.5	88.6
363	Dothan	97.7	76.1	88.2
364	Evergreen	97.3	73.6	86.9
365-366	Mobile	98.3	76.6	88.7
367	Selma	97.4	74.9	87.5
368	Phenix City	98.2	77.2	89.0
369	Butler	97.6	76.7	88.4
ALASKA				
995-996	Anchorage	121.7	115.8	119.1
997	Fairbanks	124.2	116.9	120.9
998	Juneau	123.9	115.8	120.3
999	Ketchikan	134.9	115.8	126.5
ARIZONA				
850,853	Phoenix	100.5	73.0	88.4
851,852	Mesa/Tempe	100.1	72.0	87.7
855	Globe	100.9	69.8	87.2
856-857	Tucson	98.9	70.8	86.5
859	Show Low	101.1	71.9	88.3
860	Flagstaff	102.7	71.9	89.2
863	Prescott	100.4	68.4	86.3
864	Kingman	98.9	71.3	86.7
865	Chambers	98.9	73.5	87.7
ARKANSAS				
716	Pine Bluff	99.6	65.2	84.5
717	Camden	97.6	56.0	79.3
718	Texarkana	98.5	58.7	80.9
719	Hot Springs	96.9	56.9	79.3
720-722	Little Rock	98.2	67.4	84.6
723	West Memphis	97.0	65.3	83.0
724	Jonesboro	97.4	62.6	82.1
725	Batesville	95.3	58.3	79.0
726	Harrison	96.6	57.6	79.4
727	Fayetteville	94.1	55.6	77.1
728	Russellville	95.3	59.7	79.7
729	Fort Smith	97.8	62.3	82.2
CALIFORNIA				
900-902	Los Angeles	100.2	117.2	107.7
903-905	Inglewood	96.0	115.0	104.4
906-908	Long Beach	97.5	115.0	105.2
910-912	Pasadena	97.5	114.8	105.1
913-916	Van Nuys	100.4	114.8	106.7
917-918	Alhambra	99.5	114.8	106.2
919-921	San Diego	101.4	109.8	105.1
922	Palm Springs	98.6	112.9	104.9
923-924	San Bernardino	96.4	112.3	103.4
925	Riverside	100.5	113.9	106.4
926-927	Santa Ana	98.3	112.8	104.7
928	Anaheim	100.6	114.0	106.5
930	Oxnard	99.9	113.5	105.9
931	Santa Barbara	99.2	114.3	105.9
932-933	Bakersfield	100.4	111.4	105.2
934	San Luis Obispo	100.2	113.0	105.9
935	Mojave	97.4	111.9	103.8
936-938	Fresno	100.5	116.5	107.5
939	Salinas	101.0	122.3	110.4
940-941	San Francisco	104.1	146.0	122.5
942,956-958	Sacramento	99.6	120.1	108.6
943	Palo Alto	98.1	133.4	113.7
944	San Mateo	100.5	134.3	115.4
945	Vallejo	99.1	128.9	112.2
946	Oakland	102.1	134.5	116.4
947	Berkeley	102.0	135.3	116.7
948	Richmond	101.3	132.2	114.9
949	San Rafael	103.2	135.4	117.4
950	Santa Cruz	105.9	122.5	113.2

STATE/ZIP	CITY	MAT.	INST.	TOTAL
CALIFORNIA (CONT'D)				
951	San Jose	103.9	136.9	118.4
952	Stockton	102.1	117.9	109.0
953	Modesto	102.0	116.6	108.4
954	Santa Rosa	102.6	136.3	117.4
955	Eureka	103.9	118.3	110.2
959	Marysville	103.1	115.9	108.8
960	Redding	112.3	117.8	114.7
961	Susanville	112.0	117.2	114.3
COLORADO				
800-802	Denver	101.6	80.4	92.3
803	Boulder	99.1	80.3	90.8
804	Golden	101.2	77.8	90.9
805	Fort Collins	102.4	77.6	91.5
806	Greeley	99.7	74.0	88.4
807	Fort Morgan	99.7	78.0	90.1
808-809	Colorado Springs	101.3	78.7	91.3
810	Pueblo	102.4	74.7	90.2
811	Alamosa	104.3	73.1	90.5
812	Salida	103.9	75.4	91.3
813	Durango	104.7	74.9	91.6
814	Montrose	103.4	76.5	91.6
815	Grand Junction	106.5	75.6	92.9
816	Glenwood Springs	104.5	76.4	92.1
CONNECTICUT				
060	New Britain	98.3	122.1	108.8
061	Hartford	99.6	122.2	109.6
062	Willimantic	98.9	122.3	109.2
063	New London	95.6	122.4	107.4
064	Meriden	97.5	122.2	108.4
065	New Haven	99.8	122.3	109.7
066	Bridgeport	99.5	122.0	109.4
067	Waterbury	99.0	121.9	109.1
068	Norwalk	99.0	129.7	112.5
069	Stamford	99.1	129.7	112.6
D.C.				
200-205	Washington	100.6	95.9	98.5
DELAWARE				
197	Newark	98.9	109.4	103.5
198	Wilmington	98.7	109.4	103.4
199	Dover	99.0	109.4	103.6
FLORIDA				
320,322	Jacksonville	97.3	69.6	85.1
321	Daytona Beach	97.5	75.5	87.8
323	Tallahassee	98.9	65.0	84.0
324	Panama City	98.8	62.0	82.6
325	Pensacola	101.4	71.0	88.0
326,344	Gainesville	99.0	68.8	85.7
327-328,347	Orlando	99.5	72.0	87.4
329	Melbourne	100.1	76.6	89.8
330-332,340	Miami	98.9	72.3	87.2
333	Fort Lauderdale	97.8	71.2	86.1
334,349	West Palm Beach	96.9	69.9	85.0
335-336,346	Tampa	99.3	72.8	87.6
337	St. Petersburg	101.8	70.2	87.9
338	Lakeland	98.8	72.7	87.3
339,341	Fort Myers	98.1	68.8	85.1
342	Sarasota	100.0	73.8	88.5
GEORGIA				
300-303,399	Atlanta	98.1	76.5	88.6
304	Statesboro	98.2	61.0	81.8
305	Gainesville	96.9	63.6	82.2
306	Athens	96.3	65.2	82.6
307	Dalton	98.0	63.0	82.6
308-309	Augusta	96.8	69.9	85.0
310-312	Macon	95.8	66.2	82.7
313-314	Savannah	97.8	67.6	84.5
315	Waycross	97.1	64.4	82.7
316	Valdosta	97.0	62.3	81.7
317,398	Albany	96.9	63.1	82.0
318-319	Columbus	96.8	66.1	83.3

STATE/ZIP	CITY	MAT.	INST.	TOTAL
HAWAII				
967	Hilo	120.2	119.7	120.0
968	Honolulu	125.0	119.7	122.7
STATES & POSS.				
969	Guam	143.0	57.6	105.4
IDAHO				
832	Pocatello	102.5	77.2	91.3
833	Twin Falls	103.8	66.3	87.3
834	Idaho Falls	101.1	74.3	89.3
835	Lewiston	109.4	85.0	98.6
836-837	Boise	101.5	77.8	91.1
838	Coeur d'Alene	109.3	82.2	97.4
ILLINOIS				
600-603	North Suburban	99.4	136.8	115.9
604	Joliet	99.4	139.0	116.8
605	South Suburban	99.4	137.0	115.9
606-608	Chicago	99.9	140.4	117.8
609	Kankakee	96.0	131.6	111.7
610-611	Rockford	99.3	123.5	109.9
612	Rock Island	97.4	101.6	99.2
613	La Salle	98.6	123.4	109.5
614	Galesburg	98.4	107.5	102.4
615-616	Peoria	100.3	109.8	104.5
617	Bloomington	97.7	110.0	103.1
618-619	Champaign	101.2	108.4	104.4
620-622	East St. Louis	95.9	108.4	101.4
623	Quincy	98.1	101.9	99.8
624	Effingham	97.3	106.6	101.4
625	Decatur	98.8	105.8	101.9
626-627	Springfield	99.3	106.9	102.6
628	Centralia	94.9	108.3	100.8
629	Carbondale	94.6	106.7	99.9
INDIANA				
460	Anderson	97.1	83.6	91.1
461-462	Indianapolis	99.0	85.2	93.0
463-464	Gary	98.4	110.0	103.5
465-466	South Bend	97.8	84.1	91.8
467-468	Fort Wayne	97.6	78.6	89.3
469	Kokomo	95.3	81.5	89.2
470	Lawrenceburg	93.8	78.2	86.9
471	New Albany	94.9	74.1	85.7
472	Columbus	97.1	81.3	90.1
473	Muncie	97.3	83.4	91.2
474	Bloomington	98.8	81.2	91.0
475	Washington	95.7	82.6	89.9
476-477	Evansville	96.6	86.6	92.2
478	Terre Haute	97.4	86.6	92.6
479	Lafayette	96.8	81.9	90.2
IOWA				
500-503,509	Des Moines	98.4	83.9	92.0
504	Mason City	97.0	64.4	82.6
505	Fort Dodge	97.2	57.6	79.8
506-507	Waterloo	98.4	73.7	87.5
508	Creston	97.5	78.3	89.1
510-511	Sioux City	99.6	72.3	87.6
512	Sibley	98.7	56.0	79.9
513	Spencer	100.5	56.0	80.9
514	Carroll	97.4	71.0	85.8
515	Council Bluffs	100.6	79.0	91.1
516	Shenandoah	97.8	73.5	87.1
520	Dubuque	99.0	80.6	90.9
521	Decorah	98.5	66.6	84.4
522-524	Cedar Rapids	99.9	82.6	92.3
525	Ottumwa	98.3	74.7	87.9
526	Burlington	97.5	76.9	88.4
527-528	Davenport	99.0	93.6	96.6
KANSAS				
660-662	Kansas City	99.6	95.9	98.0
664-666	Topeka	99.8	69.8	86.6
667	Fort Scott	98.3	73.7	87.5
668	Emporia	98.3	70.1	85.9
669	Belleville	100.1	66.5	85.3
670-672	Wichita	101.0	66.1	85.6
673	Independence	101.6	70.2	87.8
674	Salina	101.8	68.9	87.3
675	Hutchinson	97.1	65.2	83.1
676	Hays	101.2	66.9	86.1
677	Colby	101.8	68.0	86.9

STATE/ZIP	CITY	MAT.	INST.	TOTAL
KANSAS (CONT'D)				
678	Dodge City	103.0	66.1	86.8
679	Liberal	101.1	65.6	85.5
KENTUCKY				
400-402	Louisville	95.1	84.3	90.4
403-405	Lexington	95.5	82.2	89.7
406	Frankfort	96.1	80.2	89.1
407-409	Corbin	93.3	71.9	83.9
410	Covington	95.8	91.3	93.8
411-412	Ashland	94.5	99.8	96.8
413-414	Campton	95.6	77.6	87.7
415-416	Pikeville	96.8	89.3	93.5
417-418	Hazard	95.1	78.6	87.8
420	Paducah	93.7	85.5	90.1
421-422	Bowling Green	95.7	84.1	90.6
423	Owensboro	95.8	88.0	92.4
424	Henderson	93.5	86.3	90.3
425-426	Somerset	93.0	81.7	88.0
427	Elizabethtown	92.5	82.2	87.9
LOUISIANA				
700-701	New Orleans	99.2	70.7	86.6
703	Thibodaux	97.6	65.6	83.5
704	Hammond	95.2	61.0	80.1
705	Lafayette	97.0	66.6	83.6
706	Lake Charles	97.2	69.9	85.2
707-708	Baton Rouge	99.5	69.8	86.4
710-711	Shreveport	100.9	60.7	83.2
712	Monroe	99.6	59.3	81.8
713-714	Alexandria	99.7	63.3	83.7
MAINE				
039	Kittery	94.5	87.4	91.4
040-041	Portland	101.6	89.4	96.3
042	Lewiston	98.5	89.4	94.5
043	Augusta	101.2	81.0	92.3
044	Bangor	97.9	88.8	93.9
045	Bath	96.9	83.3	90.9
046	Machias	96.3	86.9	92.2
047	Houlton	96.5	86.9	92.3
048	Rockland	95.6	86.9	91.7
049	Waterville	96.8	81.0	89.8
MARYLAND				
206	Waldorf	97.1	82.6	90.7
207-208	College Park	97.1	83.8	91.2
209	Silver Spring	96.4	84.0	90.9
210-212	Baltimore	100.9	83.4	93.2
214	Annapolis	101.0	82.7	92.9
215	Cumberland	97.3	84.1	91.5
216	Easton	98.9	71.2	86.7
217	Hagerstown	97.7	85.2	92.2
218	Salisbury	99.4	61.8	82.8
219	Elkton	96.3	84.2	91.0
MASSACHUSETTS				
010-011	Springfield	99.1	110.3	104.0
012	Pittsfield	98.5	109.7	103.4
013	Greenfield	96.9	110.0	102.6
014	Fitchburg	95.6	124.5	108.3
015-016	Worcester	99.0	124.5	110.2
017	Framingham	94.9	132.9	111.6
018	Lowell	98.4	132.3	113.3
019	Lawrence	99.2	132.6	113.9
020-022, 024	Boston	100.6	139.1	117.5
023	Brockton	98.8	125.1	110.4
025	Buzzards Bay	93.7	125.9	107.9
026	Hyannis	95.9	125.9	109.1
027	New Bedford	97.9	126.3	110.4
MICHIGAN				
480,483	Royal Oak	95.1	106.1	99.9
481	Ann Arbor	97.0	107.4	101.6
482	Detroit	98.2	108.2	102.6
484-485	Flint	96.7	95.6	96.2
486	Saginaw	96.3	90.8	93.9
487	Bay City	96.4	90.3	93.7
488-489	Lansing	98.7	94.4	96.8
490	Battle Creek	97.5	86.0	92.4
491	Kalamazoo	97.8	83.9	91.7
492	Jackson	96.0	94.2	95.2
493,495	Grand Rapids	98.8	83.9	92.3
494	Muskegon	96.3	83.2	90.5

STATE/ZIP	CITY	MAT.	INST.	TOTAL
MICHIGAN (CONT'D)				
496	Traverse City	95.6	80.9	89.1
497	Gaylord	96.7	83.6	90.9
498-499	Iron Mountain	98.7	85.8	93.0
MINNESOTA				
550-551	Saint Paul	100.7	120.9	109.6
553-555	Minneapolis	100.7	119.9	109.1
556-558	Duluth	101.3	110.5	105.3
559	Rochester	100.3	106.3	103.0
560	Mankato	97.5	102.7	99.8
561	Windom	96.2	93.5	95.0
562	Willmar	95.7	98.8	97.1
563	St. Cloud	96.8	114.0	104.4
564	Brainerd	97.3	100.5	98.7
565	Detroit Lakes	99.4	96.1	98.0
566	Bemidji	98.6	99.1	98.8
567	Thief River Falls	98.2	93.3	96.0
MISSISSIPPI				
386	Clarksdale	96.6	55.3	78.4
387	Greenville	100.3	65.1	84.8
388	Tupelo	98.2	55.6	79.4
389	Greenwood	97.9	52.0	77.7
390-392	Jackson	99.8	68.4	86.0
393	Meridian	97.1	67.9	84.2
394	Laurel	98.8	55.4	79.7
395	Biloxi	98.8	60.5	81.9
396	McComb	97.1	51.5	77.0
397	Columbus	98.6	53.4	78.7
MISSOURI				
630-631	St. Louis	100.2	104.3	102.0
633	Bowling Green	99.1	96.0	97.7
634	Hannibal	97.9	92.5	95.6
635	Kirksville	101.5	89.7	96.3
636	Flat River	100.0	95.1	97.9
637	Cape Girardeau	99.6	92.8	96.6
638	Sikeston	98.4	90.5	94.9
639	Poplar Bluff	97.9	89.7	94.3
640-641	Kansas City	100.2	104.8	102.2
644-645	St. Joseph	99.4	96.7	98.2
646	Chillicothe	96.8	98.6	97.6
647	Harrisonville	96.3	101.0	98.4
648	Joplin	98.2	81.0	90.7
650-651	Jefferson City	99.6	98.4	99.1
652	Columbia	100.3	99.7	100.0
653	Sedalia	100.7	93.9	97.7
654-655	Rolla	98.3	96.7	97.6
656-658	Springfield	101.8	83.3	93.7
MONTANA				
590-591	Billings	103.7	78.8	92.7
592	Wolf Point	103.8	74.2	90.8
593	Miles City	101.6	74.0	89.4
594	Great Falls	105.2	76.8	92.7
595	Havre	102.7	73.0	89.6
596	Helena	102.8	73.6	90.0
597	Butte	103.7	77.1	92.0
598	Missoula	101.1	75.7	89.9
599	Kalispell	100.4	75.5	89.5
NEBRASKA				
680-681	Omaha	99.1	79.5	90.5
683-685	Lincoln	99.8	76.9	89.7
686	Columbus	98.1	77.5	89.0
687	Norfolk	99.6	76.0	89.2
688	Grand Island	99.3	76.6	89.3
689	Hastings	99.1	78.9	90.2
690	McCook	99.0	70.5	86.4
691	North Platte	98.9	75.9	88.7
692	Valentine	101.3	71.8	88.3
693	Alliance	101.2	70.5	87.7
NEVADA				
889-891	Las Vegas	103.6	109.4	106.1
893	Ely	102.4	100.6	101.6
894-895	Reno	102.2	86.5	95.3
897	Carson City	101.8	86.7	95.1
898	Elko	101.1	81.4	92.4
NEW HAMPSHIRE				
030	Nashua	99.3	93.6	96.8
031	Manchester	99.9	93.6	97.1

STATE/ZIP	CITY	MAT.	INST.	TOTAL
NEW HAMPSHIRE (CONT'D)				
032-033	Concord	100.0	93.8	97.3
034	Keene	96.2	75.6	87.1
035	Littleton	96.3	77.6	88.0
036	Charleston	95.7	70.3	84.5
037	Claremont	94.9	70.4	84.1
038	Portsmouth	96.5	93.7	95.3
NEW JERSEY				
070-071	Newark	100.3	129.3	113.1
072	Elizabeth	97.8	128.8	111.5
073	Jersey City	96.7	128.9	110.9
074-075	Paterson	98.5	129.4	112.1
076	Hackensack	96.5	129.0	110.8
077	Long Branch	96.0	124.7	108.6
078	Dover	96.7	129.2	111.0
079	Summit	96.7	128.8	110.9
080,083	Vineland	96.4	123.6	108.4
081	Camden	98.4	124.7	110.0
082,084	Atlantic City	97.2	125.3	109.6
085-086	Trenton	100.2	124.9	111.1
087	Point Pleasant	98.2	124.0	109.6
088-089	New Brunswick	98.8	128.7	112.0
NEW MEXICO				
870-872	Albuquerque	100.3	72.1	87.9
873	Gallup	100.7	72.1	88.1
874	Farmington	101.0	72.1	88.2
875	Santa Fe	101.1	72.1	88.3
877	Las Vegas	99.0	72.1	87.1
878	Socorro	98.6	72.1	86.9
879	Truth/Consequences	98.3	68.8	85.3
880	Las Cruces	97.5	68.1	84.6
881	Clovis	99.9	72.0	87.6
882	Roswell	101.4	72.1	88.5
883	Carrizozo	102.1	72.1	88.9
884	Tucumcari	100.6	72.0	88.0
NEW YORK				
100-102	New York	100.7	169.8	131.1
103	Staten Island	96.8	167.5	128.0
104	Bronx	95.2	166.0	126.4
105	Mount Vernon	95.5	143.1	116.5
106	White Plains	95.2	143.1	116.3
107	Yonkers	99.4	146.3	120.1
108	New Rochelle	95.9	143.0	116.7
109	Suffern	95.5	131.9	111.5
110	Queens	100.3	166.2	129.3
111	Long Island City	101.9	166.2	130.2
112	Brooklyn	102.3	167.9	131.2
113	Flushing	102.4	166.2	130.5
114	Jamaica	100.7	166.2	129.6
115,117,118	Hicksville	100.3	152.7	123.4
116	Far Rockaway	102.5	166.2	130.6
119	Riverhead	101.1	150.9	123.0
120-122	Albany	98.3	104.6	101.1
123	Schenectady	98.1	104.5	101.0
124	Kingston	101.8	117.1	108.6
125-126	Poughkeepsie	101.0	135.2	116.1
127	Monticello	100.3	115.3	106.9
128	Glens Falls	93.3	101.1	96.7
129	Plattsburgh	98.4	96.7	97.7
130-132	Syracuse	98.5	98.2	98.3
133-135	Utica	96.5	97.5	97.0
136	Watertown	98.0	96.7	97.4
137-139	Binghamton	98.2	98.4	98.3
140-142	Buffalo	100.3	106.4	103.0
143	Niagara Falls	98.0	107.2	102.1
144-146	Rochester	100.8	100.0	100.5
147	Jamestown	97.0	93.0	95.3
148-149	Elmira	96.9	98.3	97.6
NORTH CAROLINA				
270,272-274	Greensboro	99.2	61.7	82.7
271	Winston-Salem	98.9	62.1	82.7
275-276	Raleigh	97.7	59.7	81.0
277	Durham	100.7	61.7	83.5
278	Rocky Mount	96.8	61.8	81.4
279	Elizabeth City	97.8	54.5	78.7
280	Gastonia	99.8	62.4	83.3
281-282	Charlotte	99.1	63.8	83.6
283	Fayetteville	102.2	60.6	83.9
284	Wilmington	98.5	61.5	82.2
285	Kinston	97.0	63.0	82.0

STATE/ZIP	CITY	MAT.	INST.	TOTAL
NORTH CAROLINA (CONT'D)				
286	Hickory	97.3	61.8	81.7
287-288	Asheville	99.0	62.4	82.9
289	Murphy	98.2	51.0	77.4
NORTH DAKOTA				
580-581	Fargo	101.1	75.0	89.6
582	Grand Forks	101.9	65.1	85.7
583	Devils Lake	101.7	64.8	85.5
584	Jamestown	101.9	64.7	85.5
585	Bismarck	102.1	70.0	87.9
586	Dickinson	102.6	65.9	86.4
587	Minot	101.9	76.9	90.9
588	Williston	101.0	66.2	85.7
OHIO				
430-432	Columbus	98.2	87.4	93.5
433	Marion	95.1	84.6	90.5
434-436	Toledo	98.3	97.6	98.0
437-438	Zanesville	95.6	86.3	91.5
439	Steubenville	96.7	93.9	95.5
440	Lorain	98.9	92.2	96.0
441	Cleveland	99.2	99.4	99.3
442-443	Akron	99.9	93.9	97.3
444-445	Youngstown	99.2	88.1	94.4
446-447	Canton	99.3	89.3	94.9
448-449	Mansfield	97.1	86.2	92.3
450	Hamilton	97.8	84.4	91.9
451-452	Cincinnati	98.2	83.7	91.8
453-454	Dayton	97.9	83.9	91.7
455	Springfield	97.8	84.1	91.8
456	Chillicothe	97.2	92.0	94.9
457	Athens	100.2	83.1	92.6
458	Lima	100.5	86.3	94.3
OKLAHOMA				
730-731	Oklahoma City	99.8	71.3	87.3
734	Ardmore	98.6	64.9	83.7
735	Lawton	100.7	67.5	86.1
736	Clinton	99.8	66.2	85.0
737	Enid	100.2	65.2	84.8
738	Woodward	98.6	68.4	85.3
739	Guymon	99.7	69.1	86.2
740-741	Tulsa	98.3	64.5	83.4
743	Miami	95.2	72.1	85.0
744	Muskogee	97.5	62.9	82.3
745	McAlester	94.8	65.3	81.8
746	Ponca City	95.5	67.6	83.2
747	Durant	95.5	65.9	82.5
748	Shawnee	97.0	65.3	83.0
749	Poteau	94.7	66.4	82.2
OREGON				
970-972	Portland	99.7	101.1	100.3
973	Salem	100.9	99.2	100.2
974	Eugene	99.5	99.2	99.3
975	Medford	101.0	96.9	99.2
976	Klamath Falls	101.3	97.1	99.4
977	Bend	100.3	99.2	99.8
978	Pendleton	96.2	101.4	98.5
979	Vale	94.1	89.1	91.9
PENNSYLVANIA				
150-152	Pittsburgh	100.2	104.6	102.1
153	Washington	97.5	103.6	100.2
154	Uniontown	97.7	102.8	99.9
155	Bedford	98.7	95.3	97.2
156	Greensburg	98.7	102.8	100.5
157	Indiana	97.6	101.4	99.2
158	Dubois	99.1	97.5	98.4
159	Johnstown	98.7	98.1	98.5
160	Butler	92.3	103.2	97.1
161	New Castle	92.4	100.3	95.9
162	Kittanning	92.8	103.1	97.3
163	Oil City	92.3	99.7	95.6
164-165	Erie	94.0	97.5	95.5
166	Altoona	94.1	93.5	93.9
167	Bradford	95.6	97.4	96.4
168	State College	95.1	93.0	94.2
169	Wellsboro	96.2	96.6	96.3
170-171	Harrisburg	98.5	96.5	97.6
172	Chambersburg	95.8	91.2	93.8
173-174	York	95.9	96.8	96.3
175-176	Lancaster	94.6	98.3	96.2

STATE/ZIP	CITY	MAT.	INST.	TOTAL
PENNSYLVANIA (CONT'D)				
177	Williamsport	93.2	96.2	94.5
178	Sunbury	95.3	97.3	96.2
179	Pottsville	94.4	99.1	96.5
180	Lehigh Valley	95.7	112.3	103.0
181	Allentown	97.3	110.6	103.2
182	Hazleton	95.2	100.3	97.5
183	Stroudsburg	95.1	105.0	99.4
184-185	Scranton	98.1	99.8	98.8
186-187	Wilkes-Barre	94.9	101.0	97.6
188	Montrose	94.6	98.7	96.4
189	Doylestown	94.8	127.4	109.1
190-191	Philadelphia	99.5	134.7	115.0
193	Westchester	96.3	126.3	109.5
194	Norristown	95.4	131.2	111.2
195-196	Reading	97.1	104.5	100.4
PUERTO RICO				
009	San Juan	122.0	24.5	79.0
RHODE ISLAND				
028	Newport	97.2	116.8	105.8
029	Providence	99.8	116.8	107.3
SOUTH CAROLINA				
290-292	Columbia	98.2	68.3	85.0
293	Spartanburg	98.3	67.3	84.6
294	Charleston	99.8	66.1	84.9
295	Florence	98.0	67.7	84.7
296	Greenville	98.1	66.0	83.9
297	Rock Hill	98.0	56.5	79.7
298	Aiken	98.9	66.8	84.7
299	Beaufort	99.6	52.5	78.8
SOUTH DAKOTA				
570-571	Sioux Falls	99.0	60.5	82.0
572	Watertown	99.9	49.9	77.9
573	Mitchell	98.6	48.1	76.4
574	Aberdeen	101.0	53.7	80.1
575	Pierre	102.2	59.3	83.3
576	Mobridge	99.4	48.7	77.0
577	Rapid City	100.8	59.6	82.7
TENNESSEE				
370-372	Nashville	98.2	74.1	87.6
373-374	Chattanooga	99.9	66.6	85.2
375,380-381	Memphis	98.9	72.5	87.3
376	Johnson City	99.9	56.4	80.7
377-379	Knoxville	96.5	66.8	83.5
382	McKenzie	97.7	59.4	80.8
383	Jackson	99.0	63.5	83.4
384	Columbia	96.2	67.2	83.4
385	Cookeville	97.5	60.0	81.0
TEXAS				
750	McKinney	98.2	66.9	84.4
751	Waxahackie	98.3	65.4	83.8
752-753	Dallas	99.2	67.9	85.4
754	Greenville	98.4	67.0	84.6
755	Texarkana	97.6	63.6	82.6
756	Longview	98.4	64.3	83.4
757	Tyler	98.4	65.2	83.8
758	Palestine	94.8	64.4	81.4
759	Lufkin	95.6	67.7	83.3
760-761	Fort Worth	98.8	65.7	84.2
762	Denton	99.2	66.5	84.8
763	Wichita Falls	96.5	64.6	82.5
764	Eastland	95.6	64.1	81.8
765	Temple	94.1	61.6	79.8
766-767	Waco	95.9	63.7	81.7
768	Brownwood	99.1	60.6	82.1
769	San Angelo	98.7	59.9	81.6
770-772	Houston	100.7	69.8	87.1
773	Huntsville	99.7	66.9	85.3
774	Wharton	100.6	68.7	86.5
775	Galveston	98.7	68.5	85.4
776-777	Beaumont	98.9	68.9	85.7
778	Bryan	96.4	68.4	84.1
779	Victoria	100.7	66.1	85.5
780	Laredo	99.4	63.5	83.6
781-782	San Antonio	100.3	64.1	84.4
783-784	Corpus Christi	102.0	63.2	84.9
785	McAllen	102.9	59.9	83.9
786-787	Austin	100.0	63.4	83.9

STATE/ZIP	CITY	MAT.	INST.	TOTAL
TEXAS (CONT'D)				
788	Del Rio	102.5	63.0	85.1
789	Giddings	99.3	63.8	83.7
790-791	Amarillo	99.6	62.3	83.1
792	Childress	99.7	64.0	84.0
793-794	Lubbock	101.5	63.6	84.8
795-796	Abilene	100.0	62.3	83.4
797	Midland	102.1	64.6	85.6
798-799,885	El Paso	98.3	61.6	82.2
UTAH				
840-841	Salt Lake City	103.8	71.2	89.4
842,844	Ogden	99.3	71.2	86.9
843	Logan	101.3	71.2	88.0
845	Price	102.0	65.3	85.8
846-847	Provo	101.8	71.1	88.2
VERMONT				
050	White River Jct.	98.5	77.3	89.2
051	Bellows Falls	96.9	91.7	94.6
052	Bennington	97.3	92.3	95.1
053	Brattleboro	97.7	91.6	95.0
054	Burlington	102.2	82.5	93.5
056	Montpelier	100.6	81.8	92.3
057	Rutland	99.0	82.5	91.7
058	St. Johnsbury	98.7	76.8	89.1
059	Guildhall	97.2	76.8	88.2
VIRGINIA				
220-221	Fairfax	100.7	89.2	95.6
222	Arlington	101.5	89.4	96.2
223	Alexandria	100.6	89.4	95.6
224-225	Fredericksburg	99.4	87.7	94.2
226	Winchester	100.0	85.2	93.5
227	Culpeper	99.9	85.5	93.6
228	Harrisonburg	100.2	66.9	85.5
229	Charlottesville	100.5	67.9	86.1
230-232	Richmond	99.8	68.7	86.1
233-235	Norfolk	100.1	68.3	86.1
236	Newport News	99.8	68.3	85.9
237	Portsmouth	99.3	65.4	84.4
238	Petersburg	99.8	68.7	86.1
239	Farmville	99.0	52.1	78.3
240-241	Roanoke	101.0	66.6	85.8
242	Bristol	99.4	54.8	79.7
243	Pulaski	99.0	57.5	80.7
244	Staunton	99.7	61.7	83.0
245	Lynchburg	99.9	66.2	85.1
246	Grundy	99.2	52.6	78.7
WASHINGTON				
980-981,987	Seattle	102.3	103.8	103.0
982	Everett	102.2	99.0	100.8
983-984	Tacoma	102.6	99.6	101.3
985	Olympia	101.0	99.7	100.4
986	Vancouver	103.4	94.4	99.4
988	Wenatchee	102.8	85.9	95.3
989	Yakima	102.8	94.9	99.3
990-992	Spokane	104.8	82.6	95.0
993	Richland	104.4	90.7	98.4
994	Clarkston	103.8	81.9	94.1
WEST VIRGINIA				
247-248	Bluefield	98.2	93.6	96.2
249	Lewisburg	99.8	92.8	96.7
250-253	Charleston	99.7	97.4	98.7
254	Martinsburg	100.4	87.0	94.5
255-257	Huntington	101.4	97.8	99.8
258-259	Beckley	98.8	95.1	97.2
260	Wheeling	100.8	94.5	98.0
261	Parkersburg	99.6	92.5	96.5
262	Buckhannon	99.6	94.7	97.4
263-264	Clarksburg	100.2	94.7	97.7
265	Morgantown	100.2	94.8	97.8
266	Gassaway	99.5	94.4	97.3
267	Romney	99.4	89.8	95.1
268	Petersburg	99.3	87.7	94.2
WISCONSIN				
530,532	Milwaukee	98.9	106.9	102.4
531	Kenosha	99.2	101.7	100.3
534	Racine	98.7	102.0	100.2
535	Beloit	98.6	99.8	99.1
537	Madison	99.3	99.1	99.2

STATE/ZIP	CITY	MAT.	INST.	TOTAL
WISCONSIN (CONT'D)				
538	Lancaster	96.8	93.7	95.4
539	Portage	95.3	96.4	95.8
540	New Richmond	97.1	93.8	95.6
541-543	Green Bay	101.2	97.5	99.5
544	Wausau	96.5	96.1	96.3
545	Rhinelander	99.8	91.8	96.3
546	La Crosse	97.6	94.8	96.4
547	Eau Claire	99.2	95.5	97.6
548	Superior	96.8	97.7	97.2
549	Oshkosh	97.2	91.1	94.5
WYOMING				
820	Cheyenne	102.3	66.5	86.5
821	Yellowstone Nat'l Park	100.2	71.5	87.6
822	Wheatland	101.4	65.5	85.6
823	Rawlins	103.1	71.2	89.0
824	Worland	101.0	67.4	86.2
825	Riverton	102.1	66.4	86.4
826	Casper	102.5	63.8	85.5
827	Newcastle	100.8	72.4	88.3
828	Sheridan	103.7	66.5	87.3
829-831	Rock Springs	104.9	68.4	88.8
CANADIAN FACTORS (reflect Canadian currency)				
ALBERTA				
	Calgary	123.8	93.4	110.4
	Edmonton	123.9	93.4	110.5
	Fort McMurray	124.5	91.7	110.0
	Lethbridge	119.1	91.2	106.8
	Lloydminster	113.9	87.9	102.5
	Medicine Hat	114.1	87.2	102.2
	Red Deer	114.6	87.2	102.5
BRITISH COLUMBIA				
	Kamloops	115.0	89.3	103.7
	Prince George	116.0	88.6	104.0
	Vancouver	117.7	89.2	105.1
	Victoria	116.6	86.7	103.4
MANITOBA				
	Brandon	126.4	74.8	103.7
	Portage la Prairie	113.9	73.1	95.9
	Winnipeg	122.6	70.8	99.8
NEW BRUNSWICK				
	Bathurst	111.8	66.3	91.8
	Dalhousie	112.6	66.5	92.3
	Fredericton	118.8	71.5	97.9
	Moncton	112.0	70.7	93.8
	Newcastle	111.9	67.0	92.1
	St. John	114.4	77.8	98.3
NEWFOUNDLAND				
	Corner Brook	131.2	66.6	102.7
	St. John's	123.9	83.9	106.3
NORTHWEST TERRITORIES				
	Yellowknife	136.7	82.9	113.0
NOVA SCOTIA				
	Bridgewater	113.4	73.6	95.9
	Dartmouth	127.5	73.6	103.7
	Halifax	118.2	79.9	101.3
	New Glasgow	125.5	73.6	102.6
	Sydney	123.4	73.6	101.4
	Truro	113.0	73.6	95.6
	Yarmouth	125.4	73.6	102.6
ONTARIO				
	Barrie	118.3	91.7	106.6
	Brantford	115.5	95.5	106.7
	Cornwall	115.4	92.0	105.1
	Hamilton	116.9	95.6	107.5
	Kingston	116.5	92.1	105.7
	Kitchener	109.9	91.8	101.9
	London	116.3	92.2	105.7
	North Bay	128.6	89.8	111.5
	Oshawa	112.5	95.3	104.9
	Ottawa	117.6	92.9	106.7
	Owen Sound	118.4	89.9	105.8
	Peterborough	115.5	91.7	105.0
	Sarnia	115.3	96.1	106.9

STATE/ZIP	CITY	MAT.	INST.	TOTAL
ONTARIO (CONT'D)				
	Sault Ste. Marie	110.7	91.1	102.1
	St. Catharines	107.8	93.3	101.5
	Sudbury	107.6	92.1	100.8
	Thunder Bay	109.1	92.3	101.7
	Timmins	115.6	89.8	104.2
	Toronto	116.9	100.9	109.9
	Windsor	108.3	92.3	101.3
PRINCE EDWARD ISLAND				
	Charlottetown	120.4	61.9	94.6
	Summerside	128.1	62.0	99.0
QUEBEC				
	Cap-de-la-Madeleine	112.8	86.4	101.1
	Charlesbourg	112.8	86.4	101.1
	Chicoutimi	111.7	92.0	103.0
	Gatineau	112.5	86.1	100.9
	Granby	112.7	86.1	101.0
	Hull	112.6	86.1	100.9
	Joliette	113.0	86.4	101.3
	Laval	112.8	86.1	101.1
	Montreal	117.9	92.4	106.7
	Quebec City	117.9	92.8	106.8
	Rimouski	112.3	92.0	103.4
	Rouyn-Noranda	112.5	86.1	100.9
	Saint-Hyacinthe	112.0	86.1	100.6
	Sherbrooke	112.8	86.1	101.1
	Sorel	113.0	86.4	101.2
	Saint-Jerome	112.5	86.1	100.9
	Trois-Rivieres	126.1	86.4	108.6
SASKATCHEWAN				
	Moose Jaw	111.2	68.5	92.4
	Prince Albert	110.3	66.8	91.1
	Regina	123.0	89.0	108.0
	Saskatoon	112.3	89.0	102.0
YUKON				
	Whitehorse	137.1	67.5	106.4

L1010-101 Minimum Design Live Loads in Pounds per S.F. for Various Building Codes

Occupancy or Use	Uniform (psf)	Concentrated (lbs)
1. Access floor systems		
Office use	50	2000
Computer use	100	2000
2. Armories and drill rooms	100	-
3. Assembly areas		-
Fixed seats (fastened to floor)	60	
Follow spot, projections and control rooms	50	
Lobbies	100	
Movable seats	100	
Stage floors	150	
Platforms (assembly)	100	
Other assembly areas	100	
4. Balconies and decks	Same as occupancy served	-
5. Catwalks	40	300
6. Comices	60	-
7. Corridors		-
First floor	100	
Other floors	Same as occupancy served except as indicated	
8. Dining rooms and restaurants	100	-
9. Elevator machine room grating (on an area of 2 in by 2 in)	-	300
10. Finish light floorplate construction (on area of 1 in by 1 in)	-	200
11. Fire escapes	100	-
On single-family dwellings only		
12. Garages (passenger vehicles only)	40	Note a
13. Hospitals		
Corridors above first floor	80	1000
Operating rooms, laboratories	60	1000
Patient rooms	40	1000
14. Libraries		
Corridors above first floor	80	1000
Reading rooms	60	1000
Stacks	150, Note b	1000
15. Manufacturing		
Heavy	250	3000
Light	125	2000
16. Marquees	75	-
17. Office buildings		
Corridors above first floor	80	2000
Lobbies and first-floor corridors	100	2000
Offices	50	2000
18. Penal institutions		
Cell blocks	40	
Corridors	100	
19. Recreational uses:		
Bowling alleys, poolrooms, and similar uses	75	
Dance halls and ballrooms	100	
Gymnasiums	100	
Reviewing stands, grandstands and bleachers	100	
Stadiums and arenas with fixed seats (fastened to floor)	60	
20. Residential		
One- and two- family dwellings		
Uninhabitable attics without storage, Note c	10	
Uninhabitable attics with storage, Note c, d, e	20	
Habitable attics and sleeping areas, Note e	30	
All other areas	0	
Hotels and multifamily dwellings		
Private rooms and corridors serving them	40	
Public rooms and corridors serving them	100	

L1010-101 Minimum Design Live Loads in Pounds per S.F. for Various Building Codes (cont.)

Occupancy or Use	Uniform (psf)	Concentrated (lbs)
21. Roofs		
All roof surfaces subject to maintenance workers		300
Awnings and canopies:		
Fabric construction supported by a skeleton structure	5	-
All other construction	20	
Ordinary flat, pitched and curved roofs (not occupiable)	20	
Where primary roof members are exposed to a work floor, at single panel point of lower cord of roof trusses or any point along primary structural members supporting roofs:		
Over manufacturing, storage warehouses, and repair garages		2000
All other primary roof members		300
Occupiable roofs:		
Roof gardens	100	
Assembly areas	100	
All other similar areas	Note f	Note f
22. Schools		
Classrooms	40	1000
Corridors	80	1000
First-floor corridors	100	1000
23. Scuttles, skylight ribs and accessible ceilings	-	200
24. Sidewalks, vehicular driveways and yards, subject to trucking	250	8000, Note g
25. Stairs and exits		
One- and two- family dwellings	40	300, Note h
All others	100	3000, Note h
26. Storage warehouses		
Heavy	250	-
Light	125	
27. Stores		
Retail		
First floor	100	1000
Upper floors	75	1000
Wholesale, all floors	125	1000
28. Walkways and elevated platforms (other than exitways)	60	-
29. Yards and terraces, pedestrians	100	-

Notes:

a. Floors in garages or portions of buildings used for the storage of motor vehicles shall be designed for the uniformly distributed live loads of Table 1607.1 or the following concentrated loads: (1) for garages restricted to passenger vehicles accommodating not more than nine passengers, 3,000 pounds acting on an area of 4.5 in. by 4.5 in.; (2) for mechanical parking structures without slab or deck that are used for storing passenger vehicles only, 2,250 pounds per wheel.

b. The loading applies to stack room floors that support nonmobile, double-faced library books stacks, subject to the following limitations:
1. The nominal bookstack unit height shall not exceed 90 inches;
2. The nominal shelf depth shall not exceed 12 inches for each face; and
3. Parallel rows of double-faced book stacks shall be separated by aisles not less than 36 inches wide.

c. Uninhabitable attics without storage are those where the maximum clear height between the joists and rafters is less than 42 inches, or where there are not two or more adjacent trusses with web configurations capable of accommodating an assumed rectangle 42 inches in height by 24 inches in width, or greater, within the plane of the trusses. This live load need not be assumed to act concurrently with any other live load requirements.

d. Uninhabitable attics with storage are those where the maximum clear height between the joists and rafters is 42 inches or greater, or where there are two or more adjacent trusses with web configurations capable of accommodating an assumed rectangle 42 inches in height by 24 inches in width, or greater, within the plane of the trusses.

The live load need only be applied to those portions of the joists or truss bottom chords where both of the following conditions are met:
i. The attic area is accessible from an opening not less than 20 inches in width by 30 inches in length that is located where the clear height in the attic is a minimum of 30 inches; and
ii. The slopes of the joists or truss bottom chords are no greater than two units vertical in 12 units horizontal.

The remaining portions of the joists or truss bottom chords shall be designed for a uniformly distributed concurrent live load of not less than 10 lb/ft^2.

e. Attic spaces served by stairways other than the pull-down type shall be designed to support the minimum live load specified for habitable attics and sleeping rooms.

f. Areas of occupiable roofs, other than roof gardens and assembly areas, shall be designed for approporate loads as approved by the building official.

g. The concentrated wheel load shall be applied on an area of 4.5 inches by 4.5 inches.

h. The minimum concentrated load on stair treads shall be applied on an area of 2 in by 2 in. This load need not be assumed to act concurrently with the uniform load.

Table L1010-201 Design Weight Per S.F. for Walls and Partitions

Type	Wall Thickness	Description	Weight Per S.F.
Brick	4″	Clay brick, high absorption Clay brick, medium absorption	34 lb. 39
		Clay brick, low absorption	46
		Sand-lime brick	38
		Concrete brick, heavy aggregate Concrete brick, light aggregate	46 33
	8″	Clay brick, high absorption Clay brick, medium absorption	69 79
		Clay brick, low absorption	89
		Sand-lime brick	74
		Concrete brick, heavy aggregate Concrete brick, light aggregate	89 68
	12″	Common brick Pressed brick	120 130
		Sand-lime brick	105
		Concrete brick, heavy aggregate	130
		Concrete brick, light aggregate	98
	16″	Clay brick, high absorption	134
		Clay brick, medium absorption	155
		Clay brick, low absorption	173
		Sand-lime brick	138
		Concrete brick, heavy aggregate	174
		Concrete brick, light aggregate	130
Concrete block	4″	Solid conc. block, stone aggregate Solid conc. block, lightweight	45 34
		Hollow conc. block, stone aggregate	30
		Hollow conc. block, lightweight	20
	6″	Solid conc. block, stone aggregate Solid conc. block, lightweight	50 37
		Hollow conc. block, stone aggregate	42
		Hollow conc. block, lightweight	30
	8″	Solid conc. block, stone aggregate Solid conc. block, lightweight	67 48
		Hollow conc. block, stone aggregate	55
		Hollow conc. block, lightweight	38
	10″	Solid conc. block, stone aggregate	84
		Solid conc. block, lightweight	62
		Hollow conc. block, stone aggregate Hollow conc. block, lightweight	55 38
	12″	Solid conc. block, stone aggregate Solid conc. block, lightweight	108 72
		Hollow conc. block, stone aggregate Hollow conc. block, lightweight	85 55
Drywall	6″	Drywall on wood studs	10

Type	Wall Thickness	Description	Weight Per S.F.
Clay tile	2″	Split terra cotta furring Non load bearing clay tile	10 lb. 11
	3″	Split terra cotta furring	12
		Non load bearing clay tile	18
	4″	Non load bearing clay tile Load bearing clay tile	20 24
	6″	Non load bearing clay tile Load bearing clay tile	30 36
	8″	Non load bearing clay tile	36
		Load bearing clay tile	42
	12″	Non load bearing clay tile Load bearing clay tile	46 58
Gypsum block	2″	Hollow gypsum block Solid gypsum block	9.5 12
	3″	Hollow gypsum block	10
		Solid gypsum block	18
	4″	Hollow gypsum block	15
		Solid gypsum block	24
	5″	Hollow gypsum block	18
	6″	Hollow gypsum block	24
Structural facing tile	2″	Facing tile	15
	4″	Facing tile	25
	6″	Facing tile	38
Glass	4″	Glass block	18
	1″	Structural glass	15
Plaster	1″	Gypsum plaster (1 side)	5
		Cement plaster (1 side) Gypsum plaster on lath	10 8
		Cement plaster on lath	13
Plaster partition (2 finished faces)	2″	Solid gypsum on metal lath Solid cement on metal lath Solid gypsum on gypsum lath Gypsum on lath & metal studs	18 25 18 18
	3″	Gypsum on lath & metal studs	19
	4″	Gypsum on lath & metal studs	20
	6″	Gypsum on lath & wood studs	18
Concrete	6″	Reinf concrete, stone aggregate Reinf. concrete, lightweight	75 36-60
	8″	Reinf. concrete, stone aggregate Reinf. concrete, lightweight	100 48-80
	10″	Reinf. concrete, stone aggregate Reinf. concrete, lightweight	125 60-100
	12″	Reinf concrete stone aggregate Reinf concrete, lightweight	150 72-120

Table L1010-202 Design Weight per S.F. for Roof Coverings

Type	Description		Weight lb. Per S.F.
Sheathing	Gypsum	1″ thick	4
	Wood	¾″ thick	3
Insulation	per 1″	Loose	.5
		Poured in place Rigid	2 1.5
Built-up	Tar & gravel	3 ply felt 5 ply felt	5.5 6.5

Type	Wall Thickness	Description	Weight lb. Per S.F.
Metal	Aluminum	Corr. & ribbed, .024″ to .040″	.4-.8
	Copper	or tin	1.5-2.5
	Steel	Corrugated, 29 ga. to 12 ga.	.6-5.0
Shingles	Asphalt	Strip shingles	1.7-2.8
	Clay	Tile	8-16
	Slate	¼″ thick	9.5
	Wood		2-3

Table L1010-203 Design Weight per Square Foot for Floor Fills and Finishes

Type		Description	Weight lb. per S.F.	Type		Description	Weight lb. per S.F.
Floor fill	per 1″	Cinder fill	5	Wood	Single 7/8″	On sleepers, light concrete fill	16
		Cinder concrete	9			On sleepers, stone concrete fill	25
		Lightweight concrete	3-9		Double	On sleepers, light concrete fill	19
		Stone concrete	12		7/8″	On sleepers, stone concrete fill	28
		Sand	8		3″	Wood block on mastic, no fill	15
		Gypsum	6			Wood block on ½″ mortar	16
Terrazzo	1″	Terrazzo, 2″ stone concrete	25		per 1″	Hardwood flooring (25/32″)	4
Marble	and mortar	on stone concrete fill	33			Underlayment (Plywood per 1″)	3
Resilient	1/16″-1/4″	Linoleum, asphalt, vinyl tile	2	Asphalt	1-1/2″	Mastic flooring	18
Tile	3/4″	Ceramic or quarry	10		2″	Block on ½″ mortar	30

Table L1010-204 Design Weight Per Cubic Foot for Miscellaneous Materials

Type		Description	Weight lb. per C.F.	Type		Description	Weight lb. per C.F.
Bituminous	Coal, piled	Anthracite	47-58	Masonry	Ashlar	Granite	165
		Bituminous	40-54			Limestone, crystalline	165
		Peat, turf, dry	47			Limestone, oolitic	135
		Coke	75			Marble	173
	Petroleum	Unrefined	54			Sandstone	144
		Refined	50		Rubble, in mortar	Granite	155
		Gasoline	42			Limestone, crystalline	147
	Pitch		69			Limestone, oolitic	138
	Tar	Bituminous	75			Marble	156
Concrete	Plain	Stone aggregate	144			Sandstone & Bluestone	130
		Slag aggregate	132		Brick	Pressed	140
		Expanded slag aggregate	100			Common	120
		Haydite (burned clay agg.)	90			Soft	100
		Vermiculite & perlite, load bearing	70-105		Cement	Portland, loose	90
		Vermiculite & perlite, non load bear	35-50			Portland set	183
	Reinforced	Stone aggregate	150		Lime	Gypsum, loose	53-64
		Slag aggregate	138		Mortar	Set	103
		Lightweight aggregates	30-120	Metals	Aluminum	Cast, hammered	165
Earth	Clay	Dry	63		Brass	Cast, rolled	534
		Damp, plastic	110		Bronze	7.9 to 14% Sn	509
		and gravel, dry	100		Copper	Cast, rolled	556
	Dry	Loose	76		Iron	Cast, pig	450
		Packed	95			Wrought	480
	Moist	Loose	78		Lead		710
		Packed	96		Monel		556
	Mud	Flowing	108		Steel	Rolled	490
		Packed	115		Tin	Cast, hammered	459
	Riprap	Limestone	80-85		Zinc	Cast rolled	440
		Sandstone	90	Timber	Cedar	White or red	24.2
		Shale	105		Fir	Douglas	23.7
	Sand & gravel	Dry, loose	90-105			Eastern	25
		Dry, packed	100-120		Maple	Hard	44.5
		Wet	118-120			White	33
Gases	Air	0C., 760 mm.	.0807		Oak	Red or Black	47.3
	Gas	Natural	.0385			White	47.3
Liquids	Alcohol	100%	49		Pine	White	26
	Water	4°C., maximum density	62.5			Yellow, long leaf	44
		Ice	56			Yellow, short leaf	38
		Snow, fresh fallen	8		Redwood	California	26
		Sea water	64		Spruce	White or black	27

Table L1010-225 Design Weight Per S.F. for Structural Floor and Roof Systems

Type Slab		Description	Weight in Pounds per S.F. Slab Depth in Inches											
Concrete Slab	Reinforced	Stone aggregate	1″	12.5	2″	25	3″	37.5	4″	50	5″	62.5	6″	75
		Lightweight sand aggregate		9.5		19		28.5		38		47.5		57
		All lightweight aggregate		9.0		18		27.0		36		45.0		54
	Plain, nonreinforced	Stone aggregate		12.0		24		36.0		48		60.0		72
		Lightweight sand aggregate		9.0		18		27.0		36		45.0		54
		All lightweight aggregate	↓	8.5	↓	17	↓	25.5	↓	34	↓	42.5	↓	51
Concrete Waffle	19″ x 19″	5″ wide ribs @ 24″ O.C.	6+3	77	8+3	92	10+3	100	12+3	118				
			6+4 ½	96	8+4½	110	10+4½	119	12+4½	136				
	30″ x 30″	6″ wide ribs @ 36″ O.C.	8+3	83	10+3	95	12+3	109	14+3	118	16+3	130	20+3	155
			8+4½	101	10+4½	113	12+4½	126	14+4½	137	16+4½	149	20+4½	173
Concrete Joist	20″ wide form	5″ wide rib	8+3	60	10+3	67	12+3	74	14+3	81				
		6″ wide rib	↓	63	↓	70	↓	78	↓	86	16+3	94	20+3	111
		7″ wide rib									↓	99	↓	118
		5″ wide rib	8+4½	79	10+4½	85	12+4½	92	14+4½	99				
		6″ wide rib	↓	82	↓	89	↓	97	↓	104	16+4½	113	20+4½	130
		7″ wide rib									↓	118	↓	136
	30″ wide form	5″ wide rib	8+3	54	10+3	58	12+3	63	14+3	68				
		6″ wide rib	↓	56	↓	61	↓	67	↓	72	16+3	78	20+3	91
		7″ wide rib									↓	83	↓	96
		5″ wide rib	8+4½	72	10+4½	77	12+4½	82	14+4½	87				
		6″ wide rib	↓	75	↓	80	↓	85	↓	91	16+4½	97	20+4½	109
		7″ wide rib									↓	101	↓	115
Wood Joists	Incl. Subfloor	12″ O.C.	2x6	6	2x8	6	2x10	7	2x12	8	3x8	8	3x12	11
		16″ O.C.	↓	5	↓	6	↓	6	↓	7	↓	7	↓	9

Table L1010-226 Superimposed Dead Load Ranges

Component	Load Range (PSF)
Ceiling	5-10
Partitions	20-30
Mechanical	4-8

Table L1010-301 Design Loads for Structures for Wind Load

Wind Loads: Structures are designed to resist the wind force from any direction. Usually 2/3 is assumed to act on the windward side, 1/3 on the leeward side.

For more than 1/3 openings, add 10 psf for internal wind pressure or 5 psf for suction, whichever is critical.

For buildings and structures, use psf values from Table L1010-301.

For glass over 4 S.F. use values in Table L1010-302 after determining 30′ wind velocity from Table L1010-303.

Type Structure	Height Above Grade	Horizontal Load in Lb. per S.F.
Buildings	Up to 50 ft. 50 to 100 ft. Over 100 ft.	15 20 20 + .025 per ft.
Ground signs & towers	Up to 50 ft. Over 50 ft.	15 20
Roof structures		30
Glass	See Table below	

Table L1010-302 Design Wind Load in PSF for Glass at Various Elevations

Height From Grade	Velocity in Miles per Hour and Design Load in Pounds per S.F.																					
	Vel.	PSF	Vel.	PSF	Vel.	PSF	Vel.	PSF	Vel.	PSF	Vel.	PSF	Vel.	PSF	Vel.	PSF	Vel.	PSF	Vel.	PSF	Vel.	PSF
To 10 ft.	42	6	46	7	49	8	52	9	55	10	59	11	62	12	66	14	69	15	76	19	83	22
10-20	52	9	58	11	61	11	65	14	70	16	74	18	79	20	83	22	87	24	96	30	105	35
20-30*	60	12	67	14	70	16	75	18	80	20	85	23	90	26	95	29	100	32	110	39	120	46
30-60	66	14	74	18	77	19	83	22	88	25	94	28	99	31	104	35	110	39	121	47	132	56
60-120	73	17	82	21	85	12	92	27	98	31	104	35	110	39	116	43	122	48	134	57	146	68
120-140	81	21	91	26	95	29	101	33	108	37	115	42	122	48	128	52	135	48	149	71	162	84
240-480	90	26	100	32	104	35	112	40	119	45	127	51	134	57	142	65	149	71	164	86	179	102
480-960	98	31	110	39	115	42	123	49	131	55	139	62	148	70	156	78	164	86	180	104	197	124
Over 960	98	31	110	39	115	42	123	49	131	55	139	62	148	70	156	78	164	86	180	104	197	124

*Determine appropriate wind at 30′ elevation Fig. L1010-303 below.

Table L1010-303 Design Wind Velocity at 30 Ft. Above Ground

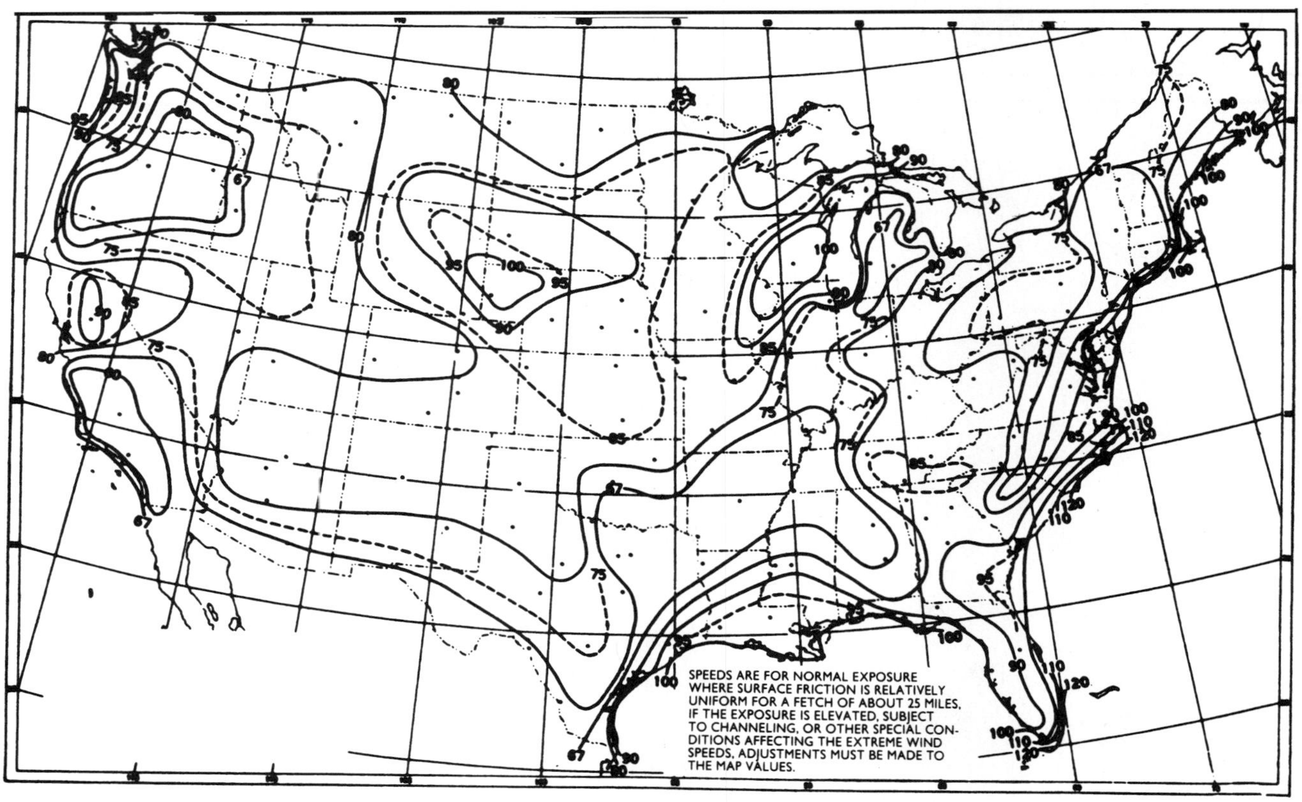

Table L1010-401 Snow Load in Pounds Per Square Foot on the Ground

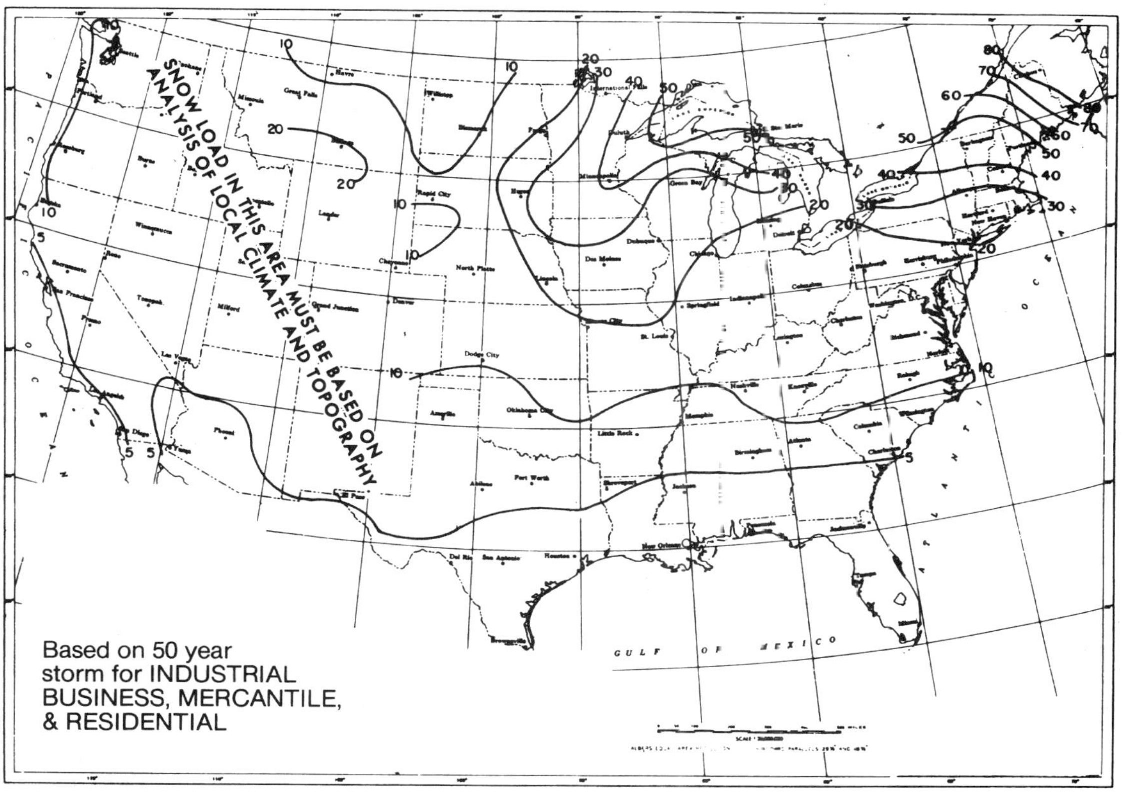

Based on 50 year storm for INDUSTRIAL BUSINESS, MERCANTILE, & RESIDENTIAL

Table L1010-402 Snow Load in Pounds Per Square Foot on the Ground

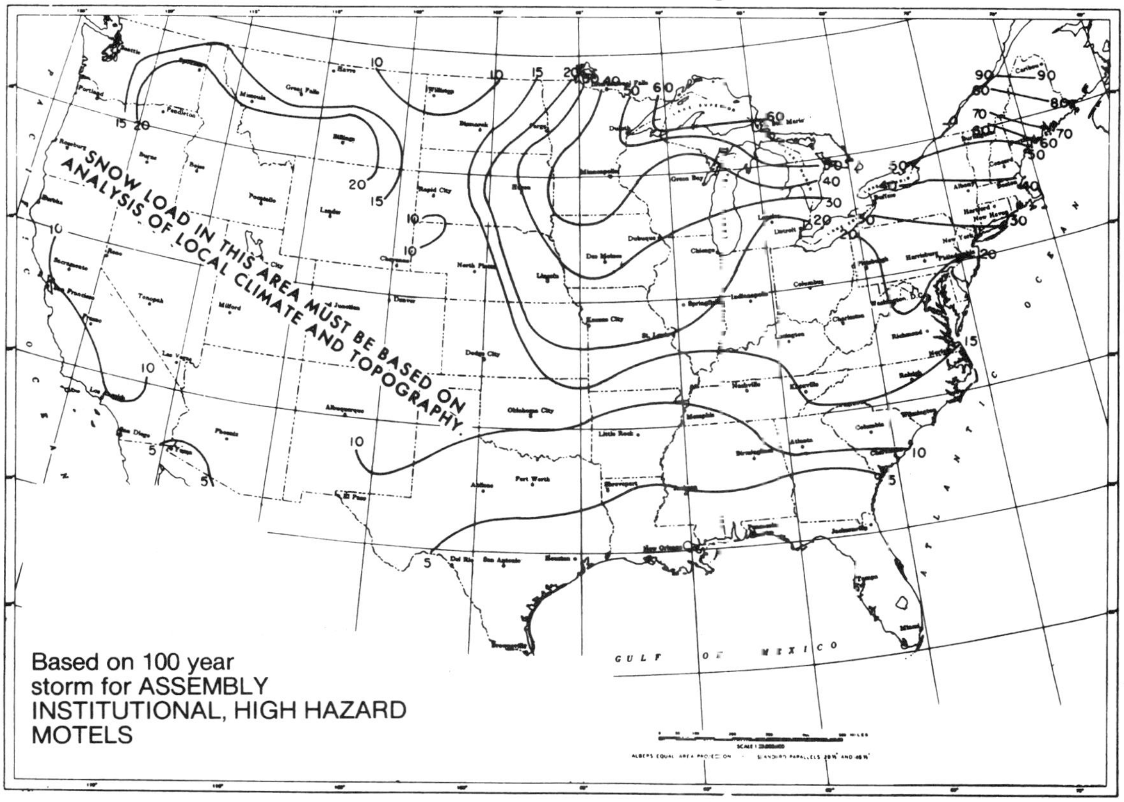

Based on 100 year storm for ASSEMBLY INSTITUTIONAL, HIGH HAZARD MOTELS

Table L1010-403 Snow Loads

To convert the ground snow loads on the previous page to roof snow loads, the ground snow loads should be multiplied by the following factors depending upon the roof characteristics.

Note, in all cases $\frac{\alpha - 30}{50}$ is valid only for > 30 degrees.

Description	Sketch	Formula	Angle	Conversion Factor = C_f Sheltered	Exposed
Simple flat and shed roofs	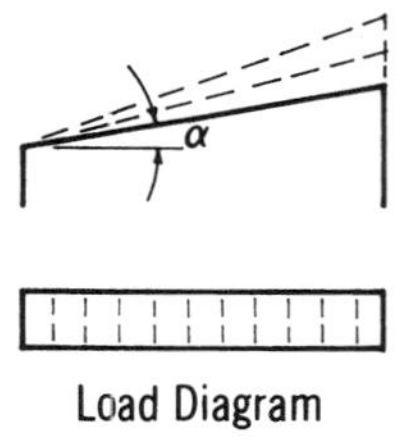	For α > 30° $C_f = 0.8 - \frac{\alpha - 30}{50}$	0° to 30° 40° 50° 60° 70° to 90°	0.8 0.6 0.4 0.2 0	0.6 0.45 0.3 0.15 0
				Case I	**Case II**
Simple gable and hip roofs	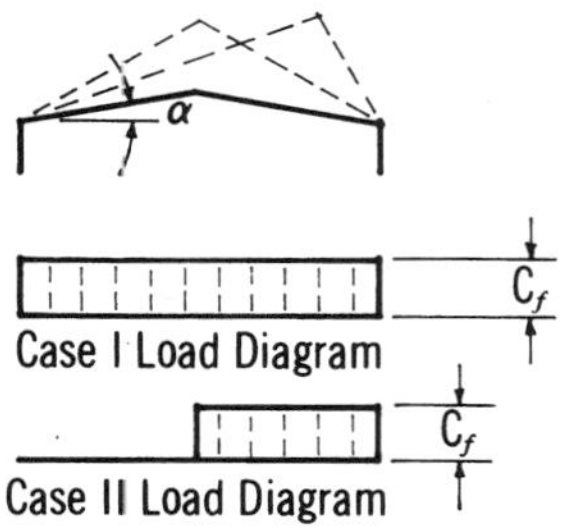	$C_f = 0.8 - \frac{\alpha - 30}{50}$ $C_f = 1.25\left(0.8 - \left(\frac{\alpha - 30}{50}\right)\right)$	10° 20° 30° 40° 50° 60°	0.8 0.8 0.8 0.6 0.4 0.2	— — 1.0 0.75 0.5 0.25
Valley areas of Two span roofs	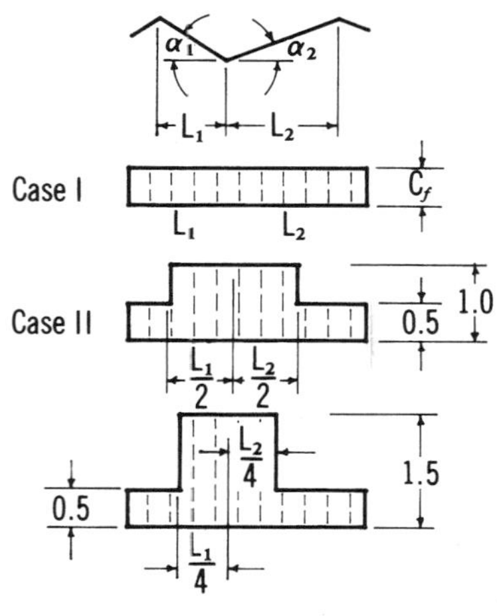	$\beta = \frac{\alpha_1 + \alpha_2}{2}$ $C_f = 0.8 - \frac{\alpha - 30}{50}$	β ≤ 10° use Case I only β > 10° β < 20° use Case I & II β ≥ 20° use Case I, II & III		
Lower level of multi level roofs (or on an adjacent building not more than 15 ft. away)	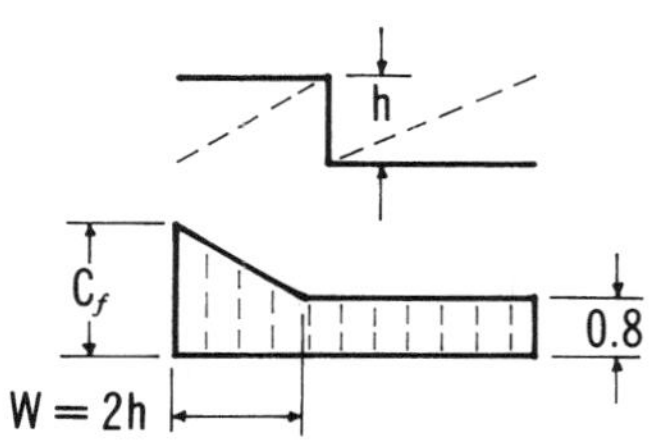	$C_f = 15\frac{h}{g}$ h = difference in roof height in feet g = ground snow load in psf w = width of drift For h < 5, w = 10 h >15, w = 30	When $15\frac{h}{g} < .8$, use 0.8 When $15\frac{h}{g} > 3.0$, use 3.0		

Summary of above:

1. For flat roofs or roofs up to 30°, use 0.8 x ground snow load for the roof snow load.
2. For roof pitches in excess of 30°, conversion factor becomes lower than 0.8.
3. For exposed roofs there is a further 25% reduction of conversion factor.
4. For steep roofs a more highly loaded half span must be considered.
5. For shallow roof valleys conversion factor is 0.8.
6. For moderate roof valleys, conversion factor is 1.0 for half the span.
7. For steep roof valleys, conversion factor is 1.5 for one quarter of the span.
8. For roofs adjoining vertical surfaces the conversion factor is up to 3.0 for part of the span.
9. If snow load is less than 30 psf, use water load on roof for clogged drain condition.

Table L1020-101 Floor Area Ratios

The table below lists commonly used gross to net area and net to gross area ratios expressed in % for various building types.

Building Type	Gross to Net Ratio	Net to Gross Ratio	Building Type	Gross to Net Ratio	Net to Gross Ratio
Apartment	156	64	School Buildings (campus type)		
Bank	140	72	Administrative	150	67
Church	142	70	Auditorium	142	70
Courthouse	162	61	Biology	161	62
Department Store	123	81	Chemistry	170	59
Garage	118	85	Classroom	152	66
Hospital	183	55	Dining Hall	138	72
Hotel	158	63	Dormitory	154	65
Laboratory	171	58	Engineering	164	61
Library	132	76	Fraternity	160	63
Office	135	75	Gymnasium	142	70
Restaurant	141	70	Science	167	60
Warehouse	108	93	Service	120	83
			Student Union	172	59

The gross area of a building is the total floor area based on outside dimensions.

The net area of a building is the usable floor area for the function intended and excludes such items as stairways, corridors, and mechanical rooms.

In the case of a commercial building, it might be considered the "leasable area."

Table L1020-201 Partition/Door Density

Building Type	Stories	Partition/Density	Doors	Description of Partition
Apartments	1 story	9 SF/LF	90 SF/door	Plaster, wood doors & trim
	2 story	8 SF/LF	80 SF/door	Drywall, wood studs, wood doors & trim
	3 story	9 SF/LF	90 SF/door	Plaster, wood studs, wood doors & trim
	5 story	9 SF/LF	90 SF/door	Plaster, metal studs, wood doors & trim
	6-15 story	8 SF/LF	80 SF/door	Drywall, metal studs, wood doors & trim
Bakery	1 story	50 SF/LF	500 SF/door	Conc. block, paint, door & drywall, wood studs
	2 story	50 SF/LF	500 SF/door	Conc. block, paint, door & drywall, wood studs
Bank	1 story	20 SF/LF	200 SF/door	Plaster, wood studs, wood doors & trim
	2-4 story	15 SF/LF	150 SF/door	Plaster, metal studs, wood doors & trim
Bottling Plant	1 story	50 SF/LF	500 SF/door	Conc. block, drywall, metal studs, wood trim
Bowling Alley	1 story	50 SF/LF	500 SF/door	Conc. block, wood & metal doors, wood trim
Bus Terminal	1 story	15 SF/LF	150 SF/door	Conc. block, ceramic tile, wood trim
Cannery	1 story	100 SF/LF	1000 SF/door	Drywall on metal studs
Car Wash	1 story	18 SF/LF	180 SF/door	Concrete block, painted & hollow metal door
Dairy Plant	1 story	30 SF/LF	300 SF/door	Concrete block, glazed tile, insulated cooler doors
Department Store	1 story	60 SF/LF	600 SF/door	Drywall, metal studs, wood doors & trim
	2-5 story	60 SF/LF	600 SF/door	30% concrete block, 70% drywall, wood studs
Dormitory	2 story	9 SF/LF	90 SF/door	Plaster, concrete block, wood doors & trim
	3-5 story	9 SF/LF	90 SF/door	Plaster, concrete block, wood doors & trim
	6-15 story	9 SF/LF	90 SF/door	Plaster, concrete block, wood doors & trim
Funeral Home	1 story	15 SF/LF	150 SF/door	Plaster on concrete block & wood studs, paneling
	2 story	14 SF/LF	140 SF/door	Plaster, wood studs, paneling & wood doors
Garage Sales & Service	1 story	30 SF/LF	300 SF/door	50% conc. block, 50% drywall, wood studs
Hotel	3-8 story	9 SF/LF	90 SF/door	Plaster, conc. block, wood doors & trim
	9-15 story	9 SF/LF	90 SF/door	Plaster, conc. block, wood doors & trim
Laundromat	1 story	25 SF/LF	250 SF/door	Drywall, wood studs, wood doors & trim
Medical Clinic	1 story	6 SF/LF	60 SF/door	Drywall, wood studs, wood doors & trim
	2-4 story	6 SF/LF	60 SF/door	Drywall, metal studs, wood doors & trim
Motel	1 story	7 SF/LF	70 SF/door	Drywall, wood studs, wood doors & trim
	2-3 story	7 SF/LF	70 SF/door	Concrete block, drywall on metal studs, wood paneling
Movie Theater 200-600 seats	1 story	18 SF/LF	180 SF/door	Concrete block, wood, metal, vinyl trim
601-1400 seats		20 SF/LF	200 SF/door	Concrete block, wood, metal, vinyl trim
1401-22000 seats		25 SF/LF	250 SF/door	Concrete block, wood, metal, vinyl trim
Nursing Home	1 story	8 SF/LF	80 SF/door	Drywall, metal studs, wood doors & trim
	2-4 story	8 SF/LF	80 SF/door	Drywall, metal studs, wood doors & trim
Office	1 story	20 SF/LF	200-500 SF/door	30% concrete block, 70% drywall on wood studs
	2 story	20 SF/LF	200-500 SF/door	30% concrete block, 70% drywall on metal studs
	3-5 story	20 SF/LF	200-500 SF/door	30% concrete block, 70% movable partitions
	6-10 story	20 SF/LF	200-500 SF/door	30% concrete block, 70% movable partitions
	11-20 story	20 SF/LF	200-500 SF/door	30% concrete block, 70% movable partitions
Parking Ramp (Open)	2-8 story	60 SF/LF	600 SF/door	Stair and elevator enclosures only
Parking Garage	2-8 story	60 SF/LF	600 SF/door	Stair and elevator enclosures only
Pre-engineered Steel	1 story	0		
Store	1 story	60 SF/LF	600 SF/door	Drywall on metal studs, wood doors & trim
Office	1 story	15 SF/LF	150 SF/door	Concrete block, movable wood partitions
Shop	1 story	15 SF/LF	150 SF/door	Movable wood partitions
Warehouse	1 story	0		
Radio & TV Broadcasting	1 story	25 SF/LF	250 SF/door	Concrete block, metal and wood doors
& TV Transmitter	1 story	40 SF/LF	400 SF/door	Concrete block, metal and wood doors
Self Service Restaurant	1 story	15 SF/LF	150 SF/door	Concrete block, wood and aluminum trim
Cafe & Drive-in Restaurant	1 story	18 SF/LF	180 SF/door	Drywall, metal studs, ceramic & plastic trim
Restaurant with seating	1 story	25 SF/LF	250 SF/door	Concrete block, paneling, wood studs & trim
Supper Club	1 story	25 SF/LF	250 SF/door	Concrete block, paneling, wood studs & trim
Bar or Lounge	1 story	24 SF/LF	240 SF/door	Plaster or gypsum lath, wooded studs
Retail Store or Shop	1 story	60 SF/LF	600 SF/door	Drywall metal studs, wood doors & trim
Service Station Masonry	1 story	15 SF/LF	150 SF/door	Concrete block, paint, door & drywall, wood studs
Metal panel	1 story	15 SF/LF	150 SF/door	Concrete block, paint, door & drywall, wood studs
Frame	1 story	15 SF/LF	150 SF/door	Drywall, wood studs, wood doors & trim
Shopping Center (strip)	1 story	30 SF/LF	300 SF/door	Drywall, metal studs, wood doors & trim
(group)	1 story	40 SF/LF	400 SF/door	50% concrete block, 50% drywall, wood studs
	2 story	40 SF/LF	400 SF/door	50% concrete block, 50% drywall, wood studs
Small Food Store	1 story	30 SF/LF	300 SF/door	Concrete block drywall, wood studs, wood trim
Store/Apt. above Masonry	2 story	10 SF/LF	100 SF/door	Plaster, metal studs, wood doors & trim
Frame	2 story	10 SF/LF	100 SF/door	Plaster, metal studs, wood doors & trim
Frame	3 story	10 SF/LF	100 SF/door	Plaster, metal studs, wood doors & trim
Supermarkets	1 story	40 SF/LF	400 SF/door	Concrete block, paint, drywall & porcelain panel
Truck Terminal	1 story	0		
Warehouse	1 story	0		

Table L1020-301 Occupancy Determinations

Function of Space	SF/Person Required
Accessory storage areas, mechanical equipment rooms	300
Agriculture Building	300
Aircraft Hangars	500
Airport Terminal	
Baggage claim	20
Baggage handling	300
Concourse	100
Waiting areas	15
Assembly	
Gaming floors (Keno, slots, etc.)	11
Exhibit gallery and museum	30
Assembly w/ fixed seats	load determined by seat number
Assembly w/o fixed seats	
Concentrated (chairs only-not fixed)	7
Standing space	5
Unconcentrated (tables and chairs)	15
Bowling centers, allow 5 persons for each lane including 15 feet of runway, and for additional areas	7
Business areas	100
Courtrooms-other than fixed seating areas	40
Day care	35
Dormitories	50
Educational	
Classroom areas	20
Shops and other vocational room areas	50
Exercise rooms	50
Fabrication and manufacturing areas where hazardous materials are used	200
Industrial areas	100
Institutional areas	
Inpatient treatment areas	240
Outpatient areas	100
Sleeping areas	120
Kitchens commercial	200
Library	
Reading rooms	50
Stack area	100
Mercantile	
Areas on other floors	60
Basement and grade floor areas	30
Storage, stock, shipping areas	300
Parking garages	200
Residential	200
Skating rinks, swimming pools	
Rink and pool	50
Decks	15
Stages and platforms	15
Warehouses	500

Table L1020-302 Length of Exitway Access Travel (ft.)

Occupancy Type	Without Sprinkler System (feet)	With Sprinkler System (feet)
A, E, F-1, M, R, S-1	200	250
I-1	Not Permitted	250
B	200	300
F-2, S-2, U	300	400
H-1	Not Permitted	75
H-2	Not Permitted	100
H-3	Not Permitted	150
H-4	Not Permitted	175
H-5	Not Permitted	200
I-2, I-3, I-4	Not Permitted	200

Note:

Refer to the 2012 *International Building Code* Section 1016 Exit Access Travel Distance for any exceptions or additions to the information above.

Table L1020-303 Capacity Per Unit Egress Width*

Use Group	Without Fire Suppression System (Inches per Person)*		With Fire Suppression System (Inches per Person)*	
	Stairways	Doors, Ramps and Corridors	Stairways	Doors, Ramps and Corridors
Assembly, Business, Educational, Factory Industrial, Mercantile, Residential, Storage	0.3	0.2	0.2	0.15
Institutional—1	0.3	0.2	0.2	0.15
Institutional—2	—	0.7	0.3	0.2
Institutional—3	0.3	0.2	0.2	0.15
High Hazard	0.7	0.4	0.3	0.2

* 1″=25.4 mm

Table L1030-101 "U" Values for Type "A" Buildings

Type A buildings shall include:

A1 Detached one and two family dwellings

A2 All other residential buildings, three stories or less, including but not limited to: multi-family dwellings, hotels and motels.

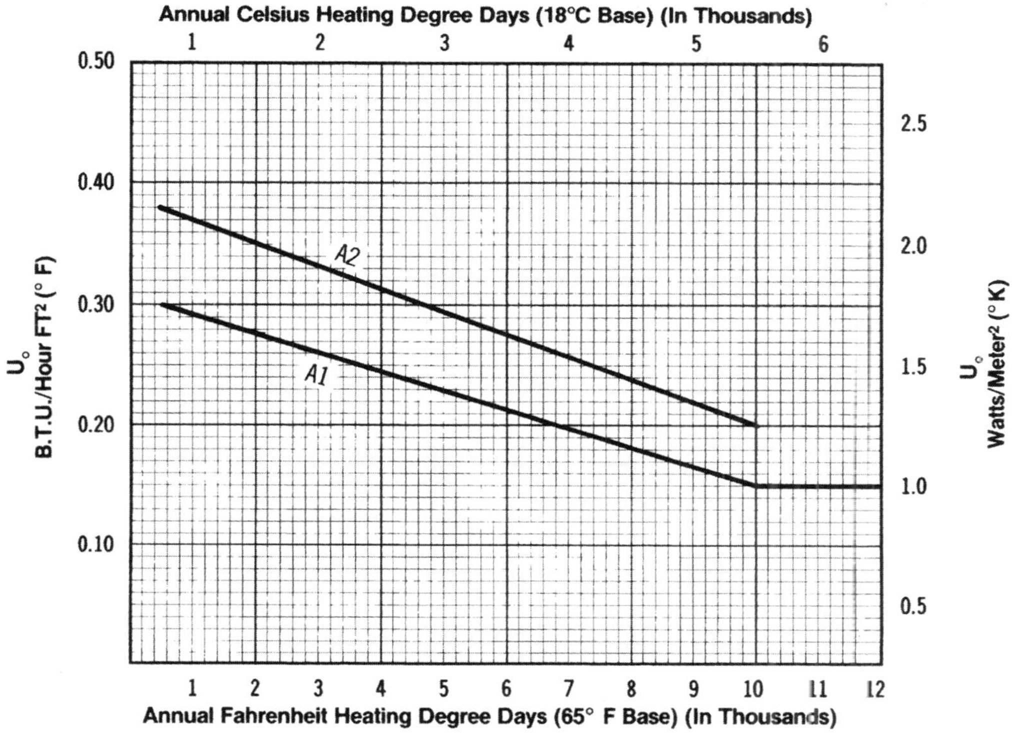

Table L1030-102 "U" Values for Type "B" Buildings

For all buildings not classified Type "A"

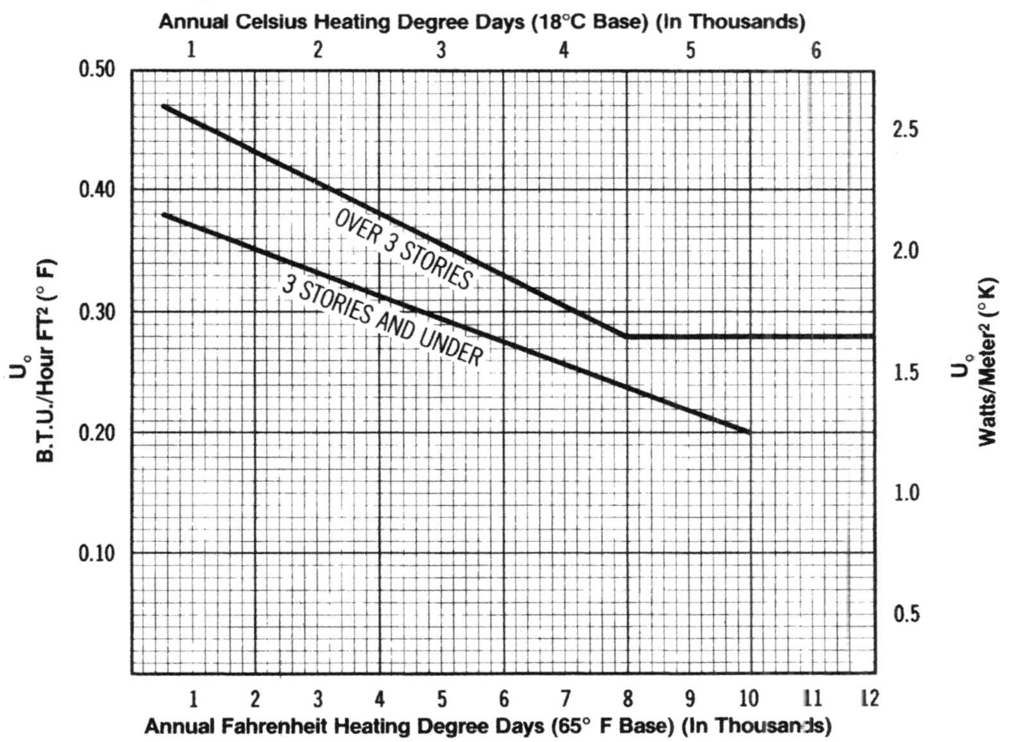

Table L1030-201 Combinations of Wall and Single-Glazed Openings (For Use With ASHRAE 90-75)

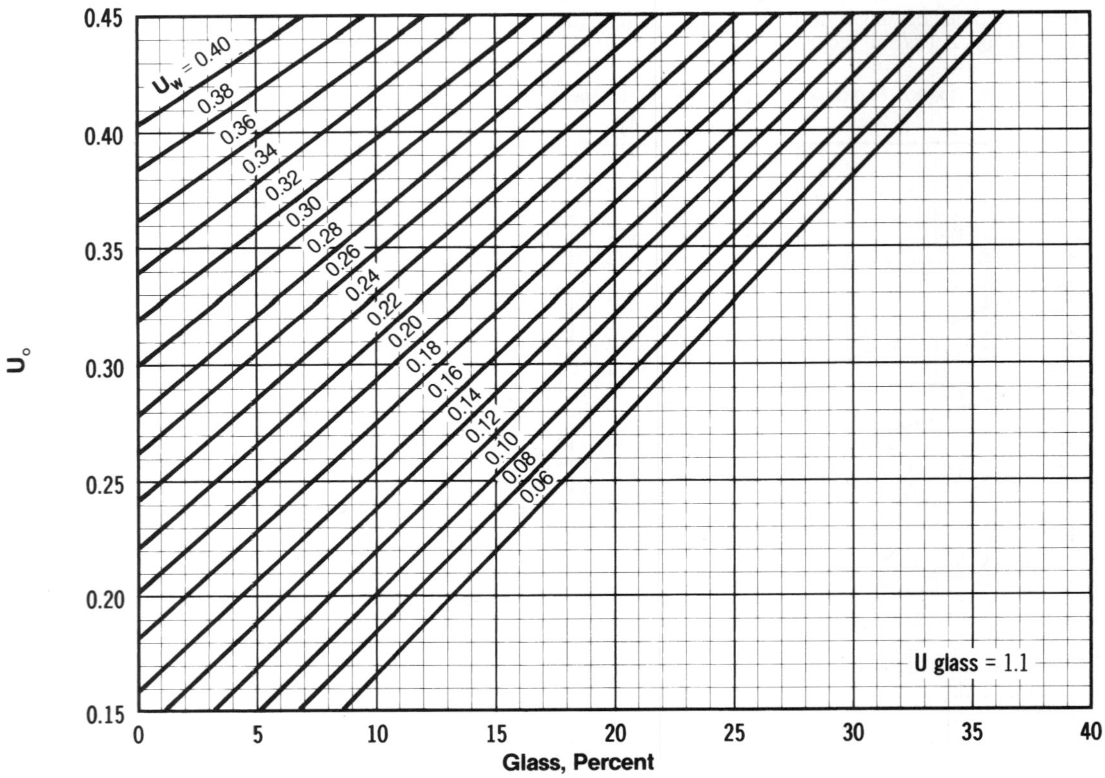

Table L1030-202 Combinations of Wall and Double-Glazed Openings (For Use With ASHRAE 90-75)

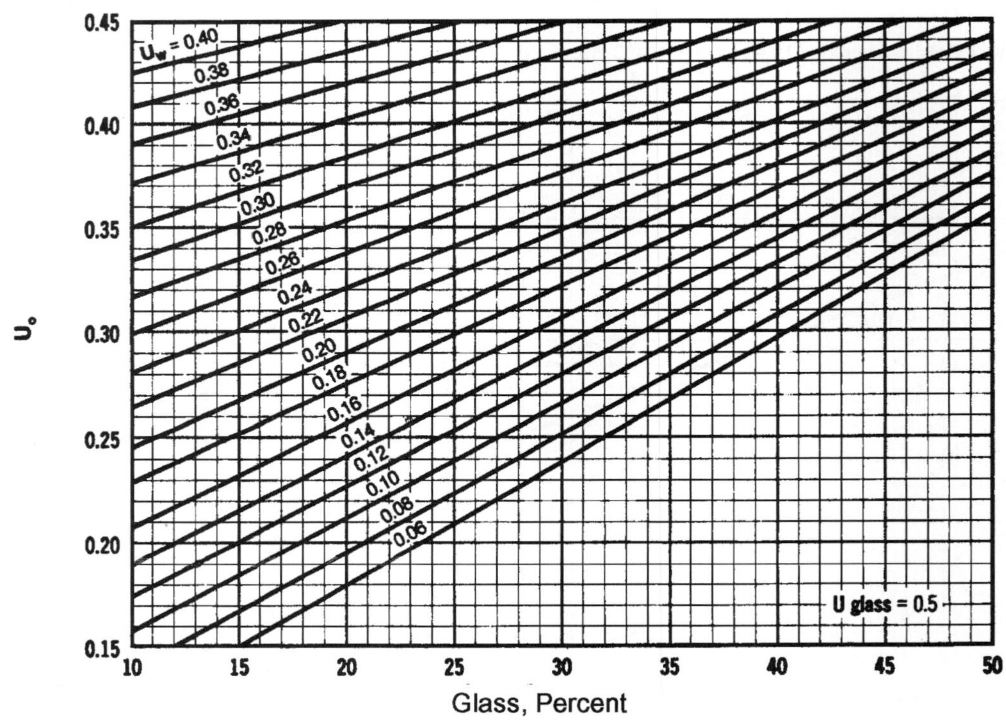

Table L1030-203 Influence of Wall and Roof Weight on "U" Value Correction for Heating Design

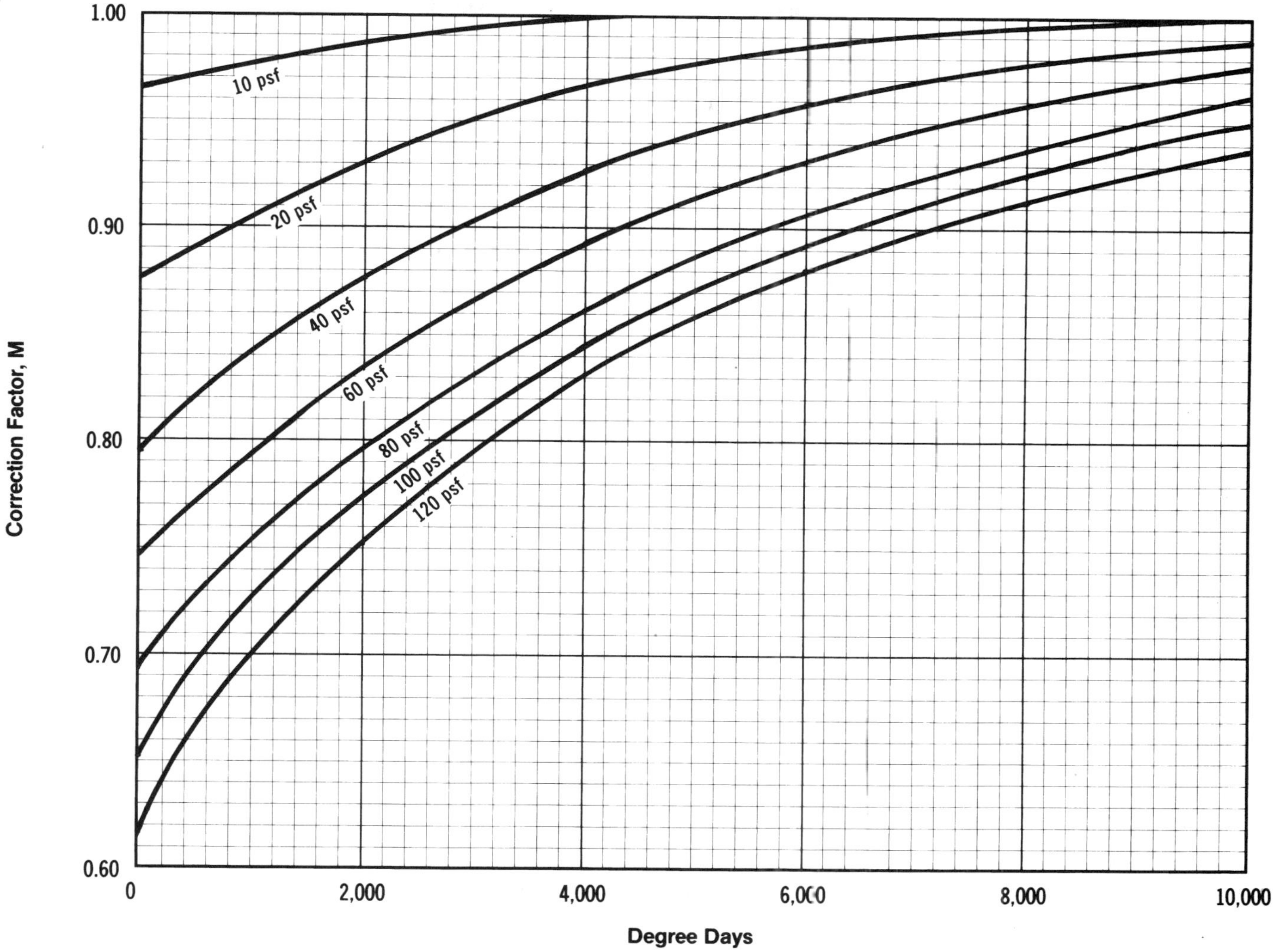

Effective "U" for walls: Uw = Uw_{ss} x M (similarly, Ur for roofs)

where Uw = effective thermal transmittance of opaque wall area BTU/h x Ft.^{2} x ° F

Uw_{ss} = steady state thermal transmittance of opaque wall area BTU/h x Ft.^{2} x ° F (steady state "U" value)

M = weight correction factor

Example: Uw_{ss} = 0.20 with wall weight = 120 psf in Providence, R.I. (6000 degree days)

Enter chart on bottom at 6000, go up to 120 psf curve, read to the left .88 M = .88

and Uw = 0.20 x 0.88 = 0.176

Table L1030-204 "U" Values for Type "A" Buildings

One and two family dwellings and residential buildings of three stories or less shall have an overall U value as follows:

A. For roof assemblies in which the finished interior surface is essentially the underside of the roof deck, such as exposed concrete slabs, joist slabs, cored slabs, or cathedral ceilings or wood-beam construction.

U_{or} = 0.08 BTU/hr./ft.^{2}/° F. (all degree days)

B. For roof-ceiling assemblies, such as roofs with finished ceilings attached to or suspended below the roof deck.

U_{or} = 0.05 for below 8,000 degree days
= 0.04 for 8,000 degree days and greater.

Table L1030-205 "U" Values for Type "B" Buildings*

For all buildings not classified type "A".

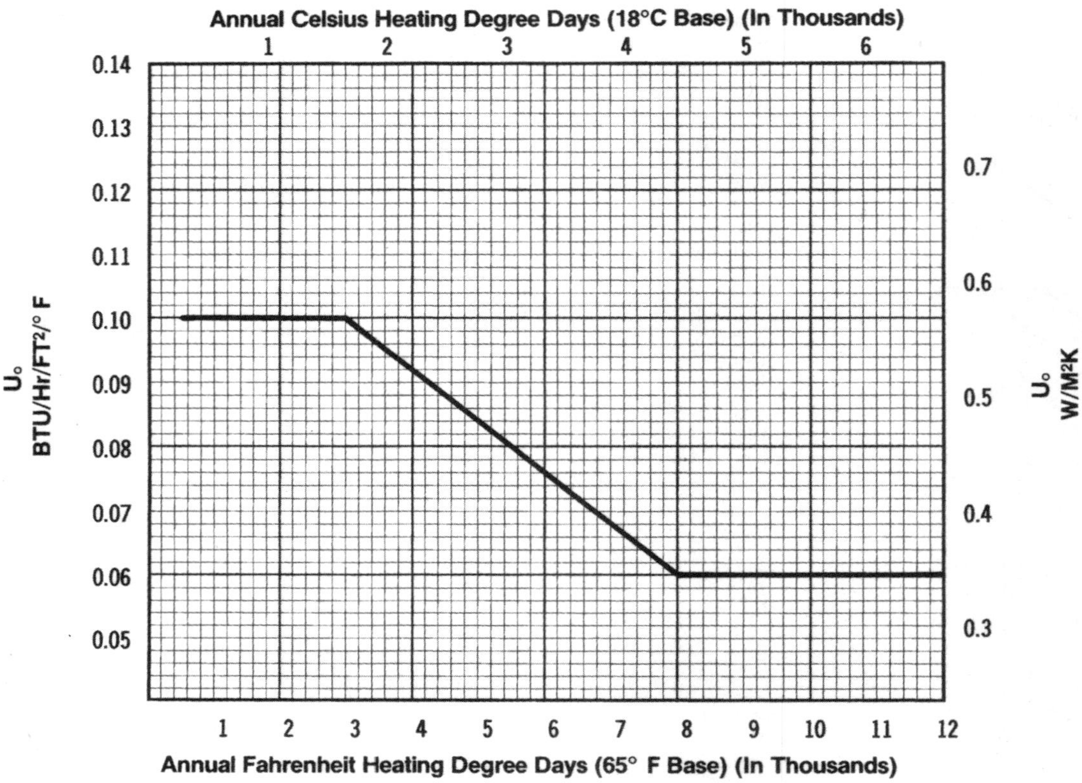

*Minimum requirements for thermal design, ASHRAE Standard 90-75

Table L1030-301 "U" Values for Floors Over Unheated Spaces

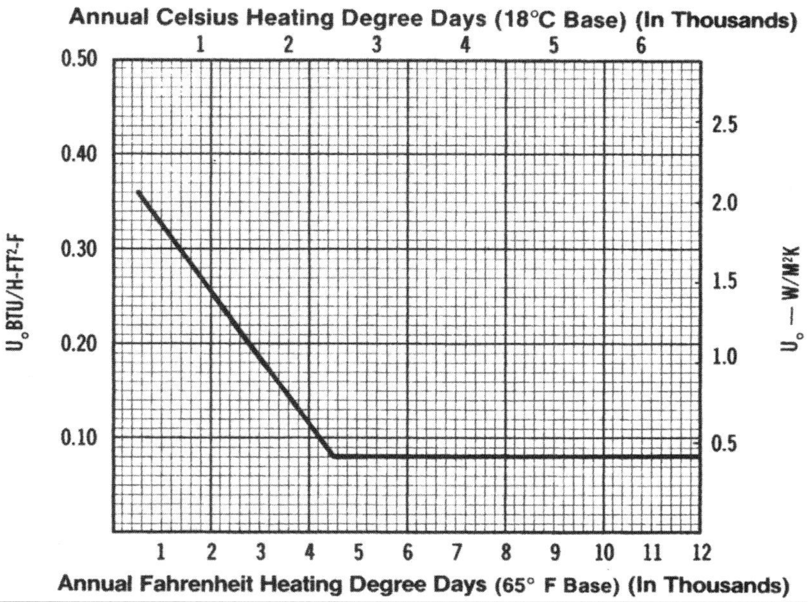

Table L1030-302 "R" Values for Slabs on Grade

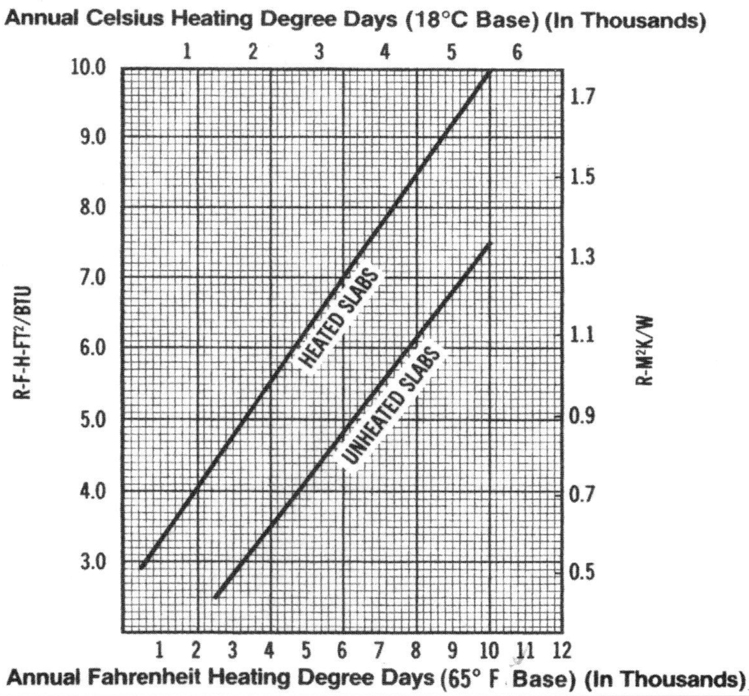

Table L1030-303 Insulation Requirements for Slabs on Grade for all Buildings

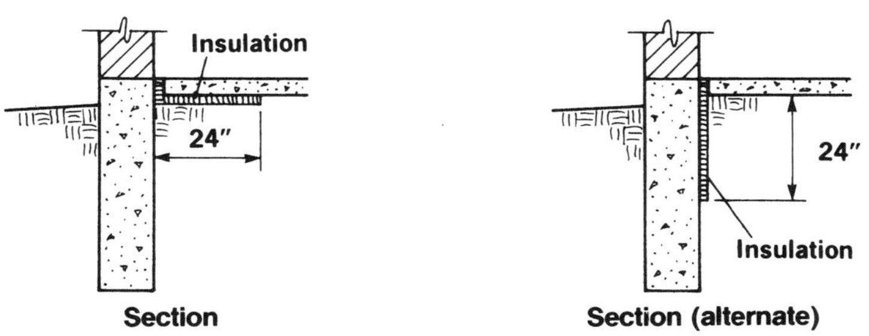

Table L1030-401 Resistances ("R") of Building and Insulating Materials

Material	Wt./Lbs. per C.F.	R per Inch	R Listed Size
Air Spaces and Surfaces			
Enclosed non-reflective spaces, E=0.82			
50° F mean temp., 30°/10° F diff.			
.5"			.90/.91
.75"			.94/1.01
1.50"			.90/1.02
3.50"			.91/1.01
Inside vert. surface (still air)			0.68
Outside vert. surface (15 mph wind)			0.17
Building Boards			
Asbestos cement, 0.25" thick	120		0.06
Gypsum or plaster, 0.5" thick	50		0.45
Hardboard regular	50	1.37	
Tempered	63	1.00	
Laminated paper	30	2.00	
Particle board	37	1.85	
	50	1.06	
	63	0.85	
Plywood (Douglas Fir), 0.5" thick	34		0.62
Shingle backer, .375" thick	18		0.94
Sound deadening board, 0.5" thick	15		1.35
Tile and lay-in panels, plain or acoustical, 0.5" thick	18		1.25
Vegetable fiber, 0.5" thick	18		1.32
	25		1.14
Wood, hardwoods	48	0.91	
Softwoods	32	1.25	
Flooring Carpet with fibrous pad			2.08
With rubber pad			1.23
Cork tile, 1/8" thick			0.28
Terrazzo			0.08
Tile, resilient			0.05
Wood, hardwood, 0.75" thick			0.68
Subfloor, 0.75" thick			0.94
Glass			
Insulation, 0.50" air space			2.04
Single glass			0.91
Insulation Blanket or Batt, mineral, glass or rock fiber, approximate thickness			
3.0" to 3.5" thick			11
3.5" to 4.0" thick			13
6.0" to 6.5" thick			19
6.5" to 7.0" thick			22
8.5" to 9.0" thick			30
Boards			
Cellular glass	8.5	2.63	
Fiberboard, wet felted			
Acoustical tile	21	2.70	
Roof insulation	17	2.94	
Fiberboard, wet molded			
Acoustical tile	23	2.38	
Mineral fiber with resin binder	15	3.45	
Polystyrene, extruded,			
cut cell surface	1.8	4.00	
smooth skin surface	2.2	5.00	
	3.5	5.26	
Bead boards	1.0	3.57	
Polyurethane	1.5	6.25	
Wood or cane fiberboard, 0.5" thick			1.25

Material	Wt./Lbs. per C.F.	R per Inch	R Listed Size
Insulation Loose Fill			
Cellulose	2.3	3.13	
	3.2	3.70	
Mineral fiber, 3.75" to 5" thick	2-5		11
6.5" to 8.75" thick			19
7.5" to 10" thick			22
10.25" to 13.75" thick			30
Perlite	5-8	2.70	
Vermiculite	4-6	2.27	
Wood fiber	2-3.5	3.33	
Masonry Brick, Common	120	0.20	
Face	130	0.11	
Cement mortar	116	0.20	
Clay tile, hollow			
1 cell wide, 3" width			0.80
4" width			1.11
2 cells wide, 6" width			1.52
8" width			1.85
10" width			2.22
3 cells wide, 12" width			2.50
Concrete, gypsum fiber	51	0.60	
Lightweight	120	0.19	
	80	0.40	
	40	0.86	
Perlite	40	1.08	
Sand and gravel or stone	140	0.08	
Concrete block, lightweight			
3 cell units, 4"-15 lbs. ea.			1.68
6"-23 lbs. ea.			1.83
8"-28 lbs. ea.			2.12
12"-40 lbs. ea.			2.62
Sand and gravel aggregates,			
4"-20 lbs. ea.			1.17
6"-33 lbs. ea.			1.29
8"-38 lbs. ea.			1.46
12"-56 lbs. ea.			1.81
Plastering Cement Plaster,			
Sand aggregate	116	0.20	
Gypsum plaster, Perlite aggregate	45	0.67	
Sand aggregate	105	0.18	
Vermiculite aggregate	45	0.59	
Roofing			
Asphalt, felt, 15 lb.			0.06
Rolled roofing	70		0.15
Shingles	70		0.44
Built-up roofing .375" thick	70		0.33
Cement shingles	120		0.21
Vapor-permeable felt			0.06
Vapor seal, 2 layers of mopped 15 lb. felt			0.12
Wood, shingles 16"-7.5" exposure			0.87
Siding			
Aluminum or steel (hollow backed)			
oversheathing			0.61
With .375" insulating backer board			1.82
Foil backed			2.96
Wood siding, beveled, ½" x 8"			0.81

Table L1030-501 Weather Data and Design Conditions

City	Latitude (1) °	Latitude (1) 1′	Winter Temperatures (1) Med. of Annual Extremes	Winter Temperatures (1) 99%	Winter Temperatures (1) 97½%	Winter Degree Days (2)	Summer (Design Dry Bulb) Temperatures and Relative Humidity 1%	2½%	5%
UNITED STATES									
Albuquerque, NM	35	0	5.1	12	16	4,400	96/61	94/61	92/61
Atlanta, GA	33	4	11.9	17	22	3,000	94/74	92/74	90/73
Baltimore, MD	39	2	7	14	17	4,600	94/75	91/75	89/74
Birmingham, AL	33	3	13	17	21	2,600	96/74	94/75	92/74
Bismarck, ND	46	5	-32	-23	-19	8,800	95/68	91/68	88/67
Boise, ID	43	3	1	3	10	5,800	96/65	94/64	91/64
Boston, MA	42	2	-1	6	9	5,600	91/73	88/71	85/70
Burlington, VT	44	3	-17	-12	-7	8,200	88/72	85/70	82/69
Charleston, WV	38	2	3	7	11	4,400	92/74	90/73	87/72
Charlotte, NC	35	1	13	18	22	3,200	95/74	93/74	91/74
Casper, WY	42	5	-21	-11	-5	7,400	92/58	90/57	87/57
Chicago, IL	41	5	-8	-3	2	6,600	94/75	91/74	88/73
Cincinnati, OH	39	1	0	1	6	4,400	92/73	90/72	88/72
Cleveland, OH	41	2	-3	1	5	6,400	91/73	88/72	86/71
Columbia, SC	34	0	16	20	24	2,400	97/76	95/75	93/75
Dallas, TX	32	5	14	18	22	2,400	102/75	100/75	97/75
Denver, CO	39	5	-10	-5	1	6,200	93/59	91/59	89/59
Des Moines, IA	41	3	-14	-10	-5	6,600	94/75	91/74	88/73
Detroit, MI	42	2	-3	3	6	6,200	91/73	88/72	86/71
Great Falls, MT	47	3	-25	-21	-15	7,800	91/60	88/60	85/59
Hartford, CT	41	5	-4	3	7	6,200	91/74	88/73	85/72
Houston, TX	29	5	24	28	33	1,400	97/77	95/77	93/77
Indianapolis, IN	39	4	-7	-2	2	5,600	92/74	90/74	87/73
Jackson, MS	32	2	16	21	25	2,200	97/76	95/76	93/76
Kansas City, MO	39	1	-4	2	6	4,800	99/75	96/74	93/74
Las Vegas, NV	36	1	18	25	28	2,800	108/66	106/65	104/65
Lexington, KY	38	0	-1	3	8	4,600	93/73	91/73	88/72
Little Rock, AR	34	4	11	15	20	3,200	99/76	96/77	94/77
Los Angeles, CA	34	0	36	41	43	2,000	93/70	89/70	86/69
Memphis, TN	35	0	10	13	18	3,200	98/77	95/76	93/76
Miami, FL	25	5	39	44	47	200	91/77	90/77	89/77
Milwaukee, WI	43	0	-11	-8	-4	7,600	90/74	87/73	84/71
Minneapolis, MN	44	5	-22	-16	-12	8,400	92/75	89/73	86/71
New Orleans, LA	30	0	28	29	33	1,400	93/78	92/77	90/77
New York, NY	40	5	6	11	15	5,000	92/74	89/73	87/72
Norfolk, VA	36	5	15	20	22	3,400	93/77	91/76	89/76
Oklahoma City, OK	35	2	4	9	13	3,200	100/74	97/74	95/73
Omaha, NE	41	2	-13	-8	-3	6,600	94/76	91/75	88/74
Philadelphia, PA	39	5	6	10	14	4,400	93/75	90/74	87/72
Phoenix, AZ	33	3	27	31	34	1,800	109/71	107/71	105/71
Pittsburgh, PA	40	3	-1	3	7	6,000	91/72	88/71	86/70
Portland, ME	43	4	-10	-6	-1	7,600	87/72	84/71	81/69
Portland, OR	45	4	18	17	23	4,600	89/68	85/67	81/65
Portsmouth, NH	43	1	-8	-2	2	7,200	89/73	85/71	83/70
Providence, RI	41	4	-1	5	9	6,000	89/73	86/72	83/70
Rochester, NY	43	1	-5	1	5	6,800	91/73	88/71	85/70
Salt Lake City, UT	40	5	0	3	8	6,000	97/62	95/62	92/61
San Francisco, CA	37	5	36	38	40	3,000	74/63	71/62	69/61
Seattle, WA	47	4	22	22	27	5,200	85/68	82/66	78/65
Sioux Falls, SD	43	4	-21	-15	-11	7,800	94/73	91/72	88/71
St. Louis, MO	38	4	-3	3	8	5,000	98/75	94/75	91/75
Tampa, FL	28	0	32	36	40	680	92/77	91/77	90/76
Trenton, NJ	40	1	4	11	14	5,000	91/75	88/74	85/73
Washington, DC	38	5	7	14	17	4,200	93/75	91/74	89/74
Wichita, KS	37	4	-3	3	7	4,600	101/72	98/73	96/73
Wilmington, DE	39	4	5	10	14	5,000	92/74	89/74	87/73
ALASKA									
Anchorage	61	1	-29	-23	-18	10,800	71/59	68/58	66/56
Fairbanks	64	5	-59	-51	-47	14,280	82/62	78/60	75/59
CANADA									
Edmonton, Alta.	53	3	-30	-29	-25	11,000	85/66	82/65	79/63
Halifax, N.S.	44	4	-4	1	5	8,000	79/66	76/65	74/64
Montreal, Que.	45	3	-20	-16	-10	9,000	88/73	85/72	83/71
Saskatoon, Sask.	52	1	-35	-35	-31	11,000	89/68	86/66	83/65
St. John, Nwf.	47	4	1	3	7	8,600	77/66	75/65	73/64
Saint John, N.B.	45	2	-15	-12	-8	8,200	80/67	77/65	75/64
Toronto, Ont.	43	4	-10	-5	-1	7,000	90/73	87/72	85/71
Vancouver, B.C.	49	1	13	15	19	6,000	79/67	77/66	74/65
Winnipeg, Man.	49	5	-31	-30	-27	10,800	89/73	86/71	84/70

(1) Handbook of Fundamentals, ASHRAE, Inc., NY 1989
(2) Local Climatological Annual Survey, USDC Env. Science Services Administration, Asheville, NC

Table L1030-502 Maximum Depth of Frost Penetration in Inches

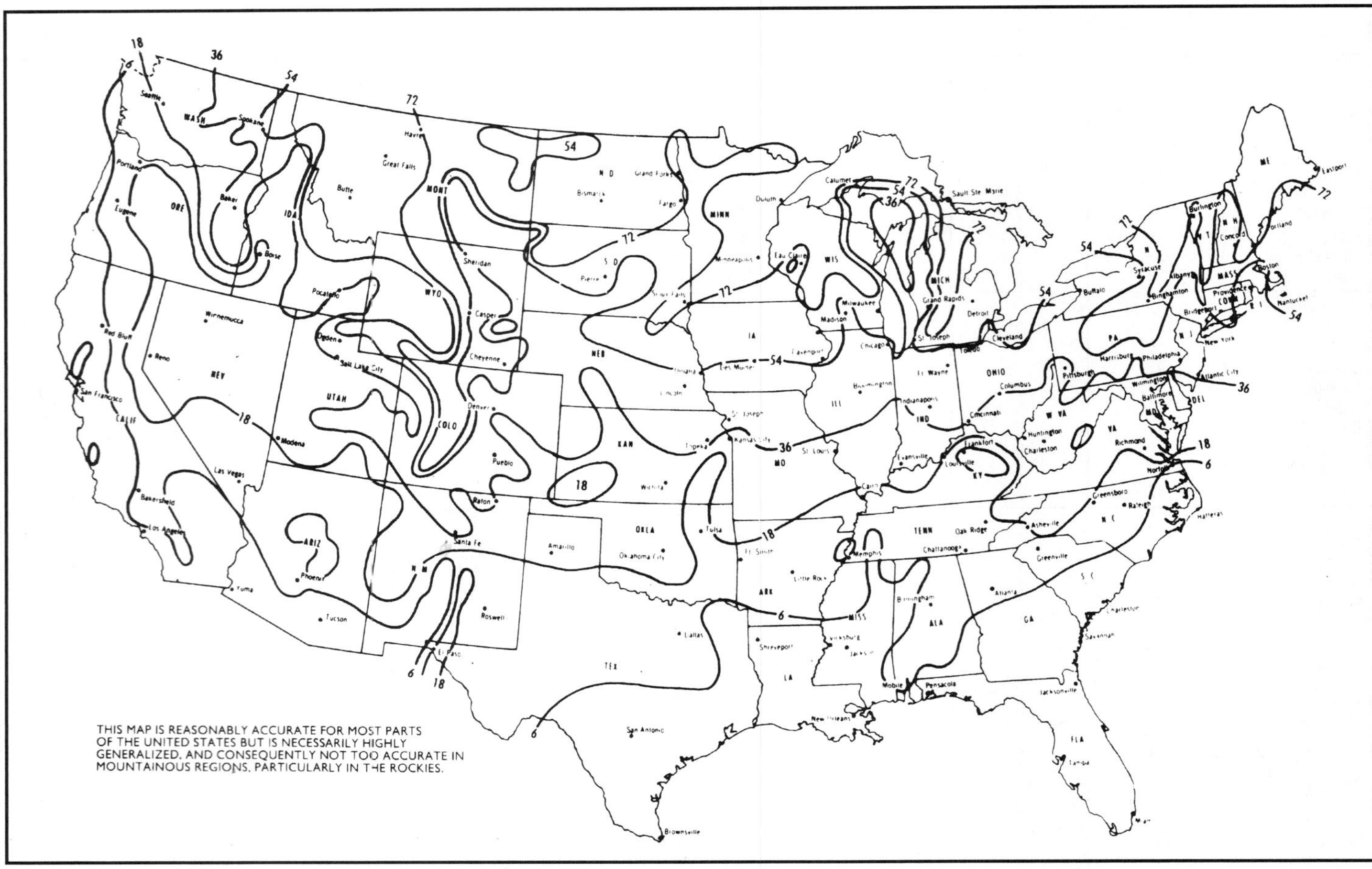

Table L1040-101 Fire-Resisting Ratings of Structural Elements (in hours)

Description of the Structural Element	Type of Construction									
	No. 1 Fireproof		No. 2 Non Combustible			No. 3 Exterior Masonry Wall			No. 4 Frame	
			Protected		Unprotected	Heavy Timber	Ordinary		Pro-tected	Unpro-tected
							Pro-tected	Unpro-tected		
	1A	1B	2A	2B	2C	3A	3B	3C	4A	4B
Exterior, Bearing Walls Nonbearing Walls	4 2	3 2	2 1½	1½ 1	1 1	2 2	2 2	2 2	1 1	1 1
Interior Bearing Walls and Partitions	4	3	2	1	0	2	1	0	1	0
Fire Walls and Party Walls	4	3	2	2	2	2	2	2	2	2
Fire Enclosure of Exitways, Exit Hallways and Stairways	2	2	2	2	2	2	2	2	1	1
Shafts other than Exitways, Hallways and Stairways	2	2	2	2	2	2	2	2	1	1
Exitway access corridors and Vertical separation of tenant space	1	1	1	1	0	1	1	0	1	0
Columns, girders, trusses (other than roof trusses) and framing: Supporting more than one floor Supporting one floor only	 4 3	 3 2	 2 1½	 1 1	 0 0	 — —	 1 1	 0 0	 1 1	 0 0
Structural members supporting wall	3	2	1½	1	0	1	1	0	1	0
Floor construction including beams	3	2	1½	1	0	—	1	0	1	0
Roof construction including beams, trusses and framing arches and roof deck 15′ or less in height to lowest member	2	1½	1	1	0	—	1	0	1	0

Note: a. Codes include special requirements and exceptions that are not included in the table above.
b. Each type of construction has been divided into sub-types which vary according to the degree of fire resistance required. Sub-types (A) requirements are more severe than those for sub-types (B).
c. Protected construction means all structural members are chemically treated, covered or protected so that the unit has the required fire resistance.

Type No. 1, Fireproof Construction — Buildings and structures of fireproof construction are those in which the walls, partitions, structural elements, floors, ceilings, roofs, and the exitways are protected with approved noncombustible materials to afford the fire-resistance rating specified in Table L1040-101; except as otherwise specifically regulated. Fire-resistant treated wood may be used as specified.

Type No. 2, Noncombustible Construction — Buildings and structures of noncombustible construction are those in which the walls, partitions, structural elements, floors, ceilings, roofs and the exitways are approved noncombustible materials meeting the fire-resistance rating requirements specified in Table L1040-101; except as modified by the fire limit restrictions. Fire-retardant treated wood may be used as specified.

Type No. 3, Exterior Masonry Wall Construction — Buildings and structures of exterior masonry wall construction are those in which the exterior, fire and party walls are masonry or other approved noncombustible materials of the required fire-resistance rating and structural properties. The floors, roofs, and interior framing are wholly or partly wood or metal or other approved construction. The fire and party walls are ground-supported; except that girders and their supports, carrying walls of masonry shall be protected to afford the same degree of fire-resistance rating of the supported walls. All structural elements have the required fire-resistance rating specified in Table L1040-101.

Type No. 4, Frame Construction — Buildings and structures of frame construction are those in which the exterior walls, bearing walls, partitions, floor and roof construction are wholly or partly of wood stud and joist assemblies with a minimum nominal dimension of two inches or of other approved combustible materials. Fire stops are required at all vertical and horizontal draft openings in which the structural elements have required fire-resistance ratings specified in Table L1040-101.

Table L1040-201 Fire Resistance Ratings

Fire Hazard for Fire Walls

The degree of fire hazard of buildings relating to their intended use is defined by "Fire Rating" the occupancy type. Such a rating system is listed in Table L1040-201 below. This type of rating determines the requirements for fire walls and fire separation walls (exterior fire exposure). For mixed use occupancy, use the higher Fire Rating requirement of the components.

Group	Fire-Resistance Rating (hours)
A, B, E, H-4[1], I, R-1, R-2, U	3
F-1[2], H-3[1], H-5[1], M, S-1	3
H-1[1], H-2[1]	4
F-2[3], S-2, R-3, R-4	2

Note: *The difference in "Fire Hazards" is determined by their occupancy and use.

1. High Hazard: Industrial and storage buildings in which the combustible contents might cause fires to be unusually intense or where explosives, combustible gases or flammable liquids are manufactured or stored.
2. Moderate Hazard: Mercantile buildings, industrial and storage buildings in which combustible contents might cause fires of moderate intensity.
3. Low Hazard: Business buildings that ordinarily do not burn rapidly.

Note:
In Type II or V construction, walls shall be permitted to have a 2-hour fire-resistance rating.

Table 706.4
Excerpted from the 2012 *International Building Code*, Copyright 2011. Washington, D.C.: International Code Council. Reproduced with permission.

Table L1040-202 Interior Finish Classification

Flame Spread for Interior Finishes

The flame spreadability of a material is the burning characteristic of the material relative to the fuel contributed by its combustion and the density of smoke developed. The flame spread classification of a material is based on a ten minute test on a scale of 0 to 100. Cement asbestos board is assigned a rating of 0 and select red oak flooring a rating of 100.

The three classes are listed in Table L1040-202.

The flame spread ratings for interior finish walls and ceilings shall not be greater than the Class listed in Table L1040-203.

Finish Class	Flame Spread Index	Smoke Developed Index
A	0-25	0-450
B	26-75	0-450
C	76-200	0-450

Section 803.1.1 Interior wall and ceiling finish materials

Excerpted from the 2012 *International Building Code*, Copyright 2011. Washington, D.C.: International Code Council. Reproduced with permission.

Table L1040-203 Interior Finish Requirements by Class

Group	Sprinklered			Nonsprinklered		
	Interior exit stairways, interior exit ramps and exit passageways Note a, b	Corridors and enclosure for exit access stairways and exit access ramps	Rooms and enclosed spaces Note c	Interior exit stairways, interior exit ramps and exit passageways Note a, b	Corridors and enclosure for exit access stairways and exit access ramps	Rooms and enclosed spaces Note c
A-1 & A-2	B	B	C	A	A (d)	B (e)
A-3 (f), A-4, A-5	B	B	C	A	A (d)	C
B, E, M, R-1	B	C	C	A	B	C
R-4	B	C	C	A	B	B
F	C	C	C	B	C	C
H	B	B	C	A	A	B
I-1	B	C	C	A	B	B
I-2	B	B	B (h, i)	A	A	B
I-3	A	A (j)	C	A	A	B
I-4	B	B	B (h, i)	A	A	B
R-2	C	C	C	B	B	C
R-3	C	C	C	C	C	C
S	C	C	C	B	B	C
U	No Restrictions			No Restrictions		

Notes:

a. Class C interior finish materials shall be permitted for wainscotting or paneling of not more than 1,000 square feet of applied surface area in the grade lobby where applied directly to a noncombustible base or over furring strips applied to a noncombustible base and fireblocked as required by IBC Section 803.11.1.

b. In other Group I-2 occupancies in buildings less than three stories above grade plane of other than Group I-3, Class B interior finish for nonsprinklered buildings and Class C interior finish for sprinklered buildings shall be permitted in interior exit stairways and ramps.

c. Requirements for rooms and enclosed spaces shall be based upon spaces enclosed by partitions Where a fire-resistance rating is required for structural elements, the enclosing partitions shall extend from the floor to the ceiling. Partitions that do not comply with this shall be considered enclosed spaces and the rooms or spaces on both sides shall be considered one. In determining the applicable requirements for rooms and enclosed spaces, the specific occupancy thereof shall be the governing factor regardless of the group classification of the building or structure.

d. Lobby areas in Group A-1, A-2, and A-3 occupancies shall not be less than Class B materials.

e. Class C interior finish materials shall be permitted in places of assembly with an occupant load of 300 persons or less.

f. For places of religious worship, wood used for ornamental purposes, trusses, paneling or chancel furnishing shall be permitted.

g. Class B material is required where the building exceeds two stories.

h. Class C interior finish materials shall be permitted in administrative spaces.

i. Class C interior finish materials shall be permitted in rooms with a capacity of four persons or less.

j. Class B materials shall be permitted as wainscotting extending not more than 48 inches above the finished floor in corridors and exit access stairways and ramps.

Section 803.9

Reference Aids — RL1090-100 Conversion Factors

Description: This table is primarily for converting customary U.S. units in the left hand column to SI metric units in the right hand column. In addition, conversion factors for some commonly encountered Canadian and non-SI metric units are included.

Table L1090-101 Metric Conversion Factors

	If You Know		Multiply By		To Find
Length	Inches	x	25.4[a]	=	Millimeters
	Feet	x	0.3048[a]	=	Meters
	Yards	x	0.9144[a]	=	Meters
	Miles (statute)	x	1.609	=	Kilometers
Area	Square inches	x	645.2	=	Square millimeters
	Square feet	x	0.0929	=	Square meters
	Square yards	x	0.8361	=	Square meters
Volume (Capacity)	Cubic inches	x	16,387	=	Cubic millimeters
	Cubic feet	x	0.02832	=	Cubic meters
	Cubic yards	x	0.7646	=	Cubic meters
	Gallons (U.S. liquids)[b]	x	0.003785	=	Cubic meters[c]
	Gallons (Canadian liquid)[b]	x	0.004546	=	Cubic meters[c]
	Ounces (U.S. liquid)[b]	x	29.57	=	Milliliters[c, d]
	Quarts (U.S. liquid)[b]	x	0.9464	=	Liters[c, d]
	Gallons (U.S. liquid)[b]	x	3.785	=	Liters[c, d]
Force	Kilograms force[d]	x	9.807	=	Newtons
	Pounds force	x	4.448	=	Newtons
	Pounds force	x	0.4536	=	Kilograms force[d]
	Kips	x	4448	=	Newtons
	Kips	x	453.6	=	Kilograms force[d]
Pressure, Stress, Strength (Force per unit area)	Kilograms force per square centimeter[d]	x	0.09807	=	Megapascals
	Pounds force per square inch (psi)	x	0.006895	=	Megapascals
	Kips per square inch	x	6.895	=	Megapascals
	Pounds force per square inch (psi)	x	0.07031	=	Kilograms force per square centimeter[d]
	Pounds force per square foot	x	47.88	=	Pascals
	Pounds force per square foot	x	4.882	=	Kilograms force per square meter[d]
Flow	Cubic feet per minute	x	0.4719	=	Liters per second
	Gallons per minute	x	0.0631	=	Liters per second
	Gallons per hour	x	1.05	=	Milliliters per second
Bending Moment Or Torque	Inch-pounds force	x	0.01152	=	Meter-kilograms force[d]
	Inch-pounds force	x	0.1130	=	Newton-meters
	Foot-pounds force	x	0.1383	=	Meter-kilograms force[d]
	Foot-pounds force	x	1.356	=	Newton-meters
	Meter-kilograms force[d]	x	9.807	=	Newton-meters
Mass	Ounces (avoirdupois)	x	28.35	=	Grams
	Pounds (avoirdupois)	x	0.4536	=	Kilograms
	Tons (metric)	x	1000	=	Kilograms
	Tons, short (2000 pounds)	x	907.2	=	Kilograms
	Tons, short (2000 pounds)	x	0.9072	=	Megagrams[e]
Mass per Unit Volume	Pounds mass per cubic foot	x	16.02	=	Kilograms per cubic meter
	Pounds mass per cubic yard	x	0.5933	=	Kilograms per cubic meter
	Pounds mass per gallon (U.S. liquid)[b]	x	119.8	=	Kilograms per cubic meter
	Pounds mass per gallon (Canadian liquid)[b]	x	99.78	=	Kilograms per cubic meter
Temperature	Degrees Fahrenheit		(F-32)/1.8	=	Degrees Celsius
	Degrees Fahrenheit		(F+459.67)/1.8	=	Degrees Kelvin
	Degrees Celsius		C+273.15	=	Degrees Kelvin

[a]The factor given is exact
[b]One U.S. gallon = 0.8327 Canadian gallon
[c]1 liter = 1000 milliliters = 1000 cubic centimeters
1 cubic decimeter = 0.001 cubic meter
[d]Metric but not SI unit
[e]Called "tonne" in England and "metric ton" in other metric countries

Table L1090-201 Metric Equivalents of Cement Content for Concrete Mixes

94 Pound Bags per Cubic Yard	Kilograms per Cubic Meter	94 Pound Bags per Cubic Yard	Kilograms per Cubic Meter
1.0	55.77	7.0	390.4
1.5	83.65	7.5	418.3
2.0	111.5	8.0	446.2
2.5	139.4	8.5	474.0
3.0	167.3	9.0	501.9
3.5	195.2	9.5	529.8
4.0	223.1	10.0	557.7
4.5	251.0	10.5	585.6
5.0	278.8	11.0	613.5
5.5	306.7	11.5	641.3
6.0	334.6	12.0	669.2
6.5	362.5	12.5	697.1

a. If you know the cement content in pounds per cubic yard, multiply by .5933 to obtain kilograms per cubic meter.

b. If you know the cement content in 94 pound bags per cubic yard, multiply by 55.77 to obtain kilograms per cubic meter.

Table L1090-202 Metric Equivalents of Common Concrete Strengths (to convert other psi values to megapascals, multiply by .006895)

U.S. Values psi	SI Value Megapascals	Non-SI Metric Value kgf/cm^2*
2000	14	140
2500	17	175
3000	21	210
3500	24	245
4000	28	280
4500	31	315
5000	34	350
6000	41	420
7000	48	490
8000	55	560
9000	62	630
10,000	69	705

* kilograms force per square centimeter

Table L1090-203 Comparison of U.S. Customary Units and SI Units for Reinforcing Bars

U.S. Customary Units							
		Nominal Dimensions[a]			Deformation Requirements, in.		
Bar Designation No.[b]	Nominal Weight, lb/ft	Diameter in.	Cross Sectional Area, $in.^2$	Perimeter in.	Maximum Average Spacing	Minimum Average Height	Maximum Gap (Chord of 12-1/2% of Nominal Perimeter)
3	0.376	0.375	0.11	1.178	0.262	0.015	0.143
4	0.668	0.500	0.20	1.571	0.350	0.020	0.191
5	1.043	0.625	0.31	1.963	0.437	0.028	0.239
6	1.502	0.750	0.44	2.356	0.525	0.038	0.286
7	2.044	0.875	0.60	2.749	0.612	0.044	0.334
8	2.670	1.000	0.79	3.142	0.700	0.050	0.383
9	3.400	1.128	1.00	3.544	0.790	0.056	0.431
10	4.303	1.270	1.27	3.990	0.889	0.064	0.487
11	5.313	1.410	1.56	4.430	0.987	0.071	0.540
14	7.65	1.693	2.25	5.32	1.185	0.085	0.648
18	13.60	2.257	4.00	7.09	1.58	0.102	0.864
SI UNITS							
		Nominal Dimensions[a]			Deformation Requirements, mm		
Bar Designation No.[b]	Nominal Weight kg/m	Diameter, mm	Cross Sectional Area, cm^2	Perimeter, mm	Maximum Average Spacing	Minimum Average Height	Maximum Gap (Chord of 12-1/2% of Nominal Perimeter)
3	0.560	9.52	0.71	29.9	6.7	0.38	3.5
4	0.994	12.70	1.29	39.9	8.9	0.51	4.9
5	1.552	15.88	2.00	49.9	11.1	0.71	6.1
6	2.235	19.05	2.84	59.8	13.3	0.96	7.3
7	3.042	22.22	3.87	69.8	15.5	1.11	8.5
8	3.973	25.40	5.10	79.8	17.8	1.27	9.7
9	5.059	28.65	6.45	90.0	20.1	1.42	10.9
10	6.403	32.26	8.19	101.4	22.6	1.62	11.4
11	7.906	35.81	10.06	112.5	25.1	1.80	13.6
14	11.384	43.00	14.52	135.1	30.1	2.16	16.5
18	20.238	57.33	25.81	180.1	40.1	2.59	21.9

[a]Nominal dimensions of a deformed bar are equivalent to those of a plain round bar having the same weight per foot as the deformed bar.

[b]Bar numbers are based on the number of eighths of an inch included in the nominal diameter of the bars.

Table L1090-204 Metric Rebar Specification - ASTM A615-81

Grade 300 (300 MPa* = **43,560 psi; +8.7% vs. Grade 40)**				
Grade 400 (400 MPa* = **58,000 psi; –3.4% vs. Grade 60)**				
Bar No.	**Diameter mm**	**Area mm^2**	**Equivalent $in.^2$**	**Comparison with U.S. Customary Bars**
10	11.3	100	.16	Between #3 & #4
15	16.0	200	.31	#5 (.31 $in.^2$)
20	19.5	300	.47	#6 (.44 $in.^2$)
25	25.2	500	.78	#8 (.79 $in.^2$)
30	29.9	700	1.09	#9 (1.00 $in.^2$)
35	35.7	1000	1.55	#11 (1.56 $in.^2$)
45	43.7	1500	2.33	#14 (2.25 $in.^2$)
55	56.4	2500	3.88	#18 (4.00 $in.^2$)

* MPa = megapascals

Grade 300 bars are furnished only in sizes 10 through 35

PRELIMINARY ESTIMATE

PROJECT	TOTAL SITE AREA
BUILDING TYPE	OWNER
LOCATION	ARCHITECT
DATE OF CONSTRUCTION	ESTIMATED CONSTRUCTION PERIOD
BRIEF DESCRIPTION	
TYPE OF PLAN	TYPE OF CONSTRUCTION
QUALITY	BUILDING CAPACITY

Floor		Wall Areas			
Below Grade Levels		Foundation Walls	L.F.	Ht.	S.F.
Area	S.F.	Frost Walls	L.F.	Ht.	S.F.
Area	S.F.	Exterior Closure		Total	S.F.
Total Area	S.F.	Comment			
Ground Floor		Fenestration		%	S.F.
Area	S.F.			%	S.F.
Area	S.F.	Exterior Wall		%	S.F.
Total Area	S.F.			%	S.F.
Supported Levels		**Site Work**			
Area	S.F.	Parking		S.F. (For	Cars)
Area	S.F.	Access Roads		L.F. (X	Ft. Wide)
Area	S.F.	Sidewalk		L.F. (X	Ft. Wide)
Area	S.F.	Landscaping		S.F. (	% Unbuilt Site)
Area	S.F.	**Building Codes**			
Total Area	S.F.	City		County	
Miscellaneous		National		Other	
Area	S.F.	**Loading**			
Area	S.F.	Roof	psf	Ground Floor	psf
Area	S.F.	Supported Floors	psf	Corridor	psf
Area	S.F.	Balcony	psf	Partition, allow	psf
Total Area	S.F.	Miscellaneous			psf
Net Finished Area	S.F.	Live Load Reduction			
Net Floor Area	S.F.	Wind			
Gross Floor Area	S.F.	Earthquake		Zone	
Roof		Comment			
Total Area	S.F.	Soil Type			
Comments		Bearing Capacity			K.S.F.
		Frost Depth			Ft.
Volume		**Frame**			
Depth of Floor System		Type		Bay Spacing	
Minimum	In.	Foundation, Standard			
Maximum	In.	Special			
Foundation Wall Height	Ft.	Substructure			
Floor to Floor Height	Ft.	Comment			
Floor to Ceiling Height	Ft.	Superstructure, Vertical		Horizontal	
Subgrade Volume	C.F.	Fireproofing		☐ Columns	Hrs.
Above Grade Volume	C.F.	☐ Girders	Hrs.	☐ Beams	Hrs.
Total Building Volume	C.F.	☐ Floor	Hrs.	☐ None	

Estimating Form

Systems Costs

ASSEMBLY NUMBER	DESCRIPTION	QTY	UNIT	TOTAL COST		COST PER S.F.
				UNIT	TOTAL	
A10	**Foundations**					
A20	**Basement Construction**					
B10	**Superstructure**					
B20	**Exterior Closure**					
B30	**Roofing**					

Estimating Form

Systems Costs

ASSEMBLY NUMBER	DESCRIPTION	QTY	UNIT	TOTAL COST		COST PER S.F.
				UNIT	TOTAL	
C	**Interior Construction**					
D10	**Conveying**					
D20	**Plumbing**					

Estimating Form

Systems Costs

ASSEMBLY NUMBER	DESCRIPTION	QTY	UNIT	TOTAL COST		COST PER S.F.
				UNIT	TOTAL	
D30	**HVAC**					
D40	**Fire Protection**					
D50	**Electrical**					

ASSEMBLY NUMBER	DESCRIPTION	QTY	UNIT	TOTAL COST		COST PER S.F.
				UNIT	TOTAL	
E	Equipment & Furnishings					
F	Special Construction					
G	Sitework					

Preliminary Estimate Cost Summary

PROJECT	TOTAL AREA	SHEET NO.
LOCATION	TOTAL VOLUME	ESTIMATE NO.
ARCHITECT	COST PER S.F.	DATE
OWNER	COST PER C.F.	NO OF STORIES
QUANTITIES BY	EXTENSIONS BY	CHECKED BY

DIV	DESCRIPTION	SUBTOTAL COST	COST/S.F.	PERCENTAGE
A	SUBSTRUCTURE			
B10	SHELL: SUPERSTRUCTURE			
B20	SHELL: EXTERIOR CLOSURE			
B30	SHELL: ROOFING			
C	INTERIOR CONSTRUCTION			
D10	SERVICES: CONVEYING			
D20	SERVICES: PLUMBING			
D30	SERVICES: HVAC			
D40	SERVICES: FIRE PROTECTION			
D50	SERVICES: ELECTRICAL			
E	EQUIPMENT & FURNISHINGS			
F	SPECIAL CONSTRUCTION			
G	SITEWORK			
	BUILDING SUBTOTAL	$ -		

Sales Tax __ % x Subtotal /2		$ -
General Conditions ___% x Subtotal		$ -
	Subtotal "A"	$ -
Overhead ___% x Subtotal "A"		$ -
	Subtotal "B"	$ -
Profit ___ % x Subtotal "B"		$ -
	Subtotal "C"	$ -
Location Factor ____ % x Subtotal "C" (Boston, MA)	Localized Cost	$ -
Architects Fee ___ x Localized Cost =		$ -
Contingency ____ x Localized Cost =		$ -
	Project Total Cost	**$ -**
Square Foot Cost $ / S.F. =	**S.F. Cost**	**$ -**
Cubic Foot Cost $ / C.F. =	**C.F. Cost**	**$ -**

Abbreviations

A	Area Square Feet; Ampere
AAFES	Army and Air Force Exchange Service
ABS	Acrylonitrile Butadiene Stryrene; Asbestos Bonded Steel
A.C., AC	Alternating Current; Air-Conditioning; Asbestos Cement; Plywood Grade A & C
ACI	American Concrete Institute
ACR	Air Conditioning Refrigeration
ADA	Americans with Disabilities Act
AD	Plywood, Grade A & D
Addit.	Additional
Adj.	Adjustable
af	Audio-frequency
AFUE	Annual Fuel Utilization Efficiency
AGA	American Gas Association
Agg.	Aggregate
A.H., Ah	Ampere Hours
A hr.	Ampere-hour
A.H.U., AHU	Air Handling Unit
A.I.A.	American Institute of Architects
AIC	Ampere Interrupting Capacity
Allow.	Allowance
alt., alt	Alternate
Alum.	Aluminum
a.m.	Ante Meridiem
Amp.	Ampere
Anod.	Anodized
ANSI	American National Standards Institute
APA	American Plywood Association
Approx.	Approximate
Apt.	Apartment
Asb.	Asbestos
A.S.B.C.	American Standard Building Code
Asbe.	Asbestos Worker
ASCE	American Society of Civil Engineers
A.S.H.R.A.E.	American Society of Heating, Refrig. & AC Engineers
ASME	American Society of Mechanical Engineers
ASTM	American Society for Testing and Materials
Attchmt.	Attachment
Avg., Ave.	Average
AWG	American Wire Gauge
AWWA	American Water Works Assoc.
Bbl.	Barrel
B&B, BB	Grade B and Better; Balled & Burlapped
B&S	Bell and Spigot
B.&W.	Black and White
b.c.c.	Body-centered Cubic
B.C.Y.	Bank Cubic Yards
BE	Bevel End
B.F.	Board Feet
Bg. cem.	Bag of Cement
BHP	Boiler Horsepower; Brake Horsepower
B.I.	Black Iron
bidir.	bidirectional
Bit., Bitum.	Bituminous
Bit., Conc.	Bituminous Concrete
Bk.	Backed
Bkrs.	Breakers
Bldg., bldg	Building
Blk.	Block
Bm.	Beam
Boil.	Boilermaker
bpm	Blows per Minute
BR	Bedroom
Brg., brng.	Bearing
Brhe.	Bricklayer Helper
Bric.	Bricklayer
Brk., brk	Brick
brkt	Bracket
Brs.	Brass
Brz.	Bronze
Bsn.	Basin
Btr.	Better
BTU	British Thermal Unit
BTUH	BTU per Hour
Bu.	Bushels
BUR	Built-up Roofing
BX	Interlocked Armored Cable
°C	Degree Centigrade
c	Conductivity, Copper Sweat
C	Hundred; Centigrade
C/C	Center to Center, Cedar on Cedar
C-C	Center to Center
Cab	Cabinet
Cair.	Air Tool Laborer
Cal.	Caliper
Calc	Calculated
Cap.	Capacity
Carp.	Carpenter
C.B.	Circuit Breaker
C.C.A.	Chromate Copper Arsenate
C.C.F.	Hundred Cubic Feet
cd	Candela
cd/sf	Candela per Square Foot
CD	Grade of Plywood Face & Back
CDX	Plywood, Grade C & D, exterior glue
Cefi.	Cement Finisher
Cem.	Cement
CF	Hundred Feet
C.F.	Cubic Feet
CFM	Cubic Feet per Minute
CFRP	Carbon Fiber Reinforced Plastic
c.g.	Center of Gravity
CHW	Chilled Water; Commercial Hot Water
C.I., CI	Cast Iron
C.I.P., CIP	Cast in Place
Circ.	Circuit
C.L.	Carload Lot
CL	Chain Link
Clab.	Common Laborer
Clam	Common Maintenance Laborer
C.L.F.	Hundred Linear Feet
CLF	Current Limiting Fuse
CLP	Cross Linked Polyethylene
cm	Centimeter
CMP	Corr. Metal Pipe
CMU	Concrete Masonry Unit
CN	Change Notice
Col.	Column
CO_2	Carbon Dioxide
Comb.	Combination
comm.	Commercial, Communication
Compr.	Compressor
Conc.	Concrete
Cont., cont	Continuous; Continued, Container
Corr.	Corrugated
Cos	Cosine
Cot	Cotangent
Cov.	Cover
C/P	Cedar on Paneling
CPA	Control Point Adjustment
Cplg.	Coupling
CPM	Critical Path Method
CPVC	Chlorinated Polyvinyl Chloride
C.Pr.	Hundred Pair
CRC	Cold Rolled Channel
Creos.	Creosote
Crpt.	Carpet & Linoleum Layer
CRT	Cathode-ray Tube
CS	Carbon Steel, Constant Shear Bar Joist
Csc	Cosecant
C.S.F.	Hundred Square Feet
CSI	Construction Specifications Institute
CT	Current Transformer
CTS	Copper Tube Size
Cu	Copper, Cubic
Cu. Ft.	Cubic Foot
cw	Continuous Wave
C.W.	Cool White; Cold Water
Cwt.	100 Pounds
C.W.X.	Cool White Deluxe
C.Y.	Cubic Yard (27 cubic feet)
C.Y./Hr.	Cubic Yard per Hour
Cyl.	Cylinder
d	Penny (nail size)
D	Deep; Depth; Discharge
Dis., Disch.	Discharge
Db	Decibel
Dbl.	Double
DC	Direct Current
DDC	Direct Digital Control
Demob.	Demobilization
d.f.t.	Dry Film Thickness
d.f.u.	Drainage Fixture Units
D.H.	Double Hung
DHW	Domestic Hot Water
DI	Ductile Iron
Diag.	Diagonal
Diam., Dia	Diameter
Distrib.	Distribution
Div.	Division
Dk.	Deck
D.L.	Dead Load; Diesel
DLH	Deep Long Span Bar Joist
dlx	Deluxe
Do.	Ditto
DOP	Dioctyl Phthalate Penetration Test (Air Filters)
Dp., dp	Depth
D.P.S.T.	Double Pole, Single Throw
Dr.	Drive
DR	Dimension Ratio
Drink.	Drinking
D.S.	Double Strength
D.S.A.	Double Strength A Grade
D.S.B.	Double Strength B Grade
Dty.	Duty
DWV	Drain Waste Vent
DX	Deluxe White, Direct Expansion
dyn	Dyne
e	Eccentricity
E	Equipment Only; East; Emissivity
Ea.	Each
EB	Encased Burial
Econ.	Economy
E.C.Y	Embankment Cubic Yards
EDP	Electronic Data Processing
EIFS	Exterior Insulation Finish System
E.D.R.	Equiv. Direct Radiation
Eq.	Equation
EL	Elevation
Elec.	Electrician; Electrical
Elev.	Elevator; Elevating
EMT	Electrical Metallic Conduit; Thin Wall Conduit
Eng.	Engine, Engineered
EPDM	Ethylene Propylene Diene Monomer
EPS	Expanded Polystyrene
Eqhv.	Equip. Oper., Heavy
Eqlt.	Equip. Oper., Light
Eqmd.	Equip. Oper., Medium
Eqmm.	Equip. Oper., Master Mechanic
Eqol.	Equip. Oper., Oilers
Equip.	Equipment
ERW	Electric Resistance Welded

Abbreviations

E.S.	Energy Saver
Est.	Estimated
esu	Electrostatic Units
E.W.	Each Way
EWT	Entering Water Temperature
Excav.	Excavation
excl	Excluding
Exp., exp	Expansion, Exposure
Ext., ext	Exterior; Extension
Extru.	Extrusion
f.	Fiber Stress
F	Fahrenheit; Female; Fill
Fab., fab	Fabricated; Fabric
FBGS	Fiberglass
F.C.	Footcandles
f.c.c.	Face-centered Cubic
f'c.	Compressive Stress in Concrete; Extreme Compressive Stress
F.E.	Front End
FEP	Fluorinated Ethylene Propylene (Teflon)
F.G.	Flat Grain
F.H.A.	Federal Housing Administration
Fig.	Figure
Fin.	Finished
FIPS	Female Iron Pipe Size
Fixt.	Fixture
FJP	Finger jointed and primed
Fl. Oz.	Fluid Ounces
Flr.	Floor
FM	Frequency Modulation; Factory Mutual
Fmg.	Framing
FM/UL	Factory Mutual/Underwriters Labs
Fdn.	Foundation
FNPT	Female National Pipe Thread
Fori.	Foreman, Inside
Foro.	Foreman, Outside
Fount.	Fountain
fpm	Feet per Minute
FPT	Female Pipe Thread
Fr	Frame
F.R.	Fire Rating
FRK	Foil Reinforced Kraft
FSK	Foil/Scrim/Kraft
FRP	Fiberglass Reinforced Plastic
FS	Forged Steel
FSC	Cast Body; Cast Switch Box
Ft., ft	Foot; Feet
Ftng.	Fitting
Ftg.	Footing
Ft lb.	Foot Pound
Furn.	Furniture
FVNR	Full Voltage Non-Reversing
FVR	Full Voltage Reversing
FXM	Female by Male
Fy.	Minimum Yield Stress of Steel
g	Gram
G	Gauss
Ga.	Gauge
Gal., gal.	Gallon
Galv., galv	Galvanized
GC/MS	Gas Chromatograph/Mass Spectrometer
Gen.	General
GFI	Ground Fault Interrupter
GFRC	Glass Fiber Reinforced Concrete
Glaz.	Glazier
GPD	Gallons per Day
gpf	Gallon per Flush
GPH	Gallons per Hour
gpm, GPM	Gallons per Minute
GR	Grade
Gran.	Granular
Grnd.	Ground
GVW	Gross Vehicle Weight
GWB	Gypsum Wall Board
H	High Henry
HC	High Capacity
H.D., HD	Heavy Duty; High Density
H.D.O.	High Density Overlaid
HDPE	High Density Polyethylene Plastic
Hdr.	Header
Hdwe.	Hardware
H.I.D., HID	High Intensity Discharge
Help.	Helper Average
HEPA	High Efficiency Particulate Air Filter
Hg	Mercury
HIC	High Interrupting Capacity
HM	Hollow Metal
HMWPE	High Molecular Weight Polyethylene
HO	High Output
Horiz.	Horizontal
H.P., HP	Horsepower; High Pressure
H.P.F.	High Power Factor
Hr.	Hour
Hrs./Day	Hours per Day
HSC	High Short Circuit
Ht.	Height
Htg.	Heating
Htrs.	Heaters
HVAC	Heating, Ventilation & Air-Conditioning
Hvy.	Heavy
HW	Hot Water
Hyd.; Hydr.	Hydraulic
Hz	Hertz (cycles)
I.	Moment of Inertia
IBC	International Building Code
I.C.	Interrupting Capacity
ID	Inside Diameter
I.D.	Inside Dimension; Identification
I.F.	Inside Frosted
I.M.C.	Intermediate Metal Conduit
In.	Inch
Incan.	Incandescent
Incl.	Included; Including
Int.	Interior
Inst.	Installation
Insul., insul	Insulation/Insulated
I.P.	Iron Pipe
I.P.S., IPS	Iron Pipe Size
IPT	Iron Pipe Threaded
I.W.	Indirect Waste
J	Joule
J.I.C.	Joint Industrial Council
K	Thousand; Thousand Pounds; Heavy Wall Copper Tubing, Kelvin
K.A.H.	Thousand Amp. Hours
kcmil	Thousand Circular Mils
KD	Knock Down
K.D.A.T.	Kiln Dried After Treatment
kg	Kilogram
kG	Kilogauss
kgf	Kilogram Force
kHz	Kilohertz
Kip	1000 Pounds
KJ	Kilojoule
K.L.	Effective Length Factor
K.L.F.	Kips per Linear Foot
Km	Kilometer
KO	Knock Out
K.S.F.	Kips per Square Foot
K.S.I.	Kips per Square Inch
kV	Kilovolt
kVA	Kilovolt Ampere
kVAR	Kilovar (Reactance)
KW	Kilowatt
KWh	Kilowatt-hour
L	Labor Only; Length; Long; Medium Wall Copper Tubing
Lab.	Labor
lat	Latitude
Lath.	Lather
Lav.	Lavatory
lb.; #	Pound
L.B., LB	Load Bearing; L Conduit Body
L. & E.	Labor & Equipment
lb./hr.	Pounds per Hour
lb./L.F.	Pounds per Linear Foot
lbf/sq.in.	Pound-force per Square Inch
L.C.L.	Less than Carload Lot
L.C.Y.	Loose Cubic Yard
Ld.	Load
LE	Lead Equivalent
LED	Light Emitting Diode
L.F.	Linear Foot
L.F. Nose	Linear Foot of Stair Nosing
L.F. Rsr	Linear Foot of Stair Riser
Lg.	Long; Length; Large
L & H	Light and Heat
LH	Long Span Bar Joist
L.H.	Labor Hours
L.L., LL	Live Load
L.L.D.	Lamp Lumen Depreciation
lm	Lumen
lm/sf	Lumen per Square Foot
lm/W	Lumen per Watt
LOA	Length Over All
log	Logarithm
L-O-L	Lateralolet
long.	Longitude
L.P., LP	Liquefied Petroleum; Low Pressure
L.P.F.	Low Power Factor
LR	Long Radius
L.S.	Lump Sum
Lt.	Light
Lt. Ga.	Light Gauge
L.T.L.	Less than Truckload Lot
Lt. Wt.	Lightweight
L.V.	Low Voltage
M	Thousand; Material; Male; Light Wall Copper Tubing
M^{2}CA	Meters Squared Contact Area
m/hr.; M.H.	Man-hour
mA	Milliampere
Mach.	Machine
Mag. Str.	Magnetic Starter
Maint.	Maintenance
Marb.	Marble Setter
Mat; Mat'l.	Material
Max.	Maximum
MBF	Thousand Board Feet
MBH	Thousand BTU's per hr.
MC	Metal Clad Cable
MCC	Motor Control Center
M.C.F.	Thousand Cubic Feet
MCFM	Thousand Cubic Feet per Minute
M.C.M.	Thousand Circular Mils
MCP	Motor Circuit Protector
MD	Medium Duty
MDF	Medium-density fibreboard
M.D.O.	Medium Density Overlaid
Med.	Medium
MF	Thousand Feet
M.F.B.M.	Thousand Feet Board Measure
Mfg.	Manufacturing
Mfrs.	Manufacturers
mg	Milligram
MGD	Million Gallons per Day
MGPH	Million Gallons per Hour
MH, M.H.	Manhole; Metal Halide; Man-Hour
MHz	Megahertz
Mi.	Mile
MI	Malleable Iron; Mineral Insulated
MIPS	Male Iron Pipe Size
mj	Mechanical Joint
m	Meter
mm	Millimeter
Mill.	Millwright
Min., min.	Minimum, Minute

Abbreviations

Misc.	Miscellaneous
ml	Milliliter, Mainline
M.L.F.	Thousand Linear Feet
Mo.	Month
Mobil.	Mobilization
Mog.	Mogul Base
MPH	Miles per Hour
MPT	Male Pipe Thread
MRGWB	Moisture Resistant Gypsum Wallboard
MRT	Mile Round Trip
ms	Millisecond
M.S.F.	Thousand Square Feet
Mstz.	Mosaic & Terrazzo Worker
M.S.Y.	Thousand Square Yards
Mtd., mtd., mtd	Mounted
Mthe.	Mosaic & Terrazzo Helper
Mtng.	Mounting
Mult.	Multi; Multiply
M.V.A.	Million Volt Amperes
M.V.A.R.	Million Volt Amperes Reactance
MV	Megavolt
MW	Megawatt
MXM	Male by Male
MYD	Thousand Yards
N	Natural; North
nA	Nanoampere
NA	Not Available; Not Applicable
N.B.C.	National Building Code
NC	Normally Closed
NEMA	National Electrical Manufacturers Assoc.
NEHB	Bolted Circuit Breaker to 600V.
NFPA	National Fire Protection Association
NLB	Non-Load-Bearing
NM	Non-Metallic Cable
nm	Nanometer
No.	Number
NO	Normally Open
N.O.C.	Not Otherwise Classified
Nose.	Nosing
NPT	National Pipe Thread
NQOD	Combination Plug-on/Bolt on Circuit Breaker to 240V.
N.R.C., NRC	Noise Reduction Coefficient/ Nuclear Regulator Commission
N.R.S.	Non Rising Stem
ns	Nanosecond
nW	Nanowatt
OB	Opposing Blade
OC	On Center
OD	Outside Diameter
O.D.	Outside Dimension
ODS	Overhead Distribution System
O.G.	Ogee
O.H.	Overhead
O&P	Overhead and Profit
Oper.	Operator
Opng.	Opening
Orna.	Ornamental
OSB	Oriented Strand Board
OS&Y	Outside Screw and Yoke
OSHA	Occupational Safety and Health Act
Ovhd.	Overhead
OWG	Oil, Water or Gas
Oz.	Ounce
P.	Pole; Applied Load; Projection
p.	Page
Pape.	Paperhanger
P.A.P.R.	Powered Air Purifying Respirator
PAR	Parabolic Reflector
P.B., PB	Push Button
Pc., Pcs.	Piece, Pieces
P.C.	Portland Cement; Power Connector
P.C.F.	Pounds per Cubic Foot
PCM	Phase Contrast Microscopy
PDCA	Painting and Decorating Contractors of America
P.E., PE	Professional Engineer; Porcelain Enamel; Polyethylene; Plain End
P.E.C.I.	Porcelain Enamel on Cast Iron
Perf.	Perforated
PEX	Cross Linked Polyethylene
Ph.	Phase
P.I.	Pressure Injected
Pile.	Pile Driver
Pkg.	Package
Pl.	Plate
Plah.	Plasterer Helper
Plas.	Plasterer
plf	Pounds Per Linear Foot
Pluh.	Plumber Helper
Plum.	Plumber
Ply.	Plywood
p.m.	Post Meridiem
Pntd.	Painted
Pord.	Painter, Ordinary
pp	Pages
PP, PPL	Polypropylene
P.P.M.	Parts per Million
Pr.	Pair
P.E.S.B.	Pre-engineered Steel Building
Prefab.	Prefabricated
Prefin.	Prefinished
Prop.	Propelled
PSF, psf	Pounds per Square Foot
PSI, psi	Pounds per Square Inch
PSIG	Pounds per Square Inch Gauge
PSP	Plastic Sewer Pipe
Pspr.	Painter, Spray
Psst.	Painter, Structural Steel
P.T.	Potential Transformer
P. & T.	Pressure & Temperature
Ptd.	Painted
Ptns.	Partitions
Pu	Ultimate Load
PVC	Polyvinyl Chloride
Pvmt.	Pavement
PRV	Pressure Relief Valve
Pwr.	Power
Q	Quantity Heat Flow
Qt.	Quart
Quan., Qty.	Quantity
Q.C.	Quick Coupling
r	Radius of Gyration
R	Resistance
R.C.P.	Reinforced Concrete Pipe
Rect.	Rectangle
recpt.	Receptacle
Reg.	Regular
Reinf.	Reinforced
Req'd.	Required
Res.	Resistant
Resi.	Residential
RF	Radio Frequency
RFID	Radio-frequency Identification
Rgh.	Rough
RGS	Rigid Galvanized Steel
RHW	Rubber, Heat & Water Resistant; Residential Hot Water
rms	Root Mean Square
Rnd.	Round
Rodm.	Rodman
Rofc.	Roofer, Composition
Rofp.	Roofer, Precast
Rohe.	Roofer Helpers (Composition)
Rots.	Roofer, Tile & Slate
R.O.W.	Right of Way
RPM	Revolutions per Minute
R.S.	Rapid Start
Rsr	Riser
RT	Round Trip
S.	Suction; Single Entrance; South
SBS	Styrene Butadiere Styrene
SC	Screw Cover
SCFM	Standard Cubic Feet per Minute
Scaf.	Scaffold
Sch., Sched.	Schedule
S.C.R.	Modular Brick
S.D.	Sound Deadening
SDR	Standard Dimension Ratio
S.E.	Surfaced Edge
Sel.	Select
SER, SEU	Service Entrance Cable
S.F.	Square Foot
S.F.C.A.	Square Foot Contact Area
S.F. Flr.	Square Foot of Floor
S.F.G.	Square Foot of Ground
S.F. Hor.	Square Foot Horizontal
SFR	Square Feet of Radiation
S.F. Shlf.	Square Foot of Shelf
S4S	Surface 4 Sides
Shee.	Sheet Metal Worker
Sin.	Sine
Skwk.	Skilled Worker
SL	Saran Lined
S.L.	Slimline
Sldr	Solder
SLH	Super Long Span Bar Joist
S.N.	Solid Neutral
SO	Stranded with oil resistant inside insulation
S-O-L	Socketolet
sp	Standpipe
S.P.	Static Pressure; Single Pole; Self-Propelled
Spri.	Sprinkler Installer
spwg	Static Pressure Water Gauge
S.P.D.T.	Single Pole, Double Throw
SPF	Spruce Pine Fir; Sprayed Polyurethane Foam
S.P.S.T.	Single Pole, Single Throw
SPT	Standard Pipe Thread
Sq.	Square; 100 Square Feet
Sq. Hd.	Square Head
Sq. In.	Square Inch
S.S.	Single Strength; Stainless Steel
S.S.B.	Single Strength B Grade
sst, ss	Stainless Steel
Sswk.	Structural Steel Worker
Sswl.	Structural Steel Welder
St.; Stl.	Steel
STC	Sound Transmission Coefficient
Std.	Standard
Stg.	Staging
STK	Select Tight Knot
STP	Standard Temperature & Pressure
Stpi.	Steamfitter, Pipefitter
Str.	Strength; Starter; Straight
Strd.	Stranded
Struct.	Structural
Sty	Story
Subj.	Subject
Subs.	Subcontractors
Surf.	Surface
Sw.	Switch
Swbd.	Switchboard
S.Y.	Square Yard
Syn.	Synthetic
S.Y.P.	Southern Yellow Pine
Sys.	System
t.	Thickness
T	Temperature; Ton
Tan	Tangent
T.C.	Terra Cotta
T & C	Threaded and Coupled
T.D.	Temperature Difference
TDD	Telecommunications Device for the Deaf
T.E.M.	Transmission Electron Microscopy
temp	Temperature, Tempered, Temporary
TFFN	Nylon Jacketed Wire

Abbreviations

TFE	Tetrafluoroethylene (Teflon)
T. & G.	Tongue & Groove; Tar & Gravel
Th., Thk.	Thick
Thn.	Thin
Thrded	Threaded
Tilf.	Tile Layer, Floor
Tilh.	Tile Layer, Helper
THHN	Nylon Jacketed Wire
THW.	Insulated Strand Wire
THWN	Nylon Jacketed Wire
T.L., TL	Truckload
T.M.	Track Mounted
Tot.	Total
T-O-L	Threadolet
tmpd	Tempered
TPO	Thermoplastic Polyolefin
T.S.	Trigger Start
Tr.	Trade
Transf.	Transformer
Trhv.	Truck Driver, Heavy
Trlr	Trailer
Trlt.	Truck Driver, Light
TTY	Teletypewriter
TV	Television
T.W.	Thermoplastic Water Resistant Wire
UCI	Uniform Construction Index
UF	Underground Feeder
UGND	Underground Feeder
UHF	Ultra High Frequency
U.I.	United Inch
U.L., UL	Underwriters Laboratory
Uld.	Unloading
Unfin.	Unfinished
UPS	Uninterruptible Power Supply
URD	Underground Residential Distribution
US	United States
USGBC	U.S. Green Building Council
USP	United States Primed
UTMCD	Uniform Traffic Manual For Control Devices
UTP	Unshielded Twisted Pair
V	Volt
VA	Volt Amperes
VAT	Vinyl Asbestos Tile
V.C.T.	Vinyl Composition Tile
VAV	Variable Air Volume
VC	Veneer Core
VDC	Volts Direct Current
Vent.	Ventilation
Vert.	Vertical
V.F.	Vinyl Faced
V.G.	Vertical Grain
VHF	Very High Frequency
VHO	Very High Output
Vib.	Vibrating
VLF	Vertical Linear Foot
VOC	Volatile Organic Compound
Vol.	Volume
VRP	Vinyl Reinforced Polyester
W	Wire; Watt; Wide; West
w/	With
W.C., WC	Water Column; Water Closet
W.F.	Wide Flange
W.G.	Water Gauge
Wldg.	Welding
W. Mile	Wire Mile
W-O-L	Weldolet
W.R.	Water Resistant
Wrck.	Wrecker
WSFU	Water Supply Fixture Unit
W.S.P.	Water, Steam, Petroleum
WT., Wt.	Weight
WWF	Welded Wire Fabric
XFER	Transfer
XFMR	Transformer
XHD	Extra Heavy Duty
XHHW	Cross-Linked Polyethylene Wire Insulation
XLPE	Cross-linked Polyethylene
XLP	
Xport	Transport
Y	Wye
yd	Yard
yr	Year
Δ	Delta
%	Percent
~	Approximately
Ø	Phase; diameter
@	At
#	Pound; Number
<	Less Than
>	Greater Than
Z	Zone

S

T

Notes

Notes

Notes

Notes

Notes

Notes

Notes

Cost Data Selection Guide

The following table provides definitive information on the content of each cost data publication. The number of lines of data provided in each unit price or assemblies division, as well as the number of crews, is listed for each data set. The presence of other elements such as reference tables, square foot models, equipment rental costs, historical cost indexes, and city cost indexes, is also indicated. You can use the table to help select the RSMeans data set that has the quantity and type of information you most need in your work.

Unit Cost Divisions	Building Construction	Mechanical	Electrical	Commercial Renovation	Square Foot	Site Work Landsc.	Green Building	Interior	Concrete Masonry	Open Shop	Heavy Construction	Light Commercial	Facilities Construction	Plumbing	Residential
1	562	383	400	504		498	205	302	445	561	496	245	1040	393	177
2	776	279	85	732		991	206	399	213	775	733	480	1220	286	273
3	1691	340	230	1084		1483	986	354	2037	1691	1693	482	1791	316	389
4	960	21	0	920		725	180	615	1158	928	615	533	1175	0	445
5	1893	158	155	1090		844	1799	1098	720	1893	1037	979	1909	204	746
6	2453	18	18	2111		110	589	1528	281	2449	123	2141	2125	22	2661
7	1594	215	128	1633		580	763	531	523	1591	26	1328	1695	227	1048
8	2111	81	44	2645		257	1136	1757	105	2096	0	2219	2890	0	1534
9	2004	86	45	1828		310	451	2088	391	1944	15	1668	2252	54	1437
10	1049	17	10	645		216	27	864	157	1049	29	539	1141	233	224
11	1096	206	166	551		129	54	933	28	107[illegible]	0	231	1116	169	110
12	530	0	2	285		205	133	1533	14	500	0	259	1554	23	217
13	744	149	158	253		367	122	252	78	720	244	108	759	115	103
14	273	36	0	221		0	0	257	0	273	0	12	293	16	6
21	91	0	16	37		0	0	250	0	9[illegible]	0	73	535	540	220
22	1168	7548	160	1201		1572	1066	848	20	1157	1681	874	7450	9370	718
23	1178	6975	580	929		157	903	778	38	117[illegible]	110	879	5213	1922	469
26	1380	458	10150	1033		793	630	1144	55	1372	562	1310	10099	399	619
27	72	0	339	34		13	0	71	0	72	39	52	321	0	22
28	96	58	136	71		0	21	78	0	99	0	40	151	44	25
31	1510	733	610	806		3263	288	7	1217	1455	3277	604	1569	660	613
32	813	49	8	881		4409	353	405	291	784	1826	421	1687	133	468
33	529	1076	538	252		2173	41	0	236	522	2157	128	1697	1283	154
34	107	0	47	4		190	0	0	31	62	2[illegible]4	0	136	0	0
35	18	0	0	0		327	0	0	0	18	442	0	84	0	0
41	60	0	0	33		8	0	22	0	61	31	0	68	14	0
44	75	79	0	0		0	0	0	0	0	0	0	75	75	0
46	23	16	0	0		274	261	0	0	23	264	0	33	33	0
48	12	0	25	0		0	25	0	0	12	17	12	12	0	12
Totals	24868	18981	14050	19783		19894	10239	16114	8038	24440	15631	15617	50090	16531	12690

Assem Div	Building Construction	Mechanical	Elec-trical	Commercial Renovation	Square Foot	Site Work Landscape	Assemblies	Green Building	Interior	Concrete Masonry	Heavy Construction	Light Commercial	Facilities Construction	Plumbing	Asm Div	Residential
A		15	0	188	150	577	598	0	0	536	571	154	24	0	1	378
B		0	0	848	2506	0	5666	56	329	1975	368	2098	174	0	2	211
C		0	0	647	928	0	1309	0	1628	146	0	818	249	0	3	588
D		1067	941	712	1859	72	2538	330	825	0	0	1345	1105	1088	4	851
E		0	0	86	261	0	301	0	5	0	0	258	5	0	5	390
F		0	0	0	114	0	114	0	0	0	0	114	3	0	6	357
G		527	447	318	312	3364	792	0	0	534	1349	205	293	677	7	307
															8	760
															9	80
															10	0
															11	0
															12	0
Totals		1609	1388	2799	6130	4013	11318	386	2787	3191	2288	4992	1853	1765		3922

Reference Section	Building Construction Costs	Mechanical	Electrical	Commercial Renovation	Square Foot	Site Work Landscape	Assem.	Green Building	Interior	Concrete Masonry	Open Shop	Heavy Construction	Light Commercial	Facilities Construction	Plumbing	Resi.
Reference Tables	yes	yes	yes	yes	no	yes	yes	yes	yes	yes	yes	yes	yes	yes	yes	yes
Models					111			25					50			28
Crews	575	575	575	553		575		575	575	575	551	575	551	553	575	551
Equipment Rental Costs	yes	yes	yes	yes		yes		yes	yes	yes	yes	yes	yes	yes	yes	yes
Historical Cost Indexes	yes	yes	yes	yes	yes	yes	yes	yes	yes	yes	yes	yes	yes	yes	yes	no
City Cost Indexes	yes	yes	yes	yes	yes	yes	yes	yes	yes	yes	yes	yes	yes	yes	yes	yes

Online Book CD eBook Online add-on available for additional fee

For more information visit our website at www.RSMeans.com

RSMeans Building Construction Cost Data 2016

Offers you unchallenged unit price reliability in an easy-to-use format. Whether used for verifying complete, finished estimates or for periodic checks, it supplies more cost facts better and faster than any comparable source. More than 24,700 unit prices have been updated for 2016. The City Cost Indexes and Location Factors cover more than 930 areas for indexing to any project location in North America. Order and get *RSMeans Quarterly Update Service* FREE.

RSMeans Green Building Cost Data 2016

Estimate, plan, and budget the costs of green building for both new commercial construction and renovation work with this sixth edition of *RSMeans Green Building Cost Data*. More than 10,000 unit costs for a wide array of green building products plus assemblies costs. Easily identified cross references to LEED and Green Globes building rating systems criteria.

RSMeans Mechanical Cost Data 2016

Total unit and systems price guidance for mechanical construction—materials, parts, fittings, and complete labor cost information. Includes prices for piping, heating, air conditioning, ventilation, and all related construction.

Plus new 2016 unit costs for:

- Thousands of installed HVAC/controls, sub-assemblies and assemblies
- "On-site" Location Factors for more than 930 cities and towns in the U.S. and Canada
- Crews, labor, and equipment

RSMeans Facilities Construction Cost Data 2016

For the maintenance and construction of commercial, industrial, municipal, and institutional properties. Costs are shown for new and remodeling construction and are broken down into materials, labor, equipment, and overhead and profit. Special emphasis is given to sections on mechanical, electrical, furnishings, site work, building maintenance, finish work, and demolition.

More than 49,800 unit costs, plus assemblies costs and a comprehensive Reference Section are included.

RSMeans Square Foot Costs 2016

Accurate and Easy-to-Use

- Updated price information based on nationwide figures from suppliers, estimators, labor experts, and contractors.
- Green building models
- Realistic graphics, offering true-to-life illustrations of building projects
- Extensive information on using square foot cost data, including sample estimates and alternate pricing methods

RSMeans Commercial Renovation Cost Data 2016

Commercial/Multi-family Residential

Use this valuable tool to estimate commercial and multi-family residential renovation and remodeling.

Includes: Updated costs for hundreds of unique methods, materials, and conditions that only come up in repair and remodeling, PLUS:

- Unit costs for more than 19,600 construction components
- Installed costs for more than 2,700 assemblies
- More than 930 "on-site" localization factors for the U.S. and Canada

RSMeans Electrical Cost Data 2016

Pricing information for every part of electrical cost planning. More than 13,800 unit and systems costs with design tables; clear specifications and drawings; engineering guides; illustrated estimating procedures; complete labor-hour and materials costs for better scheduling and procurement; and the latest electrical products and construction methods.

- A variety of special electrical systems, including cathodic protection
- Costs for maintenance, demolition, HVAC/mechanical, specialties, equipment, and more

RSMeans Electrical Change Order Cost Data 2016

RSMeans Electrical Change Order Cost Data provides you with electrical unit prices exclusively for pricing change orders. Analyze and check your own change order estimates against those prepared when using RSMeans cost data. It also covers productivity analysis and change order cost justifications. With useful information for calculating the effects of change orders and dealing with their administration.

RSMeans Assemblies Cost Data 2016

RSMeans Assemblies Cost Data takes the guesswork out of preliminary or conceptual estimates. Now you don't have to try to calculate the assembled cost by working up individual component costs. We've done all the work for you.

Presents detailed illustrations, descriptions, specifications, and costs for every conceivable building assembly—over 350 types in all—arranged in the easy-to-use UNIFORMAT II system. Each illustrated "assembled" cost includes a complete grouping of materials and associated installation costs, including the installing contractor's overhead and profit.

RSMeans Open Shop Building Construction Cost Data 2016

The latest costs for accurate budgeting and estimating of new commercial and residential construction, renovation work, change orders, and cost engineering.

RSMeans Open Shop "BCCD" will assist you to:

- Develop benchmark prices for change orders.
- Plug gaps in preliminary estimates and budgets.
- Estimate complex projects.
- Substantiate invoices on contracts.
- Price ADA-related renovations.

RSMeans Residential Cost Data 2016

Contains square foot costs for 28 basic home models with the look of today, plus hundreds of custom additions and modifications you can quote right off the page. Includes more than 3,900 costs for 89 residential systems. Complete with blank estimating forms, sample estimates, and step-by-step instructions.

Contains line items for cultured stone and brick, PVC trim, lumber, and TPO roofing.

RSMeans Site Work & Landscape Cost Data 2016

Includes unit and assemblies costs for earthwork, sewerage, piped utilities, site improvements, drainage, paving, trees and shrubs, street openings/repairs, underground tanks, and more. Contains more than 60 types of assemblies costs for accurate conceptual estimates.

Includes:

- Estimating for infrastructure improvements
- Environmentally-oriented construction
- ADA-mandated handicapped access
- Hazardous waste line items

RSMeans Facilities Maintenance & Repair Cost Data 2016

RSMeans Facilities Maintenance & Repair Cost Data gives you a complete system to manage and plan your facility repair and maintenance costs and budget efficiently. Guidelines for auditing a facility and developing an annual maintenance plan. Budgeting is included, along with reference tables on cost and management, and information on frequency and productivity of maintenance operations.

The only nationally recognized source of maintenance and repair costs. Developed in cooperation with the Civil Engineering Research Laboratory (CERL) of the Army Corps of Engineers.

RSMeans Light Commercial Cost Data 2016

Specifically addresses the light commercial market, which is a specialized niche in the construction industry. Aids you, the owner/designer/contractor, in preparing all types of estimates—from budgets to detailed bids. Includes new advances in methods and materials.

The Assemblies section allows you to evaluate alternatives in the early stages of design/planning.

More than 15,500 unit costs ensure that you have the prices you need, when you need them.

RSMeans Concrete & Masonry Cost Data 2016

Provides you with cost facts for virtually all concrete/masonry estimating needs, from complicated formwork to various sizes and face finishes of brick and block—all in great detail. The comprehensive Unit Price Section contains more than 8,000 selected entries. Also contains an Assemblies [Cost] Section, and a detailed Reference Section that supplements the cost data.

RSMeans Labor Rates for the Construction Industry 2016

Complete information for estimating labor costs, making comparisons, and negotiating wage rates by trade for more than 300 U.S. and Canadian cities. With 46 construction trades in each city, and historical wage rates included for comparison. Each city chart lists the county and is alphabetically arranged with handy visual flip tabs for quick reference.

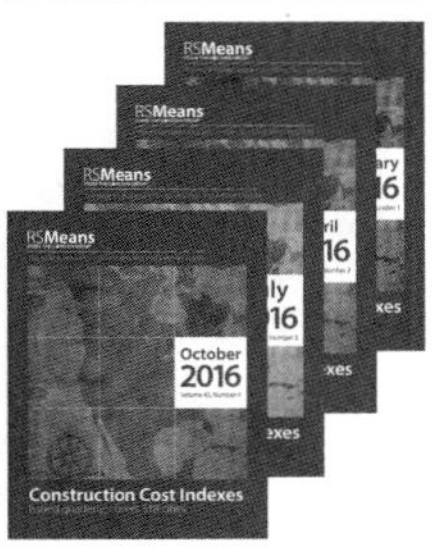

RSMeans Construction Cost Indexes 2016

What materials and labor costs will change unexpectedly this year? By how much?

- Breakdowns for 318 major cities
- National averages for 30 key cities
- Expanded five major city indexes
- Historical construction cost indexes

RSMeans Interior Cost Data 2016

Provides you with prices and guidance needed to make accurate interior work estimates. Contains costs on materials, equipment, hardware, custom installations, furnishings, and labor for new and remodel commercial and industrial interior construction, including updated information on office furnishings, as well as reference information.

RSMeans Heavy Construction Cost Data 2016

A comprehensive guide to heavy construction costs. Includes costs for highly specialized projects such as tunnels, dams, highways, airports, and waterways. Information on labor rates, equipment, and materials costs is included. Features unit price costs, systems costs, and numerous reference tables for costs and design.

RSMeans Plumbing Cost Data 2016

Comprehensive unit prices and assemblies for plumbing, irrigation systems, commercial and residential fire protection, point-of-use water heaters, and the latest approved materials. This publication and its companion, *RSMeans Mechanical Cost Data*, provide full-range cost estimating coverage for all the mechanical trades.

Contains updated costs for potable water, radiant heat systems, and high efficiency fixtures.

RSMeans Online

Competitive Cost Estimates Made Easy

RSMeans Online is a web-based service that provides accurate and up-to-date cost information to help you build competitive estimates or budgets in less time.

Quick, intuitive, easy to use, and automatically updated, RSMeans Online gives you instant access to hundreds of thousands of material, labor, and equipment costs from RSMeans' comprehensive database, delivering the information you need to build budgets and competitive estimates every time.

With RSMeans Online, you can perform quick searches to locate specific costs and adjust costs to reflect prices in your geographic area. Tag and store your favorites for fast access to your frequently used line items and assemblies and clone estimates to save time. System notifications will alert you as updated data becomes available. RSMeans Online is automatically updated throughout the year.

RSMeans Online's visual, interactive estimating features help you create, manage, save and share estimates with ease! It provides increased flexibility with customizable advanced reports. Easily edit custom report templates and import your company logo onto your estimates.

	Standard	Advanced	Professional
Unit Prices	✓	✓	✓
Assemblies	X	✓	✓
Sq Ft Models	X	X	✓
Editable Sq Foot Models	X	X	✓
Editable Assembly Components	X	✓	✓
Custom Cost Data	X	✓	✓
User Defined Components	X	✓	✓
Advanced Reporting & Customization	X	✓	✓
Union Labor Type	✓	✓	✓
Open Shop*	Add on for additional fee	Add on for additional fee	✓
History (3-year lookback)**	Add on for additional fee	Add on for additional fee	Add on for additional fee
Full History (2007)	Add on for additional fee	Add on for additional fee	Add on for additional fee

*Open shop included in Complete Library & Bundles

** 3-year history lookback included with complete library

Estimate with Precision

Find everything you need to develop complete, accurate estimates.

- Verified costs for construction materials
- Equipment rental costs
- Crew sizing, labor hours and labor rates
- Localized costs for U.S. and Canada

Save Time & Increase Efficiency

Make cost estimating and calculating faster and easier than ever with secure, online estimating tools.

- Quickly locate costs in the searchable database
- Create estimates in minutes with RSMeans cost lines
- Tag and store favorites for fast access to frequently used items

Improve Planning & Decision-Making

Back your estimates with complete, accurate and up-to-date cost data for informed business decisions.

- Verify construction costs from third parties
- Check validity of subcontractor proposals
- Evaluate material and assembly alternatives

Increase Profits

Use RSMeans Online to estimate projects quickly and accurately, so you can gain an edge over your competition.

- Create accurate and competitive bids
- Minimize the risk of cost overruns
- Reduce variability
- Gain control over costs

30 Day Free Trial

Register for a free trial at www.RSMeansOnline.com

- Access to unit prices, building assemblies and facilities repair and remodeling items covering every category of construction
- Powerful tools to customize, save and share cost lists, estimates and reports with ease
- Access to square foot models for quick conceptual estimates
- Allows up to 10 users to evaluate the full range of collaborative functionality used items.

2016 RSMeans Seminar Schedule

Note: call for exact dates and details.

Location	Dates	Location	Dates
Seattle, WA	January and August	Bethesda, MD	June
Dallas/Ft. Worth, TX	January	El Segundo, CA	August
Austin, TX	February	Jacksonville, FL	September
Anchorage, AK	March and September	Dallas, TX	September
Las Vegas, NV	March	Philadelphia, PA	October
New Orleans, LA	March	Houston, TX	October
Washington, DC	April and September	Salt Lake City, UT	November
Phoenix, AZ	April	Baltimore, MD	November
Kansas City, MO	April	Orlando, FL	November
Toronto	May	San Diego, CA	December
Denver, CO	May	San Antonio, TX	December
San Francisco, CA	June	Raleigh, NC	December

877-620-6245

RSMeans beginning early 2016 will be offering a suite of online self paced offerings. Check our website at www.RSMeans.com for more information.

Professional Development

eLearning Training Sessions

Learn how to use *RSMeans Online*® or the *RSMeans CostWorks*® CD from the convenience of your home or office. Our eLearning training sessions let you join a training conference call and share the instructors' desktops, so you can view the presentation and step-by-step instructions on your own computer screen. The live webinars are held from 9 a.m. to 4 p.m. or 11 a.m. to 6 p.m. eastern standard time, with a one-hour break for lunch.

For these sessions, you must have a computer with high speed Internet access and a compatible Web browser. Learn more at www.RSMeansOnline.com or call for a schedule: 781-422-5115. Webinars are generally held on selected Wednesdays each month.

RSMeans Online	RSMeans CostWorks CD
$299 per person	$299 per person

RSMeans Online Training

Construction estimating is vital to the decision-making process at each state of every project. RSMeansOnline works the way you do. It's systematic, flexible and intuitive. In this one day class you will see how you can estimate any phase of any project faster and better.

Some of what you'll learn:
- Customizing RSMeansOnline
- Making the most of RSMeans "Circle Reference" numbers
- How to integrate your cost data
- Generate reports, exporting estimates to MS Excel, sharing, collaborating and more

Also available: RSMeans Online training webinar

Maintenance & Repair Estimating for Facilities

This *two-day* course teaches attendees how to plan, budget, and estimate the cost of ongoing and preventive maintenance and repair for existing buildings and grounds.

Some of what you'll learn:
- The most financially favorable maintenance, repair, and replacement scheduling and estimating
- Auditing and value engineering facilities
- Preventive planning and facilities upgrading
- Determining both in-house and contract-out service costs
- Annual, asset-protecting M&R plan

Who should attend: facility managers, maintenance supervisors, buildings and grounds superintendents, plant managers, planners, estimators, and others involved in facilities planning and budgeting.

Facilities Construction Estimating

In this *two-day* course, professionals working in facilities management can get help with their daily challenges to establish budgets for all phases of a project.

Some of what you'll learn:
- Determining the full scope of a project
- Identifying the scope of risks and opportunities
- Creative solutions to estimating issues
- Organizing estimates for presentation and discussion
- Special techniques for repair/remodel and maintenance projects
- Negotiating project change orders

Who should attend: facility managers, engineers, contractors, facility tradespeople, planners, and project managers.

Practical Project Management for Construction Professionals

In this *two-day* course, acquire the essential knowledge and develop the skills to effectively and efficiently execute the day-to-day responsibilities of the construction project manager.

Some of what you'll learn:
- General conditions of the construction contract
- Contract modifications: change orders and construction change directives
- Negotiations with subcontractors and vendors
- Effective writing: notification and communications
- Dispute resolution: claims and liens

Who should attend: architects, engineers, owners' representatives, and project managers.

Construction Cost Estimating: Concepts and Practice

This introductory course to improve estimating skills and effectiveness starts with the details of interpreting bid documents and ends with the summary of the estimate and bid submission.

Some of what you'll learn:
- Using the plans and specifications for creating estimates
- The takeoff process—deriving all tasks with correct quantities
- Developing pricing using various sources; how subcontractor pricing fits in
- Summarizing the estimate to arrive at the final number
- Formulas for area and cubic measure, adding waste and adjusting productivity to specific projects
- Evaluating subcontractors' proposals and prices
- Adding insurance and bonds
- Understanding how labor costs are calculated
- Submitting bids and proposals

Who should attend: project managers, architects, engineers, owners' representatives, contractors, and anyone who's responsible for budgeting or estimating construction projects.

Mechanical & Electrical Estimating

This *two-day* course teaches attendees how to prepare more accurate and complete mechanical/electrical estimates, avoid the pitfalls of omission and double-counting, and understand the composition and rationale within the RSMeans mechanical/electrical database.

Some of what you'll learn:
- The unique way mechanical and electrical systems are interrelated
- M&E estimates—conceptual, planning, budgeting, and bidding stages
- Order of magnitude, square foot, assemblies, and unit price estimating
- Comparative cost analysis of equipment and design alternatives

Who should attend: architects, engineers, facilities managers, mechanical and electrical contractors, and others who need a highly reliable method for developing, understanding, and evaluating mechanical and electrical contracts.

Unit Price Estimating

This interactive *two-day* seminar teaches attendees how to interpret project information and process it into final, detailed estimates with the greatest accuracy level.

The most important credential an estimator can take to the job is the ability to visualize construction and estimate accurately.

Some of what you'll learn:

- Interpreting the design in terms of cost
- The most detailed, time-tested methodology for accurate pricing
- Key cost drivers—material, labor, equipment, staging, and subcontracts
- Understanding direct and indirect costs for accurate job cost accounting and change order management

Who should attend: corporate and government estimators and purchasers, architects, engineers, and others who need to produce accurate project estimates.

RSMeans CostWorks CD Training

This *one-day* course helps users become more familiar with the functionality of the *RSMeans CostWorks* program. Each menu, icon, screen, and function found in the program is explained in depth. Time is devoted to hands-on estimating exercises.

Some of what you'll learn:

- Searching the database using all navigation methods
- Exporting RSMeans data to your preferred spreadsheet format
- Viewing crews, assembly components, and much more
- Automatically regionalizing the database

This training session requires you to bring a laptop computer to class.

When you register for this course you will receive an outline for your laptop requirements.

Also offering web training for the RSMeans CostWorks CD!

Facilities Estimating Using the RSMeans CostWorks CD

Combines hands-on skill building with best estimating practices and real-life problems. Brings you up-to-date with key concepts and provides tips, pointers, and guidelines to save time and avoid cost oversights and errors.

Some of what you'll learn:

- Estimating process concepts
- Customizing and adapting RSMeans cost data
- Establishing scope of work to account for all known variables
- Budget estimating: when, why, and how
- Site visits: what to look for—what you can't afford to overlook
- How to estimate repair and remodeling variables

This training session requires you to bring a laptop computer to class.

Who should attend: facility managers, architects, engineers, contractors, facility tradespeople, planners, project managers and anyone involved with JOC, SABRE, or IDIQ.

Conceptual Estimating Using the RSMeans CostWorks CD

This *two-day* class uses the leading industry data and a powerful software package to develop highly accurate conceptual estimates for your construction projects. All attendees must bring a laptop computer loaded with the current year *Square Foot Costs* and the *Assemblies Cost Data* CostWorks titles.

Some of what you'll learn:

- Introduction to conceptual estimating
- Types of conceptual estimates
- Helpful hints
- Order of magnitude estimating
- Square foot estimating
- Assemblies estimating

Who should attend: architects, engineers, contractors, construction estimators, owners' representatives, and anyone looking for an electronic method for performing square foot estimating.

Assessing Scope of Work for Facility Construction Estimating

This *two-day* practical training program addresses the vital importance of understanding the scope of projects in order to produce accurate cost estimates for facility repair and remodeling.

Some of what you'll learn:

- Discussions of site visits, plans/specs, record drawings of facilities, and site-specific lists
- Review of CSI divisions, including means, methods, materials, and the challenges of scoping each topic
- Exercises in scope identification and scope writing for accurate estimating of projects
- Hands-on exercises that require scope, take-off, and pricing

Who should attend: corporate and government estimators, planners, facility managers, and others who need to produce accurate project estimates.

Unit Price Estimating Using the RSMeans CostWorks CD

Step-by-step instructions and practice problems to identify and track key cost drivers—material, labor, equipment, staging, and subcontractors—for each specific task. Learn the most detailed, time-tested methodology for accurately "pricing" these variables, their impact on each other and on total cost.

Some of what you'll learn:

- Unit price cost estimating
- Order of magnitude, square foot, and assemblies estimating
- Quantity takeoff
- Direct and indirect construction costs
- Development of contractors' bill rates
- How to use *RSMeans Building Construction Cost Data*

This training session requires you to bring a laptop computer to class.

Who should attend: architects, engineers, corporate and government estimators, facility managers, and government procurement staff

For more information visit our website at www.RSMeans.com

Registration Information

Register early and save up to $100!

Register 30 days before the start date of a seminar and save $100 off your total fee. Note: This discount can be applied only once per order. It cannot be applied to team discount registrations or any other special offer.

How to register

Register by phone today! The RSMeans toll-free number for making reservations is 781-422-5115.

Two-day seminar registration fee - $935. One-day RSMeans CostWorks® or RSMeans Online training registration fee - $375.

To register by mail, complete the registration form and return, with your full fee, to: RSMeans Seminars, 1099 Hingham Street, Suite 201, Rockland, MA 02370.

One-day Construction Cost Estimating - $575.

Government pricing

All federal government employees save off the regular seminar price. Other promotional discounts cannot be combined with the government discount.

Team discount program

For over five attendee registrations. Call for pricing: 781-422-5115

Multiple course discounts

When signing up for two or more courses, call for pricing.

Refund policy

Cancellations will be accepted up to ten business days prior to the seminar start. There are no refunds for cancellations received later than ten working days prior to the first day of the seminar. A $150 processing fee will be applied for all cancellations. Written notice of the cancellation is required. Substitutions can be made at any time before the session starts. No-shows are subject to the full seminar fee.

AACE approved courses

Many seminars described and offered here have been approved for 14 hours (1.4 recertification credits) of credit by the AACE International Certification Board toward meeting the continuing education requirements for recertification as a Certified Cost Engineer/Certified Cost Consultant.

AIA Continuing Education

We are registered with the AIA Continuing Education System (AIA/CES) and are committed to developing quality learning activities in accordance with the CES criteria. Many seminars meet the AIA/CES criteria for Quality Level 2. AIA members may receive 14 learning units (LUs) for each two-day RSMeans course.

Daily course schedule

The first day of each seminar session begins at 8:30 a.m. and ends at 4:30 p.m. The second day begins at 8:00 a.m. and ends at 4:00 p.m. Participants are urged to bring a hand-held calculator since many actual problems will be worked out in each session.

Continental breakfast

Your registration includes the cost of a continental breakfast and a morning and afternoon refreshment break. These informal segments allow you to discuss topics of mutual interest with other seminar attendees. (You are free to make your own lunch and dinner arrangements.)

Hotel/transportation arrangements

RSMeans arranges to hold a block of rooms at most host hotels. To take advantage of special group rates when making your reservation, be sure to mention that you are attending the RSMeans seminar. You are, of course, free to stay at the lodging place of your choice. (Hotel reservations and transportation arrangements should be made directly by seminar attendees.)

Important

Class sizes are limited, so please register as soon as possible.

Note: Pricing subject to change.

Registration Form

ADDS-1000

Call 781-422-5115 to register or fax this form to 800-632-6732. Visit our website: www.RSMeans.com

Please register the following people for the RSMeans construction seminars as shown here. We understand that we must make our own hotel reservations if overnight stays are necessary.

❑ Full payment of $__________ enclosed.

❑ Bill me.

Please print name of registrant(s).

(To appear on certificate of completion)

____

____

____

____

P.O. #: __________

GOVERNMENT AGENCIES MUST SUPPLY PURCHASE ORDER NUMBER OR TRAINING FORM.

Please mail check to: 1099 Hingham Street, Suite 201, Rockland, MA 02370 USA

Firm name __________

Address __________

City/State/Zip __________

Telephone no. __________ Fax no. __________

E-mail address __________

Charge registration(s) to: ☐ MasterCard ☐ VISA ☐ American Express

Account no. __________ Exp. date __________

Cardholder's signature __________

Seminar name __________

Seminar City __________

Installing Contractor's Overhead & Profit

Below are the average installing contractor's percentage markups applied to base labor rates to arrive at typical billing rates.

Column A: Labor rates are based on union wages averaged for 30 major U.S. cities. Base rates including fringe benefits are listed hourly and daily. These figures are the sum of the wage rate and employer-paid fringe benefits such as vacation pay, employer-paid health and welfare costs, and pension costs, plus appropriate training and industry advancement funds costs.

Column B: Workers' compensation rates are the national average of state rates established for each trade.

Column C: Column C lists average fixed overhead figures for all trades. Included are federal and state unemployment costs set at 7.8%; social security taxes (FICA) set at 7.65%; builder's risk insurance costs set at 0.50%; and public liability costs set at 2.02%. All the percentages except those for social security taxes vary from state to state as well as from company to company.

Columns D and E: Percentages in Columns D and E are based on the presumption that the installing contractor has annual billing of $4,000,000 and up. Overhead percentages may increase with smaller annual billing. The overhead percentages for any given contractor may vary greatly and depend on a number of factors such as the contractor's annual volume, engineering and logistical support costs, and staff requirements. The figures for overhead and profit will also vary depending on the type of job, the job location, and the prevailing economic conditions. All factors should be examined very carefully for each job.

Column F: Column F lists the total of Columns B, C, D, and E.

Column G: Column G is Column A (hourly base labor rate) multiplied by the percentage in Column F (O&P percentage).

Column H: Column H is the total of Column A (hourly base labor rate) plus Column G (Total O&P).

Column I: Column I is Column H multiplied by eight hours.

		A		B	C	D	E	F	G	H	I
		Base Rate Incl. Fringes		Workers' Comp. Ins.	Average Fixed Overhead	Overhead	Profit	Total Overhead & Profit		Rate with O & P	
Abbr.	Trade	Hourly	Daily					%	Amount	Hourly	Daily
Skwk	Skilled Workers Average (35 trades)	$49.90	$399.20	13.0%	18.0%	13.0%	10%	54.0%	$26.95	$ 76.85	$614.80
	Helpers Average (5 trades)	36.00	288.00	15.8		11.0		54.8	19.75	55.75	446.00
	Foreman Average, Inside ($.50 over trade)	50.40	403.20	13.0		13.0		54.0	27.20	77.60	620.80
	Foreman Average, Outside ($2.00 over trade)	51.90	415.20	13.0		13.0		54.0	28.05	79.95	639.60
Clab	Common Building Laborers	37.90	303.20	14.4		11.0		53.4	20.25	58.15	465.20
Asbe	Asbestos/Insulation Workers/Pipe Coverers	53.40	427.20	11.6		16.0		55.6	29.70	83.10	664.80
Boil	Boilermakers	61.65	493.20	8.0		16.0		52.0	32.05	93.70	749.60
Bric	Bricklayers	46.25	370.00	13.7		11.0		52.7	24.35	70.60	564.80
Brhe	Bricklayer Helpers	38.00	304.00	13.7		11.0		52.7	20.05	58.05	464.40
Carp	Carpenters	48.45	387.60	14.4		11.0		53.4	25.85	74.30	594.40
Cefi	Cement Finishers	45.65	365.20	9.3		11.0		43.3	22.05	67.70	541.60
Elec	Electricians	55.10	440.80	5.6		16.0		49.6	27.35	82.45	659.60
Elev	Elevator Constructors	79.10	632.80	5.2		16.0		49.2	38.90	118.00	944.00
Eqhv	Equipment Operators, Crane or Shovel	52.25	418.00	9.3		14.0		51.3	26.80	79.05	632.40
Eqmd	Equipment Operators, Medium Equipment	51.10	408.80	9.3		14.0		51.3	26.20	77.30	618.40
Eqlt	Equipment Operators, Light Equipment	49.15	393.20	9.3		14.0		51.3	25.20	74.35	594.80
Eqol	Equipment Operators, Oilers	46.05	368.40	9.3		14.0		51.3	23.60	69.65	557.20
Eqmm	Equipment Operators, Master Mechanics	52.50	420.00	9.3		14.0		51.3	26.95	79.45	635.60
Glaz	Glaziers	46.70	373.60	13.1		11.0		52.1	24.35	71.05	568.40
Lath	Lathers	47.30	378.40	8.6		11.0		47.6	22.50	69.80	558.40
Marb	Marble Setters	45.90	367.20	13.7		11.0		52.7	24.20	70.10	560.80
Mill	Millwrights	51.10	408.80	7.6		11.0		46.6	23.80	74.90	599.20
Mstz	Mosaic & Terrazzo Workers	44.60	356.80	8.8		11.0		47.8	21.30	65.90	527.20
Pord	Painters, Ordinary	40.35	322.80	11.7		11.0		50.7	20.45	60.80	486.40
Psst	Painters, Structural Steel	41.30	330.40	32.7		11.0		71.7	29.60	70.90	567.20
Pape	Paper Hangers	40.60	324.80	11.7		11.0		50.7	20.60	61.20	489.60
Pile	Pile Drivers	48.40	387.20	14.6		16.0		58.6	28.35	76.75	614.00
Plas	Plasterers	44.90	359.20	10.5		11.0		49.5	22.25	67.15	537.20
Plah	Plasterer Helpers	38.05	304.40	10.5		11.0		49.5	18.85	56.90	455.20
Plum	Plumbers	59.20	473.60	6.9		16.0		50.9	30.15	89.35	714.80
Rodm	Rodmen (Reinforcing)	53.00	424.00	12.9		14.0		54.9	29.10	82.10	656.80
Rofc	Roofers, Composition	41.70	333.60	31.4		11.0		70.4	29.35	71.05	568.40
Rots	Roofers, Tile & Slate	41.75	334.00	31.4		11.0		70.4	29.40	71.15	569.20
Rohe	Roofers, Helpers (Composition)	31.35	250.80	31.4		11.0		70.4	22.05	53.40	427.20
Shee	Sheet Metal Workers	57.25	458.00	8.8		16.0		52.8	30.25	87.50	700.00
Spri	Sprinkler Installers	57.60	460.80	7.6		16.0		51.6	29.70	87.30	698.40
Stpi	Steamfitters or Pipefitters	60.70	485.60	6.9		16.0		50.9	30.90	91.60	732.80
Ston	Stone Masons	47.05	376.40	13.7		11.0		52.7	24.80	71.85	574.80
Sswk	Structural Steel Workers	53.20	425.60	27.4		14.0		69.4	36.90	90.10	720.80
Tilf	Tile Layers	44.50	356.00	8.8		11.0		47.8	21.25	65.75	526.00
Tilh	Tile Layers Helpers	34.80	278.40	8.8		11.0		47.8	16.65	51.45	411.60
Trlt	Truck Drivers, Light	41.95	335.60	12.0		11.0		51.0	21.40	63.35	506.80
Trhv	Truck Drivers, Heavy	43.20	345.60	12.0		11.0		51.0	22.05	65.25	522.00
Sswl	Welders, Structural Steel	53.20	425.60	27.4		14.0		69.4	36.90	90.10	720.80
Wrck	*Wrecking	37.90	303.20	22.9		11.0		61.9	23.45	61.35	490.80

*Not included in averages

RSMeans Assemblies Cost Data 2016

Your 2016 cost guide features

- Fully updated cost data, with more than 11,200 building assembly and component costs
- Expanded exterior wall and framing systems giving you more options to choose from
- Revised pedestal access floor assemblies

Plus:

- Receive access to free quarterly cost index and key material price updates online for a full year

Assemblies

A Substructure
B Shell
C Interiors
D Services
E Equipment & Furnishings
F Special Construction
G Building Sitework

RSMeans Online

Quick, intuitive, easy to use, and automatically updated, RSMeans Online gives you instant access to hundreds of thousands of material, labor, and equipment costs from RSMeans' comprehensive database, delivering the information you need to build budgets and competitive estimates every time.

ACCURATE

Visit www.RSMeans.com/Online to get started.

RSMeans Catalog No. 60066

THE GORDIAN GROUP®